Charles Winston

ACOUSTIC WAVES: DEVICES, IMAGING, AND ANALOG SIGNAL PROCESSING

PRENTICE-HALL SIGNAL PROCESSING SERIES

Alan V. Oppenheim, Editor

ANDREWS and HUNT *Digital Image Restoration*
BRIGHAM *The Fast Fourier Transform*
BURDIC *Underwater Acoustic System Analysis*
CASTLEMAN *Digital Image Processing*
COWAN and GRANT *Adaptive Filters*
CROCHIERE and RABINER *Multirate Digital Signal Processing*
DUDGEON and MERSEREAU *Multidimensional Digital Signal Processing*
HAMMING *Digital Filters, 2e*
HAYKIN, ED. *Array Signal Processing*
JAYANT and NOLL *Digital Coding of Waveforms*
KINO *Acoustic Waves: Devices, Imaging and Analog Signal Processing*
LEA, ED. *Trends in Speech Recognition*
LIM, ED. *Speech Enhancement*
MARPLE *Digital Spectral Analysis with Applications*
MCCLELLAN and RADER *Number Theory in Digital Signal Processing*
MENDEL *Lessons in Digital Estimation Theory*
OPPENHEIM, ED. *Applications of Digital Signal Processing*
OPPENHEIM, WILLSKY, with YOUNG *Signals and Systems*
OPPENHEIM and SCHAFER *Digital Signal Processing*
RABINER and GOLD *Theory and Applications of Digital Signal Processing*
RABINER and SCHAFER *Digital Processing of Speech Signals*
ROBINSON and TREITEL *Geophysical Signal Analysis*
STEARNS and DAVID *Signal Processing Algorithms*
TRIBOLET *Seismic Applications of Homomorphic Signal Processing*
WIDROW and STEARNS *Adaptive Signal Processing*

ACOUSTIC WAVES: DEVICES, IMAGING, AND ANALOG SIGNAL PROCESSING

Professor Gordon S. Kino
Stanford University

PRENTICE-HALL, INC.
Englewood Cliffs, New Jersey 07632

Library of Congress Cataloging-in-Publication Data

KINO, GORDON S.
   Acoustic waves.

   Bibliography: p.
     1. Acoustical engineering. 2. Sound—Equipment
and supplies. 3. Piezoelectric devices. I. Title.
TA365.K53 1987   620.2   86–25223
ISBN 0-13-003047-3

Editorial/production supervision and
  interior design: Richard Woods
Cover design: Ben Santora
Manufacturing buyer: Gordon Osbourne

©1987 by Prentice-Hall, Inc.
A Division of Simon & Schuster
Englewood Cliffs, New Jersey 07632

All rights reserved. No part of this book may be
reproduced, in any form or by any means, without
permission in writing from the publisher.

Printed in the United States of America

10 9 8 7 6 5 4 3 2 1

ISBN 0-13-003047-3 025

PRENTICE-HALL INTERNATIONAL (UK) LIMITED, *London*
PRENTICE-HALL OF AUSTRALIA PTY. LIMITED, *Sydney*
PRENTICE-HALL CANADA INC., *Toronto*
PRENTICE-HALL HISPANOAMERICANA, S.A. *Mexico*
PRENTICE-HALL OF INDIA PRIVATE LIMITED, *New Delhi*
PRENTICE-HALL OF JAPAN, INC., *Tokyo*
PRENTICE-HALL OF SOUTHEAST ASIA PTE. LTD., *Singapore*
EDITORA PRENTICE-HALL DO BRASIL, LTDA., *Rio de Janeiro*

To Carol and Dorothy

Contents

Preface   xix

Chapter 1   Sound Wave Propagation   1

1.1 INTRODUCTION   1

     1.1.1 Sound Waves in Nonpiezoelectric Materials: One-Dimensional Theory,   2
        Stress,   3
        Displacement and strain,   4
        Hooke's law and elasticity,   4
        Equation of motion,   4
        Conservation of mass,   5
        The wave equation and definition of propagation constant,   5
        Energy,   6
        Poynting's theorem,   7
        Acoustic losses,   8
        Acoustic impedance,   9
        Shear waves,   13
        Extensional waves,   14
     Problem Set 1.1,   15

1.2 PIEZOELECTRIC MATERIALS   17

     1.2.1 Constitutive Relations,   17
        Poling and domains in ferroelectric materials,   19

1.2.2 Effect of Piezoelectric Coupling on Wave Propagation in a Medium of Infinite Extent, 21
Piezoelectrically stiffened elastic constant, 21
Stress-free dielectric constant, 22
Problem Set 1.2, 22

1.3 ENERGY CONSERVATION IN PIEZOELECTRIC MEDIA 24

Problem Set 1.3, 26

1.4 PIEZOELECTRIC TRANSDUCERS 27

1.4.1 Introduction, 27
1.4.2 The Transducer as a Three-Port Network, 29
Example: electrical input impedance of a transducer, 32
1.4.3 Mason Equivalent Circuit, 33
Example: clamped transducer, 34
1.4.4 Redwood Equivalent Circuit, 35
Example: open-circuited pulse-excited acoustically matched receiving transducer, 35
1.4.5 Impedance of an Unloaded Transducer, 35
Equivalent circuit of an unloaded transducer, 37
1.4.6 Broadband Operation of Transducers into an Acoustic Medium: The KLM Model, 41
Effect of reflections on the response of a transducer to a sinusoidal signal, 45
Half-wave and quarter-wave resonators, 47
1.4.7 Pulse Response of a Transducer with Arbitrary Terminations, 47
Air-backed transducer and unmatched front surface, 50
Example: PZT transducer, 50
1.4.8 Electrical Input Impedance of a Loaded Transducer, 51
Constructional techniques, 51
Electrical impedance at resonance, 52
Terminated transducer ($Z_1 = Z_2 = Z_C$) or $Z_1 = 0$, $Z_2 = Z_C$, 52
1.4.9 Efficiency of Power Transfer to Transducer, 53
Example: zinc oxide transducer on sapphire, 54
1.4.10 Electrical Matching of a Loaded Transducer for Optimum Bandwidth and Efficiency, 55
Transducer acoustically matched at each end, $Z_1 = Z_2 = Z_C$, 56
Air-backed transducer matched at output, $Z_1 = 0$, $Z_2 = Z_C$, 56
Example: acoustic emission transducer, 57
$Z_1 < Z_C$ and $Z_2 < Z_C$, 57
Optimum value of the acoustic load with the source resistance $R_0 = 0$, 58
Optimum value of the acoustic load with electrical tuning and source impedance $R_0 = R_{a0}$, 60
Tuned and untuned transducer, $Z_1 = Z_2 = Z_C$, 60
Tuned and untuned air-backed transducer with $Z_2 = Z_C$, 60
1.4.11 Reeder–Winslow Design Method for Air-Backed Transducers, 61
Example: $Z_1 = Z_2 \gg Z_C$, 65
1.4.12 Some Examples of Transducer Design, 66
Problem Set 1.4, 71

1.5 TENSOR NOTATION AND CONSTITUTIVE
    RELATIONS FOR PIEZOELECTRIC AND
    NONPIEZOELECTRIC MATERIALS   75

   1.5.1 Introduction, 75
   1.5.2 Mathematical Treatment, 75
      Displacement and strain, 75
      Stress, 76
      Hooke's law and elasticity, 77
      Tensor notation, 77
      Reduced subscript notation, 78
   Problem Set 1.5, 83

Chapter 2 Waves in Isotropic Media   85

2.1 INTRODUCTION   85

2.2 BASIC THEORY FOR WAVES IN ISOTROPIC
    MEDIA   86

   2.2.1 Mathematical Formalism for Waves in Isotropic Media, 86
      Lamé constants, 86
      Dilation and Hooke's law, 86
      Energy, 87
      Ideal fluid, 87
      Young's modulus and the extensional wave velocity, 88
      Poisson's ratio, 88
      Shear, longitudinal, extensional, and strip guide wave velocities, 89
   2.2.2 Equations of Motion for Solids and Fluids, 90
   2.2.3 Wave Equation in an Isotropic Medium, 92
      Longitudinal waves, 93
      Shear waves, 94
   2.2.4 Plane Wave Reflection and Refraction, 94
      Incident longitudinal wave, 96
      Conversion of longitudinal to shear waves and shear to longitudinal waves, 98
      Reflection and refraction, 100
   Problem Set 2.2, 104

2.3 ACOUSTIC WAVEGUIDES   105

   2.3.1 Introduction, 105
   2.3.2 Shear Horizontal Modes, 106
   2.3.3 Lamb Waves, 108
   2.3.4 Surface Waves, 109
      Rayleigh waves in an isotropic medium, 110
   Problem Set 2.3, 114

2.4 INTERDIGITAL TRANSDUCERS 117

2.4.1 Introduction, 117
2.4.2 Delta-Function Model of the Transducer, 119
  Example: uniform transducer, 120
2.4.3 Network Theory of the Transducer, 123
Problem Set 2.4, 130

2.5 NORMAL-MODE THEORY AND PERTURBATION THEORY 131

2.5.1 Introduction, 131
2.5.2 Power Flow Concepts, 133
  Normalization, 134
2.5.3 Excitation of a Surface Acoustic Wave, 134
  Physical implications of the excitation equations, 135
  Quantitative evaluation of $g_n(z)$, 135
2.5.4 Perturbation Theory of the Interdigital Transducer, 137
2.5.5 Leaky Waves, 141
  Example: aluminum substrate loaded by water, 144
2.5.6 Wedge Transducer, 144
  Finite attenuation, 146
  Efficiency variation with incident angle or velocity, 147
  Example: excitation of surface acoustic waves in aluminum from water, 149
  Reflected wave, 149
  Example: reflection of waves from aluminum in water, 149
  Experimental results, 150
Problem Set 2.5, 150

Chapter 3 Wave Propagation with Finite Exciting Sources 154

3.1 DIFFRACTION AND NONUNIFORM EXCITATION 154

3.1.1 Introduction, 154
3.1.2 Spherical Waves in a Liquid or Solid, 156
  Vibrating sphere, 156
  Liquid medium, 157
3.1.3 Green's Function, 158
  Helmholtz's theorem, 159
  Sommerfeld radiation condition, 160
  Kirchhoff formula, 160
  Rayleigh–Sommerfeld formula, 161
  Rigid baffle, 162
  Pressure release baffle, 162
  Transient source, 163
Problem Set 3.1, 163

3.2 PLANE PISTON TRANSDUCERS 164

3.2.1 Fields on the Axis, 164
Fresnel or paraxial approximation, 165
Fraunhofer and Fresnel zone, 165
Fraunhofer approximation, 167
3.2.2 Radial Variation of the Field Using a Hankel Transform and Spatial Frequency Concepts, 171
Example: piston transducer for probing the human body, 175
3.2.3 Diffraction from Rectangular Transducers, 175
Fraunhofer diffraction field by the method of stationary phase, 178
3 dB points, 180
Problem Set 3.2, 180

3.3 FOCUSED TRANSDUCERS 182

3.3.1 Field of a Focused Spherical Transducer, 182
Transverse definition, 185
3-dB definition, F number, and lens aperture, 185
Sidelobes, 185
Coherent and incoherent imaging, 186
Rayleigh two-point definition, 187
Sparrow two-point definition, 189
Speckle, 189
Depth of focus, 190
Geometrical concepts for depth of focus, 191
Example: focused transducer for medical applications, 194
Reflection from a plane reflector, 194
3-dB resolution for V(z), 196
3.3.2 Scanned Acoustic Microscope, 197
3.3.3 Gaussian Beams and the Paraxial Equation, 206
Wave equation for paraxial beams, 206
Gaussian beam, 207
Apodization, 209
Problem Set 3.3, 210

3.4 PULSED EXCITATION OF TRANSDUCERS 212

Transient response on- and off-axis of a piston resonator, 213
Fresnel region, 214
Fraunhofer region, 215
Excitation with a tone burst, 216
Problem Set 3.4, 218

3.5 LENSLESS ACOUSTIC IMAGING 218

3.5.1 Introduction, 218
A. Applications of Acoustic Imaging, 218
B. A-Scan, B-Scan, and C-Scan Imaging, 220
A-scan, 220
B-scan, 222
C-scan, 223

C. Focusing Systems and High-Speed Scanning, 225
    Elimination of physical lenses, 226
    Physical lens, 227
    Time delay focusing, 227
    Phase delay focusing, 228
3.5.2 Basic Imaging Theory, 229
  A. Matched Filter Concepts, 229
  B. Paraxial Approximation, 231
    Line spread function, 232
    3-dB definition of a paraxial rectilinear system, 232
  C. A Radial Sector Scan System, 232
    Rayleigh criterion, 234
    Sparrow criterion, 235
    Range definition, 235
  D. Generalization of the Matched Filter Theory, 237
  E. Sidelobes and Grating Lobes, 238
    Grating lobes, 239
    Examples of apodization and grating lobes, 241
  F. Effect of Missing Elements (Amplitude Errors), 243
3.5.3 Fresnel Lenses and Digital Sampling, 245
  A. Basic System, 245
  B. Fresnel Lens Sidelobes and Phase Sampling in Digital Systems, 248
    Sidelobe level, 249
3.5.4 Chirp-Focused System, 251
  A. A Basic System, 251
  B. Various Forms of Chirp-Focused Systems, 255
3.5.5 Time-Delay and Tomographic Systems, 258
  A. Introduction, 258
  B. TV Display, 265
  C. Sidelobes, Grating Lobes, and Sampling Lobes in Time Delay Systems, 265
    Range resolution, 266
    Transverse definition, 266
    Grating lobes, 267
  D. Examples of Synthetic Aperture Imaging, 270
    Imaging modes, 270
    Surface acoustic wave imaging, 272
  E. Tomographic Imaging Systems, 275
3.5.6 Acoustic Holography, 277
  A. Introduction, 277
  B. Holographic Reconstruction with Spherical Reference Waves, 279
    Spherical wave sources, 281
    Reconstruction of the image, 282
  C. Water-Air Surface as an Acoustic Imaging Intensity Detector, 285
    Smith and Brenden holographic technique, 286
  D. Scanned Holographic Imaging, 288
  E. The Scanning Laser Acoustic Microscope, 291
  F. Holographic Imaging of Vibrating Objects, 294
Problem Set 3.5, 296

3.6 REFLECTION AND SCATTERING BY SMALL AND LARGE OBJECTS 300

    3.6.1 Introduction, 300
    3.6.2 Scattering by Large Objects (Physical Concepts), 301
    3.6.3 General Scattering Theory, 303
        Surface integral formulation, 303
        Volume integral formulation, 304
        Born approximation, 305
        Quasistatic approximation, 306
        Example: scattering from a rigid sphere in a liquid, 307
        Example: air bubbles and sand grains in water, 308
        Kirchhoff approximation for scattering from a large sphere ($ka \gg 1$), 308
        Comparison with exact results and the concept of creeping waves, 309
        Generalizations of the theory, 310
    Problem Set 3.6, 312

Chapter 4 Transversal Filters 318

4.1 INTRODUCTION 318

4.2 LINEAR PASSIVE SURFACE WAVE DEVICES 322

    4.2.1 Interdigital Transducer, 322
        Digitally coded devices, 322
        Mathematical model of the transducer, 326
    4.2.2 Bandpass Filter, 329
        Uniform transducer, 329
        Apodized transducer, 330
    4.2.3 FM Chirp Analog Filter, 332
    4.2.4 Resonators, 334
    4.2.5 RAC Filters, 335
    4.2.6 Stabilized SAW Oscillators, 336
    Problem Set 4.2, 337

4.3 CHARGE-TRANSFER DEVICES 339

    4.3.1 Introduction, 339
    4.3.2 Bucket Brigade Devices, 341
        Performance limitation of the BBD, 342
    4.3.3 Charge-Coupled Devices, 347
    4.3.4 Effect of Imperfect Charge-Transfer Efficiency, 352
    4.3.5 Single Transfer Devices (Switched Capacitors), 356
    Problem Set 4.3, 358

4.4 MATCHED FILTERS 359

    4.4.1 Introduction, 359
    4.4.2 Matched Filter Theory, 361

        Example: A matched filter for a square pulse, 364
        Matched filter in the frequency domain, 365
    4.4.3 Pulse Compression, 366
    4.4.4 Matched Filters for Complex Signals, 367
    4.4.5 Range and Velocity Accuracy of a Radar System, 368
        Simple physical treatment, 368
        More rigorous treatment, 369
        Accuracy of frequency measurement, 372
        Uncertainty relation, 372
        Proof of the inequality, 373
        Narrowband waveforms, 373
    4.4.6 Ambiguity Function, 376
    Problem Set 4.4, 382

4.5 FM CHIRP FILTERS AND CHIRP TRANSFORM PROCESSORS 383

    4.5.1 Introduction, 383
    4.5.2 Mathematical Treatment of FM Chirp Filter, 384
        Width of main lobe, 385
        Sidelobes, 385
        Physics of FM Chirp Filter, 386
    4.5.3 Filter Weighting for Sidelobe Reduction, 387
        Effect of apodization on signal-to-noise ratio, 390
    4.5.4 Fourier Transform Operations with FM Chirp Filters, 391
        Compressive receiver, 395
        Variable bandpass–bandstop filters and bandshape filters, 395
        Variable-time-delay filters, 396
        Correlation or convolution of two signals, 396
        Spectral whitening and nonlinear processing, 399
    4.5.5 The Implementation of the Chirp z Transforms with SAW Devices, 399
        Experimental SAW FFT processor realization, 404
    4.5.6 Chirp z Transform with a CCD, 406
        Sliding transform, 408
    4.5.7 Chirp Transforms with Superconductive Delay Lines, 409
    Problem Set 4.5, 413

4.6 BANDPASS FILTERS 415

    4.6.1 Introduction, 415
    4.6.2 Strip Coupler, 416
    4.6.3 Amplitude-Weighted Bandpass Filters, 421
    4.6.4 Phase-Weighted Transducers, 426
    4.6.5 Building Block Design Technique, 427
    4.6.6 Recursive Filters and Comb Filters, 432
    A. Introduction, 432
    B. Recursive Filter with Feedback, 434
        Example of an SAW recursive comb filter, 436
    C. Fiber-Optic Recursive Filters, 439
    Problem Set 4.6, 443

4.7 CONVOLVERS 448

4.7.1 Introduction, 448
4.7.2 Acoustic Convolver, 449
4.7.3 Spread-Spectrum Communications, 454
  Spread-spectrum systems for communications, 458
  Range finding, 459
4.7.4 Waveguide Convolvers, 459
4.7.5 Semiconductor convolver, 461
4.7.6 Acoustic Convolver Using External Mixers, 464
4.7.7 CCD Convolver or Correlator, 467
4.7.8 SAW Storage Correlator, 469
  Tapped-delay-line correlator, 470
  Monolithic and air gap storage correlators, 471
  Input correlation mode, 474
  Spread-spectrum communications, 475
Problem Set 4.7, 478

4.8 ADAPTIVE FILTERS 480

4.8.1 Introduction, 480
4.8.2 Removing Phase Errors with an Adpative Matched Filter, 481
4.8.3 Inverse Filter, 482
4.8.4 Wiener Filter, 488
Pseudo-inverse filter, 491
4.8.5 Storage Correlator as an Adaptive Wiener Filter, 492
4.8.6 Elimination of an Interfering Signal with an SAW Correlator, 495
  Other uses for adaptive filters, 497
Problem Set 4.8, 498

4.9 ACOUSTO-OPTIC FILTERS 501

4.9.1 Introduction, 501
4.9.2 Photoelastic Effect, 501
  Example: Piezo-optic constant of water, 503
4.9.3 Light Diffraction by Sound, 503
  Acousto-optic figure of merit, 505
  Example: calculation of $v(w)$ for water, 506
  Limitations of Rahman–Nath theory, 508
  Example: Rahman–Nath criterion for water, 509
4.9.4 Schlieren Effect, 509
4.9.5 Bragg Diffraction by Sound, 511
  Diffraction theory, 513
  Rahman–Nath interaction $Q \ll \pi/2$, 513
  Bragg regime $Q \gg \pi/2$, 514
  Response as a function of acoustic beam profile, 514
  Response as a function of input angle θ_i, 515
  Response as a function of input frequency, 515
4.9.6 Application of Bragg Deflection Systems, 517
4.9.7 Acousto-Optic Convolver, 518

4.9.8 Time-Integrating Correlator, 521
One-dimensional transforms, 524
Two-dimensional transforms, 526
Problem Set 4.9, 528

Appendix A Stress, Strain, and the Reduced Notation   537

STRESS AND STRAIN VECTORS   537

Strain Tensor, 537
Change in Volume, 540
Stress Tensor, 540
Equation of Motion, 542

SYMBOLIC NOTATION AND ABBREVIATED SUBSCRIPTS   542

Strain Tensor, 542
Stress Tensor, 544
Elasticity, 544
Example: A Cubic Crystal, 545
Example: Isotropic Material, 546

PIEZOELECTRIC TENSORS   546

REFERENCE   547

Appendix B Acoustic Parameters of Common Materials   548

TABLE B.1: BULK MATERIAL CONSTANTS FOR SELECTED SOLIDS   549

TABLE B.2: MATERIAL CONSTANTS FOR LIQUIDS   552

TABLE B.3: ACOUSTIC CONSTANTS FOR GASES   553

TABLE B.4: PROPERTIES OF COMMONLY USED PIEZOELECTRIC CERAMICS   554

TABLE B.5: PROPERTIES OF SELECTED TRANSDUCER MATERIALS   557

TABLE B.6: PROPERTIES OF SOME COMMON ACOUSTO-OPTIC MATERIALS   560

TABLE B.7: MATERIAL CONSTANTS OF SOME COMMON ACOUSTO-OPTIC MATERIALS 560

REFERENCES 561

Appendix C Poynting's Theorem in Piezoelectric Media 563

Appendix D Determination of the Impedance Z_o in Terms of $\Delta V/V$ 566

Appendix E A Rigorous Derivation of Normal-Mode Theory 569

Normal-mode expansion, 571

Appendix F Transducer Admittance Matrix 574

IN-LINE MODEL 576

CROSSED-FIELD MODEL 577

Appendix G Method of Stationary Phase 578

Appendix H Quasistatic Theory for Fields Inside a Sphere 580

Appendix I Rate of Movement of Charge in a CCD Due to Diffusion and Space Charge 583

Appendix J Acousto-Optic Effect in Anisotropic Crystals 588

Index 591

Preface

The purpose of this book is to introduce the reader to the acoustic wave concepts required for the design of a wide range of acoustic devices, and to discuss some of these devices in detail. My aim has been to describe a wide range of classical and more recently developed theoretical methods, such as perturbation and coupled wave theories, and to show how these concepts are applied to understanding the wide range of acoustic devices now being used in practical applications. The subjects covered in this book reflect my own research interests during the last ten years. Many of the new results are due to the original work of myself and my colleagues Bert Auld, Marvin Chodorow, Pierre Khuri-Yakub, John Shaw, and Cal Quate, and our research associates and students in the Ginzton Laboratory at Stanford University. Most of the text was first developed for class notes and taught by me, at Stanford, in two separate one-quarter graduate courses, "Theory of Acoustic Devices" and "Analog Signal Processing."

Chapter 1 introduces the reader to the basic concepts of wave propagation in piezoelectric and nonpiezoelectric media, and deals with the theory of piezoelectric transducers in some detail. To keep the theory simple, I have used one-dimensional theories and theories for isotropic media wherever possible. When necessary, it is shown how the theory must be modifed to take account of anisotropic piezoelectric media and to relate the constants used in the one-dimensional theory to the properties of anisotropic materials.

Chapter 2 is concerned with the theory of wave propagation in finite media, waveguide and surface waves. The basic theory of surface wave transducers is dealt with there, as well as concepts of leaky waves and wedge transducers. Again, the theory deals mainly with isotropic media, and it is shown how the results can be modified to take account of anisotropic piezoelectric materials. In this chapter, the reader is introduced to several important and powerful theoretical techniques, such as perturbation, normal mode theory, and the network theory for interdigital transducers.

Chapter 3 deals with diffraction and imaging. The basic theory of diffraction is developed and applied to piston transducers and focused beams, with the acoustic microscope as the prime example. The effects of using short pulses are dealt with, and it is shown how short-pulse-excited systems can be extremely useful in medical and nondestructive testing applications. Electronic focusing with transducer arrays, holography, and tomography are discussed in the last part of the chapter.

Chapter 4 is application-oriented to all kinds of analog signal processing devices. Various ways in which surface acoustic wave devices can be used in practice as filters for taking Fourier transforms in real time, for correlation of signals, for pulse compression, and so on, are discussed. Basic concepts of radar signal processing and spread spectrum communications are also described. It is shown how the same concepts cna be applied for use with other transversal filters such as charge-coupled devices and fiber-optic and superconductive delay lines. Considerable attention is paid to applications of charge-coupled devices. The last part of the chapter is devoted to acousto-optic devices and shows how these devices can be used to realize many of the concepts developed earlier in the chapter.

Many have helped me with this book and deserve thanks for doing so. I began teaching the first course on which this book is based in 1976; since then, many of my students have pointed out errors and made helpful suggestions. There have been so many, in fact, that by singling out some for praise, I am bound to neglect others who deserve equal mention, and so I thank all of them. Those whom I particularly remember as having made major contributions are Rick Baer, Tim Corle, Mindy Garber, Jonathan Green, Steve Hessel, Didier Husson, Robert Joly, Lou Lome, Kent Peterson, Alan Selfridge, and Fred Stanke.

I would also like to thank my friends and colleagues Eric Ash, Bert Auld, Marvin Chodorow, John Shaw, and Cal Quate for many stimulating discussions on acoustic concepts, as well as Simon Bennett, whose detailed comments enhanced my discussion of imaging and microscopy. I owe special thanks to Pierre Khuri-Yakub for the many research ideas we have generated together over the years, and for his helpful suggestions at all stages of the text.

I am grateful to Ching-Hua Chou, who calculated several of the graphs in this book, and to Alan Selfridge, who provided the material for several of the tables in Appendix B.

I would like to thank Melinda Sterling, who typed some of the early drafts, as well as Ingrid Tarien, who has spent many hours, in the years she has worked for me, occupied with the book manuscript and class notes.

I would like to thank my wife, Dorothy, for patiently bearing with me for the several years that it took to create this book.

Finally, I am extremely grateful to my daughter, Carol, who delayed her own career plans to edit the book and put it in final form for the publishers. She made the text far more readable, caught numerous errors, and despite my many distractions, kept me working hard on the manuscript. Without her, I doubt that this book would ever have been completed.

Gordon S. Kino
Stanford, California

Chapter 1

Sound Wave Propagation

1.1 INTRODUCTION

This book covers the propagation of acoustic waves in solids and liquids. We want to know what kinds of waves can propagate in various types of materials, how they are reflected from and transmitted through the boundaries between two materials, how they are excited, and how their properties are related to the mass, density, and elasticity of the materials involved.

We shall be particularly concerned with the properties of piezoelectric transducers, which are the most efficient, and hence most widely used ultrasonic transducers for converting electrical into mechanical energy. Therefore, we will have to derive formulas for their conversion efficiency and electrical impedance as a function of frequency, and for their pulse response characteristics. To do this, we need to know how waves propagate in piezoelectric and nonpiezoelectric materials. We start with the simpler case of nonpiezoelectric materials and then extend the analysis to piezoelectric materials.

In general, the treatment of acoustic wave propagation in solids is complicated by the fact that solids are not always isotropic. Thus the parameters of the acoustic wave must be expressed in terms of tensor quantities and the relations between them.

For simplicity, we shall assume that the waves of interest are either pure longitudinal or pure shear waves, and that all physical quantities (particle displacement, particle velocity, stress, strain, elasticity, and the piezoelectric coupling constant) can be expressed in one-dimensional form. In addition, we restrict our analysis in this chapter to plane wave propagation, which reduces the problem from three dimensions to one. The results we obtain are identical in form to a general, more complete treatment for isotropic materials, and are valid even for propagation of waves along an axis of a crystalline material, provided that the elastic constants and coupling coefficients are defined correctly. The required formalism will be described briefly in Sec. 1.1.1; Sec. 1.5 has a more detailed treatment and Appendix A contains more rigorous derivations of some of the essential formulas. Later, in Sec. 2.2, we generalize the theory to account for propagation in an arbitrary direction in isotropic media; in Sec. 2.3 we deal with waves in materials of finite cross section.

1.1.1 Sound Waves in Nonpiezoelectric Materials: One-Dimensional Theory

We must first define the two basic types of waves that are important in acoustic wave propagation. The first is a *longitudinal wave*, in which the motion of a particle in the acoustic medium is only in the direction of propagation. Thus when a force is applied to the acoustic medium, the medium expands or contracts in the z direction, as shown in Fig. 1.1.1(a). The second type of wave is a *shear wave*, in which the motion of a particle in the medium is transverse to the direction of propagation, as illustrated in Fig. 1.1.1(b). Shear waves are associated with the flexing or bending of a material (e.g., twisting a rod). There is no change in volume or density of the material in a shear wave mode, as shown in Fig. 1.1.1(b).

In general, the acoustic waves that can propagate through a solid medium may combine shear and longitudinal motion. However, in a crystalline medium

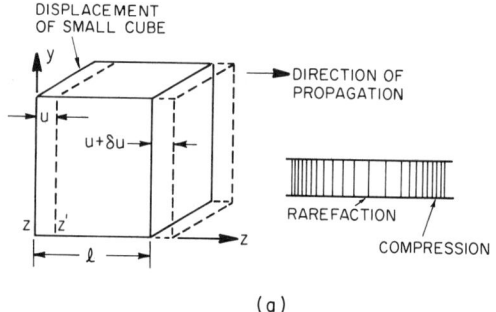

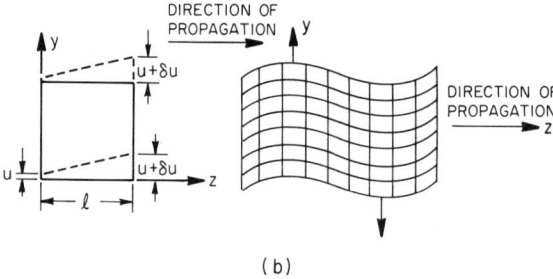

Figure 1.1.1 (a) Longitudinal wave propagation; (b) shear wave propagation (full cube not shown).

with anisotropic elastic properties, the direction of propagation can be chosen to be along one of the principal axes of the crystal; in this case the basic modes can be purely longitudinal or purely shear waves. In acoustic transducers, microwave delay lines, amplifiers, and most other acoustic devices, propagation is usually chosen to be along one of the principal axes of the material.

We shall define the basic wave equation for acoustic propagation in the longitudinal case. The results obtained are identical in form to those for shear wave propagation. A more complicated treatment is needed only when considering propagation at an angle to one of the principal axes of the crystal. Therefore, we will carry out the initial derivations in a one-dimensional form. The reader is referred to Sec. 1.5, Chapter 2, and Appendix A for more general derivations of these relations with a tensor formalism [1].

Stress. The force per unit area applied to a solid is called the *stress*. In the one-dimensional case, we shall denote it by the symbol T. A force, applied to a solid, stretches or compresses it. We first consider a slab of material of infinitesimal length l, as shown in Fig. 1.1.2. Figure 1.1.2(a) illustrates the application of a longitudinal stress, and Fig. 1.1.2(b) illustrates the application of shear stress. The stress $T(z)$ is defined as the force per unit area on particles to the left of the plane z. We note that the longitudinal stress is defined as positive if the external stress applied to the right-hand side of the slab is in the $+z$ direction, while the external stress applied to the left-hand side of the slab is $-T$ in the $-z$ direction. If the stress is taken to be positive in the $+x$ or $+y$ directions, these

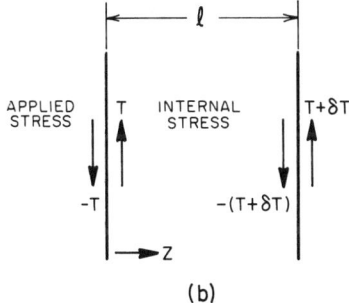

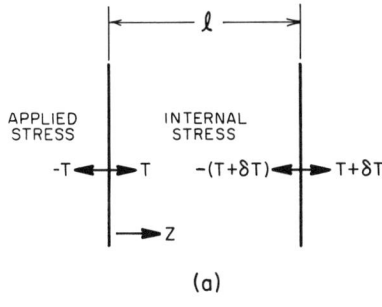

Figure 1.1.2 (a) Stress in the longitudinal direction for a slab of length l; (b) stress in the shear direction.

definitions also apply to shear stress. The net difference between the external stresses applied to each side of the slab is $l(\partial T/\partial z)$. Thus the net force applied to move a unit volume of the material relative to its center of mass is $\partial T/\partial z$.

Displacement and strain. Suppose that in the one-dimensional case, the plane z in the material is displaced in the z direction by longitudinal stress to a plane $z' = z + u$, as shown in Fig. 1.1.1(a). The parameter u is called the *displacement* of the material and in general is a function of z. At some other point in the material $z + l$, the displacement u changes to $u + \delta u$. If the displacement u is a constant throughout the entire material, the material has simply undergone a bulk translation. Such gross movements are of no interest to us here. We are interested in the variation of particle displacement as a function of z.

We can use a Taylor expansion to show that, to first order, the change in u in a length l is δu, where

$$\delta u = \frac{\partial u}{\partial z} l = Sl \tag{1.1.1}$$

The fractional extension of the material is defined as

$$S = \frac{\partial u}{\partial z} \tag{1.1.2}$$

The parameter S is called the *strain*.

We may also consider the one-dimensional case of shear motion where the material is displaced in the y direction by a wave propagating in the z direction. Then, as shown in Fig. 1.1.1(b), a particle in the position $\mathbf{a}_z z$ is displaced to a point $\mathbf{a}_z z + \mathbf{a}_y u$, where $\mathbf{a}_y$ and $\mathbf{a}_z$ are unit vectors in the y and z directions, respectively. The same treatment used for longitudinal motion holds for the shear wave case. We define a shear strain as $S = \delta u/l = \delta u/\delta z$. Here the only difference is that the displacement u is in the y direction. The diagram shows that there is no change in the area of the rectangle as shear motion distorts it. Longitudinal motion, however, changes the cube volume by $\delta u A$, where A is the area of the x, y face. Thus the relative change in volume is $\delta u/l = S$. The reader is referred to Appendix A for a more general definition of strain.

Hooke's law and elasticity. Hooke's law states that for small stresses applied to a one-dimensional system, the stress is proportional to the strain, or

$$T = cS \tag{1.1.3}$$

where c is the elastic constant of the material. The parameters T and c would be tensors in the general system, but can be represented by one component for one-dimensional longitudinal or shear wave propagation. Because it is easier to bend a solid than to stretch it, the shear elastic constant is normally smaller than the longitudinal elastic constant.

Equation of motion. Consider now the equation of motion of a point in the material when a small time-variable stress is applied to it. From Newton's second law, the net translational force per unit area applied to the material is $l\, \partial T/\partial z$, so the equation of motion must be

$$\frac{\partial T}{\partial z} = \rho_{m0}\ddot{u} = \rho_{m0}\dot{v} \qquad (1.1.4)$$

Here v is defined as the particle velocity of the material and ρ_{m0} is its mass density in the stationary state.

Conservation of mass. The particle velocity of the material is $v = \dot{u}$, so that in a small length l in the one-dimensional case, the change in velocity is

$$\delta v = \frac{\partial v}{\partial z} l \qquad (1.1.5)$$

However, from Eq. (1.1.1),

$$\delta v = \frac{\partial}{\partial t}\delta u = l\frac{\partial S}{\partial t} \qquad (1.1.6)$$

We can therefore combine Eqs. (1.1.5) and (1.1.6) to show that

$$\frac{\partial S}{\partial t} = \frac{\partial v}{\partial z} \qquad (1.1.7)$$

Equation (1.1.7) is essentially another way of writing the equation of conservation of mass of the material for longitudinal waves. The same equation holds for shear waves; however, as there is no change of density with shear motion, the equation of conservation of mass is not implied. Suppose that we write $\rho_m = \rho_{m0} + \rho_{m1}$, where ρ_{m0} is the unperturbed density and ρ_{m1} is the perturbation in the density. We can write the one-dimensional equation of conservation of mass for longitudinal waves in the form

$$\frac{\partial}{\partial z}\rho_m v + \frac{\partial \rho_m}{\partial t} = 0 \qquad (1.1.8)$$

Assuming that v and ρ_{m1} are first-order perturbations, and keeping only first-order terms, it follows that

$$\rho_{m0}\frac{\partial v}{\partial z} + \frac{\partial \rho_{m1}}{\partial t} = 0 \qquad (1.1.9)$$

where $\rho_m = \rho_{m0} + \rho_{m1} = \rho_{m0}/(1 + S) \approx \rho_{m0}(1 - S)$. Thus

$$\frac{\partial v}{\partial z} = -\frac{\partial}{\partial t}\left(\frac{\rho_{m1}}{\rho_{m0}}\right) = \frac{\partial S}{\partial t} \qquad (1.1.10)$$

The wave equation and definition of propagation constant. We may now use Eqs. (1.1.3), (1.1.4), and (1.1.7), and put $v = \partial u/\partial t$, to obtain the small signal wave equation for sound wave propagation in the material:

$$\frac{\partial^2 T}{\partial z^2} = \rho_{m0}\frac{\partial^2 S}{\partial t^2} = \frac{\rho_{m0}}{c}\frac{\partial^2 T}{\partial t^2} \qquad (1.1.11)$$

The solutions of this wave equation for the stress are of the form $F(t \pm z/V_a)$. For a wave of radian frequency ω, with all field quantities varying as $\exp(j\omega t)$, the solutions are of the form $\exp[j(\omega t \pm \beta_a z)]$, where the negative sign in the exponential corresponds to a forward wave, the positive sign corresponds to a backward wave, and

$$\beta_a = \omega \left(\frac{\rho_{m0}}{c}\right)^{1/2} = \frac{\omega}{V_a} \qquad (1.1.12)$$

and $V_a = (c/\rho_{m0})^{1/2}$ is the acoustic wave velocity. The parameter β_a is called the *propagation constant* of the acoustic wave.

For propagation in the forward direction, it follows that

$$v = -V_a S = -\frac{V_a}{c} T \qquad (1.1.13)$$

The magnitude of the strain S is also the ratio of the particle velocity of the medium to the sound wave velocity of the medium.

Tables of the relevant parameters for longitudinal and shear wave propagation for various types of commonly used materials are given in Appendix B. A liquid such as water, for example, is relatively easy to compress and has a small mass density (1000 kg/m³). Its elastic constant is 2.25×10^9 N/m², so the acoustic wave velocity in water is 1.5 km/s. Sapphire has a larger mass density than water (3990 kg/m³), but it is a rigid material with a relatively large longitudinal elastic coefficient (4.92×10^{11} N/m²). Therefore, it has a high longitudinal wave velocity along the z axis of 11.1 km/s. Most metals have longitudinal wave velocities on the order of 5 km/s, with shear wave velocities approximately half this value. There are exceptions, such as beryllium, which is extremely light (1870 kg/m³) and rigid and therefore has a longitudinal wave velocity of 12.9 km/s. Lead, which is very heavy (11,400 kg/m³), has a low longitudinal wave velocity of 1.96 km/s. However, the acoustic wave velocity in a gas is small because it is relatively easy to compress. Thus the acoustic wave velocity in air is 330 m/s. As liquids and gases cannot support shear stresses, shear waves cannot propagate through them.

Energy. The total stored energy per unit volume in the medium is the sum of two components: (1) the elastic energy per unit volume due to the force applied to displace the material, $W_c = \frac{1}{2} TS$; and (2) the kinetic energy per unit volume due to the motion of the medium, $W_v = \frac{1}{2}\rho_{m0} v^2$. For a propagating plane wave whose components vary as $\exp(j\omega t)$, following the analogy for electromagnetic (EM) waves, the average elastic energy per unit volume is

$$W_c = \tfrac{1}{4}\operatorname{Re}(TS^*) = \tfrac{1}{4}\operatorname{Re}(cSS^*) \qquad (1.1.14)$$

and the average kinetic energy per unit volume is†

$$W_v = \tfrac{1}{4}\operatorname{Re}(\rho_{m0} vv^*) \qquad (1.1.15)$$

†Suppose that we put $A = A_0 \exp(j\omega t)$ and $B = B_0 \exp[j(\omega t + \phi)]$. Then Re $A \times$ Re B is the average over one radio-frequency (RF) cycle of $\langle A_0 B_0 \cos \omega t \cos(\omega t + \phi)\rangle = \tfrac{1}{2}(A_0 B_0 \cos \phi)$. This quantity is just $\tfrac{1}{2}\operatorname{Re}(AB^*)$.

It follows from Eqs. (1.1.12)–(1.1.15) that $W_c = W_v$ and that the total energy per unit volume in an acoustic wave is

$$W_a = \tfrac{1}{4}\text{Re}\,(\rho_{m0}vv^* + TS^*) = \tfrac{1}{2}\text{Re}\,(TS^*) \tag{1.1.16}$$

Similarly, the power flow per unit area in the acoustic wave may be defined as the product of the force per unit area $-T$ applied by the material on the left-hand side of the plane z, and the material velocity (i.e., the average value of $-vT$ during the RF cycle). The complex power flow through an area A may therefore be defined as

$$P_a = -\tfrac{1}{2}(v^*T)A \tag{1.1.17}$$

For a propagating wave in a lossless medium, v and T are in phase, so P_a is real and

$$P_a = V_a W_a A = -\tfrac{1}{2}(v^*T)A \tag{1.1.18}$$

Poynting's theorem. The power and energy in an acoustic wave obey conservation laws similar to those of Poynting's theorem in EM theory [1, 2]. To prove the conservation laws, we write Eq. (1.1.4) in the form

$$\frac{\partial T}{\partial z} = j\omega\rho_{m0}v \tag{1.1.19}$$

and Eq. (1.1.10) in the form

$$\frac{\partial v}{\partial z} = j\omega S = \frac{j\omega T}{c} \tag{1.1.20}$$

By multiplying Eq. (1.1.19) by v^* and the complex conjugate of Eq. (1.1.20) by T, and adding the results, it follows that

$$-2\frac{\partial}{\partial z}\text{Re}\,(P_a) = A\frac{\partial}{\partial z}\text{Re}\,(Tv^*) = (j\omega\rho_{m0}vv^* - j\omega TS^*)A \tag{1.1.21}$$

Writing $T = cS$, and assuming that c is real, we find that

$$-2\frac{\partial}{\partial z}\text{Re}\,(P_a) = (j\omega\rho_{m0}vv^* - j\omega cSS^*)A \tag{1.1.22}$$

As both terms on the right-hand side of Eq. (1.1.22) are imaginary, it follows that

$$\frac{\partial}{\partial z}\text{Re}\,(P_a) = 0 \tag{1.1.23}$$

This relation is entirely analogous to Poynting's theorem in EM theory and shows that Re (P_a) is constant (i.e., the power in the wave is conserved). For a propagating wave, since P_a itself is real, the average stored kinetic energy per unit volume is equal to the average elastic energy, or

$$\rho_{m0}vv^* = cSS^* \tag{1.1.24}$$

Acoustic losses. When the system can dissipate energy so that it is not purely elastic, and c is complex, there is a loss term just like that due to conduction current in EM theory. The viscous forces between neighboring particles with different velocities are a major cause of acoustic wave attenuation in solids and liquids. These are additional viscous stresses T_η on the particles in a medium through which a plane wave is propagating of the form

$$T_\eta = \eta \frac{\partial v}{\partial z} = \eta \frac{\partial S}{\partial t} \tag{1.1.25}$$

where the coefficient of viscosity is called η. It follows from Eqs. (1.1.3) and (1.1.25) that the total stress is

$$T = cS + \eta \frac{\partial S}{\partial t} \tag{1.1.26}$$

The equation of motion has the same form as for a lossless medium:

$$\frac{\partial T}{\partial z} = \rho_{m0} \frac{\partial v}{\partial t} \tag{1.1.27}$$

Just as in EM theory, the attenuation due to loss can be calculated approximately by using Poynting's theorem. Following the derivation of Eqs. (1.1.22) and (1.1.23), and assuming that all field quantities vary as $\exp(j\omega t)$, Poynting's theorem in the complex form now has an additional term associated with viscosity:

$$\frac{\partial}{\partial z} \text{Re}(P_a) = -\tfrac{1}{2}\eta\omega^2 SS^* A \tag{1.1.28}$$

where A is the area of the beam. Thus viscosity gives rise to a loss term that varies as the square of the frequency. If we regard this loss term as small, it follows from Eqs. (1.1.16) and (1.1.18) that

$$\tfrac{1}{2}cSS^* \approx W_a \approx \frac{P_a}{V_a A} \tag{1.1.29}$$

Hence, after substitution in Eq. (1.1.28), writing P_a for $\text{Re}(P_a)$, we see that

$$\frac{1}{P_a}\frac{\partial P_a}{\partial z} \approx \frac{-\eta\omega^2}{V_a c} = \frac{-\eta\omega^2}{V_a^3 \rho_{m0}} \tag{1.1.30}$$

The solution of this equation is $P_a = P_0 \exp(-2\alpha z)$, where α is the attenuation constant of the wave and P_0 is a constant. The attenuation constant is given by the relation

$$\alpha = \frac{\eta\omega^2}{2V_a^3 \rho_{m0}} \tag{1.1.31}$$

Thus the attenuation of the wave due to viscous losses varies as the square of the frequency and inversely as the cube of the velocity. As shear waves typically have velocities of the order of half those of longitudinal waves in the same material, we might expect the shear wave attenuation per unit length to be considerably larger than the longitudinal wave attenuation per unit length, although this is not always so in practice. In water at room temperature, the attenuation is 0.22 dB/m at 1 MHz; thus low-frequency waves can propagate over long distances in water. At 1 GHz, however, the attenuation is estimated to be 2.2×10^5 dB/m, or 0.22 dB/μm. Therefore, for such applications as the acoustic microscope, it is important to limit the propagation path of the waves in water to the order of tens of micrometers.

There are many other sources of loss in real materials. One is thermal conduction. When a material is compressed adiabatically, its temperature increases; its temperature decreases when the material expands. Since thermal conduction causes the process to be nonadiabatic and contributes to a loss of energy, it tends to give higher attenuations in metals than in insulators. The attenuation due to thermal conduction also varies as the square of the frequency. In addition, attenuation may exist because waves have been scattered by finite-size grains or by dislocations in a solid; another cause of attenuation is the unequal thermal conduction and expansion of neighboring grains due to their axes being rotated with respect to each other. For these reasons, high-quality single-crystal materials exhibit much lower sound attenuation at high frequencies than do the same materials in a polycrystalline form.

Loss mechanisms also result from sound-induced changes of state in a solid or a liquid such as water. Water molecules are thought to exhibit a partial crystalline structure. A sound wave loses energy when it disturbs this structure, which creates a further loss mechanism. Such mechanisms can be a powerful tool for measuring chemical changes, and the relaxation times associated with them, in solids and liquids. A further discussion of grain scattering losses is given in Sec. 3.6.

The acoustic attenuation in common materials varies over a very wide range. Thus a hard, high-quality, single-crystalline material such as sapphire can be used for acoustic delay lines at frequencies up to 10 GHz. The attenuation at this frequency may be as much as 40 dB/cm. However, the wavelength at this frequency is of the order of 1 μm. Thus the attenuation per wavelength is 4×10^{-3} dB, a lower loss than is typical for EM waves in a waveguide. On the other hand, viscous materials such as rubber, exhibit relatively high losses at frequencies greater than a few kilohertz, and thus make good sound absorbers.

Acoustic impedance. In analogy to EM theory or transmission line theory, we can define an acoustic impedance Z, which we more formally call the *specific acoustic impedance*. We write

$$Z = -\frac{T}{v} \tag{1.1.32}$$

For a plane wave traveling in the forward direction, which we denote by subscript F, we define the *characteristic impedance* Z_0 as

$$Z_0 = -\frac{T_F}{v_F} = (\rho_{m0} c)^{1/2} = V_a \rho_{m0} \tag{1.1.33}$$

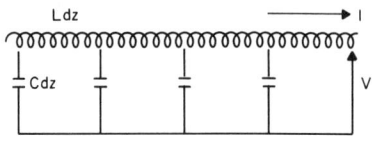

Figure 1.1.3 Transmission line with a series inductance L per unit length and shunt capacitance C per unit length.

while the impedance Z for a wave traveling in the backward direction, denoted by subscript B, (for which β_a is negative) is defined as

$$Z = -\frac{T_B}{v_B} = -V_a \rho_{m0} = -Z_0 \qquad (1.1.34)$$

Both the impedances defined here have the dimensions of

$$\frac{\text{pressure}}{\text{velocity}} = \frac{\text{N/m}^2}{\text{m/s}} = \frac{\text{N}}{\text{m}^3/\text{s}} = \frac{\text{kg}}{\text{m}^2\text{-s}}$$

With the use of Eq. (1.1.32), Eqs. (1.1.19) and (1.1.20) take the standard form of the transmission line equations and may be be used in the same way.

To see this result, we compare Eqs. (1.1.19) and (1.1.20) with the transmission-line equations for the voltage and current along a transmission line, which are

$$\frac{\partial V}{\partial z} = -j\omega L I \qquad (1.1.35)$$

and

$$\frac{\partial I}{\partial z} = -j\omega C V \qquad (1.1.36)$$

where L is the series inductance per unit length and C is the shunt capacitance per unit length, as illustrated in Fig. 1.1.3. A wave propagating along this line has a propagation constant $\beta = \omega\sqrt{LC}$ and an impedance $Z_0 = \sqrt{L/C}$.

Suppose that we replace V by $-T$ and I by v. Then Eqs. (1.1.19) and (1.1.20) are equivalent to Eqs. (1.1.35) and (1.1.36), respectively, provided that we replace L with ρ_{m0} and C with $1/c$ and define a characteristic acoustic impedance as in Eq. (1.1.33). It follows that we can use these equations just as we did in EM theory [1, 2].

It is convenient to define the reflection coefficient of an acoustic wave in terms of the stress. This gives a relation exactly equivalent to the one used in electromagnetic theory. As we shall see, such relations are useful in defining equivalent circuits, because piezoelectric transducers convert electromagnetic energy to acoustic energy.

If we consider a wave reflected normally from the interface between two media of differing characteristic impedances Z_{01} and Z_{02}, as shown in Fig. 1.1.4(a), then T and v will be continuous at the interface. We therefore write the stress $\hat{T}_1(z)$ and velocity $\hat{v}_1(z)$ of the left-hand side of the interface $z = 0$ as

$$\hat{T}_1 = T_{F1}e^{-j\beta_1 z} + T_{B1}e^{j\beta_1 z} \qquad (1.1.37)$$

and

$$\hat{v}_1 = v_{F1}e^{-j\beta_1 z} + v_{B1}e^{j\beta_1 z} \qquad (1.1.38)$$

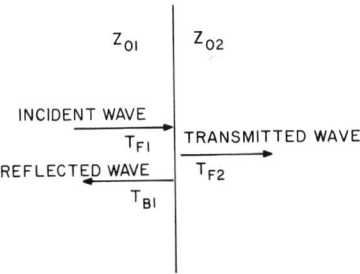

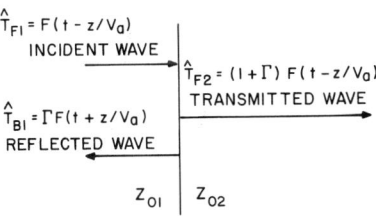

Figure 1.1.4 (a) Reflection of a continuous wave at the interface between two boundaries; (b) reflection of a pulsed signal at the interface of the two boundaries. The symbol ^ represents a quantity that varies with z.

where T_{F1} and T_{B1} are the amplitudes of the forward and backward waves on the left-hand side of $z = 0$, respectively, and the symbol ^ denotes a quantity that varies with z. We may also define the reflection coefficient Γ of the wave at the plane $z = 0$ as the ratio of the amplitude of the backward to the forward stress components, or as

$$\Gamma = \frac{T_{B1}}{T_{F1}} \quad (1.1.39)$$

Furthermore, using Eqs. (1.1.33) and (1.1.34) to write the velocities in terms of the stress, we can substitute into Eqs. (1.1.37) and (1.1.38) and write

$$\hat{T}_1 = T_{F1}(e^{-j\beta_1 z} + \Gamma e^{j\beta_1 z}) \quad (1.1.40)$$

and

$$\hat{v}_1 = -\frac{T_{F1}}{Z_{01}}(e^{-j\beta_1 z} - \Gamma e^{j\beta_1 z}) \quad (1.1.41)$$

respectively.

In the region to the right of the interface, there is only an excited wave propagating in the forward direction. Thus we can write

$$\hat{T}_2 = T_{F2} e^{-j\beta_2 z} \quad (1.1.42)$$

and

$$\hat{v}_2 = -\frac{T_{F2}}{Z_{02}} e^{-j\beta_2 z} \quad (1.1.43)$$

where T_{F2} is the amplitude of the stress on the right-hand side of $z = 0$. The boundary condition at the plane $z = 0$ is that the stress T and the velocity v must be continuous. This leads to the result

$$\Gamma = \frac{Z_{02} - Z_{01}}{Z_{02} + Z_{01}} \tag{1.1.44}$$

The stress transmission coefficient $\mathcal{T}_T$ is defined as

$$\mathcal{T}_T = \frac{T_{F2}}{T_{F1}} = 1 + \Gamma = \frac{2Z_{02}}{Z_{02} + Z_{01}} \tag{1.1.45}$$

and the power transmission coefficient $\mathcal{T}_P$ is defined as

$$\mathcal{T}_P = \frac{P_{F2}}{P_{F1}} = \frac{T_{F2}^2/Z_{02}}{T_{F1}^2/Z_{01}} = 1 - |\Gamma|^2 \tag{1.1.46}$$

This parameter determines how efficiently power is transmitted from one medium to another. Thus it is normally desirable to keep the reflection coefficient Γ as small as possible.

Note that the mismatches that can occur between different acoustic media are generally much more severe than they are in the electromagnetic case. For example, a rigid material such as sapphire has an impedance of $Z_0 = 44.3 \times 10^6$ kg/m²-s for longitudinal waves, while a heavy material such as lead, with a large mass density ρ_{m0}, also tends to have a relatively high impedance. On the other hand, water has an impedance of $Z_0 = 1.5 \times 10^6$ kg/m²-s, and air has an extremely low impedance, which for our present purposes can be considered to be zero.

It is obviously important to provide good matches between different media. Quarter-wave matching sections can be used to match a material of one impedance to another, just as they are with optical lenses or microwave transmission lines.

Let us consider the general case when a layer of impedance Z_0, propagation constant β, and thickness l is placed in contact with a medium that presents a load impedance Z_L. If this medium were semi-infinite, Z_L would in fact be the impedance $Z_L = Z_{02}$ of the medium. It follows from Eqs. (1.1.40) and (1.1.41) that if the load Z_L is at $z = 0$, the value of the input impedance $Z_{\text{in}} = -T(-l)/v(-l)$ of the layer of length l is given by the expression

$$Z_{\text{in}} = Z_0 \frac{e^{j\beta l} + \Gamma e^{-j\beta l}}{e^{j\beta l} - \Gamma e^{-j\beta l}} \tag{1.1.47}$$

where β is the propagation constant in the layer. It follows from Eqs. (1.1.44) and (1.1.47) that

$$Z_{\text{in}} = Z_0 \frac{Z_L \cos \beta l + jZ_0 \sin \beta l}{Z_0 \cos \beta l + jZ_L \sin \beta l} \tag{1.1.48}$$

By using an intermediate layer, the input impedance can be changed to a different value. In particular, if the matching layer is $\lambda/4$ thick (i.e., $\beta l = \pi/2$), then

$$Z_{\text{in}} = \frac{Z_0^2}{Z_L} \tag{1.1.49}$$

Thus matching can be obtained between two media of widely different impedances by choosing an intermediate matching layer properly, although perfect matching occurs only at one frequency. Typically, the larger the ratio of impedances Z_{in}/Z_L or Z_L/Z_{in}, the narrower the bandwidth of the impedance match.

Techniques also exist for matching with multiple quarter-wavelength layers to improve the bandwidth. The reader is referred to the considerable literature on this subject which has been developed to deal with the problem of matching transmission lines [2–5].

Another important case is the reflection of an acoustic pulse by an interface. We consider a stress pulse of value $\hat{T}_{F1}(t, z) = F(t - z/V_a)$ incident on the interface $z = 0$, as illustrated in Fig. 1.1.4(b). This will give rise to a reflected pulse of the same form, $\hat{T}_{B1}(t, z) = \Gamma F(t + z/V_a)$. Thus we can write the total stress T_1 in medium 1 as

$$\hat{T}_1 = F\left(t - \frac{z}{V_a}\right) + \Gamma F\left(t + \frac{z}{V_a}\right) \tag{1.1.50}$$

with a corresponding velocity v_1 of value

$$\hat{v}_1 = -\frac{1}{Z_{01}}\left[F\left(t - \frac{z}{V_a}\right) - \Gamma F\left(t + \frac{z}{V_a}\right)\right] \tag{1.1.51}$$

By following exactly the same procedure as before, we arrive at the same value of the reflection coefficient given by Eq. (1.1.44).

Suppose that the incident stress is in the form of a square pulse, as illustrated in Fig. 1.1.5. If $Z_{01} > Z_{02}$, as illustrated in Fig. 1.1.5(a), the reflection coefficient is negative, so the reflected stress pulse has the opposite sign from that of the incident pulse. In particular, at an air interface with $Z_{02} = 0$, the total stress at the interface $z = 0$ will be zero, and $\Gamma = -1$. At all events, the total stress at the interface will be less than the incident stress if $Z_{01} > Z_{02}$. On the other hand, it follows from Eq. (1.1.34) that because the velocity associated with the return echo is of the same sign as the velocity of the stress, the total velocity at the interface $z = 0$ is increased and doubled at an air interface. In this case, the reflected velocity pulse then has the same sign as the incident velocity pulse. On the other hand, if $Z_{01} < Z_{02}$, as illustated in Fig. 1.1.5(b), the reflected stress pulse has the same sign as the incident pulse but the velocity is reversed in sign. Furthermore, as might be expected, the stress at the interface is larger than that of the incident wave; in fact, if the second layer is perfectly rigid so that $Z_{02} = \infty$, then $\Gamma = 1$ and the stress is doubled in value at the time the pulse reaches the interface.

Shear waves. The analysis for shear wave propagation is exactly the same as that for longitudinal waves. Now the stress may be taken to have components in both the y and z directions, with propagation in the z direction. As an example, an isotropic material may be displaced in the y direction, and the displacement u_y will be a function of z. The shear motion of particles in the material is thus like a rotation about the x axis. Now the shear strain t is defined by the one-dimensional

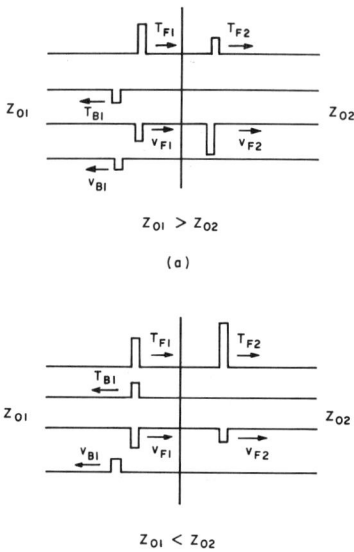

Figure 1.1.5 Acoustic pulse incident on the interface between two media: (a) $Z_{01} > Z_{02}$; (b) $Z_{01} < Z_{02}$.

parameter $S = \partial u_y/\partial z$. The derivation of the wave equation for shear wave propagation follows as before, the only difference being that the appropriate elastic coefficient relating shear stress and strain must be used. This leads to sound wave velocities and wave impedances, which are typically smaller than for longitudinal waves in the same material, as discussed in Sec. 2.2. With these restrictions, the energy conservation theorem and expression for attenuation may be derived for shear waves and will have the same forms as the ones for longitudinal waves [1].

Extensional waves. Another type of wave of great practical importance, which can be treated by the simple one-dimensional theories given here, is the *extensional wave* in a thin rod. In this mode, illustrated in Fig. 1.1.6, it is assumed that the cross-sectional dimensions of the rod, or in the case of a cylindrical rod, its diameter, are much smaller than the wavelength of the wave of interest. In this case, the only stress component present is a longitudinal one in the direction of propagation along the rod. There can be no stress components within the rod perpendicular to its axis, because the normal stress must be zero at its surface. When a wave propagates along the rod, there is particle motion both along the rod and perpendicular to it, for the rod is free to expand its cross-sectional area when it is compressed in the z direction, and vice versa. However, the component of strain $S = \partial u_z/\partial z$ associated with the motion in the z direction is directly related by Hooke's law [Eq. (1.1.3)] to the stress T in the z direction, but with a different effective coefficient of elasticity. The relevant coefficient of elasticity is called *Young's modulus*, which we denote by the symbol E, thus writing $T = ES$. Young's modulus is the coefficient of elasticity normally measured when a long wire or rod is stretched by a static force. In all cases Young's modulus is smaller than the longitudinal coefficient of elasticity, because a larger force is required to stretch a

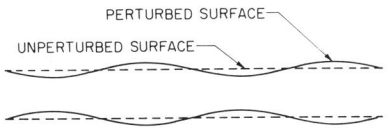

Figure 1.1.6 Extensional wave in a rod or strip.

slab of material with the same strain S when the slab is not free to expand or contract in the transverse direction.

The general case is far more complicated, for the stress and strain must be tensors of rank 2 and may contain both shear and longitudinal components. In general, the solution of acoustic problems tends to be more complicated than for the equivalent electromagnetic problems, because the stress and strain terms are tensors rather than vectors. The formalism required to deal with the general case of propagation of an acoustic wave in an arbitrary direction in an anisotropic material is given in Appendix A. The techniques developed there are applied to propagation of waves in finite isotropic media in Chapter 2. Fortunately, in isotropic materials, it is always possible to resolve any waves present into plane shear wave and longitudinal wave components. In a finite isotropic medium, however, both types of fields are required to satisfy the boundary conditions.

PROBLEM SET 1.1

1. (a) The correct formula for attenuation due to viscosity in water was derived by Lord Rayleigh, who showed that the result in Eqs. (1.1.26) and (1.1.28) should have η replaced by $\frac{4}{3}\eta_s$, where η_s is the shear viscosity of water. Assuming that $\eta_s = 0.01$ poise (dyn-s/cm²) for water at room temperature, calculate the attenuation at 1 MHz, and compare your result with the measured value of 0.0022 dB/cm at room temperature. Your results for power attenuation will be lower than the measured result because not all contributions to loss have been taken into account.
   (b) Suppose that we want to construct an acoustic microscope working at 1 GHz to examine a biological specimen suspended in water. Find the wavelength of sound in water ($V_a = 1.5$ km/s) at 1 GHz. From the experimental value of loss at 1 MHz, estimate the maximum path length that could be used at 1 GHz if the attenuation had to be less than 80 dB.

2. A longitudinal wave is excited in a sapphire rod that is immersed in water. Use the data of the tables in Appendix B. Assuming that the water is lossless, find the reflection coefficient at the sapphire–water interface. What proportion of the power is reflected and transmitted at the sapphire–water interface? Despite the mismatch, this combination of materials is convenient to use in an acoustic microscope. Suppose that you wanted to improve the efficiency by using a quarter-wavelength matching section of impedance $Z = (Z_{01}Z_{02})^{1/2}$, where Z_{01} is the impedance of the sapphire and Z_{02} is that of water. What would the impedance of this matching material have to be? As you will see from the tables in Appendix B, very few, if any, materials with this impedance exist. Can you suggest a compromise solution or solutions using two layers, one of the layers having a higher impedance than sapphire, that might at least improve the transmission efficiency? Basically, the lowest-impedance common material that can be deposited in the form of a thin film is glass, with an impedance of the order of 12×10^6

kg/m²-s. What are the advantages and disadvantages of such a two-layer solution compared to using a single matching layer of impedance $(Z_{01}Z_{02})^{1/2}$?

3. (a) We want to design an acoustic transducer that can emit a pulsed signal with as great a transmission efficiency as possible into water. Because the signal required is a short pulse, matching over a broad frequency band is desirable. Thus a single quarter-wavelength matching layer might not be the best choice. Another possible choice is to use a long buffer rod between the transducer and the water. Suppose that the transducer were made of PZT with an impedance of 34×10^6 kg/m²-s. Then the signal emitted from the transducer into the rod would suffer a certain transmission loss, and there would be further transmission loss from the rod to the water. What would be the best choice of impedance for optimum transmission efficiency from the transducer to the water? Ignore multiple echoes.

   (b) Consider a second situation where a matching layer a quarter-wavelength long at the center frequency is employed. In this case, assume that a short tone burst $A(t)$ sin ωt, two RF cycles long, is employed [$A(t) = 0$ when $\omega t < 0$, $A(t) = 0$ when $\omega t > 4\pi$, and $A(t) = 1$ when $0 < \omega t < 4\pi$]. The return echoes from the two interfaces tend to cancel out if the RF frequency is chosen correctly. Sketch the form of the return echo for a sinusoidal pulse two RF cycles long, chosen with its frequency to be at the center frequency of the quarter-wavelength matching layer. Choose the impedance of the matching layer to be $Z_0 = (Z_{01}Z_{02})^{1/2}$, where Z_{01} and Z_{02} are the impedances of the transducer material and the load, which are both regarded as semi-infinite in extent.

4. (a) An acoustic transducer emits a power of 1 W/cm² into water at a frequency of 1 MHz. Determine the peak RF stress or pressure in the acoustic beam and the maximum displacement of the water. Consider what happens when the RF pressure is negative (i.e., during the negative half of the RF cycle). In this case, a vacancy or a bubble would form. Normally, this does not occur at low acoustic levels of power because of the finite pressure due to gravity and the atmosphere. Consider a laboratory water tank with a transducer located near its top surface, so that only atmospheric pressure need be considered. At what power density would this effect, known as *cavitation*, occur?

   Note: In comparison with experiments, your result will predict very low power for the onset of cavitation; neither will it vary with frequency, as it does in practice. This is because viscosity and surface tension have been neglected in this example.

   (b) Now consider sapphire, a solid material with an acoustic impedance of 44.3×10^6 kg/m²-s. Suppose that its breaking strain is $S = 5 \times 10^{-3}$. Estimate the power density required to fracture it. Your estimate in part (a) was really equivalent to the one for a liquid.

   Note: This effect can cause damage by a high-power laser beam passing through a transparent solid, due to the excitation of acoustic waves (phonons) by the laser beam (the Raman effect).

5. Consider the effect of a thin bond on the transmission of an acoustic wave between two identical materials of impedance Z_0. Work out an approximate formula for the reflection coefficient through a bond material of impedance Z_1 and length l such that $\beta l \ll 1$. Show that if $Z_1 \ll Z_0$, even though $\beta l \ll 1$, there can be a serious reflection due to the bond. As an example, consider two layers of PZT-5A with $Z_0 = 34 \times 10^6$ kg/m²-s, bonded with an epoxy layer of impedance $Z_1 = 3.24 \times 10^6$ kg/m²-s and $V_a = 2.67$ km/s that is 2 μm thick. At what frequency does the magnitude of the reflection coefficient become 0.5? What is the ratio of the bond length to the wavelength in the bond at this frequency?

6. Suppose that a uniform, static tensile stress of 300 MPa (1 MPa = 10^6 N/m²-s) were applied to an aluminum rod. What would be the strain? How much does the effective density of the material decrease? Assuming that there is no change in the elastic constant, estimate the relative change in acoustic wave velocity along the rod. Suppose that the rod were 5 cm long. What would be the relative change in phase? How much phase change would there be in a 10-MHz wave passing through the total expanded length of the rod? This idea forms the basis of an acoustic technique for measuring stress, although the relative change of the effective elastic constant may be two or three times larger than the relative change in density.

1.2 PIEZOELECTRIC MATERIALS

1.2.1 Constitutive Relations

A piezoelectric material has an asymmetric atomic lattice. When an electric field is applied to such a material, it changes its mechanical dimensions. Conversely, an electric field is generated in a piezoelectric material that is strained. All ferroelectric materials are piezoelectric. In the ferroelectric state, the center of positive charge of the crystal does not coincide with its center of negative charge, as illustrated in Fig. 1.2.1. In such a material two neighboring atoms that are not identical will not move the same distances in an applied electric field. Therefore, the dimensions of a ferroelectric material will change (i.e., the material will be strained when an electric field is applied to it).

Consider the periodic atomic system illustrated in Fig. 1.2.1, in which the equilibrium spacings between neighboring rows of atoms in the z direction are a_1 and a_2, and the spacings between neighboring rows of atoms in the x and y directions

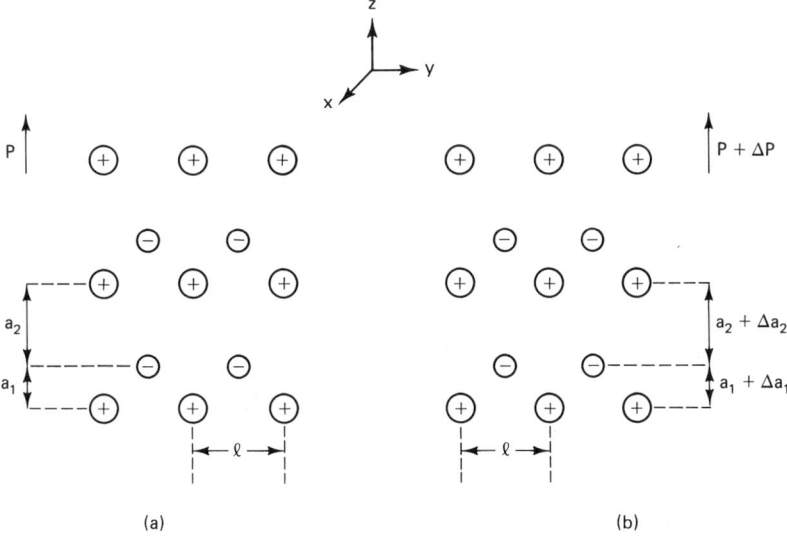

Figure 1.2.1 Ferroelectric crystals: (a) unstrained; (b) strained.

is l. The dipole moment per unit volume of these atoms is

$$P = \frac{q(a_2 - a_1)}{l^2(a_2 + a_1)} = \frac{\text{dipole strength of unit cell}}{\text{volume of unit cell}} \quad (1.2.1)$$

where the charges on the atoms are q and $-q$, respectively.

We now consider the effect of strain on the polarization of the material. When a_1 changes to $a_1 + \Delta a_1$ and a_2 changes to $a_2 + \Delta a_2$, the polarization changes by ΔP, and we can write $\Delta a_1 = a_1 S$ and $\Delta a_2 = a_2 S$. It follows from Eq. (1.2.1) that to first order in strain,

$$\Delta P = PS = eS \quad (1.2.2)$$

where the parameter e is called the piezoelectric stress constant, and S is the macroscopic strain in the material.

The total change in electric displacement in the presence of an electric field is

$$D = \varepsilon E + \Delta P \quad (1.2.3)$$

We can also write

$$D = \varepsilon^S E + eS \quad (1.2.4)$$

where the dielectric constant is the permittivity with zero or constant strain and is thus denoted by ε^S. The electric displacement in a piezoelectric material depends on the strain as well as the electric field.

We shall now determine the stress in a piezoelectric medium due to an electric field E. The forces per unit area on the positive and negative atoms are qE/l^2 and $-qE/l^2$, respectively. The stresses in the regions of length a_2 and a_1 are therefore

$$T_2 = \frac{qE}{l^2} \quad (1.2.5)$$

and

$$T_1 = \frac{-qE}{l^2} \quad (1.2.6)$$

respectively. Therefore, the average stress in the medium due to the electric field is

$$T_E = \frac{a_1 T_1 + a_2 T_2}{a_1 + a_2} = eE \quad (1.2.7)$$

The total stress applied to the medium is the sum of the externally applied stress T and the internal stress T_E due to the electric field. The application of Hooke's law leads to the result

$$T + T_E = c^E S \quad (1.2.8)$$

or

$$T = c^E S - eE \quad (1.2.9)$$

where we have defined the elastic constant in the presence of a constant or zero E field to be c^E. Equations (1.2.4) and (1.2.9) are known as the *piezoelectric constitutive relations*.

We can estimate the value of e by taking $a_1 = 2$ Å, $a_2 = 4$ Å, and $l = 3$ Å, and taking q to be the charge of a single electron; these assumptions yield a value of $e = 0.6$ C/m². Because the atoms can have multiple charges, this figure is considerably lower than the largest measured values of e in strongly piezoelectric materials such as the lead zirconium titanate (PZT) ceramics ($e_{z3} = 23.3$ C/m² in PZT-5H) or lithium niobate crystals, but it is comparable to the values observed in many other piezoelectric materials.

Poling and domains in ferroelectric materials. Ferroelectric materials such as the PZT ceramics and lithium niobate, which are important piezoelectric materials, exhibit a dipole moment. Above a certain temperature, known as the *Curie point*, the dipole directions have random orientations. The dipoles may be aligned by applying a strong electric field at a temperature near the Curie point; this process is known as *poling*. As illustrated in Fig. 1.2.2, an unpoled ferroelectric material is normally divided into macroscopic regions, called *domains*, in which the dipoles are aligned. The alignment of the dipoles in these separate domains is, however, random with respect to those in the other domains. Ideally, after poling, all the domains are aligned with each other.

A crystal such as quartz can be piezoelectric without being ferroelectric. For example, a crystal with a threefold symmetry axis, as illustrated in Fig. 1.2.3, represents three dipoles aligned at 120° to each other. The sum of the dipole moments at each vertex is zero. When an electric field is applied in the z direction, the three dipoles tend to expand or contract by different distances in the z direction; thus the crystal can develop a net stress. Similarly, when the material is strained in the z direction, it can develop a net dipole moment, so it is a piezoelectric

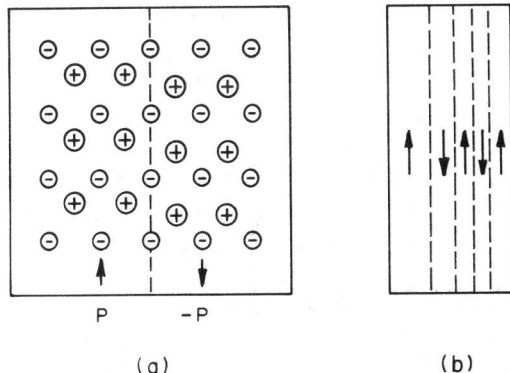

Figure 1.2.2 Displacement of the atoms on either side of a domain boundary: (a) unpolarized medium with domains on either side of the boundary polarized in opposite directions; (b) domain structure showing neighboring domains polarized in opposite directions.

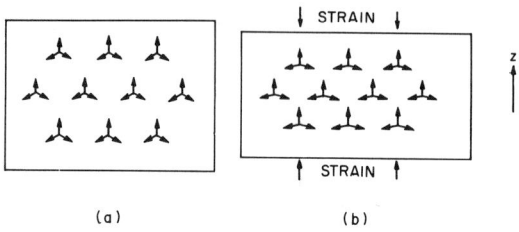

Figure 1.2.3 Piezoelectric crystal that is not ferroelectric. (a) The unstressed crystal has a threefold symmetry axis. The three arrows represent a planar group of ions with a triple charge, a positive charge at each vertex, and single negative charges at the arrowheads. The sum of the three dipole moments is zero. (b) When the sample is stressed, so that the sum of the three dipole moments is finite, there is no longer threefold symmetry.

material. Such piezoelectric materials have much smaller piezoelectric stress constants than do ferroelectric materials.

The argument we have given here also applies in the presence of shear strain, which means that an electric field applied in the x or y directions would create shear stress. We shall see in Sec. 1.5 that it is possible to generalize the definitions of the elastic constants and piezoelectric coupling parameters to take both longitudinal and shear stress and strain components into account.

There are other possible ways in which an electric field can generate stress. In any dielectric, there is an electrostrictive stress applied to the material. In a liquid, its value is $\frac{1}{2}(\varepsilon - \varepsilon_0 + S \, \partial\varepsilon/\partial S)E^2$ [6]. For simplicity, we shall neglect the third term in this expression. When a dc field E_0 is applied to a dielectric, a first-order perturbing field E_1 yields $E^2 = (E_0 + E_1)^2 \approx E_0^2 + 2E_0 E_1$. So a small first-order component of stress will be generated, of value $T_1 \approx (\varepsilon - \varepsilon_0)E_0 E_1$. Taking fused quartz as an example, if we use fields comparable to the breakdown strength of quartz (i.e., 10^8 V/m), by comparing the electrostrictive component of stress to Eq. (1.2.9), we find the effective value of e to be 5×10^{-3} C/m². In an unpoled ferroelectric material such as barium titanate, with a relative permittivity of 5000 and $E_0 = 10^6$ V/m, the effective value of e can be quite large, of the order of 0.1 C/m. Thus in most materials, except those with a very high dielectric constant, electrostriction is a very weak effect in comparison to the piezoelectric effect. This is basically because the internal fields resulting from the asymmetry of the lattice are many orders of magnitude larger than the fields that can be applied externally. The same arguments applied to deduce the electrostrictive effect lead us to this conclusion.

Note that for one-dimensional shear or longitudinal wave propagation, the **E** field must be along the direction of propagation, at least to the limit of the assumption that $\mathbf{E} = -\nabla\phi$, because the only variation of potential is in this direction. Because of symmetry, this usually implies that **D**, if finite, is also in this direction.

1.2.2 Effect of Piezoelectric Coupling on Wave Propagation in a Medium of Infinite Extent

We can now consider wave propagation in a piezoelectric medium. If the medium is infinite in extent, the wave motion is one-dimensional, there is no free charge within the medium, and **D** is in the z direction, then

$$\frac{\partial D_z}{\partial z} = 0 \tag{1.2.10}$$

This implies that D_z = constant (i.e., D does not vary with z), although it may vary with time. So the total displacement current density in the medium is

$$i_D = \frac{\partial D}{\partial t} \tag{1.2.11}$$

The current i_D must either be uniform with z or zero. In a piezoelectric transducer with metal electrodes on each surface, a displacement current passes between the electrodes, as it does in any capacitor. In a piezoelectric medium of infinite extent, however, we would expect the displacement current to be zero; hence $D = 0$ in the medium.

Piezoelectrically stiffened elastic constant. Let us now solve for the effective value of the elastic constant c with $D = 0$, or c^D, and determine the propagation constant of a wave in the infinite piezoelectric medium. Writing $D = 0$ in Eq. (1.2.4) and substituting the result in Eq. (1.2.9), we find that

$$E = -\frac{eS}{\varepsilon^S} \tag{1.2.12}$$

and

$$T = c^E\left(1 + \frac{e^2}{c^E \varepsilon^S}\right)S = c^D S \tag{1.2.13}$$

Thus it is as if the piezoelectric medium has an effective elasticity

$$c^D = c^E(1 + K^2) \tag{1.2.14}$$

where we define the parameter c^D as the *stiffened elastic constant*, and call the parameter K, defined by the relation

$$K^2 = \frac{e^2}{c^E \varepsilon^S} \tag{1.2.15}$$

the *piezoelectric coupling constant*. Values of K^2 and the acoustic wave velocity in some of the more common piezoelectric materials are given in Appendix B. The energy definition for K^2 will be discussed in Sec. 1.3. Section 1.5 gives various definitions of K^2 for wave fields applied in arbitrary directions in anisotropic piezoelectric materials.

Now it is easy to define the propagation constant in the piezoelectric medium for a wave of frequency ω. Using the notations $\overline{\beta}_a$ for the stiffened propagation constant and $\overline{V}_a$ for the stiffened acoustic velocity, we see that

$$\overline{\beta}_a = \omega\left(\frac{c^D}{\rho_{m0}}\right)^{-1/2} = \frac{\omega}{\overline{V}_a} \tag{1.2.16}$$

and

$$\overline{V}_a = V_a(1 + K^2)^{1/2} \tag{1.2.17}$$

The piezoelectrically stiffened velocity is always larger than the equivalent velocity in a nonpiezoelectric medium or a piezoelectric medium with $E = 0$ (i.e., a perfectly conducting medium). In materials such as indium antimonide or gallium arsenide, K^2 can have values of the order of 10^{-4}; in a piezoelectric ceramic such as PZT, or a crystal such as lithium niobate, it can be as large as 0.5.

Stress-free dielectric constant. Let us consider the properties of a finite-length medium of infinite cross section. In this case, D may be finite and we can define an effective dielectric constant of the medium. We have seen that if the strain is zero, the ratio D/E is ε^S, the strain-free dielectric constant. In the same way, we define an effective permittivity under stress-free conditions ε^T. Putting $T = 0$ in Eq. (1.2.9), we see that

$$S = \frac{e}{c^E}E \tag{1.2.18}$$

Thus, on substituting Eq. (1.2.18) into Eq. (1.2.4), we find that

$$D = \varepsilon^T E = \varepsilon^S(1 + K^2)E \tag{1.2.19}$$

Therefore,

$$\varepsilon^T = \varepsilon^S(1 + K^2) \tag{1.2.20}$$

Thus the stress-free dielectric constant is larger than the strain-free dielectric constant.

The definitions given here can and must be generalized to define ε^T correctly. In using our definitions, we must know whether all components of T or D in a finite piezoelectric medium are zero, or whether only one of them is.

PROBLEM SET 1.2

1. (a) A simple piezoelectric transducer consists of a thin piezoelectric layer, of thickness l, with thin metal film electrodes. The transducer is backed by a material of the same acoustic impedance $\overline{Z}_0$, and placed against a material of impedance Z_1, as illustrated below. Assume that $l \ll \lambda$, the acoustic wavelength, so that the RF field E and the strain S are essentially uniform within the transducer.

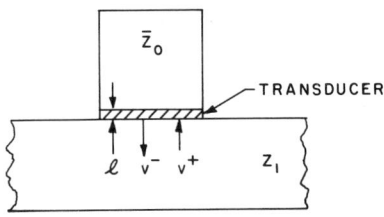

Suppose that a wave with an RF velocity v^+ is incident from the material of impedance Z_1 on the transducer, and passes through the transducer. Taking the transducer to be open-circuited ($i_D = j\omega D = 0$), use Eq. (1.2.4) to find the voltage induced across the transducer.

    (b) Find the minimum detectable velocity v and displacement u at the front surface of a 100-μm-thick PZT-5H transducer, at center frequencies of 100 kHz and 1 MHz, respectively, when the usable sensitivity of the input amplifier is 5×10^{-4} V. Assume that the transducer is thin compared to a wavelength. Over this frequency range, assume that the induced fields in the material are uniform. Assume that the transducer has the same impedance as the material with which it is in contact, and that it is terminated on its other side with a matched impedance. Use e_{z3} for e.

2. **(a)** Consider a capacitive transducer consisting of two metal electrodes spaced by an air gap of width l. Assume that a dc field E_0 is applied to the transducer and induces charges ρ_{s0} and $-\rho_{s0}$ per unit area on each electrode. Show that if the charge on the electrodes is kept constant, the induced transducer voltage is proportional to the RF displacement u at its front surface.

    (b) Suppose that a capacitive transducer is used to detect the emission of acoustic waves incident on a metal surface. In this case, the metal surface is one of the capacitor plates. Such acoustic emissions occur when a crack spreads under stress, the way wood creaks as it is strained. Take the air gap l to be 2 μm and the applied static potential to be 50 V. Assume that a forward wave of displacement amplitude u^+ and velocity v^+ is normally incident at the surface. Suppose that the detector has a usable sensitivity of 5×10^{-4} V. Find the minimum detectable displacement u^+ and the minimum detectable RF velocity at 100 kHz and 1 MHz. Take account of the reflection of the wave at the metal surface to find the total displacement u, which excites the transducer.

    (c) Compare this result to your result for Prob. 1.

3. **(a)** Consider the propagation of an acoustic wave in a conducting piezoelectric medium with a conductivity σ, such as cadmium sulfide. In this case, Eq. (1.2.11) becomes $i = \partial D/\partial t + \sigma E = 0$. Work out an expression for the propagation constant of waves in this medium. Assume that $K^2 \ll 1$. Show that for very low frequencies, the attenuation varies as the square of the frequency, but for very high frequencies, reaches a constant value.

    (b) Taking $K = 0.15$, $\varepsilon^S/\varepsilon_0 = 9.5$, and $\bar{V}_a = 4.45$ km/s, find the value of resistivity of the material for which the attenuation is a maximum at 50 MHz. Find the value of resistivity when the attenuation is a maximum at 1 GHz. Find the attenuation per unit length in each case.

4. Consider what occurs when a dc field is applied to a piezoelectric semiconductor such as the one described in Prob. 1, so that the carriers drift with a velocity v_0. The conduction

current density associated with the carriers is

$$i_C = qn\mu E$$

We put

$$i_C = i_{C0} + i_{C1}$$
$$v_0 = \mu E_0$$
$$E = E_0 + E_1$$

and

$$n = n_0 + n_1$$

where E_0 is the applied dc field, E_1 is the RF field, μ is the carrier mobility, q is the carrier charge (the conductivity $\sigma = \mu q n_0$), n_0 is the static carrier density, n_1 is the RF carrier density, and it is assumed that $E_1 \ll E_0$ (i.e., $i_{C1} \ll i_{C0}$). Then

$$i_{C1} = q(n_1 v_0 + \mu n_0 E_1)$$

to first order in the RF quantities, with

$$v_0 = \mu E_0$$

The carriers obey the equation of continuity of charge

$$\frac{\partial i_{C1}}{\partial z} + q\frac{\partial n_1}{\partial t} = 0$$

The total first-order current density is

$$i = i_{C1} + \frac{\partial D}{\partial t} = 0$$

Following the method of Prob. 3, solve for the propagation constant β of the wave. Find a simple formula for β by writing $\beta = \beta_0 + K^2\beta_1 + K^4\beta_2 \cdots$, and keep only the first-order terms in K^2. Show that when $K^2 \ll 1$, the wave grows with distance if $\overline{V}_a < v_0$, and is attenuated if $v_0 < \overline{V}_a$. This is the basic theory of the acoustoelectric amplifier. In practice, the gain or growth rate of the device does not remain large at very high frequencies because carrier diffusion effects at such levels cause it to decrease.

1.3 ENERGY CONSERVATION IN PIEZOELECTRIC MEDIA

It is possible to obtain the equivalent of Poynting's theorem for piezoelectric media by generalizing the derivation of Eq. (1.1.22), the Poynting's theorem for nonpiezoelectric media. This is done for the general case in Appendix C.

For a plane wave of frequency ω, like the one treated in Sec. 1.2, it follows from Eq. (C.11) that the generalized one-dimensional Poynting's theorem can be written in the form

$$\frac{\partial}{\partial z}[(-v^*T) + (\mathbf{E} \times \mathbf{H}^*)_z]$$

$$= j\omega c^E SS^* + j\omega \varepsilon^S \mathbf{E} \cdot \mathbf{E}^* - j\omega p_{m0}\mathbf{v} \cdot \mathbf{v}^* - j\omega \mu \mathbf{H} \cdot \mathbf{H}^* - \mathbf{i}_C^* \cdot \mathbf{E} \quad (1.3.1)$$

where $\mathbf{i}_C$ is the conduction current in the medium.

The left-hand side of Eq. (1.3.1) is associated with the total power flow density in the medium. If the permittivity, elasticity, and permeability are real, then

$$\frac{\partial}{\partial z} \text{Re}\, (P_a + P_e) = -\tfrac{1}{2} \text{Re} \int_A \mathbf{i}_C^* \cdot \mathbf{E}\, ds \qquad (1.3.2)$$

where the integral is taken over the cross section A of the system. Thus we have generalized the usual complex Poynting's theorem to find that the only change required is that the total power must now include the electromagnetic power flow

$$P_e = \tfrac{1}{2} \text{Re} \int_A (\mathbf{E} \times \mathbf{H}^*) \cdot ds \qquad (1.3.3)$$

Equation (1.3.1) shows that in addition to the stored elastic energy W_S per unit volume, defined as

$$W_S = \tfrac{1}{4} S^* c^E S \qquad (1.3.4)$$

and the stored kinetic energy W_v per unit volume, defined as

$$W_v = \tfrac{1}{4} \rho_{m0} \mathbf{v} \cdot \mathbf{v}^* \qquad (1.3.5)$$

there is stored magnetic energy W_H per unit volume, defined as

$$W_H = \tfrac{1}{4} \mu \mathbf{H} \cdot \mathbf{H}^* \qquad (1.3.6)$$

The stored magnetic energy is usually negligible for an acoustic wave. The stored electric energy W_E per unit volume is

$$W_E = \tfrac{1}{4} \varepsilon^S \mathbf{E} \cdot \mathbf{E}^* \qquad (1.3.7)$$

Let us consider the ratio of the electrical energy to the acoustic energy, defined as $W_E/(W_S + W_v) = W_E/W_a$, where we have taken the acoustic energy per unit volume to be

$$W_a = W_S + W_v \qquad (1.3.8)$$

In the one-dimensional case, as $D = 0$, it follows from Eq. (1.2.12) that

$$E = \frac{-eS}{\varepsilon^S} \qquad (1.3.9)$$

The electrical energy per unit volume is

$$\begin{aligned} W_E &= \frac{1}{4} \varepsilon^S |E|^2 \\ &= \frac{1}{4} \frac{e^2 |S|^2}{\varepsilon^S} \qquad (1.3.10) \\ &= \frac{1}{4} \frac{e^2}{c^E \varepsilon^S} TS = K^2 W_S \end{aligned}$$

Thus K^2 can be defined as the ratio of the stored electrical energy to the stored elastic energy. This ratio is usually much less than unity. It also follows from

Sec. 1.3   Energy Conservation in Piezoelectric Media

Eq. (1.3.1), with $\mathbf{i}_C = 0$ and $\mathbf{H} = 0$ (the magnetic field can usually be neglected), that

$$W_E + W_S - W_v = 0 \tag{1.3.11}$$

Therefore, Eqs. (1.3.10) and (1.3.11) lead to the result

$$W_v = (1 + K^2)W_S = \frac{1 + K^2}{K^2} W_E \tag{1.3.12}$$

or

$$\frac{W_E}{W} = \frac{K^2}{2(1 + K^2)} \tag{1.3.13}$$

where the total stored energy per unit volume is $W = W_S + W_v + W_E$. Because the square of the piezoelectric coupling constant K^2 is normally small, the stored electrical energy is usually much less than the total stored energy in a piezoelectric medium.

The same results also follow simply from the one-dimensional static equations. If we consider a material in which the electric field is zero, then under static conditions, the stored elastic energy W_S is

$$W_S = \tfrac{1}{2}TS = \tfrac{1}{2}cS^2 \text{ per unit volume} \tag{1.3.14}$$

Suppose that the electric field is finite and $D = 0$ (i.e., no charge flows into the system). The stored energy is now

$$\begin{aligned} W &= \tfrac{1}{2}TS = \tfrac{1}{2}c^E S^2 + \tfrac{1}{2}\varepsilon^s E^2 \\ &= \tfrac{1}{2}c^E(1 + K^2)S^2 \text{ per unit volume} \end{aligned} \tag{1.3.15}$$

Thus the increase in stored energy due to the piezoelectric effect is $K^2 W_S$.

To put it another way, it follows from Eq. (1.2.9) that the static energy per unit volume is

$$\tfrac{1}{2}TS = \tfrac{1}{2}c^E S^2 - \tfrac{1}{2}eES \tag{1.3.16}$$

The elastic energy is $\tfrac{1}{2}c^E S^2$ per unit volume, and the energy resulting from coupling between electrical and mechanical quantities is $\tfrac{1}{2}eES$ per unit volume. The ratio of the mutual coupling to the self-energy term is $-eE/c^E S$. If $D = 0$, this quantity is equal to K^2. Thus K^2 is also the ratio of the mutual coupling energy to the stored energy.

PROBLEM SET 1.3

1. Consider the effect of resistive loss in a piezoelectric medium. Suppose that the conductivity of the material is σ. Using Poynting's theorem [Eq. (1.3.2)] and the Maxwell equation

$$\nabla \times \mathbf{H}^* = -j\omega \mathbf{D}^* + \mathbf{i}_C^*$$

and assuming that for an acoustic beam of area A with no attenuation, the power in the acoustic beam is $P_a = W\overline{V}_a A$, find the attenuation of a plane wave due to resistive loss. Carry out a treatment such as the one to derive viscous losses in Sec. 1.1 and the derivation leading to Eq. (1.1.31), and use the values of E, S, T, and P_a for the perfectly insulating medium.

2. Derive the one-dimensional form of Poynting's theorem, using the one-dimensional piezoelectric constitutive relations and the one-dimensional equation of motion for a plane wave.

1.4 PIEZOELECTRIC TRANSDUCERS

1.4.1 Introduction

Electromechanical transducers convert electrical energy into mechanical energy, and vice versa, in low-frequency applications; microphones and loudspeakers are well-known examples of such transducers used at frequencies below 20 kHz. But for higher-frequency applications, piezoelectric transducers are useful. Their high Q as mechanical resonators means that they can be used at medium frequencies (1 to 50 MHz) in quartz crystal resonators and filters, and also at much higher frequencies as bulk and surface wave electromechanical transducers, from a few kilohertz up into the microwave range.

A simple configuration for a delay line operating at frequencies above 100 MHz, which might be made of sapphire, for example, is shown in Fig. 1.4.1(a). A metal film is deposited on either end of the delay-line rod to create two metal counter electrodes, one at each end. Then a film of a piezoelectric material, such as zinc oxide, is deposited on each counter electrode by sputtering or evaporation in a vacuum; typically, this layer is chosen to be between a quarter- and a half-wavelength thick. Finally, a metal film is laid down on the surface of the piezoelectric material to form a metal top electrode; this second metal electrode is normally a small fraction of a wavelength thick. We now have a transducer at either end of the rod, comprised of a metal counter electrode, a piezoelectric film, and a metal top electrode. At one end of the rod, a potential is applied between the two metal electrodes on either side of the piezoelectric film to excite a longitudinal acoustic wave in the delay line. After the wave has traveled through the delay line, it is detected on the transducer at the end of the rod. The electrical impedance at the input terminals will depend on the thickness and acoustic impedance of the electrodes and the piezoelectric material, and on the nature of the substrate material. To make a broadband UHF delay line, it is often necessary to use a quarter-wavelength-thick matching layer of material between the counter electrode and the delay-line rod, to match the impedance of the piezoelectric material to the impedance of the delay line.

Another type of application, illustrated in Fig. 1.4.1(b), uses an air-backed piezoelectric transducer to excite a wave in water. In a low-frequency device, using a PZT transducer ($Z_0 = 34 \times 10^6$ kg/m^2-s) to excite a wave in water ($Z_0 = 1.5 \times 10^6$ kg/m^2-s), there is a 22-to-1 mismatch in impedance. This leads to a resonant characteristic with an acoustic Q of the order of 30. But using a quarter-

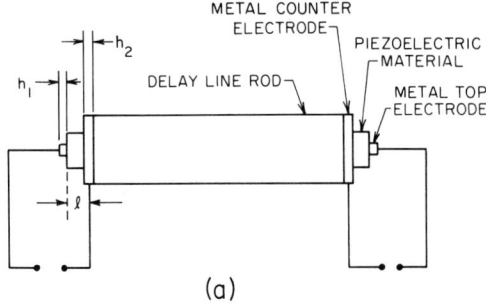

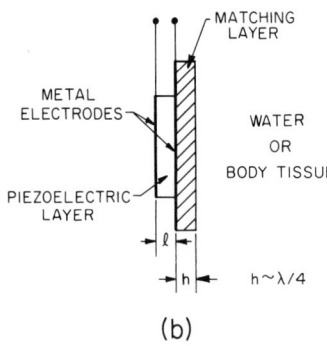

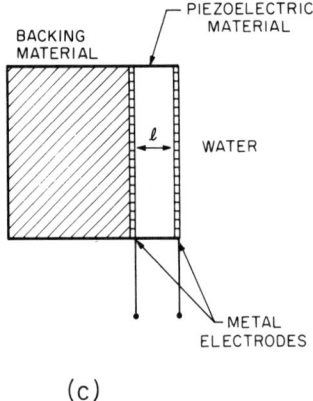

Figure 1.4.1 (a) Delay line used at UHF frequencies. All materials are typically deposited by vacuum deposition on the delay-line rod. (b) Air-backed piezoelectric transducer with a matching layer on its front surface, for exciting a wave in water or body tissue. (c) Piezoelectric transducer loaded on its back surface, for exciting a wave in water.

wavelength-thick intermediate layer, with an impedance $Z_0 \sim \sqrt{34 \times 1.5} \times 10^6$ = 7.1×10^6 kg/m²-s, between the ceramic and the water will yield a broader-band characteristic with better power transmission. The resonance can also be broadened by using acoustically lossy material with an impedance comparable to that of the PZT ($Z_0 = 34 \times 10^6$ kg/m²-s) bonded to the other side of the transducer, as illustrated in Fig. 1.4.1(c).

First we will carry out a general derivation to determine the properties of a transducer with arbitrary acoustic impedances at each surface. The piezoelectric transducer can be regarded as a "black box" having one or more mechanical ports and one electrical port.

1.4.2 The Transducer as a Three-Port Network

We shall consider a uniform transducer or resonator with cross-sectional dimensions of many wavelengths, and electrodes on opposite surfaces normal to the z direction, as shown in Figs. 1.4.1 and 1.4.2. Because the electrodes short out the field, it is reasonable to assume that $E_x = 0$ and $E_y = 0$. The symmetry means that if the transducer is designed to operate with longitudinal waves, there will be no motion in the x and y directions. In this case, the parameters S, E, D, v, u, and T have components only in the z direction, and Eqs. (1.2.4) and (1.2.9) are the natural ones to use with the appropriate values of c^E, e, and ε^S. Similarly, for shear wave resonators or transducers in which S, E, D, v, u, and T have only one component, Eqs. (1.2.4) and (1.2.9) can be used with the appropriate shear wave constants c^E and e. The definitions of these parameters are discussed in more detail in Sec. 1.5.

We now regard the transducer as a three-port black box. We define the force F at the surface of the transducer as we do voltage in electrical circuits, and the particle velocity v as we do current in electrical circuits. Using the notation shown in Fig. 1.4.3(a) for the three-port network, and that shown in Fig. 1.4.3(b) for the physical transducer, we can find an equivalent circuit for this black box.

The external force applied to the piezoelectric material at the surface of the resonator is

$$F = -AT \quad (1.4.1)$$

where A is the area of the transducer and T is the internal stress.

The equivalent circuit definitions used in the theory of the piezoelectric transducer are based on the idea that particle velocity is equivalent to current and

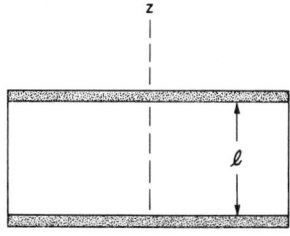

Figure 1.4.2 Piezoelectric resonator of length l with electrodes on opposite surfaces.

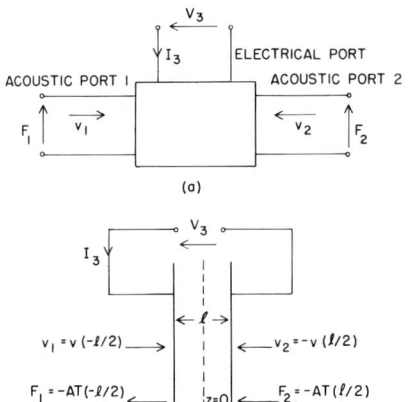

Figure 1.4.3 (a) Transducer, regarded as a three-port black box; (b) relation of three-port notation to the physical parameters of the transducer.

stress is equivalent to voltage, as we discussed in the transmission-line analogy in Sec. 1.1.

The definition for velocity, as it applies to the two acoustic ports or terminals, follows the definition for current at the ports of an electrical two-port network. Thus the particle velocity is positive *inward* to the piezoelectric material. It follows that the boundary conditions at the acoustic ports are

$$F_1 = -AT\left(\frac{-l}{2}\right)$$

$$F_2 = -AT\left(\frac{l}{2}\right)$$

$$v_1 = v\left(\frac{-l}{2}\right)$$

$$v_2 = -v\left(\frac{l}{2}\right)$$

(1.4.2)

respectively, where $v(-l/2)$ and $v(l/2)$ are the velocity components at the surface of the piezoelectric material.

The relation between T and v within the material of the transducer is

$$\frac{dT}{dz} = j\omega \rho_{m0} v \qquad (1.4.3)$$

and

$$\frac{dv}{dz} = j\omega S \qquad (1.4.4)$$

The total current through the transducer is

$$I_3 = j\omega AD \qquad (1.4.5)$$

With the sign convention shown in Fig. 1.4.3, the voltage across the transducer is

$$V_3 = \int_{-l/2}^{l/2} E \, dz \tag{1.4.6}$$

Since current is conserved, D must be uniform with z. Eliminating E from Eqs. (1.2.4) and (1.2.9), the generalization of Eq. (1.2.13) with D finite is

$$T = c^D S - hD \tag{1.4.7}$$

where h is known as the *transmitting constant*, defined as

$$h = \frac{e}{\varepsilon^S} \tag{1.4.8}$$

with

$$c^D = c^E \left(1 + \frac{e^2}{c^E \varepsilon^S}\right) = c^E(1 + K^2) \tag{1.4.9}$$

If we eliminate T and S from Eqs. (1.4.3), (1.4.5), and (1.4.7), then v obeys the wave equation

$$\frac{d^2v}{dz^2} + \frac{\omega^2 \rho_{m0}}{c^D} v = 0 \tag{1.4.10}$$

This has the solutions

$$v = v_F e^{-j\bar{\beta}_a z} + v_B e^{j\bar{\beta}_a z} \tag{1.4.11}$$

and

$$T = T_F e^{-j\bar{\beta}_a z} + T_B e^{j\bar{\beta}_a z} - hD \tag{1.4.12}$$

where the subscripts F and B denote forward and backward propagating waves, respectively. We define the following parameters:

$$\bar{\beta}_a = \omega \left(\frac{\rho_{m0}}{c^D}\right)^{1/2} \tag{1.4.13}$$

and

$$\bar{Z}_0 = (\rho_{m0} c^D)^{1/2} \tag{1.4.14}$$

with

$$T_F = -\bar{Z}_0 v_F \tag{1.4.15}$$

and

$$T_B = \bar{Z}_0 v_B \tag{1.4.16}$$

Using the boundary conditions of Eq. (1.4.2) in Eq. (1.4.11), we see that

$$v = \frac{-v_2 \sin[\bar{\beta}_a(z + l/2)] + v_1 \sin[\bar{\beta}_a(l/2 - z)]}{\sin \bar{\beta}_a l} \tag{1.4.17}$$

Substituting this result in Eqs. (1.4.2) and (1.4.4)–(1.4.7), it follows, after some algebra, that

$$\begin{bmatrix} F_1 \\ F_2 \\ V_3 \end{bmatrix} = -j \begin{bmatrix} Z_C \cot \overline{\beta}_a l & Z_C \operatorname{cosec} \overline{\beta}_a l & \dfrac{h}{\omega} \\ Z_C \operatorname{cosec} \overline{\beta}_a l & Z_C \cot \overline{\beta}_a l & \dfrac{h}{\omega} \\ \dfrac{h}{\omega} & \dfrac{h}{\omega} & \dfrac{1}{\omega C_0} \end{bmatrix} \begin{bmatrix} v_1 \\ v_2 \\ I_3 \end{bmatrix} \qquad (1.4.18)$$

where the clamped (zero strain) capacitance of the transducer is

$$C_0 = \frac{\varepsilon^S A}{l} \qquad (1.4.19)$$

Consistent with our definition of electrical impedance, we define the acoustic impedance of an area A of piezoelectric material as

$$Z_C = \overline{Z}_0 A \qquad (1.4.20)$$

where the parameter Z_C has the dimensions force/velocity or kg/s and $\overline{Z}_0$ has the dimensions pressure/velocity or kg/m²-s. Impedances with the dimensions of force/velocity are sometimes called *radiation impedances*.

Example: Electrical Input Impedance of a Transducer

We can determine the electrical input impedance of a transducer terminated by acoustic load impedances Z_1 and Z_2 by using the matrix formula (1.4.18). We define the radiation impedances of the loads (looking outward from the transducer) as

$$Z_1 = -\frac{F_1}{v_1} = \frac{AT(-l/2)}{v(-l/2)} \qquad (1.4.21)$$

and

$$Z_2 = -\frac{F_2}{v_2} = \frac{-AT(l/2)}{v(l/2)} \qquad (1.4.22)$$

Using these relations in Eq. (1.4.18) yields the electrical input impedance of the transducer in the form

$$Z_3 = \frac{V_3}{I_3} = \frac{1}{j\omega C_0}\left[1 + k_T^2 \frac{j(Z_1 + Z_2)Z_C \sin \overline{\beta}_a l - 2Z_C^2(1 - \cos \overline{\beta}_a l)}{[(Z_C^2 + Z_1 Z_2) \sin \overline{\beta}_a l - j(Z_1 + Z_2)Z_C \cos \beta_a l]\overline{\beta}_a l}\right]$$

$$(1.4.23)$$

where

$$k_T^2 = \frac{K^2}{1 + K^2} \qquad (1.4.24)$$

and

$$\frac{c^D}{c^E} = 1 + K^2 = \frac{1}{1 - k_T^2} \qquad (1.4.25)$$

For longitudinal waves, the parameter k_T is often defined as the piezoelectric coupling constant for a transversely clamped material, for it is the effective piezoelectric constant used when there is no motion transverse to the electric field. For $K^2 \ll 1$, k_T^2 and K^2 are essentially identical. For materials such as quartz, cadmium sulfide, and zinc oxide, this assumption is adequate. It tends to break down, however, with piezoelectric ceramics; with PZT-5A, for instance, $K^2 = 0.5$ and $k_T^2 = 0.33$.

The definitions of K^2 and k_T^2 are useful only if the material can be regarded as being transversely clamped. A piezoelectric transducer with a cross section of many wavelengths has very little transverse motion, so k_T is the effective coupling constant. Such transducers are commonly employed in UHF applications, where the wavelengths are less than 100 μm, and often in low-frequency applications, where the wavelengths are of the order of 1 mm and the cross-sectional dimensions of the transducer are 1 or more centimeters. Similar relations hold for a shear wave transducer with a cross section of many wavelengths.

1.4.3 Mason Equivalent Circuit

We show here that the matrix formula (1.4.18) results in the Mason equivalent circuit of Fig. 1.4.4. First we consider the value of F_1:

$$F_1 = -jZ_C v_1 \cot \overline{\beta}_a l - jZ_C v_2 \csc \overline{\beta}_a l + \frac{hI_3}{j\omega} \quad (1.4.26)$$

When $I_3 = 0$, the first two terms can be written in the form of an impedance matrix of the type shown in Fig. 1.4.5(b), where

$$Z_{11} = -jZ_C \cot \overline{\beta}_a l \quad (1.4.27)$$

$$Z_{12} = -jZ_C \csc \overline{\beta}_a l \quad (1.4.28)$$

and

$$Z_{11} - Z_{12} = jZ_C(\csc \overline{\beta}_a l - \cot \overline{\beta}_a l) = jZ_C \tan \frac{\overline{\beta}_a l}{2} \quad (1.4.29)$$

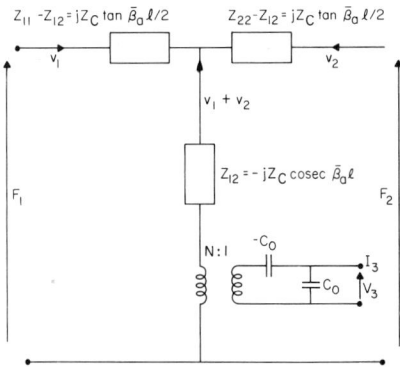

Figure 1.4.4 Mason series equivalent circuit for the in-line transducer with the E field in the same direction as the material velocity. In this circuit the transformer ratio is $N = hC_0 = eC_0/\varepsilon^S = eA/l$, and $Z_C = \overline{Z}_0 A$.

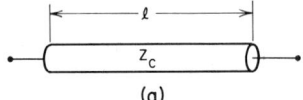

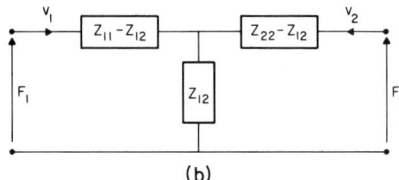

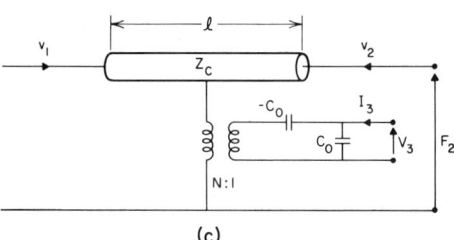

Figure 1.4.5 Redwood equivalent circuit. (a) Coaxial transmission line of impedance Z_C. In this circuit, $Z_{11} - Z_{12} = Z_{22} - Z_{12} = jZ_C \tan \bar{\beta}_a l/2$, $Z_{12} = -jZ_C \csc \bar{\beta}_a l$. (b) T network equivalent of the coaxial line. (c) Redwood equivalent circuit, derived from the Mason model of Fig. 1.4.4. In this circuit the transformer ratio is $N = hC_0 = eC_0/\varepsilon^S = eA/l$, and $Z_C = \bar{Z}_0 A$.

The right-hand side of the network has the same form as the left-hand side. In general, there is an extra potential of value $hI_3/j\omega$ in series with the potentials generated by v_1 and v_2. Before treating this source, let us consider the value of V_3:

$$V_3 = \frac{h}{j\omega}(v_1 + v_2) + \frac{I_3}{j\omega C_0} \qquad (1.4.30)$$

The last term in this equation is merely a voltage across the capacitor C_0. The first term is a voltage proportional to the total equivalent current $v_1 + v_2$ flowing into the transformer. A perfect transformer of value N to 1, where $N = hC_0 = eC_0/\varepsilon^S = eA/l$, would introduce a current of value $(v_1 + v_2)N$ into the right-hand side of the circuit, and the current would then develop a potential $(h/j\omega)(v_1 + v_2)$ across the capacitor C_0.

The transformed potential V_3 does not appear across terminals 1 or 2. A potential $-(h/j\omega)/(v_1 + v_2)$ is developed across the negative capacitor $-C_0$ in the circuit of Fig. 1.4.4, which just cancels out the potential across C_0. This equivalent circuit also gives the last term in the expression for F_1. Thus the final Mason equivalent circuit is the one shown in Fig. 1.4.4.

Example: Clamped Transducer

If the transducer is rigidly held (clamped) so that $v_1 = v_2 = 0$, the mechanical terminating impedances are infinite and it follows from Eq. (1.4.18), or the Mason equivalent circuit of Fig. 1.4.4, that

$$\frac{F_1}{V_3} = \frac{F_2}{V_3} = N = \frac{eA}{l} \qquad (1.4.31)$$

There is no motion in a rigidly clamped system. Thus with $v(z) = 0$ and $S = 0$, it follows from Eqs. (1.2.4) and (1.2.9) that $T = -eE$ and $D = \varepsilon^s E$. So if $F = -AT$ and $V_3 = El$, then $F_1/V_3 = F_2/V_3 = eA/l$, which is the result of Eq. (1.4.31).

1.4.4 Redwood Equivalent Circuit

An alternative equivalent circuit, the *Redwood equivalent circuit* [7], can be derived from the Mason model of Fig. 1.4.4. The T network Z_{11}, Z_{12}, Z_{22} of the Mason equivalent circuit can be represented by a transmission line of impedance Z_C, as illustrated in Fig. 1.4.5(a) and (b). This means that we can write the Mason model in the form shown in Fig. 1.4.5(c), which is the Redwood equivalent circuit. The transmission line in this new circuit can be regarded as a coaxial line whose outer shield is connected to the transformer. The Redwood equivalent circuit is particularly useful for dealing with a short pulse excitation of the transducer, especially when the pulse length is less than the delay time of an acoustic wave passing through the transducer. In this case, the input impedance of the coaxial line appears to be Z_C, and it is very easy to determine the pulse response of the system.

Example: Open-Circuited Pulse-Excited Acoustically Matched Receiving Transducer

We assume that the transducer is electrically open-circuited (i.e., $I_3 = 0$). In this case, the voltage across the transformer is zero because the two capacitors C_0 and $-C_0$, in series, form a short circuit. Suppose that the transducer is excited at its left-hand side with a velocity pulse $v_1(t)$. The pulse propagates along the transmission line and gives rise to a pulse $v_2(t)$, where

$$v_2(t) = -v_1(t - T) \tag{1.4.32}$$

and $T = l/\overline{V_a}$ is its transit time along the line.

The current flowing into the capacitor C_0 is

$$I = N(v_1 + v_1)$$
$$= N[v_1(t) - v_1(t - T)] \tag{1.4.33}$$

Thus the output voltage V_3 is defined as

$$V_3 = \frac{N}{C_0} \int_0^t [v_1(t) - v_1(t - T)] \, dt \tag{1.4.34}$$

or

$$V_3 = h \int_0^t [v_1(t) - v_1(t - T)] \, dt \tag{1.4.35}$$

If $v_1(t)$ has the form of a δ function velocity pulse (a very short pulse), $V_3(t)$ will be a square-topped pulse of length T, as illustrated in Fig. 1.4.6.

1.4.5 Impedance of an Unloaded Transducer

First we consider an unloaded transducer in air (e.g., a ceramic or quartz plate with thin electrodes). In this case, $F_1 = F_2 = 0$ or $Z_1 = Z_2 = 0$.

From Eq. (1.4.23) we find that the RF input impedance $Z_3 = V_3/I_3$ of the

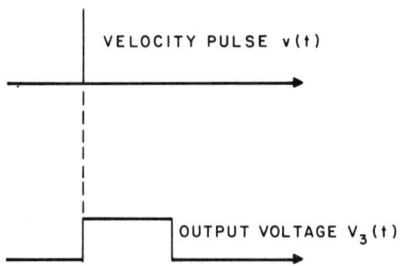

Figure 1.4.6 Output voltage of an open-circuited receiving transducer, terminated in a matching acoustic impedance when excited by a δ-function velocity pulse.

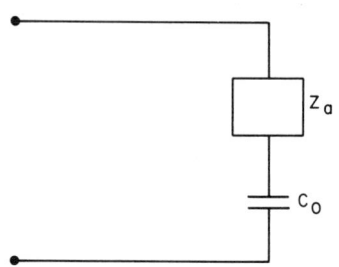

Figure 1.4.7 Equivalent circuit of a piezoelectric transducer.

transducer is

$$Z_3 = \frac{V_3}{I_3} = \frac{1}{j\omega C_0}\left(1 - k_T^2 \frac{\tan \overline{\beta}_a l/2}{\overline{\beta}_a l/2}\right) \quad (1.4.36)$$

Equation (1.4.36) shows that the equivalent circuit of the transducer shown in Fig. 1.4.7 can be represented by the clamped capacity of the transducer in series with the *motional impedance* Z_a (i.e., where Z_a is the acoustic contribution to the electrical impedance), defined by the relation

$$Z_a = -\frac{k_T^2}{j\omega C_0} \frac{\tan \overline{\beta}_a l/2}{\overline{\beta}_a l/2} \quad (1.4.37)$$

The transducer exhibits a parallel resonance with an infinite electrical impedance. Thus the transducer impedance is like that of an inductance and capacitance in parallel, at frequencies where the transducer is an odd number of half-wavelengths long [i.e., where $\beta_a l = (2n + 1)\pi$]. The corresponding resonant frequencies ω_{0n} are given by the relation

$$\omega_{0n} = \frac{\pi(2n + 1)\overline{V}_a}{l} \quad (1.4.38)$$

For simplicity, we shall call the lowest-order parallel resonance ($n = 0$) ω_0. Resonances also exist when the transducer length is an even multiple of a half-wavelength, but because the RF electric field associated with these modes has an odd symmetry about the center of the resonator, there is no net potential across the resonance. Thus there is no electrical coupling to even modes.

The transducer exhibits zero electrical impedance at a frequency ω_1 near the $n = 0$ parallel resonance. Near this frequency ω_1, the transducer behaves like an inductance and capacitance in series, and so exhibits a series resonance. At this frequency ω_1, the transducer impedance is $Z_3 = 0$, where

$$\frac{\tan \overline{\beta}_a l/2}{\overline{\beta}_a l/2} = \frac{1}{k_T^2} \quad (1.4.39)$$

It follows from Eqs. (1.4.38) and (1.4.39) that

$$\frac{\tan(\pi\omega_1/2\omega_0)}{\pi\omega_1/2\omega_0} = \frac{1}{k_T^2} \qquad (1.4.40)$$

We can, in principle, determine k_T from Eq. (1.4.40) by measuring ω_1 and ω_0.

Equivalent circuit of an unloaded transducer. It is convenient to express the impedance Z_a in the form of equivalent lumped circuits that correspond to the fundamental resonance and higher-order resonances of the resonator. The function $\tan x$ has poles at $x = (2n + 1)\pi/2$. This allows us to obtain a partial fraction, or a Mittag Leffler expansion, for $\tan x$ in the form [2]

$$\tan x = \sum_{n=0}^{\infty} \frac{2x}{[(2n+1)\pi/2]^2 - x^2} \qquad (1.4.41)$$

We can express the motional impedance Z_a in a similar form, as

$$Z_a = -\frac{1}{j\omega C_0} \sum_n \frac{k_{\text{eff},n}^2}{1 - \omega^2/\omega_{0n}^2} \qquad (1.4.42)$$

where $k_{\text{eff},n}$ is an effective coupling coefficient for the nth mode, defined by the relation

$$k_{\text{eff},n}^2 = \frac{8}{[(2n+1)\pi]^2} k_T^2 \qquad (1.4.43)$$

The coupling coefficients to the higher-order modes of the resonator $k_{\text{eff},n}$ fall off with n. The value of $\int_{-l/2}^{l/2} E\, dz$ falls off with n for a given maximum value of E, because the negative contributions to the integral tend to cancel out the positive contributions. This implies that it is possible to excite a resonator at an odd harmonic of its fundamental resonant frequency, although the effective coupling coefficient for this higher-order mode is smaller than for the fundamental mode. Coupling to higher-order modes is often very convenient; for example, quartz or lithium niobate resonators are often used as narrowband transducers at frequencies many times their fundamental resonant frequency. This makes it possible, for instance, to work at a frequency of 200 MHz with a quartz or lithium niobate resonator whose fundamental frequency is in the 20-MHz range. In this case, the transducer material has a thickness on the order of 0.1 to 0.2 mm, making it easier to handle than a fundamental mode resonator, which would be only 15 μm thick. This technique also tends to give a higher Q or a narrower bandwidth than a thin fundamental mode resonator, whose lapped surfaces might contribute some power loss.

Let us derive a lumped equivalent circuit for the transducer. Using Eq. (1.4.42), we write the impedance Z_3 of the resonator as

$$Z_3 = \frac{1}{j\omega C_0} \left[1 - \sum_n \frac{k_{\text{eff},n}^2}{1 - \omega^2/\omega_{0n}^2} \right] \qquad (1.4.44)$$

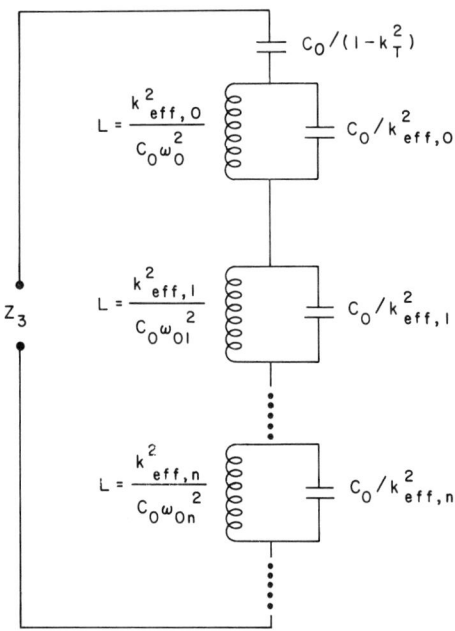

Figure 1.4.8 Complete series equivalent circuit including all resonances. Each parallel inductance–capacitance combination has a resonant frequency ω_{0n}, and the total series capacitance is $C_0/(1 - k_T^2)$, the unclamped parallel-plate capacitance. A similar parallel equivalent circuit can also be found.

It is convenient to write this relation in the form

$$Z_3 = \frac{1}{j\omega C_0}\left[1 - \sum_n k_{\text{eff},n}^2 + \sum_n \frac{j\omega k_{\text{eff},n}^2}{(\omega_{0n}^2 - \omega^2)C_0}\right] \quad (1.4.45)$$

$$= \frac{1 - k_T^2}{j\omega C_0} + \sum_n \frac{j\omega k_{\text{eff},n}^2}{(\omega_{0n}^2 - \omega^2)C_0}$$

This impedance can be represented exactly by the equivalent circuit shown in Fig. 1.4.8. It may be approximated by ignoring the higher-order terms and taking only the fundamental resonance into account, which leads us to the equivalent circuit shown in Fig. 1.4.9(a), with

$$Z_3 \approx \frac{1 - k_T^2}{j\omega C_0} + \frac{8}{\pi^2}\frac{j\omega k_T^2/C_0}{\omega_0^2 - \omega^2} \quad (1.4.46)$$

This equivalent circuit has both a series resonance at ω_1 and a parallel resonance at ω_0. The ratio of these frequencies is given by the relation

$$\frac{\omega_1}{\omega_0} = \left(1 + \frac{8}{\pi^2}\frac{k_T^2}{1 - k_T^2}\right)^{-1/2} = \left(1 + \frac{8K^2}{\pi}\right)^{-1/2} \quad (1.4.47)$$

To second order in k_T^2, it follows that

$$\frac{\omega_1}{\omega_0} \approx \left(1 - \frac{8k_T^2}{\pi^2}\right)^{1/2} \quad (1.4.48)$$

When $k_T^2 = 0.5$, the approximate formula of Eq. (1.4.47) has an error of only 0.15% from the exact formula of Eq. (1.4.40). Figure 1.4.10 shows that even the

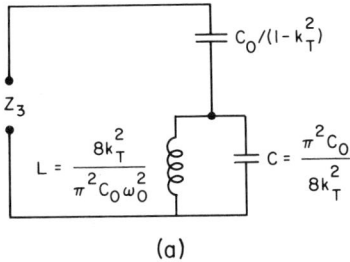

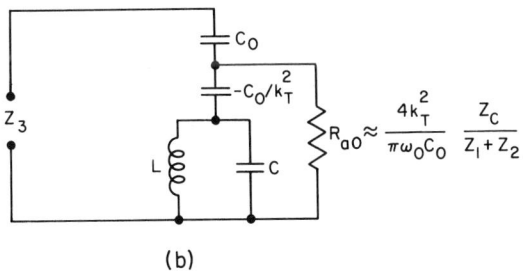

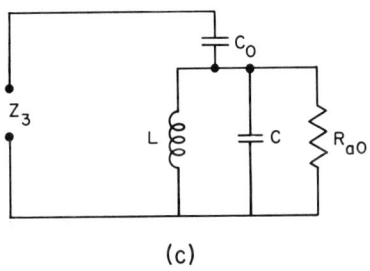

Figure 1.4.9 (a) Series equivalent circuit for Z_3. The paralllel inductance–capacitance combination has a resonant frequency ω_0. (b) Transducer loaded by impedances Z_1 and Z_2. It is assumed that $|(Z_1 + Z_2)| \ll Z_C$. (c) Usually, in the neighborhood of resonance, the reactance $-C_0/k_T^2$ can be ignored in comparison to the impedance of the shunt resonant circuit.

approximate formula of Eq. (1.4.48) agrees with the exact formula Eq. (1.4.40) for all reasonable values of k_T^2. For example, the piezoelectric ceramic PZT-5H has a relatively large k_T^2 value of 0.25, while a material such as zinc oxide has a k_T^2 value of 0.078. In both cases the errors in either of the approximate formulas (1.4.47) and (1.4.48) are negligible.

Let us estimate the effect on Z_a, at the shunt resonance frequency $\omega = \omega_0$, of load impedances Z_1 and Z_2 at each end of the transducer. We use Eq. (1.4.23) with Z_1 and Z_2 finite to show that

$$Z_a = R_{a0} = \frac{4k_T^2 Z_C}{(Z_1 + Z_2)\pi\omega_0 C_0} \qquad (1.4.49)$$

Thus it is as if a capacity C_0 is placed in series with a resistance $R_{a0} = 4k_T^2 Z_C/(Z_1 + Z_2)\pi\omega_0 C_0$; R_{a0} itself is placed in parallel with the lumped resonant circuit. The reactance of the capacitor $-C_0/k_T^2$ is small compared to that of the shunt resonator in the neighborhood of resonance. We will discuss the implications of this circuit

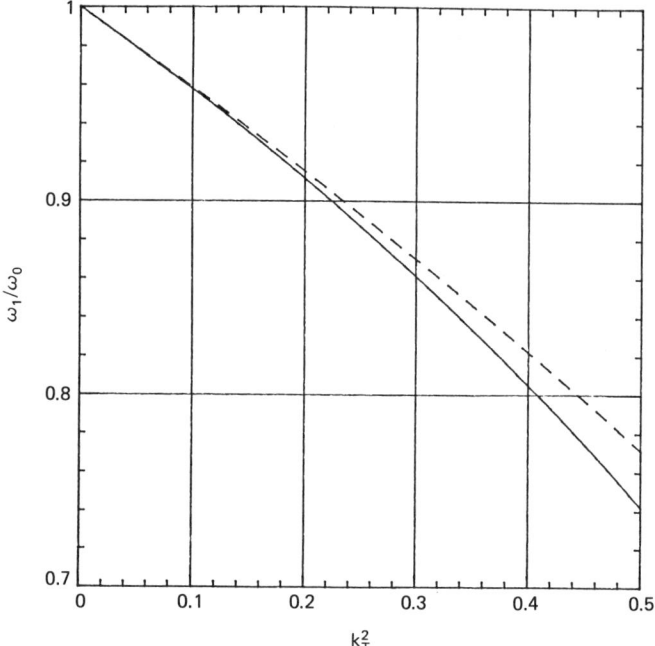

Figure 1.4.10 Plot of ω_1/ω_0 as a function of k_T^2: solid line, plot of Eq. (1.4.40) or Eq. (1.4.47); dashed line, plot of Eq. (1.4.48). The results of Eqs. (1.4.40) and (1.4.47) are within 0.15% at $k_T^2 = 0.5$.

on matching to an external electrical source and to bandwidth throughout this section.

We can also use Eq. (1.4.46), or the Mason equivalent circuit of Fig. 1.4.4, to derive an alternative parallel equivalent circuit for the admittance $Y_3 = 1/Z_3$. This circuit is convenient for cases where we need to operate near the series resonance into a real input impedance that is relatively small compared to the impedance at shunt resonance. The result obtained is shown in Fig. 1.4.11(a). The series and parallel resonances are at frequencies ω_1 and ω_0, respectively, as we might expect.

A similar derivation from the Mason equivalent circuit, which takes account of the loads Z_1 and Z_2, can be used to derive the equivalent circuit shown in Fig. 1.4.11(b). Here the direct use of the Mason equivalent circuit is the most convenient approach (see Prob. 10). Such a circuit is most accurate when $(Z_1 + Z_2) \ll Z_C$. Using the series resonance is helpful if an input circuit with a relatively low input impedance is required.

We have seen that the value k_T^2 of a piezoelectric material can, in principle, be measured by determining the frequencies ω_0 and ω_1 of the parallel and series resonances of an unloaded piezoelectric resonator. This technique of measurement works well, again in principle, with a plate of piezoelectric material on which thin electrodes have been deposited. With piezoelectric ceramics, however, other resonances may be associated with shear waves, and with other transverse modes in which stress and strain variations occur in directions parallel, as well as perpen-

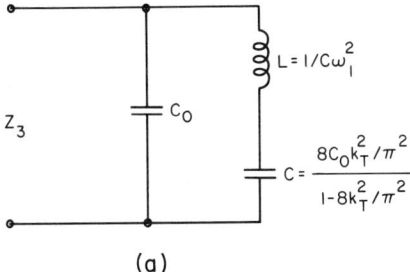

(a)

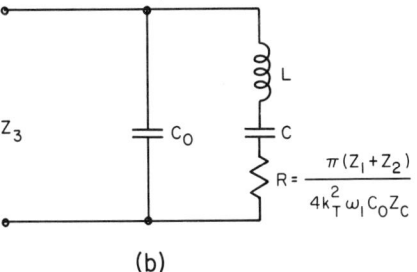

(b)

Figure 1.4.11 (a) Equivalent circuit for an unloaded resonator with an inductance and capacitance in series, and a series resonance at $\omega = \omega_1$. Note that C is equal in value to a capacity $-C_0$ in series with a capacity $8C_0 k_T^2/\pi^2$. (b) Equivalent circuit for a resonator loaded with impedances Z_1 and Z_2 at each end.

dicular, to the electrodes. Such resonances can sometimes hinder good readings of ω_1, making it difficult to determine k_T^2 accurately. However, Onoe et al. [8] have shown that measurement of the points where the conductance of a transducer goes through maxima and minima will normally give accurate readings of ω_1 and ω_0, respectively. Furthermore, difficulties in determining the resonant frequency ω_0, due to the presence of other modes, can often be circumvented by using a higher-order resonance, usually the third harmonic.

1.4.6 Broadband Operation of Transducers into an Acoustic Medium: The KLM Model

We now consider the operation of a transducer used to excite a wave in an acoustic medium. Normally, it is difficult to design transducers for broadband operation and predict their behavior without resorting to numerical computation. But we can obtain some physical insight into the behavior of such transducers by deducing yet another equivalent circuit, due to Krimholtz, Leedom, and Matthaei [9], which we shall refer to as the *KLM model*.

We regard the piezoelectric material as capable of propagating two waves or modes, one in the forward direction, denoted by a subscript F, the other in the backward direction, denoted by a subscript B. These modes are continually excited along the length of the transducer by the displacement current $j\omega DA$ that passes through it.

In the Redwood model of Fig. 1.4.5, the continuous excitation is replaced by two electrical sources, one at each end of the transmission line. In the KLM model, a more conventional circuit is derived, which has an electrical source at the

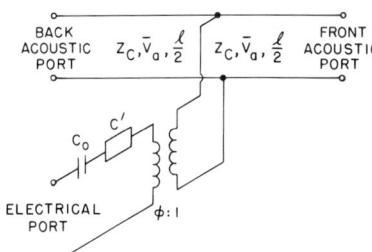

Figure 1.4.12 KLM model of a piezoelectric transducer: $\phi = k_T(\pi/\omega_0 C_0 Z_C)^{1/2}$ sinc $(\omega/2\omega_0)$ and $C' = -C_0/[k_T^2 \text{ sinc}(\omega/\omega_0)]$. (From Krimholtz et al. [9].)

center of a transmission line. Its disadvantage is that the ratio of the required transformer varies with frequency.

The KLM model is represented by the three-port network shown in Fig. 1.4.12. This equivalent circuit relates the voltage V_3 and the current I_3 at the electrical port to the forces $F_1 = -AT(-l/2)$ and $F_2 = -AT(l/2)$ and the velocities $v_1 = v(-l/2)$ and $v_2 = -v(l/2)$. It is convenient because it expresses the acoustic parameters in terms of an equivalent transmission line, regarding the stress T as equivalent to a voltage $-V$, and the velocity v as equivalent to a current I. This makes it easier to design multiple matching layers [10]. On the other hand, the electrical terminal parameters are expressed in terms of an equivalent lumped circuit, which is convenient for the design of electrical matching networks at commonly used frequencies. Using the relations $\bar{\beta}_a = \omega/\bar{V}_a$, $\bar{V}_a = (c^D/\rho_{m0})^{1/2}$, $\bar{Z}_0 = (c^D \rho_{m0})^{1/2}$, $Z_C = A\bar{Z}_0$, and $I_3 = j\omega AD$ and substituting Eq. (1.4.7) in Eq. (1.4.4), then Eqs. (1.4.3) and (1.4.4) can be written as

$$\frac{dT}{dz} - j\bar{\beta}_a \bar{Z}_0 v = 0 \tag{1.4.50}$$

and

$$\frac{dv}{dz} - j\left(\frac{\bar{\beta}_a}{\bar{Z}_0}\right)T = \frac{hI_3}{Z_C \bar{V}_a} \tag{1.4.51}$$

To obtain a transmission-line representation, it is convenient to define parameters proportional to the amplitudes of the forward and backward waves. Again, we denote these forward and backward waves propagating in the transducer by subscripts F and B, respectively. Thus the total stress is

$$T = T_F(z) + T_B(z) \tag{1.4.52}$$

and the total velocity is

$$v = v_F(z) + v_B(z) \tag{1.4.53}$$

We write $T_F(z) = -\bar{Z}_0 v_F(z)$ for the forward wave and $T_B(z) = \bar{Z}_0 v_B(z)$ for the backward wave. (This concept of forward and backward waves is slightly different to the one described in Sec. 1.4.1.) We substitute Eqs. (1.4.52) and (1.4.53) into Eqs. (1.4.50) and (1.4.51). After performing the necessary addition and subtraction, and leaving the z dependences of $v_F(z)$ and $v_B(z)$ implicit, v_F and v_B obey the following relations:

$$\frac{dv_F}{dz} + j\bar{\beta}_a v_F = \frac{hI_3}{2Z_C \overline{V}_a} \tag{1.4.54}$$

and

$$\frac{dv_B}{dz} - j\bar{\beta}_a v_B = \frac{hI_3}{2Z_C \overline{V}_a} \tag{1.4.55}$$

When $I_3 = 0$ and there is no external excitation, the solutions of Eqs. (1.4.54) and (1.4.55) are the normal modes or propagating waves in the piezoelectric medium: thus these two equations vary as $v_F \sim \exp(-j\bar{\beta}_a z)$ and $v_B \sim \exp(j\bar{\beta}_a z)$, respectively. More generally, Eqs. (1.4.54) and (1.4.55) express the excitation of the modes of the system by external signals. As we shall see in Sec. 2.5, this normal-mode method can be generalized fairly easily and used in the same form for systems of finite cross section. We show in Sec. 2.5 that the normal-mode formulation is particularly useful for surface acoustic wave interdigital and wedge transducers.

If the center of the transducer is at $z = 0$, we can integrate Eq. (1.4.54) to yield the result

$$v_F(z) = v_F\left(\frac{-l}{2}\right) e^{-j\bar{\beta}_a(z + l/2)} + \frac{h}{2Z_C \overline{V}_a} e^{-j\bar{\beta}_a z} \int_{-l/2}^{z} e^{j\bar{\beta}_a z} I_3 \, dz \tag{1.4.56}$$

Equation (1.4.56) shows that the amplitude of the forward wave at the plane z depends on the wave initially propagating through the medium from the plane $z = -l/2$ and on an additional term due to the cumulative excitation by the current I_3 along the transducer. A similar result can be obtained for the excitation of the wave propagating in the $-z$ direction.

When I_3 is uniform, as it is in the transducer considered here, Eq. (1.4.56) can be integrated and written in the form

$$v_F\left(\frac{l}{2}\right) = v_F\left(\frac{-l}{2}\right) e^{-j\bar{\beta}_a l} + \frac{hI_3}{Z_C \omega} e^{-j\bar{\beta}_a l/2} \sin\frac{\bar{\beta}_a l}{2} \tag{1.4.57}$$

with

$$v_B\left(\frac{-l}{2}\right) = v_B\left(\frac{l}{2}\right) e^{-j\bar{\beta}_a l} - \frac{hI_3}{Z_C \omega} e^{-j\bar{\beta}_a l/2} \sin\frac{\bar{\beta}_a l}{2} \tag{1.4.58}$$

The first term of Eq. (1.4.57) corresponds to the wave that is excited at the left-hand side of the transducer and propagates through it. The second term indicates that the current I_3 behaves as though it is exciting waves at the center of the transducer. These waves propagate along the transducer with amplitudes varying as $\exp(-j\bar{\beta}_a|z|)$ and reach the ends $z = \pm l/2$ with a phase delay $\bar{\beta}_a l/2$. We can formalize this concept by defining the velocities just to the right of the center of the transducer as v_F^+ and v_B^+, and the velocities just to the left of its center as v_F^- and v_B^-. Hence we write

$$v_F\left(\frac{l}{2}\right) = v_F^+ e^{-j\bar{\beta}_a l/2}$$

$$v_B\left(\frac{l}{2}\right) = v_B^+ e^{j\bar{\beta}_a l/2}$$

$$v_F\left(\frac{-l}{2}\right) = v_F^- e^{j\bar{\beta}_a l/2} \qquad (1.4.59)$$

$$v_B\left(\frac{-l}{2}\right) = v_B^- e^{-j\bar{\beta}_a l/2}$$

It follows from Eqs. (1.4.57)–(1.4.59) that if $I_3 = 0$, then $v_F^+ = v_F^-$ and $v_B^+ = v_B^-$. When I_3 is finite, then

$$v_B^+ - v_B^- = v_F^+ - v_F^- = \frac{hI_3}{Z_C \omega} \sin \frac{\bar{\beta}_a l}{2} \qquad (1.4.60)$$

The total change in velocity induced by the current at the center of the transducer is therefore

$$v^+ - v^- = (v_F^+ + v_B^+) - (v_F^- + v_B^-) = \frac{2hI_3}{Z_C \omega} \sin \frac{\bar{\beta}_a l}{2} \qquad (1.4.61)$$

The electrical current I_3 flowing into the transducer excites the backward and forward waves equally, so the ratio of induced acoustic velocity (or equivalent current) to the electrical current can be represented by a perfect transformer with a ratio 1 to ϕ, defined by the relation

$$\phi = \frac{2h \sin \bar{\beta}_a l/2}{Z_C \omega} = k_T \left(\frac{\pi}{\omega_0 C_0 Z_C}\right)^{1/2} \operatorname{sinc} \frac{\omega}{2\omega_0} \qquad (1.4.62)$$

where $\operatorname{sinc} x = \sin \pi x / \pi x$ and the center frequency of the transducer ω_0 is defined from the relation $\bar{\beta}_a l = \pi$. When $\omega = n\omega_0$ and n is even, the transformer ratio is zero, resulting in zero excitation. At these frequencies, the transducer is an integral number of wavelengths thick and the generated waves cancel exactly at the output terminals. As $\omega \to 0$, ϕ increases by $\pi/2$ from its value at $\omega = \omega_0$.

To complete the equivalent circuit, we also need to know the relationship between T_F, T_B, and V_3. It follows from Eq. (1.4.18) or Eq. (1.4.30) that

$$V_3 = \frac{I_3}{j\omega C_0} - \frac{h}{j\omega}\left[v\left(\frac{l}{2}\right) - v\left(\frac{-l}{2}\right)\right] \qquad (1.4.63)$$

Substituting Eq. (1.4.59) into Eq. (1.4.63), we write

$$V_3 = \frac{I_3}{j\omega C_0} - \frac{h}{j\omega}\left[(v_F^+ - v_F^- + v_B^+ - v_B^-)\cos\frac{\bar{\beta}_a l}{2}\right.$$

$$\left. - j(v_F^+ + v_F^- - v_B^+ - v_B^-)\sin\frac{\bar{\beta}_a l}{2}\right] \qquad (1.4.64)$$

From Eq. (1.4.60), we see that
$$v_F^+ - v_B^+ = v_F^- - v_B^- \tag{1.4.65}$$
Because we can write
$$T = T_F^+ + T_B^+ = T_F^- + T_B^- = (v_B^+ - v_F^+)\overline{Z}_0 = (v_B^- - v_F^-)\overline{Z}_0 \tag{1.4.66}$$
it follows that the input stress T is continuous through the center terminal and is analogous to voltage. Therefore,
$$V_3 = \frac{I_3}{j\omega C_0} - \frac{h^2 I_3}{j\omega^2 Z_C} \sin \overline{\beta}_a l - \frac{2hT}{\omega Z_C} \sin \frac{\overline{\beta}_a l}{2} \tag{1.4.67}$$

The first term of Eq. (1.4.67) is identified with the clamped capacitance of the transducer. The second and third terms are associated with acoustic wave excitation. The second term is a reactance $X = (h^2/\omega^2 Z_C) \sin \overline{\beta}_a l$, that is, a capacitance of value
$$C' = -\frac{C_0}{k_T^2} \frac{1}{\text{sinc}(\omega/\omega_0)} \tag{1.4.68}$$

This capacitor is in series with the transducer capacity C_0; it is negative for frequencies where $\omega < \omega_0$ and is normally much larger than C_0. At resonance when $\omega = \omega_0$, $C' = \infty$; when $\omega \to 0$, then $C' \to -C_0/k_T^2$. The last term of Eq. (1.4.67) contains the transformer ratio already derived from the relationship between the electric and the equivalent acoustic currents. Thus the model is consistent for both voltage and current.

The KLM model of Fig. 1.4.12 retains close ties with the actual physical processes in an acoustic transducer. Its advantage is the use of a single central coupling point between electrical and acoustic quantities, gained at the cost of a variable series reactance and a variable transformer ratio.

Effect of reflections on the response of a transducer to a sinusoidal signal. We shall now use the KLM model of Fig. 1.4.12 to consider the effect of different terminations at the acoustic ports when the transducer is excited by a sinusoidal signal of frequency ω. We see from Eq. (1.4.57), or from the KLM equivalent circuit, that the forward wave reaching $z = l/2$ has stress and velocity components

$$-Z_0 v_F\left(\frac{l}{2}\right) = T_F\left(\frac{l}{2}\right) = -\frac{hI_3}{\omega A} \sin \frac{\overline{\beta}_a l}{2} e^{-j\overline{\beta}_a l/2}$$
$$= \frac{jhI_3}{\omega A} \frac{1 - e^{-j\overline{\beta}_a l}}{2} \tag{1.4.69}$$

In Sec. 1.4.7 we show that it is better to use the second form of this equation when considering the pulse response of the transducer. But here, as there is no external acoustic excitation, we have assumed that $v_F(-l/2) = 0$.

We have written the expression for $v_F(l/2)$ on the first line of Eq. (1.4.69) as if the current I_3 generates a wave at the center of the transducer ($z = 0$), which

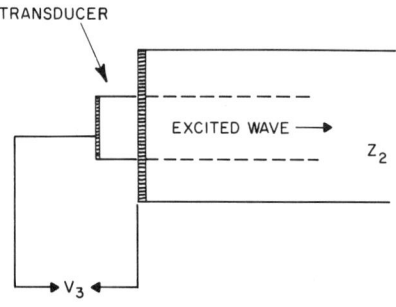

Figure 1.4.13 Transducer used to excite a wave in a medium of impedance Z_2. The metal electrodes are regarded as being infinitesimally thin.

is delayed in phase by $\bar{\beta}_a l/2$ by the time it reaches $z = l/2$. This form is more convenient than, although equivalent to, writing $v_F(l/2) \propto I_3[1 - \exp(-j\bar{\beta}_a l)]$ and assuming that the wave is generated at two points, $z = -l/2$ and $z = l/2$, as it is in the Redwood model of Fig. 1.4.5.

We now consider what happens if the terminations are imperfect. In this case, it is as if the waves are produced at the center of the transducer and suffer reflections at $z = -l/2$ and $z = l/2$; thus the total stress or velocity depends on the reflection coefficients at each surface.

Suppose that we consider a transducer of the type shown in Fig. 1.4.13, with the surface $z = -l/2$ open to air (i.e., $Z_1 = 0$ at $z = -l/2$). In this case, two waves appear to be generated *at the center of the transducer $z = 0$*, as shown in Fig. 1.4.14. The velocity and stress components of the forward wave reaching $z = l/2$ are given by Eq. (1.4.69). The velocity and stress components of the backward wave reaching $z = -l/2$ are

$$Z_0 v_B\left(\frac{-l}{2}\right) = T_B\left(\frac{-l}{2}\right) = -\frac{hI_3}{\omega A} \sin \frac{\bar{\beta}_a l}{2} e^{-j\bar{\beta}_a l/2} \tag{1.4.70}$$

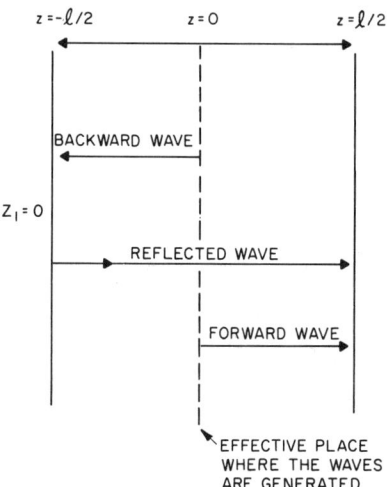

Figure 1.4.14 Waves generated in a transducer with $Z_1 = 0$ at $z = -l/2$.

The wave is reflected at $z = -l/2$. Because of the zero-impedance boundary condition, the reflected wave has its stress T of opposite sign to the incident wave. This wave suffers a further phase change $\bar{\beta}_a l$ by the time it reaches $z = l/2$. This implies that the wave suffers a total phase delay of $3\bar{\beta}_a l/2 + \pi$ from its effective point of generation to the output plane $z = l$. We therefore expect that when both waves add in phase, the output stress will be double the stress of a transducer that is perfectly matched at both ends. This occurs at $\bar{\beta}_a l = \pi$. On the other hand, because the reflected wave changes phase with frequency more rapidly than the forward wave, the bandwidth will tend to be narrower than for a transducer that is perfectly matched at both ends.

We can obtain this result in a more mathematical form by evaluating the amplitude and phase of the backward wave, after reflection and a total phase delay of $3\beta_a l/2$. In accordance with the sign convention used (where T reverses in sign after reflection at $z = -l/2$), Eq. (1.4.70) yields the result

$$T_B\left(\frac{l}{2}\right) = \frac{hI_3}{\omega A} \sin \frac{\bar{\beta}_a l}{2} e^{-3j\bar{\beta}_a l/2} \tag{1.4.71}$$

Adding Eqs. (1.4.69) and (1.4.71), we see that the total stress at $z = l/2$ is

$$T\left(\frac{l}{2}\right) = -2j\frac{hI_3}{\omega A} e^{-j\bar{\beta}_a l} \sin^2 \frac{\bar{\beta}_a l}{2} \tag{1.4.72}$$

Thus $|T(l/2)|$ varies as $(\omega_0/\omega) \sin^2 (\pi\omega/2\omega_0)$. This result should be compared with Eq. (1.4.57), where $|v_F(l/2)|$ and hence $|T_F(l/2)|$ vary as $(\omega_0/\omega) |\sin (\pi\omega/2\omega_0)|$, that is, less rapidly than when the wave reflected at $z = -l/2$ is added to the forward wave component.

Half-wave and quarter-wave resonators. In general, if there are reflections at both surfaces of the transducer, then T, v, and E will tend to be maximum when the forward and reflected waves add constructively. This occurs near a resonant condition. If $Z_1 \ll Z_C$ and $Z_2 \ll Z_C$, then v and T build up in amplitude when $l \approx (2n + 1)\lambda/2$, so for the strongest excitation, the resonator should be approximately a half-wavelength long. Similarly, if $Z_1 \gg Z_C$ and $Z_2 \ll Z_C$, or if $Z_1 \ll Z_C$ and $Z_2 \gg Z_C$, the reflections add up when $l \approx \lambda/4$; this implies that a rigidly backed resonator will show the strongest excitation near a frequency where $l \approx \lambda/4$. Thus by carefully choosing the backing and matching impedance into the medium of interest, the frequency response of the transducer can be varied and optimized at particular frequencies.

1.4.7 Pulse Response of a Transducer with Arbitrary Terminations

We can also consider what occurs when a pulse rather than a continuous-wave (CW) RF signal is injected into the transducer. We shall first assume that the transducer is perfectly terminated at both acoustic ports. In this case,

$v_F(-l/2) = 0$, and we can write Eq. (1.4.57) in the form

$$v_F\left(\omega, \frac{l}{2}\right) = \frac{hI_3}{2jZ_C\omega}(1 - e^{-j\bar{\beta}_a l}) \tag{1.4.73}$$

If the signal varies as exp $(j\omega t)$, the charge on the right-hand electrode is $Q = I_3/j\omega$ and that on the left-hand electrode, $-Q$. Thus, in the frequency domain, we can write

$$\frac{-T_F(\omega, l/2)}{Z_0} = v_F\left(\omega, \frac{l}{2}\right) = \frac{hQ}{2Z_C}(1 - e^{-j\bar{\beta}_a l}) \tag{1.4.74}$$

We put $\bar{\beta}_a l = \omega l/\bar{V}_a = \omega T$, where T is the transit time of an acoustic wave through the transducer. Then, taking the Fourier transform of Eq. (1.4.74), we see that the stress or velocity, as a function of time at $z = l/2$, is

$$\frac{-T_F(t, l/2)}{Z_0} = v_F\left(t, \frac{l}{2}\right) = \frac{h}{2Z_C}[Q(t) - Q(t - T)] \tag{1.4.75}$$

This equation leads to an equally valid viewpoint of transducer operation in the time domain. A stress or velocity component proportional to the charge on the right-hand electrode is excited. The charge on the left-hand electrode excites a signal of opposite sign, which suffers a time delay T before reaching $z = l/2$. If we assume in this model that $k_T^2 \ll 1$, the transducer behaves like a capacitor with a voltage source $V(t)$ across it. This voltage $V(t)$ is linearly proportional to the charges $Q(t)$ and $-Q(t)$ at each electrode; the transducer therefore emits signals proportional to $V(t)$. We can write

$$\frac{-T_F(t, l/2)}{Z_0} = v_F\left(t, \frac{l}{2}\right) \approx \frac{hC_0}{2Z_C}[V(t) - V(t - T)] \tag{1.4.76}$$

or

$$T_F\left(t, \frac{l}{2}\right) = -\frac{hC_0}{2A}[V(t) - V(t - T)] \tag{1.4.77}$$

It is instructive to obtain this same result using the Redwood model of Fig. 1.4.5 for the equivalent circuit. If the transducer is terminated at each end, the Redwood circuit becomes the one shown in Fig. 1.4.15. When the coupling coefficient is sufficiently small (i.e., when h is small), very little current flows through the negative capacitor. Thus we can ignore its presence. In this case, the equivalent voltage developed across the output of the transformer is hC_0V, and it is divided

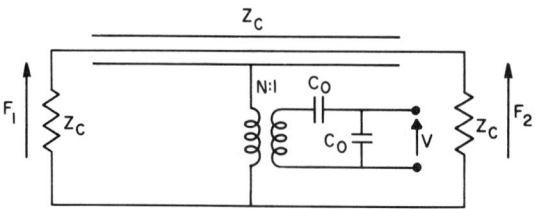

Figure 1.4.15 Redwood circuit with terminations Z_C and $N = hC_0$.

equally between the load impedance Z_C and the impedance Z_C of the coaxial line. Consequently, the equivalent voltages initially developed across the two loads (i.e., the forces in acoustic terms) are given by the simple relation

$$F_1(t) = F_2(t) = \frac{hC_0V(t)}{2} \qquad (1.4.78)$$

The voltages developed across the coaxial line and the load are of opposite sign, as defined in Fig. 1.4.15. Consequently, the induced voltage or stress at the left-hand end of the coaxial line will propagate along it, giving rise to a stress pulse, of opposite sign to the initial stress pulse and at a later time T, across the right-hand load impedance Z_C. Using the relations $F_1 = -AT(-l/2)$ and $F_2 = -AT(l/2)$, the Redwood equivalent circuit results in Eq. (1.4.77). Thus the Redwood model demonstrates clearly that pulsed excitation makes the transducer behave as if excited by equal sources of opposite sign at each end, as shown in Fig. 1.4.16(a) and (b).

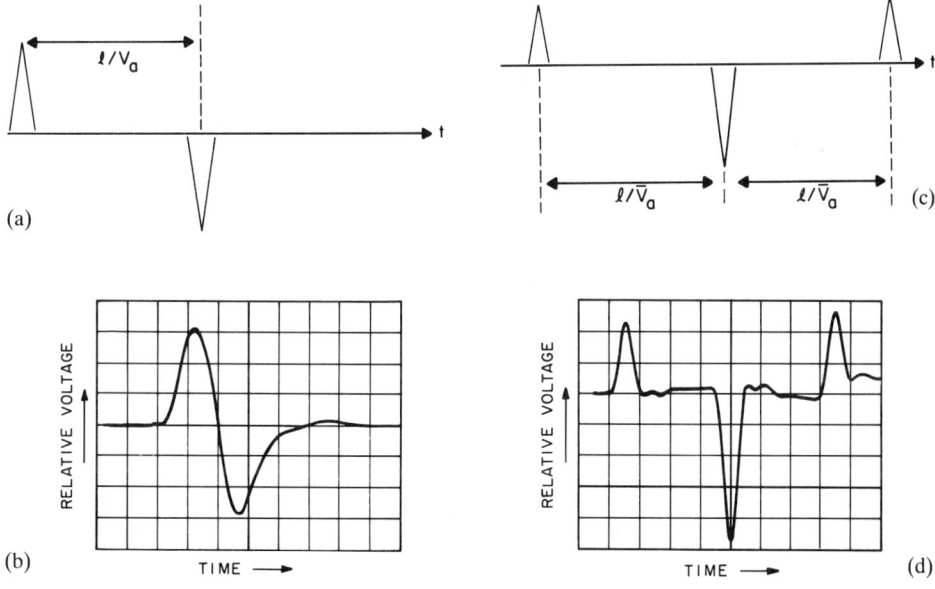

Figure 1.4.16 Pulses generated by a transducer excited by a voltage spike. The signals are detected by a long transducer working into a low-impedance electrical load [see Eq. (1.4.63), which yields $I_3 = -hC_0v(-l/2)$ for the initial signal]. (a) Transducer terminated by a matched load at $z = 0$. (b) Corresponding experimental result obtained in the author's laboratory for a 5-MHz commercial transducer with a matched backing (Panametrics) excited by a short pulse. (c) Transducer terminated by a matched load at $z = l$ and with zero impedance at $z = 0$. (d) Corresponding result obtained in the author's laboratory with an unbacked, 2-mm-thick PZT-5A transducer excited by a short electrical pulse. Note that the experimental results tend to give pulses that are somewhat less sharp than indicated by simple theory. This is because the theory neglects finite coupling, and hence loading.

Sec. 1.4   Piezoelectric Transducers

Air-backed transducer and unmatched front surface. An air-backed transducer terminated with an impedance $Z_2 = Z_C$ at $z = l/2$ will emit a series of three pulses, as illustrated in Fig. 1.4.16(c) and (d). The first pulse comes from the plane $z = l/2$. The second pulse has twice the amplitude of the first and is emitted from $z = -l/2$, arriving at a time $T = l/\overline{V}_a$. It is doubled in size because it corresponds to the sum of the forward and backward waves excited at the left-hand electrode. The third pulse is emitted in the backward direction from $z = l/2$; it arrives back at $z = l/2$ at a time $2l/\overline{V}_a$ after the first pulse.

Terminating the back surface of the transducer with a matched load $Z_1 = Z_C$ is a good technique for obtaining the shortest possible pulse because there is no reflection of a wave from the back surface. This is true even if the front surface of the transducer is not terminated with a matched load ($Z_2 \neq Z_C$). In this case, the Redwood equivalent circuit of Fig. 1.4.5 shows that

$$T_F(t) = -\frac{h}{A} \frac{Z_2/Z_C}{1 + Z_2/Z_C} [Q(t) - Q(t - T)] \quad (1.4.79)$$

Thus if a short voltage pulse is applied to the transducer, the output will be similar to the one shown in Fig. 1.4.16(a) and (b), differing only in amplitude from the situation when $Z_2 = Z_C$.

In practice, because the impedance of the transducer is not purely real, the transducer cannot be driven with a simple voltage spike. However, the stress excited by a transducer that is perfectly terminated at its back surface can approximate a single cycle of a sinusoidal wave when it is driven by a short pulse of current or voltage. In the same way, a transducer that is well terminated at its front surface ($Z_2 = Z_C$), but unmatched at the back surface, will give an output much like that of Fig. 1.4.16(c) and (d), although the relative amplitudes and signs of the three spikes may differ.

Matching at the back surface of the transducer tends to give a broader bandwidth but not the best efficiency. In this case, even if the front surface of the transducer is matched, half the power is lost at $z = -l/2$, and there is at least a 3-dB power loss in converting electrical to mechanical energy, or vice versa. Worse still, if the back surface is terminated with an impedance Z_C and the front surface is badly mismatched, most of the electrical power entering the transducer will be dissipated in the backing. The Redwood or KLM equivalent circuits (Figs. 1.4.5 and 1.4.12, respectively) show that when a transducer is excited at the center frequency ($\omega = \omega_0$), where each half of the transducer is $\lambda/4$ long, the fraction of the input power f reaching the acoustic load Z_2 is

$$f = \frac{Z_2}{Z_2 + Z_1} \quad (1.4.80)$$

Example: PZT Transducer

As an example, a PZT transducer of unit area with a matched backing ($Z_1 = Z_C = 34 \times 10^6$ kg/m²-s), radiating into water, gives a fractional efficiency of $f = 0.042$, or -13.7 dB, when the electrical terminal is perfectly matched. If the backing impedance is reduced to 17×10^6 kg/m²-s, the bandwidth decreases, and the efficiency increases to $f = 0.08$, or -10.9 dB.

More generally, if there are reflections at both surfaces of the transducer, it will "ring" with an output that looks like a decaying sinusoidal wave, with a period $2l/V_a$, whose decay rate depends on the acoustic mismatch and electrical loading of the transducer.

1.4.8 Electrical Input Impedance of a Loaded Transducer

We now consider the input impedance of a transducer used to excite a wave in a medium of impedance Z_2. When such a transducer is employed at UHF frequencies for a delay line, it is typically used in the configuration shown in Fig. 1.4.1(a). More generally, the back surface loading by the electrode and the backing materials, and the front surface loading by the impedances of intermediate layers between the transducer and the medium in which a wave is being excited, must be taken into account. Careful choice of these layers can optimize the bandwidth and efficiency of the transducer, although numerical calculations are normally required for a complete design of an optimized broadband transducer.

Constructional techniques. The transducers used in ultrahigh-frequency (UHF) delay lines (i.e., for frequencies over 100 MHz) are typically 1 to 30 μm thick. Therefore, a piezoelectric layer is normally deposited on the delay-line substrate; at the moment, the method of deposition most commonly used is sputter deposition of a material such as zinc oxide. Because the basic zinc oxide crystals have hexagonal symmetry, the individual crystallites deposited by this process need have only their z axes aligned normal to the surface of the substrate to obtain strong longitudinal wave coupling. Rotation of the individual crystallites about their axes makes no difference. In practice, with a value of k_T^2, at least 90% of the value for a single crystal can be obtained in high-quality deposited layers. This means that the substrate must be chosen carefully and that its surface conditions must be well controlled.

Electrodes must of course be employed on either side of the piezoelectric layer. Unless the substrate is itself a conductor, a metal film such as gold, typically 1000 Å thick, is deposited on the substrate before the deposition of the piezoelectric layer. A similar film is deposited through a mask on the top surface of the transducer to form a top dot electrode, typically with a diameter of less than 1 mm. This top dot defines the effective diameter of the transducer and hence of the emitted acoustic beam.

Broadband impedance matching considerations may require that additional layers of materials, such as various metals or glasses, be deposited on the substrate before the electrode is deposited on the substrate side of the transducer. In practice, multiple layers of matching materials have only in rare instances been deposited on UHF transducers. But it is important to take account of the effect of the metal electrode on the acoustic matching.

Low-frequency broadband transducers used in the frequency range from 0.1 to 10 MHz are often made of high dielectric ($\varepsilon^S > 500\varepsilon_0$) and high-coupling-constant PZT ceramics so that they can be designed with electrical impedances of the order of 50 Ω. The full diameter of the ceramic is electroded. To obtain broadband or short-pulse operation, the back side may be bonded to a lossy high-impedance material such as epoxy filled with tungsten, as illustrated in Fig. 1.4.1(b). Because tungsten has a very high impedance ($Z_0 = 103 \times 10^6$ kg/m²-s for longitudinal waves) and epoxy has a low impedance ($Z_0 = 3.4 \times 10^6$ kg/m²-s), the impedance of the composite material can be varied from 3.5×10^6 to 40×10^6 kg/m²-s by changing the proportions of tungsten and epoxy. At the same time, such composites have relatively high attenuation coefficients, due to acoustic scattering by the tungsten powder and the high viscosity of the medium. Rubber-like media, such as urethane or vinyl filled with tungsten powder, provide still higher losses.

Such transducers have an excellent pulse response but are inefficient for exciting waves in a low-impedance medium such as water. To obtain high efficiency, however, it is better to use multiple quarter-wave matching layers between a PZT ceramic ($\overline{Z}_0 = 34 \times 10^6$ kg/m²-s) and water ($Z_0 = 1.5 \times 10^6$ kg/m²-s), employing for the layers such materials as glass ($Z_0 = 9$ to 13×10^6 kg/m²-s), epoxy ($Z_0 = 3.4 \times 10^6$ kg/m²-s), or filled epoxies ($Z_0 = 3$ to 12×10^6 kg/m²-s). Either no backing or a low-impedance backing is used. These materials are either epoxy-bonded to each other or cast in place.

Electrical impedance at resonance. The impedance of the transducer at its center frequency ω_0 can be found easily using the KLM equivalent circuit of Fig. 1.4.12. When $\omega = \omega_0$, the KLM circuit becomes a half-wavelength-long transmission line terminated by the impedances Z_1 and Z_2 at each end. We take Z_1 and Z_2 to be real. The effective impedance at the center point is calculated by transforming Z_1 and Z_2 through the two quarter-wavelength-long transmission lines to the center tap. At the center frequency ω_0, as follows from Eq. (1.1.49), their effective values are Z_C^2/Z_1 and Z_C^2/Z_2, respectively, yielding a total shunt resistance of $R = Z_C^2/(Z_1 + Z_2)$.

The electrical radiation resistance of the transducer at its center frequency is determined by finding the electrical resistance seen on the other side of the transformer. This is

$$R_{a0} = [\phi(\omega_0)]^2 \frac{Z_C}{Z_1 + Z_2} = \frac{4k_T^2}{\pi\omega_0 C_0} \frac{Z_C^2}{Z_1 + Z_2} \qquad (1.4.81)$$

This result can also be obtained directly from Eq. (1.4.23), with $\omega = \omega_0$ and $\overline{\beta}_a l = \pi$.

Because of the quarter-wavelength transformation along the transmission lines, the electrical radiation resistance R_{a0} increases as Z_1 and Z_2 are decreased. When the ends are short-circuited, the radiation resistance becomes infinite, as we would expect for an unloaded loss-free transducer at the center frequency.

Terminated transducer ($Z_1 = Z_2 = Z_C$) or $Z_1 = 0, Z_2 = Z_C$. When the transducer is perfectly terminated at each end, the radiation resistance is $R_{a0} = 2k_T^2/\pi\omega_0 C_0$. Thus when k_T is small, the radiation resistance typically is less than the series reactance $X = 1/\omega_0 C_0$ of the transducer. When the left-hand side of the transducer is air-backed ($Z_1 = 0$), the input resistance at resonance becomes

$$R_{a0} = \frac{4k_T^2}{\pi\omega_0 C_0} \frac{Z_C}{Z_2} \qquad (1.4.82)$$

Thus with $Z_2 = Z_C$, the input resistance is double that of a transducer terminated on both sides ($Z_1 = Z_2 = Z_C$) and the total input power, as opposed to half the input power, goes into the acoustic load. As we have seen, however, the bandwidth is narrower in this case.

1.4.9 Efficiency of Power Transfer to Transducer

We now consider frequency response and efficiency, and how matching is affected by the impedance variation of the transducer. When a transducer is terminated by a matched load $Z_2 = \overline{Z}_0$, for example, and air-backed by an impedance $Z_1 = 0$, all the electrical power entering it must appear as acoustic power in the acoustic load, which is the only resistive load present. At the center frequency $\omega = \omega_0$, it follows from Eq. (1.4.23) that the transducer exhibits zero motional reactive impedance. If $k_T \ll 1$, however, the electrical radiation resistance R_{a0} presented by the transducer is much less than the capacitive reactance $1/\omega C_0$ in series with R_{a0}. Thus the transducer tends to present a highly reactive load to the input source. Typically, for most efficient operation, the series capacitance can be tuned out with an inductance, in which case the main source of loss is the resistance of the contacts, while the bandwidth may be seriously limited by the Q of this tuning circuit. Often, because the timing inductance may itself be lossy, it is better to choose the source impedance to obtain maximum power into the acoustic load with a given input voltage, instead of tuning the transducer electrically. Typically, this means that the resistance of the source is chosen to equal the reactive impedance of the load.

We can derive the result by considering the equivalent circuit shown in Fig. 1.4.17. If we take the impedance of the power supply to be R_0, then the maximum power P_{in} available from the power supply of voltage V, when terminated by an impedance R_0, is

$$P_{in} = \frac{V^2}{8R_0} \qquad (1.4.83)$$

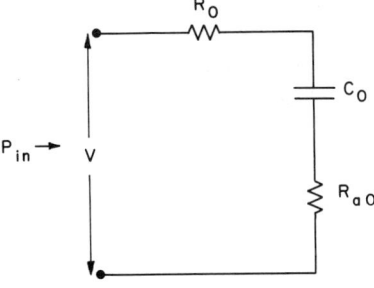

Figure 1.4.17 Equivalent circuit of a piezoelectric transducer at its resonant frequency ω_0.

The power delivered to the transducer with the same input voltage V is that developed across the load R_{a0}. This is $P_L = R_{a0}I^2/2$, where I is the current flowing in the circuit, or

$$P_L = \frac{V^2 R_{a0}}{2[(R_0 + R_{a0})^2 + (1/\omega C_0)^2]} \tag{1.4.84}$$

Thus the available input power is P_{in}, and the power dissipated in the radiation resistance is P_L. The efficiency η_T is

$$\eta_T = \frac{P_L}{P_{in}} = \frac{4 R_0 R_{a0}}{(R_0 + R_{a0})^2 + (1/\omega C_0)^2} \tag{1.4.85}$$

If we differentiate Eq. (1.4.85) with respect to R_0, then η_T is maximum when $R_0 = \sqrt{R_{a0}^2 + 1/(\omega C_0)^2}$. When $R_{a0} \ll (1/\omega C_0)^2$, as it is when $k_T^2 \ll 1$, then η_T is maximum when $R_0 \approx 1/\omega C_0$.

The maximum power transfer is

$$\eta_T\,(\text{max}) = \frac{2 R_{a0}}{R_{a0} + R_0} = \frac{2 R_{a0}}{R_{a0} + \sqrt{R_{a0}^2 + 1/(\omega C_0)^2}}$$
$$\approx 2 R_{a0} \omega C_0 \quad \left(R_{a0} \ll \frac{1}{\omega C_0} \right) \tag{1.4.86}$$

The mismatch is severe, however, with $R_{a0} \ll 1/\omega C_0$ and $R_0 \approx 1/\omega C_0$, the maximum efficiency condition when $k_T^2 \ll 1$.

Example: Zinc Oxide Transducer on Sapphire

These considerations apply to both transmitting and receiving transducers in microwave systems. As an example, we consider a half-wavelength-thick, 1-GHz zinc oxide transducer on a sapphire substrate. For longitudinal waves, $Z_C = 36 \times 10^6\,A$ kg/m²-s and $Z_2 = 44.3 \times 10^6\,A$ kg/m²-s, where A is the area of the transducer. In this case, $l = 3.1\,\mu\text{m}$. If the transducer is designed to have a 50-Ω reactive impedance so that it gives best efficiency with a 50-Ω source, its capacity must be 3.2 pF. With a permittivity of $\varepsilon^S = 8.8\varepsilon_0$, the area of the transducer or the metal top dot electrode is 0.13 mm², and the diameter of the circular dot is 0.4 mm. Thus the beam diameter must be fairly small, although still many wavelengths in diameter ($\lambda = 11\,\mu\text{m}$ in sapphire). The radiation resistance of the transducer at this frequency is given by Eq. (1.4.82). When $k_T = 0.28$, $R_{a0} = 4\,\Omega$ and $P_L/P_{in} = 0.16$ (i.e., there is an 8-dB input loss).

Series inductance tuning of the transducer using a lossless inductance, which is equivalent to removing the $1/\omega C_0$ term in Eq. (1.4.85), improves its acoustic conversion efficiency by 3 dB. Efficiency may be improved further by using more complicated tuning circuits or transformers. However, if the radiation resistance is only a few ohms, there will always tend to be some loss associated with the resistance of the metal films and leads. It is important to realize that in this example, $Z_2 > Z_C$. Hence, as we have already discussed in Sec. 1.4.6 and will discuss further in Sec. 1.4.11, the value of R_a will tend to peak at a lower frequency, where the piezoelectric layer is a quarter-wavelength thick.

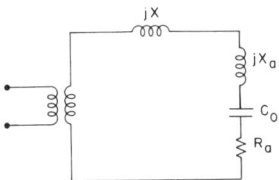

Figure 1.4.18 Matching circuit for an acoustic transducer.

1.4.10 Electrical Matching of a Loaded Transducer for Optimum Bandwidth and Efficiency

We shall consider placing the transducer in an inductive circuit that can tune out its series capacity at the center frequency. In addition, we shall consider the situation when a transformer is employed for impedance matching, that is, to match R_{a0} to the impedance of the input or output circuit, as shown in Fig. 1.4.18. Under these circumstances, the ratio of the electrical power transferred to the transducer to the available input power is

$$\eta_T = \frac{4R_0 R_a}{(R_0 + R_a)^2 + (X + X_a - 1/\omega C_0)^2} \tag{1.4.87}$$

where X is the reactance of the inductor, R_0 is the transformed value of the input resistance, and $Z_a = R_a + jX_a$ is the motional impedance of the transducer at an arbitrary frequency.

We first consider how the bandwidth is determined by the acoustic response of the circuit (i.e., how R_a varies with frequency). In the simplest case, when $R_0 \gg R_a$ and $X_a \ll 1/\omega C_0$, as occurs when the coupling coefficient k_T^2 is small and no electrical matching is used, then

$$\eta_T = \frac{4R_0 R_a}{R_0^2 + (1/\omega C_0)^2} \tag{1.4.88}$$

Here the acoustic output of the transducer is proportional only to $R_a(\omega)$, that is, it depends only on the acoustic response as a function of frequency.

We now consider the response of a transducer placed in a matching circuit, such as the one shown with the equivalent circuit (impedance jX_a, R_a, and $1/j\omega C_0$ in series) of the transducer in Fig. 1.4.18. Its bandwidth will be limited for two reasons. First, the value of R_a, the radiation resistance, varies with frequency due to the acoustic properties of the transducer. This gives rise to an effective acoustic Q, defined as $Q = f_0/\Delta f$, where f_0 is the center frequency of the response and Δf is the 3-dB bandwidth. We call the acoustic Q, Q_a. A transducer with poor acoustic matching at each end has a high value of Q_a; one that is well matched at each end has a low Q_a. The value of Q_a is derived below for various types of acoustic loads. Second, the circuit used to tune out the transducer capacity, usually a series or parallel inductance, has an effective circuit Q called Q_e (the electrical Q) of value $Q_e \approx 1/\omega_0 C_0 R_{a0}$. Thus Q_e tends to be small if $R_{a0} \sim 1/\omega_0 C_0$ (i.e., if the coupling coefficient k_T is large).

These considerations make the PZT ceramics good choices for broadband, high-efficiency, low-frequency transducers. As a class they have a very high k_T^2, approximately 0.25, which makes R_{a0} relatively high. They also have high dielectric constants ($\varepsilon^S = 1300\varepsilon_0$). Thus transducers designed to operate at frequencies of a few megaherz with diameters of the order of 1 cm can have capacitive reactances and input resistances of the order of 50 Ω. As this impedance is typical for most power sources, such a design leads to a system that can be reasonably well matched over a wide frequency range.

We shall now use the KLM circuit model of Fig. 1.4.12 to obtain a physical understanding of the matching problem. In Sec. 1.4.11 we shall also deal with a technique known as the Reeder–Winslow mathematical design method, illustrating it, for simplicity, only by reference to air-backed transducers [11]. The powerful and simple Reeder–Winslow formalism provides further insight into the design criteria for optimum efficiency and bandwidth.

Transducer acoustically matched at each end, $Z_1 = Z_2 = Z_C$. As a simple example of the KLM model, we consider both sides of the transducer to be terminated with a matching load so that $Z_1 = Z_2 = Z_C$. In this case, the acoustic response of the circuit is due only to the response of the transformer. It therefore follows from Eq. (1.4.23), or from the KLM model, that we can define a normalized radiation resistance $M_r(f)$ ($f = \omega/2\pi$ and $f_0 = \omega_0/2\pi$) as

$$M_r(f) = \frac{R_a}{R_{a0}} = \left(\frac{f_0}{f}\right)^2 \sin^2 \frac{\pi f}{2f_0} \qquad (1.4.89)$$
$$= \left(\frac{\pi}{2} \operatorname{sinc} \frac{f}{f_0}\right)^2$$

where $\operatorname{sinc} x = (\sin \pi x)/\pi x$, and

$$R_{a0} = \frac{k_T^2}{\pi^2 f_0 C_0} \qquad (1.4.90)$$

This function is plotted in Fig. 1.4.19. From the solid-line plot, we see that $M_r(0) = 2.47$ and $M_r(0.89 f_0) = 1.23$. Hence the 3-dB acoustic bandwidth of the transducer lies in the frequency range from zero to $0.89 f_0$, and the transducer acoustic response is like that of a low-pass filter because of the $(f_0/f)^2$ term in $M_r(f)$.

A transducer terminated by finite impedances on both sides has a finite response down to zero frequency. Physically, this is because the back of the transducer can push against an infinitely long backing, even at zero frequency, and thus move its front surface. An air-backed transducer cannot do this; thus at zero frequency, it cannot deliver power into the medium at its front surface. In practice, with a finite-length backing held in a rigid mount that is attached to the load, there will be a finite response at zero frequency.

Air-backed transducer matched at output, $Z_1 = 0$, $Z_2 = Z_C$. In this case, the KLM model of Fig. 1.4.12, or the use of Eq. (1.4.23), indicates that the

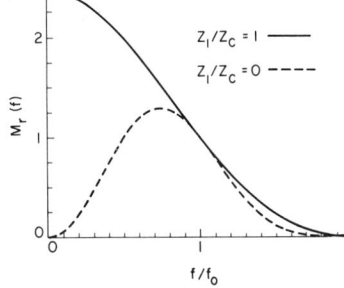

Figure 1.4.19 Plots of $M_r(f)$ for a transducer matched at its front surface ($Z_2 = Z_C$). Solid line: transducer with a matched backing ($Z_1 = Z_C$); dashed line: the real part of the normalized input radiation resistance for an air-backed transducer ($Z_1 = 0$).

real part of the normalized input radiation resistance is

$$M_r(f) = \frac{R_a}{R_{a0}} = \left(\frac{f_0}{f}\right)^2 \sin^4 \frac{\pi f}{2f_0} \tag{1.4.91}$$

where

$$R_{a0} = \frac{2k_T^2}{\pi^2 f_0 C_0} \tag{1.4.92}$$

This normalized function is plotted as the dashed line in Fig. 1.4.19. Note that an air-backed transducer has twice the value of R_{a0} of a transducer with $Z_1 = Z_C$.

The $(f_0/f)^2$ term makes the response maximum with a value of $R_a = 1.30 R_{a0}$ at $f = 0.74 f_0$. The response drops to $R_a = 0.65 R_{a0}$ where $f = 0.365 f_0$ and $f = 1.16 f_0$. Thus the bandwidth is 107% of the frequency of maximum response. (*Note:* In both examples we have ignored the series reactance of the transducer.)

Example: Acoustic Emission Transducer

These results have important implications for the design of transducers with a broad frequency response. A good example is an acoustic emission receiving transducer. When solid materials are highly stressed, small cracks and defects develop. As these defects appear, they emit sound. Creaking wood is a familiar example of this phenomenon. Such acoustic emission occurs over a very wide frequency range and can be detected by a small transducer placed in contact with the cracked material. A wideband transducer is obviously desirable. Therefore, a transducer with a solid rather than an air backing must be used to give an almost constant response down to very low frequencies. The large series reactance of the transducer at low frequencies requires an extremely high impedance electrical load on the transducer.

$Z_1 < Z_C$ and $Z_2 < Z_C$. We now consider the situation where Z_1 and Z_2 are smaller than Z_C and purely resistive, and the input circuit is inductively tuned. The KLM equivalent circuit of Fig. 1.4.12 becomes a center-tapped transmission line terminated by a load Z_1 at one end and a load Z_2 at the other. When Z_1 and $Z_2 < Z_C$, the transmission line has a resonance close to ω_0. The approximate acoustic Q of the resonator, which we call Q_a, can be calculated by transforming the impedances Z_1 and Z_2 through quarter-wave transmission lines to the center tap, where they have effective values of Z_C^2/Z_1 and Z_C^2/Z_2, respectively, at the center frequency. The total resistance is $R_{sh} = Z_C^2/(Z_1 + Z_2)$. Two shorted quarter-wave transmission lines, with a total reactive impedance $X = (Z_C/2) \tan (\pi f/2f_0)$, are placed in parallel with this resistance. Thus the shunt reactance is infinite at $f = f_0$. The resistive load due to this parallel circuit is

$$R = \frac{R_{sh}}{1 + R_{sh}^2/X^2} \qquad (1.4.93)$$

We can define the effective values of Q and Q_a by the 3-dB bandwidth of these points (i.e., the points where $X = R_{sh}$) to yield the result for $(Z_1 + Z_2) \ll Z_C$ that

$$Q_a \approx \frac{\pi}{2} \frac{Z_C}{Z_1 + Z_2} \qquad (1.4.94)$$

When $Z_1 \approx Z_C$ and $Z_2 \approx Z_C$, the value of the acoustic Q, Q_a, is lower than this approximate formula implies, as we have already seen for the extreme cases of transducers with $Z_1 = 0$, $Z_2 = Z_C$, and $Z_1 = Z_2 = Z_C$.

A more rigorous treatment follows along similar lines to the derivation used to obtain the equivalent circuits of Figs. 1.4.8 and 1.4.9. Equation (1.4.23) gives the exact impedance of the transducer; this reduces to the model where $(Z_1 + Z_2) \ll Z_C$ and $|f - f_0| \ll f_0$, which we have already derived from the KLM equivalent circuit of Fig. 1.4.12. The negative series capacity C' in the KLM equivalent circuit varies with frequency; it must therefore be treated as a variable reactance X, which we write as $X = -(k_T^2/j\omega C_0)$ sinc (ω/ω_0). This reactance can be expressed in the form of a series resonator, using the relation

$$\text{sinc } z = \frac{\sin \pi z}{\pi z} = \prod_1^\infty \left(1 - \frac{z^2}{n^2}\right) \qquad (1.4.95)$$

and keeping only the $n = 1$ term. The properties of the transmission line can be represented by a series expansion, as we did in Sec. 1.4.3. Keeping only one term, the transmission line is replaced by a parallel resonant circuit with a shunt resistance R_{sh} across it. If we ignore the variation with frequency of the turns ratio of the transformer, we obtain the equivalent circuit for the parameters on the input side of the transformer, shown in Fig. 1.4.20(a).†

When k_T^2 is small, the series resonant circuit can usually be ignored and the equivalent circuit reduces to that of Fig. 1.4.20(b). This equivalent circuit is identical to the one derived earlier shown in Fig. 1.4.9(c).

Optimum value of the acoustic load with the source resistance $R_0 = 0$. The electrical Q, Q_e, of the tuning circuit increases as the acoustic load of a transducer is decreased, while the acoustic Q, Q_a, decreases. Therefore, there is an optimum acoustic load for maximum bandwidth, which we shall derive here.

As we have seen, the electrical radiation resistance of the air-backed trans-

---

†If the variation of the transformer turns ratio with frequency in the KLM model is taken into account, the effective value of R_a is multiplied by $(\omega_0/\omega)^2$ near the center frequency. This makes the maximum value of R_a occur at a frequency lower than ω_0.

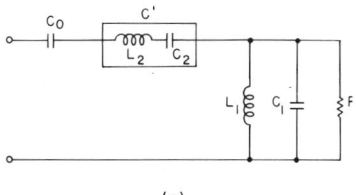

(a)

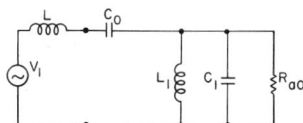

Figure 1.4.20 (a) Equivalent circuit for a loaded transducer. A similar circuit is shown in Fig. 1.4.9(b). $C_1 = C_0\pi^2/8k_T^2$, $L_1 = 1/\omega_0^2 C_1$, $C_2 = -C_0/k_T^2$, $L_2 = 1/\omega_0^2 C_2$, and $R_{a0} = (4k_T^2/\pi\omega_0 C_0)[Z_C/(Z_1 + Z_2)]$. (b) Simplified form of the equivalent circuit shown in part (a), which is identical to the one shown in Fig. 1.4.9(c). It is illustrated with an external series tuning inductance L.

ducer at its center frequency is

$$R_{a0} = \frac{4k_T^2}{\pi\omega_0 C_0} \frac{Z_C}{Z_1 + Z_2} \qquad (1.4.96)$$

Suppose that the circuit is supplied by a constant voltage source ($R_0 = 0$) through an inductance L, as illustrated in Fig. 1.4.20(b). Then, near the center frequency, the series resonant circuit has zero impedance, the parallel resonant circuit has infinite impedance, and $R_a = R_{a0}$. Thus the electrical Q at the center frequency of the transducer (i.e., of the series part of the circuit) is $Q = Q_e$, where

$$Q_e = \frac{1}{\omega_0 C_0 R_{a0}} = \frac{\pi}{4k_T^2} \frac{Z_1 + Z_2}{Z_C} \qquad (1.4.97)$$

We note from Eq. (1.4.94) that as $Z_1 + Z_2$ is decreased, the acoustic Q, Q_a, increases and the electrical Q, Q_e, decreases. Thus the optimum bandwidth of the transducer is obtained with $Q_e = Q_a$. This yields the relations

$$\frac{Z_1 + Z_2}{Z_C} = k_T \sqrt{2} \qquad (1.4.98)$$

and

$$Q_e = Q_a = \frac{\pi}{2\sqrt{2} k_T} \qquad (1.4.99)$$

The effective Q of the tuned system is approximately $Q_e \sqrt{2}$. Thus

$$\frac{\Delta f \ (3 \text{ dB})}{f_0} \approx \frac{2k_T}{\pi} \qquad (1.4.100)$$

So for a low value of k_T, an optimum bandwidth with the highest efficiency possible can be obtained with a low value of $(Z_1 + Z_2)/Z_C$.

Optimum value of the acoustic load with electrical tuning and source impedance $R_0 = R_{a0}$. In this case, the electrical Q, Q_e, is half the value given by Eq. (1.4.97), or

$$Q_e = \frac{\pi}{8k_T^2} \frac{Z_1 + Z_2}{Z_C} \quad (1.4.101)$$

Putting $Q_e = Q_a$, it follows that

$$\frac{\Delta f \text{ (3 dB)}}{f_0} \approx \frac{4k_T}{\pi} \quad (1.4.102)$$

with the optimum value of $Z_1 + Z_2$ increased by a factor of $\sqrt{2}$ from that given in Eq. (1.4.98). Thus the bandwidth of the transducer, under optimum matching conditions for high-efficiency operation, is controlled by the coupling coefficient k_T. For low-efficiency operation with a mismatched electrical circuit, we can obtain the largest bandwidth by using a low value for the acoustic Q, Q_a, and hence relatively large values for $(Z_1 + Z_2)/Z_C$. On the other hand, with a tuned input circuit, we can use high-coupling-coefficient materials, such as PZT or lithium niobate, that have $k_T = 0.5$. Then the optimum load is given by the condition $(Z_1 + Z_2)/Z_C = 0.7$, and we can obtain bandwidths of nearly one octave with high efficiency.

Tuned and untuned transducer, $Z_1 = Z_2 = Z_C$. As we have seen in Sec. 1.4.9, if the device is untuned, the minimum power loss will occur with a generator impedance of $R_0 \approx \sqrt{R_{a0}^2 + (1/\omega C_0)^2}$. For $Z_1 = Z_2 = Z_C$, it follows from Eqs. (1.4.86) and (1.4.90) that this condition corresponds to a power transfer efficiency η from the generator to the load at the center frequency f_0 of

$$\eta = \frac{1}{1 + (1 + \pi^2/4k_T^4)^{1/2}} \quad (1.4.103)$$

This formula takes into account the 3-dB loss due to power being emitted from both acoustic ports. Therefore, to obtain high efficiency with an untuned transducer, we must use as high a coupling coefficient k_T as possible. In such a case, the best one-way efficiency we can expect from a PZT ceramic with $k_T^2 = 0.25$ is 13.6%; this transducer has a minimum round-trip insertion loss of 17.3 dB and a low-pass bandwidth characteristic.

Tuning further lowers the minimum round-trip loss, ideally to 6 dB. With $k_T^2 = 0.25$, it follows from Eqs. (1.4.94) and (1.4.97) that when $Z_1 + Z_2 = Z_C$, the resultant electrical bandwidth of 32% is the controlling factor. From Eq. (1.4.101), a constant voltage source lowers the bandwidth to 16%. Therefore, we want as large a coupling coefficient as possible, to increase $R\omega_0 C_0$ and thus lower the electrical Q of the tuning circuit.

Tuned and untuned air-backed transducer with $Z_2 = Z_C$. We can make the transducer more efficient by leaving its left-hand side air-backed, that is, short-circuiting the left-hand transmission line in the KLM model of Fig. 1.4.12. In this case, if the right-hand transmission line is terminated by an impedance $Z_2 = Z_C$, the effective resistance at the center terminals of the KLM circuit is doubled and the one-way optimized untuned efficiency at the center frequency is

$$\frac{2}{1 + (1 + \pi^2/16k_T^4)^{1/2}} \qquad (1.4.104)$$

For $k_T^2 = 0.25$, this corresponds to 46.5% efficiency. The electrical bandwidth of the untuned system is now approximately 100%. The series-tuned system with $R_0 = R_{a0}$ will have an electrical bandwidth (the controlling bandwidth) of approximately 64%.

1.4.11 Reeder–Winslow Design Method for Air-Backed Transducers

A more direct mathematical formulation for piezoelectric transducer design has been given by Reeder and Winslow [11]. Their technique has been used extensively as a guide to designing very high frequency transducers laid down on a solid substrate. It involves calculating the real and imaginary parts of the input impedance of an air-backed piezoelectric resonator for different values of Z_2/Z_C ($Z_1 = 0$), and using the resultant curves as a basis for design. This procedure can be generalized for Z_1 finite, but it is not so convenient to work with the added variable this introduces. In such cases, it is better to use the KLM model of Fig. 1.4.12, or the Mason model of Fig. 1.4.4, first, to obtain some physical insight into the problem before resorting to numerical procedures.

Reeder and Winslow considered an air-backed transducer ($Z_1 = 0$) in contact with a solid substrate (Z_2 finite). They expressed the acoustic component Z_a of the input impedance of the transducer in a normalized form, symmetric about the center frequency f_0, as follows:

$$Z_a = R_{a0}\left(\frac{f_0}{f}\right)^2 H_a(f) \qquad (1.4.105)$$

We shall also find it convenient to define a second normalized quantity $M_a(f)$ that completely takes account of the variation of Z_a with frequency. Therefore, we write

$$Z_a = R_{a0} M_a(f) \qquad (1.4.106)$$

where

$$M_a(f) = \left(\frac{f_0}{f}\right)^2 H_a(f) \qquad (1.4.107)$$

The radiation resistance at the resonant frequency $f = f_0$ ($\bar{\beta}_a l = \pi$) is defined by the relation

$$R_{a0} = \frac{2k_T^2 Z_C}{\pi^2 f_0 C_0 Z_2} \qquad (1.4.108)$$

The parameters $H_a(f)$ and $M_a(f)$ are complex quantities. Thus we write

$$H_a(f) = H_r(f) + jH_i(f) \tag{1.4.109}$$

and

$$M_a(f) = M_r(f) + jM_i(f) \tag{1.4.110}$$

From Eq. (1.4.23), with $Z_1 = 0$,

$$H_r(f) = \frac{1}{4}\left(\frac{Z_2}{Z_C}\right)^2 \frac{[1 - \cos(\pi f/f_0)]^2}{1 + [(Z_2/Z_C)^2 - 1]\cos^2(\pi f/f_0)} \tag{1.4.111}$$

and

$$H_i(f) = \frac{1}{2}\frac{\sin(\pi f/f_0)\{1 + [1/2(Z_2/Z_C)^2 - 1]\cos(\pi f/f_0)\}}{1 + [(Z_2/Z_C)^2 - 1]\cos^2(\pi f/f_0)} \tag{1.4.112}$$

Figure 1.4.21 gives plots of $H_r(f)$ and $H_i(f)$ as functions of f/f_0 for different values of Z_2/Z_C. This normalization procedure is convenient because $H_r(f)$ and $H_i(f)$ are symmetric functions of $(f - f_0)$. Plots of $M_r(f)$ and $M_i(f)$ are given in Fig. 1.4.22.

When $Z_2 = Z_C$, then

$$H_r(f) = \sin^4\frac{\bar{\beta}_a l}{2} = \sin^4\left(\frac{\pi}{2}\frac{f}{f_0}\right) \tag{1.4.113}$$

Thus the 3-dB power points of $H_r(f)$ are at $f = 0.64f_0$ and $f = 1.36f_0$. The bandwidth is $\Delta f/f_0 = 0.72$. This result may be compared to the value of $\Delta f/f_0 = 0.79$, calculated for the bandwidth of M_r with a frequency of maximum response at $f_M = 0.74f_0$ and with $\Delta f/f_M = 1.07$. Therefore, an estimate of the effective bandwidth of $H_r(f)$ gives a conservative estimate of the actual bandwidth, and a somewhat pessimistic estimate of Q_a, the acoustic Q.

By choosing $Z_2 \ll Z_C$, the bandwidth decreases when Z_2/Z_C is decreased, as we have shown in Eq. (1.4.96) and can observe from Fig. 1.4.21(a). In this case,

$$H_r(f) \approx \left[\left(\frac{Z_C}{Z_2}\right)^2 \sin^2 \pi\left(\frac{f - f_0}{f_0}\right) + 1\right]^{-1} \tag{1.4.114}$$

or

$$H_r(f) \approx \left[1 + \left(\frac{\pi(f - f_0)}{f_0}\frac{Z_C}{Z_2}\right)^2\right]^{-1} \tag{1.4.115}$$

Defining Q as $f/\Delta f$(3 dB), the spacing Δf(3 dB) between the 3-dB points of $H_r(f)$ gives the effective acoustic Q, Q_a, of the resonator as

$$Q_a = \frac{1}{2}\left|\frac{f_0}{f(3\text{ dB}) - f_0}\right| = \frac{\pi Z_C}{2Z_2} \tag{1.4.116}$$

This result is the same as Eq. (1.4.94), which we derived using the KLM equivalent

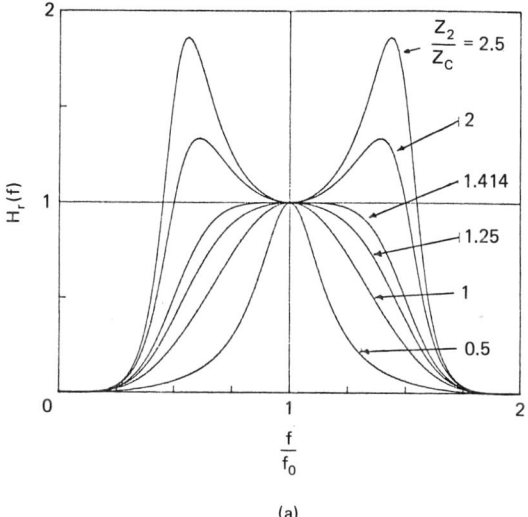

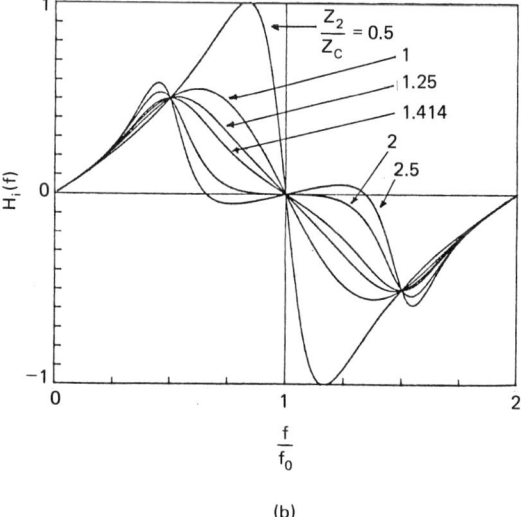

Figure 1.4.21 Acoustic bandshape function $H_a(f)$ for a thin disk unbacked transducer, plotted as a function of normalized frequency. Z_C and Z_2 are the characteristic mechanical impedances of the transducer medium and the medium in which the wave is being excited, respectively. (a) $H_r(f)$, real part of the bandshape function; (b) $H_i(f)$, imaginary part of the bandshape function.

circuit of Fig. 1.4.12. Note that this result also applies to $M_r(f)$, and hence $R_a(f)$, when $Z_2 \ll Z_C$. For $Z_2/Z_C = \sqrt{2}$, the exact acoustic response function $H_r(f)$ is maximally flat. Beyond this point, when $Z_2/Z_C > \sqrt{2}$, there are peaks on both sides of f_0.

The basic physical reason for the sharp response near $f/f_0 = 1$ is that when $Z_2 \ll Z_C$, the transducer exhibits a strong resonance when $l \approx \lambda/2$. When Z_2/Z_C is very large, the peaks in $H_r(f)$ tend to occur at $f = 0.5f_0$ and $f = 1.5f_0$ because the reflection at $z = l$ tends to make the transducer resonant when its thickness is one quarter-wavelength or three-quarters of a wavelength.

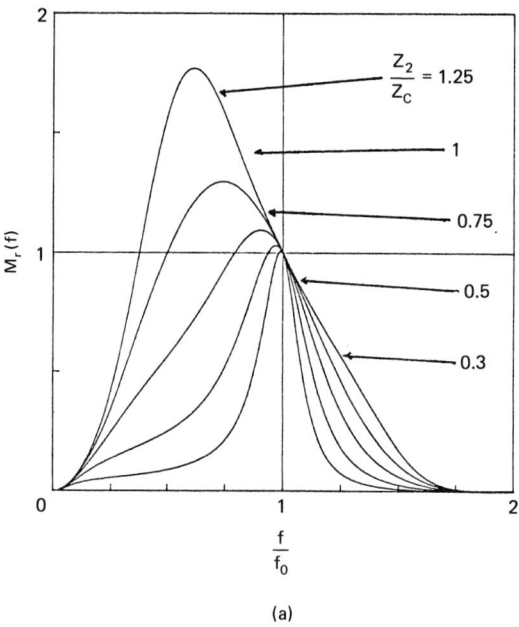

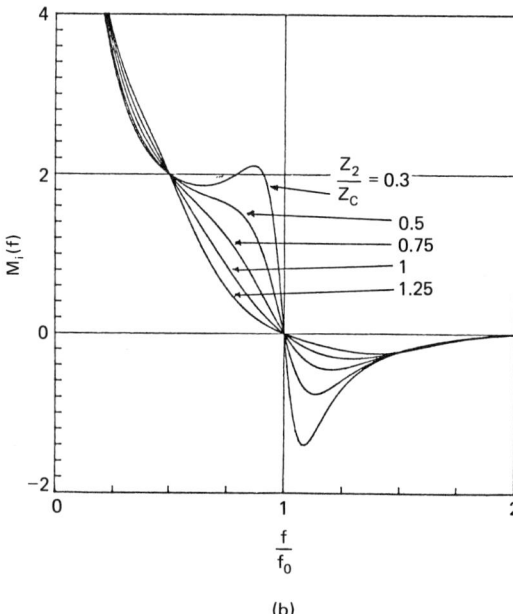

Figure 1.4.22 Acoustic bandshape $M_a(f)$ function for a thin disk unbacked transducer plotted as a function of normalized frequency: (a) $M_r(f)$, real part of the bandshape function; (b) $M_i(f)$, imaginary part of the bandshape function.

Example: Z_1 or $Z_2 \gg Z_C$

A zinc oxide thin-film air-backed transducer on a high-impedance substrate, such as sapphire, tends to resonate nearer $l = \lambda/4$ than $l = \lambda/2$. This gives broadband characteristics and leads to the use of thinner films, which is convenient because of the long sputter deposition times such films often require.

Furthermore, it follows from Eq. (1.4.112) or Figs. 1.4.21(b) and 1.4.22(b) that when $f = f_0/2$, then $H_i(f) = 0.5$ and $M_i(f) = 2$. So, at this frequency, the motional reactance is inductive and tends to cancel out the reactance of the series capacity. When $k_T^2/\pi = 1$ or $k_T^2 = 0.39$, the input impedance of the transducer is purely real at a frequency $f = f_0/2$. It may therefore be possible to make a PZT transducer, without external tuning, that can operate in an extensional mode ($k_{z3}^2 \approx 0.5$) and is well matched to a source with a real input impedance. At the same time, it is useful to choose the impedance Z_2 so that the real part of the transducer impedance is maximum at this frequency $f = f_0/2$.

It is also interesting to consider some practical examples for the situation when either $Z_1 > Z_C$ and $Z_2 < Z_C$ or $Z_1 < Z_C$ and $Z_2 > Z_C$. If a rigid backing is employed with a relatively low impedance front layer, the transmission line in the KLM model of Fig. 1.4.12 tends to resonate when it is a quarter-wavelength or three-quarters of a wavelength long and, by symmetry, the input impedance is the same as when $Z_1 = 0$ and $Z_2 > Z_C$. Thus the resonant frequency is halved. If the backing is perfectly rigid ($Z_1 = \infty$), the value of R_{a0}, with a given Z_2 at a center frequency $f = f_0/2$, is halved from the R_{a0} value of the equivalent half-wavelength-long air-backed resonator. In this case, however, the frequency response is broader-band and the pulse response can be as narrow as that of a half-wavelength-long resonator with $Z_1 = Z_C$, like the one shown in Fig. 1.4.16(a) and (b).

A very important practical example of this phenomenon is the plastic piezoelectric material polyvinylidene difluoride (PVF_2), which has a longitudinal wave velocity of 2.2 km/s and an impedance of 3.92×10^6 kg/m²-s, and thus can match reasonably well to water if used with a rigid backing such as brass, which has an impedance of 31×10^6 kg/m²-s. This material has a low value of k_T ($k_T = 0.11$), so its electrical efficiency is poor. However, very little power is lost at the acoustic mismatch into water. Thus it has outstanding broadband short-pulse performance as a receiving transducer exciting a high-impedance unmatched electrical load (see Probs. 6 and 7).

We can calculate the true response of a transducer by multiplying the response function $H_a(f)$ by $(f_0/f)^2$. This yields the true skewed response function $M_a(f)$, which is plotted in Fig. 1.4.22. It follows from Eqs. (1.4.107) and (1.4.115) that if $Z_2 < Z_C$, then

$$M_r(f) \approx \left(\frac{f_0}{f}\right)^2 \left[1 + \left(\frac{\pi(f - f_0)}{f_0} \frac{Z_C}{Z_2}\right)^2\right]^{-1} \quad (1.4.117)$$

Thus the response function for $M_r(f)$ or $R_a(f)$ is skewed and tends to peak at a lower frequency. We can obtain a similar result with the KLM model of Fig. 1.4.12 if we take account of how the transformer turns ratio varies with frequency.

It can be shown from Eq. (1.4.117) that if $(Z_2/Z_C\pi)^2 \ll 1$, the maximum value of R_a occurs when $f = f_M$, where f_M is given by the relation

$$\frac{f_M}{f_0} \approx 1 - \frac{Z_2^2}{Z_C^2 \pi^2} \tag{1.4.118}$$

It follows from Eqs. (1.4.98) and (1.4.118), or from the derivation given in this section, that for optimum electrical matching from a constant-voltage zero-impedance source, $Z_2 = Z_C k_T \sqrt{2}$. This implies that

$$\frac{f_M - f_0}{f_0} = 1 - \frac{2k_T^2}{\pi^2} \tag{1.4.119}$$

at the point where R_a is maximum.

Thus, when the criteria for electrical matching are satisfied for $k_T^2 = 0.25$ (PZT-5A) and $Z_2 = 0.7 Z_C$, R_a is maximum at 95% of the resonant frequency of the transducer. The result given in Eq. (1.4.118) is apparent in the theoretical curve of Fig. 1.4.22(a). Equation (1.4.119) implies that when a quarter-wavelength matching layer is designed for a transducer, it should be optimized for the frequency f_M, rather than for the shunt resonant frequency f_0 of the transducer.

The results of this analysis are identical to those that can be obtained with the KLM equivalent circuit of Fig. 1.4.12, in which the transformer response skews the response function. On the other hand, the KLM model provides a more general physical picture of the frequency behavior of the transducer which allows us to consider terminations at both its ends. We conclude that both the KLM model and the Reeder–Winslow model provide useful criteria for choosing the design parameters of a broadband transducer. In general, however, it is still necessary to make a final check of the design with numerical procedures. This is particularly important when electrical matching conditions must also be considered.

1.4.12 Some Examples of Transducer Design

The design criteria we have given in Secs. 1.4.10 and 1.4.11 lead to a good understanding of the optimum parameters required for efficient operation of a transducer with a large bandwidth. In practice, it is not always easy to obtain materials with the optimum values of Z_1 or Z_2, although this can usually be achieved by using one or more acoustic matching layers, each approximately a quarter-wavelength thick at the center frequency.

As we saw in Sec. 1.4.7, we expect optimum pulse responses when the back of the transducer is well terminated. More generally, Fourier transform theory leads to the conclusion that we can obtain a Gaussian-shaped pulse if the amplitude response of the transducer, as a function of frequency, is Gaussian, while at the same time its phase response is flat or varies linearly with frequency. A distortion in the phase response gives severe ringing in the pulse response. Such effects tend to show up when multiple matching layers are employed.

Figures 1.4.23–1.4.27 illustrate some examples of experimental results obtained with PZT-5A transducers. Figure 1.4.23 shows a longitudinal wave PZT-4 transducer with no matching layers, a lossy tungsten–epoxy backing of impedance 26×10^6 kg/m²-s, and an acoustic Q of value $Q_a = 2.05$. The transducer disk

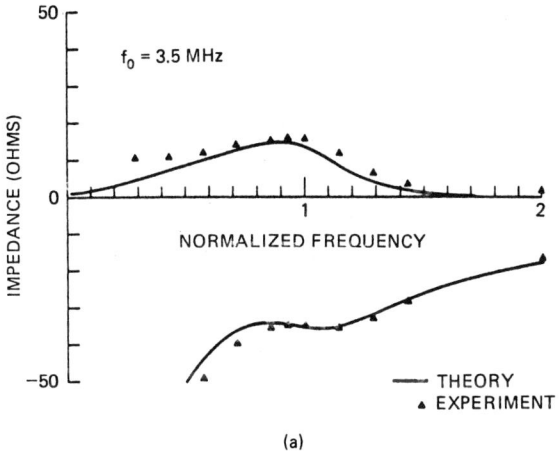
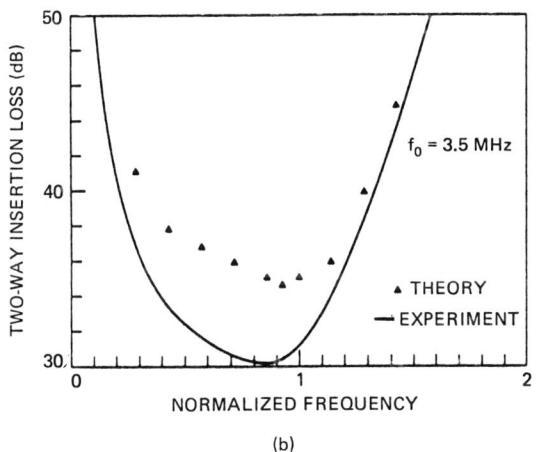
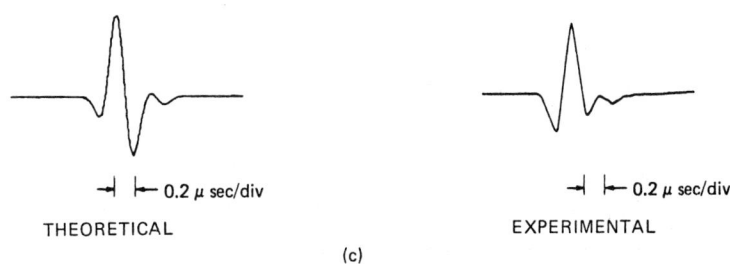

Figure 1.4.23 Comparison of experimental results with theory for a PZT-4 transducer matched with a high-loss, high-impedance backing (solid curve, theory; dots, experimental results): (a) electrical impedance; (b) two-way insertion loss; (c) theoretical and experimental impulse responses. (After DeSilets et al. [10].)

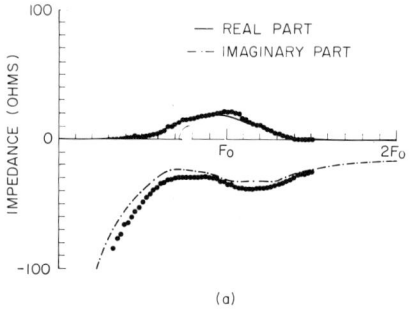

(a)

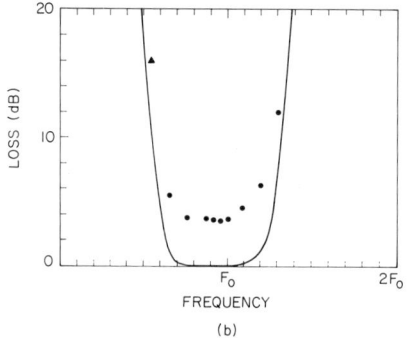

(b)

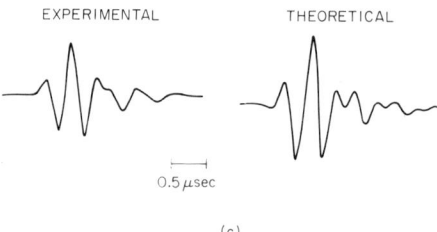

(c)

Figure 1.4.24 Comparison of experiment with theory for a 19-mm-diameter, 3.4-MHz PZT transducer with matching layers of impedance, 2.45×10^6 kg/m^2-s and 8.91×10^6 kg/m^2-s, and no backing (solid curve, theory; dots, experimental results): (a) electrical impedance; (b) two-way insertion loss; (c) theoretical and experimental impulse responses.

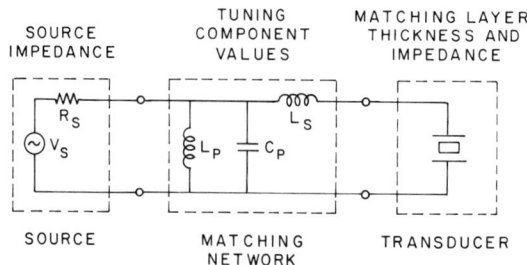

Figure 1.4.25 Transducer and driving network to be optimized. (After Selfridge et al. [12].)

68 Sound Wave Propagation   Chap. 1

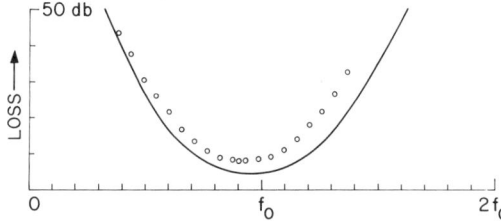

Figure 1.4.26 Measured insertion loss compared with theory: solid curve, theory; circles, experimental results. (After Selfridge et al. [12].)

was 12.7 mm in diameter with a resonant frequency of 3.5 MHz; its important material parameters were $k_T^2 = 0.24$, $\varepsilon = 730\varepsilon_0$, and $Z_0 = 34 \times 10^6$ kg/m^2-s. The impedance, insertion loss, and impulse response were measured by determining the power into a matched 50-Ω electrical load after reflection from a perfect acoustic reflector. No electrical tuning was employed and a 50-Ω source was used. The distance between the transducer and the reflector was 3.0 cm.

In the impedance plot of Fig. 1.4.23(a), note the small value of the resistance compared to the reactance at the center frequency. (This would have led to a high Q, as well as a narrow bandwidth, if electrical tuning had been used.) The insertion-loss plot of Fig. 1.4.23(b) shows the very high loss and the broad, smooth

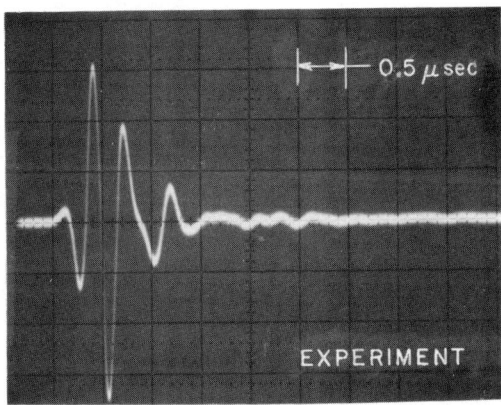

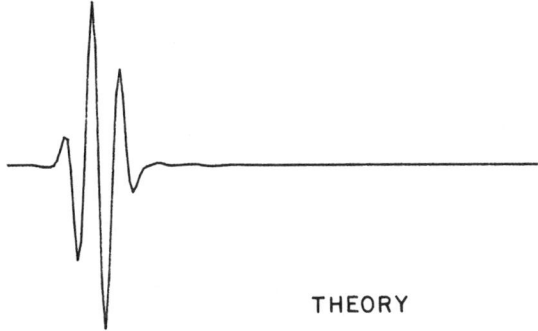

Figure 1.4.27 Measured impulse response compared with theory. (After Selfridge et al. [12].)

bandshape characteristic of such transducers. The measured two-way insertion loss of 34.5 dB was about 3.3 dB worse than the theoretical value of 30.2 dB for a lossless transducer material; this loss can be attributed to the finite loss of the transducer material. Calculations of commercial transducers that account for loss, when present, in the transducer material and matching layers usually give excellent agreement with theory. The 3-dB bandwidth is 105%, so the very narrow impulse response that results remains the outstanding property. Note that the effective center frequency f_M, where R_a is maximum, is lower than the resonance frequency f_0. With $Z_1 = 26 \times 10^6$ kg/m²-s, we expect, from the simple formula of Eq. (1.4.118) with Z_2 replaced by Z_1, that $f_M/f_0 \approx 0.94$.

Figure 1.4.24 shows results for a transducer with two quarter-wave matching layers on an air-backed 19 mm PZT disk whose resonant frequency is 2.3 MHz. The transducer material parameters were $k_T^2 = 0.25$, $\varepsilon = 830\varepsilon_0$, and $\overline{Z}_0 = 31.6 \times 10^6$ kg/m²-s. Two matching layers were made, one of silicon carbide–loaded epoxy with an impedance of 8.91×10^6 kg/m²-s, the other of an epoxy with an impedance of 2.45×10^6 kg/m²-s. These matching layers were designed, using criteria of Riblet [4] and DeSilets et al. [10], to give an approximate maximally flat response. Calculations indicated that the impedance presented by the matching layers at the transducer surface was 19.8×10^6 kg/m²-s, leading to $Q_a = 2.5$. From Eq. (1.4.118), the frequency for R_a to be maximum was reduced by 4%. These matching layers were therefore chosen to be a quarter-wave thick at 4% below the center frequency of the transducer. However, in the actual plates produced, the equivalent number for the epoxy layer was 5%. This 5% change in the thickness of the epoxy plate amounted to 12 μm, which illustrates the difficulty of fabricating these devices. A series inductor was used to tune the impedance to a real value at the center frequency, and an autotransformer was employed to obtain a value of 50 Ω.

The electrical impedance of the transducer was measured as a function of frequency; this is compared with the calculated values in Fig. 1.4.24(a). After electrical matching, the two-way insertion loss, when a signal is reflected from a perfect reflection as a function of frequency and impulse response, is compared to theory in Fig. 1.4.24(b) and (c). The measured tuned insertion loss was 3.5 dB at 2.3 MHz, and the 3-dB bandwidth was 79%, a considerably higher value than would be expected from the approximate value of Q_a, the acoustic Q. The impulse response was not as narrow as desirable, but as the calculation shows, we might have expected this from the rather square bandshape of this transducer design. We can attribute half the measured insertion loss of 3.5 dB to losses in the electrical components and diffraction of the acoustic beam from a piston radiator, as discussed in Chapter 3; the rest is due to loss in the ceramic material.

In practice it is difficult to design a transducer for optimum pulse response on the basis of a desired frequency response. As we saw in Fig. 1.4.24, the optimum frequency response is more like a Gaussian than a maximally flat response. Ideally, it is desirable to optimize the thickness and impedance of one or more matching layers, as well as the parameters of the electrical matching circuit, to obtain as short a pulse response as possible simultaneously with the maximum peak response.

Iterative techniques for such a design have been worked out by Chou et al. [13] and by Selfridge et al. [12]. The components to be used are specified and then adjusted iteratively in the theoretical design to optimize the length of the pulse response. A transducer made of high-quality Murata PZT ceramic, with $Q \approx 4000$, $\overline{V}_a = 4.72$ km/s, $k_T^2 = 0.24$, and impedance 39.6×10^6 kg/m^2-s, which uses one matching layer of epoxy to water and an epoxy backing of impedance 4×10^6 kg/m^2-s, has been designed by this technique. The matching circuit is shown in Fig. 1.4.25 [12]; its adjustable parameters are the matching layer thickness, the area of the transducer (or the input impedance), and the three tuning components L_p, C_p, and L_s. Figures 1.4.26 and 1.4.27 demonstrate the excellent agreement between theory and experiment, showing the smooth frequency response and the short pulse, respectively; the tuning circuit used values of $L_p = 5.7$ µH, $L_s = 4.1$ µH, $C_p = 332$ µF, and $R_s = 86$ Ω. As in the previous designs, there is a two-way loss in excess of theory, due to loss in the PZT, of approximately 2.5 dB. Note in Fig. 1.4.26 that the response in the frequency domain has approximately a Gaussian form. Interestingly, the optimum impedance and thickness of the matching layer were not those that we would arrive at by simple theoretical considerations; instead, they were 3.45×10^6 kg/m^2-s and 0.243λ, respectively, at the center frequency. This impedance is far lower than the value we expect from a design for a maximally flat response (i.e., with $Z \sim 7 \times 10^6$ kg/m^2-s). The iterative technique provides a much-improved pulse response in comparison to the simpler concepts of quarter-wavelength matching discussed earlier.

PROBLEM SET 1.4

1. A ceramic transducer made of PZT-5A, 1 cm square and 1 mm thick, is coated with a thin silver film on each side. The transducer is tested in air, and its impedance as a function of frequency is measured.
   (a) Work out the frequencies f_0 and f_1, for maximum and minimum impedance, respectively.
   (b) The transducer is now used to excite a longitudinal wave in an infinitely long aluminum bar by bonding it to the bar. Work out the impedance, including the series capacity at the resonant frequency f_0, and the efficiency of conversion of power from a generator with a 50-Ω impedance. Estimate the acoustic bandwidth of the transducer. Use the values of c_{33}, k_T, and ε_{zz}^s given in Appendix B.

2. Suppose that the bar in Prob. 1 is not infinitely long but has a length of 2.5 mm. What would be the impedance looking into the transducer at a frequency f_0 under CW conditions? This impedance is purely reactive.

3. The transducer described in Prob. 1 is placed with its front surface in water.
   (a) Work out its impedance, including the series capacitance at the resonant frequency f_0, and the efficiency of conversion of power from a generator with a 50-Ω impedance.
   (b) Estimate the acoustic bandwidth and acoustic Q of the transducer.

4. Consider a transducer terminated on its back surface by a perfect match $Z_1 = Z_c$. Suppose that the transducer is excited by a voltage pulse of the shape shown below, which increases in amplitude for a time T_1.

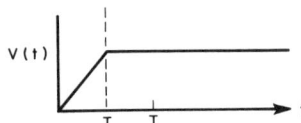

(a) Take the transit time through the transducer to be $T = l/V_a$ and l to be the length of the transducer. Assume that the coupling coefficient $k_T^2 \ll 1$, so that the current I passing through the transducer is $I \approx C_0 \partial V/\partial t$. By using the treatment of Sec. 1.4.7 for pulsed transducers, show that when $T_1 = T$, the transducer excites a triangular pulse of pressure or stress of base length $2T$, as shown below.

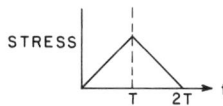

(b) Sketch the form of the output when $T_1 < T$.
(c) Sketch the form of the output when $T_1 > T$.

5. (a) Consider a receiving transducer terminated on its back surface by its own impedance. Suppose that the transducer is excited by an acoustic signal on its front surface ($z = -l/2$), whose velocity is $v_0 \exp(j\omega t)$. Find the output into an electrical load R_L. Assume that $k_T^2 \ll 1$. Thus the wave in the medium of the transducer is unaffected by the induced current in the electrical load.
(b) Now, with $R_L \gg 1/\omega C_0$ for all frequency components of interest, use a Fourier or Laplace transform to derive the form of the output when the input signal is a very short velocity pulse of length $\tau \ll T$ (a δ function).
(c) What will the form of the output be with the same excitation when $R_L \ll 1/\omega C_0$?

6. Consider a transducer made of the plastic piezoelectric transducer material PVF_2 (polyvinylidene difluoride). This material is a plastic much like Teflon, and can be cut easily with a pair of scissors. It has a density of 1.78×10^3 kg/m³, a longitudinal wave acoustic velocity of 2.2 km/s, and an acoustic impedance of 3.92×10^6 kg/m²-s. Therefore, its impedance is far closer to the impedance of water (1.5×10^6 kg/m²-s) than to that of ceramics ($\sim 30 \times 10^6$ kg/m²-s), so the impedance mismatch into water is not too severe. This transducer normally shows a very broadband characteristic, especially when used as a receiving transducer.
(a) Using the equivalent circuit or the expressions of Eq. (1.4.23), consider a PVF_2 transducer of length l bonded to a brass backing material that you may regard as having infinite impedance (i.e., being perfectly rigid), which means that $v_1 = 0$ when the other side is terminated by a medium of impedance Z_2. Determine the input impedance of the transducer, and show that the motional impedance or acoustic impedance Z_a is purely real when the material is a quarter-wavelength thick.
(b) Taking $Z_2 = 0$ (i.e., with the front surface looking into air), find an expression for the frequency ω_1 where the electrical impedance of the transducer is zero. Following the derivation of Eq. (1.4.46), work out a lumped equivalent circuit like that of Fig. 1.4.9 for the rigidly backed transducer.

Note. The change in result for part (b) from Eq. (1.4.46) is minor. Basically, it requires changes by factors of 2.

7. For the brass-backed PVF$_2$ transducer described and illustrated in Prob. 6, work out the ratio $R = V_3/T_{F1}$, an important parameter when the device is used as a receiving transducer operating into a very high impedance circuit for which $I_3 = 0$. This result gives the voltage output due to a stress wave of stress T_{F1} incident on the transducer. Find $R(\omega = 0)/R(\omega_0)$, where ω_0 is the resonant frequency corresponding to the quarter-wavelength-long transducer. What is this ratio for a transducer operating in water?

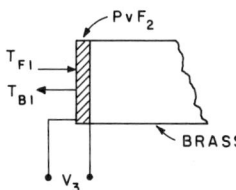

Note. In this situation, $|R|$ does not change very much with frequency, so the device is a broadband receiving transducer.

Hint. For this problem, consider the piezoelectric coupling to be small, so that the wave approaching the transducer has an amplitude T_{F1}. The reflection coefficient $\Gamma = T_{B1}/T_{F1}$ of this wave is dictated by the transducer length and the impedance of the piezoelectric material. Find the value of $v_1 = v_{F1} + v_{B1}$ at the H$_2$O–PVF$_2$ interface, and hence V_3/T_{F1}.

8. Consider a multiple transducer array terminated by an acoustic impedance Z_C at each end, that is, the medium is infinite on each side and of the same impedance as the transducer medium, as illustrated below.

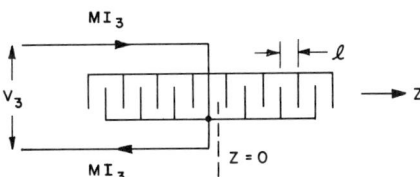

Suppose that each individual transducer is of length l. It has been shown that at $z > l/2$ ($z = 0$ at the center),

$$T_F(z) = -\frac{eI_3}{\varepsilon^S \omega A} e^{-j\bar{\beta}_a z} \sin \frac{\bar{\beta}_a l}{2}$$

(a) Now suppose that there are M transducers stacked in a row, which are interdigitally connected so that the currents flowing into neighboring transducers are flowing in opposite directions, and that the total current flowing into this M-section transducer is MI_3. By using superposition (i.e., adding the contribution from each transducer in the correct phase when all transducers are excited by a current I_3 or $-I_3$), find the total stress at a plane z where $z > Ml$. Add up the sum of the contributions to $T(z)$ by using the relations

$$\sum_{0}^{M-1} e^{-jnx} = \frac{e^{-jMx} - 1}{e^{-jx} - 1}$$

$$= e^{-j(M-1)x/2} \frac{\sin(Mx/2)}{\sin(x/2)}$$

Determine the real power in the forward traveling wave excited at a plane z outside the interdigital transducer by the total current I, by using the relation

$$P_F = A \, \text{Re}\left(\frac{TT^*}{2Z_0}\right)$$

where A is the area of the transducer.

Assuming, by symmetry, that the same amount of power is radiated in the backward direction from the left-hand side of the transducer, find the total power radiated by the transducer in terms of I. We may now determine the radiation resistance of this multiple element transducer as follows. If the current entering the transducer is I and the voltage across it is V, the radiated power must be

$$P = \tfrac{1}{2} \, \text{Re}(VI^*)$$

However, if R_a is the radiation resistance and jX_a is the reactance of the transducer,

$$P = \tfrac{1}{2} \, \text{Re}(R_a + jX_a)II^*$$

$$= \tfrac{1}{2} R_a II^*$$

It follows that $R_a = 2P/II^*$. Using the expression for P that you have already found, determine the radiation resistance of the transducer (the real part of its impedance). Show that R_a is maximum when each transducer is a half-wavelength long.

You have now worked out a simple theory for the radiation resistance of an interdigital transducer, the type most often used for surface acoustic wave devices. Note that in most interdigital transducer theories (see Secs. 2.4, 2.5, and 4.2), the number of transducers is defined by the number of transducer pairs N. Here $M = 2N$; because M can be an even or an odd integer, however, N will not be an integer if M is odd.

(b) From your result, consider an M-element transducer with M large, so that you can write $\sin x \approx x$ but cannot approximate $\sin Mx$. Since for $y = \pi/2$, $(\sin y)/y$ corresponds to a 4-dB change from the maximum value of 1, find the bandwidth between 4-dB points of an M-element transducer.

9. Derive Eq. (1.4.23) from the KLM equivalent circuit.

10. (a) Derive the circuit of Fig. 1.4.11(a) directly from the Mason equivalent circuit of Fig. 1.4.4. You will need to write an expression for an admittance term Y_a in the form of a series such as Eq. (1.4.42), keeping only one term corresponding to the lowest-order resonance.

(b) Assuming that $(Z_1 + Z_2) \ll Z_C$, derive the circuit of Fig. 1.4.11(b) from the Mason equivalent circuit of Fig. 1.4.4.

11. Derive the equivalent circuit of Fig. 1.4.20(a) using the methods described in the discussion of the schematic in Sec. 1.4.10 [see Eq. (1.4.95)].

12. Derive Eq. (1.4.118) from Eq. (1.4.117).

13. Use the equivalent circuit of Fig. 1.4.20(a) to prove Eq. (1.4.118). You will find it helpful to expand your expressions to first order in $\omega - \omega_0$.

1.5 TENSOR NOTATION AND CONSTITUTIVE RELATIONS FOR PIEZOELECTRIC AND NONPIEZOELECTRIC MATERIALS

1.5.1 Introduction

So far we have dealt mainly with one-dimensional forms for stress, strain, and the equation of motion in piezoelectric and nonpiezoelectric materials; we have also derived the theory of the thickness expander mode for a piezoelectric plate. The piezoelectric constitutive relations we have used throughout this chapter are particularly convenient when reduced to their one-dimensional forms [Eqs. (1.2.4) and (1.2.9)], because only one component of strain, the longitudinal component in the z direction, is finite in a piezoelectric plate. For certain other situations, however, this form of the constitutive relations may be less convenient. Thus, in this section, we describe some alternative formulations [14, 15].

In all cases, whether referring to shear or longitudinal waves, we will describe wave interactions in a one-dimensional form. This is actually a correct and rigorous approach, provided that the propagation of the wave of interest is along an axis of symmetry of a crystal. However, to carry out quantitative calculations, we must state the equation of motion, Hooke's law, and the elastic and piezoelectric parameters of the crystal, and then reduce them to one-dimensional terms. *Tensor notation* will be introduced; to simplify the resulting equations, we also introduce *reduced subscript notation*. A detailed treatment of some of the notation used is given in Appendix A.

1.5.2 Mathematical Treatment

Displacement and strain. In general, the displacement is a vector **u** with three Cartesian components u_x, u_y, and u_z, each of which can be a function of the three Cartesian components x, y, and z of the vector **r**. Thus, in general, **S** will be a tensor with nine components. For example,

$$S_{xx} = \frac{\partial u_x}{\partial x} \tag{1.5.1}$$

and

$$S_{xy} = \frac{1}{2}\left(\frac{\partial u_x}{\partial y} + \frac{\partial u_y}{\partial x}\right) \tag{1.5.2}$$

with S_{xx}, S_{xy}, S_{yz}, S_{yx}, S_{yy}, S_{yz}, S_{zx}, S_{zy}, and S_{zz} defined similarly. For pure one-dimensional motion, however, we can represent **S** by only one component (e.g., S_{xx} for one-dimensional longitudinal strain and S_{xy} for one-dimensional shear strain). The symmetry of Eq. (1.5.2) shows that $S_{xy} = S_{yx}$.

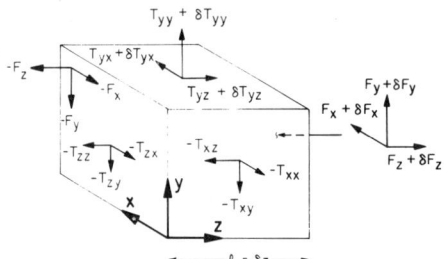

Figure 1.5.1 Application of general stress components.

Stress. We now consider the stress on a cube of volume $\delta x\, \delta y\, \delta z$, illustrated in Fig. 1.5.1. The force applied to the left-hand surface $\delta x\, \delta y$ is called the traction **F**. The traction applied to this surface has three components: $-F_x$ and $-F_y$, both parallel to the surface, and $-F_z$, perpendicular to the surface. The traction applied to the opposite surface has components $F_x + \delta F_x$, $F_y + \delta F_y$, and $F_z + \delta F_z$. By carrying out a Taylor expansion to first order in δz, we see that this traction has components $F_x + (\partial F_x/\partial z)\,\delta z$, $F_y + (\partial F_y/\partial z)\,\delta z$, and $F_z + (\partial F_z/\partial z)\,\delta z$, where δz is equivalent to the parameter l used in the one-dimensional formulation of Sec. 1.1. We define the stresses on a surface perpendicular to the z axis as follows:

$$\text{Shear stress:} \quad T_{zx} = \frac{F_x}{\delta x\, \delta y} \quad (1.5.3)$$

$$\text{Shear stress:} \quad T_{zy} = \frac{F_y}{\delta x\, \delta y} \quad (1.5.4)$$

$$\text{Longitudinal stress:} \quad T_{zz} = \frac{F_z}{\delta x\, \delta y} \quad (1.5.5)$$

The first subscript of the tensor **T** denotes the coordinate axis normal to a given plane; the second subscript denotes the axis to which the traction is parallel. There are nine possible stress components:

$$\mathbf{T} = \begin{bmatrix} T_{xx} & T_{xy} & T_{xz} \\ T_{yx} & T_{yy} & T_{yz} \\ T_{zx} & T_{zy} & T_{zz} \end{bmatrix} \quad (1.5.6)$$

The terms T_{xx}, T_{yy}, and T_{zz} are longitudinal stress components, while the terms $T_{xy} = T_{yx}$, $T_{xz} = T_{zx}$, and $T_{yz} = T_{yz}$ are shear stress components that are euqal in pairs, because internal stresses can give no net rotation of the body.

By analogy to our earlier simple derivation in Sec. 1.1, the force per unit volume in the z direction is the net resultant of the forces per unit volume applied to an infinitesimal cube, that is,

$$f_z = \frac{\partial T_{zz}}{\partial z} + \frac{\partial T_{xz}}{\partial x} + \frac{\partial T_{yz}}{\partial y} \quad (1.5.7)$$

Hooke's law and elasticity. We write, for example, the relation between T_{xx} and the applied strains S_{xx} and S_{yy} as

$$T_{xx} = c_{xxxx}S_{xx} + c_{xxyy}S_{yy} \qquad (1.5.8)$$

and the relation between T_{xz} and the applied strain S_{xz} as

$$T_{xz} = c_{xzxz}S_{xz} + c_{xzzx}S_{zx} \qquad (1.5.9)$$
$$= 2c_{xzxz}S_{xz}$$

where $S_{xz} = S_{zx}$ and $c_{xzxz} = c_{xzzx}$, due to symmetry. In each case, the first two subscripts of the elastic tensor correspond to the subscripts for the stress tensor, and the last two subscripts correspond to those for the strain tensor.

It follows from Eq. (1.5.7) that the equation of motion in the z direction is

$$\rho_{m0}\ddot{u}_z = \rho_{m0}\dot{v}_z = \frac{\partial T_{zz}}{\partial z} + \frac{\partial T_{xz}}{\partial x} + \frac{\partial T_{yz}}{\partial y} \qquad (1.5.10)$$

with corresponding equations for the other components of $\ddot{\mathbf{u}}$ and $\dot{\mathbf{v}}$.

Tensor notation. A simpler notation, described more fully in Appendix A, can be used to denote the components of the vectors and tensors involved. We shall use the subscripts i, j, and k to denote any one of the x, y, and z axes. For example, E_i is the ith component of the E field. We write the electric displacement field in terms of the E fields in the form

$$D_i = \varepsilon_{ij}E_j \qquad (1.5.11)$$

More fully, this means that

$$D_i = \sum_j \varepsilon_{ij}E_j \qquad (1.5.12)$$

or

$$\begin{aligned} D_x &= \sum_j \varepsilon_{xj}E_j = \varepsilon_{xx}E_x + \varepsilon_{xy}E_y + \varepsilon_{xz}E_z \\ D_y &= \sum_j \varepsilon_{yj}E_j = \varepsilon_{yx}E_x + \varepsilon_{yy}E_y + \varepsilon_{yz}E_z \\ D_z &= \sum_j \varepsilon_{zj}E_j = \varepsilon_{zx}E_x + \varepsilon_{zy}E_y + \varepsilon_{zz}E_z \end{aligned} \qquad (1.5.13)$$

Here j is a floating subscript over which summation is automatically implied, while the subscript i denotes the required component of interest on the left-hand side of the equation.

As a further example, we write the electric field in terms of the scalar electric potential as follows:

$$E_i = \frac{-\partial \phi}{\partial x_i} \qquad (1.5.14)$$

Here there is no floating subscript. Similarly, the tensors denoting stress and strain are T_{ij} and S_{ij}, respectively. We write Hooke's law in the form

$$T_{ij} = c_{ijkl} S_{kl} \tag{1.5.15}$$

where k and l are floating subscripts indicating summation over k and l, while i and j indicate the stress components required. Similarly, the equation of motion in tensor form is

$$\frac{\partial T_{ij}}{\partial x_j} = \rho_{m0} \ddot{u}_i \tag{1.5.16}$$

where j is the floating subscript.

Reduced subscript notation. Because $T_{ij} = T_{ji}$ and $S_{ij} = S_{ji}$, there are actually only six independent tensor quantities. We can therefore simplify the notation even further. The subscripts i, j, and k will be used to denote the tensor components of interest in what is called the *reduced form*. Hence T_I is a component of the stress tensor, which replaces the longer unreduced notation T_{ij}, and S_I is a component of the strain tensor, which replaces the longer unreduced notation S_{ij}.

Table 1.5.1 summarizes how this notation is used, taking the E field vector, the strain tensor, and the stress tensor as examples. It also gives notation describing the stress and strain tensors in nonreduced form.

We can now write the full tensor form of the constitutive relations for a

TABLE 1.5.1 TENSOR AND VECTOR COMPONENTS

Stress or vector component	Example	Meaning
Electric field, E_i	E_x ($i = x$)	E field in x direction
	E_y ($i = y$)	E field in y direction
	E_z ($i = z$)	E field in z direction
Strain, S_I	S_1 ($I = 1$) $= S_{xx}$	Longitudinal strain in x direction
	S_2 ($I = 2$) $= S_{yy}$	Longitudinal strain in y direction
	S_3 ($I = 3$) $= S_{zz}$	Longitudinal strain in z direction
	S_4 ($I = 4$) $= 2S_{yz}$	Shear strain, motion about x axis; shear in y and z directions
	S_5 ($I = 5$) $= 2S_{zx}$	Shear strain, motion about y axis; shear in x and z directions
	S_6 ($I = 6$) $= 2S_{xy}$	Shear strain, motion about z axis; shear in x and y directions
Stress, T_I	T_1 ($I = 1$) $= T_{xx}$	Longitudinal stress in x direction
	T_2 ($I = 2$) $= T_{yy}$	Longitudinal stress in y direction
	T_3 ($I = 3$) $= T_{zz}$	Longitudinal stress in z direction
	T_4 ($I = 4$) $= T_{yz}$	Shear stress about x axis
	T_5 ($I = 5$) $= T_{zx}$	Shear stress about y axis
	T_6 ($I = 6$) $= T_{xy}$	Shear stress about z axis

TABLE 1.5.2 EXAMPLES OF THE REDUCED TENSOR NOTATION

Parameter		
Reduced notation	Standard notation	Meaning
Elastic constant		
c_{IJ}	c_{ijkl}	The ratio of the Ith stress component to the Jth strain component
c_{11}	c_{1111}	The longitudinal elastic constant relating longitudinal stress and strain components in the x direction
c_{44}	c_{2323}	The shear elastic constant relating shear stress and strain components in the 4-direction (motion about x axes)
$c_{12} = c_{21}$	$c_{1122} = c_{2211}$	$c_{IJ} = c_{JI}$
Dielectric constant		
ε_{ij}		The ratio of the ith component of displacement density to the jth component electric field
	ε_{xx}	The dielectric constant relating D_x and E_x
	ε_{xy}	Note: ε_{ij} is generally finite; $\varepsilon_{ij} = \varepsilon_{ji} = 0$ if i, j, and k are along symmetry axes of a crystal
Piezoelectric stress constant		
e_{Ij}	e_{ikj}	The ratio of the Ith stress component to the jth vector component of electric field
e_{3z}	e_{333}	The ratio of the longitudinal stress in the z direction to the E field in the z direction
e_{iJ}	e_{ijk}	The ratio of the i vector component of electric displacement density to the Jth component of strain
e_{Ij}, e_{Ji}	$e_{ijk} = e_{jki}$	Note that e_{Ij} is the transposed form of e_{jI}

piezoelectric material, which were given in one-dimensional form as Eqs. (1.2.4) and (1.2.9):

$$T_I = c_{IJ}^E S_J - e_{Ij} E_j \qquad (1.5.17)$$
$$D_i = \varepsilon_{ij}^S E_j + e_{iJ} S_J$$

To clarify the meaning of these expressions further, some more examples of the notation are given in Table 1.5.2.

When the material is not piezoelectric, we can write

$$T_I = c_{IJ} S_J \qquad (1.5.18)$$

or, more fully,

$$\begin{bmatrix} T_1 \\ T_2 \\ T_3 \\ T_4 \\ T_5 \\ T_6 \end{bmatrix} = \begin{bmatrix} c_{11} & c_{12} & c_{13} & c_{14} & c_{15} & c_{16} \\ c_{21} & c_{22} & c_{23} & c_{24} & c_{25} & c_{26} \\ c_{31} & c_{32} & c_{33} & c_{34} & c_{35} & c_{36} \\ c_{41} & c_{42} & c_{43} & c_{44} & c_{45} & c_{46} \\ c_{51} & c_{52} & c_{53} & c_{54} & c_{55} & c_{56} \\ c_{61} & c_{62} & c_{63} & c_{64} & c_{65} & c_{66} \end{bmatrix} \begin{bmatrix} S_1 \\ S_2 \\ S_3 \\ S_4 \\ S_5 \\ S_6 \end{bmatrix} \quad (1.5.19)$$

Generally, there are 21 independent elastic constants ($c_{IJ} = c_{JI}$), but these reduce to far fewer independent terms in crystals with certain symmetries. In a cubic crystal, for instance, $c_{11} = c_{22} = c_{33}$, $c_{12} = c_{21} = c_{31} = c_{13} = c_{23} = c_{32}$, and $c_{14} = c_{15} = c_{16} = c_{24} = c_{25} = c_{26} = c_{34} = c_{35} = c_{36} = 0$. Thus there are only three independent constants: c_{11}, c_{44}, and c_{12}. If the material is isotropic, it can be shown that

$$c_{11} - c_{12} = 2c_{44} \quad (1.5.20)$$

With a piezoelectric material such as PZT, which is symmetrical about its z axis, or isotropic in the x and y directions, we find that only e_{z3}, e_{z1}, e_{z2}, and $e_{x5} = e_{y4}$ are finite with $c_{44} = c_{55}$, $c_{11} = c_{22}$, $c_{11} - c_{12} = 2c_{66}$, and $\varepsilon_{xx} = \varepsilon_{yy}$. In this case, if we consider the longitudinal component of stress in the z direction, in a large area plate in which there is no motion in the x and y directions, we see that $S_1 = S_2 = 0$.

It follows from Eq. (1.5.17) that†

$$\begin{aligned} T_3 &= c_{33}^E S_3 - e_{3z} E_z \\ D_z &= \varepsilon_{zz}^S E_z + e_{z3} S_3 \end{aligned} \quad (1.5.21)$$

Thus the values of c and e that we must use for longitudinal waves correspond to c_{33} and e_{3z}, respectively.

In the same way, if we consider shear wave propagation along the x axis of the material, we see that

$$\begin{aligned} T_5 &= c_{55}^E S_5 - e_{5x} E_x \\ D_x &= \varepsilon_{xx}^S E_x + e_{x5} S_5 \end{aligned} \quad (1.5.22)$$

because E_x couples only to T_5.

Thus we can excite a shear wave in the 5-direction (motion about the y axis or shear in the x and z directions) by using electrodes placed at $x = 0$ and $x = l$. This means that a piezoelectric transducer for shear waves can be constructed using a ceramic material "poled" in the z direction. In this case, the appropriate elastic and piezoelectric stress constants in the one-dimensional theory are c_{55} and e_{x5}, respectively.

Consider the expansion of a thin bar along its length, with the E field in the same direction. As the bar is compressed in the z direction, it will bulge in the x

---

†In the literature, the subscripts 1, 2, and 3 are often used instead of x, y, and z, respectively. For example, ε_{zz}^S would be written as ε_{33}^S; e_{z3} would be e_{33} [16].

or y direction; the restoring force in those directions is negligible. Thus we assume that T_3 is the only finite component of stress. In this case, it is more convenient to use constitutive relations stated with stress as the independent variable. These are written in the reduced subscript form as follows:

$$S_I = d_{Ij}E_j + s^E_{IJ}T_J \qquad (1.5.23)$$
$$D_i = \varepsilon^T_{ij}E_j + d_{iJ}T_J$$

Both c^E_{IJ} and s^E_{IJ} are reciprocal matrices; so are c^D_{IJ} and s^D_{IJ}. The parameter s is called the *compliance* and is often quoted in preference to c. The customary names and definitions of these parameters are given in Table 1 of a paper by Jaffe and Berlincourt [15].

Leaving out the subscripts, Eq. (1.5.23) reduces to a set of one-dimensional equations that are equivalent to Eqs. (1.4.4) and (1.4.9):

$$S = dE + s^E T \qquad (1.5.24)$$
$$D = \varepsilon^T E + dT$$

For a thin piezoelectric rod transducer of PZT poled in the z direction, for example, this set of equations would be more convenient than Eqs. (1.4.4) and (1.4.9).

We define a piezoelectric coupling coefficient k_{z3} as follows [16]:

$$k^2_{z3} = \frac{d^2_{z3}}{s^E_{33}\varepsilon^T_{zz}} \qquad (1.5.25)$$

In general, d_{iJ} is called the *transmitting constant* or *piezoelectric stress constant*. Often, in the literature, the parameter k_{z3} in Eq. (1.5.25) is called k_{33}, the two nomenclatures being interchangeable (see the footnote on p. 80).

The acoustic velocity of a wave along a thin bar is normally lower than it is along a plate. When $E = 0$, for example, the acoustic velocity of an acoustic wave along a thin rod of PZT-5A is 2.62 km/s, while when $D = 0$, the stiffened velocity is increased by a factor of $(1 - k^2_{z3})^{1/2}$, to 3.71 km/s. Equivalently, we can write

$$s^D = s^E(1 - k^2_{z3}) \qquad (1.5.26)$$

The k_{z3} term defined here is used in the expressions for the impedance of a piezoelectric transducer [see, e.g., Eq. (1.4.23)], just as the parameter k_T is used for a piezoelectric plate in Sec. 1.4.

Let us consider some other useful forms of the constitutive relations. When the transducer is used as a transmitter, we can obtain a good idea of its characteristics by using the constitutive relations given above. Alternatively, if it is used as a receiving transducer operating into a very high impedance (a common situation in low-frequency devices), we can assume that $D = 0$ (i.e., that the load impedance is much higher than the capacitive reactance of the transducer).

When the acoustic load at one surface of the transducer is specified and the impedance of the acoustic source is known, we can find the stress and strain component within the transducer in terms of the driving velocity or strain. However, it is often easier to specify the input stress to a receiving transducer. When a severe mismatch exists, as it would using a piezoelectric receiving transducer in water, the stress or pressure at the transducer's surface may be a known or measurable quantity, for the transducer will tend to be rigid relative to the water, and the velocity at its surface will not be easily measurable or calculable. Thus, for low-frequency work, it is often convenient to write the constitutive relations with the known quantities as the dependent variables in the reduced subscript notation, that is,

$$E_i = \beta_{ij}^T D_j - g_{iJ} T_J$$
$$S_I = g_{Ij} D_j + s_{IJ}^D T_J \qquad (1.5.27)$$

If the current is small, as it would be with a high-impedance receiver, then $D = 0$, and the one-dimensional longitudinal wave equations for a plate transducer become

$$E = gT$$
$$S = s^D T \qquad (1.5.28)$$

The parameter g is often called the *receiver constant* and is the only one required to determine the voltage output of the transducer with a given applied stress. By comparison with Eq. (1.5.24), we see that

$$g = \frac{d}{\varepsilon^T} \qquad (1.5.29)$$

The constitutive relations can be written in yet another convenient form:

$$E_i = \beta_{ij}^S D_j - h_{iJ} S_J$$
$$T_i = -h_{Ij} D_j + c_{IJ}^D S_J \qquad (1.5.30)$$

In this case, the one-dimensional relations become

$$E = -hS + \beta^S D$$
$$T = c^D S - hD \qquad (1.5.31)$$

The parameter h is called the *transmitting constant*; it determines the voltage required across the transducer to produce a given strain. We see from Eqs. (1.4.7), (1.4.8), and (1.5.31) that

$$h = \frac{e}{\varepsilon^S} \qquad (1.5.32)$$

as we have shown in Sec. 1.4.2. Similarly, comparing the one-dimensional forms, we see that

$$\beta^S = \frac{1}{\varepsilon^S} \qquad (1.5.33)$$

In practice, any one of these constitutive relations can be used; the choice entirely a matter of convenience and custom. When one complete set of parameters is given, the others can be determined in terms of them, but to do so may require considerable algebraic manipulations.

PROBLEM SET 1.5

1. (a) We have worked out the equivalent circuit of the in-line transducer by using the fact that D is uniform in x, y, and z, while E varies with z. Another important case is the *crossed-field transducer*, also known as the *length expander bar*, in which a thin strip of ceramic is coated with electrodes, with the crystal axes chosen so that when a potential is applied between the electrodes, the piezoelectric material expands parallel to them. This transducer is used to excite longitudinal or shear waves in hollow cylinders or thin strips. Analyze it, taking the D and E fields to be in the z direction (i.e., perpendicular to the electrodes) and $v = u_x$ and $T = T_1$ to be in the x direction. In this case, we assume that $T_3 = 0$ and $S_2 = 0$.

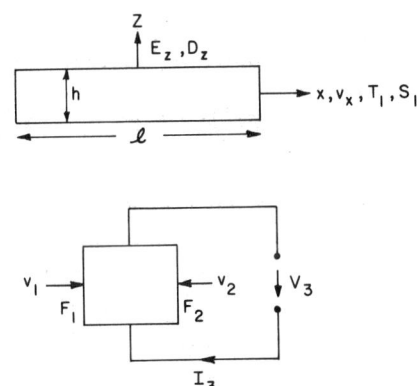

We assume that T_1, like S_1, varies with x, but not with y or z. The length of the transducer is l, and the electrode spacing is h.

Hint: You will find it convenient to work in terms of E rather than D. Take $E_x = 0$ so that as $\nabla \times \mathbf{E} = 0$, $\partial E_z/\partial x = 0$ (i.e., E_z and ϕ are independent of x). Now, however, D_z will vary with x.

(b) Show that the Mason equivalent circuit of the transducer is just like that of the in-line transducer illustrated in Fig. 1.4.4, except that it lacks a negative capacity. Using the three-port concept, state what coefficients with what subscripts are required to specify the parameters of the transducer when it is a ceramic material such as PZT-5H poled in the z direction.

2. Work out the KLM equivalent circuit for the crossed-field transducer described in Prob. 1.

REFERENCES

1. B. A. Auld, *Acoustic Fields and Waves in Solids*, Vol. 1. New York: John Wiley & Sons, Inc., 1973.
2. S. Ramo, J. R. Whinnery, and T. Van Duzer, *Fields and Waves in Communication Electronics*. New York: John Wiley & Sons, Inc., 1965.
3. R. E. Collin, "Theory and Design of Wide-Band Multisection Quarter-Wave Transformers," *Proc. IRE*, 43, No. 2 (Feb. 1955), 179–85.
4. H. J. Riblet, "General Synthesis of Quarter-Wave Impedance Transformers," *IRE Trans. Microwave Theory Tech.*, MTT-5, No. 1 (Jan. 1957), 36–43.
5. J. L. Altman, *Microwave Circuits*. Princeton, N.J.: D. Van Nostrand Company, 1964.
6. J. A. Stratton, *Electromagnetic Theory*. New York: McGraw-Hill Book Company, 1941.
7. M. Redwood, "Transient Performance of a Piezoelectric Transducer," *J. Acoust. Soc. Am.*, 33, No. 4 (Apr. 1961), 527–36.
8. M. Onoe, H. F. Tiersten, and A. H. Meitzler, "Shift in the Location of Resonant Frequencies Caused by Large Electromechanical Coupling in Thickness-Mode Resonators," *J. Acoust. Soc. Am.*, 35, No. 1 (Jan. 1963), 36–42.
9. R. Krimholtz, D. A. Leedom, and G. L. Matthaei, "New Equivalent Circuits for Elementary Piezoelectric Transducers," *Electron. Lett.*, 6, No. 13 (June 25, 1970), 398–99.
10. C. S. DeSilets, J. D. Fraser, and G. S. Kino, "The Design of Efficient Broad-Band Piezoelectric Transducers," *IEEE Trans. Sonics Ultrason.*, SU-25, No. 3 (May 1978), 115–25.
11. T. M. Reeder and D. K. Winslow, "Characteristics of Microwave Acoustic Transducers for Volume Wave Excitation," *IEEE Trans. Microwave Theory Tech.*, MTT-17, No. 11 (Nov. 1969), 927–41.
12. A. R. Selfridge, R. Baer, B. T. Khuri-Yakub, and G. S. Kino, "Computer-Optimized Design of Quarter-Wave Acoustic Matching and Electrical Matching Networks for Acoustic Transducers," *1981 Ultrason. Symp. Proc.*, (IEEE), 81-CH1689-9, Vol. 2, 644–48.
13. C. H. Chou, J. E. Bowers, A. R. Selfridge, B. T. Khuri-Yakub, and G. S. Kino, "The Design of Broadband and Efficient Acoustic Wave Transducers," *1980 Ultrason. Symp. Proc.*, (IEEE), 80CH1602-2, Vol. 2, 984–88.
14. J. F. Nye, *Physical Properties of Crystals: Their Representation by Tensors and Matrices*, corr. rpt. of 1st ed. (1957: rpt. Oxford: Clarendon Press, 1960).
15. H. Jaffe and D. A. Berlincourt, "Piezoelectric Transducer Materials," *Proc. IEEE*, 53, No. 10 (Oct. 1965), 1372–86.
16. "IRE Standards on Piezoelectric Crystals: Measurements of Piezoelectric Ceramics, 1961," *Proc. IRE*, 49, No. 7 (July 1961), 1161–69.

Chapter 2

Waves in Isotropic Media

2.1 INTRODUCTION

In this chapter we describe the various types of waves that can exist in isotropic media. In Sec. 2.2 we introduce the commonly used notation for the elastic constants in isotropic media and then derive the wave equations for propagation of longitudinal and shear waves. These will be stated in terms of potential functions, in a similar manner as in electromagnetic (EM) theory. We also discuss the basic concepts of reflection and refraction at the boundary between two media, stressing their importance to various types of mode conversion schemes.

Section 2.3 carries these ideas further by giving a short description of some of the various kinds of waveguide modes that can exist in bounded media. We then consider the important topic of surface acoustic waves and give the basic derivation for Rayleigh waves in an isotropic medium.

We describe the properties of the interdigital transducer in Sec. 2.4, and the delta function model is given, together with the network theory for determining its impedance.

In Sec. 2.5, we derive a perturbation theory or *normal-mode theory* that is used to calculate excitation and perturbation of waves in a waveguide. First we show how this theory may be applied to various problems; then we use it to determine the impedance of an interdigital transducer. Next we introduce the concept of *leaky waves* and determine the attenuation of leaky Rayleigh waves propagating on a surface loaded by water. Finally, we employ the leaky wave theory to derive the properties of the wedge transducer.

2.2 BASIC THEORY FOR WAVES IN ISOTROPIC MEDIA

2.2.1 Mathematical Formalism for Waves in Isotropic Media

Here we show how the tensor formalism for acoustic waves given in Sec. 1.5 and Appendix A can be simplified for isotropic media [1–6]. We use the condition that only two independent elastic constants can exist in an isotropic medium; these determine the ratio between the shear wave velocity and the longitudinal wave velocity. As we show, this ratio must be between 0 and 0.707; for many solids it is approximately 0.5. Because liquids cannot support shear stresses, their shear wave velocities must always be zero. In liquids, the stress components in all directions are equal in magnitude, which means that the stress can be expressed in terms of pressure, a scalar quantity. Materials such as polyethylene or rubber have very small ratios of shear to longitudinal wave velocity and thus behave like viscous liquids. We also define *Poisson's ratio* [2], a parameter used frequently in solid mechanics.

We shall find it convenient to derive the reciprocal relations to the elastic constitutive equations so that we can determine strain in terms of stress. The expressions obtained will be stated in terms of Poisson's ratio and an elasticity parameter known as *Young's modulus*. Such relations are particularly convenient for dealing with wave propagation along thin rods for which the applied stresses normal to the surface of the rod are zero. As the frequency is reduced to zero and the wavelength becomes much larger than the cross-sectional dimensions of the rod, these relations reduce to simple forms used in statics.

Lamé constants. We have shown in Appendix A that, due to symmetry, the number of independent elastic constants in an isotropic medium reduces to two. In the literature, these two independent constants are called the Lamé constants, λ and μ [1–6]. These parameters are useful for determining the total stored energy of the system [see Eq. (2.2.11)] and are related to the elastic constants already defined, as follows:

$$c_{11} = c_{22} = c_{33} = \lambda + 2\mu \qquad (2.2.1)$$

$$c_{12} = c_{13} = c_{23} = c_{21} = c_{31} = c_{32} = \lambda \qquad (2.2.2)$$

and

$$c_{44} = c_{55} = c_{66} = \mu = \frac{c_{11} - c_{12}}{2} \qquad (2.2.3)$$

All the other off-diagonal terms are zero. The parameter μ is known as the *shear modulus*, or the modulus of rigidity.

Dilation and Hooke's law. The relation between stress and strain can be written in several equivalent forms. One way to write it is in the reduced tensor notation introduced in Sec. 1.5 and Appendix A:

$$T_I = c_{IJ} S_J \qquad (2.2.4)$$

A second convenient way to state this relation is in a dyadic form:
$$\mathbf{T} = \mathbf{c} : \mathbf{S} \tag{2.2.5}$$
For isotropic materials, the relation between longitudinal stress and strain is
$$T_1 = c_{11}S_1 + c_{12}S_2 + c_{13}S_3$$
$$= \lambda(S_1 + S_2 + S_3) + 2\mu S_1 = \lambda\Delta + 2\mu S_1 \tag{2.2.6}$$
or, in general,
$$T_I = \lambda\Delta + 2\mu S_I \quad (I = 1, 2, 3) \tag{2.2.7}$$
where we have defined Δ, the *dilation*, as
$$\Delta = S_1 + S_2 + S_3 \tag{2.2.8}$$
or the fractional change in volume of the material [see Eq. (A.16)], that is, the sum of the fractional extensions of a cube along the x, y, and z axes.

The similar shear relations are
$$T_4 = \mu S_4$$
$$T_5 = \mu S_5 \tag{2.2.9}$$
$$T_6 = \mu S_6$$

Energy. The stored elastic energy W_c is defined as
$$W_c = \tfrac{1}{2} T_I S_I = \tfrac{1}{2}\mathbf{T} : \mathbf{S} \tag{2.2.10}$$
For an isotropic material this relationship may be written in the following form:
$$W_s = \tfrac{1}{2}\lambda\Delta^2 + \mu(S_1^2 + S_2^2 + S_3^2) + \frac{\mu}{2}(S_4^2 + S_5^2 + S_6^2) \tag{2.2.11}$$

Ideal fluid. An ideal fluid cannot support shear stresses. Hence $\mu = 0$, $c_{11} = c_{12} = \lambda$, $c_{44} = 0$, and, from Eq. (2.2.7), $T_1 = T_2 = T_3 = \lambda D$. Thus we can define the pressure p in a fluid in terms of the stress within a small volume, as follows:
$$p = -T_1 = -T_2 = -T_3 = -\kappa\Delta \tag{2.2.12}$$
The parameter $\kappa = T$ is called *the bulk elastic modulus*. More generally, in a solid, when μ is finite, Eq. (2.2.7) shows that we can still write $p = -\kappa\Delta$, provided that we now define p in terms of the average stress as
$$p = \frac{-(T_1 + T_2 + T_3)}{3} \tag{2.2.13}$$
and put
$$\kappa = \left(\frac{\lambda + 2}{3\mu}\right) \tag{2.2.14}$$

A perfectly incompressible material would have $\kappa = \infty$, whereas an easily compressible medium such as air has a very small κ compared to that of water [κ(water)/κ(air) = 19,000]. This relation is important when considering the propagation of waves through water containing air bubbles.

Young's modulus and the extensional wave velocity. Certain other notations are useful when dealing with stress in a thin rod. It is a simple matter to express the strain in terms of the stress. Equation (2.2.7) implies that

$$S_1 = \frac{\lambda + \mu}{\mu(3\lambda + 2\mu)} T_1 - \frac{\lambda}{2\mu(3\lambda + 2\mu)} (T_2 + T_3) \qquad (2.2.15)$$

with symmetric expressions for S_2 and S_3.

Consider a thin rod stressed in the z direction so that the only finite component of stress is T_3. In this case the rod may bulge in the x and y directions so that S_1 and S_2 are finite. Putting $T_1 = T_2 = 0$, we find that

$$S_3 = \frac{T_3}{E} \qquad (2.2.16)$$

where

$$E = \frac{1}{s_{11}} = \frac{\mu(3\lambda + 2\mu)}{\lambda + \mu} = c_{11} - \frac{2c_{12}^2}{c_{11} + c_{12}} \qquad (2.2.17)$$

The parameter E is called *Young's modulus*; it is the elastic constant normally measured when a rod is stretched in a testing machine. The parameter s_{11} is defined in Sec. 1.5 and Appendix A. The longitudinal wave that propagates along a thin rod is called an *extensional wave*. It follows from Eq. (2.2.17) and the one-dimensional wave equation (1.1.11) that we can find the velocity of this wave by replacing c with E. The extensional wave velocity is therefore

$$V_e = \sqrt{\frac{E}{\rho_{m0}}} \qquad (2.2.18)$$

Equations (2.2.17) and (2.2.18) imply that the extensional wave velocity V_e is always less than the longitudinal velocity V_l.

Poisson's ratio. Putting $T_1 = T_2 = 0$ in Eq. (2.2.15) and using Eqs. (2.2.16) and (2.2.17), we find that

$$S_1 = -\sigma S_3 \qquad (2.2.19)$$

where the parameter σ is called *Poisson's ratio*, defined as

$$\sigma = \frac{\lambda}{2(\lambda + \mu)} = \frac{c_{12}}{c_{11} + c_{12}} = -\frac{s_{12}}{s_{11}} \qquad (2.2.20)$$

Poisson's ratio is the ratio of the transverse compression to the longitudinal expansion of a thin rod to which a static longitudinal axial stress is applied. For a liquid, Poisson's ratio is 0.5; for a solid, it must be between 0 and 0.5. For materials with a small modulus of rigidity, such as polyethylene or rubber, it is close to 0.5 ($\sigma = 0.46$ for polyethylene). For most metals it is of the order of 0.3; for beryllium, however, it is 0.05. For a perfect fluid, $\mu = 0$, $E = 0$, and $\sigma = 0.5$. Thus, if we ignore surface tension which may make T_1 and T_2 finite, a stream of liquid cannot support an extensional wave, so that $V_e = 0$. In seismic work, for reasons of mathematical convenience, it is sometimes assumed that $\lambda = \mu$. In this case, $E = 5/2\lambda$ and $\sigma = \frac{1}{4}$ with $\kappa = 5\lambda/3$; this value of σ is typical for ceramic materials and many types of rock.

It also follows from these relations that, in general,

$$S_1 = \frac{T_1}{E} - \frac{\sigma}{E}(T_2 + T_3) \qquad (2.2.21)$$

with symmetric expressions for S_2 and S_3, and that

$$S_4 = \frac{2(1 + \sigma)}{E} T_4 = \frac{T_4}{\mu} \qquad (2.2.22)$$

with similar expressions for S_5 and S_6. In mechanics texts, the constitutive relations are usually expressed in these forms because they are convenient to use for statics when there is no longitudinal stress variation over the cross section of a rod.

Suppose that we consider a thin rod of rubber or a thin stream of liquid, each aligned in the z direction. As the stresses T_1 and T_2 normal to the axis are zero, and as $\sigma = 0.5$, then $S_1 = S_2 = -S_3/2$. Thus the material expands in the transverse direction by half the amount it is compressed in the longitudinal direction. This is just what we would expect from conservation of mass, for when there is no restoring force on the outside of the rod, the volume of the material cannot change.

Shear, longitudinal, extensional, and strip guide wave velocities. The shear wave velocity V_s determined from the effective elastic constant for shear waves, $c_{44} = E$, is defined as $V_s = \sqrt{c_{44}/\rho_{m0}}$. The ratio of the shear wave velocity to the longitudinal wave velocity V_l is defined as

$$\frac{V_s}{V_l} = \sqrt{\frac{c_{44}}{c_{11}}} = \sqrt{\frac{\mu}{\lambda + 2\mu}} = \sqrt{\frac{0.5 - \sigma}{1 - \sigma}} \qquad (2.2.23)$$

Similarly, the ratios of the extensional wave velocity V_e to the longitudinal and shear wave velocities are

$$\frac{V_e}{V_l} = \sqrt{\frac{E}{c_{11}}} = \sqrt{\frac{(1 + \sigma)(1 - 2\sigma)}{1 - \sigma}} \qquad (2.2.24)$$

and

$$\frac{V_s}{V_e} = \sqrt{\frac{\mu}{E}} = \sqrt{\frac{1}{2(1 + \sigma)}} \qquad (2.2.25)$$

respectively.

Finally, we give the results for an extensional wave propagating along a strip that is infinitesimally thin in the y direction and infinitely wide in the x direction. We call this wave a *strip guide mode*; for it, we assume that $T_2 = 0$ and $S_1 = 0$. We shall call its velocity V_L; the ratio of this velocity to the longitudinal wave velocity V_l is

$$\frac{V_L}{V_l} = \sqrt{1 - \left(\frac{c_{12}}{c_{11}}\right)^2} = \sqrt{\frac{1 - 2\sigma}{(1 - \sigma)^2}} \qquad (2.2.26)$$

Plots of the values of V_s/V_l, V_e/V_l, V_s/V_e, and V_L/V_l as functions of Poisson's ratio are given in Fig. 2.2.1. For $\sigma = 0$, $V_s/V_l = 0.707$. As might be expected, rubber-like materials such as polyethylene ($\sigma = 0.46$) tend to have relatively large values of Poisson's ratio; because they behave somewhat like liquids, they have a small ratio of V_s/V_l. Harder materials such as ceramics and glasses tend to have smaller values of σ, and hence a larger ratio of V_s/V_l. Most metals have Poisson's ratios of the order of 0.3, with $V_s/V_l \sim 0.6$ and $V_e/V_l \sim 0.9$. Note that V_e/V_l, V_L/V_l, and V_s/V_l decrease monotonically with σ.

2.2.2 Equations of Motion for Solids and Fluids

It is shown in Sec. 1.5 and Appendix A that the force in the x direction on a cube of volume $dx\,dy\,dz$ due to a stress $T_1 = T_{xx}$ is $(\partial T_1/\partial x)\,dx\,dy\,dz$, or $\partial T_1/\partial x$ per unit volume. This is illustrated in Fig. 2.2.2(a). Similarly, as illustrated in Fig. 2.2.2(b), the force in the x direction on a cube of volume $dx\,dy\,dz$ due to a shear stress $T_6 = T_{xy}$ is $(\partial T_6/\partial y)\,dx\,dy\,dz$, or $\partial T_6/\partial y$ per unit volume [7]. A similar relation can be obtained for the force due to a shear stress about the y axis (a shear stress in the xz direction). Thus, as discussed more fully in Sec. 1.5 and

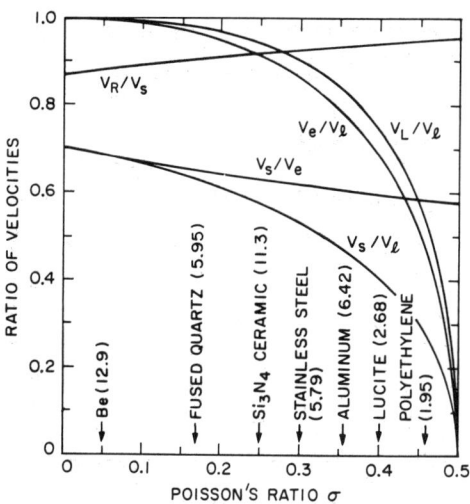

Figure 2.2.1 Relationships of velocity ratios to Poisson's ratio. The longitudinal wave velocities in km/s are given in brackets for various materials. V_l, Longitudinal wave velocity; V_e, extensional wave velocity; V_s, shear wave velocity; V_R, Rayleigh wave velocity; V_L, strip guide mode velocity.

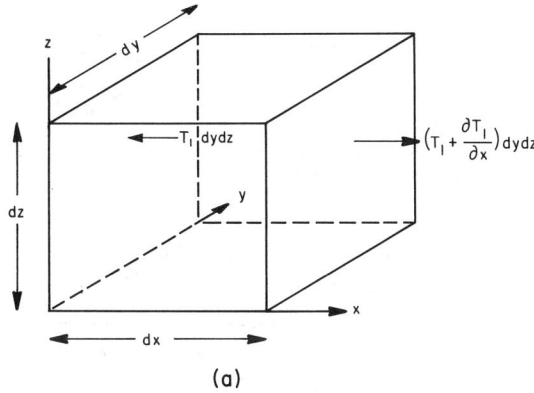

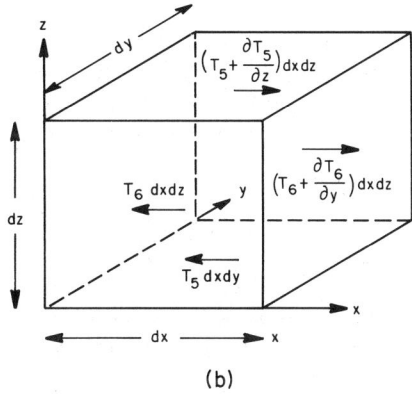

Figure 2.2.2 Stresses applied to a cube with sides dx, dy, and dz.
(a) Longitudinal stress T_1; (b) shear stresses $T_5 = T_{xz}$ and $T_6 = T_{xy}$. The longitudinal stress applies forces to the faces x and $x + dx$ of the cube, while the shear stresses T_5 and T_6 apply forces to the faces z and $z + dz$, and y and $y + dy$, respectively.

Appendix A, after adding all the force terms, the equation of motion in the x direction is

$$\rho_{m0}\frac{\partial^2 u_x}{\partial t^2} = \frac{\partial T_1}{\partial x} + \frac{\partial T_6}{\partial y} + \frac{\partial T_5}{\partial z} \tag{2.2.27}$$

In full tensor notation, Eq. (2.2.27) may be written in the form

$$\rho_{m0}\frac{\partial^2 u_x}{\partial t^2} = \frac{\partial T_{xx}}{\partial x} + \frac{\partial T_{xy}}{\partial y} + \frac{\partial T_{xz}}{\partial z} \tag{2.2.28}$$

Similar relations for the equations of motion in the y and z directions can be obtained.

Following Auld [4], it is often convenient to use the symbolic notation, which is discussed more fully in Appendix A, and write

$$\nabla \cdot \mathbf{T} = \rho_{m0}\frac{\partial^2 \mathbf{u}}{\partial t^2} \tag{2.2.29}$$

Gauss's divergence theorem can often be used on this divergence term and physically meaningful formulas for surface stress can easily be derived from it.

In the full tensor notation, Eq. (2.2.29) is equivalent to

$$\frac{\partial T_{ij}}{\partial x_j} = \rho_{m0} \frac{\partial^2 u_i}{\partial t^2} \qquad (2.2.30)$$

2.2.3 Wave Equation in an Isotropic Medium

We can substitute Eq. (2.2.6) into Eq. (2.2.29) or Eq. (2.2.30) to obtain an equation that relates the displacement in the x direction and the strain, as follows:

$$\rho_{m0} \frac{\partial^2 u_x}{\partial t^2} = \frac{\partial}{\partial x}(\lambda \Delta + 2\mu S_1) + \mu \left(\frac{\partial S_6}{\partial y} + \frac{\partial S_5}{\partial z}\right) \qquad (2.2.31)$$

Similar expressions hold for $\partial^2 u_y/\partial t^2$ and $\partial^2 u_z/\partial t^2$. Note that the dilation Δ of the material, or the relative volume expansion, is defined as

$$\Delta = S_1 + S_2 + S_3 = \nabla \cdot \mathbf{u} \qquad (2.2.32)$$

To obtain the wave equation for the displacement, we must use an additional relation between strain and displacement. As far as the longitudinal components of strain are concerned, this relation is the same as that derived in Eq. (1.1.2):

$$S_1 = \frac{\partial u_x}{\partial x} \qquad (2.2.33)$$

Similar expressions exist for the other longitudinal components of strain.

To determine the shear strain component in the y-z plane, we must take account of the displacements in both the y and z directions. Generalizing the one-dimensional formulation in Sec. 1.1, we write

$$S_4 = \frac{\partial u_z}{\partial y} + \frac{\partial u_y}{\partial z} \qquad (2.2.34)$$

with similar expressions for the other shear components of strain. These relations are derived in more detail in Appendix A. Following Auld [4], we can summarize them, with the symbolic notation defined in Appendix A, as

$$\mathbf{S} \approx \nabla_s \mathbf{u} \qquad (2.2.35)$$

where $\nabla_s \mathbf{u}$ denotes the symmetric part of the dyadic $\nabla \mathbf{u}$. Alternatively, using the full tensor notation, we can write

$$S_{ij} = \frac{1}{2}\left(\frac{\partial u_i}{\partial x_j} + \frac{\partial u_j}{\partial x_i}\right) \qquad (2.2.36)$$

Note the $\frac{1}{2}$ used in the definition of S_{ij} (see Sec. 1.5, Table 1.5.1, and Appendix A). By substituting for S_5 and S_6 from Eq. (2.2.35) or Eq. (2.2.36) into Eq. (2.2.31), we obtain (after some algebra) the relatively simple relation

$$\rho_{m0} \frac{\partial^2 \mathbf{u}}{\partial t^2} = (\lambda + 2\mu) \nabla(\nabla \cdot \mathbf{u}) - \mu \nabla \times \nabla \times \mathbf{u} \qquad (2.2.37)$$

At this point, it is convenient to define the infinitesimal rotation of a rigid body w_{ij} as

$$w_{ij} = \frac{1}{2}\left(\frac{\partial u_j}{\partial x_i} - \frac{\partial u_i}{\partial x_j}\right) \tag{2.2.38}$$

or

$$\mathbf{w} = \tfrac{1}{2} \nabla \times \mathbf{u} \tag{2.2.39}$$

We note that $\mathbf{w}$ can be finite when the strain is zero, and vice versa. By substituting Eqs. (2.2.32) and (2.2.39) into Eq. (2.2.37) and using the relation $\nabla \cdot \mathbf{w} = \tfrac{1}{2}\nabla \cdot (\nabla \times \mathbf{u}) = 0$, we obtain the following separate wave equations for Δ and $\mathbf{w}$:

$$(\lambda + 2\mu)\nabla^2 \Delta = \rho_{m0} \frac{\partial^2 \Delta}{\partial t^2} \tag{2.2.40}$$

and

$$\mu \nabla^2 \mathbf{w} = \rho_{m0} \frac{\partial^2 \mathbf{w}}{\partial t^2} \tag{2.2.41}$$

respectively. These expressions are the general forms of the longitudinal and shear wave equations, respectively. Therefore, the longitudinal wave equation can be expressed entirely in terms of the dilation Δ, and the shear wave equation can be expressed entirely in terms of the rotation $\mathbf{w}$.

At this point it is convenient to use the fact that any vector can be written in terms of a vector potential and scalar potential. Thus the displacement vector $\mathbf{u}$ can be written in the form

$$\mathbf{u} = \nabla \phi + \nabla \times \mathbf{\psi} \tag{2.2.42}$$

where ϕ is a scalar potential and $\mathbf{\psi}$ is a vector potential. We shall show that the longitudinal wave solution can be stated entirely in terms of ϕ and that the shear wave solution can be stated entirely in terms of $\mathbf{\psi}$. All displacement components, as well as the stress and strain, can be determined in terms of ϕ and $\mathbf{\psi}$.

Longitudinal waves. We substitute Eqs. (2.2.32) and (2.2.42) into Eq. (2.2.40) and use the identity $\nabla \cdot (\nabla \times \mathbf{\psi}) = 0$ to obtain

$$\nabla^2 \left[\rho_{m0} \frac{\partial^2 \phi}{\partial t^2} - (\lambda + 2\mu)\nabla^2 \phi \right] = 0 \tag{2.2.43}$$

This expression is satisfied if the terms in brackets are zero. Hence we can take the longitudinal wave equation to be

$$\rho_{m0} \frac{\partial^2 \phi}{\partial t^2} - (\lambda + 2\mu)\nabla^2 \phi = 0 \tag{2.2.44}$$

For waves that vary as $\exp[j(\omega t - \mathbf{k}_l \cdot \mathbf{r})]$, either Eq. (2.2.40) or Eq. (2.2.44) leads to the result

$$k_l^2 = \frac{\omega^2 \rho_{m0}}{\lambda + 2\mu} = \frac{\omega^2}{V_l^2} \qquad (2.2.45)$$

Thus the longitudinal wave velocity V_l is given by the relation

$$V_l = \sqrt{\frac{\lambda + 2\mu}{\rho_{m0}}} = \sqrt{\frac{c_{11}}{\rho_{m0}}} \qquad (2.2.46)$$

By now this is a familiar result, but here we have proved that it holds for longitudinal wave propagation in an arbitrary direction in an isotropic solid, and that a pure longitudinal wave, uncoupled to a shear wave, can exist in an isotropic solid.

Shear waves. We now consider shear waves in an isotropic solid. Following a similar procedure to that used for longitudinal waves, we substitute Eqs. (2.2.39) and (2.2.42) into Eq. (2.2.41), and use the identity $\nabla \times \nabla \phi = 0$, to obtain the relation

$$\nabla \times \nabla \times \left(\mu \nabla^2 \boldsymbol{\psi} - \rho_{m0} \frac{\partial^2 \boldsymbol{\psi}}{\partial t^2} \right) = 0 \qquad (2.2.47)$$

Again, by assuming that the terms in parentheses are zero, we see that the shear wave equation is

$$\rho_{m0} \frac{\partial^2 \boldsymbol{\psi}}{\partial t^2} = \mu \nabla^2 \boldsymbol{\psi} \qquad (2.2.48)$$

We obtain a one-dimensional wave equation from Eq. (2.2.41) or Eq. (2.2.48), with a solution of the form $\exp[j(\omega t - \mathbf{k}_s \cdot \mathbf{r})]$, for a wave with velocity

$$V_s = \sqrt{\frac{\mu}{\rho_{m0}}} = \sqrt{\frac{c_{44}}{\rho_{m0}}} \qquad (2.2.49)$$

2.2.4 Plane Wave Reflection and Refraction

We now consider the reflection of an infinite plane wave at a free surface. If a plane longitudinal or shear wave is normally incident on a free surface, the reflected wave will also be a wave of the same type and of equal amplitude. The boundary condition is that the normal component of stress at the surface is zero. It is easy to apply this condition in this simple case. More generally, if a longitudinal wave is incident in the x-y plane on a free surface $y = 0$ at an angle θ_{li} to the normal, as illustrated in Fig. 2.2.3, it will give rise to a reflected longitudinal wave at an angle θ_{lr} to the normal and a reflected shear wave at an angle θ_{sr} to the normal. The displacement associated with this shear wave will have a component in the vertical direction, so we call it a *shear vertical* wave.

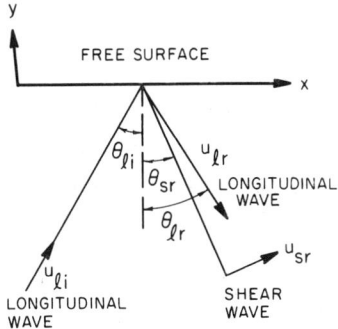

Figure 2.2.3 Longitudinal wave incident on a free surface. There are two reflected waves: a longitudinal wave and a shear wave.

The longitudinal wave can be represented in terms of a potential ϕ, which varies as $\exp(-j\mathbf{k}_l \cdot \mathbf{r})$. The incident longitudinal wave potential ϕ_{li}, in the coordinate system of Fig. 2.2.3, is defined as

$$\phi_{li} = A_{li} e^{-jk_l(x \sin \theta_{li} + y \cos \phi_{li})} \tag{2.2.50}$$

In the same way, there must be a reflected longitudinal wave ϕ_{lr} of the form

$$\phi_{lr} = A_{lr} e^{-jk_l(x \sin \theta_{lr} - y \cos \theta_{lr})} \tag{2.2.51}$$

and, in general, a reflected shear wave ψ_{sr} of the form

$$\psi_{sr} = A_{sr} e^{-jk_s(x \sin \theta_{sr} - y \cos \theta_{sr})} \tag{2.2.52}$$

where the potential ψ_{sr} is a vector in the z direction. Any other components in ψ_{sr} would give rise to additional components of $\mathbf{T}$ and $\mathbf{u}$, and thus would not satisfy the boundary conditions.

The boundary condition at the surface is that the total normal component of stress must be zero. Therefore, as shown in Sec. 1.5 and Appendix A, the stress components at the surface are

$$T_2 = T_{yy} = 0 \tag{2.2.53}$$

and

$$T_6 = T_{xy} = 0 \tag{2.2.54}$$

These stress components are derived as sums of the components of the longitudinal and shear waves. Thus, to satisfy the boundary conditions at any point along the surface $y = 0$, all components of the longitudinal and shear waves must have the same phase variation along the surface. This implies, from Eqs. (2.2.50)–(2.2.52), that

$$k_l \sin \theta_{li} = k_l \sin \theta_{lr} = k_s \sin \theta_{sr} \tag{2.2.55}$$

We conclude, just as we do for reflection of electromagnetic (EM) waves, that the angle of incidence equals the angle of reflection for the longitudinal waves, or

$$\theta_{li} = \theta_{lr} \tag{2.2.56}$$

We also arrive at the condition

$$\frac{\sin \theta_{sr}}{\sin \theta_{li}} = \frac{k_l}{k_s} = \frac{V_s}{V_l} = \sqrt{\frac{\mu}{\lambda + 2\mu}} \qquad (2.2.57)$$

This condition is like Snell's law for refraction of EM waves and is based on the same considerations. As $V_s < V_l$ in all isotropic solids, the reflected shear wave propagates at an angle closer to the normal than does the reflected longitudinal wave.

If we consider the opposite case of excitation by a shear vertical wave, we conclude, using the same type of notation, that

$$\frac{\sin \theta_{lr}}{\sin \theta_{si}} = \frac{k_s}{k_l} = \frac{V_l}{V_s} \qquad (2.2.58)$$

as illustrated in Fig. 2.2.4. Now the possibility exists that the angle θ_{lr} can become purely imaginary and that only a shear wave will be reflected if the angle of incidence of the shear wave is large enough. In this case, there will actually be a finite component of the reflected longitudinal wave, but its amplitude will fall off exponentially from the surface [4].

Incident longitudinal wave. We may solve for the amplitudes of the reflected waves by deriving the displacement and the stresses from the corresponding expressions for the potentials. For instance, for a longitudinal wave, incident on the plane $y = 0$, with amplitude varying as $\exp(j\omega t)$, we can write $\mathbf{u} = -j\mathbf{k}_l A_{li} \exp(-j\mathbf{k}_l \cdot \mathbf{r})$. The two components of displacement are

$$u_x = \frac{\partial \phi}{\partial x} = -jk_l A_{li} \sin \theta_l e^{-jk_l(x \sin \theta_l + y \cos \theta_l)} \qquad (2.2.59)$$

and

$$u_y = \frac{\partial \phi}{\partial y} = -jk_l A_{li} \cos \theta_l e^{-jk_l(x \sin \theta_l + y \cos \theta_l)} \qquad (2.2.60)$$

It is also convenient, following Auld [7], to define an amplitude coefficient in terms of the displacement

$$\mathbf{u} = \frac{-\mathbf{k}_l}{k_l} A'_{li} e^{-j\mathbf{k}_l \cdot \mathbf{r}} \qquad (2.2.61)$$

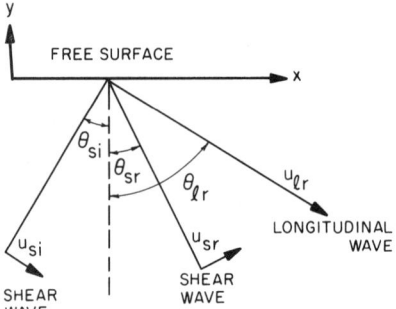

Figure 2.2.4 Shear wave incident on a free surface.

where the exp $(j\omega t)$ variation is understood. Similarly, the reflected shear wave is generated from a potential $\boldsymbol{\psi}$ in the z direction, and we can write $\mathbf{u} = \nabla \times \boldsymbol{\psi}$. The two components of displacement are

$$u_x = \frac{\partial \psi_z}{\partial y} = jk_s A_{sr} \cos \theta_s e^{-jk_s(x \sin \theta_s - y \cos \theta_s)} \qquad (2.2.62)$$

and

$$u_y = -\frac{\partial \psi_z}{\partial x} = jk_s A_{sr} \sin \theta_s e^{-jk_s(x \sin \theta_s - y \cos \theta_s)} \qquad (2.2.63)$$

with

$$\mathbf{u} = j(\mathbf{k}_s \times \mathbf{a}_z) A'_{sr} e^{j\mathbf{k}_s \cdot \mathbf{r}} \qquad (2.2.64)$$

and the reflected longitudinal wave is of the form $\mathbf{u} = jk_l A_{lr} \exp(j\mathbf{k}_l \cdot \mathbf{r})$. It has components

$$u_x = -jk_l A_{lr} \sin \theta_l e^{-jk_l(x \sin \theta_l - y \cos \theta_l)} \qquad (2.2.65)$$

and

$$u_y = jk_l A_{lr} \cos \theta_l e^{-jk_l(x \sin \theta_l - y \cos \theta_l)} \qquad (2.2.66)$$

with

$$\mathbf{u} = \frac{\mathbf{k}_l}{k_l} A'_{lr} e^{j\mathbf{k}_l \cdot \mathbf{r}} \qquad (2.2.67)$$

Writing $\mathbf{S} = \nabla_s u$ and $\mathbf{T} = \mathbf{c}:\mathbf{S}$, we can now derive the stresses and strains corresponding to these waves. In general, by differentiating, we can show that both the longitudinal potential ϕ and the shear potential ψ give rise, in the x, y coordinate system, to longitudinal and shear components of stress. For instance, the longitudinal component of strain $S_1 = \partial u_x/\partial x$ has contributions from both ϕ and ψ, so at $x = 0$ and $y = 0$, we can write

$$S_1 = -k_l^2 (A_{li} + A_{lr}) \sin^2 \theta_l + k_s^2 A_{sr} \sin \theta_s \cos \theta_s \qquad (2.2.68)$$

Following Auld's treatment of the problem [7], we shall define reflection coefficients here in terms of the magnitude of the total displacements $\mathbf{u}$ or velocity $\mathbf{v} = j\omega\mathbf{u}$ of the waves, rather than in terms of the stress, as we did in Sec. 1.1. We write

$$\Gamma_{ll} = \frac{A'_{lr}}{A'_{li}}$$

$$\Gamma_{sl} = \frac{A'_{sr}}{A'_{si}} \qquad (2.2.69)$$

To solve for Γ_{ll} and Γ_{sl}, we employ the boundary conditions $T_2 = 0$ and $T_6 = 0$. We can use Eqs. (2.2.7) and (2.2.9) to write $T_2 = \lambda(S_1 + S_2) + 2\mu S_1$ and $T_6 = \mu S_6$. This gives two boundary conditions for two unknowns. Thus, for

Sec. 2.2   Basic Theory for Waves in Isotropic Media   **97** longitudinal wave incidence, we can show (after some algebra) [7] that

$$\Gamma_{ll} = \frac{\sin 2\theta_s \sin 2\theta_l - (V_l/V_s)^2 \cos^2 2\theta_s}{\sin 2\theta_s \sin 2\theta_l + (V_l/V_s)^2 \cos^2 2\theta_s} \tag{2.2.70}$$

and

$$\Gamma_{sl} = \frac{2(V_l/V_s) \sin 2\theta_l \cos 2\theta_s}{\sin 2\theta_s \sin 2\theta_l + (V_l/V_s)^2 \cos^2 2\theta_s} \tag{2.2.71}$$

A curve, given by Auld, for a vertically polarized longitudinal wave reflected at the surface of fused quartz is shown in Fig. 2.2.5. Several interesting features of these results are apparent from the curves and the formulas used. As we might expect, at normal incidence ($\theta_l = 0$), $\Gamma_{ll} = 1$ and $\Gamma_{sl} = 0$.

Conversion of longitudinal to shear waves and shear to longitudinal waves. As the angle of incidence is increased, the longitudinal wave reflection coefficient Γ_{ll} decreases and a reflected shear wave begins to be excited, with Γ_{ll} initially increasing linearly with θ_l. At a certain point, the excitation of the reflected longitudinal wave may become zero; for fused quartz this occurs at $\theta_l = 42°$. Thus, for this condition, there is complete conversion from a longitudinal to a shear wave. This result is important because the phenomenon makes it possible to convert a longitudinal wave to a shear wave without losing power. As longitudinal wave transducers are often easier to construct than shear wave transducers, this is an extremely convenient method for obtaining shear waves. All that is required, as illustrated in Fig. 2.2.6, is a piece of fused quartz or another suitable material (see Prob. 11), cut to the correct angle.

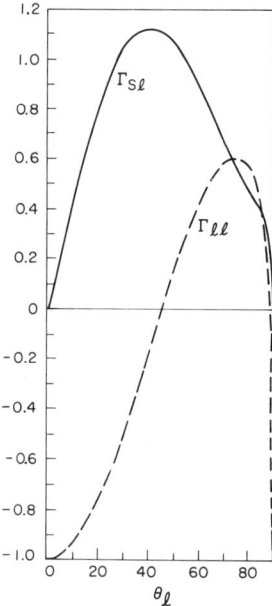

Figure 2.2.5 Reflection of a longitudinal wave at a stress-free boundary in fused silica. The vertical axis is the amplitude reflection coefficient defined in terms of the particle velocity. For fused quartz, $\sigma = 0.17$ and $V_s/V_l = 0.63$. (After Auld [7].)

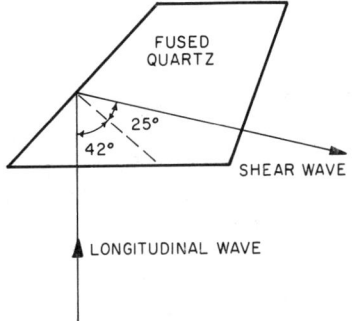

Figure 2.2.6 Longitudinal-to-shear wave converter using fused quartz. For fused quartz, $\sigma = 0.17$ and $V_s/V_l = 0.63$.

Note that the plot of Fig. 2.2.5 gives a maximum value of Γ_{sl} greater than unity. Because power flow normal to the surface is conserved, both the change in angle of incidence and reflection and the change in impedance for longitudinal and shear waves require that $\Gamma_{sl} \neq 1$ when Γ_{ll} is imaginary (see Prob. 10).

Because of reciprocity, the device also works in the opposite direction to convert a shear wave to a longitudinal one. A curve giving these results is shown in Fig. 2.2.7; note that at $\theta_s = 25°$, $\Gamma_{ss} = 0$. This phenomenon is, of course, closely analogous to the Brewster angle phenomenon in optics, although here it is used to convert one type of wave to another.

Another phenomenon we have already discussed is that when the angle of incidence of a shear wave is large enough, the angle of reflection of the longitudinal wave becomes greater than 90°. In this case, any longitudinal wave excited falls off exponentially in amplitude from the surface. Figure 2.2.7 shows that a shear wave incident at a fused quartz surface has a critical angle $\theta_{si} = \theta_{cr} = 41°$. Beyond this cutoff point, all incident shear wave power is converted to a reflected shear wave, so that $|\Gamma_{ss}| = 1$, although there is a π phase shift of Γ_{ss} as θ_s passes through the critical angle.

Beyond the longitudinal wave cutoff point, the amplitudes of the longitudinal wave components at the surface are indeed finite, but there is no real power

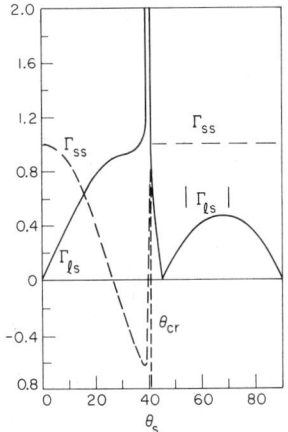

Figure 2.2.7 Reflection of a vertically polarized shear vertical wave at a stress-free boundary in fused silica. The vertical axis is the amplitude reflection coefficient defined in terms of the particle velocity. For fused quartz, $\sigma = 0.17$ and $V_s/V_l = 0.63$. (After Auld [7].)

Sec. 2.2   Basic Theory for Waves in Isotropic Media

associated with them. This implies that although $|\Gamma_{ls}|$ is finite, only decaying fields are associated with the longitudinal wave components near the surface.

Reflection and refraction. The same type of analysis can be carried out for refraction and reflection at the surface between two neighboring materials I and II, as illustrated in Fig. 2.2.8. Following the same arguments as before, the propagation constants of the reflected and refracted waves along the surface of the interface must be equal. This leads to a generalized Snell's law,

$$\sin \theta_{li} = \frac{V_{lI}}{V_{sI}} \sin \theta_{sr} = \frac{V_{lI}}{V_{lII}} \sin \theta_{lt} = \frac{V_{lI}}{V_{sII}} \sin \theta_{st} \qquad (2.2.72)$$

which relates the angles of the various reflected and refracted rays.

The phenomenon of reflection and refraction is far more complicated for acoustic waves than for EM waves. The only simple case is that for an incident *shear horizontal* wave, a wave whose particle motion is parallel to a horizontal surface; as illustrated in Fig. 2.2.8, the particle motion is into the paper the figure is drawn on. In this case, all the boundary conditions are satisfied by stress components that arise only from shear horizontal waves; therefore, the reflected wave is also an shear horizontal wave. In the general case, the boundary conditions are more complicated, requiring several components of the relevant fields to be continuous across the boundary. The continuous components are: (1) the normal component of longitudinal stress; (2) the transverse component of shear stress; and (3) both the normal and transverse components of displacement. The amplitudes of the two transmitted and the two reflected waves can then be determined from the resulting equations.

This general case is similar in many ways to the simpler one of reflection at a free surface, which we discussed earlier in this section. Again, because of Snell's law, at certain critical angles no longitudinal wave is transmitted or reflected, depending on the wave velocities in the two media. As an example, consider the

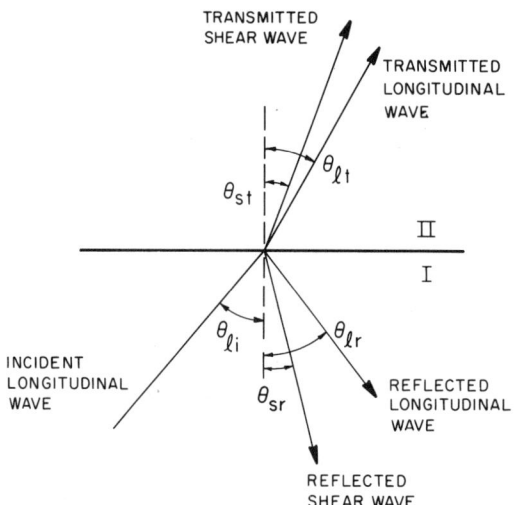

Figure 2.2.8 Refraction and reflection at the interface between two media.

longitudinal wave transmitted into aluminum from an incident longitudinal wave in water. The boundary conditions are simplified in this case because a shear wave cannot exist in water; thus there can be no reflected shear wave. Since the wave velocity in water is 1.5 km/s and the longitudinal wave velocity in aluminum is 6.42 km/s, there is no longer a refracted longitudinal wave in the aluminum if the incident wave in water reaches an angle θ_w greater than θ_{cr}, where $\sin \theta_{cr}$ = 1.5/6.42 or $\theta_w \geq 13.5°$.

This phenomenon is extremely useful for looking at flaws in metals in nondestructive testing applications. To introduce a shear wave into the metal, the sample and the transducer are immersed in water. The transducer excites a longitudinal wave in the water, which, in turn, generally excites a longitudinal wave and a shear wave in the metal. If a flaw is present, a wave will be reflected from it. To obtain good definition, it is desirable to work with the shortest wavelength possible; thus shear waves are preferable to longitudinal ones because the shear wave velocity is approximately half the longitudinal wave velocity in most metals. Furthermore, some types of flaws, such as closed cracks, may show up better with shear wave excitation, which may cause one face of the crack to slip past the other. Using a beam incident from water to metal, we can excite a shear wave in a metal relatively easily; if we make the incident angle large enough, only a shear wave will be excited. Such techniques are commonly used, for instance, in nondestructive testing of nuclear reactor walls.

In Sec. 2.5 we show that this technique is also useful for exciting surface waves on metal. In this case, the wave of interest propagates along the surface (i.e., at 90° to the normal). Of course, we must know the propagation velocity of this wave along the surface to choose the incident angle correctly. Experimentally, the problem is relatively simple: It is a matter of rotating the transducer in the water until a wave of the desired type is excited.

Since both the shear and longitudinal wave velocities in aluminum (V_s = 3.04 km/s and V_l = 6.42 km/s) are larger than the longitudinal wave velocity in water (V_l = 1.5 km/s), there are two critical angles, one for each transmitted wave. Thus for angles of incidence larger than the shear critical angle (i.e., with $\theta_w \gg \theta_{cr} = 29.6°$), the water–aluminum interface is a perfect reflector or mirror. The results of numerical calculations carried for this case are shown in Fig. 2.2.9.

Because power is conserved, it is often convenient to use transmission and reflection coefficients defined in terms of power rather than displacement, velocity, or stress. We have therefore used power reflection and transmission coefficients in Figs. 2.2.9–2.2.12. These coefficients are expressed in terms of the power density normal to the interface between the media (i.e., $P = Z|v|^2 \cos \theta/2$), where the parameters Z, θ, and v are appropriate to the wave of interest. For example, $T_{lw} = P_l/P_w$, where P_l is the power density normal to the interface associated with the longitudinal wave in aluminum and P_w is the power density normal to the interface of the incident wave in water. At normal incidence from the water, the power transmission coefficient to a longitudinal wave is $T_{lw} = 0.29$, while the power reflection coefficient is $R_{ww} = 0.71$ and the transmission coefficient to a shear wave is $T_{sw} = 0$. As the angle of incidence θ_w increases, T_{sw} increases and T_{lw} decreases to become zero at the critical angle. At an angle of incidence θ_w of approximately

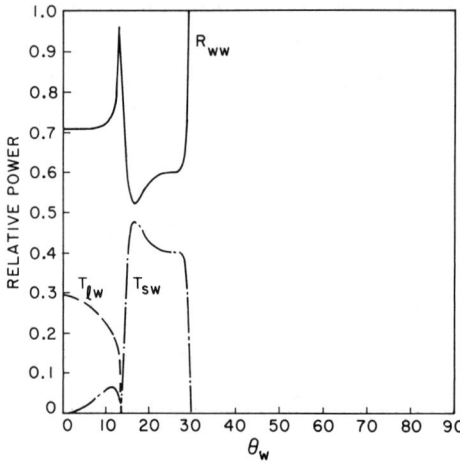

Figure 2.2.9 Plot of transmission and reflection coefficients as a function of the angle of incidence θ_w for a wave in water incident on aluminum. The transmission and reflection coefficients are defined in terms of power. For aluminum, $V_l = 6.42$ km/s, $V_s = 3.04$ km/s, and $\sigma = 0.355$.

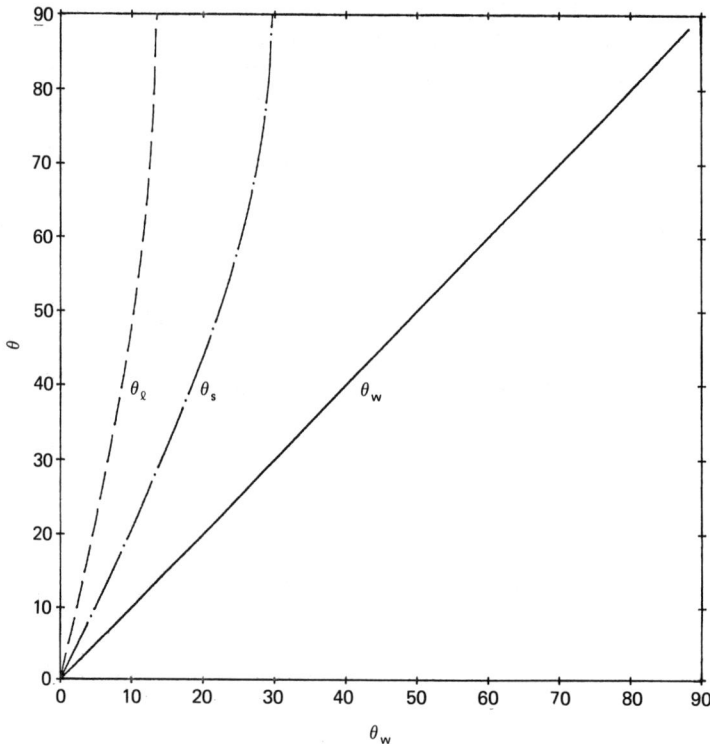

Figure 2.2.10 Plot of the relations of θ_w, θ_l, and θ_s for water and aluminum.

17°, almost half the incident power is converted to shear waves. Thus conversion to shear waves can be more efficient than conversion to longitudinal waves, because shear waves have a lower wave impedance. The relation between θ_s, θ_l, and θ_w is given in Fig. 2.2.10. The reverse situation, for conversion of longitudinal waves in aluminum to longitudinal waves in water, is shown in Fig. 2.2.11.

Another interesting use for a shear wave transducer is to excite a shear wave in a solid material such as aluminum, and to convert this wave, in turn, to a longitudinal wave in water. The curves in Fig. 2.2.12 show that approximately 40% of the incident energy can be converted to a longitudinal wave in water at incident angles for which the reflected longitudinal wave is cut off. Just as in the reverse case, transmission of a wave from water, the conversion ratio is better than for longitudinal wave excitation in the solid because the wave impedance of the shear wave is lower than that of a longitudinal wave.

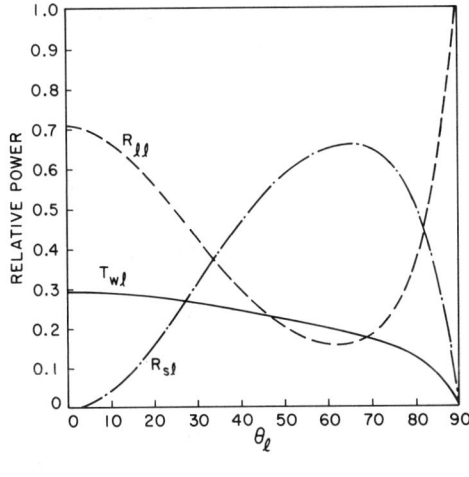

Figure 2.2.11 Plot of transmission and reflection coefficients as a function of incident angle θ_l for a longitudinal wave incident on a solid–water interface.

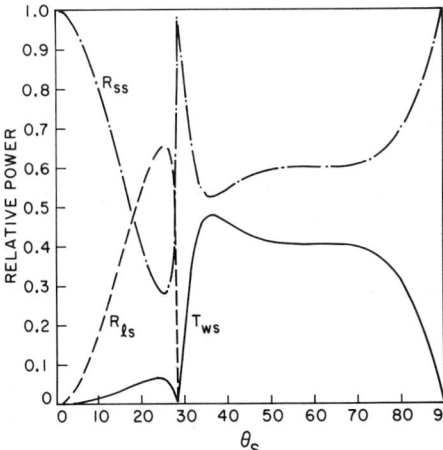

Figure 2.2.12 Plot of transmission and reflection coefficients as a function of incident angle θ_s for a shear wave incident on a solid–water interface.

Sec. 2.2   Basic Theory for Waves in Isotropic Media   103

PROBLEM SET 2.2

1. Consider the strip waveguide of isotropic material illustrated in the figure below:

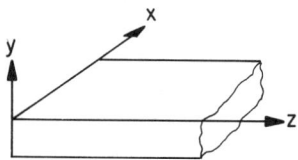

   Regarding the guide as infinitesimally thin in the y direction and infinitely wide in the x direction, we expect that $T_2 = 0$ with $S_1 = 0$ and $u_x = 0$. Under these conditions, prove Eq. (2.2.26). Compare V_E/V_l and V_L/V_l for aluminum and crown glass.

2. Consider a PZT imaging array transducer element made of a thin strip regarded as infinitely long in the x direction, infinitesimally thin in the y direction, and finite in the z direction. This material is poled in the z direction and electrodes are deposited on its top and bottom surfaces.

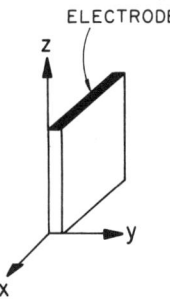

   The transducer is designed to vibrate only in the z direction, and to have transverse resonances in the y direction of much higher frequency than the fundamental resonance in the z direction.
   (a) Using the boundary conditions $T_2 = 0$, $S_1 = 0$, and $E_z = 0$ (i.e., there is no piezoelectric coupling), find the unstiffened acoustic velocity V_L^E in the z direction (see Prob. 1).
   (b) Using the boundary condition $D_z = 0$, find the stiffened acoustic velocity V_L^D in the z direction and find an expression for the effective values of K^2 and k_T^2. Work out these values for PZT-5A.

   Note. In this case, the material is anisotropic. You must assume that E_z and D_z are uniform (or zero) over any x-y plane, while D_z is independent of z. It will also be convenient to define a current per unit width, $I = j\omega \int D_z \, dy$.

3. Find c_{11}, c_{12}, and c_{44} in terms of s_{11}, s_{12}, and s_{44}.
4. Prove Eq. (2.2.37) by carrying out the complete derivation.
5. Consider a liquid in which $T_1 = T_2 = T_3 = -p$, where p is the pressure and $T_4 = T_5 = T_6 = 0$.
   (a) Show that
   $$\nabla p = -\rho_{m0} \ddot{\mathbf{u}}$$
   (b) Find a wave equation for p, as well as the relation between p and the scalar longitudinal wave potential ϕ.

6. Derive Eqs. (2.2.70) and (2.2.71). Remember that $(V_l/V_s)^2 = (\lambda + 2\mu)/\mu$.
7. Derive the equivalents of Eqs. (2.2.70) and (2.2.71) for shear wave incidence.
8. We define a shear horizontal mode as one whose particle motion v_x is parallel to a surface of interest (see Sec. 2.3.2). Consider the transmission and reflection of a horizontal shear wave, with its propagation vector in the y-z plane, incident from an angle θ_i to the normal, to the interface $y = 0$ between two media with Lamé constants λ_1, μ_1 and λ_2, μ_2 and densities ρ_{m1} and ρ_{m2}, respectively. Show that the boundary conditions at the interface can be satisfied by shear horizontal modes in both media. Find the transmission and reflection coefficients for particle velocity v_x at the interface as a function of the incident angle θ_i, and the shear wave impedances of the two media.
9. Suppose that an incident longitudinal wave is converted completely to a shear wave, as illustrated in Fig. 2.2.6 and described by Eqs. (2.2.70) and (2.2.71). Suppose, in addition, that we require the exit wave to be at right angles to the incident wave (i.e., $\theta_s + \theta_l = \pi/2$). Find the condition on V_l/V_s for this to occur. What value of Poisson's ratio would the material need?
10. Power is conserved when an incident wave is reflected at an interface into shear and longitudinal wave components. Show that Eqs. (2.2.70) and (2.2.71) are consistent with this fact. You will need to consider carefully the power per unit area transmitted in the direction normal to the interface. Work out expressions for the power reflection coefficients R_{ll} and R_{sl}, using Eqs. (2.2.70) and (2.2.71), and define these parameters in terms of the power density normal to the interface.
11. Suppose that an incident longitudinal wave is converted completely to a shear wave, as illustrated in Fig. 2.2.6 and described by Eqs. (2.2.70) and (2.2.71).
    (a) By solving for V_l/V_s in terms of θ_s and then finding Poisson's ratio σ, show numerically that the range $V_l/V_s = 1.414$ to $V_l/V_s = 1.76$, with Poisson's ratio varying from 0 to 0.263, can satisfy the condition $\Gamma_{sl} = 0$. If $\sigma > 0.263$, however, it is not possible to satisfy the condition $\Gamma_{sl} = 0$.
    (b) Show that as $\sigma \to 0$, there are two possible solutions: $\theta_s \to 45°$ and $\theta_l \to 90°$ or $\theta_s \to 25.9°$ and $\theta_l \to 38.1°$.

2.3 ACOUSTIC WAVEGUIDES

2.3.1 Introduction

In this section we discuss the waves that can propagate in an acoustic waveguide. The modes involved in this case are more complicated than in the analogous electromagnetic (EM) waveguide, for even in a simple acoustic strip waveguide that has finite thickness in the y direction and is infinite in extent in the x and z directions, several different types of modes can exist. The simplest, which will be described in Sec. 2.3.2, is the *shear horizontal* mode, in which all particle motion is in the x direction and the wave is a pure shear wave. This mode is analogous to an EM waveguide mode. In Sec. 2.3.3 we describe Lamb waves, which have particle motion in both the y and z directions. Both shear and longitudinal wave components are needed to satisfy the boundary conditions of this mode. Finally, in Sec. 2.3.4 we describe surface acoustic waves, or Rayleigh waves, which propagate along the surface of a semi-infinite substrate. These waves have fields that

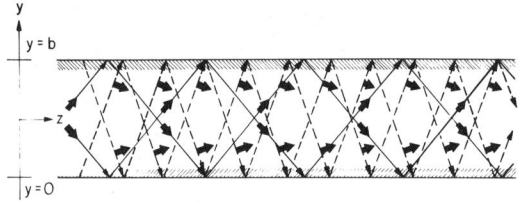

——— INCIDENT AND REFLECTED LONGITUDINAL PARTIAL WAVES
— — — INCIDENT AND REFLECTED VERTICALLY POLARIZED SHEAR PARTIAL WAVES

Figure 2.3.1 Partial wave pattern for transverse resonance analysis of Lamb wave propagation on an isotropic plate with free boundaries. (After Auld [7].)

fall off exponentially in amplitude away from the surface, and which need both shear and longitudinal wave components to satisfy the boundary conditions.

Consider a sheet of material of finite thickness b and infinite extent in the x and z directions, as illustrated in Fig. 2.3.1. If a shear wave is incident on the top surface from the inside of the material, it generally gives rise to both reflected shear and longitudinal waves, which, in turn, are reflected from the lower surface. Thus, in general, components of both shear and longitudinal waves will exist in the "waveguide," all of them with the same propagation constant β in the z direction, so as to satisfy the boundary conditions at all points on both surfaces. We will assume that the shear wave components are associated plane shear waves propagating at an angle to the z axis in the y-z plane. In the y, z coordinate system they must have fields that vary as $\exp(\pm j k_{sy} y) \exp(-j\beta z)$. By symmetry, the amplitudes of the waves incident on and reflected from the top surface must be equal. Consequently, the total shear wave potential must vary as $\psi = A \exp(-j\beta z) \sin(k_{sy} y + \alpha)$. Similarly, with the same assumptions, the longitudinal wave component must have a potential that varies as $\phi = B \sin(k_{ly} y + \gamma) \exp(-j\beta z)$, where α and γ are constants. In both cases we have assumed, for simplicity, that the field components have no variation in the x direction.

It follows from the wave equations for ϕ and ψ that

$$\beta^2 + k_{sy}^2 = k_s^2 \tag{2.3.1}$$

and

$$\beta^2 + k_{ly}^2 = k_l^2 \tag{2.3.2}$$

We must find the possible values of β for a given frequency ω (i.e., the eigenvalues of the waveguide modes). To do this, we must write the components of displacement, strain, and stress in the material in terms of the assumed forms of ϕ and ψ, and satisfy the boundary conditions at each surface.

2.3.2 Shear Horizontal Modes

We first consider the shear horizontal (SH) modes; these modes have only a u_x component, with ψ in the y direction and $\phi = 0$. As we assume only one component of ψ, ψ_y, we will, for simplicity, drop the subscript y and write

$$\psi = A e^{-j\beta z} \cos(k_{sy} y + \alpha) \tag{2.3.3}$$

with

$$u_x = -\frac{\partial \psi_y}{\partial z} = j\beta A e^{-j\beta z} \cos(k_{sy}y + \alpha) \tag{2.3.4}$$

$$u_y = u_z = 0 \tag{2.3.5}$$

$$T_5 = T_{xz} = \mu \frac{\partial u_x}{\partial z} = \beta^2 A \mu e^{-j\beta z} \cos(k_{sy}y + \alpha) \tag{2.3.6}$$

and

$$T_6 = T_{xy} = \mu \frac{\partial u_x}{\partial y} = j\beta k_{sy} \mu A e^{-j\beta z} \sin(k_{sy}y + \alpha) \tag{2.3.7}$$

The boundary condition is zero normal stress at the surface, that is, $T_6 = 0$ or $\sin(k_{sy}y + \alpha) = 0$ at $y = 0$, $y = b$. This implies that $\alpha = 0$ and

$$k_{sy} = \frac{n\pi}{b} \tag{2.3.8}$$

with

$$\psi = A e^{-j\beta z} \cos \frac{n\pi y}{b} \tag{2.3.9}$$

Thus there is a series of possible solutions or waveguide modes for different values of n, with corresponding values of β given by Eq. (2.3.1). In this case, we see that

$$\beta^2 = k_s^2 - \left(\frac{n\pi}{b}\right)^2 = \left(\frac{\omega}{V_s}\right)^2 - \left(\frac{n\pi}{b}\right)^2 \tag{2.3.10}$$

which is identical in form to the result for EM waveguides.

For β to be real for the nth mode, $\omega > n\pi V_s/b$ (i.e., there is a low-frequency cutoff for all but the $n = 0$ mode). If the frequency is less than this value, then β for the nth mode is imaginary and the wave does not propagate along the guide. Dispersion curves (plots of ω with respect to β) are given in Fig. 2.3.2. The phase velocity of the wave is

$$V_p = \frac{\omega}{\beta} \tag{2.3.11}$$

It follows from Eq. (2.3.10) that $V_p > V_s$ and that $V_p \to \infty$ at the cutoff frequency $\omega_{cn} = n\beta V_s/b$. Note also that the lowest mode $n = 0$ has a component of displacement only in the x direction and that the only finite component of stress is $T_5 = T_{xz}$ with no variation across the guide. In this case, the mode can propagate at all frequencies down to zero and its field components are like those of a plane shear wave mode propagating in the z direction. As we might expect, an SH plane wave propagating in the z direction in an infinite medium is not affected when the cross section of the medium is made finite in the y direction.

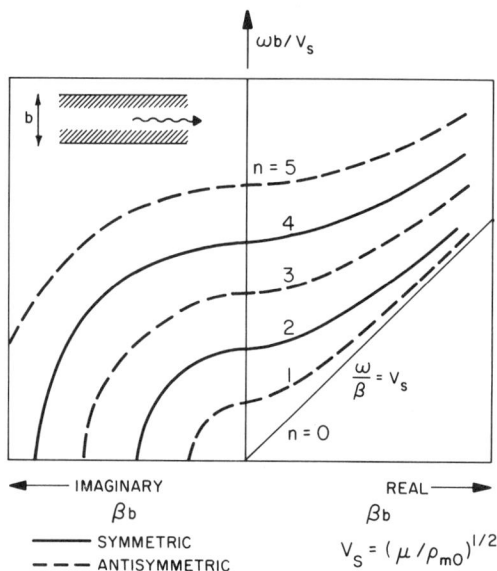

Figure 2.3.2 Dispersion curves for SH modes on an isotropic plate with free boundaries. (After Auld [7].)

2.3.3 Lamb Waves

A case of great practical interest is when the displacement is in both the y and z directions, so that ψ is in the x direction, ϕ is finite, and there are both shear and longitudinal components. The waves that exist in this case are known as *Lamb waves*. Some are primarily shear and others, primarily longitudinal, but both components are required to solve for the boundary conditions. In Sec. 2.2.1 we have already solved for one of these waves propagating in a very thin sheet; in that case, we assumed that the particle motion was essentially in the z direction with zero stress in the y direction. Our solution, with no variation of the displacement u_z across the sheet, was what is called the strip extensional mode of the system, that is, a mode that exists down to zero frequency with a phase velocity somewhat less than that of a longitudinal wave in an infinite medium. An analogous mode that exists down to zero frequency is a shear wave, called the flexural mode of the thin sheet. Higher-order modes, with several maxima and minima in the amplitude of variation of the field components across the guide, have a correspondingly higher frequency cutoff. We will not deal with the details of these waves here (see Problems 2.3.2 and 2.3.3). However, it is important to be aware of their existence and of the fact that their velocities can be calculated and measured experimentally. Furthermore, these waves can be excited from water or other media using an incident plane wave beam, incident at the correct angle, to excite the particular Lamb wave of interest. By measuring this critical angle for excitation of the Lamb wave, we can determine β, because for strong excitation, $\beta = k_{li} \sin \theta_{li}$ when k_{li} is the propagation constant of the incident wave. With this experimental technique, we can usually get a fairly good idea of the actual phase velocity of the Lamb wave $V_p = \omega/\beta$ along the waveguide.

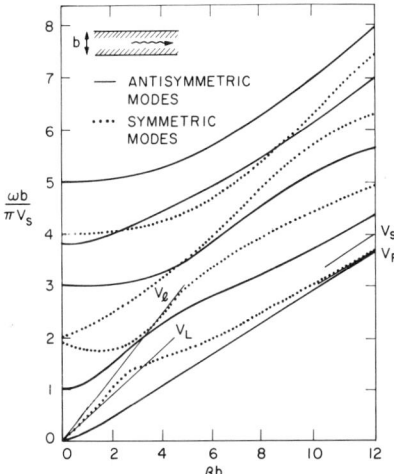

Figure 2.3.3 Lamb wave dispersion curves for the lower-order modes of an isotropic free plate with $V_l/V_s = 1.9056$. (After Auld [7].)

A plot of the dispersion curves for Lamb waves in a material such as aluminum is given in Fig. 2.3.3. The lowest-order symmetric mode (u_z is symmetric and u_y is antisymmetric about the central axis) extends down to zero frequency (see Probs. 2 and 3). This mode, as we have seen, has a phase velocity V_L that is lower than that for a longitudinal wave as $\omega \to 0$. As the frequency is increased, the phase velocity increases and approaches the longitudinal wave velocity. It then decreases again until it approaches the Rayleigh wave velocity, which is a little less than the shear wave velocity. When $\beta b \gg 1$, the value of u_z at the center of the guide becomes less than at the edges. Thus, at very high frequencies, the mode looks like two surface acoustic waves, or Rayleigh waves (which will be discussed in Sec. 2.3.4), propagating along the boundary surfaces of the guide.

There is also a lowest-order antisymmetric mode or *flexural mode* (u_z is antisymmetric and u_y is antisymmetric about the central axis) extending down to zero frequency. The velocity of this flexural mode is usually lower than the shear wave velocity and the Rayleigh wave velocity.

2.3.4 Surface Waves

We have now briefly discussed the different types of waves that can exist in a solid medium of finite width. In this section we devote considerable attention to Rayleigh waves. These are waves of great technical importance that exist only near the surface of a semi-infinite medium. For this reason such waves are called surface waves. A familiar example of a surface wave is one that propagates along the surface of water. In this case, the wave motion is strong at the surface of the water and falls off very rapidly into its interior. In a water wave, the inertial forces are associated with the mass of the water and the restoring forces are due to gravity, rather than Hooke's law.

When scientists first began to analyze seismic motions of the earth, they noted that when a distant disturbance occurs, an observer notes three distinct events.

The first is a result of longitudinal waves propagating through the interior of the earth. The second is due to shear waves, which, because they propagate at a slower velocity than longitudinal waves, reach the observer at a later time. Finally, a third disturbance, due to a wave propagating along the curved surface of the earth, reaches the observer; this surface wave disturbance is the strongest of the three.

Lord Rayleigh proposed a theory for the surface wave, which shows that it consists of a mixture of shear and longitudinal stress components [8]. Because there is no restoring force at the surface of a solid medium, any force normal to the surface must be zero. Thus the boundary condition at the surface is that the normal components of stress must be zero. If a wave propagating in the z direction exists in a semi-infinite medium, the total energy per unit length in the wave must be finite. This, in turn, implies that the field components associated with the wave will fall off exponentially into the interior of the medium. As we shall see, we can indeed obtain a mathematical solution that satisfies the boundary conditions.

Surface acoustic waves are technically important because their energy is concentrated in a relatively small region, approximately one wavelength deep, near the surface. The waves are therefore accessible from the surface. Thus Rayleigh waves produced by seismic disturbances are the ones most easily detected by sensors on the surface of the earth [3, 6, 8–13].

If the medium is piezoelectric, the electric fields associated with the wave should be stronger near the surface. Consequently, by depositing electrodes on the surface of a piezoelectric material, a surface wave can be excited or detected relatively easily. Furthermore, it is also easy to sample the wave along its path (i.e., to make taps for a surface wave delay line). The technology is inexpensive and convenient because the electrodes required to excite or detect these waves can be deposited by the standard techniques used for photolithography of integrated circuits.

Rayleigh waves in an isotropic medium. We consider a medium that is semi-infinite in the $-y$ direction with a free surface at $y = 0$ and particle displace-

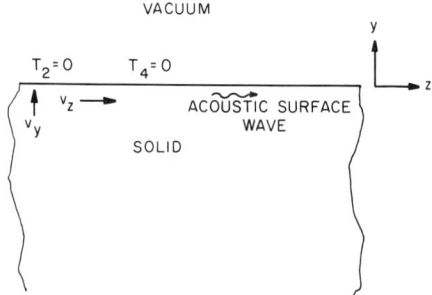

Figure 2.3.4 Configuration for acoustic surface wave analysis.

ment only along the y and z axes, as illustrated in Fig. 2.3.4. We use Eqs. (2.2.44) and (2.2.48) and write the wave equations for the potentials in the following forms:

$$\frac{\partial^2 \phi}{\partial z^2} + \frac{\partial^2 \phi}{\partial y^2} + k_l^2 \phi = 0 \qquad (2.3.12)$$

and

$$\frac{\partial^2 \psi}{\partial z^2} + \frac{\partial^2 \psi}{\partial y^2} + k_s^2 \psi = 0 \qquad (2.3.13)$$

where

$$k_l = \omega \sqrt{\frac{\rho_{m0}}{\lambda + 2\mu}} \qquad (2.3.14)$$

$$k_s = \omega \sqrt{\frac{\rho_{m0}}{\mu}}$$

The motion is assumed to be independent of the coordinate x. Thus only the component of the vector potential $\boldsymbol{\psi}$ along the x axis will be finite, and we write $\boldsymbol{\psi} = \mathbf{a}_x \psi$. If, as in Sec. 2.2.3, we associate the potential ϕ with the longitudinal component of motion, and the potential ψ with the shear waves, we can write the following relations:

$$u_z = \frac{\partial \phi}{\partial z} - \frac{\partial \psi}{\partial y} \qquad (2.3.15)$$

$$u_y = \frac{\partial \phi}{\partial y} + \frac{\partial \psi}{\partial z} \qquad (2.3.16)$$

$$T_1 = \lambda \left(\frac{\partial^2 \phi}{\partial y^2} + \frac{\partial^2 \phi}{\partial z^2} \right) \qquad (2.3.17)$$

$$T_2 = \lambda \left(\frac{\partial^2 \phi}{\partial y^2} + \frac{\partial^2 \phi}{\partial z^2} \right) + 2\mu \left(\frac{\partial^2 \phi}{\partial y^2} + \frac{\partial^2 \psi}{\partial y \partial z} \right) \qquad (2.3.18)$$

$$T_3 = \lambda \left(\frac{\partial^2 \phi}{\partial y^2} + \frac{\partial^2 \phi}{\partial z^2} \right) + 2\mu \left(\frac{\partial^2 \phi}{\partial z^2} - \frac{\partial^2 \psi}{\partial y \partial z} \right) \qquad (2.3.19)$$

and

$$T_4 = \mu \left(2 \frac{\partial^2 \phi}{\partial y \partial z} + \frac{\partial^2 \psi}{\partial z^2} - \frac{\partial^2 \psi}{\partial y^2} \right) \qquad (2.3.20)$$

We seek solutions for both the shear and longitudinal terms, which have the same phase variation in the z direction. Thus we can write ϕ as

$$\phi = F(y) e^{j(\omega t - \beta z)} \qquad (2.3.21)$$

and ψ as

$$\psi = G(y)e^{j(\omega t - \beta z)} \quad (2.3.22)$$

Substituting Eqs. (2.3.21) and (2.3.22) into Eqs. (2.3.12) and (2.3.13), respectively, leads to the following results:

$$\frac{d^2F}{dy^2} = (\beta^2 - k_l^2)F(y) \quad (2.3.23)$$

and

$$\frac{d^2G}{\partial y^2} = (\beta^2 - k_s^2)G(y) \quad (2.3.24)$$

We look for surface wave solutions where the components fall off exponentially to $y = -\infty$ so that they have finite stored energy per unit length. We take $F(y)$ to vary as $\exp(\gamma_l y)$ and $G(y)$ to vary as $\exp(\gamma_s y)$. From Eqs. (2.3.23) and (2.3.24), it follows that

$$\gamma_l^2 = \beta^2 - k_l^2 \quad (2.3.25)$$

and

$$\gamma_s^2 = \beta^2 - k_s^2 \quad (2.3.26)$$

The solutions for ϕ and ψ must take the forms

$$\phi = Ae^{-j\beta z}e^{\gamma_l y} \quad (2.3.27)$$

and

$$\psi = Be^{-j\beta z}e^{\gamma_s y} \quad (2.3.28)$$

respectively. The boundary condition of the problem is that the normal component of stress at the surface must be zero. In turn, this implies that T_2 and T_4 are zero at the surface. We substitute Eqs. (2.3.27) and (2.3.28) into Eqs. (2.3.18) and (2.3.20) to write

$$T_4 = \mu[-2j\beta\gamma_l Ae^{\gamma_l y} - (\beta^2 + \gamma_s^2)Be^{\gamma_s y}]e^{-j\beta z} \quad (2.3.29)$$

and

$$T_2 = \mu[A(\gamma_s^2 + \beta^2)e^{\gamma_l y} - 2j B\beta\gamma_s e^{\gamma_s y}]e^{-j\beta z} \quad (2.3.30)$$

The condition that $T_4 = 0$ and $T_2 = 0$ at $y = 0$ imply that

$$B = -\frac{2j\beta\gamma_l A}{\beta^2 + \gamma_s^2} \quad (2.3.31)$$

and

$$A = \frac{2j\beta\gamma_s B}{\beta^2 + \gamma_s^2} \quad (2.3.32)$$

Substituting Eq. (2.3.32) into Eq. (2.3.31), we obtain one form of the Rayleigh wave dispersion relation:

$$4\beta^2 \gamma_l \gamma_s - (\beta^2 + \gamma_s^2)^2 = 0 \qquad (2.3.33)$$

Substituting Eqs. (2.3.25) and (2.3.26) into Eq. (2.3.33), we can also write the Rayleigh wave dispersion relation as a cubic equation in β^2. We then write the dispersion relation in terms of the Rayleigh wave velocity $V_R = \omega/\beta$, in the form

$$\left(\frac{V_R}{V_s}\right)^6 - 8\left(\frac{V_R}{V_s}\right)^4 + 8\left[3 - 2\left(\frac{V_s}{V_l}\right)^2\right]\left(\frac{V_R}{V_s}\right)^2 - 16\left(1 - \frac{V_s}{V_l}\right)^2 = 0 \qquad (2.3.34)$$

where V_s, V_l, and V_R are the shear wave velocity, the longitudinal wave velocity, and the Rayleigh wave velocity, respectively. This dispersion relation has a real root, the *Rayleigh root*, which can be stated in the approximate form [7, 10] (see Prob. 6) as

$$\frac{k_s}{\beta} = \frac{V_R}{V_s} \approx \frac{0.87 + 1.12\sigma}{1 + \sigma} \qquad (2.3.35)$$

The Rayleigh wave is a nondispersive wave, with V_R varying from $0.87V_s$ to $0.95V_s$ as Poisson's ratio varies from 0 to 0.5. A plot of V_R/V_s as a function of V_s/V_l is compared with the approximate solution in Fig. 2.3.5.

Note that we have neglected two other roots of the dispersion relation. These correspond to waves propagating from $y = -\infty$, which are reflected from the surface (see Prob. 7).

The Rayleigh wave velocity V_R is always less than the shear wave velocity V_s or the longitudinal wave velocity V_l. This is necessary because for the waves to fall off in amplitude exponentially into the interior of the medium, γ_l and γ_s must be real, and hence $k_r > k_l$ and $k_r > k_s$. We have already seen that the shear wave velocity is always less than the longitudinal wave velocity. Consequently, the Rayleigh wave velocity must always be considerably less than the longitudinal wave velocity but only slightly less than the shear wave velocity. Because the Rayleigh wave velocity is closest to the shear wave velocity, most of the stored energy in the medium is associated with shear wave components, rather than the longitudinal ones; thus in many respects the Rayleigh wave behaves like a shear wave. So

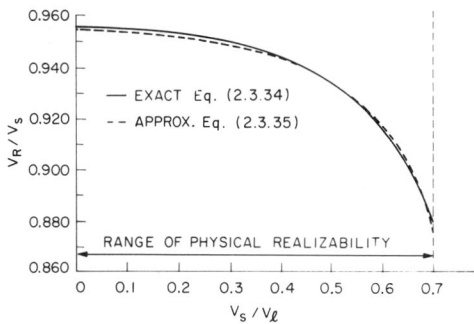

Figure 2.3.5 Isotropic Rayleigh wave velocity V_R as a function of the bulk shear wave velocity V_s and the bulk longitudinal wave velocity V_l. (After Auld [7].)

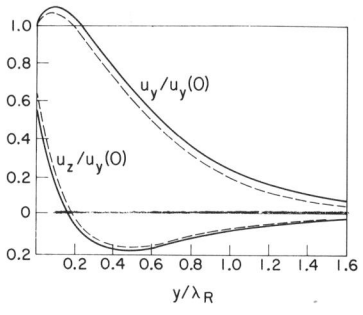

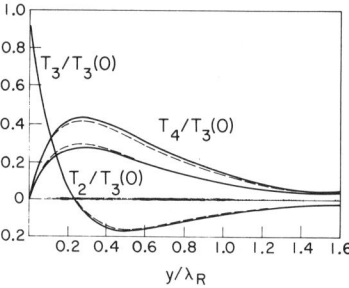

Figure 2.3.6 Plot of the normalized values of stress and displacement components of a Rayleigh wave: solid lines, $\sigma = 0.34$; dashed lines, $\sigma = 0.25$.

when surface roughness or air loading are not dominant effects, the Rayleigh wave attenuation is comparable to that of shear wave loss in most materials. The way the Rayleigh wave velocity varies with the angle of propagation relative to the crystal axis in an anisotropic crystal also follows fairly closely to the way the shear wave velocity varies with direction.

The displacement and stress components of the wave do not fall off as simple exponentials into the interior of the medium because both the longitudinal and shear potentials ϕ and ψ contribute to these components. Thus it follows from Eqs. (2.3.27)–(2.3.32) that

$$T_4 = -2jA\gamma_l\mu A(e^{\gamma_l y} - e^{\gamma_s y}) \tag{2.3.36}$$

and

$$T_2 = \mu A(\gamma_s^2 + \beta^2)(e^{\gamma_l y} - e^{\gamma_s y}) \tag{2.3.37}$$

Hence T_4 and T_2 increase from an initial zero to fall off finally as $e^{\gamma_s y}$ into the interior. A plot of how some of the important components vary with y is given in Fig. 2.3.6.

PROBLEM SET 2.3

1. Consider a layered material with a shear wave elastic constant μ_1, density ρ_{m1}, and a thickness h, laid down on a semi-infinite substrate with a shear wave elastic constant μ_2 and density ρ_{m2}. A type of SH wave known as a *Love* wave can propagate in this

configuration. To have finite energy, its fields must fall off exponentially in the $-y$ direction in the semi-infinite substrate. This wave is dispersive (i.e., the phase velocity varies with frequency).

Take u_x to vary as $\cos(k_y y + \alpha) \exp(-j\beta z)$ in the upper layer and as $\exp(\gamma y) \exp(-j\beta z)$ in the substrate, the interface between the layers to be at $y = 0$, and all field components to be invariant with x, with

$$k_y^2 = k_{s1}^2 - \beta^2$$

$$\gamma^2 = \beta^2 - k_{s2}^2$$

and

$$k_{s1}^2 = \frac{\omega^2 \rho_{m1}}{\mu_1}$$

$$k_{s2}^2 = \frac{\omega^2 \rho_{m2}}{\mu_2}$$

In this case, unlike that of the simple SH mode guide, note that k_y is a variable that changes with frequency.

(a) By matching the boundary conditions at the interface and using the boundary conditions at the top surface, find a transcendental relation between γ and k_y (the dispersion relation) from which, with the relations already given, β may be found. Show that the media must be chosen so that $k_{s2} < k_{s1}$.

(b) Show that for the lowest mode the fields extend to $y = -\infty$ as $\omega \to 0$. What is the relation between β and k_{s2} and hence the phase velocity ω/β of the wave at this point? Find the high-frequency limit of this mode where $\beta \to \infty$ and $\gamma \to \infty$. What is the relation between β and k_{s1} at this point? What is the phase velocity of this mode at the high-frequency limit? Sketch how $k_y b$ varies with frequency, as well as the ω–β relation for the lowest-order mode. You will find it convenient to consider $\tan(k_y b)$ and its limits as $k_y b \to 0$ and $k_y b \to \pi/2$.

(c) Find the value of $k_y b$ when $\gamma = 0$ for the next mode of the Love wave, as well as an expression for the low-frequency cutoff of this wave.

2. (a) Assume that a symmetrical Lamb wave has ϕ, an even function of y, and ψ_x, an odd function of y, about the center of a strip of thickness b. Find a transcendental expression from which the propagation constant β may be found. Show that as $\omega \to 0$, the transverse propagation constants in the y direction of the lowest order mode also approach zero. Show that in this case the phase velocity of the wave V_p approaches V_L, which is defined in Eq. (2.2.26), or by Prob. 2.2.1. Assume that both ϕ and ψ vary as $\exp(-j\beta z)$. Follow the methods suggested in Prob. 1.

Note. As $\omega \to 0$, $k_s \to 0$, $k_l \to 0$, and $\beta \to 0$, so $k_{sy} \to 0$ and $k_{1y} \to 0$.

Answer.

$$\frac{\tan(k_{sy}b/2)}{\tan(k_{1y}b/2)} = -\frac{4\beta^2 k_{sy} k_{ly}}{(k_{sy}^2 - \beta^2)^2}$$

where $k_{sy}^2 = k_s^2 - \beta^2$, $k_{ly}^2 = k_l^2 - \beta^2$, $k_s^2 = \omega^2 \rho_{m0}/\mu$, and $k_l^2 = \omega^2 \rho_{m0}/(\lambda + 2\mu)$.

3. Consider the antisymmetric Lamb wave with ϕ, an odd function of y, and ψ_x, an even function of y, about the center of a strip of thickness b, using the methods suggested in Prob. 2.

(a) Find a transcendental expression from which the propagation constant may be found.

Answer.

$$\frac{\tan(k_{sy}b/2)}{\tan(k_{ly}b/2)} = -\frac{(k_{sy}^2 - \beta^2)^2}{4\beta^2 k_{sy}k_{ly}}$$

with the same definitions as in Prob. 2.

(b) Find the phase velocity of the antisymmetric mode as $\omega \to 0$. Assume that both ϕ and ψ vary as $\exp(-j\beta z)$. Show that as $\omega \to 0$, the group velocity is twice the phase velocity, where the phase velocity is $V_p = \omega/\beta$ and the group velocity is $V_g = d\omega/d\beta$.

Answer.

$$(\beta b)^4 = \frac{3k_s^2 b^2}{1 - (k_l/k_s)^2} \qquad V_p = \frac{\omega}{\beta}$$

Note. You will need to expand the trigonometric functions to third order in the argument to find this relation. For a thin strip, look for a solution where $\beta \to 0$ less slowly than k_l or k_s, as can be seen from final result (i.e., neglect k_l^2 or k_s^2 compared to β^2). Do not neglect k_{sy}^2 relative to β^2 as $(\beta b)^2 \to 0$.

4. (a) Show from general considerations based on Eq. (2.2.21) that at $y = 0$, the Rayleigh wave components satisfy the relation $T_1(0) = \sigma T_3(0)$. Using Eq. (2.2.21) and the symmetric expressions for S_2 and S_3, find $S_2(0)$ and $S_3(0)$ in terms of $T_3(0)$.

(b) Using Eqs. (2.3.17) and (2.3.19), find expressions for $T_1(y)$ and $T_3(y)$ and show that your results also yield $T_1(0) = \sigma T_3(0)$.

5. Find the total power flow in the z direction associated with the propagation of a Rayleigh wave. Show that if v_y is the RF velocity at $y = 0$, then for a beam of width w,

$$\frac{v_y v_y^*}{P} = \frac{f_y \omega}{w \rho_{m0} V_s^2}$$

where

$$f_y = \frac{4\gamma^2 [1 - (V_R/V_s)^2]^{3/2}}{3\gamma - 2\gamma(V_R/V_s)^2 - 1} \left(\frac{V_s}{V_R}\right)^2$$

and

$$\gamma^2 = \frac{1 - (V_R/V_l)^2}{1 - (V_R/V_s)^2}$$

and V_R, V_s, and V_l are the Rayleigh wave, shear wave, and longitudinal wave velocities, respectively.

6. Prove the approximate result of Eq. (2.3.35) by putting $\beta = k_s(1 + \Delta)$ in Eq. (2.3.33) and keeping only first-order terms in Δ.

Answer.

$$\frac{k_s}{\beta} = \frac{0.875 + 1.125\sigma}{1 + \sigma}$$

7. Consider the solutions of Eq. (2.3.34) with $\sigma = 0.25$. In this case show that $(V_l/V_s)^2 = 3$ and that there are three real solutions of the Rayleigh wave equations: $(V_R/V_s)^2 = $

4, $2 + 2/\sqrt{3}$, and $2 - 2/\sqrt{3}$. The last root is the normal Rayleigh root with $V_R/V_s = 0.9194$.

(a) Discuss the first root, which corresponds to a velocity $V_R > V_l$. Is it a true root of the equation? Find the values of γ_l and γ_s that satisfy Eq. (2.3.33). Compare your results with the situation for a longitudinal wave incident at an angle θ_l to the normal and giving rise to only a reflected shear wave. Find θ_l and θ_s, the reflected wave angle. You may find it useful to make use of Eqs. (2.2.70) and (2.2.71) in your discussion.

(b) Find θ_l and θ_s for the second root of Eq. (2.3.33).

2.4 INTERDIGITAL TRANSDUCERS

2.4.1 Introduction

The major advantage of surface acoustic waves is simply that they are accessible at the surface; thus surface acoustic wave (SAW) devices can be easily adapted to the technology developed for creating microcircuits in thin, flat structures. In typical applications, most of the acoustic energy is contained within 1 to 100 μm from the surface. A surface acoustic wave can be easily excited anywhere on the surface and readily received elsewhere on the same "chip." Thus it is easy, as well as desirable, to construct a delay line in which an acoustic wave travels along the surface of the crystal rather than through its interior. Because the wave is so easily accessible, signals with different delay times can be picked up at various points along their path; thus the delay line can be tapped at several intermediate points to create a *transversal filter*. Chapter 4 deals with the theory of such filters in detail.

In recent years, the technology of acoustic wave devices has expanded rapidly, due to the development of *interdigital transducers*, which convert an electrical signal into a surface acoustic wave and reconvert it into an electrical signal. This type of transducer, illustrated in Fig. 2.4.1, is normally used to excite a surface acoustic wave on a piezoelectric material.

We can produce an electric field at the surface of a piezoelectric crystal by applying an electrical potential to two parallel metal electrodes deposited on it, as shown in Fig. 2.4.1(a). This field excites a surface acoustic wave that can be reconverted to an electrical signal at a second similar pair of electrodes laid down on the piezoelectric substrate. But a single pair of electrodes cannot excite surface acoustic waves very efficiently, so instead, it is customary to use an interdigital transducer consisting of several pairs of electrodes or fingers, placed one after the other in an interdigital pattern, as shown in Fig. 2.4.1(b). Thus each pair of electrodes excites a Rayleigh wave, and the transducer is designed so that these separately excited waves reinforce one another and give rise to a usefully large acoustic signal. This is accomplished by choosing the spacing between each finger pair so that a Rayleigh wave travels that distance in exactly the time required for the exciting signal to repeat itself (i.e., the finger pair spacing is one wavelength).

If the frequency of the wave is altered from this ideal value, the individual excitations from each pair of fingers will have a tendency to cancel each other out.

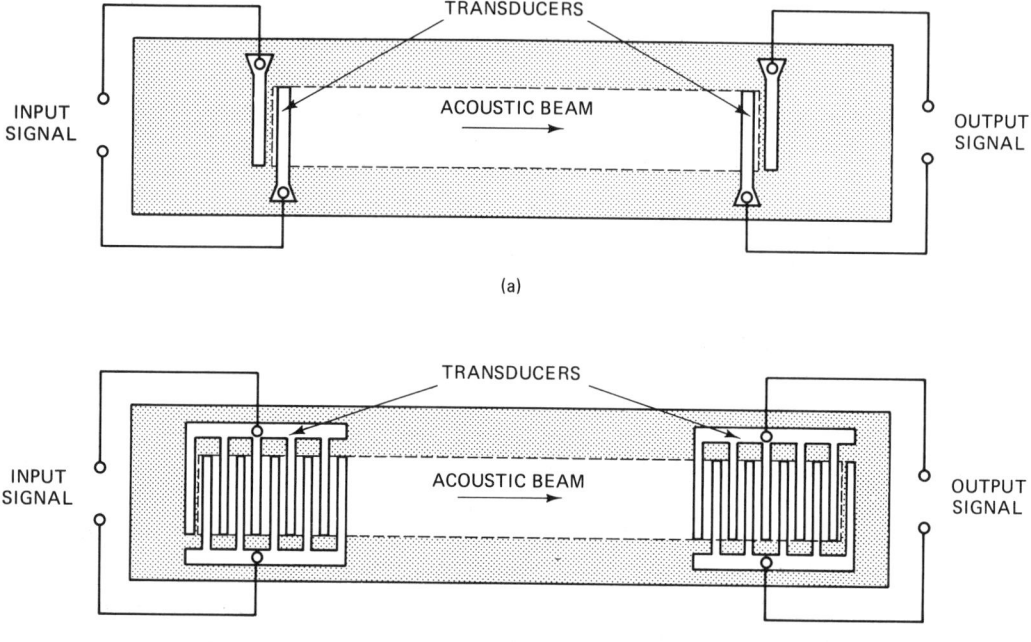

Figure 2.4.1 Excitation and detection of Rayleigh waves: (a) by a simple interdigital transducer consisting of two metal electrodes deposited on a piezoelectric crystal; (b) by a multiple-finger interdigital transducer. (After Kino and Shaw [14].)

The longer the transducer, and hence the more fingers it has, the more easily a slight change in frequency will cause the signal from one end of the transducer to be out of phase with the signal excited at the other end. Thus a long transducer with many fingers tends to be efficient for exciting and receiving signals only over a narrow frequency range and can also act as a filter to differentiate signals of one frequency from signals of another. Conversely, a short transducer with only a few fingers can be used to excite signals over a wider frequency range.

Figure 2.4.2 shows the performance of one of the first efficient SAW delay lines, of the type shown in Fig. 2.4.1(b), fabricated on the surface of a lithium niobate crystal. Each interdigital transducer consists of five identical finger pairs with finger widths and spacings of 8 μm; this gives a center frequency of operation of 105 MHz. These dimensions are scaled inversely with frequency.

Filters of this kind are commonly used in communication systems. A television receiver, for example, must be able to switch among several channels, each of which is at a different frequency; thus it requires several individual filters corresponding to the frequency of each channel. The signal from the chosen channel is then shifted to an "intermediate" frequency and passed through another filter that separates the picture and sound information, as well as strongly rejecting signals from the adjacent channels. SAW filters are now widely used as intermediate-frequency (IF) filters in this application.

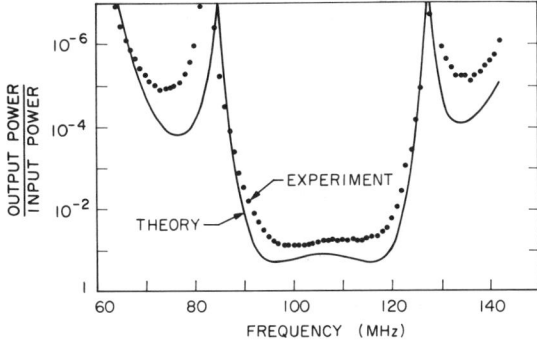

Figure 2.4.2 Bandpass frequency response of an early two-port SAW delay line. (After Smith et al. [11].)

The interdigital transducers used to generate and detect surface acoustic waves are extremely small. For example, if we want to excite waves with a frequency of 40 MHz, then each finger can be no more than about 20 μm wide and the spacing between them must be of comparable dimensions, since the finger pair spacing should be one wavelength. Lithium niobate, a typical material used in this application, has a Rayleigh velocity of 3.3 km/s. Thus for 40 MHz, λ ∼ 80 μm. A frequency near 40 MHz is a common intermediate frequency in television applications. To excite waves with a frequency of 1 GHz, the fingers must be spaced by one twenty-fifth of this distance, or 0.8 μm. Transducers with such dimensions can be produced by the well-developed photolithographic techniques commonly used in the semiconductor industry. In practice, the metal fingers are deposited on the crystal by evaporating, in a high vacuum, a metal such as gold or aluminum through a suitably exposed photoresist mask. The mask through which the photoresist layer is exposed is itself made photographically, by reducing a large-scale reproduction of the interdigital transducer by a factor on the order of 100. This means that many identical filters can be produced by relatively simple and inexpensive photographic techniques.

2.4.2 Delta-Function Model of the Transducer

We can use a simple mathematical model to represent the excitation of a Rayleigh wave by an SAW transducer. We consider the excitation by an individual finger to be proportional to the total charge Q on the finger. Suppose that all the fingers are of equal length w in the x direction across the acoustic beams, and that the charge density per unit length in the direction of acoustic propagation z at z' on the surface of the piezoelectric material is $\sigma(z')$, as shown in Fig. 2.4.3. Because the system is linear, the SAW signal excited by the charge in a length dz' is proportional to $\sigma(z')dz'$. This surface acoustic wave has a propagation constant $k = \omega/V_R$, where ω is the radian frequency, V_R is the Rayleigh wave velocity, $k = 2\pi/\lambda$ is the propagation constant, and λ is the wavelength.

Let the SAW signal at the plane z have an amplitude $A(z, w)$. In a linear system, $A(z, w)$ varies linearly with the amplitude of stress, strain, or particle velocity at any cross section of the plane z. Commonly, $|A(z, w)|$ is taken to be the square root of the power at the plane z. The elementary signal of frequency

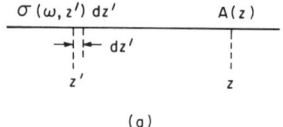

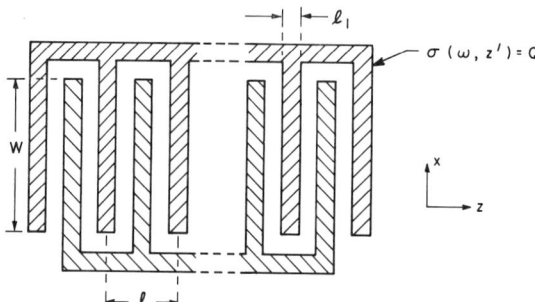

Figure 2.4.3 Notation used (a) to estimate the response of a transducer, and (b) for the finger pair spacing and width.

ω that reaches the plane z from the element between z' and $z' + dz'$ is of amplitude $dA(z, z', \omega)$. We can write

$$dA(z, z', \omega) = \alpha\sigma(\omega, z') e^{-jk(z-z')} dz' \qquad (2.4.1)$$

where α is a coupling factor between the charge and the acoustic excitation and $k(z - z')$ is the phase shift of the wave between the excitation point z' and the observation point z. The signal amplitude $A(z)$ induced at z must be the sum of the contributions from all the elements of length dz' that are to the left of z. The total signal $A(z, \omega)$ reaching z is therefore

$$A(z, \omega) = \alpha \int_{-\infty}^{z} \sigma(\omega, z') e^{-jk(z-z')} dz' \qquad (2.4.2)$$

If z is beyond the end of the exciting transducer, then $\sigma(z) = 0$, and we can write

$$A(z, \omega) = \alpha \int_{-\infty}^{\infty} \sigma(\omega, z') e^{-jk(z-z')} dz' \qquad (2.4.3)$$

Thus *the output amplitude as a function of frequency is the Fourier transform of the charge on the fingers*. Since the frequency response of a transducer is the Fourier transform of its impulse response, it follows that the impulse response of a transducer is the same as the spatial charge distribution along its length. A different proof of this result based on time domain concepts is given in Sec. 4.2.

Example: Uniform Transducer

Consider the simple transducer with constant finger length and uniform finger pair spacing shown in Fig. 2.4.3. Suppose that the fingers have a *pair spacing l* and an individual width l_1, and that the charges on them are Q and $-Q$, in series.

We assume, for simplicity, that the charge is uniformly distributed on the fingers,† and that the amplitudes of all fields vary at a radian frequency ω. Hence, for a single finger (the right-hand one of a pair), we can write

$$\sigma = \frac{Q}{l_1} \tag{2.4.4}$$

If the center of this finger is at $z' = l/4$, its contribution to the field at the plane z is

$$A(z) \text{ [single finger]} = \frac{\alpha Q}{l_1} e^{-jk(z - l/4)} \int_{-l_1/2}^{l_1/2} e^{jkz'} dz'$$

$$= \alpha Q e^{-jk(z - l/4)} \frac{\sin (kl_1/2)}{kl_1/2} \tag{2.4.5}$$

The contribution of the left-hand finger of this pair, whose center is at $z' = -l/4$ and whose charge is $-Q$, can be added to this term to give

$$A(z) \text{ [finger pair]} = \alpha Q e^{-jkz} e^{jkl/4} - e^{-jkl/4} \frac{\sin (kl_1/2)}{kl_1/2}$$

$$= 2\alpha j Q e^{-jkz} \sin \frac{kl}{4} \frac{\sin (kl_1/2)}{kl_1/2} \tag{2.4.6}$$

We can now add the contributions from N finger pairs whose centers are at $z' = nl$ ($n = 0$ to $N - 1$), using the appropriate delays to obtain

$$A(z) = 2j\alpha Q e^{-jkz} \sin \frac{kl}{4} \operatorname{sinc} \frac{l_1}{\lambda} \int_0^{N-1} e^{jknl}$$

$$= 2j\alpha Q \sin \frac{kl}{4} e^{-jkz} \frac{\sin (kNl/2)}{\sin (kl/2)} \operatorname{sinc} \frac{l_1}{\lambda} e^{jk(N-1)l/2} \tag{2.4.7}$$

$$= j\alpha Q e^{-jkz} \frac{\sin (kNl/2)}{\cos (kl/4)} \operatorname{sinc} \frac{l_1}{\lambda} e^{jk(N-1)l/2}$$

where $\operatorname{sinc} x = \sin (\pi x)/(\pi x)$. Equation (2.4.7) is, of course, the Fourier transform of the charge distribution on the fingers. When N is large, the most rapidly varying function in Eq. (2.4.7) is the term $\sin (kNl/2)/\cos(kl/4)$. Using L'Hospital's rule, it can be shown that as $kl \to 2\pi$, this function has a peak value of $\pm 2N$ and is zero where

$$kl(\text{zero response}) = 2\pi \left(1 \pm \frac{1}{N}\right) \tag{2.4.8}$$

We conclude that as k is varied (the frequency is proportional to k, for $k = \omega/V_r$), the response is maximum where $kl = 2\pi$. Therefore, this value of k corresponds to the center frequency, or the *synchronous frequency*, $\omega = \omega_0$. At this point, $|A(z)| \to 2\alpha QN \operatorname{sinc} (l_1/l)$. It follows from Eq. (2.4.8) that the bandwidth between the zeros in the response is

$$\frac{(\Delta\omega) \text{ (zeros)}}{\omega_0} = \frac{2}{N} \tag{2.4.9}$$

†In practice the charge is peaked at the edges of the fingers: a single finger of width l_1 in free space has a charge distribution of the form $\rho_s = K/\sqrt{l_1^2/4 - x^2}$, where K is a constant. A more exact treatment of the problem uses a quasistatic solutions for a periodically spaced set of fingers.

If N is large, it is convenient to write

$$\frac{k - k_0}{k_0} = \frac{\omega - \omega_0}{\omega_0} = \frac{x}{N\pi} \qquad (2.4.10)$$

In this case, it can be shown that for N large,

$$|A(z)| \approx 2\gamma QN \left|\frac{\sin x}{x}\right| \qquad (2.4.11)$$

where

$$\gamma = \alpha \operatorname{sinc} \frac{l_1}{\lambda} \qquad (2.4.12)$$

and sinc (l_1/λ) is regarded as constant and of value sinc (l_1/l) over the bandwidth of the transducer. This is equivalent to assuming that $l_1 \ll Nl$, which is usually true. It follows that the bandwidth to the 3−dB points where $x = \pm .89\pi/2$ is

$$\frac{\Delta\omega}{\omega_0}(3 \text{ db}) = \frac{0.89}{N} \qquad (2.4.13)$$

This behavior is illustrated in Figs. 2.4.2 and 2.4.4. We note that the points where $x = \pm \pi/2$ actually correspond to $(\sin x)/x = 2/\pi$, which is equal to 4 dB.

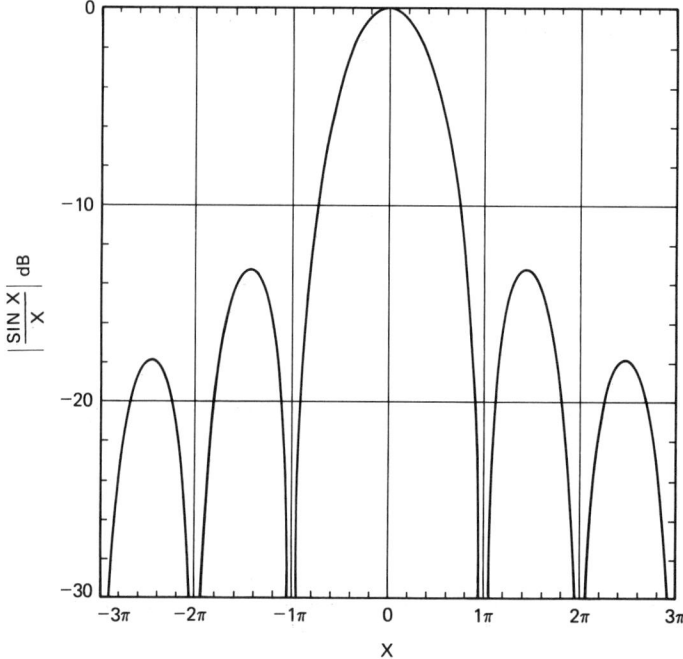

Figure 2.4.4 Plot of $|(\sin x)/x|$ as a function of x. (From Smith et al. [11].)

2.4.3 Network Theory of the Transducer

Several theories have been used to determine the electrical input impedance and frequency response of an interdigital transducer. The first, the *network theory*, is an equivalent-circuit method based on the idea that the response of a single finger pair is like that of a bulk wave transducer, so that the parameters of a single finger pair can be derived as if the finger pair were a bulk wave transducer (see Prob. 1.4.8). The N finger pairs are cascaded and connected together appropriately to establish the impedance matrix of the three-port network, comprised of the two acoustic ports, one at either end of the transducer, and the one electrical port.

A more fundamental approach to the theory is to use the *normal-mode formalism*, which is based on the conservation of power. While the equivalent-circuit method requires the use of empirically fitted parameters, this method allows us to obtain a solution for the electrical impedance of the transducer with no fitted parameters.

A third technique, developed by Ingebrigtsen [12], is based on wave impedance concepts. It, too, is useful and relatively simple; furthermore, it is often equivalent to, and sometimes more powerful than, normal-mode theory. But as it is a rather specialized method, we shall not deal with it here.

Another approach is a direct solution of the field theory [13]. This may be the most exact technique, but it is extremely complicated and inflexible and must be resolved for each new case. Generally, it is better to use a more physical theory, with some simplifying approximations, to get a physical understanding of what is happening in the system. The three approximate theories mentioned above have this advantage.

We shall first deal with the network theory, which was developed by Smith et al. [11]. They considered the system shown in Figs. 2.4.5 and 2.4.6(a), and based their theory of the transducer on the analogous one-dimensional models shown in Fig. 2.4.6(b) and (c). Following their method, we assume that a pair of bulk wave transducers is arranged acoustically in cascade and electrically in parallel so that the necessary electric field reversal takes place. For the configuration shown in Fig. 2.4.6(c), we assume that the component of applied electric field parallel to the direction of acoustic wave propagation is the most important. This is called the *in-line* model; it corresponds to a bulk wave transducer, of the type dealt with in Sec. 1.4, used with an equivalent circuit (the Mason equivalent circuit) of the type shown in Fig. 2.4.7, which we derived earlier (see Sec. 1.4.3). Alter-

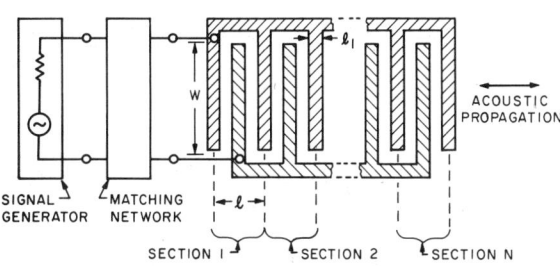

Figure 2.4.5 Interdigital transducer with its external circuit. (From Smith et al. [11].)

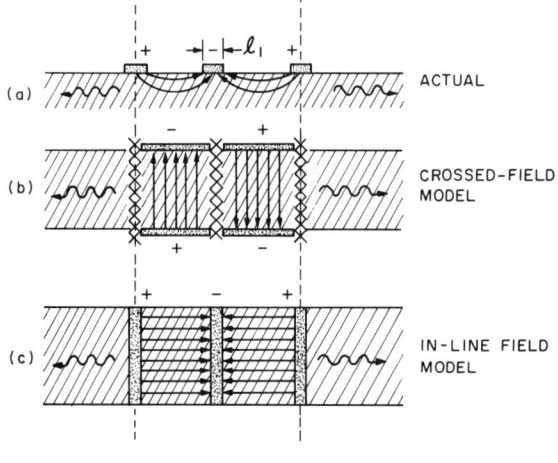

Figure 2.4.6 Side view of the interdigital transducer, showing field patterns: (a) actual field pattern; (b) crossed-field approximation; (c) in-line field approximation. (From Smith et al. [11].)

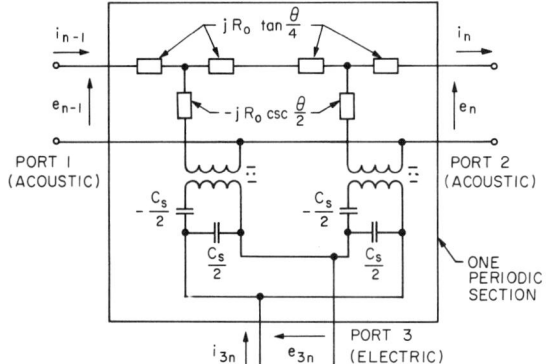

Figure 2.4.7 Mason equivalent circuit for one periodic section. The negative capacitors are short circuited for the crossed-field model: l, Periodic length; A, cross-sectional area; v, sound velocity; ρ, density; Z_0, $A\rho v$; h, piezo constant; f_0, v/L = synchronism frequency; θ, $2\pi(\omega/\omega_0)$ = periodic section transit angle; R_0, electrical equivalent of Z_0; C_s, electrode capacitance per section; K, electromechanical coupling constant. (From Smith et al. [11].)

natively, as shown in Fig. 2.4.6(b), we consider the field component normal to the surface to be the most important. We call this the *crossed-field* model; it is represented by the equivalent circuit shown in Fig. 2.4.7 (see Prob. 1.5.1). Its one difference from the in-line model is that it lacks a negative capacitor.

We define electrical equivalents to the acoustic force F_i and the terminal velocity V_i by setting

$$V_i = \frac{F_i}{\phi} \tag{2.4.14}$$

and

$$i_i = v_i \phi \tag{2.4.15}$$

where ϕ is the transformer ratio, defined as[†]

$$\phi = \frac{hC_s}{2} \tag{2.4.16}$$

[†] Note that here we use ϕ instead of N to avoid conflict with our definition for the number of finger pairs.

These definitions allow the mechanical impedance Z_0 of the substrate to be expressed in electrical ohms by the formula

$$R_0 = \frac{Z_0}{\phi^2} = \frac{2\pi}{\omega_0 C_s K^2} \tag{2.4.17}$$

where C_s is the capacitance of one periodic section and $\omega_0 = 2\pi f_0$ is the synchronous frequency defined by the formula $f_0 = V/l$, where l is the periodic length of the system, V is the acoustic velocity, and K is the effective electromechanical coupling constant for a surface acoustic wave.

The most difficult parameter to determine is the value of K, for the fields are no longer uniform within the transducer; thus an *effective* value of K is the best we can expect. Obviously, we cannot easily base our definition of K on the same type of field theory analysis used directly in the one-dimensional theory of the bulk wave transducer. It is possible to define K as the ratio of the stored electrical energy to the stored mechanical energy, but this definition is unreliable because in an interdigital transducer, not all the electric field lines intercept a single electrode pair, as they do in a bulk wave transducer.

A better definition is based on the velocity change of the wave in the medium when the RF field between the electrodes is shorted out, for it will give the same result, at least as far as the external electrical impedance is concerned, for bulk and SAW transducers. We know that the ratio of the stiffened to unstiffened velocity of a piezoelectric medium is $1 + \Delta V/V = (1 + K^2)^{1/2}$. Therefore, the fractional change in velocity of a bulk wave transducer when the E field within the medium (i.e., between the electrodes) is made zero is

$$\frac{\Delta V}{V} = 1 - (1 + K^2)^{1/2} \approx \frac{-K^2}{2} \tag{2.4.18}$$

Now suppose that we place a metal film on the surface of the substrate. This is equivalent to shorting out the E field between the electrodes. In this case we measure a fractional change in velocity $\Delta V/V$. We then define the effective value of K on this basis, using Eq. (2.4.18). Thus we can relate the effective value of K to a parameter that can be experimentally measured or calculated from a field theory for the Rayleigh waves in a piezoelectric medium. This is a better procedure than the direct use of field theory for an interdigital transducer, because the boundary conditions are simpler.

In practice, we add a correction factor to Eq. (2.4.18) to compensate for the fact that this theory is heuristic, and for the effect of changing the finger width (the finger itself partially shorts out the fields). Thus, more generally, we write

$$K^2 = 2F \left| \frac{\Delta V}{V} \right| \tag{2.4.19}$$

where F is a "filling factor," which, as we discuss in Sec. 2.5 [see Eq. (2.5.43)], turns out to be about 1.12 when the strips and gaps between them are equal in size.

With the equivalent circuit for one periodic section given in Fig. 2.4.7, we can now calculate the admittance matrix for the connected N sections, as illustrated

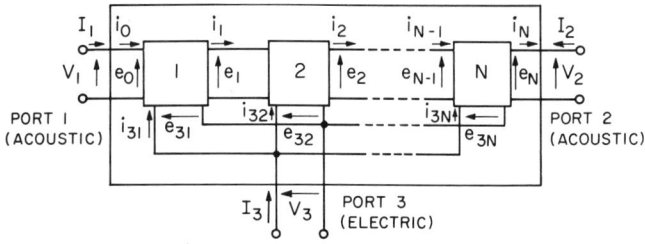

Figure 2.4.8 Transducer composed of N periodic sections, acoustically in cascade and electrically in parallel. (From Smith et al. [11].)

in Fig. 2.4.8. The three-port transducer equation is written like the equation for any three-port network, in the form

$$\begin{bmatrix} I_1 \\ I_2 \\ I_3 \end{bmatrix} = [Y] \begin{bmatrix} V_1 \\ V_2 \\ V_3 \end{bmatrix} \qquad (2.4.20)$$

where the admittance matrix, because of the network's symmetry, can be written in the form

$$[Y] = \begin{bmatrix} Y_{11} & Y_{12} & Y_{13} \\ Y_{12} & -Y_{11} & -Y_{13} \\ Y_{13} & -Y_{13} & Y_{33} \end{bmatrix} \qquad (2.4.21)$$

It is shown in Appendix F that for the crossed-field model, the admittances are

$$\begin{aligned} Y_{11} &= -jG_0 \cot N\theta \\ Y_{12} &= jG_0 \operatorname{cosec} N\theta \\ Y_{13} &= -jG_0 \tan \frac{\theta}{4} \\ Y_{33} &= j\omega C_T + 4jNG_0 \tan \frac{\theta}{4} \end{aligned} \qquad (2.4.22)$$

For the in-line model, the admittances are

$$\begin{aligned} Y_{11} &= \frac{-S_{11}}{S_{12}} \\ Y_{12} &= \frac{1}{S_{12}} \\ Y_{13} &= \frac{-jG_0 \tan(\theta/4)}{1 - 2x \tan(\theta/4)} \\ Y_{33} &= \frac{j\omega C_T}{1 - 2x \tan(\theta/4)} \end{aligned} \qquad (2.4.23)$$

where $G_0 = (R_0)^{-1}$, $C_T = NC_s$ is the capacity of the transducer, $\theta = 2\pi f/f_0 = kl$, and the matrix S is a complex function of N, θ, G_0, and C_s. Thus we have a complete three-port network model that can be used to calculate reflections from the transducer, its input admittance with different terminations, and so on.

The crossed-field model yields a simplified form of the admittance near synchronism. Writing $\theta = 2\pi + \delta$ and assuming that δ is small, we can show by the methods given in Appendix F that

$$[Y] = \frac{jG_0}{\delta} \begin{bmatrix} -\frac{1}{N} & \frac{1}{N} & 4 \\ \frac{1}{N} & -\frac{1}{N} & -4 \\ 4 & -4 & -16N + \frac{\delta\omega C_T}{G_0} \end{bmatrix} \quad (2.4.24)$$

Similar results for the synchronous case $\theta = 2\pi$ of the in-line model are given in Appendix F as Eq. (F.14).

We may determine the input impedance or admittance of the transducer by assuming that ports 1 and 3 are terminated with the matching admittance G_0. For the crossed-field model, the equivalent input circuit has the form

$$Y_3 = Y_a + j\omega C_T \quad (2.4.25)$$

where the acoustic contribution to the shunt admittance is given by the relation

$$Y_a = G_a(\omega) + jB_a(\omega) \quad (2.4.26)$$

where

$$G_a(\omega) = 2G_0 \left(\tan \frac{\theta}{4} \sin \frac{N\theta}{2} \right)^2 \quad (2.4.27)$$

and

$$B_a(\omega) = G_0 \tan \frac{\theta}{4} \left(4N + \tan \frac{\theta}{4} \sin N\theta \right) \quad (2.4.28)$$

as shown in Fig. 2.4.9(b).

For frequencies near synchronism, the resultant admittance can be stated in the simple forms, on which most design work is based, as follows:

$$G_a(\omega) = G_{a0} \left(\frac{\sin x}{x} \right)^2 \quad (2.4.29)$$

$$B_a(\omega) = G_{a0} \frac{\sin 2x - 2x}{2x^2} \quad (2.4.30)$$

where

$$x = \frac{N\pi(\omega - \omega_0)}{\omega_0} = \frac{N\delta}{2} \quad (2.4.31)$$

Sec. 2.4   Interdigital Transducers

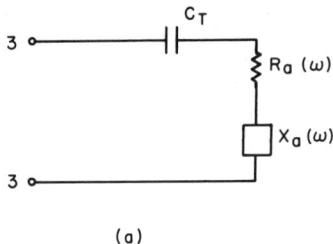

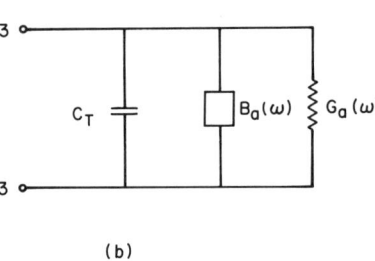

Figure 2.4.9 Transducer electrical input admittance: (a) series representation; (b) shunt representation.

and where

$$G_{a0} = \frac{4}{\pi} K^2 \omega_0 C_s N^2 F \qquad (2.4.32)$$

is the synchronous conductance and $K^2 \approx 2|\Delta V/V|$.

Similar formulas can be obtained for the in-line circuit, which looks like a capacitance in series with the acoustic impedance. For this model, we write

$$Z_3 = Z_a + \frac{1}{j\omega C_T} \qquad (2.4.33)$$

as illustrated in Fig. 2.4.9(a).

For frequencies near synchronism, we can now show that

$$Z_a = R_a + jX_a \qquad (2.4.34)$$

with

$$R_a = R_{a0} \left(\frac{\sin x}{x}\right)^2 \left(\frac{\omega_0}{\omega}\right)^2 \qquad (2.4.35)$$

and

$$X_a = R_{a0} \frac{\sin 2x - 2x}{2x^2} \left(\frac{\omega_0}{\omega}\right)^2 \qquad (2.4.36)$$

where

$$R_{a0} = \frac{4}{\pi} \frac{K^2 F}{\omega_0 C_s} \qquad (2.4.37)$$

where $K^2 = 2|\Delta V/V|$.

The series equivalent circuit predicts that the acoustic impedance Z_a is independent of N, although C_T is of course proportional to N. On the other hand, Y_a in the crossed-field model is proportional to N^2. Assuming that the capacitive reactance is much larger than the series impedance, we can show that the crossed-field result follows logically from the series formulation. We write

$$Y_3 \approx \frac{1}{Z_3} \approx \frac{1}{Z_a + (1/j\omega C_T)} = \frac{Z_a - (1/j\omega C_T)}{Z_a^2 + (1/\omega^2 C_T^2)}$$
$$\approx Z_a \omega^2 C_T^2 + j\omega C_T \qquad (2.4.38)$$

This result is identical to that which follows from Eqs. (2.4.25)–(2.4.32).

Using the crossed-field result, we see that the input radiation conductance G_a becomes zero at $x = \pi$, or where $(\omega - \omega_0)/\omega_0 = 1/N$. The acoustic Q, Q_a, is defined from the frequencies where the acoustic response is 3 dB down from the center frequency, that is,

$$Q_a = \frac{\omega_0}{\Delta\omega(3 \text{ dB})} \qquad (2.4.39)$$

where $\Delta\omega(3 \text{ dB})$ is the frequency difference between the two -3-dB response points. However, it is more convenient to use the -4-dB points and ignore the slight error this yields because the resulting expression from the $(\sin x/x)$ function, taken where $x = \pi/2$, is so simple. We therefore write

$$Q_a \approx \frac{\omega_0}{\Delta\omega(4 \text{ dB})} = N \qquad (2.4.40)$$

Thus the larger the number of fingers, the higher the value of Q_a and the lower the bandwidth.

If we try to tune the transducer with a parallel inductance to eliminate the effect of the transducer capacity, the electrical Q, Q_e, will be

$$Q_e = \frac{\omega_0 C_T}{G_{a0}} = \frac{\pi}{4K^2 NF} \qquad (2.4.41)$$

When the number of fingers is smaller, Q_e increases because the capacity decreases. To obtain the lowest total Q, and hence the greatest bandwidth, we choose $Q_e = Q_a$. The condition for this to occur is

$$N^2 = \frac{\pi}{4K^2 F} \qquad (2.4.42)$$

Thus the maximum bandwidth is determined by the effective value of K. For lithium niobate, with an effective K^2 value of 0.046 and $F = 1.12$ [see discussion

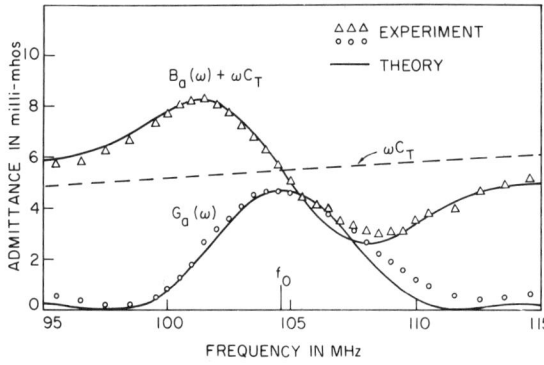

Figure 2.4.10 Measured radiation admittance for an $N = 15$ transducer on YZ lithium niobate, compared with theoretical curves calculated from the crossed-field model. (After Smith et al. [11].)

after Eq. (2.5.42)], the optimum number of fingers is 4 and the bandwidth is of the order of 20%.

A similar treatment for the in-line model gives the same result. Experiments confirm the theoretical results very accurately. An early experimental result is compared to theory in Fig. 2.4.10. This network theory has been widely used in the design of transducers and, in practice, provides accurate, comprehensible results.

PROBLEM SET 2.4

1. The assumption that the charge on the interdigital fingers is uniform is a crude approximation. It is better to write

$$\sigma = \sigma_0 l_1 \left(\frac{l_1^2}{4 - z^2} \right)^{-1/2}$$

for each finger.
   (a) Find σ_0 in terms of Q. Find $A(Z, \omega)$ for an N-finger pair with a period l.

   Hint. Put $z = (l_1/2) \sin \theta$ and use the Bessel function formula

$$J_0(x) = \frac{1}{\pi} \int_0^\pi e^{jx \sin \theta} \, d\theta$$

   (b) Consider an interdigital transducer for which $l_1 = l/4$. What is the ratio of the excitation amplitude to the total charge at the center frequency? Compare your result to that obtained with the use of the uniform charge assumption.

2. A transducer uses three fingers for each section, as illustrated below. These three fingers are supplied by a three-phase system, with signals $\exp[j(\omega t - \pi/3)]$, $-\exp(j\omega t)$, and $\exp[j(\omega t + \pi/3)]$. The most convenient way to obtain a three-phase excitation is to connect one set of electrodes to ground, to make the voltage between the input grounded electrode and the input terminals $\exp(j\omega t - \pi/3)$.

   Work out the excitation of waves in the forward and backward directions, respectively. Take the fingers to have a width l_1 and the period of the transducer to be l.

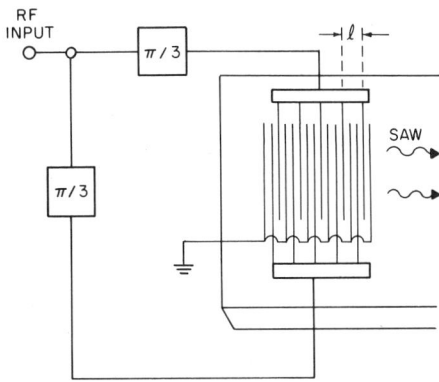

Assume that $l_1 \ll l$, that the charges on the fingers are of magnitude Q, and that the charge density is uniform on each finger.

3. (a) Consider the network circuit theory of an interdigital transducer. Use the KLM model, as illustrated in Fig. 1.4.12, to write down a circuit for the in-line field model that is equivalent to the one obtained using the Mason model, as illustrated in Fig. 2.4.7. Do not work out the theory of this transducer.

   (b) How would you modify the circuit to take account of a finite finger width l_1? The region under the finger can be regarded as a transmission line of slightly different impedance from the region between the fingers (the gap).

4. Prove Eqs. (2.4.29) and (2.4.30) from Eqs. (2.4.27) and (2.4.28).

2.5 NORMAL-MODE THEORY AND PERTURBATION THEORY

2.5.1 Introduction

The problem of determining the excitation of surface acoustic waves or other types of waveguide modes by a transducer is difficult to tackle directly. In the same way, it is difficult to determine the effects of small perturbations of the medium directly. For instance, if a metal strip is placed on the surface of the medium on which a surface acoustic wave is propagating, the wave will be perturbed. Generally, part of the wave will be reflected by the obstruction, part transmitted past it, and part may be converted to a bulk shear or longitudinal wave.

Problems of this type can sometimes be solved exactly by carrying out a detailed field theory analysis and taking account of the boundary conditions at the perturbation. But as discussed in Sec. 2.4.3, it is extremely difficult to apply such techniques, except in the simplest cases, because it is hard to satisfy the boundary conditions for all the field components of interest. Even when the exact field theory can be solved, it tends to be extremely complicated in mathematical form and provides very little intuition about the basic parameters governing the solution. Furthermore, again as mentioned in Sec. 2.4.3, field theories must be solved anew for each individual case.

For example, when a thin metal film a fraction of a wavelength in thickness is evaporated onto a substrate, the solution for the surface acoustic wave must obviously be very close to the solution for the original unperturbed substrate. But to determine the perturbation by an exact field theory would mean solving the problem again completely. Instead, however, we can find the required solution by using a perturbation theory, which will provide more physical insight into the solution. Field theory is an even less desirable choice when a thin film covers only a short region of the substrate, for the solution then becomes so complicated that we tend to lose physical insight into the problem.

Here we discuss a perturbation theory based on the idea that the total field in a perturbed medium is very similar to a field of a wave in the unperturbed medium. In Appendix E, this theoretical approach is justified on a more rigorous basis by expressing the total field of a single mode in the perturbed medium as a weighted sum of the original modes in the unperturbed medium. When the perturbation is small, only one of these unperturbed modes is strongly excited.

This mode expansion technique is similar to Fourier analysis or, more generally, the expansion of a function in terms of a sum of orthogonal functions. For instance, if $F(x)$ is a periodic function with a period 2π, we can write

$$F(x) = \sum_n A_n e^{jnx} \qquad (2.5.1)$$

and use the orthogonality relation

$$\frac{1}{2\pi} \int_0^{2\pi} e^{j(m-n)x} \, dx = \delta_{nm} \qquad (2.5.2)$$

where $\delta_{nm} = 1$ when $n = m$ and $\delta_{nm} = 0$ when $n \neq m$, to show that

$$A_n = \frac{1}{2\pi} \int_0^{2\pi} F(x) e^{-jnx} \, dx \qquad (2.5.3)$$

If $F(x)$ is a function very close in form to $\exp(jNx)$, the dominant term in the expression of Eq. (2.5.1) is that for $n = N$. This analogy holds exactly for rectangular electromagnetic waveguides, where any arbitrary cross-sectional variation of the field in the waveguide can be expressed as the sum of a set of trigonometric functions (the wave solutions for the individual modes of the waveguide). More generally, however, the same kinds of analogies still hold, and we can work in terms of the wave solutions of the unperturbed system to determine the nature of the waves in the perturbed system.

One advantage of this method is it allows us to find the amplitude of a mode that is excited in a waveguide without detailed knowledge of the fields in the exciting region. For instance, we need only a relatively crude estimate of the charge distribution on the metal electrodes of an interdigital transducer to determine the amplitudes of the excited waves and the transducer input impedance. Here we use the procedure of Sec. 2.4.2 to obtain quantitative results.

We state the amplitude of an excited wave in terms of the power flow associated with it. Because power flow is a one-dimensional concept, the final results are often very simple in form and provide a great deal of physical insight into the interactions involved. The concept is particularly useful for dealing with piezoelectric materials because power has a meaning for both electromagnetic and acoustic fields. Thus, to determine how a piezoelectric transducer is excited by a surface acoustic wave, it is useful to be able to work in terms of power concepts. [9].

2.5.2 Power Flow Concepts

Before beginning a detailed analysis for the excitation of a surface acoustic wave, we must express the surface acoustic wave (SAW) amplitude in terms of the power flow. Suppose that the amplitude of any component of the wave is $A(z)$. In general, if the amplitude is associated with stress, strain, velocity, or some other component of the wave, it has a fixed cross-sectional variation. The power in the wave, however, varies as the square of the amplitude. Thus it is convenient to choose the amplitude $A(z)$ to be directly associated with the total power of the wave, which is a one-dimensional parameter [9, 15]. We call the power of the wave $A(z)$ propagating in the forward direction P_F, and define it as

$$P_F = AA^* \tag{2.5.4}$$

We use an amplitude parameter $B(z)$ for a wave propagating in the backward direction and take the power P_B associated with it to be

$$P_B = -BB^* \tag{2.5.5}$$

We can connect this concept with the stress and strain components in a general way because the stress, strain, velocity, and displacement vary linearly in amplitude with $A(z)$. Thus if we know $A(z)$, in principle we also know the amplitude of the stress at any point $\mathbf{T}(x, y, z)$ and the amplitudes of all the other field components.

We shall find it convenient to express the cross-sectional variations of the fields in a normalized form so that they correspond to unit power flow in the guide. When the nth mode is excited by the fields at a perturbation, its amplitude does not necessarily vary as $\exp(-jk_n z)$. Thus for the nth mode with a propagation constant k_n, we more generally write

$$\hat{v}_n(x,y,z) = A_n(z)\mathbf{v}_n(x, y) \tag{2.5.6}$$

$$\hat{\mathbf{T}}_n(x,y,z) = A_n(z)\mathbf{T}_n(x,y) \tag{2.5.7}$$

$$\hat{\phi}_n(x,y,z) = A_n(z)\phi_n(x, y) \tag{2.5.8}$$

and

$$\hat{\mathbf{D}}_n(x,y,z) = A_n(z)\mathbf{D}_n(x,y) \tag{2.5.9}$$

where $\hat{\phi}_n$ is the electric potential and $\mathbf{T}_n(x, y)$, $\mathbf{v}_n(x, y)$, $\phi_n(x, y)$, and $\mathbf{D}_n(x, y)$ are the cross-sectional variations of the components of the nth mode. The symbol ˆ is associated with the field variation in the z direction. For an unperturbed mode (i.e., a mode in the unperturbed guide), $A_n(z) = \exp(-jk_n z)$.

The power flow P_n associated with this propagating wave is the sum of the electromagnetic and acoustic power flows (see Sec. 1.2 and Appendix E), where [4, 9]

$$P_n = \frac{1}{2} \operatorname{Re} \left(A_n A_n^* \int_s (\mathbf{E}_n^* \times \mathbf{H}_n - \mathbf{v}_n^* \cdot \mathbf{T}_n) \mathbf{a}_z \, ds \right) \quad (2.5.10)$$

The phase velocity of a surface acoustic wave is typically several orders of magnitude less than that of light. This means that we can neglect RF magnetic fields, put $\nabla \times \mathbf{E} = 0$, and write the electric field in terms of a quasistatic electric potential ϕ, with $\mathbf{E} = -\nabla \phi$. In this case (see Prob. 1),

$$P_n = \frac{1}{2} \operatorname{Re} \left(A_n A_n^* \int_s (j\omega \mathbf{D}_n \phi_n^* - \mathbf{v}_n^* \cdot \mathbf{T}_n) \cdot \mathbf{a}_z \, ds \right) \quad (2.5.11)$$

where $j\omega \mathbf{D}_n \cdot \mathbf{a}_z$ is the displacement current density in the z direction and s is the cross-sectional area of the system.

Normalization. We choose the amplitudes of the normalized fields $\mathbf{v}_n$, $\mathbf{T}_n$, ϕ_n, and $\mathbf{D}_n$, so that for a propagating wave,

$$\frac{1}{2} \operatorname{Re} \int_s (j\omega \mathbf{D}_n \phi_n^* - \mathbf{v}_n^* \cdot \mathbf{T}_n) \cdot \mathbf{a}_z \, ds = \pm 1 \quad (2.5.12)$$

where the $+$ and $-$ signs denote waves propagating in the forward $(+z)$ and backward $(-z)$ directions, respectively. Thus we have chosen the amplitudes of the normalized fields $\mathbf{v}_n$, $\mathbf{T}_n$, ϕ_n, and $\mathbf{D}_n$ to correspond to unit power flow. In this case, it follows that the power flow P_{Fn} associated with the nth forward wave mode is

$$P_{Fn} = A_n A_n^* \quad (2.5.13)$$

while that associated with the nth backward wave mode, P_{Bn}, is

$$P_{Bn} = -A_{-n} A_{-n}^* = -B_n B_n^* \quad (2.5.14)$$

We have used the symbol B_n, rather than A_{-n}, to indicate the nth backward wave mode. Therefore, the total real power flow P_n in the nth forward and backward modes is

$$P_n = A_n A_n^* - B_n B_n^* \quad (2.5.15)$$

2.5.3 Excitation of a Surface Acoustic Wave

Here we give a heuristic derivation of the perturbation theory for the excitation of the nth mode. A more rigorous derivation of the results, using mode orthogonality, is given in Appendix E.

Suppose that the nth forward wave mode is excited from an external source of amplitude $f(z') \, dz'$ in an element of length dz'. Typical examples are the charges on a set of electrodes, which couple from another acoustic beam or scatter from an inhomogeneity. As we have already seen in Sec. 2.5.2, because the system is linear and the guide is taken to be uniform, the signal reaching the plane z must be of the form

$$dA_n(z) = \alpha_n f(z') e^{-jk_n(z-z')} \, dz' \tag{2.5.16}$$

where we have used the subscript n to denote the nth mode and α_n is a coupling parameter from the external source to the nth mode. If $f(z')$ is doubled in amplitude, so is $A(z)$, as it should be in a linear system. If there is a distributed excitation, we can write the *excitation equation* in the following integral form:

$$A_n(z) = \alpha_n \int_{-\infty}^{z} f(z') e^{-jk_n(z-z')} \, dz' = \int_{-\infty}^{z} g_n(z') e^{-jk_n(z-z')} \, dz' \tag{2.5.17}$$

where now we have combined $\alpha_n f(z')$ into one parameter, $g_n(z')$.

By taking the $\exp(-jk_n z)$ term outside the integral, Eq. (2.5.17) may be differentiated with respect to z to obtain the differential form of the *excitation equation*:

$$\frac{dA_n}{dz} + jk_n A_n = g_n(z) = \alpha_n f(z) \tag{2.5.18}$$

Equations (2.5.17) and (2.5.18) imply that if we know the excitation as a function of z and can determine the coupling coefficient α or the parameter $g_n(z)$, we can determine the amplitude $A_n(z)$ and hence the power of the wave excited at any plane z. Outside the excitation region $[g_n(z) = 0]$, the solution is of the form $A_n(z) \sim \exp(-jk_n z)$. A similar derivation can be carried out for the backward wave or other modes of the system.

Physical implications of the excitation equations. Using the simple form of Eq. (2.5.18), we can usually determine how the distribution of the excitation will affect the amplitude of the final wave, even without a quantitative knowledge of α_n.

Suppose that the excitation is of the form $f(z') = f_0(z') \exp(-j\beta z')$ and $\beta \neq k$. From Eq. (2.5.17) or Eq. (2.5.18), if $f(z')$ is finite over a region many wavelengths long, the net excitation of the surface acoustic wave of amplitude $A_n(z)$ tends to be small, because if $\beta \neq k$, the waves excited at planes $z' < z$ arrive at the plane z out of phase.

Now suppose that $f_0(z')$ is constant from $z' = 0$ to $z' = L$, and that $\beta = k_n$. The excitation is then cumulative at the plane z, so that $A_n(L) = \alpha_n f_0 L$ and $A_n(L)$ increases linearly with L. Thus, even without knowing the coupling coefficient α_n, we can gather much information about the nature of the excitation.

Quantitative evaluation of $g_n(z)$. We can derive the value of $g_n(z)$ in terms of the exciting fields at the surface of a SAW waveguide by regarding these fields as perturbations at the waveguide surface that change the power flow along the guide. Let us consider a piezoelectric waveguide, as illustrated in Fig. 2.5.1,

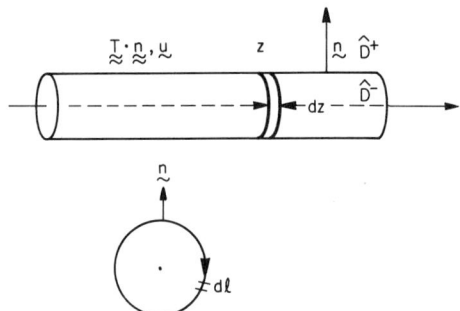

Figure 2.5.1 Notation used in the normal mode theory for a cylindrical system that is uniform in the z direction. The shape of the guide is arbitrary.

for which the boundary conditions for the nth mode are such that the normal stress is zero, that is

$$\mathbf{T}_n \cdot \mathbf{n} = 0 \qquad (2.5.19)$$

and there is no surface charge, that is, the boundary condition on the normal electric displacement densities of the nth mode $\mathbf{D}_n$ is

$$(\mathbf{D}_n^+ - \mathbf{D}_n^-) \cdot \mathbf{n} = 0 \qquad (2.5.20)$$

where $\mathbf{D}_n^+$ and $\mathbf{D}_n^-$ are the electric displacement densities just outside and just inside the surface, respectively, and $\mathbf{n}$ is the outward unit vector normal to the surface.

We derive the excitation parameter $g_n(z)$ by assuming that there is a finite normal stress $\hat{\mathbf{T}} \cdot \mathbf{n}$ and surface charge $\hat{\rho}_s$ at the boundary surface of the guide. Thus the boundary conditions for an unperturbed mode are no longer satisfied. The required derivation is carried out rigorously in Appendix E. Here we will carry out the derivation heuristically by using the concept of conservation of real power.

The velocity due to the nth mode at the surface of the guide is $A_n v_n(x, y)$, while the potential is $A_n \phi_n(x, y)$. Therefore, the power delivered to the nth mode by an outward normal applied stress $\hat{\mathbf{T}} \cdot \mathbf{n}$ and an electric surface charge density $\hat{\rho}_s = (\hat{\mathbf{D}}^+ - \hat{\mathbf{D}}^-) \cdot \mathbf{n}$ is $\frac{1}{2} \operatorname{Re} [j\omega \hat{\rho}_s \phi_n^* + \hat{\mathbf{T}} \cdot \mathbf{v}_n^*] A_n^*$. Here the $+$ and $-$ superscripts are associated with the fields just outside and just inside the surface, respectively. Thus the rate of change of real power in the nth forward mode is

$$\frac{dP_n}{dz} = \frac{d}{dz}(A_n A_n^*) = \frac{1}{2} \int_l A_n^* [(\hat{\mathbf{T}} \cdot \mathbf{v}_n^*) \cdot \mathbf{n} + j\omega \hat{\rho}_s \phi_n^*] \, dl \qquad (2.5.21)$$

where the integral is taken around the periphery of the guide at a plane z and the displacement current per unit area flowing into the guide is $j\omega \hat{\rho}_s$.

Multiplying Eq. (2.5.18) by A_n^* and adding the result to its complex conjugate, we can show that

$$A_n^* \frac{dA_n}{dz} + A_n \frac{dA_n^*}{dz} = A_n^* g_n + A_n g_n^* \qquad (2.5.22)$$

This relation can be written in the form

$$\frac{d}{dz}(A_n A_n^*) = 2(A_n^* g_n) \qquad (2.5.23)$$

Comparing this expression with Eq. (2.5.21), we can write

$$g_n(z) = \tfrac{1}{4} \int_l [(\hat{\mathbf{T}} \cdot \mathbf{v}_n^*) \cdot \mathbf{n} + j\omega \hat{\rho}_s \phi_n^*] \, dl \tag{2.5.24}$$

Thus

$$\frac{dA_n}{dz} + jk_n A_n = \tfrac{1}{4} \int_l [(\hat{\mathbf{T}} \cdot \mathbf{v}_n^*) \cdot \mathbf{n} + j\omega \hat{\rho}_s \phi_n^*] \, dl \tag{2.5.25}$$

This is the relation required for the excitation of the nth mode of the system.

2.5.4 Perturbation Theory of the Interdigital Transducer

We now use the results of Secs. 2.5.2 and 2.5.3 to determine how surface acoustic waves are excited by an interdigital transducer extending from $z = 0$ to $z = L$. We derive the input impedance of the transducer, dropping the use of the subscript n, as we are interested only in the surface acoustic wave mode, and finding the amplitude of the forward wave $A(z)$ at points $z > L$ and of the backward wave $B(z)$ at points $z < 0$. Because of symmetry, we expect that $|A(L)| = |B(0)|$. Thus when a transducer converts an electrical signal to an acoustic wave traveling in one direction, and vice versa, there will be a 3-dB loss.

We determine the input impedance of the transducer by finding the total acoustic power P excited by the transducer. Because of conservation of power [Poynting's theorem; Eq. (1.3.2)], we can define the electrical input resistance R_a of the transducer as

$$R_a = \frac{2P}{II^*} \tag{2.5.26}$$

where I is the current into the transducer. We must now derive A more quantitatively than in Sec. 2.5.2.

Consider the piezoelectric substrate, illustrated in Fig. 2.5.2, in which waves are excited by metal electrodes or by charges in a semiconductor. Initially, we suppose that there is a charge per unit length $\sigma(z') = \rho_s(z')w$ at any point along the substrate within the acoustic beam width w, where the charge per unit area $P_s(z')$ is assumed to be uniform over a width w. From Eq. (2.5.25),

$$\frac{dA}{dz} + jkA = \frac{j\omega}{4}\sigma \phi_0^* \tag{2.5.27}$$

where we have used the subscript 0 to indicate that ϕ_n is associated with the principal mode, and have dropped the subscript n on A and k.

We may treat the excitation of the backward traveling wave in a similar way, writing

$$\frac{dB}{dz} - jkB = -\frac{j\omega}{4}\sigma \phi_0^* \tag{2.5.28}$$

where B denotes the wave traveling in the backward direction, as discussed in

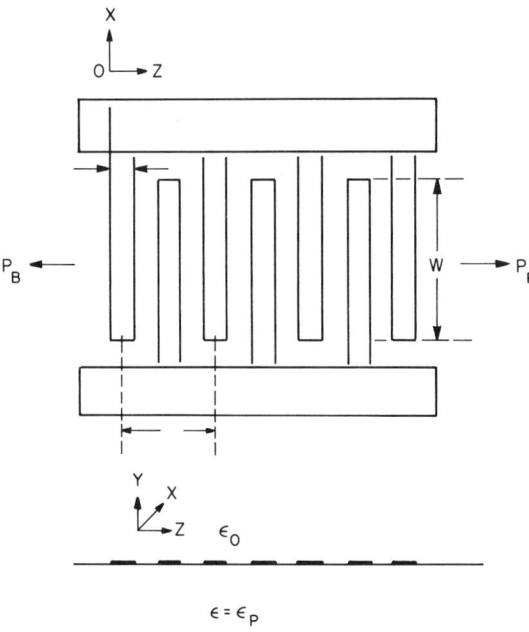

Figure 2.5.2 Interdigital transducer.

Appendix E, and we take $\phi_0 = -\phi_{-0}$ and $P_{-0} = -P_0$. When there is no excitation $[\sigma(z) = 0]$, the solutions of Eqs. (2.5.27) and (2.5.28) vary as $\exp(-jkz)$ and $\exp(jkz)$, respectively.

Now we can use Eqs. (2.5.27) and (2.5.28) to determine the input impedance of the interdigital transducer as a function of frequency. We will not do this in the most general way, but will instead determine only the real part of the input impedance of the interdigital transducer. From this result, we can determine the reactive part of the input impedance by using a Hilbert transform [15, 16]. The result is equivalent to the series in-line model of the circuit theory, which was described in Sec. 2.4.3. Following very similar procedures, a result equivalent to that of the crossed-field model can be obtained. The reader is referred to the paper by Auld and Kino [9] for further details of the full theory.

Suppose that we consider an interdigital structure with N finger pairs, each finger of length w, and width l_1; the periodic length is l and the transducer extends from $z = 0$ to L for a total length L. We can assume that the surface charge $\sigma(z)$ on each finger is uniform and of value Q/l_1. A schematic of this system is shown in Fig. 2.5.2.

We use Eq. (2.5.27) to replace α in Eqs. (2.4.3), (2.4.7), and (2.5.18) and write

$$\alpha = \frac{j\omega \phi_0^*}{4} \qquad (2.5.29)$$

From Eq. (2.4.7),

$$|A(L)| = \frac{\omega \phi_0^* Q}{4} \left| \frac{\sin (Nkl/2)}{\cos (kl/4)} \frac{\sin (kl_1/2)}{kl_1/2} \right| \qquad (2.5.30)$$

Similarly,

$$|B(0)| = \frac{\omega \phi_0^* Q}{4} \left| \frac{\sin (Nkl/2)}{\cos (kl/4)} \frac{\sin (kl_1/2)}{(kl_1/2)} \right| \qquad (2.5.31)$$

The total electrical power input to the transducer is

$$P = |A(L)|^2 + |B(0)|^2 \qquad (2.5.32)$$

The input current to the transducer is

$$I = j\omega NQ \qquad (2.5.33)$$

We have defined the electrical radiation resistance R_a of the transducer as

$$R_a = \frac{2P}{II^*} \qquad (2.5.34)$$

Substituting Eqs. (2.5.30)–(2.5.33) into Eq. (2.5.34) results in

$$R_a = \frac{Z_0}{2} \left(\frac{\sin (Nkl/2)}{N \cos (kl/4)} \frac{\sin (kl_1/2)}{(kl_1/2)} \right)^2 \qquad (2.5.35)$$

where we define the wave impedance Z_0 as the

$$Z_0 = \frac{\phi_0 \phi_0^*}{2} \qquad (2.5.36)$$

The normalized potential ϕ_0, and hence the wave impedance Z_0, can be evaluated directly from the exact field theory for a Rayleigh wave propagating on the unperturbed substrate.

We can also evaluate the impedance Z_0 another way. First we use the field theory to solve a simpler problem, determining the propagation constant of a surface acoustic wave on a piezoelectric substrate with a uniform perfect conductor deposited on it. Then we solve the same problem by perturbation theory and compare the two results.

As an example, we can begin by finding the change in velocity when an infinitesimally thin metal conductor is deposited on the piezoelectric substrate. This problem can be solved by field theory. In this case, the potential ϕ at the surface is forced to become zero, which changes the propagation constant from k to k' and the wave velocity from V to $V + \Delta V$. As we might expect, $|\Delta V/V|$ is directly proportional to the coupling coefficient α_0 or impedance Z_0, for if the electrical coupling is zero, an infinitesimally thin metal conductor cannot interact with the wave. On the other hand, if the value of $|\Delta V/V|$ is large, the potential at the surface for a given power flow is large, as is the interaction with charges at the surface. Thus we can work in terms of the coupling coefficient α, the relative change in velocity $\Delta V/V$, or the impedance of the wave at the surface defined as $Z_0 = \Phi_0^2/2P_0 = \phi_0^2/2$, where Φ_0 is the unnormalized potential at the surface for a wave traveling on the free piezoelectric substrate and P_0 is the total power in the wave.

Appendix D shows that

$$Z_0 = \frac{2}{\omega(\varepsilon + \varepsilon_0)w} \left|\frac{\Delta V}{V}\right| \tag{2.5.37}$$

where ε is the permittivity of the medium. Thus

$$R_a = \frac{1}{\omega(\varepsilon_0 + \varepsilon)w} \left(\frac{\sin (Nkl/2)}{N \cos (kl/4)} \frac{\sin (kl_1/2)}{kl_1/2}\right)^2 \left|\frac{\Delta V}{V}\right| \tag{2.5.38}$$

We now consider the frequency response of the transducer. We assume that it is operated near the synchronous frequency $\omega = \omega_0$, so that $kl \approx 2\pi$ and $kl_1 \approx 2\pi l_1/l$. Putting $x = N\pi(\omega - \omega_0)/\omega_0$ in Eq. (2.5.38), we find that if N is large, the input resistance R_a may be written in the approximate form

$$R_a \approx \frac{\omega_0}{\omega} R_{a0} \left(\frac{\sin x}{x}\right)^2 \tag{2.5.39}$$

where

$$R_{a0} = \frac{4}{\omega_0(\varepsilon_0 + \varepsilon)w} \left(\frac{\sin (\pi l_1/l)}{(\pi l_1/l)}\right)^2 \left|\frac{\Delta V}{V}\right| \tag{2.5.40}$$

When $l_1 = l/4$ (equal strip and gap widths), R_{a0} becomes

$$R_{a0} = \frac{32}{\pi^2 \omega_0 (\varepsilon_0 + \varepsilon)w} \left|\frac{\Delta V}{V}\right| \tag{2.5.41}$$

We may compare this result to Eq. (2.4.37), obtained from the circuit theory of Sec. 2.4.3. Equation (2.4.37) has a very similar frequency dependence and a value of R_{a0} corresponding to

$$R_{a0} = \frac{8}{\pi} \frac{F}{\omega_0 C_s} \left|\frac{\Delta V}{V}\right| \tag{2.5.42}$$

where C_s is the capacity per finger pair. From electrostatic theory, if $l_1 = l/4$ (equal strip width and spacing), then $C_s = w(\varepsilon_0 + \varepsilon)$. Equation (2.5.41) yields a numerical factor, $32/\pi^2 = 3.24$; the circuit theory for an interdigital transducer gives a factor $8F/\pi$, or $2.54F$.

A more accurate derivation of the perturbation theory, based on the true electrostatic distribution of the charge on the fingers, with $l_1 = l/4$ [27], leads to the result

$$R_{a0} = \frac{2.87}{\omega_0 C_s} \left|\frac{\Delta V}{V}\right| \tag{2.5.43}$$

We conclude that the parameter F must have a value of $2.87/2.54 = 1.12$.

One more change is needed to make this model conform to reality. We must take the permittivity of a piezoelectric material to be anisotropic. This is done easily; we need only replace ε by a parameter ε_p, defined by the relation

$$\varepsilon_p = (\varepsilon_{yy}\varepsilon_{zz} - \varepsilon_{yz}^2)^{1/2} \tag{2.5.44}$$

Careful comparisons have been made between this normal mode theory and the experimental results for R_{a0} taken on Y-cut lithium niobate with propagation in the z direction and with $\varepsilon_p = 50\varepsilon_0$, which corresponds to the stress-free values of the dielectric constant ε_{ij}^T. When the value of $|\Delta V/V|$ is taken to be 0.023, the experimental and theoretical results agree to within less than 1%.

Note that here we have neither calculated the transducer capacity directly nor shown its connection with the acoustic theory. In Appendix D, however, we derive the *electrostatic field*, which results from the charge. Similarly, when there is a charge on the electrodes of the transducer, we must determine the two potentials resulting from it: the acoustic potential ϕ_a, which we determine from the perturbation theory and the electrostatic potential. The electrostatic potential is, of course, associated with the capacity of the transducer; it gives an additional potential in series with the acoustic term and hence a capacity in series with the acoustic impedance. This capacity C_T is calculable by methods very similar to those given above, or by a direct solution of electrostatic theory, as

$$C_T = N\varepsilon_p w \tag{2.5.45}$$

where $l_1 = l/4$.

2.5.5 Leaky Waves

Now consider the effect on a surface acoustic wave when the surface of the substrate on which it propagates is loaded by another semi-infinite medium such as air or water. We assume that a wave can propagate in this medium with a phase velocity lower than that of the surface acoustic wave on the free substrate. If a surface acoustic wave with a propagation constant β is then excited on the substrate, we expect it to excite a quasi-plane wave in the liquid propagating at an angle θ to the normal, as illustrated in Fig. 2.5.3, with

$$k \sin \theta = \beta \tag{2.5.46}$$

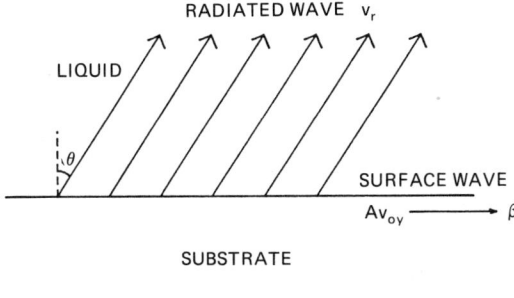

Figure 2.5.3 Leaky surface wave radiating into a liquid or into another medium that propagates a wave with a lower phase velocity than the surface wave.

or
$$V_R \sin \theta = V_L \tag{2.5.47}$$

where k is the propagation constant in the liquid. To avoid confusion, we have changed the notation for the propagation constant on the substrate. The wave velocity in the liquid is now V_L, and V_R is the Rayleigh wave phase velocity. Thus, under these conditions, power is continuously radiated into the liquid medium and the surface acoustic wave is attenuated so that its propagation constant changes from β to $\beta - j\alpha$ (i.e., the wave amplitude varies as $\exp[-(j\beta + \alpha)z]$). Such a wave is known as a *leaky wave*. In principle, it can exist as a simple outgoing wave only if the upper medium is semi-infinite. When the upper medium is finite in extent, it forms a type of waveguide and the radiated wave is reflected from the top surface of the medium. Thus energy is not radiated, although there may be several propagating modes in the liquid waveguide into which the energy is coupled, and hence several solutions for the perturbed value of β.

We assume that the perturbing medium is a liquid in which only longitudinal stress components can exist, and we take the substrate surface to be at $y = 0$. The presence of the liquid perturbs the boundary conditions at the substrate surfaces so that the normal component of stress at this surface is no longer zero. We assume the width of the acoustic beam to be w and the fields to be uniform in the x direction. In this case, Eq. (2.5.25) becomes

$$\frac{dA}{dz} + j\beta A = \frac{w}{4} v_{0y}^* \hat{T}_2 \tag{2.5.48}$$

where T_2 is the component of stress normal to the substrate and v_{0y} is the velocity normalized to unit power at the surface $y = 0$ of the unperturbed Rayleigh wave. The particle velocity in the y direction $\hat{v}_y$, associated with the Rayleigh wave, is defined as

$$\hat{v}_y = A v_{0y} \tag{2.5.49}$$

In this solution, we use the boundary condition on $\hat{v}_y$ but use Eq. (2.5.48) to account for the change in the boundary condition on T_2. We assume that the radiated wave has a particle velocity $\hat{v}$ in a direction θ to the substrate normal. The boundary conditions on continuity of v_y yield the relation

$$\hat{v} \cos \theta = A v_{0y} \tag{2.5.50}$$

The stress in a liquid is invariant with the angle. Thus

$$\hat{T}_2 = -Z_L \hat{v} = \frac{-Z_L A V_{0y}}{\cos \theta} \tag{2.5.51}$$

It follows from Eqs. (2.5.48)–(2.5.51) that

$$\frac{dA}{dz} + j\beta A = -\alpha A \tag{2.5.52}$$

where

$$\alpha = \frac{\omega v_{0y} v_{0y}^* Z_L}{4 \cos \theta} \quad (2.5.53)$$

The solution of Eq. (2.5.52) is

$$A(z) = A_0 e^{-(j\beta + \alpha)z} \quad (2.5.54)$$

where A_0 is the amplitude of the wave at $z = 0$. The value of v_{0y} can be found from the Rayleigh wave solution given in Sec. 2.3.4. This is the leaky wave we postulated.

After considerable algebra to evaluate the power flow per unit width of a surface acoustic wave, $P = -(w/2) \int (v_z T_3^* + v_y T_4^*) dy$, Auld [7] has shown that

$$v_{0y} v_{0y}^* = \frac{f_y \omega}{\rho_{m0} V_s^2 w} \quad (2.5.55)$$

where f_y is a dimensionless parameter of order unity, given by the relation

$$f_y = \left(\frac{V_s}{V_R}\right)^2 \frac{4\gamma^2 [1 - (V_R/V_s)^2]^{3/2}}{3\gamma - 2\gamma(V_R/V_s)^2 - 1} \quad (2.5.56)$$

The parameters V_s and V_l are the shear and longitudinal phase velocities in the substrate, respectively, and

$$\gamma^2 = \frac{1 - (V_R/V_l)^2}{1 - (V_R/V_s)^2} \quad (2.5.57)$$

We often need the parameter f_y, as well as another useful parameter, f_z, which is defined by the relation

$$v_{0z} v_{0z}^* = \frac{f_z \omega}{\rho_{m0} V_s^2 w} \quad (2.5.58)$$

where

$$f_z = f_z'/\gamma \quad (2.5.59)$$

Both these parameters, f_z and f_y, are plotted in Fig. 2.5.4 as a function of V_s/V_l.

We can therefore write the leakage rate α in the form

$$\alpha = \frac{Z_L \omega f_y}{4\rho_{m0} V_s^2 \cos \theta} = \frac{k_s}{4} \frac{Z_L}{Z_s} \frac{f_y}{\cos \theta} \quad (2.5.60)$$

where Z_s and k_s are the shear wave impedance and the propagation constant, respectively, in the substrate.

If $\alpha/\beta \ll 1$, the leaky wave attenuation due to radiation into a liquid is small when $Z_L \ll Z_s$ (i.e., if the impedance of the liquid is low compared to the shear wave impedance of the substrate). Even air, for which $Z_L/Z_s \ll 1$, can cause attenuation; this effect is noticeable in very long SAW delay lines and at very high frequencies because α is linearly proportional to ω.

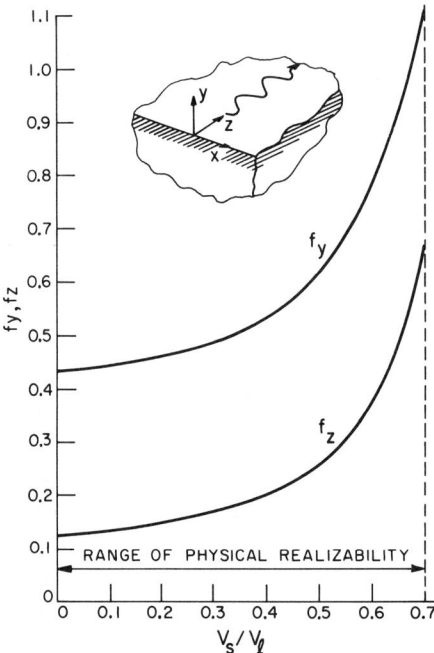

Figure 2.5.4 Field parameters f_y and f_z in Eqs. (2.5.56) and (2.6.58) as a function of the ratio of the bulk shear wave velocity V_s to the bulk longitudinal wave velocity V_l. (After Auld [7].)

This solution does not specifically depend on the fact that the substrate is isotropic. Provided that the parameter $v_{0y}v_{0y}^*$ can be calculated or measured for the substrate of interest, the leakage rate, or the attenuation α, can be determined from Eq. (2.5.53).

Example: Aluminum Substrate Loaded by Water

Consider an aluminum substrate with $V_s = 3.04$ km/s, $V_l = 6.42$ km/s, and $Z_s = 8.2 \times 10^6$ kg/m²-s, on which a surface acoustic wave leaks into water with $Z_s = 1.5 \times 10^6$ kg/m²-s and $V_L = 1.5$ km/s. We find that $Z_L/Z_s = 0.183$ and $V_s/V_l = 0.473$; thus Fig. 2.5.4 shows that $f_y = 0.58$ and Fig. 2.3.5 shows that $V_R/V_s = 0.94$ or $V_R = 2.86$ km/s. Hence $\theta = 31.6°$ and $\alpha = 0.029k_R$ or $\alpha\gamma_R = 0.18$. The distance for the fields to drop to $1/e$ of their value at $z = 0$ is therefore 5.6 wavelengths.

2.5.6 Wedge Transducer

If power can radiate from the substrate, the reverse process can also occur, with a wave incident from the liquid exciting a surface acoustic wave on the substrate. Thus if a wave is incident on the substrate at the angle θ defined by Eq. (2.5.47), as shown in Fig. 2.5.5(a), we expect part of it to be reflected directly and the rest to excite a surface acoustic wave. Furthermore, as the surface acoustic wave propagates along the substrate, it will reradiate a wave at the angle θ [17–24].

On the other hand, if the region of excitation is kept short enough that there is little reradiation, the excitation length can be optimized to convert most of the incident energy into a surface acoustic wave. Thus a bulk or volume wave can be excited in water and, if incident at the correct angle on the substrate, will excite a surface acoustic wave. Alternatively, a wedge-shaped solid material can be used, as shown in Fig. 2.5.5(b), to make an SAW transducer, known as a wedge transducer. These same principles are used to make prism couplers to optical waveguides [19].

We now consider a liquid wedge with the configuration shown in Fig. 2.5.5(b). We choose the incident longitudinal wave in the liquid medium to be at an angle θ to the normal, which excites Rayleigh waves cumulatively, as defined by Eq. (2.5.47). If the incident and reflected waves have particle velocities $\hat{v}_i$ and $\hat{v}_r$, respectively, the boundary conditions at the substrate are

$$(\hat{v}_i - \hat{v}_r) \cos \theta = -A v_{0y} \tag{2.5.61}$$

and

$$\hat{T}_2 = -Z_L(\hat{v}_i + \hat{v}_r) \tag{2.5.62}$$

where Z_L is the wave impedance of the wedge medium. Eliminating the reflected wave component $\hat{v}_r$ from Eqs. (2.5.61), (2.5.62), and (2.5.48) and assuming $T_4 = 0$ as before, we obtain the following differential equation for A:

$$\frac{dA}{dz} + j\beta A + \alpha A = -\frac{w Z_L \hat{v}_i v_{0y}^*}{2} \tag{2.5.63}$$

The parameter α corresponds to the leak rate or attenuation per unit length of the acoustic surface when the wedge material is present. From Eqs. (2.5.53) and (2.5.60), we find that

$$\alpha = \frac{w v_{0y} v_{0y}^* Z_L}{4 \cos \theta} = \frac{Z_L}{Z_s \cos \theta} \frac{k_s f_y}{4} \tag{2.5.64}$$

We consider a wedge transducer of length $l \cos \theta$, which emits a parallel beam

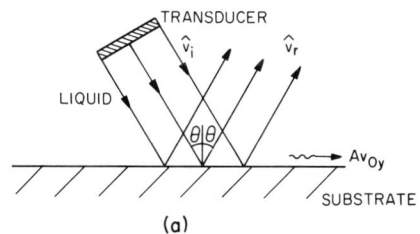

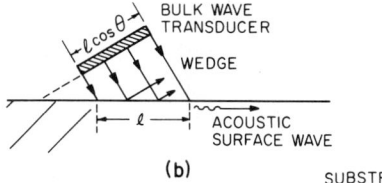

Figure 2.5.5 (a) Excitation of a surface wave from a liquid; (b) solid wedge transducer.

of length l in the z direction with a width w in the x direction. The total incident power on the substrate is

$$P_i(l) = \frac{1}{2} wl|v_i|^2 Z_L \cos\theta \qquad (2.5.65)$$

To remove the phase factors, we write

$$A(z) = A_0(z)e^{-j\beta z} \qquad (2.5.66)$$

and

$$\hat{v}_r(z) = V_r(z)e^{-j\beta z} = V_r(z)e^{-jk_L z \sin\theta} \qquad (2.5.67)$$

It is also convenient to define the power density per unit length in the incident beam in terms of an amplitude A_i. From Eq. (2.5.65), for a length z,

$$P_i(z) = |A_i^2|z \qquad (2.5.68)$$

where

$$\hat{v}_i(z) = A_i e^{-j\beta z} \frac{2}{wZ_L \cos\theta} = v_i e^{-j\beta z} \qquad (2.5.69)$$

From Eqs. (2.5.65)–(2.5.69), Eq. (2.5.63) can be written in the form

$$\frac{dA_0}{dz} + \alpha A_0 = -A_i(2\alpha)^{1/2} \qquad (2.5.70)$$

This equation can be integrated to determine $A_0(l)/A_i$. With the boundary condition $A_0(0) = 0$, the solution is

$$A_0(z) = -(1 - e^{-\alpha z}) A_i \left(\frac{2}{\alpha}\right)^{1/2} \qquad (2.5.71)$$

The surface wave power excited by a transducer of length l is $|A_o(l)|^2$. Thus, from Eqs. (2.5.68) and (2.5.71), the conversion efficiency η is defined as

$$\eta = \frac{P_o(l)}{P_i(l)} = \frac{2(1 - e^{-\alpha l})2}{\alpha l} \qquad (2.5.72)$$

The efficiency is plotted as a function of αl in the upper curve ($\gamma/\alpha = 0$) of Fig. 2.5.6. It has a maximum value of 0.815 at $\alpha l = 1.26$ and half-power points at $\alpha l = 0.27$ and 4.8; thus the tolerance on the choice of the optimum value of αl is relatively loose. Since α is proportional to frequency, this is also a plot of efficiency versus frequency.

Finite attenuation. This excitation theory has also been modified to account for finite attenuation in the wedge. Due to the magnitude of the input velocity attenuation, the excitation at the substrate decays as $\exp(-\gamma z)$, while the

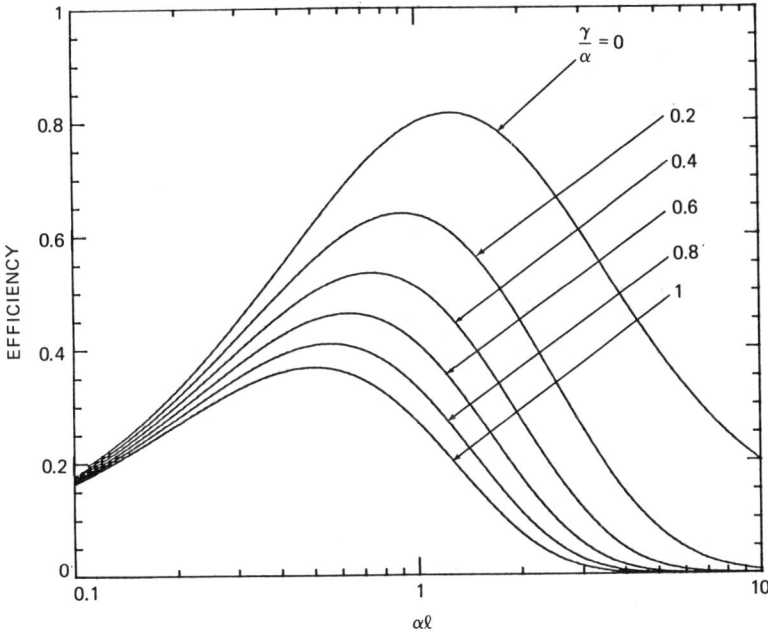

Figure 2.5.6 Wedge efficiency η versus normalized length or frequency αl for several values of wedge loss factor γ/α.

magnitude of the velocity at the transducer stays constant. The efficiency function η becomes

$$\eta = 2\alpha l e^{-2\gamma l} \left[\frac{1 - e^{-(\alpha-\gamma)l}}{(\alpha-\gamma)l} \right]^2 \quad (2.5.73)$$

Curves of η as a function of αl for different values of γ/α are plotted in Fig. 2.5.6.

Efficiency variation with incident angle or velocity. The variation of the efficiency can also be calculated as a function of the incident angle θ. If θ is not at its correct value θ_0, the incident wave will not excite a wave at the point z that is in phase with the wave traveling along the substrate that was excited at a point z'. Equation (2.5.72) can be modified to account for this effect and to solve the equations exactly. The physics of the problem lead us to expect that as the phase change in a length z is $\phi = k_L z \sin \theta$, the phase error $\Delta \phi$ due to a change in angle from the optimum value θ_0, where $k_L \sin \theta_0 = \beta$ to θ, is

$$\Delta \phi = (k_L \sin \theta - \beta) z \approx k_L z \cos \theta_0 \, \Delta \theta \quad (2.5.74)$$

where $\Delta \theta = \theta - \theta_0$. Thus, after a distance l, the phasor error $\Delta \phi$ is approximately

$$\Delta \phi \approx K \alpha l \quad (2.5.75)$$

where the velocity error is normalized and expressed by the parameter K as

$$K = \frac{k_L \sin \theta - \beta}{\alpha} \tag{2.5.76}$$

Beyond the point $\Delta\phi = \pi$, the induced signals will be out of phase with those induced earlier near $z = 0$. Thus when $\Delta\phi > \pi$, the efficiency will drop radically. Hence if $\alpha l \sim 1.3$, then $K < 4$ becomes the approximate condition for efficient power transfer.

A more complete treatment can be carried out by putting $\theta = \theta_0$ in Eqs. (2.5.61) and (2.5.62) but taking account of the phase change in the arguments of the exponentials. In this case, the efficiency is

$$\eta = \frac{2}{\alpha l} \frac{[e^{-\alpha l} - \cos(K\alpha l)]^2 + \sin^2(K\alpha l)}{1 + K^2} \tag{2.5.77}$$

This relation is plotted for different values of K in Fig. 2.5.7. These calculations show that, in agreement with the approximate physical arguments, the efficiency drops to half its synchronous value when $K \approx 3.5$. The optimum length for maximum energy transfer also decreases when the synchronous condition is not satisfied. If αl is kept at a value of 1.26, there is a 3-dB drop in output for $K \approx 2$. We observe that the 3-dB point occurs when the error in velocity along the substrate ΔV_R corresponds to

$$\frac{\Delta V_R}{V_R} = -\frac{\Delta\beta}{\beta} = \frac{K\alpha}{\beta} \tag{2.5.78}$$

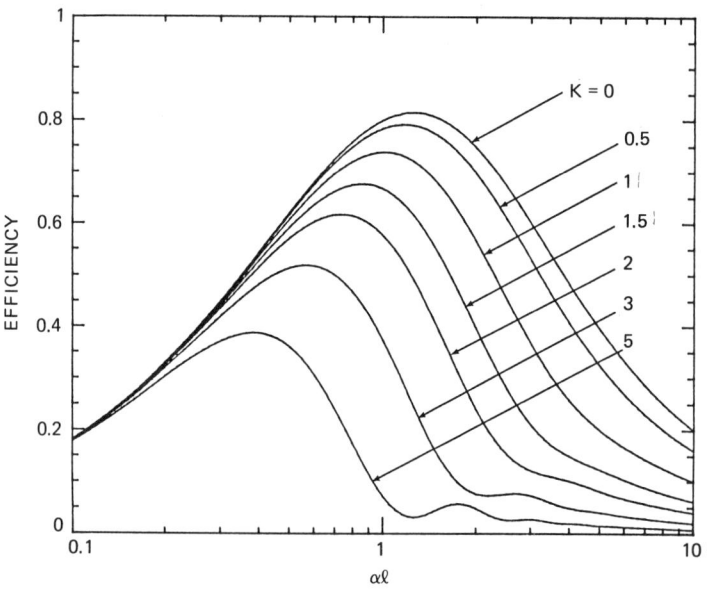

Figure 2.5.7 Plot of wedge transducer transduction efficiency versus αl for a different value of velocity mismatch parameter K, where $K = (k_L \sin \theta - \beta)/\alpha$.

Example: Excitation of Surface Acoustic Waves in Aluminum from Water

We can use the parameters already calculated in Sec. 2.5.5 for the example of leaky surface waves on aluminum in water. There we found that $\alpha = 0.18/\lambda_r$. Hence the optimum length l for transduction is 1.3/0.18, or 7.2 wavelengths. If we keep $\alpha l = 1.3$, then Fig. 2.5.7 indicates that for a K value of approximately 2, the efficiency drops by a factor of 3 dB. In turn, this implies an error in the incident angle corresponding approximately to $k_L \cos\theta \, \Delta\theta = K\alpha$ or $\Delta\theta = K\alpha\lambda_L/2\pi = K\alpha\lambda_R \sin\theta/2\pi$. This corresponds to an allowable error of $\pm 2°$ for a 3-dB drop in output or a velocity mismatch of $\pm 2\%$. The effect on the phase change of the reflected wave as the angle θ is varied may be considerably more critical, especially if $\alpha l \gg 1$.

Reflected wave. Finally, we consider the wave reflected from the substrate. Assuming that $\theta = \theta_0$, it can be shown from Eqs. (2.5.61), (2.5.67), and (2.5.69) that, for the optimum angle of incidence,

$$V_r(z) = V_i(2e^{-\alpha z} - 1) \tag{2.5.79}$$

We observe that $V_r(z)$ is positive and equal to V_i at $z = 0$. At the point where $\alpha z = \ln 2 = 0.69$, $V_r(z)$ becomes zero, reverses in sign, and increases to a magnitude V_i at large z. Thus, initially, there is a reflected beam; its amplitude then passes through zero and a second reflected beam generated by the leaky surface wave appears to be radiated from the region where $z > 0.69\alpha$. The radiated wave due to the surface acoustic wave excitation is π out of phase with the initial reflected wave and constant in amplitude for large z because the power lost is continually being made up by the excitation from the transducer.

The fact that the signal for large z suffers a π phase change is in marked contrast to the situation where no surface acoustic wave is excited. In this case there is no phase change. The implication is that if the transducer angle is changed by a small amount from the optimum, the surface acoustic wave excitation becomes weak and there is a π phase change in the reflected signal. Therefore, it is relatively easy to measure the proper angle for synchronism and, hence, the surface acoustic wave velocity of the substrate.

If the transducer excites the substrate only over a length l, the leaky wave will fall off in amplitude exponentially at points where $z \gg l$.

Example: Reflection of Waves from Aluminum in Water

It follows from Eq. (2.5.79) that the reflected beam drops in amplitude by 3 dB from its initial value at the point where $z = 0.16/\alpha$. The amplitude becomes zero at $z = 0.69/\alpha$, changes phase, and rises to within 3 dB of its maximum value where $z = \Delta = 1.92/\alpha$. For excitation of Rayleigh waves on aluminum ($\alpha\lambda_r = 0.187$), we conclude that the specularly reflected beam has its first 3-dB point at $z = 0.16/0.18\lambda_r = 0.9\lambda_r$ or $z = 1.7\lambda_L$, while the zero point is at $z = 5.6\lambda_r$ or $z = 11\lambda_L$, respectively, and the 3-djB point of the shifted beam is at $D = 10.7\lambda_r$ or $D = 20\lambda_wL$, respectively. This phenomenon was first predicted by Schock for Rayleigh waves, using a completely different method, and is known as the *Goos–Hänchen effect* in optics [20–23]. The initial theories predicted that there is a secondary beam emitted at a distance Δ along the substrate, whose value is very close to $1.92/\alpha$, as we have calculated [23].

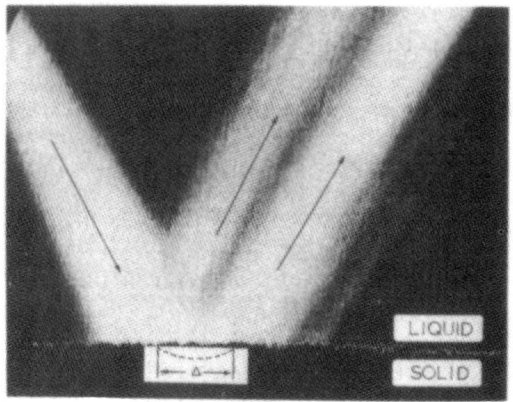

Figure 2.5.8 Schlieren photograph of an ultrasonic beam incident on a liquid–solid (aluminum) interface at the Rayleigh angle. The reflected beam, as indicated, is split into two components: a specularly reflected beam and a beam displaced a distance laterally down the interface. Secondary beams are visible at greater distances. (After Breazeale et al. [20].)

Experimental results. Wedge transducers have been made using a liquid medium and efficiencies up to 68% have been observed [24]. Thus for nondestructive testing or for SAW devices, it is obviously preferable to use a wedge of solid material and to excite and reradiate only over the optimum length l for which $\alpha l = 1.3$. The problem here is that the wedge transducer can generate both shear and longitudinal waves at the interface with the substrate, which lowers its efficiency. But choosing a wedge material that propagates only in a pure mode can circumvent this problem. One possibility is to use silicon rubber (RTV 615), which can propagate only longitudinal waves. At frequencies below 5 MHz, where this material is not too lossy, transducers with conversion efficiencies up to 35% have been constructed. At higher frequencies, polystyrene wedges have been employed to excite surface acoustic waves on ceramics with good efficiency; these are also often used on metals, because the coupling to the shear wave in the wedge is relatively weak and thus the shear wave is not strongly excited [24]. Several experiments have been carried out that clearly demonstrate the shift in the reflected beam excited by a beam of finite length. Excellent photos have been taken of this effect using Schlieren image techniques, as shown in Fig. 2.5.8.

Surface acoustic wave excitation from a liquid is important in materials testing because the change in amplitude of a surface acoustic wave as the angle of excitation by a bulk wave in water is varied, gives a very sensitive measure of the Rayleigh wave velocity. This basic phenomenon, discussed in Sec. 3.3.2, gives rise to the strong contrast effects of the scanned acoustic microscope, in which a lens in water is used to produce a highly convergent acoustic beam incident on a solid substrate. In turn, waves that are out of phase with the directly reflected waves are reradiated into the water. [25,26].

PROBLEM SET 2.5

1. (a) The power associated with the electric fields in a wave propagating through a piezoelectric material is

$$P_e = \tfrac{1}{2} \operatorname{Re} \int_s (\hat{\mathbf{E}} \times \hat{\mathbf{H}}^*) \cdot \mathbf{a}_z \, ds$$

Writing $\mathbf{E} \approx -\nabla\phi$ for a slow acoustic wave and using the vector relation

$$\nabla \times (\mathbf{A}\psi) = \psi\nabla \times \mathbf{A} + \mathbf{A} \times \nabla\psi$$

and Stoke's theorem, prove that

$$P_e = \tfrac{1}{2} \operatorname{Re} \int (j\omega \mathbf{D}\phi^*) \cdot \mathbf{a}_z \, ds$$

(b) Show that both the definitions of P_e above imply that $P_e = 0$ for a longitudinal plane acoustic wave traveling in the z direction through a piezoelectric medium poled in the z direction. Define $\mathbf{v}_0(x, y)$, $\mathbf{T}_0(x, y)$, $\phi_0(x, y)$, and $\mathbf{D}_0(x, y)$ for a forward wave.

2. A thin layer of water of thickness $h \ll \lambda$ is laid down on a surface wave substrate for which the Rayleigh wave velocity is V_r. Use the notation and parameters for a solid employed in Sec. 2.5 and take the density of the water to be ρ_w. Find the perturbed propagation constant of the surface wave when the substrate is mass loaded by the water layer.

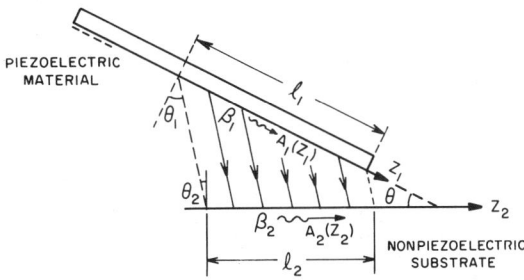

Note. The attenuation of the loaded surface wave will be zero. It is the real part of the propagation constant that is determined by the mass of the thin layer of water.

3. Acoustic waves can be excited in a conductor by electromagnetic $\mathbf{J} \times \mathbf{B}$ forces. Consider the electromagnetic acoustic transducer (EMAT) illustrated in the figure, which is used to excite Rayleigh waves. The magnet is used to excite a uniform β field parallel to the surface of the metal substrate. A meander line is placed very close to the substrate. A current I flows along the N elements of the transducer, which are spaced a distance $l/2 = \lambda/2$ apart, where λ is the Rayleigh wave wavelength. Assume that an identical current I is excited in the substrate just below each strip, and that the strips are each of length w and width l_1, spaced $l/2$ apart.

(a) By determining the total Rayleigh wave power excited, find the radiation resistance R_a (input resistance) of the meander line at the synchronous or center frequency, where $R_a = 2P_a/I^2$ and P_a is the acoustic power excited.

(b) Assume that the device is operated at a center frequency of 1 MHz, that $l_1 = l/4$, and that the wave is excited on aluminum. Suppose that the total resistance of the input circuit, including the meander line and surface resistance of the aluminum, is 1 Ω. Find the efficiency of excitation η by assuming that the inductance of the meander line is tuned out and the input is matched. In this case, the input power is $P_a = 1/2(R_a + R_0)I^2/2$, where R_a is the radiation resistance and R_0 is the resistance of the circuit (1 Ω). Thus the efficiency is

$$\eta = \frac{R_a}{2(R_a + R_0)}$$

where the factor of 2 compensates for the acoustic waves excited in each direction.

Note. The efficiency of this device is very low. Its advantage is that no contact has to be made between it and the substrate.

4. Consider the surface-to-surface (STS) wave transducer shown in the figure. A Rayleigh wave with a propagation constant $\beta_1 = \omega/V_{r1}$ is excited on a piezoelectric material. It radiates into water at an angle θ_1 to the normal. The piezoelectric substrate, which we will call substrate 1, is placed at an angle θ to a lower nonpiezoelectric substrate, which we will call substrate 2. The wave in the water between the two substrates excites a surface wave with a propagation constant $\beta_2 = \omega/V_{R2}$ on substrate 2. Determine the correct relations between the angles θ_1, θ_2, and θ, as shown in the figure, for optimum excitation. Assuming that the leak rates of waves from substrates 1 and 2 are α_1 and α_2, respectively, determine an expression for the power transferred in a length l_2 of the substrate 2. Consider the case for which $\alpha_1/\cos\theta_1 = \alpha_2/\cos\theta_2$ (you will need l'Hospital's rule). Find the optimum value of $\alpha_2 l_2$ for maximum power transfer and the corresponding theoretical maximum power transfer efficiency.

Hint. It is convenient to use coordinates z_1 and z_2 for the top and bottom substrates, respectively. Remember that $A_1(0)$ is defined and that you must find $A_2(l_2)$ in terms of $Aj_1(0)$ and determine when $|A_2(l_2)/A_1(0)|^2$ is maximum.

5. Work out a similar analysis to that of Prob. 4 for two parallel substrates of the same material of equal velocity and interaction length l which are held parallel to each other. Assume that the separating medium is thick enough that the attenuation of the wave passing through it is η_0, and large enough that the radiated wave from the lower substrate has no influence on the upper substrate. Determine the optimum coupling value of αl for maximum power transfer and the maximum power transfer efficiency for this configuration. Ignore the loss η_0 in your definition of the leak rate α for the two substrates.

REFERENCES

1. L. D. Landau and E. M. Lifshitz, *Theory of Elasticity*, Vol. 7 of *Course of Theoretical Physics*, J. B. Sykes and W. H. Reid, trans. Oxford: Pergamon Press Ltd., 1959.
2. Y. C. Fung, *Foundations of Solid Mechanics*. Englewood Cliffs, N.J.: Prentice-Hall, Inc., 1965.
3. W. M. Ewing, W. S. Jardetsky, and F. Press, *Elastic Waves in Layered Media*. New York: McGraw-Hill Book Company, 1957.
4. B. A. Auld, *Acoustic Fields and Waves in Solids*, Vol. I. New York: John Wiley & Sons, Inc., 1973.
5. J. F. Nye, *Physical Properties of Crystals: Their Representation by Tensors and Matrices*, corr. rpt. of 1st ed. (1957: rpt. Oxford: Clarendon Press, 1960).
6. R. C. McMaster, ed., *Nondestructive Testing Handbook*, 2 vols. New York: Roland Press, 1959.
7. B. A. Auld, *Acoustic Fields and Waves in Solids*, Vol. II. New York: John Wiley & Sons, Inc., 1973.
8. Lord Rayleigh, "On Waves Propagated Along the Plane Surfaces of an Elastic Solid," *Proc. London Math. Soc.*, 17 (1885), 4–11.
9. B. A. Auld and G. S. Kino, "Normal Mode Theory for Acoustic Waves and Its Application to the Interdigital Transducer," *IEEE Trans. Electron Devices*, ED-18, No. 10 (Oct. 1971), 898–908.

10. I. A. Viktorov, *Rayleigh and Lamb Waves: Physical Theory and Applications*. New York: Plenum Press, 1967.
11. W. R. Smith, H. M. Gerard, J. H. Collins, T. M. Reeder, and H. J. Shaw, "Analysis of Interdigital Surface Wave Transducers by Use of an Equivalent Circuit Model," *IEEE Trans. Microwave Theory Tech.*, MTT-17, No. 11 (Nov. 1969), 856–64.
12. K. A. Ingebrigtsen, "Surface Waves in Piezoelectrics," *J. Appl. Phys.*, 40, No. 7 (June 1969), 2681–86.
13. A. K. Ganguly and M. O. Vassell, "Frequency Response of Acoustic Surface Wave Filters," *J. Appl. Phys.*, 44, No. 3 (Mar. 1973), 1072–85.
14. G. S. Kino and J. Shaw, "Acoustic Surface Waves," *Sci. Am.*, 227, No. 4 (Oct. 1972), 50–68.
15. S. Ramo, J. R. Whinnery, and T. Van Duzer, *Fields and Waves in Communication Electronics*. New York: John Wiley & Sons, Inc., 1965.
16. R. N. Bracewell, *The Fourier Transform and Its Applications*, 2nd ed. New York: McGraw-Hill Book Company, 1978.
17. W. G. Neubauer, "Ultrasonic Reflection of a Bounded Beam at Rayleigh and Critical Angles for a Plane Liquid-Solid Interface," *J. Appl. Phys.*, 44, No. 1 (Jan. 1973), 48–55.
18. H. L. Bertoni and T. Tamir, "Characteristics of Wedge Transducers for Acoustic Surface Waves," *IEEE Trans. Sonics Ultrason.*, SU-22, No. 6 (Nov. 1975), 415–20.
19. P. K. Tien and R. Ulrich, "Theory of Prism-Film Coupler and Thin-Film Light Guides," *J. Opt. Soc. Am.*, 60, No. 10 (Oct. 1970), 1325–37.
20. M. A. Breazeale, L. Adler, and G. W. Scott, "Interaction of Ultrasonic Waves Incident at the Rayleigh Angle onto a Liquid-Solid Interface," *J. Appl. Phys.*, 48, No. 2 (Feb. 1977), 530–37.
21. A. Schoch, "Seitliche Versetzung eines total reflektierten Strahls bei Ultraschallwellen," *Acustica*, 2, No. 1 (1952), 18–19.
22. F. Goos and H. Hänchen, "Ein neuer und fundamentaler Versuch zur Totalreflexion," *Ann. Phys.* (Leipzig), 1, No. 6 (1947), 333–46.
23. L. M. Brekhovskikh, *Waves in Layered Media*, R. T. Beyer, trans., 2nd ed. New York: Academic Press, Inc., 1980.
24. J. Fraser, B. T. Khuri-Yakub, and G. S. Kino, "The Design of Efficient Broadband Wedge Transducers," *Appl. Phys. Lett.*, 32, No. 11 (June 1978), 698–700.
25. R. D. Weglein, "Metrology and Imaging in the Acoustic Microscope," in *Scanned Image Microscopy*, E. A. Ash, ed. London: Academic Press, Inc. (London) Ltd., 1980, pp. 127–36.
26. C. F. Quate, "Microwaves, Acoustics and Scanning Microscopy," in *Scanned Image Microscopy*, E. A. Ash, ed. London: Academic Press, Inc. (London) Ltd., 1980, pp. 23–55.
27. H. Engan, "Excitation of Elastic Surface Waves by Spatial Harmonics of Interdigital Transducers," *IEEE Trans. Electron Devices*, ED-16, No. 12 (Dec. 1969), 1014–17.

Chapter 3

Wave Propagation with Finite Exciting Sources

3.1 DIFFRACTION AND NONUNIFORM EXCITATION

3.1.1 Introduction

In Chapter 1 we described the one-dimensional theory of the piston transducer, on the assumption that the stress and strain fields within it are uniform. In practice, when the diameter of the transducer is finite, the stress fields may have several nodes and antinodes across the diameter, as would any waveguide or electromagnetic (EM) resonator. When the diameter of the transducer is larger than the wavelength and the Q of the system is not too high, these resonant effects tend to be washed out and the stress fields can be assumed to remain essentially uniform between the exciting electrodes.

Even if the excitation is perfectly uniform, however, the stress fields at some distance from the transducer may not be. Because of diffraction, the acoustic beam emitted from the transducer will increase its diameter with distance and the field components in the beam will exhibit fine structure variations, both along its length and across its diameter. Just as in optics, there are two distinct regions of interest, the near-field region, or the *Fresnel zone*, and the far-field region, or the *Fraunhofer zone*. Within the Fresnel zone, the outside diameter of the beam remains essentially uniform; the beam then spreads beyond this region. Within the Fresnel zone there are rapid variations of the stress fields within the beam, both along its axis and radially.

Throughout this chapter the analysis will be based, for simplicity, on the excitation of longitudinal waves in a liquid medium. In Sec. 3.1.2 we consider excitation of spherical waves in a liquid. In Sec. 3.1.3 we develop a Green's function and, from it, derive the Rayleigh Sommerfeld integral, which we will use extensively in our analyses throughout the rest of the chapter. Our results for excitation by a piston transducer in a liquid medium will be essentially the same as those we would expect for longitudinal or shear wave excitation in a solid medium [1].

In Sec. 3.2 we employ the Rayleigh–Sommerfeld integral to derive the essential features of diffraction from a plane piston transducer. Then we will use Hankel and Fourier transform methods to derive more extensive results for diffraction from a piston transducer and from its surface wave equivalent, the one-dimensional rectangular transducer.

In Sec. 3.3 we use the same techniques to analyze diffraction from a concave spherical focused transducer or lens focused system, and we will discuss excitation of and propagation of Gaussian beams. We will pay considerable attention to the concepts of transverse definition and range definition of a focused beam, to sidelobes, and to the problem of speckle, which occurs with coherent wave excitation. The example of a scanned acoustic microscope will be discussed in some detail.

In Sec. 3.4 we consider the effect of diffraction when short pulses are used to excite a transducer. In this case, much of the "ringing" phenomenon in the Fresnel zone tends to disappear.

In Sec. 3.5 we discuss concepts of acoustic imaging without the use of physical lenses. In Sec. 3.5.1 we first define the A-scan, B-scan, and C-scan with examples drawn from medical imaging and nondestructive testing; we also discuss the concepts of time-delay and phase-delay focusing.

In Sec. 3.5.2 we derive basic imaging theories for transducer arrays, using matched filter concepts. By employing the paraxial approximation, we determine sidelobe levels for various types of continuous array systems, and then show how grating lobes occur in a system with a finite number of periodically spaced transducer elements, and how sidelobe levels are affected by missing elements.

In Sec. 3.5.3 we consider the concept of the Fresnel lens and show how subsidiary foci occur. We show that the errors due to a finite number of phase samples in digitally sampled systems are very similar to those of Fresnel lenses: in both cases there are subsidiary foci, and the sidelobe levels are increased by the sampling errors.

In Sec. 3.5.4 we discuss chirp focused systems. We then consider examples of various types of transmission, imaging, reflection imaging, and two-dimensional array systems.

In Sec. 3.5.5 we discuss time-delay focused systems. We consider the use of lumped delay lines, charge-coupled-device (CCD) delay lines, and synthetic aperture digitally processed systems, and give a brief description of tomography, comparing this concept to synthetic aperture focusing.

In Sec. 3.5.6 we discuss the concepts of holographic imaging, giving examples of various types of holographic image reconstruction schemes. In addition, we describe the scanning laser acoustic microscope (SLAM), which uses the same type of technology. We also describe how the holographic methods can be very powerful for measuring the vibration amplitude of vibrating objects.

Finally, in Sec. 3.6 we use the analytical methods developed in Sec. 3.1 to analyze scattering from small and large objects. We deal with the scattering theory for objects whose size is much larger than the wavelength, as well as for objects at the opposite Rayleigh limit, whose size is much less than a wavelength. We also describe the use of quasistatic analysis and other approximate techniques based on ray-tracing concepts.

3.1.2 Spherical Waves in a Liquid or Solid

First we consider the excitation of waves by a small vibrating sphere. This derivation will show that for good excitation efficiency, the size of the sphere must be comparable to the wavelength. Furthermore, we need the basic results to derive the waves excited by various forms of transducers and to obtain a simple understanding of the scattering of waves from a small object, such as a flaw in a solid or a dust particle in water.

The displacement **u** associated with longitudinal waves in an isotropic medium can be derived from a potential ϕ, where $\mathbf{u} = \Delta\phi$ and ϕ obeys the wave equation

$$\nabla^2 \phi - \frac{1}{V^2} \frac{\partial^2 \phi}{\partial t^2} = 0 \qquad (3.1.1)$$

where $V = \sqrt{(\lambda + 2\mu)/\rho_{m0}}$ is the longitudinal wave velocity and λ and μ are the Lamé constants for an isotropic solid.

In spherical coordinates for a wave whose components vary as $\exp(j\omega t)$, this equation can be written as

$$\frac{1}{R^2} \frac{\partial}{\partial R} \left(R^2 \frac{\partial \phi}{\partial R} \right) + k^2 \phi = 0 \qquad (3.1.2)$$

where $k = \omega/V$ and we have assumed no variation of the potential with angle. The solutions of this equation are

$$\phi = \frac{A e^{\pm jkR}}{R} \qquad (3.1.3)$$

The $-$ sign corresponds to a spherical wave propagating outward from the origin and the $+$ sign corresponds to one propagating in toward it. More generally, the solution of Eq. (3.1.2) has the form

$$\phi = \frac{f(t \pm R/V)}{R} \qquad (3.1.4)$$

Vibrating sphere. Before dealing with the plane piston transducer, let us first consider the wave excited by a small spherical transducer of radius a. In this case, we use the relation $\mathbf{u} = \nabla\phi$ in Eq. (3.1.3) for a wave propagating radially outward to write

$$u_R = \frac{\partial \phi}{\partial R} = -\frac{A e^{-jkR}}{R^2} (1 + jkR) \qquad (3.1.5)$$

Thus if $u_R = u_R(a)$ at the spherical surface, then

$$\phi = -\frac{u_R(a)a^2}{R(1 + jka)} e^{-jk(R-a)} \tag{3.1.6}$$

Liquid medium. In a liquid medium, the pressure p is isotropic. Thus we can write $p = -T_1 = -T_2 = -T_3$ or $p = -T_R$. It follows that

$$-\nabla \cdot \mathbf{T} = \nabla p = \omega^2 \rho_{m0} \mathbf{u} = \omega^2 \rho_{m0} \nabla \phi \tag{3.1.7}$$

On integration, we see that, generally, for a liquid,

$$p = \omega^2 \rho_{m0} \phi \tag{3.1.8}$$

with

$$p = -\kappa \Delta = -c_{11} S_1 \tag{3.1.9}$$

where $\kappa = (\lambda + \frac{2}{3}\mu)$. As discussed in Sec. 2.1, the parameter κ is called the bulk elastic modulus and the dilation is $\Delta = S_1 + S_2 + S_3$. We note that

$$k^2 = \frac{\omega^2 \rho_{m0}}{c_{11}} \tag{3.1.10}$$

For a liquid, it follows that

$$k^2 = \frac{\omega^2 \rho_{m0}}{3\kappa} \tag{3.1.11}$$

If $v_R = j\omega u_R$ is the radial velocity, the effective wave impedance Z for a spherical wave traveling outward in an anisotropic solid or a liquid is

$$Z = -\frac{T_R}{v_R} = \frac{p}{v_R} = \frac{j\omega \rho_{m0} R}{1 + jkR}$$
$$= \frac{k^2 R^2 + jkR}{1 + k^2 R^2} Z_0 \tag{3.1.12}$$

where $Z_0 = \rho_{m0}\omega/k$ is the plane wave impedance of the medium.

We note that far from the origin, where $kR \gg 1$, the wave impedance approaches Z_0. However, for a small source near the origin ($kR \ll 1$), the wave impedance becomes imaginary, and there is very little real power flow. Because there is very little real power flow near the source, it is difficult to excite acoustic waves from a source much smaller than a wavelength in diameter.

To put it another way, suppose that the surface of a sphere of radius a is pulsating so that its surface acceleration is $\partial^2 u_R/\partial t^2$ or $-\omega^2 u_R(a)$. It follows from Eq. (3.1.5) that the radial displacement $u_R(R)$ is

$$u_R(R) = \frac{a^2 u_R(a) e^{-jk(R-a)}}{R^2} \frac{1 + jkR}{1 + jka} \tag{3.1.13}$$

The real power Re (P) radiated is therefore

$$\text{Re}(P) = 2\pi R^2 \text{Re}(Z) v_R v_R^* = 2\pi R^2 \text{Re}(Z) \omega^2 |u_R^2(R)| \tag{3.1.14}$$

It follows, by substituting Eqs. (3.1.12) and (3.1.13) in (3.1.14), that the complex power P, defined as $P = 2\pi R^2 Z v_R v_R$, is

$$P = 2\pi a^2 Z_0 \frac{k^2 a^2 + jka^2/R}{1 + k^2 a^2} |v_R^2(a)| \qquad (3.1.15)$$

This is the final formula required for determining radiation from a sphere. We conclude that:

1. When $kR \gg 1$, then $u_R \approx -jk\phi$, so that $|u_R| \propto |\phi|$ at distances more than a few wavelengths from the transducer.
2. Real power is conserved and does not change with radius. The real power emitted is

$$\text{Re}(P) = \frac{k^2 a^2}{1 + k^2 a^2} 2\pi a^2 Z_0 |v_R^2(a)| \qquad (3.1.16)$$

   and the power intensity $I(R) = \text{Re}(P)/4\pi R^2$ falls off as $1/R^2$.
3. The real power emitted for a given $v_R(a)$ does not vary with R even if $ka \gg 1$. However, if $ka \ll 1$, the power intensity at the transducer for a given $v_R(a)$ varies as $k^2 a^2$. Thus it becomes very small as $ka \to 0$.
4. As R changes, reactive power is not conserved and becomes dominant at the spherical transducer if $ka \ll 1$. When $ka \ll 1$, the pressure at the transducer is essentially $\pi/2$ out of phase with the velocity and very small, and the effective impedance is inductive of value $jkRZ_0$. Thus, in this case, there is no real load on the transducer.

These conclusions also hold for the plane transducer and other transducer shapes. The transducer size (in this case its diameter) must be of the order of half a wavelength or more to obtain efficient excitation of acoustic waves.

3.1.3 Green's Function

Here we show how to derive a Green's function, which we use in the analysis of diffraction of the waves excited by transducers. We derive this function by adding spherical wave solutions to give the potential at any point due to waves excited by, for instance, a piston transducer. The boundary conditions we will apply are most conveniently stated for pressure waves in a liquid, but the results we obtain are a good approximation to the truth for waves in isotropic solids, provided that the transducer is several wavelengths in diameter [1,2]. For very small transducers or for excitation of waves in a solid, other techniques are needed. The angular spectrum methods described in Secs. 3.2.2 and 3.2.3 are more rigorous, while still convenient to employ. In practice they yield essentially the same results as the Green's function approach given here.

The spherically symmetric solution of the wave equation of Sec. 3.1.2 implies that the potential at the point x', y', z' of a longitudinal wave due to a source at a point x, y, z can be written in the form

$$G = \frac{Ae^{-jkR}}{R} \quad (3.1.17)$$

where

$$R = (x - x')^2 + (y - y')^2 + (z - z')^2 \quad (3.1.18)$$

It is easy to show by differentiation that this solution obeys the relation $\nabla^2\phi + k^2\phi = 0$ ($G = \phi$), except where $R=0$ when $\nabla^2\phi \to \infty$. More generally, Eq. (3.1.17) is a solution of the wave equation for a source point at x, y, z, with

$$\nabla^2 G + k^2 G = \delta(x' - x, y' - y, z' - z) \quad (3.1.19)$$

where x', y', and z' are now the independent variables, so all differentiations are with respect to the variables x', y', and z'. The function $\delta(x' - x, y' - y, z' - z)$ is the three-dimensional Dirac delta function, which satisfies the equation

$$\int_{-\infty}^{\infty}\int_{-\infty}^{\infty}\int_{-\infty}^{\infty} \delta(x' - x, y' - y, z' - z) \, dx' \, dy' \, dz' = 1 \quad (3.1.20)$$

with $\delta(x' - x, y' - y, z' - z) \to \infty$ as $R \to 0$.

Equation (3.1.17) satisfies the boundary conditions at infinity and corresponds to a wave propagating outward from the point x, y, z. The function G is called the *free-space Green's function*. To find the constant A, we integrate Eq. (3.1.19) over a small sphere of radius a, enclosing the point x, y, z. Thus

$$\int_{a \to 0} (\nabla^2 G + k^2 G) \, dV = 1 \quad (3.1.21)$$

Using Gauss's theorem, it follows that the first term in the integrand becomes

$$\int_{S_a} \nabla G \cdot \mathbf{n} \, ds = -4\pi A \quad (3.1.22)$$

where s_a is the surface of the sphere. The second term in the integrand becomes

$$k^2 \int_{a \to 0} G \, dV = 4\pi k^2 A \int_{a \to 0} \frac{1}{R} R^2 \, dR = 0 \quad (3.1.23)$$

It is apparent, therefore, that as $a \to 0$, $A \to -1/4\pi$. Thus the free-space Green's function is

$$G = \frac{-e^{-jkR}}{4\pi R} \quad (3.1.24)$$

Helmholtz's theorem. We want to find a solution for the potential at a point due to excitation by a transducer. We shall show that this is mathematically equivalent to finding the general solution for ϕ at any point, given ϕ and $\nabla\phi$ on a surface surrounding the point x, y, z. To do this, we consider the solutions of the wave equation

$$\nabla^2 \phi + k^2 \phi = 0 \tag{3.1.25}$$

We now use Green's theorem: We multiply Eq. (3.1.25) by G and subtract the result from the product of Eq. (3.1.19) with ϕ. We integrate the resulting expression over an arbitrary volume V, using the relation $\int \phi(x', y', z')\delta(x' - x, y' - y, z' - z) \, dV' = \phi(x, y, z)$, to obtain the formula

$$\phi(x, y, z) = \int_{V'} (\phi \nabla'^2 G - G \nabla'^2 \phi) \, dV'$$

$$= \int_{V'} \nabla' \cdot (\phi \nabla' G - G \nabla' \phi) \, dV' \tag{3.1.26}$$

where now we have used the $'$ superscript to make it clear that we are differentiating with respect to the variables x', y', and z'. By applying Gauss's theorem to the integral of Eq. (3.1.26), we see that

$$\phi(x, y, z) = \int_{s'} (\phi \nabla' G - G \nabla' \phi) \cdot \mathbf{n} \, ds' \tag{3.1.27}$$

where s' is any surface enclosing the point x, y, z and $\mathbf{n}$ is the outward normal from the volume V. Substituting the solution we have already found for G [Eq. (3.1.24)] in Eq. (3.1.27), we see that we can determine ϕ at any point if we know ϕ and $\nabla \phi$ on the surrounding surface. This result is known as *Helmholtz's integral theorem*.

Sommerfeld radiation condition. Now consider a piston transducer of area S_1 in a baffle of area S_2, as illustrated in Fig. 3.1.1. Let the rest of the enclosing surface of the volume considered be called Σ and let $R \to \infty$ as $z \to \infty$.

When $R \to \infty$, then $\phi \to (\exp -jkR)/R$, as does G. In this case, the contribution to the integral of Eq. (3.1.27) becomes zero on Σ, because the two terms in the integrand cancel out. Therefore, we can ignore contributions to the integral from the surface at infinity; this assumption is known as the *Sommerfeld radiation condition*.

Kirchhoff formula. We assume that both ϕ and $\nabla \phi \cdot \mathbf{n}$ are zero on the baffle. This is equivalent to assuming that in a liquid, both the pressure and normal displacement are zero on the baffle. Under these conditions, we can write Eq. (3.1.27) in the form

$$\phi(x, y, z) = \int_{S_1} (\phi \nabla' G - G \nabla' \phi) \cdot \mathbf{n} \, ds' \tag{3.1.28}$$

where $\mathbf{n}$ is the normal pointing *into* the transducer and S_1 is the area of the transducer. Therefore, if both ϕ and $\nabla \phi \cdot \mathbf{n}$ are specified at the transducer, then $\phi(x, y, z)$ can be found by using the value of G given in Eq. (3.1.24).

The Kirchhoff formula of Eq. (3.1.28) is useful and often employed in prac-

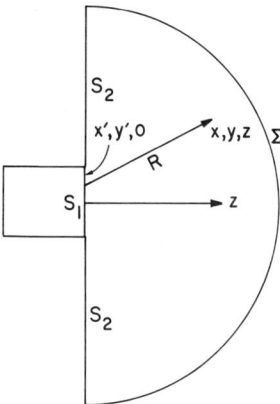

Figure 3.1.1 Piston transducer enclosed in a baffle.

tice; it is not rigorously correct, however, because it cannot be self-consistent, since ϕ and $\nabla\phi \cdot \mathbf{n}$ cannot be specified independently everywhere on the enclosing surface. For example, if $\phi = 0$ on a flat baffle at $z = 0$, and ϕ is finite on the transducer, then $\nabla\phi \cdot \mathbf{n}$ must be large near the edge of the transducer (i.e., there must be a fringing field). Mathematically, ϕ and $\nabla\phi \cdot \mathbf{n}$ cannot both be specified on all enclosing surfaces; this is an overspecification.

It is not even rigorously correct to specify ϕ and $\nabla\phi \cdot \mathbf{n}$ on the transducer. However, for transducers whose cross-sectional dimensions are several wavelengths, it is usually reasonable to employ the simple plane wave condition

$$\frac{\partial \phi}{\partial z} = -\frac{\partial \phi}{\partial z'} \approx -jk\phi \qquad (3.1.29)$$

at the transducer, and take $\phi = \partial\phi/\partial z' = 0$ on S_2. We can then specify both ϕ and $\nabla\phi \cdot \mathbf{n}$ on S_1.

We expect these approximations to be valid if the cross-sectional dimensions of the transducer are more than several wavelengths; then the boundary conditions at the baffle are not very important and the use of Eq. (3.1.29) gives results close to the truth. The boundary conditions at the baffle become important only when the transducer is relatively narrow (e.g., less than a wavelength in extent). This happens, for instance, with the small elements of a transducer array used in acoustic imaging devices (see Sec. 3.5).

Rayleigh–Sommerfeld formula. Because we cannot specify both ϕ and $\partial\phi/\partial z'$, we observe that there are certain internal inconsistencies in the direct use of the Kirchhoff formula of Eq. (3.1.28). It is better, if we can, to use a formula in which only the potential or its gradient is specified on the boundary. To do this, we choose a different Green's function in Eq. (3.1.27), such that $G = 0$ or $\partial G/\partial z = 0$ on the transducer and its baffle. Then we need specify only $\partial\phi/\partial n$ or ϕ at the transducer.

Possible Green's functions for the fields at $z \leq 0$, which satisfy one of the required boundary conditions at the plane $z = 0$, are those due to a source

$\delta(x' - x, y' - y, z' - z)$ and its image $\delta(x' - x, y' - y, z' + z)$ in the plane $z = 0$. These Green's functions must have one of the following forms:

$$G = -\frac{1}{4\pi}\left(\frac{e^{-jkR_1}}{R_1} \pm \frac{e^{-jkR_2}}{R_2}\right) \quad (3.1.30)$$

where

$$R_1 = \sqrt{(z - z')^2 + (x - x')^2 + (y - y')^2} \quad (3.1.31)$$

and

$$R_2 = \sqrt{(z + z')^2 + (x - x')^2 + (y - y')^2} \quad (3.1.32)$$

We observe that the Green's function with the positive sign in Eq. (3.1.30) obeys the boundary condition $\partial G/\partial z' = 0$ at $z' = 0$, while that with the negative sign obeys the boundary condition $G = 0$ at $z' = 0$.

Rigid baffle. The first choice of Green's function, with a positive sign, in Eq. (3.1.30) yields a solution for the potential at any point $(z \geq 0)$ in the form

$$\phi(x, y, z) = -\frac{1}{2\pi}\int_s u_z(x', y', 0)\frac{e^{-jkR}}{R}ds' \quad (3.1.33)$$

where the integral is taken over the area of the transducer and it is assumed that $u_z = 0$ outside the transducer (i.e., on the baffle). Thus this choice of Green's function is rigorously correct for a flat piston transducer surrounded by a rigid baffle.

Pressure release baffle. The second choice of Green's function, with a negative sign, in Eq. (3.1.30) yields a solution for the potential at any point $(z \geq 0)$ in the form

$$\phi(x, y, z) = \frac{jk}{2\pi}\int_s \phi(x', y', 0)\frac{e^{-jkR}(1 + 1/jkR)\cos\theta}{R}ds' \quad (3.1.34)$$

where θ is the angle between the radius vector **R** and the z axis.

Normally, we are interested in regions several wavelengths from the transducer. In this case we can assume that $kR \gg 1$ and write Eq. (3.1.34) in the Huygens–Fresnel form:

$$\phi(x, y, z) = \frac{jk}{2\pi}\int_s \phi(x', y', 0)\frac{e^{-jkR}}{R}\cos\theta\, ds' \quad (3.1.35)$$

Note that the choice of Green's function used in Eqs. (3.1.34) and (3.1.35) implies that it is convenient to assume that $\phi(x', y') = 0$ on the baffle. Thus for waves excited in a liquid, Eq. (3.1.34) or (3.1.35) apply to a transducer for which the pressure is zero on the $z = 0$ plane outside the transducer (i.e., the *pressure release baffle*). The results of Eq. (3.1.34) or (3.1.35) are not exact for waves in a solid unless it is assumed that $\phi = 0$ on the plane outside the transducer. This zero potential assumption is equivalent to assuming, for a liquid, that normal stress is zero on the plane $z = 0$ outside the transducer. But zero longitudinal wave potential does not imply zero stress in a solid, although this is probably a fairly reasonable approximation at the baffle for most practical cases.

Transient source. The similar expressions for a source $u_z(x', y', 0)$ or $\phi(x', y', 0)$, which vary arbitrarily with time, can be found by taking the Fourier or Laplace transform of either Eq. (3.1.33) or Eq. (3.1.35). The result for a transducer in a rigid baffle is

$$\phi(t, x, y, z) = -\frac{1}{2\pi} \int_s \frac{1}{R} u_z \left(t - \frac{R}{V}\right) ds' \qquad (3.1.36)$$

where u_z is given at the point $x', y', 0$.

It follows that if $u_z(x', y', 0)$ or $\phi(x', y', 0)$ is known at a transducer surface and, correspondingly, $u_z(x', y', 0)$ or $\phi(x', y', 0)$ is zero on the surrounding baffle, it is possible to find the potential at any point x, y, z.

PROBLEM SET 3.1

1. Writing the impedance Z for a spherical wave in the form $Z = R + jX$, plot R/Z_0 and X/Z_0 as a function of kR from $kR = 0$ to 10. At what values of R/λ is $|Z|$ within 10% and 1% of Z_0? You will see from your results that reactive impedance effects are important only for very small transducers or very near the edge of large transducers.

2. Use Eq. (3.1.33) to work out general integral formulas for the values of u_z, u_x, and u_y excited by a planar transducer that is surrounded by a rigid baffle.

3. Assuming that a transducer is infinite in extent in the y direction and that the fields are uniform with y, work out from Eq. (3.1.33) a general formula for the potential at x, z when $u_z(x', 0)$ is known. You will find it convenient to write $R = [r^2 + (y - y')^2]^{1/2}$ with $r = [z^2 + (x' - x)^2]^{1/2}$ and $t = (y' - y)/r$. You will need the relation

$$H_0^{(2)}(\alpha) = \frac{2j}{\pi} \int_0^\infty \frac{e^{-j\alpha(1+t^2)^{1/2}}}{(1+t^2)^{1/2}} dt$$

where $H_0^{(2)}(\alpha)$ is a zeroth-order Hankel function of the second kind. The asymptotic form of this Bessel function ($\alpha \gg 1$) is

$$H_0^{(2)}(\alpha) \sim j \frac{1-j}{\sqrt{\pi\alpha}} e^{-j\alpha}$$

Find the form of $\phi(x, z)$ for arbitrary values of kr and for the case, normally of most practical interest, when $kr \gg 1$.

Answer:

$$\phi(r) = \frac{j-1}{2\sqrt{\pi}} \int u_z(x', 0) \frac{e^{-jkr}}{\sqrt{kr}} dx'$$

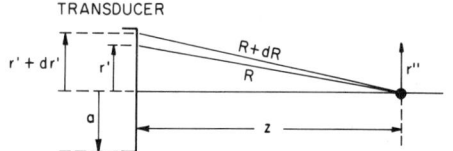

Figure 3.2.1 Notation used in the diffraction theory for the plane piston transducer.

3.2 PLANE PISTON TRANSDUCERS

3.2.1 Fields on the Axis

We will use the results of Sec. 3.1.2 to derive the field on the axis of a cylindrical plane piston transducer. We assume that the displacement at the face of the transducer $u_z(r, 0) = u_0$ to be uniform over its radius and zero outside it. We consider the piston transducer of radius a to be radiating into a liquid. Our results will also be valid for radiation into an isotropic solid if we use the same boundary conditions at $z = 0$ on the potential gradient $\partial \phi / \partial z$. The excitation at a distance R from a ring of radius r' to $r' + dr'$ on the transducer, as illustrated in Fig. 3.2.1, is given by Eq. (3.1.33). Thus the amplitude of a wave at point z on the axis is

$$\phi(0, z) = -u_0 \int_{r'=0}^{a} \frac{e^{-jkR}}{R} r' \, dr' \tag{3.2.1}$$

The assumption that $u_z = 0$ outside the radius a of the transducer is equivalent to assuming that the transducer is placed in a rigid baffle.

Using the relations $R^2 = r'^2 + z^2$ and $R \, dR = r' \, dr'$, we find that

$$\begin{aligned} \phi(0, z) &= -u_0 \int_{r'=0}^{a} e^{-jkR} \, dR \\ &= \left. \frac{u_0 e^{-jkR}}{jk} \right|_{r'=0}^{a} \end{aligned} \tag{3.2.2}$$

Thus

$$\begin{aligned} \phi(0, z) &= \frac{u_0}{jk} \left[e^{-jk(a^2+z^2)^{1/2}} - e^{-jkz} \right] \\ &= \frac{-2}{k} u_0 \sin \frac{k(\sqrt{a^2 + z^2} - z)}{2} e^{-jk(\sqrt{a^2 + z^2} + z)/2} \end{aligned} \tag{3.2.3}$$

By differentiating this equation with respect to z, we can show that

$$u_z(0, z) = u_0 e^{-jkz} - \frac{z}{(a^2 + z^2)^{1/2}} e^{-jk(a^2+z^2)^{1/2}} \tag{3.2.4}$$

This formula reduces to $u_z = u_0$ at $z = 0$, while the magnitude of $\phi(0, z)$ on the axis is $(2/k) u_z(0) |\sin(ka/2)|$. As we shall see later, $u_z(r, z)$ and $\phi(r, z)$ vary rapidly with radius near the transducer.

Fresnel or paraxial approximation. If $z^2 \gg a^2$, we can use the *Fresnel approximation* or *paraxial approximation* and expand the square roots in the phase-varying terms of Eq. (3.2.4) to first order in a^2/z^2, while keeping only the zeroth-order terms in a^2/z^2 in the amplitude. In this case, we use Eq. (3.1.8) to write Eq. (3.2.4) in the approximate form.†

$$\frac{-T_3(0, z)}{z_0} \approx v_z(0, z) \approx -v_0 e^{-jk[z + (a^2/2z)]} - e^{-jkz}$$ (3.2.5)

$$= 2jv_0 e^{-jkz} \sin \frac{\pi a^2}{2z\lambda} e^{-j\pi a^2/2z\lambda}$$

where $k = 2\pi/\lambda$, $T_3(0, z) \approx c_{11} \partial u_z(0, z)/\partial z$, $v_0 = j\omega u_0$, $v_z = j\omega u_z$ is the particle velocity, $T_3(0, z)$ is the stress on axis in the z direction, and Z_O is the impedance per unit area of the medium.

We assume that the average power intensity is $I(0)$ at the surface of the transducer and that $I(0) \approx Z_0|v_0^2|/2$. The power intensity at the axis, $I(z) = Z_0|V_z^2|/2$, can then be written in the form

$$I(z) = 4I(0) \sin^2 \frac{\pi a^2}{2\lambda z}$$
$$= 4I(0) \sin^2 \frac{\pi}{2S}$$ (3.2.6)

where we define a normalized parameter, *the Fresnel parameter*, by the relation

$$S = \frac{z\lambda}{a^2}$$ (3.2.7)

Fraunhofer and Fresnel zone. The region for which $z \gg a^2/\lambda$ or $S \gg 1$ is called the far-field region or the *Fraunhofer zone*. When $S \gg 1$, it follows from Eq. (3.2.6) that the power intensity at the axis, $I(z)$, is

$$I(z) \approx \frac{\pi a^2}{\lambda z^2} I(0) = \frac{\pi}{S^2} I(0)$$ (3.2.8)

In the Fraunhofer zone, where $S \gg 1$, $I(z)$ decreases as $1/z^2$.

The region for which $z < a^2/\lambda$ or $S < 1$ is called the *Fresnel zone*. In this case, the signal potential ϕ is maximum on the axis where the argument of the sine term in Eq. (3.2.6) or, more exactly, the argument of the sine term in Eq. (3.2.3), is $(2m + 1)\pi/2$. From Eq. (3.2.3), it follows that the value of z at the point where the potential is maximum is

$$z = z_{max} = \frac{a^2}{(2m + 1)\lambda} - \frac{(2m + 1)\lambda}{4}$$
$$\approx \frac{a^2}{(2m + 1)\lambda} \quad [ka \gg (2m + 1)\pi]$$ (3.2.9)

† Note that this approximation is valid only for $ka \gg 1$. An example illustrating this point is given in Eq. (3.2.9) and the discussion following it.

The approximate form of the second line of Eq. (3.2.9) is equivalent to using Eq. (3.2.5) or Eq. (3.2.6). So the Fresnel approximation tends to be valid for very large values of ka (i.e., when the transducer diameter is many wavelengths).

The Fresnel approximation form of Eq. (3.2.9) can be written in terms of the Fresnel parameter S, as follows:

$$S = S_{max} = \frac{1}{2m + 1} \qquad (3.2.10)$$

At the point where $z = z_{max}$, the amplitude $|v_z(z)/v_z(0)| = 2$ (i.e., v_z is double its average value at the transducer). The velocity v_z is zero at z_{min}, where

$$z = z_{min} = \frac{a^2}{2m\lambda} \qquad (ka \gg 2m\pi) \qquad (3.2.11)$$

or

$$S = S_{min} = \frac{1}{2m} \qquad (3.2.12)$$

Note that the axial fields vary rapidly with distance for small z. This is because $kR = k(r^2 + z^2)^{1/2}$ varies rapidly with radius when z is small, for the contributions from each ring element of the transducer alternately add and subtract. For large R, however, where

$$R = z\left(\frac{1 + r^2}{z^2}\right)^{1/2} \approx \frac{z + r^2}{2z} \qquad (3.2.13)$$

there is only a small variation of kR with r, so that all the contributions from the disk add. This occurs when $a^2/2z < \lambda/2$ (i.e., there is less than a half-wavelength difference in length between the rays) or $S > 1$.

The condition $z = a^2/\lambda$ or $S = 1$ corresponds exactly to the condition $m = 0$ (the last maximum) in Eq. (3.2.10). We call this point, $S = 1$, the crossover point between the Fresnel and Fraunhofer zones, or the *Fresnel limit*. Within the Fresnel zone, we expect to see the displacement vary rapidly with radius. Outside the radius a, however, the fields fall rapidly to zero (i.e., most of the energy is contained within the radius a). Thus if $z < a^2/\lambda$ or $S < 1$, the beam is essentially confined within its original diameter. Now suppose that we use another transducer of the same size to detect the transmitted signal. In this case the rapid variations of the field over the diameter of the receiving transducer wash out and the signal received by the receiving transducer varies very little within the Fresnel zone (see Fig. 3.2.7). This is equally true when one transducer is used as a receiver and transmitter of a perfectly reflected signal, provided that the total path length of the beam is less than a^2/λ.

Beyond the region $z = a^2/\lambda$ or $S = 1$, Eq. (3.2.8) shows that the intensity falls off monotonically as $1/z^2$ and, on axis, drops to the value at the transducer only when $z = \pi a^2/\lambda$ or $S = \pi$. A plot of $|u_z(0, z)|$ is given as a function of $S = z\lambda/a^2$ in Fig. 3.2.2 for a value of $ka = 100$, which is equivalent to $a/\lambda \approx 16$ or $z/\lambda \approx 16$ at $S = 1$. This plot is obtained from the exact solution of Eq. (3.2.2),

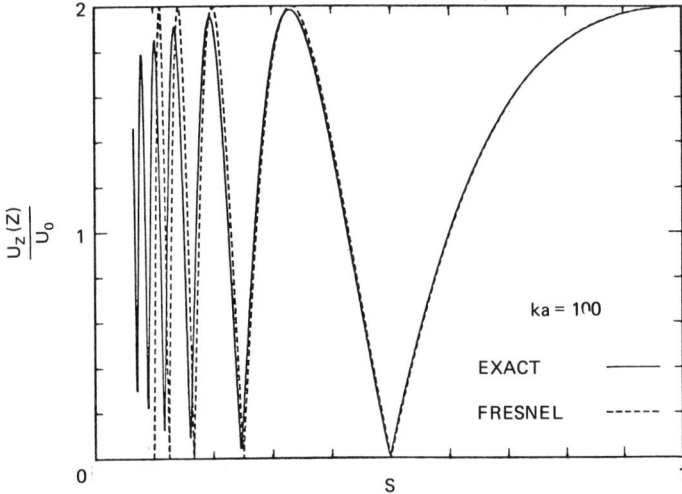

Figure 3.2.2 Plot of the variation of the normalized axial displacement field $|u_z(0, z)/u_0|$ as a function of normalized distance $S = z\lambda/a^2$.

while the dashed line in Fig. 3.2.2 is a plot for the approximate solution of Eq. (3.2.5) (the Fresnel approximation). Thus the maximum values of $|(u_z(0, z)|$ for small values of S are less than 2 and the minimum values are greater than zero but approach the appropriate paraxial or Fresnel solution as S is increased. It is convenient, and often accurate enough, to use the Fresnel approximation, because the results can be expressed in terms of only one parameter S. It is a universal solution that is valid for the values of ka used in many practical transducers employed in isotropic media, and it is easier to compute the solutions for points off-axis than to use the exact form for the solution. Further plots of the field variation on- and off-axis are given in Figs. 3.2.3 and 3.2.4. These plots are derived by the methods given in Sec. 3.2.2.

Fraunhofer approximation. In the Fraunhofer region, we assume from the start that $z \gg a$, and we can use a simple analysis appropriate to the Fraunhofer zone to find the fields off-axis [2]. As illustrated in Fig. 3.2.5, we write the coordinates of a point in spherical coordinates, as r, θ, ψ, where $r = 0$ at the center of the transducer and $\theta = 0$ on the z axis. Initially, it is convenient to work in cylindrical coordinates, with the cylindrical coordinates of a point at r'', z, ψ corresponding to a point at r, θ, ψ in spherical coordinates, and with $r', 0, \psi'$ the cylindrical coordinates of a point at the transducer surface. At a plane z, an element of area on the transducer at the point $r', 0, \psi'$ is a distance R from the point r'', z, ψ, where

$$R_1^2 = r'^2 + r''^2 - 2r'r'' \cos(\psi - \psi') \qquad (3.2.14)$$

and

$$R = \sqrt{R_1^2 + z^2} = \sqrt{z^2 + r'^2 + r''^2 - 2r'r'' \cos(\psi - \psi')} \qquad (3.2.15)$$

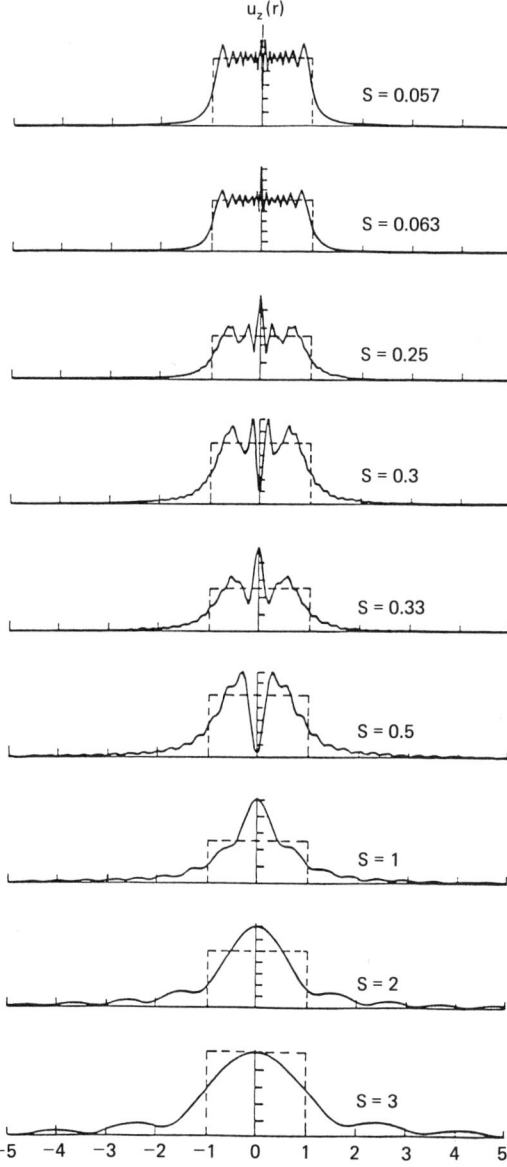

Figure 3.2.3 Radial variation of displacement $u_z(r, z)$ for $ka = 100$, plotted for different values of $S = z\lambda/a^2$ [dashed lines, value of $u_z(r, 0)$ at the transducer].

If we write $z^2 + r'^2 = r^2$ and $r'' = r \sin \theta$, Eq. (3.2.15) becomes

$$R = \sqrt{r^2 + r'^2 - 2rr' \sin \theta \cos (\psi - \psi')} \qquad (3.2.16)$$

In the Fraunhofer approximation, we assume that $r' \ll r$ and that only first order terms in r' are kept in the Taylor expansion. This yields the result

$$R \approx r - r' \sin \theta \cos (\psi - \psi') \qquad (3.2.17)$$

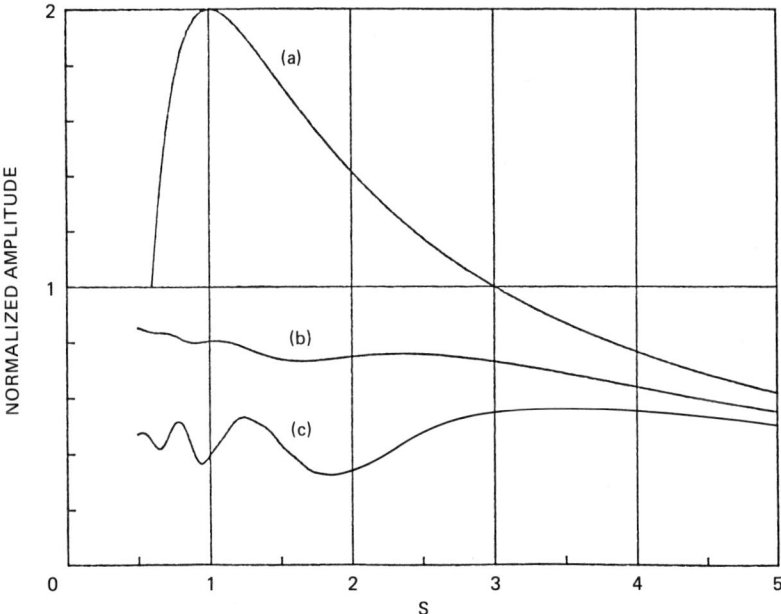

Figure 3.2.4 Plots of amplitude of the signal emitted by a transducer of radius a as a function of $S = z\lambda/a^2$: (a) ratio of the amplitude on axis $u_z(0, z)/u_z(0)$ to the average amplitude at the transducer; (b) ratio of the average amplitude at a receiving transducer, of the parameter $\bar{u}_z(r,z)/u_z(0)$ to the amplitude at the transmitter (averaged over the face of a receiving transducer of radius a); (c) ratio of the amplitude at $r = a$ to the average amplitude at the transmitter $u_z(a, z)/u_z(0)$.

From Eq. (3.1.33), the contribution to the field at $r, \theta, 0$ is

$$\phi(r, \theta) = -\frac{u_0}{2\pi} \int_{\psi=0}^{2\pi} \int_{r'=0}^{a} \frac{e^{-jkR}}{R} r'\, d\psi'\, dr' \quad (3.2.18)$$

If we take account only of the effect of the change in r on the phase term, and put $R = r$ in the denominator of Eq. (3.2.18), then

$$\phi(r, \theta) = -\frac{e^{-jkr} u_0}{2\pi r} \int_{\psi=0}^{2\pi} \int_{r'=0}^{a} e^{jkr'\sin\theta\cos(\psi-\psi')} r'\, dr'\, d\psi' \quad (3.2.19)$$

We use the Bessel function identity

$$J_0(z) = \frac{1}{2\pi} \int_0^{2\pi} e^{jz\cos\psi} d\psi \quad (3.2.20)$$

where $J_0(z)$ is a Bessel function of the first kind of zeroth order. Then it follows that

$$\phi(r, \theta) = -\frac{u_0}{r} e^{-jkr} \int_0^a r' J_0(kr' \sin\theta)\, dr' \quad (3.2.21)$$

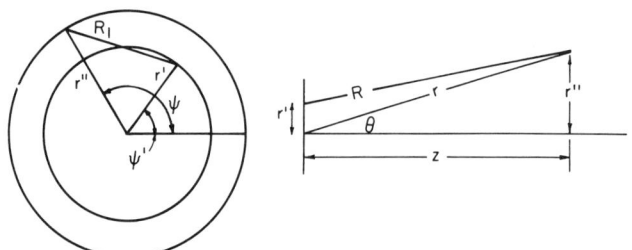

Figure 3.2.5 Notation used for Fraunhofer diffraction theory.

Using the relation $\int z J_0(z)\, dz = z J_1(z)$, where $J_1(z)$ is a first-order Bessel function, we obtain the result

$$\phi(r, \theta) = -\frac{a^2}{r} \frac{J_1(ka \sin \theta)}{ka \sin \theta} e^{-jkr} u_0 \qquad (3.2.22)$$

It follows that the power intensity $I(r, \theta)$ at r, θ, with a power intensity $I(0)$ at the transducer, is

$$I(r, \theta) \approx \left(\frac{\pi a^2}{\lambda r}\right)^2 \text{jinc}^2\left(\frac{a \sin \theta}{\lambda}\right) I(0) \qquad (3.2.23)$$

where jinc $X = J_1(2\pi X)/\pi X$. The functions jinc X and jinc^2 X are plotted in Fig. 3.2.6.†

If $\theta = 0$, then $I(r, 0) \approx (\pi a^2/\lambda z^2) I(0)$, which agrees with Eq. (3.2.8). Where $J_1(ka \sin \theta) = 0$, $I(r, \theta) = 0$. The first zero is where $ka \sin \theta = 3.83$. When $ka \gg 1$ or $a \gg \lambda$, then $\theta = \theta(\text{zero})$ is small and

$$\theta(\text{zero}) = \frac{0.61\lambda}{a} \qquad (3.2.24)$$

On the same basis, the amplitude of the signal drops by 3 dB at the point

$$\theta(3 \text{ dB}) \approx \frac{0.25\, \lambda}{a} \qquad (3.2.25)$$

This is the extent of the first lobe of the beam, which is called the *main lobe*. There are also minor sidelobes for larger values of θ.

Note that at the plane $z = a^2/\lambda$ ($S = 1$), Eq. (3.2.25) predicts that at a radius $r'' = 0.25a$, the field will be reduced in amplitude by 3 dB from the axial field. Thus, on this approximate basis at $z = a^2/\lambda$, a cylindrical beam is confined to an area smaller than its original size. More exact calculations, given in the next section and plotted in Fig. 3.2.3, indicate that at $z = a^2/\lambda$, the 3-dB point is at $r'' \approx 0.35a$.

We conclude that for $z < a^2/\lambda$, the acoustic beam tends to remain confined to its original radius. Beyond this point, its intensity varies with θ and several lobes appear in the radiation pattern. In the Fraunhofer region, the intensity $I(r)$ falls off as $1/r^2$.

† Note that there is no established convention for the normalization of jinc X. Here we have chosen a normalization that makes its value 1 at $X = 0$.

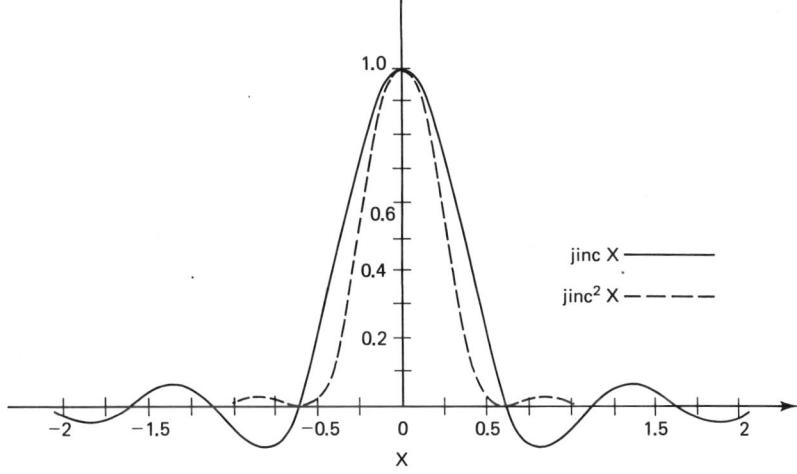

Figure 3.2.6 Plot of jinc X and jinc^2 X where jinc $X = J_1(2\pi X)/\pi X$.

If the transducer is surrounded by a pressure release baffle (zero impedance), we can find a result equivalent to Eq. (3.2.23) by using Eq. (3.1.35) to obtain

$$I(r, \theta) \approx I(0)\left[\frac{\pi a^2}{\lambda r} \text{jinc}\left(\frac{a}{\lambda}\sin\theta\right)\cos\theta\right]^2 \quad (3.2.26)$$

Thus even an infinitesimal radius transducer, with jinc $(a\sin\theta/\lambda) = 1$, in a pressure release baffle has an angular response with its 3-dB points at $\theta = 45°$.

We have now discussed the differences between the behavior of a transducer in a rigid baffle and that of one in a pressure baffle. An example of a transducer in a rigid baffle would be a transducer in water, surrounded by a metal disk whose impedance is much higher than that of water. An example of a transducer in a pressure release baffle would be a small transducer, or one element of an array of identical transducers, spaced by a distance of a wavelength or more from each other, separated from a water bath by a Mylar film, so that there is water on one side of the film and air on the other. In practice, the behavior of a transducer array depends on the spacing between the array elements. Thus the angular response of a transducer array element may lie between the response of a transducer in a pressure release baffle and that of a transducer in a rigid baffle [3].

3.2.2 Radial Variation of the Field Using a Hankel Transform and Spatial Frequency Concepts

The theory we have given so far can be extended to deal with the radial variation of the fields. However, it is often more convenient to derive the necessary results by using Fourier transform or Hankel transform methods [2].

The basic idea, stated in terms of a rectangular coordinate system, is to carry out a spatial Fourier transform in x, y coordinates of the fields at the transducer.

This determines the excitation of the plane wave components with transverse propagation constants k_x and k_y, in the x and y directions, respectively. By analogy to Fourier transformation in the time domain, k_x and k_y are called the *transverse spatial frequencies*. The contributions of the propagating plane waves are summed at the plane z and an inverse transform is carried out to determine the fields at that plane. The process is entirely equivalent in cylindrical coordinates, except that here we call the transverse propagation constant α where $k_r = \alpha$, make use of cylindrical symmetry, and employ Hankel transforms.

A spherically symmetric solution of the longitudinal wave equation in cylindrical coordinates (r, θ, z), for waves that vary as $\exp(j\omega t)$, is

$$\phi(\alpha r, \beta z) = A(\alpha)J_0(\alpha r)e^{-j\beta z} \qquad (3.2.27)$$

By analogy to Fourier transformation in the time domain, the parameters α and β are called the *radial* and *axial spatial frequencies*, respectively. By differentiating Eq. (3.2.27) with respect to z, for waves in a liquid, we can write

$$u_z(\alpha r, \beta z) = -j\beta A(\alpha)J_0(\alpha r)e^{-j\beta z} \qquad (3.2.28)$$

where $J_0(x)$ is a Bessel function of zeroth order of the first kind and

$$\beta^2 + \alpha^2 = k^2 \qquad (3.2.29)$$

In general, β may be real or imaginary. We can also use this formulation for solid materials by neglecting contributions from the shear wave potential term ψ_θ; this assumption will be discussed more fully in Sec. 3.2.3. We write the complete spatial solutions for $u_z(r, z)$ in the form

$$u_z(r, z) = -j\beta \int_0^\infty A(\alpha)e^{-j\beta z}J_0(\alpha r)\, d\alpha \qquad (3.2.30)$$

The definition of the Hankel transform of a function $g(r)$ is

$$G(\alpha) = \int_0^\infty g(r)J_0(\alpha_r)r\, dr \qquad (3.2.31)$$

where

$$g(r) = \int_0^\infty G(\alpha)J_0(\alpha r)\alpha\, d\alpha \qquad (3.2.32)$$

Therefore, at $z = 0$, it follows from Eqs. (3.2.30)–(3.2.32) that

$$\frac{-j\beta A(\alpha)}{\alpha} = \int_0^a u_z(r, 0)J_0(\alpha r)r\, dr \qquad (3.2.33)$$

where $A(\alpha)$ is the amplitude of the term with a radial spatial frequency α, and $u_z(r, 0)$ is only finite from 0 to a. For a cylindrical piston transducer, $u_z(r, 0) = u_0$ and is uniform. Integrating Eq. (3.2.33) and using the identity $xJ_1(x) = \int xJ_0(x)\, dx$ yields the result

$$-j\beta A(\alpha) = u_0 a J_1(\alpha a) \qquad (3.2.34)$$

where $J_1(x)$ is a Bessel function of the first order of the first kind. Hence, at any plane z, it follows from Eqs. (3.2.30) and (3.2.34) that

$$u_z(r, z) = au_0 \int_0^\infty J_1(\alpha a)J_0(\alpha r)e^{-j\beta z}\, d\alpha \qquad (3.2.35)$$

We can find $u_z(r, z)$ at any point r, z from this relation. We can also find the average value of $u_z(r, z)$ over a radius a, which is proportional to the signal detected by another transducer of radius a. If we call this quantity $\overline{u_z(r, z)} = (2/a^2) \int_0^a u_z(r, z)r\, dr$, the signal detected by a second identical and ideal transducer, which is perfectly matched to the propagating medium, is

$$\overline{u_z}(r, z) = 2u_0 \int_0^\infty \frac{J_1^2(\alpha a)}{\alpha} e^{-j\beta z}\, d\alpha \qquad (3.2.36)$$

It follows, by using the relation

$$\int_0^\infty \frac{J_1^2(\alpha x)}{\alpha}\, d\alpha = \frac{1}{2} \qquad (3.2.37)$$

that $\overline{u_z}(r, 0) = u_0$, as it should. Note that we can also find the longitudinal wave contribution to the pressure, as well as any other parameter of interest, by extending this formulation.

It is often convenient to normalize these formulas in terms of the Fresnel parameter $S = z\lambda/a^2$, and, in addition, to use the equivalent of the Fresnel or paraxial approximation. This makes it possible to express the results in terms of only one variable, S. We put $\alpha a = Y$ and write

$$\beta = \sqrt{k^2 - \alpha^2} \approx k - \frac{\alpha^2}{2k} \qquad (3.2.38)$$

This approximation, which is discussed more fully in Sec. 3.2.3, implies that only waves propagating at small angles to the axis contribute to the total field.

Equations (3.2.35) and (3.3.36) reduce, respectively, to the following normalized forms:

$$u_z(r, z) = u_0 e^{-jkz} \int_0^\infty J_1(Y)J_0\frac{rY}{a} e^{jy^2 S/4\pi}\, dY \qquad (3.2.39)$$

The average value of $u_z(r, z)$ in a radius a is

$$\frac{\overline{u_z(r, z)}}{u_0} = 2e^{-jkz} \int_0^\infty \frac{J_1^2(Y)}{Y} e^{jy^2 S/4\pi}\, dY \qquad (3.2.40)$$

It is not necessary, in principle, to use the assumption of Eq. (3.2.38) to obtain analogous results in Eqs. (3.2.39) and (3.2.40) (see Prob. 6), but the advantage of this procedure is that the results can be normalized and then expressed in terms of one curve. Otherwise, we must use the two parameters ka and S to specify the problem. The radial variation of $u_z(r, z)$ has been plotted in Fig. 3.2.3 from the exact solution for various values of the parameter $S = z\lambda/a^2$ and a value of $ka = 100$, which corresponds to $z/a = a/\lambda \approx 16$ when $S = 1$. This value of ka is comparable to that for a typical transducer used in medical applications (19 mm diameter for use at 2.25 MHz, giving $a/\lambda \approx 14$).

The radial variation of $u_z(r, z)$ exhibits sharp maxima and minima near the axis, as we would expect from the theory given in Sec. 3.2.1. Near the transducer, the Fresnel ripples (i.e., the radial variations) are relatively rapid and the beam is well confined to its original diameter. At $S = 1$, the effective beam diameter at the 3-dB points is reduced to approximately 0.35 of its original diameter. The beam expands beyond this point until its effective diameter reaches the original diameter, although a considerable portion of the energy remains outside the main beam near $S = 3$. The Fraunhofer theory predicts the positions of the lower-order maxima and minima in the beam even for $S = 1$, but they are not as well defined as in the simple theory.

In Figs. 3.2.4 and 3.2.7, we give plots calculated from Eqs. (3.2.39) and (3.2.40) of $u_z(0, z)/u_0$, $u_z(a, z)/u_0$, and $\overline{|u_z(r, z)|}/u_0$ as a function of S. The latter parameter gives the diffraction loss between two identical transducers, spaced a distance z apart, that is often required for calibration [4]. If the beam is reflected from a perfect plane reflector at a plane z and received by the original transmitting transducer, the diffraction loss may be determined for these curves by replacing z with $2z$. In Fig. 3.2.7, we also give the correction ϕ_R, due to diffraction, to the phase of $\overline{u_z(r, z)}$. The total phase change due to diffraction, $\phi(\text{total})$, is always

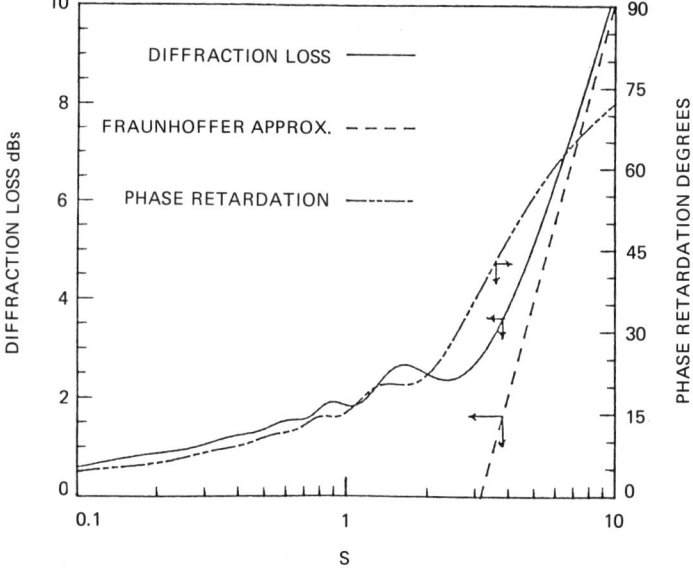

Figure 3.2.7 Diffraction loss and phase retardation between two equal-sized transducers of radius a spaced by a distance z ($S = z\lambda/a^2$). The phase retardation is $\phi_R = kz - \phi(\text{total})$, where $\phi(\text{total})$ is the total phase change between the two transducers. A comparison of the calculated loss with the simple Fraunhofer approximation (the dashed-line plot) is also given. For $S > 10$, this approximation is adequate for estimating the loss between transducers.

less than kz; ϕ_R must be known to make absolute velocity measurements based on measurement of phase.

Note that Figs. 3.2.4 and 3.2.7 are plotted using the paraxial or Fresnel approximation $\alpha^2 \ll k^2$. The use of this approximation makes it easy to calculate the result when the beam passes through several different media; the total value of S is determined by adding the values of S determined for each region, using the appropriate values of z and λ in these regions. The situation is somewhat more complicated when the interfaces occur in regions where the paraxial approximation does not hold. In this case, however, as the diffraction losses for $S < 1$ are normally small, the breakdown of the paraxial approximation in these regions may not matter too much.

Example: Piston Transducer for Probing the Human Body

Consider a transducer to be used for examining the human body by placing it in contact with the patient. The maximum depth required for examination is approximately 20 cm. As the attenuation in the body varies approximately linearly with frequency and is of the order of 1 dB/cm/MHz, the total attenuation of a reflected wave at a depth of 20 cm will be 40 dB/MHz. This tends to limit the maximum frequency employed. Typically, operating frequencies of 2 to 5 MHz are used in the diagnostic examination of adults; the higher frequencies in this range are employed for children, with frequencies as high as 20 MHz used for examination of the eye or near-surface features such as the carotid artery.

Suppose that we consider a 2.25-MHz transducer for which the acoustic wavelength is 0.67 mm ($V = 1.5 \times 10^5$ cm/s). If we choose the end of the operating range to be at $z = 3a^2/\lambda$, where the 3-dB beam diameter is comparable to that of the transducer, we find that the radius of the transducer is 0.67 cm, or that its diameter is 1.3 cm, approximately $\frac{1}{2}$ in. Thus, in medical practice, transducers of this diameter can provide a beam with a maximum excess loss, due to diffraction, of approximately 6 dB, after reflection from a flat plane and traveling a path $6a^2/\lambda$ long. The one-way diffraction loss in traveling a distance $3a^2/\lambda$ ($S = 3$) to the point of interest is, from Fig. 3.2.7, only 2.5 dB. The transverse definition of such a beam will also be better than 1.3 cm.

Note that small reflectors in the Fresnel region of the beam may not always be characterized correctly because the beam intensity is not uniform either over its cross section or along its length. For this reason, it is often advisable to sacrifice the advantage of greater beam intensity or definition obtained by operating in the Fresnel region, and to work in the Fraunhofer region instead. One or more transducers can be employed to observe specific regions, each designed with a radius such that $1 < S < 3$ where the beam has not spread too much, so that its intensity is relatively strong. This procedure is normally employed in quantitative nondestructive evaluation (NDE) of solid materials, where accurate calibrations are needed to determine the size and nature of a flaw.

3.2.3 Diffraction from Rectangular Transducers

The diffraction theory derived in the preceding section can also be adapted to rectangular systems. This is most conveniently accomplished by using a Fourier transform approach or a spatial frequency analysis that is analogous to the Hankel transform theory already given for the cylindrical transducer. In this case, there is no simple analytical solution for the field variation along the central axis of a rectangular transducer, so the more general solution is employed.

The two-dimensional solutions we shall give here are applicable to: (1) rectangular transducers of finite width in the x direction, which are assumed to be infinitely long in the y direction; and (2) surface wave transducers of finite width, whose width is much larger than the Rayleigh wave wavelength. In the latter case, we assume that the surface wave field variation in the y direction remains invariant and that the ϕ and ψ_x potential variations in the x and z directions obey the wave equation

$$\frac{\partial^2 \phi}{\partial x^2} + \frac{\partial^2 \phi}{\partial z^2} + k_r^2 \phi = 0 \tag{3.2.41}$$

where k_r is the Rayleigh wave propagation constant. Thus we use the same formalism for surface waves as for the simple rectangular transducer, only replacing k by k_r and taking the propagation constant in the z direction to be β.

For any potential function $\phi(x, z)$, we can write

$$\phi(x, z) = \int_{-\infty}^{\infty} A(\alpha) e^{-j\beta z} e^{-j\alpha x} \, d\alpha \tag{3.2.42}$$

where $A(\alpha)$ is the amplitude of the spatial frequency component α. The parameters α and $A(\alpha)$ obey the relations

$$\beta^2 + \alpha^2 = k^2 \tag{3.2.43}$$

and

$$A(\alpha) = \frac{1}{2\pi} \int_{-\infty}^{\infty} \phi(x, 0) e^{j\alpha x} \, dx \tag{3.2.44}$$

respectively. For a transducer of width w, if ϕ is a longitudinal wave potential, then

$$u_z(\alpha, x, z) = -j\beta A(\alpha) e^{-j\beta z} e^{j\alpha x} \tag{3.2.45}$$

More generally, the surface wave displacement component $u_z(x, y, z)$ depends on both ϕ and ψ_x. At any plane y below the surface we can write

$$u_z(\alpha, x, y, z) = B(\alpha, y) e^{-j\beta z} e^{-j\alpha x} \tag{3.2.46}$$

We assume that $B(\alpha, y)$ does not change with z; thus a straight crested component of the surface wave (the equivalent of a plane wave component of a bulk wave), expressed by Eq. (3.2.46), exists. On this basis, we sum the components of different α to form the total field. We assume that ϕ, ψ_x, u_x, u_y, and the other field components can be treated in the same way.

Ideally, for a rectangular bulk wave transducer on a solid material, we should take account of the contributions of the longitudinal and shear wave potential to the displacement and stress at any point. This difficulty does not occur when the propagating medium is a liquid. Even with a solid, the effect of the shear wave potential term is small for excitation from a longitudinal wave transducer. Similarly, the effect of the longitudinal wave potential term is small for excitation by a shear wave transducer. Thus we treat the longitudinal wave transducer by neglecting the shear wave potential term.

We assume that the transducer is uniformly excited by a potential ϕ_0 over its width w. From Eq. (3.2.44), the spatial frequency component $A(\alpha)$ has an amplitude

$$A(\alpha) = \frac{w}{2\pi} \phi_0 \frac{\sin(\alpha w/2)}{\alpha w/2} \qquad (3.2.47)$$

Thus

$$\phi(x, z) = \frac{w}{2\pi} \phi_0 \int_{-\infty}^{\infty} e^{-jz(k^2-\alpha^2)^{1/2}} e^{-j\alpha x} \frac{\sin(\alpha w/2)}{\alpha w/2} d\alpha \qquad (3.2.48)$$

We assume that the function $\sin(\alpha w/2)/(\alpha w/2)$ is large only for small α. This is equivalent to using the paraxial ray assumption and assuming that only rays at a small angle to the axis are important, or that $\alpha^2 \ll k^2$. We can then write Eq. (3.2.48) in the form

$$\phi(x, z) = \frac{w}{2\pi} \phi_0 e^{-jkz} \int_{-\infty}^{\infty} e^{j\alpha^2 z/2k} e^{-j\alpha x} \frac{\sin(\alpha w/2)}{\alpha w/2} d\alpha \qquad (3.2.49)$$

It is convenient to define a normalized diffraction parameter

$$S = \frac{4z\lambda}{w^2} \qquad (3.2.50)$$

with $Y = \alpha w/2$. Equation (3.2.49) can now be written in the form

$$\phi(x, z) = \frac{\phi_0}{\pi} e^{-jkz} \int_{-\infty}^{\infty} e^{-2jYx/w} e^{jY^2 S/4\pi} \frac{\sin Y}{Y} dY \qquad (3.2.51)$$

We can also find the average value of $\phi(x, z)$ over a transducer of width w to determine the diffraction loss between two transducers spaced by a distance z. The result is

$$\frac{\overline{\phi(x, z)}}{\phi_0} = \frac{e^{-jkz}}{\pi} \int_{-\infty}^{\infty} \left(\frac{\sin Y}{Y}\right)^2 e^{jk^2 S/4\pi} dY \qquad (3.2.52)$$

Figure 3.2.8 shows plots of $|\phi(0, z)/\phi_0|$ and $|\overline{\phi(x, z)}/\phi_0|$ as a function of S.

Note that now there are no deep nulls in the axial field variation with z. Complete cancellation occurs in a cylindrical system only because the contribution to the axial field by a ring of radius r and thickness dr on the transducer is proportional to the term $[\exp(-jkR)/R]r\,dr = [\exp(-jkR)]\,dR$. Thus elements of incremental length dR contribute equal amplitudes to the axial fields and can cancel each other out. This same relation does not hold for a rectangular transducer.

Note also that the last amplitude maximum of $\phi(0, z)$ is near $S = 1.2$, which is only 2.5 dB higher than the value at the transducer, rather than the 6 dB higher

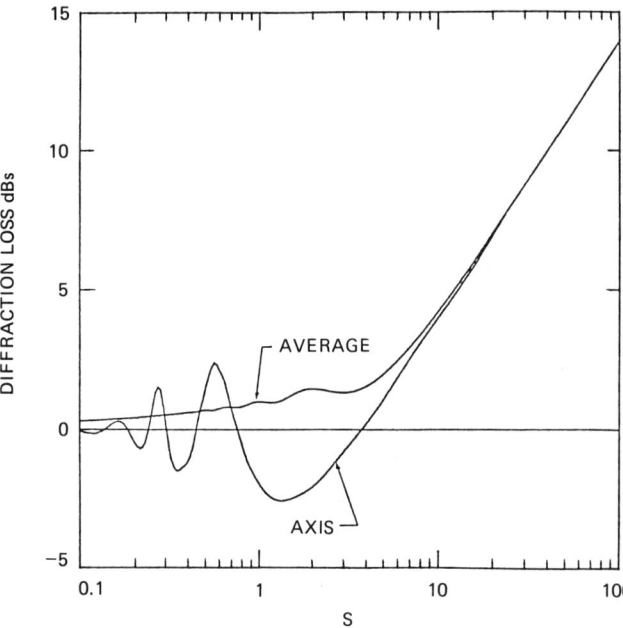

Figure 3.2.8 Plot of the average potential $|\phi(x, z)/\phi_0|$ and the potential on the axis $|\phi(0, z)/\phi_0|$ of a rectangular transducer.

obtained with a cylindrical transducer. The Fresnel length is therefore near $S \approx 1$ or $\lambda z/(w/2)^2 \approx 1$. The actual value of $S = 1.2$, the Fresnel length of a rectangular transducer, clearly corresponds to the value of $S = z\lambda/a^2 = 1$ for the Fresnel length of a cylindrical transducer.

Fraunhofer diffraction field by the method of stationary phase. Finally, we consider diffraction in the far field. First we write

$$\alpha = k \sin \theta' \qquad (3.2.53)$$

or

$$Y = \frac{kw}{2} \sin \theta' \qquad (3.2.54)$$

with

$$d\alpha = k \cos \theta' \, d\theta' \qquad (3.2.55)$$

From Eq. (3.2.47), the amplitude $A(\theta') \, d\theta'$ of the plane wave excited in the element of angle $d\theta'$ can be written in the form

$$A(\theta') \, d\theta' = kw \frac{\phi_0}{2\pi} \frac{\sin \left[(kw \sin \theta')/2\right]}{(kw \sin \theta')/2} \cos \theta' \, d\theta' \qquad (3.2.56)$$

Thus we expect the amplitude of the waves emitted from the transducer into an angular range $d\theta'$ to vary as $[\text{sinc} \, (w \sin \theta')/\lambda] \cos \theta'$.

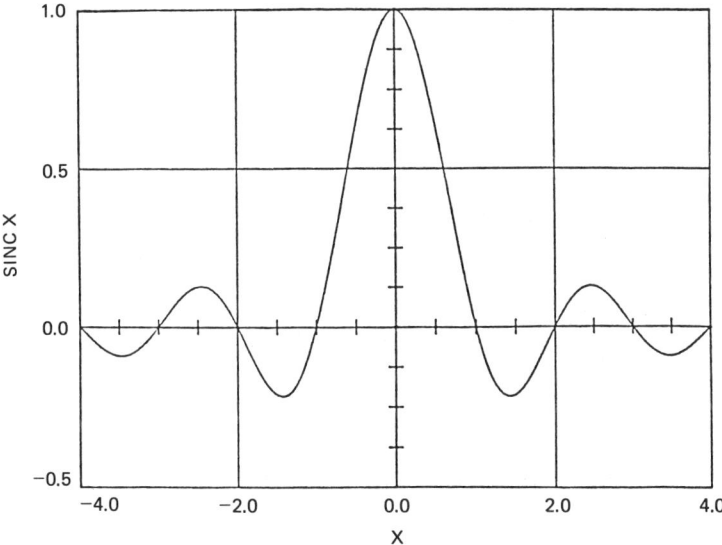

Figure 3.2.9 Plot of sinc $X = (\sin \pi X)/\pi X$.

It is convenient to write Eq. (3.2.48) in terms of θ' to obtain the result

$$\phi(x, z) = \frac{kw}{2\pi} \phi_0 \int_{-\infty}^{\infty} e^{-jkr\cos(\theta' - \theta)} \frac{\sin\left[(kw \sin \theta')/2\right]}{kw \sin (\theta'/2)} \cos \theta' \, d\theta' \qquad (3.2.57)$$

where

$$z = r \cos \theta \qquad (3.2.58)$$
$$x = r \sin \theta$$

We assume that kr is large in the Fraunhofer region. In this case we can use the *method of stationary phase*, as given in Appendix G, to evaluate the integral. This technique is based on the idea that as the phase of the exponential term in the integrand varies rapidly as θ' is varied, the main contribution to the integral must come from near the region where the rate of phase change is zero (stationary phase). This is where $\sin(\theta' - \theta) = 0$ or $\theta' = \theta$ (i.e., where the plane wave component excited by the transducer is aimed toward the observer at x, z). Then, writing $\cos(\theta' - \theta) \approx 1 - (\theta' - \theta)^2/2$ in the argument of the exponential, and putting $\sin \theta' \approx \sin \theta$ and $\cos \theta' \approx \cos \theta$, as discussed in Appendix G, we obtain the result

$$\frac{\phi(x, z)}{\phi_0} \approx \frac{w}{\sqrt{r\lambda}} \text{sinc}\left(\frac{w}{\lambda} \sin \theta\right) \cos \theta \, e^{-j(kr - \pi/4)} \qquad (3.2.59)$$

where sinc $X = \sin(\pi X)/\pi X$. The sinc X function is plotted in Fig. 3.2.9. Thus Eq. (3.2.59) leads us to conclude that $\phi(x, z)$ and u_r vary as $\cos \theta$ sinc $[(w \sin \theta)/\lambda]$.

Sec. 3.2   Plane Piston Transducers

This relation should be compared to that for a cylindrical transducer in a pressure release baffle [(Eq. (3.1.35)]. The same $\cos\theta$ term appears when the potential is specified at the plane of the transducer (see Prob. 2). Here, if we had specified $u_z = \partial\phi/\partial z = u_0$ at the plane of the transducer, with $u_z = 0$ for $|x| > w/2$, there would be no $\cos\theta$ term in the far field. An infinitely long rectangular transducer immersed in a liquid would give the same type of far-field pressure variation as a Rayleigh wave transducer.

Note that the amplitude falls off as $1/\sqrt{r}$, because the power intensity in a cylindrical spreading beam must fall off as $1/\sqrt{r}$, due to conservation of power. Therefore, at the same distance from the transducer, diffraction losses in a cylindrical beam are less severe than in a spherical beam. In practice, the diffraction loss of a rectangular transducer, of finite length h in the y direction, will tend to a $1/\sqrt{r}$ variation in the far field when $r \gg h^2/\lambda$.

3-dB points. The argument of the sinc function is $0.455\,\pi/2$ at the 3-dB points, provided that $w/\lambda \gg 1$ (i.e., provided we can ignore the $\cos\theta$ term). The spacing between 3-dB points is therefore, to this approximation,

$$\Delta x(3\text{ dB}) = \frac{0.89 r\lambda}{w} \tag{3.2.60}$$

where $x = r\sin\theta$. The zero points of the response are at

$$x(\text{zero}) = \frac{\pm r\lambda}{w} \tag{3.2.61}$$

The first sidelobe of the response occurs where $x = \pm 3r\lambda/2w$ and is reduced 13 dB in amplitude from the main lobe.

PROBLEM SET 3.2

1. Consider a cylindrical piston transducer in the form of a ring of outer radius a and inner radius b.
   (a) Find the variation of displacement and the potential field along its axis, and show that as the ring becomes infinitesimal ($b \to a$), there are no longer sharp dips in the z-directed field on the axis.
   (b) Find the radial field variation in the Fraunhofer region.
2. A rectangular piston transducer in a pressure release baffle (zero impedance baffle) is infinitely long in the y direction and finite in the x direction.
   (a) Assuming that $\phi(x', 0)$ on the transducer varies only in the x direction, use Eq. (3.1.35) to find $\phi(x, z)$ at any point by putting $t = (y' - y)/r$, where $r = [z^2 + (x - x')^2]^{1/2}$ and integrating from $t = -\infty$ to $t = \infty$. Apply the method of stationary phase given in Appendix G to the integral, to find a simple form of the integral for $kr \gg 1$. Keep the term $\cos\theta$ in Eq. (3.1.35) in that form and do not substitute for it until you have used the method of stationary phase.
   (b) Using your result for (a), find the Fraunhofer solution for the potential at a point x,

z that is due to the excitation of a rectangular transducer of width w by a uniform potential ϕ_0.

Answer:

$$\frac{\phi(x, z)}{\phi_0} = \frac{w}{\sqrt{r\lambda}} \text{sinc} \frac{w \sin \theta}{\lambda} e^{-j(kr - \pi/4)} \cos \theta$$

where $x = r \sin \theta$ and $r = \sqrt{x^2 + z^2}$. The result shows that the fields fall off as $1/\sqrt{r}$ in a cylindrical system.

3. Repeat Prob. 2 for an infinitely long rectangular piston transducer in a rigid baffle, excited by a displacement field $u_z(x', 0) = u_0$ on the transducer.
4. Repeat Prob. 3, using the Fourier transform method for a rectangular transducer.
5. Using the results of Prob. 3, and assuming that $u_z(x')$ has the form $u_z(x') = u_0 \exp(-\alpha x'^2)$ and that the transducer is infinitely wide, find an analytic expression for the fields at any point x, z. It is reasonable to employ the paraxial approximation ($x'^2 \ll z^2$); this is a valid assumption when α is so large that only regions for which x' is small contribute to the field. Find an analytic expression for the fields at any point x, z by completing the square in x' in the argument of the exponentials. You may need the result

$$\frac{\sqrt{\pi}}{\beta} = \int_{-\infty}^{\infty} e^{-\beta x^2} dx$$

6. (a) Consider the errors in the derivation of the paraxial forms of Eqs. (3.2.39) and (3.2.40). Use the exact form of Eq. (3.2.36) and expand up to fourth order in α. Write the new normalized forms of Eqs. (3.2.39) and (3.2.40) in terms of S and ka.
   (b) Suppose that the main contribution to Eq. (3.2.40) comes from the region $Y < 10$. In this case, write the condition for the maximum phase error in the exponential term (i.e., the fourth-order term) as less than $\pi/4$. Find the minimum value of S to satisfy this condition as a function of ka. What is $S(\min)$ for $ka = 100$? The result will be by no means a good estimate of the error, but this derivation should give you a feel for how the error in the paraxial theory depends on the parameters.
7. Consider the use of a rectangular transducer as a receiver. Supposes that a plane longitudinal wave of amplitude $u_0 [\exp(-jkr)]$, with the direction of propagation in the r direction at an angle θ to the z axis, is incident on a transducer of width w with its plane perpendicular to the z direction. Assume that the output of the receiving transducer is proportional to the average value of u_z over the width of the transducer, defined as

$$\bar{u}_z = \frac{1}{w} \int_{-w/2}^{w/2} u_z \, dx$$

where u_z is the displacement in the z direction due to the incident longitudinal wave. Determine how the received signal varies with θ and show that your result has the same variation with θ as Eq. (3.2.59).

8. (a) A plane wave is incident in the z direction on a thin penny-shaped crack of radius a, and is scattered by it. Determine the amplitude of the scattered wave in the forward direction (i.e., at an angle θ from the axis where $\theta < \pi/2$). Assume that just beyond the crack, $u_z = 0$ in the region $r < a$. This implies that if the incident

wave is u_z^i and the scattered wave is u_z^s, then

$$u_z = u_z^i + u_z^s$$

and

$$u_z^s = -u_z^i \quad (r < a)$$

Thus you may assume that the scattered field is due to a piston transducer with a field u_z^s at a plane $z = 0$ [just beyond the crack $(z < 0)$].

(b) Find the total field on the axis and show that in the Fresnel approximation, there is always a bright spot on the axis whose amplitude does not vary with z.

(c) Find an expression for the magnitude of the total field at a point r, θ in the Fraunhofer region of the scatterer.

3.3 FOCUSED TRANSDUCERS

3.3.1 Field of a Focused Spherical Transducer

As we discussed in Sec. 3.2, the beam emitted from a piston transducer spreads radially due to diffraction. Thus, at some distance from the transducer, the power intensity may be too low, while the beam diameter is too large to obtain good transverse definition when probing an object. Therefore, we often employ a focused acoustic beam, as in optics, to obtain good transverse definition and high acoustic beam intensity at a point of interest.

As in optics, we can use lenses to focus the acoustic beam. But often the simplest way to obtain a focused acoustic beam is to use a spherically shaped transducer. Such a transducer will produce a focused beam near its center of curvature. We normally focus the beam to a spot smaller than the transducer diameter, thus obtaining a beam intensity much higher than that at the transducer itself. So in this section we shall use the methods employed in Sec. 3.2.1 to determine the intensity of the beam near its focus in terms of the beam intensity at the transducer surface.

We consider a spherical transducer with a diameter $2a$ and a radius of curvature z_0, as illustrated in Fig. 3.3.1. We shall use a paraxial theory, or *Fresnel approximation*, in which we assume that $a^2 \ll z^2$ to treat the properties of the beam emitted from the transducer. Equations (3.1.33) and (3.2.1) for the potential of a transducer in a rigid baffle are strictly true only for a flat piston transducer in a flat baffle. To avoid the difficulty this presents, we can assume that the acoustic beam, initially, follows geometrical ray paths normal to the surface of the transducer, and that the acoustic source is at the plane of the baffle $z = z_1$. Suppose that the distance from a point on the transducer, along a ray path normal to its face, to the point of interception with the plane of the baffle, r', z_1, ψ', is R_1. We

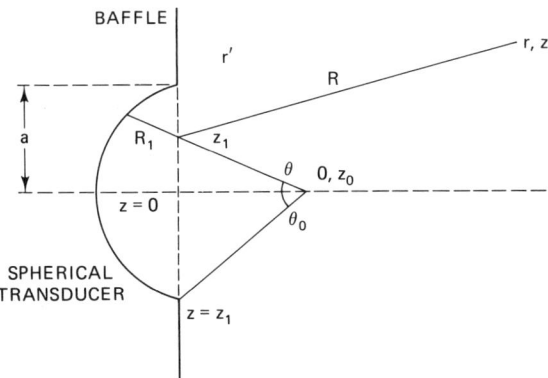

Figure 3.3.1 Notation used in the analysis of a spherical transducer.

can then take the phase of the beam emitted from this point to be $\exp(-jkR_1)$. The displacement at the plane of the baffle can be found by ray optics. We assume that the radial displacement at the transducer is $u_0(\theta)$. The displacement in the z direction at the plane of the baffle is

$$u_z(\theta) = u_0(\theta) \cos\theta \, \frac{z_0}{z_0 - R_1} \quad (3.3.1)$$

But $z_0 - R_1 = (z_0 - z_1)/\cos\theta$. Since $z_0 - z_1 = z_0 \cos\theta_0$, where θ_0 is the aperture angle of the transducer, it follows that

$$u_z(\theta) = u_0(\theta) \frac{\cos^2\theta}{\cos\theta_0} \quad (3.3.2)$$

On this basis, we can use the Rayleigh–Sommerfeld theory, in the form of Eq. (3.2.1) or (3.2.33), and treat the plane of the baffle as the source, for all we have done is to alter the phase of the source at the plane of the baffle and multiply $u_0(\theta)$ by $\cos^2\theta/\cos\theta_0$.

We now suppose that the transducer is excited by a uniform displacement u_0. The total potential at a point $r, z, 0$ from all contributions at a distance R from points $r', 0,$ and ψ' on the baffle is

$$\phi(r, z, 0) \approx -\frac{u_0}{2\pi} \iint \frac{e^{-jk(R+R_1)} \cos^2\theta}{R \cos\theta_0} r' \, dr' \, d\psi' \quad (3.3.3)$$

Referring to Fig. 3.3.1, we see that

$$R = \sqrt{r'^2 + r^2 - 2rr' \cos\psi' + (z - z_1)^2} \quad (3.3.4)$$

By simple geometry, it is easy to show, to the paraxial approximation, that

Sec. 3.3   Focused Transducers

$R_1 \approx z_1 - r'^2/2z_0$. Thus, keeping only second-order terms in r and r', we find that

$$R_1 + R \approx z + \frac{r'^2 + r^2 - 2rr'\cos\psi'}{2z} - \frac{r'^2}{2z_0} \qquad (3.3.5)$$

Here we have neglected z_1 in the denominator of the first term on the right-hand side of Eq. (3.3.5), since it is of second order in r' and would contribute only fourth-order terms in r and r' to the result. Thus if $z = z_0$, the second-order terms in r' cancel out. We will assume that as far as the amplitude variation is concerned, the R term in the denominator of Eq. (3.3.3) is equal to z, and to the paraxial or Fresnel approximation $\cos^2\theta \approx 1$ and $\cos\theta_0 \approx 1$. Then Eq. (3.3.3) can be written in the form

$$\phi(r, z) = -e^{-jk(z + r^2/2z)} \frac{u_0}{2\pi z} \int_0^a \int_0^{2\pi} e^{-(jkr'^2/2)(1/z - 1/z_0)} e^{(jkrr'/z)\cos\psi'} r'\, dr'\, d\psi' \qquad (3.3.6)$$

Integrating with respect to ψ', we find that

$$\phi(r, z) = -e^{-jk(z + r^2/2z)} \frac{u_0}{z} \int_0^a J_0\left(\frac{krr'}{z}\right) e^{-(jkr'^2/2)(1/z) - 1/z_0)} r'\, dr' \qquad (3.3.7)$$

where $J_0(x)$ is a Bessel function of the first kind and of zeroth order.

We now consider the form of the potential at the plane $z = z_0$. Ray optics indicate that this is the focal plane; thus we shall call it the *geometrical focus*. As we shall see, diffraction effects tend to bring the point of maximum axial beam intensity nearer to the lens (i.e., to a plane $z < z_0$). Hence the *true focus* is at a plane $z < z_0$.

At the geometrical focal plane, the potential on axis is

$$\phi(0, z_0) = -\frac{a^2 e^{-jkz_0}}{2z_0} u_0 \qquad (3.3.8)$$

By differentiating with respect to z, we can find $u_z(z)$ and show that

$$u_z(0, z_0) \approx \frac{j\pi a^2}{\lambda z_0} e^{-jkz_0} u_0 \qquad (kz_0 \gg 1) \qquad (3.3.9)$$

It follows that the ratio of the beam intensity $I(0, z_0)$ on axis at $z = z_0$ to the beam intensity $I(0)$ at the transducer is

$$\frac{I(0, z_0)}{I(0)} = \left(\frac{\pi a^2}{z_0 \lambda}\right)^2 = \left(\frac{\pi}{S}\right)^2 \qquad (3.3.10)$$

where, as before, $S = z_0\lambda/a^2$. Thus if the beam is focused at z_0, and if $z_0 < \pi a^2/\lambda$, the beam intensity at the geometrical focus will be larger than that at the transducer. Note that this implies that the focused transducer will normally be operated with its focal point in a region where the parameter $S = \lambda z_0/a^2$ is such that $S < \pi$.

Transverse definition. Now consider the fields at the optical focal plane $z = z_0$ for finite r. It follows from Eq. (3.3.7) that

$$u_z(r, z_0) = \frac{j\pi a^2 e^{-jkz_0} e^{-jkr^2/2z_0}}{\lambda z_0} \frac{2J_1(kra/z_0)}{kra/z_0} u_0 \qquad (3.3.11)$$

Thus the beam intensity at the plane $z = z_0$ varies as

$$\frac{I(r, z_0)}{I(0)} = \left(\frac{\pi a^2}{z_0 \lambda}\right)^2 \left(\text{jinc} \frac{ra}{\lambda z_0}\right) \qquad (3.3.12)$$

where

$$\text{jinc } X = \frac{J_1(2\pi X)}{\pi X} \qquad (3.3.13)$$

This function is plotted in Fig. 3.2.6.

The variation of the beam intensity at the focal plane is exactly the same in form as that in the Fraunhofer region of a plane piston transducer at $z = z_0$. However, the beam intensity can now be chosen at will and can also be much larger than at the transducer itself, which changes the scales of both the axial and radial variations of the field.

3-dB definitions, F number, and lens aperture. The radius at which the beam intensity becomes zero is

$$r_0(\text{zero}) = \frac{0.61 z_0 \lambda}{a} = \frac{0.61 \lambda}{\sin \theta_0} = \frac{0.61 \lambda}{\text{aperture}} \qquad (3.3.14)$$

where θ is the half-angle subtended at the geometric focus by the transducer lens, and the *aperture of the lens* is defined as $\sin \theta_0$. The diameter of the beam at the 3-dB points, which we take as a measure of its definition, is

$$d_r(3 \text{ dB}) \approx \frac{0.51 z_0 \lambda}{a} = \frac{0.51 \lambda}{\sin \theta_0} = 1.02 \lambda F \qquad (3.3.15)$$

where we call $F = z_0/2a = z_0/D$ the F number of a lens or transducer, and D its diameter. Thus

$$\frac{d_r(3 \text{ dB})}{D} = \frac{0.25 z_0 \lambda}{a^2} = 0.25 S \qquad (3.3.16)$$

So, by focusing, the effective beam diameter can be made smaller at its geometric focus than at the transducer, provided that $z_0 \ll 4a^2/\lambda$ (i.e., $S \ll 4$).

Sidelobes. Note that the beam intensity falls off radially as $(\text{jinc } X)^2$ where $X = ra/\lambda z_0$. A plot of this function is given in Fig. 3.2.6. The first sidelobe is at $kra/z_0 = 5.136$ or $r = 0.82 z_0 \lambda/a$, and is 17.6 dB lower in amplitude than the main lobe.

The sidelobe level is important because it dictates how well the transducer can respond to a wanted signal rather than to an interfering one. Suppose that a focused transducer is used to search for two point sources, A and B, and that point source B emits power at a level 30 dB higher than does A. If the transducer is focused on A, and B happens to be located at a distance $r = 0.82 z_0 \lambda / a$ from A, we see B on the first sidelobe of the transducer and are unable to detect the presence of A; therefore, we cannot be certain where the true source is located. If B is only 10 dB higher in magnitude than A, however, the resulting sidelobes do not present a major problem, because the signal received from sidelobe B is 7.6 dB less than that received by A, and we can locate both sources.

Coherent and incoherent imaging. The sidelobe problem becomes even more severe if we wish to observe a large illuminated area in which dark spots are present, or a small absorber in a bright background. In the worst cases, the sum of the signals from the higher-order sidelobes may be much larger than the signal from the main lobe.

As an example, the signal amplitude variation across a one-dimensional object with a spatial step function insonification was calculated for the one-dimensional sinc X response by Lemons and Quate [5]; it is illustrated in Fig. 3.3.2. We assume that the receiving transducer is being moved past the object. As we observe from Fig. 3.3.2(a), there is quite severe "ringing" near a sharp edge when the step object function is insonified with a coherent plane wave. This problem is often severe in coherent imaging systems. By contrast, as shown in Fig. 3.3.2(b), when we employ an incoherent imaging system (i.e., one in which a noise source or a source with a broad range of frequencies present is used), there are no phase additions of the signals. Thus only the intensities (the square of the amplitudes), rather

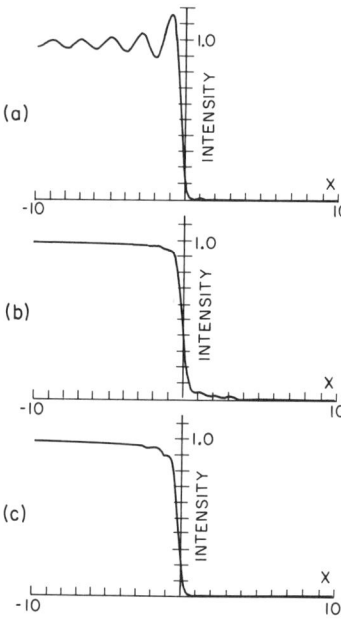

Figure 3.3.2 Calculated one-dimensional images of a step function object: (a) conventional coherent image; (b) conventional incoherent image; (c) image produced by a confocal scanning system using coherent radiation. (From Lemons and Quate [5].)

than the amplitudes of the sidelobes, add and the ringing effect virtually disappears. When two confocal lens are used, one for insonification and the other as a receiver, with a knife-edged object between them, the spatial response is squared. Hence the sidelobe level amplitudes now vary as (sinc $X)^2$ in the one-dimensional case [(jinc $X)^2$ in a cylindrically-symmetric system]; this squaring results in a clean image of a sharp edge, as shown in Fig. 3.3.2(c). It can be shown that the amplitude response of a confocal system to a step function object is identical to the intensity response of an incoherent single lens system to a step function object. Hence the confocal system gives a sharper edge response than an incoherent single lens system.

Rayleigh two-point definition. Obviously, the use of the simple 3-dB definition is not necessarily adequate. Consequently, it is more common to use the *Rayleigh definition*, which is based on the *two-point definition*. This definition uses the idea that two neighboring point sources can be distinguished from each other if the maximum response to one is located at the first zero of response to the second point. On this basis, the two points are spaced by a distance d_{cr} (Rayleigh), which is defined as

$$d_{cr}(\text{Rayleigh}) = 1.22\lambda F = \frac{0.61\lambda}{\sin \theta_0} \tag{3.3.17}$$

where the subscript c denotes coherent sources and the subscript r denotes definition in the radial direction.

The two-point response observed in practice depends on the type of insonification used. With coherent insonification, the result will depend on the relative phases of the two point sources. If the phase difference between the insonification of the two points is ϕ, the total intensity, defined as the square of the magnitude of the response to the two coherent sources, $I_c(X)$, is of the form

$$I_c(X) = |\text{jinc } (X - 0.30) + e^{j\phi} \text{ jinc } (X + 0.30)|^2 \tag{3.3.18}$$

This function is plotted in Fig. 3.3.3 for different phases ϕ.

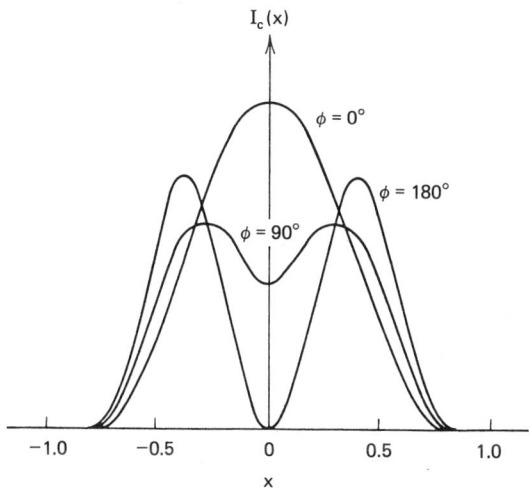

Figure 3.3.3 Image intensity for two mutually coherent point sources separated by the Rayleigh distance.

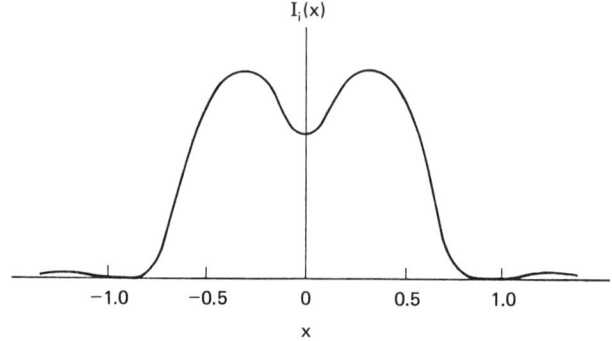

Figure 3.3.4 Image intensity for two incoherent point sources separated by the Rayleigh distance.

When the two coherent sources are 180° out of phase, there is a deep null at the point midway between them, as we might expect. However, if the two coherent sources are in phase, there is no minimum and thus the two points cannot be distinguished. When the sources are 90° out of phase, there is a minimum of intensity at the center point, which is 26.5% below the maximum value of the intensity.

On the other hand, if the sources are insonified completely incoherently, as is possible in optical systems and only possible to a limited extent, using very short pulses in acoustic systems, there is no correlation between the two sources and hence the product terms between them do not contribute to the total intensity. Thus the intensity for incoherent sources $I_i(X)$ is just the sum of the squares of the two individual intensities, or

$$I_i(X) = \text{jinc}^2 (X - 0.30) + \text{jinc}^2 (X + 0.30) \tag{3.3.19}$$

where the subscript i denotes incoherent sources. This function is plotted in Fig. 3.3.4. Because there is now a 26.5% dip at the center point, the two sources can be distinguished from each other.

Therefore, the Rayleigh criterion of definition for incoherent imaging is often stated in this form: *Two points can be distinguished if there is a 26.5% dip in intensity at the midpoint between them.* This definition corresponds to a separation between the two points by a distance $d_{ir}(\text{Rayleigh})$, defined as

$$d_{ir}(\text{Rayleigh}) = 1.22\lambda F = \frac{0.61\lambda}{\sin \theta_0} \tag{3.3.20}$$

Note that with coherent, in-phase sources, the two points must be separated by a distance $d_{cr}(\text{Rayleigh})$, defined as

$$d_{cr}(\text{Rayleigh}) = 1.64\lambda F = \frac{0.82\lambda}{\sin \theta_0} \tag{3.3.21}$$

to obtain the same 26.5% dip in level between them. Thus the use of incoherent sources improves the Rayleigh definition by a factor of 1.34 over the worst case for coherent sources.

Sparrow two-point definition. For two-point definition, we must be able to distinguish the dip in level between the images of the two points. This criterion is far more general in application than simply placing one point where the zero response to the other point occurs. It can apply equally well to coherent imaging or, for example, a Gaussian beam, where there is no sharp spatial zero in response to a point source. Detection of the minimum level in the two-point response must, in practice, depend on the noise level of the system and the properties of the display system, which means that the 26.5% dip in the Rayleigh definition is an arbitrary requirement. It is therefore worthwhile to consider yet another two-point definition, the *Sparrow criterion* [6]: *The intensity halfway between the two points is just equal to the total intensity at one point.*

Calculations define the Sparrow criterion for incoherent imaging, d_{ir}(Sparrow), as

$$d_{ir}(\text{Sparrow}) = 1.02\lambda F = \frac{0.51\lambda}{\sin \theta_0} \qquad (3.3.22)$$

while for two coherent in-phase sources, it is d_{cr}(Sparrow), defined as

$$d_{cr}(\text{Sparrow}) = 1.42\lambda F = \frac{0.76\lambda}{\sin \theta_0} \qquad (3.3.23)$$

We shall use this criterion in Sec. 3.3.3, as well as the Rayleigh criterion, to consider the scanning microscope. Note that for incoherent imaging, the Sparrow criterion is almost identical to the single-point 3-dB criterion, while for coherent imaging, it leads to a definition approximately 1.5 times worse than that for incoherent imaging.

Speckle. The *speckle* phenomenon occurs only in coherent imaging systems and often leads to images of poor quality. When an optical image is illuminated under coherent laser light rather than room light, for instance, speckle is what gives its appearance a granular character. One reason for this phenomenon is that on a rough surface (rough in the scale of wavelengths of light), light will be reflected by many points into a focused receiver lens or transducer. The transducer response to these random sources depends on the sidelobe level at the random source. When the receiver is focused on a point A, which is illuminated by a broad beam, as illustrated in Fig. 3.3.5, the contributions from all the random scatterers may add to the contribution from the main lobe, thus giving rise to a bright spot at A. On the other hand, when the receiver is focused on another point B, the reflections from the random scatterers outside the main lobe may tend to cancel it out, hence creating a dark spot at the point B in the image. As the transducer is moved from point to point, some points will be darker, and others brighter, than the average beam intensity, thus displaying the speckle phenomenon. Similar phenomena can occur in images of semitransparent media such as body tissue, where the random additions from points in front of and behind the focal point give rise to speckle. The spots observed in the image will tend to have a size of the order of the focal spot size d_r, and the image will have a granular appearance.

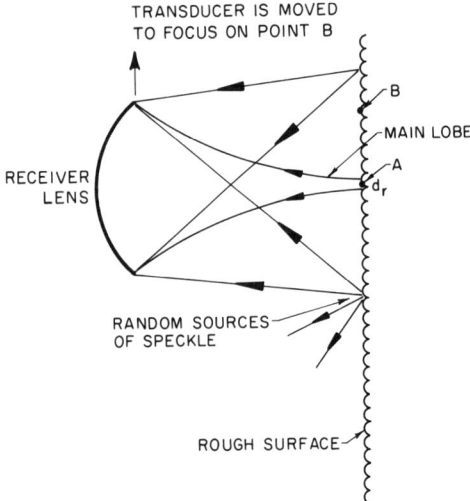

Figure 3.3.5 Speckle phenomenon. The broad insonifying beam, which is presumed to give uniform illumination over the whole area, is not shown.

Speckle is usually present in acoustic imaging systems because of the coherent sources normally used, the coherent detectors employed, and the fact that the typical media (e.g., body tissue and polycrystalline materials) are granular in nature and semitransparent. As we will discuss further in Sec. 3.5.3, the phenomenon can be minimized by using a mechanically scanned confocal imaging system where the transmitter insonifies only a small area, corresponding to the focal diameter of the receiver lens. When this is used, the point response is a (jinc X)^2 function, which is always positive; the sidelobe and speckle levels are very low and the mechanical scan gives signal averaging. Using short pulses for insonification can also minimize speckle by making the system behave as if it were partially incoherent.

Depth of focus. We now determine the variation of the acoustic potential, and hence the pressure and displacement along the axis of the lens. Just as with the plane cylindrical piston transducer, we will find that the fields along the axis of the focused transducer vary rapidly. In the limit $z_0 \ll a^2/\lambda$ or $S \ll 1$ (i.e., deep in the Fresnel zone), we can use the results to define a depth of focus d_z(3 dB) as the distance between the points where the field on axis is 3 dB less than that at the focal point.

We return to Eq. (3.3.7) and put $r = 0$ with $z \neq z_0$. The equation can then be integrated directly to yield

$$\phi(0, z) = \phi(0, z_0) Z(z) e^{-jkz} e^{(-j/2S)(z_0/z - 1)} \qquad (3.3.24)$$

where

$$Z(z) = \frac{z_0}{z} \operatorname{sinc}\left[\frac{1}{2S}\left(\frac{z_0}{z} - 1\right)\right] \qquad (3.3.25)$$

and sinc $X = (\sin \pi X)/\pi X$ and $S = z_0\lambda/a^2$.

It also follows from Eq. (3.3.24) that in addition to the phase change kz, there is a π phase change through the focal point. This is due to the term $\exp\{-j[(z_0/z) - 1]/2S\}$.

We shall assume that the beam intensity is proportional to $|\phi(z)^2|$. We may find the 3-dB range definition, deep in the Fresnel zone where $a^2/\lambda z_0 \gg 1$ (i.e., $S \ll 1$), by determining where the argument of the sinc function becomes $1/\sqrt{2}$; in this case, we can ignore the variation of z_0/z. Thus, with this approximation, the 3-dB points occur where

$$z - z_0 = \pm 0.89 \frac{z_0^2}{a^2} \lambda = \pm \frac{0.89\lambda}{\sin^2 \theta_0} \quad (3.3.26)$$

where θ is the half-angle subtended by the lens at the optical focal point 0, z_0. Thus a simple approximation for the depth of focus is

$$d_z(3 \text{ dB}) = \frac{1.8\lambda}{\sin^2 \theta_0} = \frac{1.8 z_0^2}{a^2}\lambda = 7.1\lambda F^2 = 3.5 \frac{z_0}{a} d_r(3 \text{ dB}) \quad (3.3.27)$$

Another useful way to write this relation is in the form

$$\frac{d_z(3 \text{ dB})}{z_0} = 1.8S \quad (3.3.28)$$

We see that the depth of focus is directly related to the Fresnel parameter of the lens. When the depth of focus is very small, we can write $z \approx z_0$ and use only first-order terms in $z - z_0$. In the region where the beam size is small, Eq. (3.3.25) takes the simple form

$$Z(z) = \text{sinc} \frac{z - z_0}{2S z_0} \quad (3.3.29)$$

This expression is very convenient for estimating the beam intensity along the axis of the lens, at least near the focal plane. The sinc function is plotted as the dashed line in Fig. 3.3.6(a), and also in Fig. 3.2.9.

Geometrical concepts for depth of focus. These relations for the depth of focus can be illustrated geometrically, as shown in Fig. 3.3.7. The beam initially converges along its geometrical optics path (a cone) but diverges from this path near the focus. From Eq. (3.3.14), the radius of the beam to its zero amplitude point at the focal plane is $r_0 = 0.61\lambda z_0/a$. A cylinder of this radius intersects the edge of the geometrical beam where

$$\frac{z - z_0}{z_0} = \pm \frac{r_0}{a} = \pm 0.61 \frac{z_0 \lambda}{a^2} \quad (3.3.30)$$

Thus the depth of focus estimated by this approach is

$$d_z = 1.22\lambda \left(\frac{z_0}{a}\right)^2 \quad (3.3.31)$$

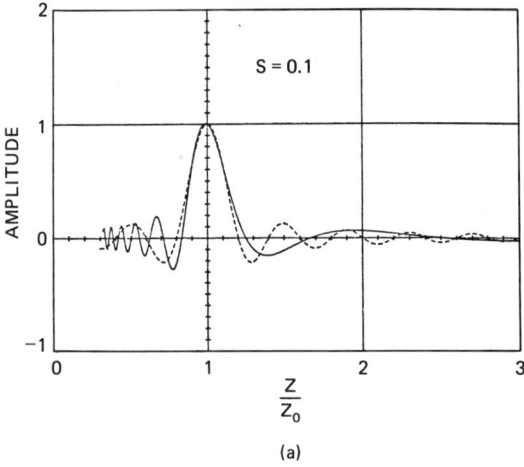

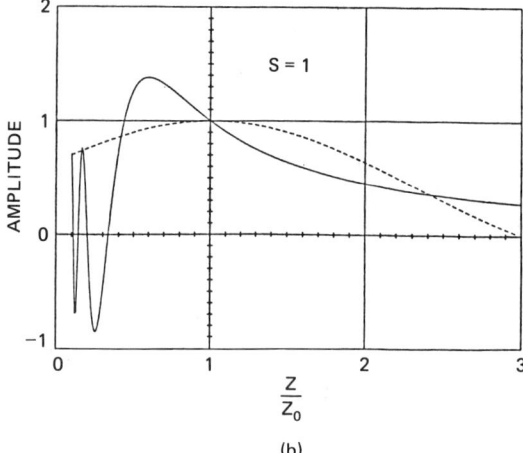

Figure 3.3.6 Normalized pressure or potential variation along the axis of a spherical lens: (a) $S = 0.1$; (b) $S = 1$. We take the amplitude to be unity at $z = z_0$. Solid line, plot of Eq. (3.3.25); dashed line, plot of Eq. (3.3.29).

This geometrical picture is a satisfying concept, but it is unreliable, for its estimate of beam intensity is, to say the least, crude.

Equations (3.3.25) and (3.3.29) show that the wave amplitude along the axis passes through several subsidiary maxima and minima. Thus simple geometrical considerations do not provide a complete picture of the behavior of the beam. In particular, if we wish to use the beam to probe for small reflectors, we may obtain inaccurate and misleading results when the reflector is outside the region of the depth of focus. This problem is illustrated in Fig. 3.3.6, which shows the multiple maxima and minima along the axis of the beam; the solid line plots are calculated from Eq. (3.3.25) and the dashed line plots, from Eq. (3.3.29). The situation illustrated in Fig. 3.3.6(a) corresponds to a value of $S = 0.1$, or a ratio of focal spot diameter d_r(3 dB) to a transducer radius of $d_r/2a = z_0\lambda/4a^2 = 0.025$.

For a weaker lens with $S = 1$ (i.e., $d_r/2a = 0.25$), as illustrated in Fig. 3.3.6(b), the formulas of Eq. (3.3.27) are not as accurate. In this case, the fields

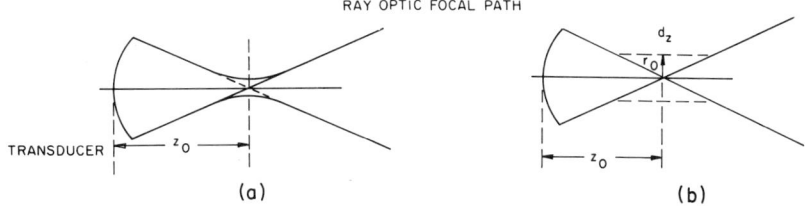

Figure 3.3.7 Focused beam from a spherical transducer. (a) The beam initially converges, as we expect from geometrical optics, but diverges from the geometrical optics path when it nears the focus. (b) Demonstration that the depth of focus can be estimated geometrically. A cylinder of radius r_0 passes through the zero amplitude points at the focal plane. The depth of focus is taken to be the length of this cylinder where it intersects the cone formed by the outside of the geometrical optic beam.

actually increase as z decreases from z_0; this is due to the variation of the z_0/z term in Eq. (3.3.25). Thus, although Eq. (3.3.25) is accurate for a wide range of the parameter S, the approximate formula of Eq. (3.3.27) for the depth of focus is completely reliable only for highly convergent lenses.

A convenient set of lens design curves is given in Fig. 3.3.8. We call the true focal length of the lens f, so that $\phi(z)$ or $u_z(z)$ is maximum on the axis at $z = f$. The parameter f/z_0 is plotted in Fig. 3.3.8; it decreases as S increases. Thus the effective focal length of a weak lens is less than its geometrical optics focal length. As we can see from Fig. 3.3.8, the true value of the depth of focus, $d_z(\text{true})$, is also less than its approximate value, $d_z(\text{approx.})$, which is given by Eq. (3.3.28). Because the focal point is nearer to the lens, the amplitude of $Z(f)$ is larger than the value of $Z(z_0)$.

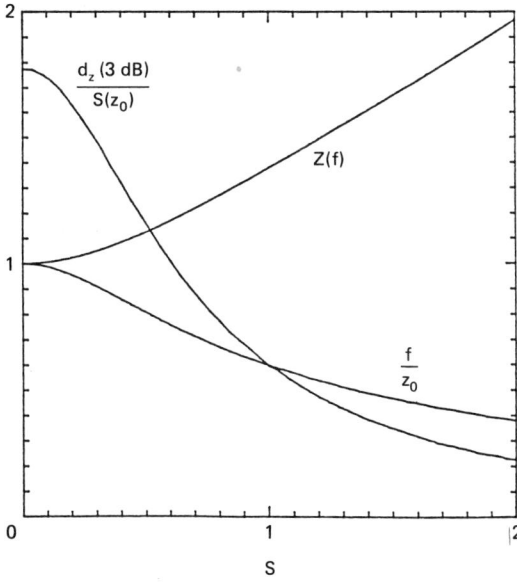

Figure 3.3.8 Plot of the effective parameters of weakly focused lenses as a function of $S = z_0\lambda/a^2$.

Example: Focused Transducer for Medical Applications

We now consider a medical example. Suppose that we require a beam diameter d_r(3 dB) = 1.5 mm, at a distance of 10 cm from the body surface, at a frequency of 2.25 MHz (λ = 0.67 mm). From Eq. (3.3.15), a = (0.5 × 10 × 0.067)/0.15 (i.e., a = 2.2 cm). In this case, S = 0.14 and Eq. (3.3.27) shows that the depth of focus is 2.5 cm. Figure 3.3.8 shows that f has, in fact, decreased to 9.6 cm and that with these parameters, d_z(true) = 2.4 cm.

Now suppose that we try to obtain a much larger depth of focus, d_z(3 dB) = 10 cm, with z_0 = 10 cm. In this case, Eq. (3.3.27) yields the result a = 1.1 cm; hence, from Eq. (3.3.15), the beam diameter at the optical focal point is d_r(3 dB) = 3 mm, with S = 0.56. But Fig. 3.3.8 shows that the focal point is actually at 8 cm and that d_z(true) = 6.8 cm. One or more iterations would, of course, let us design for the correct d_z. Even in this case, however, we can obtain a fairly good estimate of the beam parameters by using Eq. (3.3.27).

Reflection from a plane reflector. Another important criterion for judging the behavior of an imaging system is based on the reflection of a focused beam from a perfect plane reflector. We shall determine, here, how the reflected signal $V(z)$, measured with the input transducer used as the receiver, varies with the position z of a planar reflector. We will assume that the surface of the planar reflector is perpendicular to the axis of the lens.

When a perfect plane reflector is located at the geometrical focus of the lens, we expect the reflected signal to attain its maximum value. Here we define a perfect reflector as one for which the reflection coefficient $R(\theta)$ of a plane wave incident on the plane at an angle θ to the normal is such that $R(\theta)$ = 1. More generally, as we have seen in Sec. 2.5, a plane wave incident on a solid may excite Rayleigh waves, or bulk shear and longitudinal waves, in the solid. Consequently, $R(\theta)$ may, in general, have an amplitude and phase that vary with θ, and $|R(\theta)| \leq 1$.

For a perfect reflector, we will take the normalized return signal $V(z)$ to be†

$$V(z) = \frac{\iint \phi^2(x, y, z)\, dx\, dy}{\iint |\phi^2(x, y, z_0)|\, dx\, dy} \quad (3.3.32)$$

where $\phi^2(x, y, z)$ is the point spread function at the plane z of this confocal lens system used in a reflection mode. We define a two-dimensional Fourier transform of $\phi(x, y, z)$ by writing

$$\phi(k_x, k_y, z) = \iint \phi(x, y, z) e^{-j(k_x x + k_y y)}\, dx\, dy \quad (3.3.33)$$

with

$$\phi(x, y, z) = \frac{1}{4\pi^2} \iint \phi(k_x, k_y, z) e^{j(k_x x + k_y y)}\, dk_x\, dk_y \quad (3.3.34)$$

---

†Equation (3.3.32) is adequate for use with the paraxial approximation. Liang et al. [7] have shown that a more accurate formula is

$$V(z) = \frac{\int \phi(x, y, z) u_z(x, y, z)\, dx\, dy}{\int \phi(x, y, z_0) u_z(x, y, z_0)\, dx\, dy}$$

where k_x and k_y are the transverse spatial frequencies, as discussed in Secs. 3.2.2 and 3.2.3.

It follows from Parseval's theorem that for a perfect reflector, Eq. (3.3.32) can be written in the following form [8]:

$$V(z) = \frac{\iint \phi(k_x, k_y, z)\phi(-k_x, -k_y, z) \, dk_x \, dk_y}{\iint \phi(k_x, k_y, z_0)\phi(-k_x, -k_y, z_0) \, dk_x \, dk_y} \tag{3.3.35}$$

Referring to Fig. 3.3.1, we may take the displacement $u_z(x', y', z_1)$ in the z direction at the baffle to be of the form

$$u_z(x', y', z_1) = P(x', y')e^{-jkR_1(x',y')} \tag{3.3.36}$$

where we call $P(x', y')$ the *pupil function* of the lens.

Writing Eqs. (3.3.4)–(3.3.5) in rectangular coordinates, we find that

$$R = \sqrt{(x'-x)^2 + (y'-y)^2 + (z-z_1)^2} \approx z - z_1 + \frac{(x-x')^2 + (y'-y)^2}{2z} \tag{3.3.37}$$

and

$$R_1 = z_1 - \frac{x'^2 + y'^2}{2z_0} \tag{3.3.38}$$

At $z = z_0$, with the use of the paraxial approximation, we find that

$$\phi(x, y, z_0) = -\frac{1}{2\pi} \frac{e^{-jkz_0}}{z_0} e^{-jk(x^2+y^2)/2z_0} \iint P(x', y')e^{jk(xx'+yy')/z_0} \, dx' \, dy' \tag{3.3.39}$$

We assume that the focused acoustic beam at the geometrical focal plane has a relatively small cross section, so that $k(x^2 + y^2)/2z_0 \ll 1$. In this case, Eq. (3.3.39) becomes

$$\phi(x, y, z_0) = -\frac{e^{-jkz_0}}{2\pi z_0} \iint P(x', y')e^{jk(xx'+yy')/z_0} \, dx' \, dy' \tag{3.3.40}$$

From this result, we see that the potential at the geometrical focal plane is the Fourier transform of the pupil function.

Comparing Eqs. (3.3.34) and (3.3.40), we see that

$$\phi(k_x, k_y, z_0) = -\frac{z_0 e^{-jkz_0}}{2\pi k^2} P(x', y') \tag{3.3.41}$$

where

$$k_x = \frac{kx'}{z_0}$$
$$k_y = \frac{ky'}{z_0} \tag{3.3.42}$$

Thus we can regard the plane wave component k_x, k_y as being associated with the ray leaving the plane $z = z_1$ from the point x', y'.

The form of the plane wave k_x, k_y at the plane z is

$$\phi(k_x, k_y, z) = \phi(k_x, k_y, z_0) e^{-jk_z(z-z_0)} \tag{3.3.43}$$

where, from the wave equation for plane waves,

$$k_x^2 + k_y^2 + k_z^2 = k^2 \tag{3.3.44}$$

We substitute Eqs. (3.3.41), (3.3.43), and (3.3.44) in Eq. (3.3.35), to show that

$$V(z) = \frac{e^{-2jkz_0} \iint P(x', y') P(-x', -y') e^{-2jk(z-z_0)\sqrt{1-(x'^2+y'^2)/z_0^2}} \, dx' \, dy'}{\iint P(x', y') P(-x', -y') \, dx' \, dy'} \tag{3.3.45}$$

In cylindrical coordinates, this expression becomes, for a symmetric lens,

$$V(z) = \frac{e^{-2jkz_0} \int P^2(r') e^{-2jk(z-z_0)\sqrt{1-r'^2/z_0^2}} r' \, dr'}{\int P^2(r') r' \, dr'} \tag{3.3.46}$$

Let us now consider the result for a lens in which $P^2(r') = 1$ from $r' = 0$ to $r' = a$. It is convenient to write $r' = z_0 \sin \theta$ with $a = z_0 \sin \theta_0$. In this case,

$$V(z) = \frac{e^{-2jkz_0} \int_0^{\theta_0} e^{-2jk(z-z_0)\cos\theta} \sin\theta \cos\theta \, d\theta}{\int_\theta^{\theta_0} \sin\theta \cos\theta \, d\theta} \tag{3.3.47}$$

For a paraxial system, we can put $\cos \theta = 1$ in the amplitude terms, to find that

$$V(z) = \frac{e^{-2jk(z-z_0)\cos\theta_0} - e^{-2jk(z-z_0)}}{2jk(1 - \cos\theta_0)(z - z_0)} e^{-2jkz_0} \tag{3.3.48}$$

or

$$V(z) = \frac{\sin[k(z-z_0)(1-\cos\theta_0)]}{k(z-z_0)(1-\cos\theta_0)} e^{-jk(z-z_0)(1+\cos\theta_0)} e^{-2jkz_0} \tag{3.3.49}$$

This is the response $V(z)$ of a lens to a perfect plane reflector.

3-dB resolution for V(z). It follows from Eq. (3.3.48) that the 3-dB points for $V(z)$ from a plane reflector are spaced by a distance $d_{ps}(3 \text{ dB})$, where

$$d_{ps}(3 \text{ dB}) = \frac{0.45\lambda}{1 - \cos\theta_0} \tag{3.3.50}$$

For θ_0 small, Eq. (3.3.50) can be written in the form

$$d_{ps}(3 \text{ dB}) = \frac{0.9\lambda}{\theta_0^2} \tag{3.3.51}$$

Thus this result is similar to that obtained from the depth of focus (i.e., the response on the axis) of a single lens. The depth of focus for a two-way or confocal reflecting system, or the response for reflection from a point reflector, is less than this value (i.e., it is where the sinc² function drops by 3 dB). We call the confocal depth of focus $d_{zs}(3 \text{ dB})$; its value is

$$d_{zs}(3 \text{ dB}) = \frac{1.28\lambda}{\sin^2 \theta_0} \qquad (3.3.52)$$

Thus the 3-dB response points for a plane reflector are closer together than the 3-dB depth of focus for a point reflector.

The analysis we have carried out here is similar to that given by Atalar [9]. The reader is referred to his work and to that of Liang et al. [7] for more complete derivations. Liang et al. derived a more rigorous nonparaxial formulation based on the reciprocity theorem; both sets of authors derived formulas for an imperfect reflecting plane. The results they obtained for this latter case are similar to Eq. (3.3.46) and can be written, in our terms, in the form

$$V(z) = \frac{e^{-2jkz_0} \int P^2(r') R(r') e^{-2jk(z-z_0)\sqrt{1-r'^2/z_0^2}} r' \, dr'}{\int P^2(r') r' \, dr'} \qquad (3.3.53)$$

where $R(r')$ is the plane wave reflectivity function for a wave incident on the plane reflector at an angle θ, and $r' = z_0 \tan \theta$.

This result has been of great help in understanding the behavior of the acoustic microscope and its contrast mechanisms, which we discuss in the following section.

3.3.2 Scanned Acoustic Microscope

The acoustic microscope is an acoustic equivalent of the optical microscope [5, 7, 9–19]. It has been employed in both transmission and reflection modes, with water, liquid argon, liquid nitrogen, high-pressure gases, or liquid helium used as the operating medium. The acoustic wavelength in water is approximately 5000 Å at 3 GHz, which is comparable to wavelengths normally employed in the optical microscope. In practice, the device has been used over a wide range of frequencies, ranging, at the time of writing, from 2 MHz to 8 GHz, with definitions in water of better than 4000 Å at an operating frequency of 3 GHz, and in liquid helium of 300 Å at an operating frequency of 8 GHz. The upper-frequency limit, at normal temperatures, is determined by the attenuation in the operating medium; in most materials, this increases as the square of the frequency. In liquid helium, at temperatures of the order of 0.1°K, the attenuation is very low, so the upper-frequency limit is determined basically by problems of exciting and receiving acoustic waves.

Water has the dual advantages of convenience and having the lowest attenuation at room temperature (191 dB/cm at 1 GHz) of the possible operating media [5, 10]. At 60°C, its attenuation drops to 95 dB/cm, so high-frequency microscopes are often operated at this temperature using a water path in the range 30 to 100 μm.

Other operating media provide interesting alternatives to water. Liquid nitrogen and liquid argon have been used because they have low attenuations, and lower acoustic velocities than water [5]. Liquid helium, which can be used at temperatures as low as 0.1°K, has extremely low acoustic velocities and attenuation, and hence a very small wavelength and fine definition [10]. A high-pressure gas can provide yet another low-velocity operating medium with relatively low attenuation [11]. Typically, with many of these media, the acoustic matching problem is severe due to their very low acoustic impedances.

The acoustic microscope responds to different properties in the object under examination from those responded to by the optical microscope. Because acoustic waves measure mechanical properties, the contrast of the microscope is determined by variations in elasticity, density, and acoustic attenuation in the medium. Therefore, acoustic wave devices are particularly suited to observing soft tissue or biological cell structure, which tend to be transparent to light. And because acoustic waves can penetrate optically opaque materials, the acoustic microscope is also very useful for observing features in integrated-circuit structures under metal films and other such optically opaque structures. It can also be used to measure the surface structure of metals. Because the individual grains of a polycrystalline material are anisotropic, their acoustic reflectivity varies from grain to grain, according to their orientations; thus the surface structures of metals show up well acoustically.

The optical microscope has been undergoing a long period of development since it was invented 300 years ago. Extremely sophisticated compound lens designs with very little chromatic or spherical aberration are now used routinely. But it is extremely difficult to contrive such low-aberration designs for acoustic lenses, because severe off-axis aberrations result from the large changes in refractive indices between the media employed. Another difference between the two types of devices is that conventional optical microscopes are designed to use incoherent illumination with phase-insensitive detectors (the eye, a photographic film, or a photodetector), while acoustic insonification tends to be coherent and acoustic receiving transducers are phase-sensitive.

As discussed in Sec. 3.3.1, a single-lens system with a coherent source will tend to give images containing a large amount of speckle. A different approach, using *confocal scanned microscopy*, has been taken in the design of the acoustic microscope. A high-quality focus is obtained only on the axis of a lens; in a transmission system, both the transmitting and receiving lenses are focused on the same point, with either the object or the lenses moved mechanically. This produces a scanned image, in much the same manner as a TV raster scan that has very little speckle, (see the discussion of speckle in Sec. 3.3.1). The output of the scanning microscope is used to modulate the intensity of the spot on a cathode ray tube, which is itself moved synchronously over the face of the tube with the mechanical movement of the object under the microscope.

An early reflection mode version of the acoustic microscope lens system is shown in Figs. 3.3.9 and 3.3.10. A plane wave is excited by a thin film zinc oxide transducer on sapphire, which has a high longitudinal acoustic wave velocity of 11.1 km/s. A spherical depression is cut in the opposite surface of the sapphire to form a lens. A drop of water or some other fluid is placed between the lens and the flat object to be observed; the surface of the object is located close to the

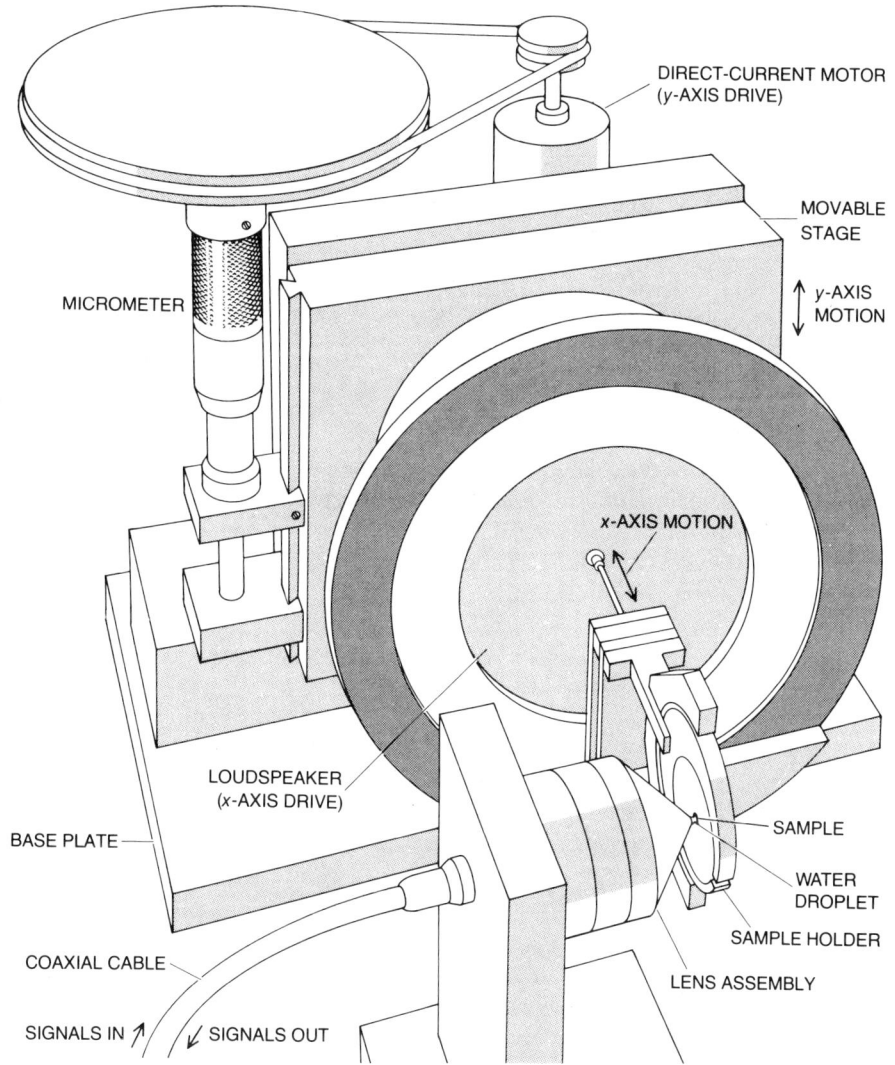

Figure 3.3.9 Scanned acoustic microscope. (After Quate [16].)

focal point of the lens. The transducer lens system is used as both transmitter and receiver for short RF tone bursts. After amplification and detection, the output from the receiving transducer is used to modulate the intensity of a cathode ray tube. At points where the effective reflectivity of the object is small, the electrical signal output is small; conversely, at points where the effective reflectivity of the object is large, the output is strong.

In the early version of the device, a picture was formed by scanning the object mechanically with a loudspeaker movement, as shown in Fig. 3.3.9; this produced the line scan. The lens system was scanned at right angles to the line scan, but much more slowly, by a piezoelectric or hydraulic movement; this formed the frame

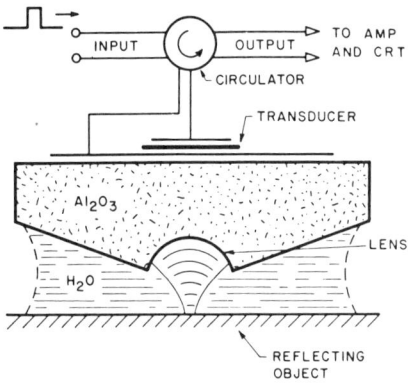

Figure 3.3.10 Lens of the acoustic microscope. (After Lemons and Quate [5].)

scan. More recent versions of the device use more sophisticated mechanical scans, with the lens scanned in both directions so that the object under observation can remain stationary. Great care is taken to ensure that the scan is repeatable.

We expect this type of lens to have a focal point on its axis but possibly to suffer from severe spherical aberrations (i.e., aberrations resulting from the inaccuracy of the paraxial approximation). Consider the ray geometry shown in Fig. 3.3.11. Ideally, if the velocity ratio between the two media were infinity, then, from Snell's law, all rays would be refracted to the center of curvature of the lens; thus there would be no spherical aberration, which would imply, in Fig. 3.3.11, that $\theta' = 0$. For a typical optical lens material in which the velocity ratio between the two media is of the order of 1.5, the aberrations are extremely severe, as illustrated in Fig. 3.3.12. In the case of sapphire and water, the ratio of the acoustic velocities is 7.4:1, which is large enough to cause very little aberration.

Figure 3.3.12 also shows calculations for materials with low and high velocity ratios, using a high ratio of acoustic velocities in the two media; in the acoustic case, the deviation from the paraxial focus is essentially undetectable. In practice, this implies that the aberrations are so weak that the diffraction limit of the lens is not affected. Thus a lens with almost perfect on-axis performance can be made for the acoustic microscope in a relatively simple manner.

The microscope illustrated in Figs. 3.3.9 and 3.3.10 has the additional ad-

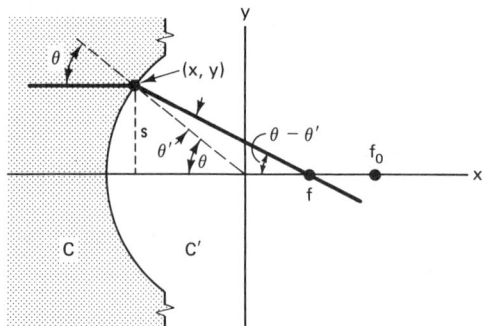

Figure 3.3.11 Geometry for the ray-tracing analysis of spherical aberrations. (After Lemons, as noted in Lemons and Quate [5, 17].)

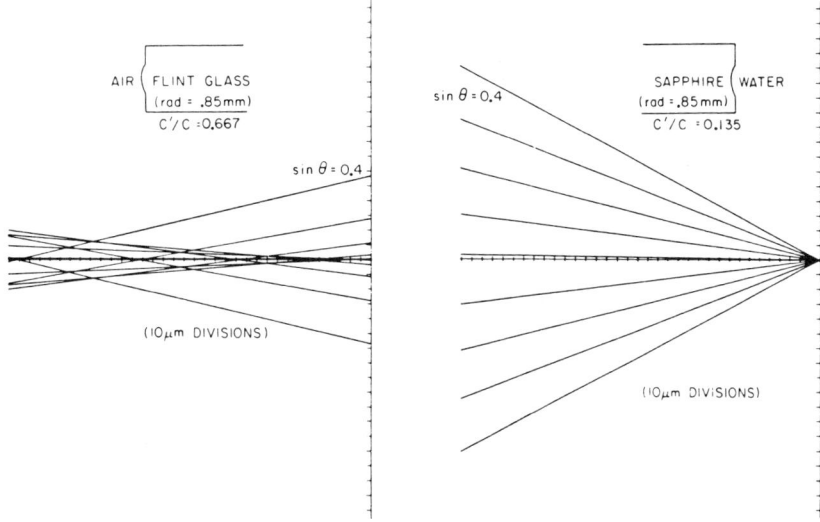

Figure 3.3.12 Ray-tracing comparison of the performance of a single-surface lens, with C' the wave velocity in the solid and C the wave velocity in the air or water medium: (a) light optical system; (b) Acoustic system. (After Lemons and Quate, as noted in Lemons and Quate [5, 18].) The paraxial focus lies at the ordinate.

vantage that the lens is used twice. Thus its amplitude response or point spread function (PSF) is the product of the response of the transmitting and receiving lenses, and the PSF is therefore (jinc $X)^2$ where $X = ra/\lambda z_0 = r/\lambda \sin \theta_0$.

Note that this function is always positive and that the amplitude away from the point $X = 0$ ($r = 0$) decreases much more rapidly than it does for the simple single lens. Consequently, the single-point 3-dB definition $d_{rs}(3 \text{ dB})$ is better for a confocal microscope than for a single lens, becoming, in this case,

$$d_{rs}(3 \text{ dB}) = \frac{0.37\lambda}{\sin \theta_0} \tag{3.3.54}$$

For a single lens, the 3-dB definition [see Eq. (3.3.15)] is

$$d_r(3 \text{ dB}) = \frac{0.51\lambda}{\sin \theta_0} \tag{3.3.55}$$

Here the subscript s denotes a confocal scanned system.

For the same reasons, the sidelobe levels for this double-lens configuration are a factor of 2 in dB lower than for a single lens. For a single lens, the first sidelobe is 17.6 dB below the main lobe amplitude, whereas for the scanning microscope double-lens system, the first sidelobe level is 35 dB below the main lobe, a radical improvement. Similarly, the total power outside the main lobe is considerably reduced from the simple lens result. This implies that when objects of finite area are insonified from a coherent focused source, there is relatively little illumination outside the main lobe, so the addition of the sidelobe amplitudes from this region does not give rise to severe spurious responses. As we have already seen in discussing Fig. 3.3.2, with incoherent insonification of a knife edge, the sidelobe intensities, rather than amplitudes, are added, so the problem is no longer serious. In the scanning microscope, the amplitudes of the sidelobes can add, but are equal in magnitude to the sidelobe intensities obtained with incoherent insonification. Thus its sidelobe problems are normally unimportant, and the ringing, speckle, and granularity of the images we normally expect with coherent imaging systems do not occur. This fact has already been illustrated from another viewpoint in our discussion of Fig. 3.3.2(c).

As we have seen throughout this section, the 3-dB definition is not always an adequate criterion for performance of an imaging system. Instead, it is better to use either the Rayleigh or Sparrow criterion for the two-point response, as described in Sec. 3.3.2. The Rayleigh definitions for the three cases of interest, incoherent imaging, coherent imaging, and confocal scanned coherent imaging, are

$$d_{ri}(\text{Rayleigh}) = \frac{0.61\lambda}{\sin \theta_0}$$

$$d_{rc}(\text{Rayleigh}) = \frac{0.82\lambda}{\sin \theta_0} \qquad (3.3.56)$$

$$d_{rs}(\text{Rayleigh}) = \frac{0.56\lambda}{\sin \theta_0}$$

respectively. Thus there is a slight improvement in the Rayleigh definition for the scanning microscopes, compared to an incoherently insonified system with a single lens of the same aperture. Similarly, we find that the Sparrow definitions for the same cases are

$$d_{ri}(\text{Sparrow}) = \frac{0.51\lambda}{\sin \theta_0}$$

$$d_{rc}(\text{Sparrow}) = \frac{0.76\lambda}{\sin \theta_0} \qquad (3.3.57)$$

$$d_{rs}(\text{Sparrow}) = \frac{0.51\lambda}{\sin \theta_0}$$

respectively.

As an example, the acoustic microscope of Lemmons and Quate typically has an acceptance angle θ_0 of 50° (i.e., an aperture of 0.77). Thus, at a frequency of 3 GHz in water, the Rayleigh two-point definition is 0.73λ or 3650 Å and the 3-dB and Sparrow definitions are 0.66λ or 3300 Å.†

The acoustic microscope has been used in transmission, for looking at thin biological samples. In the reflection mode, where the same lens and transducer

---

†With an aperture this wide, the paraxial formulation is not completely accurate. Here we have based our estimates on the value of $\sin \theta_0$. But it could equally be argued that using the F number z_0/D would be better; in this case, the estimated definitions would be slightly worse.

are used as both transmitter and receiver, the thickness of the sample is not necessarily limited, as it is in transmission microscopy, although the best definition is obtained only for objects near its surface. A reflection microscope, of course, is ideal for looking at integrated circuits.

We do not have space here to deal with the many possible applications and complexities of the acoustic microscope. It is important to realize, however, that the use of acoustics will produce images quite different from those obtained with optics. Acoustic waves respond to the elastic, rather than the optical, properties of the object under examination and these properties can often be of interest in medical applications. In addition, the contrast mechanisms of the acoustic microscope are not necessarily the same as those of the optical microscope. Attenuation plays its part, of course, but there are also phase-contrast mechanisms that occur due to the relatively large velocity changes between the media involved, and the fact that solids support more than one type of wave motion.

As an example of the contrast mechanisms which occur, we suppose, first, that the microscope is focused on the surface of a solid material under examination. As the impedance of many solids such as silicon, aluminum, and glass is very high compared to that of water, the reflection coefficient of the wave will be large and the image of the surface will be bright.

Now suppose that the lens is moved toward the solid so that the focal point is below its surface, as shown in Fig. 3.3.13. As discussed in Sec. 2.5.2, a Rayleigh wave can now be excited on the surface of the solid by rays leaving a point A on the lens at an angle of incidence θ_r, which satisfies the relation.

$$\sin \theta_r = \frac{V_w}{V_R} \tag{3.3.58}$$

where V_w is the acoustic velocity in the water medium of the miscroscope and V_R is the Rayleigh wave velocity [12–14]. The Rayleigh wave reradiates bulk waves into the water at an angle θ_r to the axis, which excite the lens at a point B, at the same distance from the axis as the point A [12].

A second near-axis ray incident on the surface of the solid, when reflected, can also reexcite the lens. The two resulting sets of reflected rays can be in or out of phase with each other; so the surface image may be dark or light depending

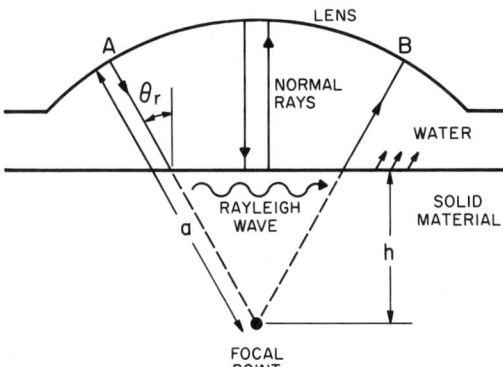

Figure 3.3.13 Rayleigh wave excitation by an acoustic microscope lens.

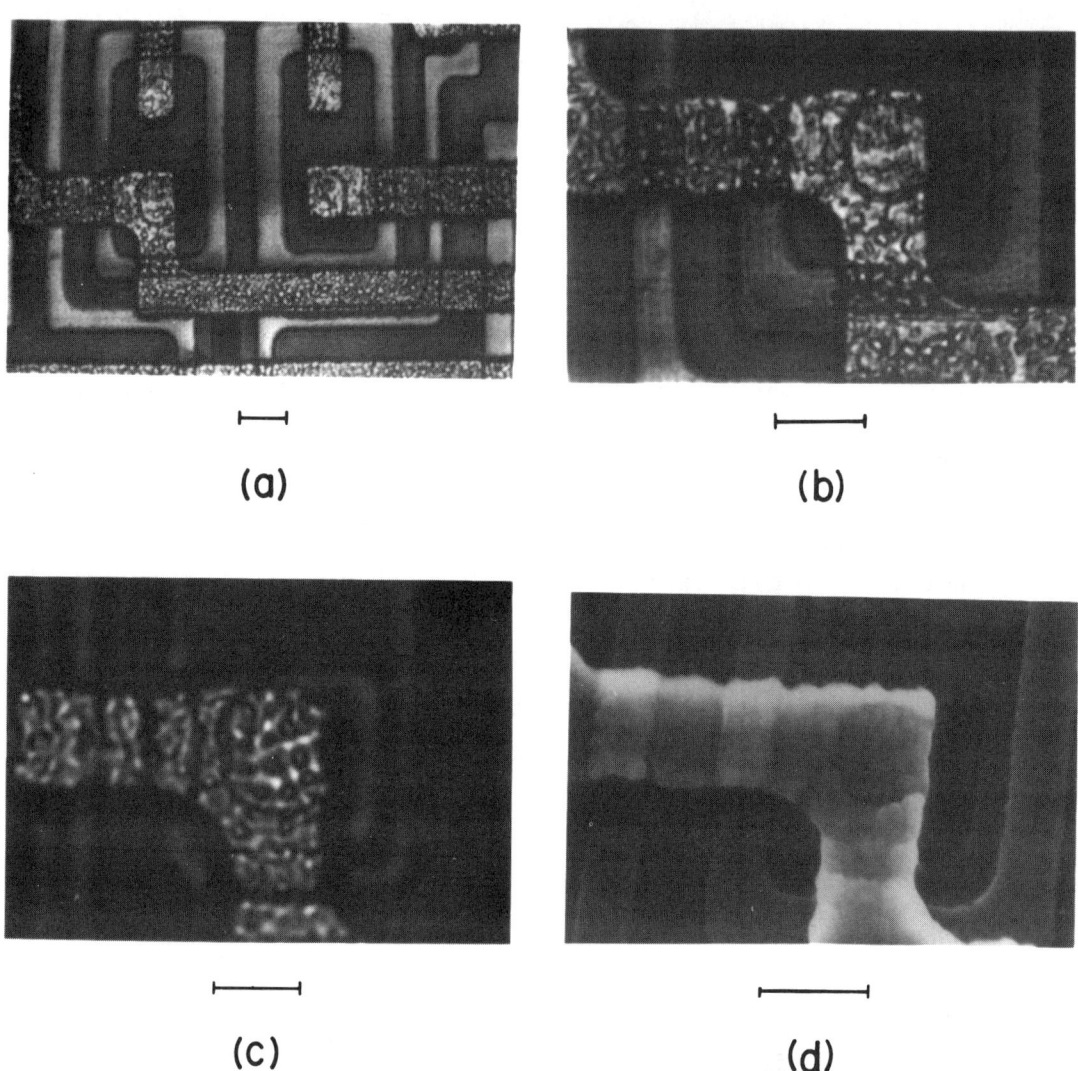

Figure 3.3.14 (a) Acoustic image of bipolar transistors on a silicon integrated circuit; (b) high-magnification acoustic image of the base contact of a transistor; (c) optical micrograph of the same area as in part (b); (d) SEM micrograph of the same area as in part (b). Scale bars are 3 μm. (After Hadimioglu and Quate [19].)

on the position $h = z_0 - z$ of the focus below the surface of the substrate. Therefore, a reversal of contrast can occur as a result of this mechanism (see Prob. 9).

By analyzing such contrast mechanisms, we can obtain a great deal of information about internal structure of solids [13–16, 19] (Problem 9). A particularly good example of this is the observation of layers underneath metal films; results of such experiments are shown in Figs. 3.3.14 and 3.3.15. In contrast to the optical situation, acoustic waves can propagate through a metal film and the amplitudes of the reflected waves will depend critically on its thickness. If there is also an oxide layer beneath the metal film, as there is in the gate structure of many integrated circuits, the reflected waves will change in amplitude and phase, and the image will be different. Thus it is often quite easy to determine the presence of an oxide layer beneath a gate, or whether there is good adhesion between a metal layer and a substrate.

The contrast mechanisms in the acoustic microscope are still being investigated and their regions of applicability determined. The importance of this device is that it can measure an entirely new range of parameters with definitions comparable to those of the best optical microscopes.

The basic principles of scanned acoustic microscopy have also led to the development of a new type of scanned coherent optical microscope with excellent definition. The scanning technique is also being employed for photoacoustic microscopy, where an optically focused scanned modulated laser beam is used to

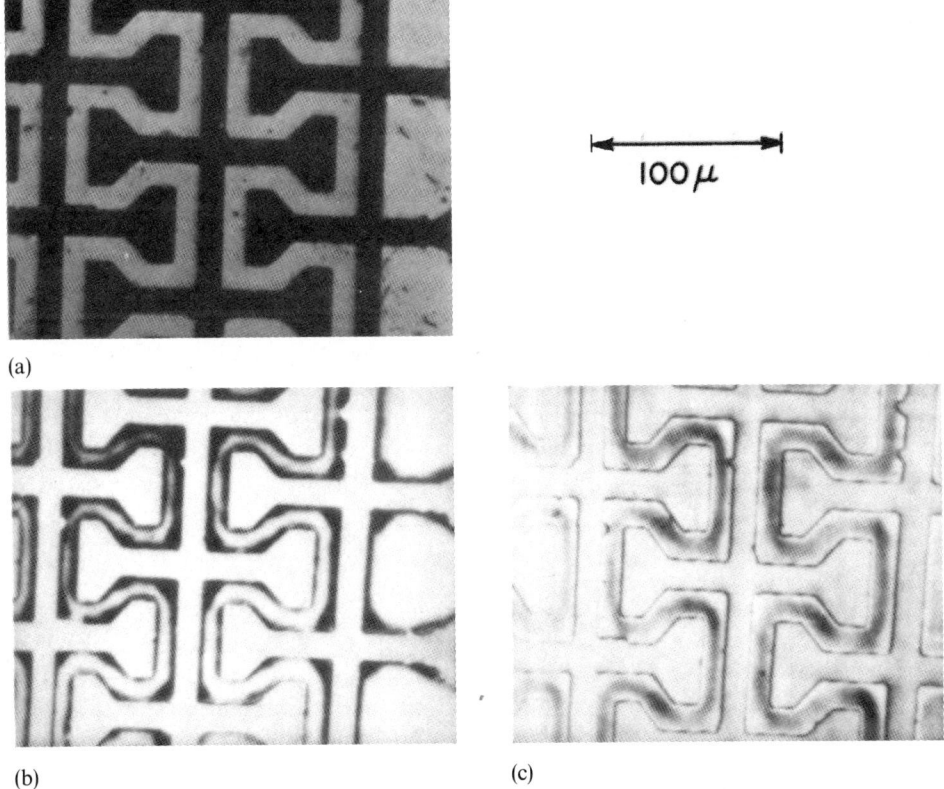

Figure 3.3.15 Optical (a) and acoustic (b) and (c) comparison of a 1000-Å layer of chrome on glass. The acoustic pictures are taken at a frequency of 2.6 GHz. In part (b), the focal spot has been moved 0.5 μm below the surface; in part (c), it has been moved −1 μm below the surface. The effects of poor adhesion can be seen clearly. (After Bray et al. [14].)

modulate the surface temperature of a solid. The consequent expansion and contraction of the solid excites acoustic waves, which can be detected by a relatively coarse acoustic transducer placed against the lower surface of the substrate. Acoustic waves can also be detected in the air or liquid medium above the substrate [20]. The images obtained are comparable in definition to those obtained with the optical beam. Similar ideas are also being applied to design a scanned x-ray microscope. The reader is referred to the proceedings of a conference on scanned image microscopy for further information on this subject [15].

3.3.3 Gaussian Beams and the Paraxial Equation

In Sec. 3.3.2 we showed that a spherical transducer produces a focused beam whose diameter at the focal plane is determined by the initial angle of convergence of the beam, and whose focal length is approximately equal to the radius of curvature of the transducer, which determines the shape of the initial phase front of the wave. The signal amplitude can vary rapidly along the axis and over radial cross sections. As we have seen, this is because of the rapidly varying phase differences between the rays emitted from neighboring points on the transducer; the more rapid the phase variation with radius, the more closely spaced are the maxima and minima. Thus at points near the transducer where the phase varies rapidly, both the axial and radial variations of the field are rapid and large.

One way to overcome this difficulty is to taper the amplitude of excitation of the transducer over its radius. It is clear that if the amplitude of the excitation falls off monotonically away from the central axis, the outer rays will not be able to cancel out the contributions from those near the axis; thus the field variation along the axis and over a radial cross section will be much less. In fact, as we shall show, a Gaussian taper gives the ideal amplitude variation, producing a beam with a Gaussian profile whose fields vary smoothly in all directions.

Wave equation for paraxial beams. Gaussian beams have been extensively studied in laser applications and are well understood [21–23]. For a longitudinal acoustic wave, we write the wave equation in the form

$$\nabla_T^2 \phi + \frac{\partial^2 \phi}{\partial x^2} + k^2 \phi = 0 \tag{3.3.59}$$

where $\nabla_T^2 \equiv \partial^2/\partial x^2 + \partial^2/\partial y^2$. We define ϕ as

$$\phi = f(r, z)e^{-jkz} \tag{3.3.60}$$

Then we make the approximation that $f(r, z)$ varies slowly enough that $|\partial f/\partial z| \ll |kf|$ and $|\partial^2 f/\partial z^2| \ll |k \partial f/\partial z|$. In this case, Eq. (3.3.59) can be written in the approximate form

$$\nabla_T^2 f - 2jk \frac{\partial f}{\partial z} = 0 \tag{3.3.61}$$

The usual treatment of laser optics is to obtain an exact solution of Eq. (3.3.59); this solution has the profile of a Gaussian beam [21–23]. Here we shall adopt a different approach, which gives the same solution. We do this to better understand the paraxial approximation and correlate the results with our earlier derivation for spherical lenses. We use Eq. (3.1.35) and write the potential at any point in the form

$$\phi(x, y, z) = -\frac{1}{2\pi} \iint u_z(x', y', 0) \frac{e^{-jkR}}{R} dx' dy' \qquad (3.3.62)$$

We now make the paraxial approximation, writing $R \approx z + [(x - x')^2 + (y - y')^2]/2z$, that is, we approximate the solution of the spherical wave equation $\phi = \exp(-jkR)/R$ by writing

$$\frac{e^{-jkR}}{R} \approx \frac{e^{-jkz}e^{-jk[(x-x')^2+(y-y')^2]/2z}}{z} \qquad (3.3.63)$$

Note that *this approximate result is an exact solution of Eq. (3.3.61)*. We can therefore use the paraxial approximation as the solution of Eq. (3.3.59), provided that the beam profiles with which we are concerned have only relatively slow variations in amplitude along the axis. In this case we can write Eq. (3.3.62) in the form

$$\phi(x, y, z) = -\frac{e^{-jkz}}{2\pi z} \iint u_z(x', y', 0) e^{-jk[(x-x')^2+(y-y')^2]/2z} dx' dy' \qquad (3.3.64)$$

Gaussian beam. We now consider a beam with a Gaussian profile at the plane $z = 0$. We shall show that the profile of this beam is Gaussian everywhere. We take the initial displacement amplitude of the beam at $z = 0$ to be

$$u_z = u_0 e^{-(r'/w_0)^2} \qquad (3.3.65)$$

where w_0 is the effective radius at the $1/e$ point of the Gaussian profile.

We substitute Eq. (3.3.65) into Eq. (3.3.64) and write

$$\phi(x, y, z) = \frac{u_0 e^{-jkz}}{2\pi z} \int_{-\infty}^{\infty}\int_{-\infty}^{\infty} [e^{-(x'^2+y'^2)/w_0^2}] \qquad (3.3.66)$$
$$\times \{e^{-jk[(x-x')^2+(y-y')^2]/2z}\} dx' dy'$$

or

$$\phi(x, y, z) = -\frac{u_0 e^{-jk(z+r^2/2z)}}{2\pi z} \int_{-\infty}^{\infty}\int_{-\infty}^{\infty} [e^{-(jk/2z+1/w_0^2)(x'^2+y'^2)}] \qquad (3.3.67)$$
$$\times [e^{-jk(xx'+yy')/z} dx' dy']$$

We complete the square in x' and y' in the argument of the exponential and carry out the infinite integrals using the relation

$$\int_{-\infty}^{\infty} e^{-\alpha x^2} dx = \sqrt{\pi/\alpha} \qquad (3.3.68)$$

We then find, after considerable algebra, that $\phi(x, y, z)$ is symmetric about the z

axis and can be written in the form

$$\phi(r, z) = \phi_0 \frac{w_0}{w(z)} e^{-j[kz - \eta(z)]} e^{-r^2[1/w^2 + jk/2R]} \quad (3.3.69)$$

This result is stated in terms of the following set of parameters:

$$z_0 = \frac{\pi w_0^2}{\lambda} \quad (3.3.70)$$

$$w^2(z) = w_0^2 \left(1 + \frac{z^2}{z_0^2}\right) \quad (3.3.71)$$

$$\eta(z) = \tan^{-1} \frac{z}{z_0} \quad (3.3.72)$$

$$R(z) = z + \frac{z_0^2}{z} \quad (3.3.73)$$

and

$$-jk\phi_0 = u_0 \quad (3.3.74)$$

The parameters $w(z)$, w_0, z, and $R(z)$ are illustrated in Fig. 3.3.16.

The potential varies smoothly along the axis and has a Gaussian amplitude variation with radius at all values of z. The radius of the beam at the $1/e$ point is $w(z)$. For $z \gg z_0$, the surfaces of constant phase are where

$$z + \frac{r^2}{2z} = \text{constant} \quad (3.3.75)$$

but the equation of a circle of radius $R \gg z_0$, centered on the origin in the region where $R \gg r$, is

$$R = \sqrt{z^2 + r^2} \approx z + \frac{r^2}{2z} \quad (3.3.76)$$

A comparison of Eq. (3.3.75) with Eq. (3.3.76) shows them to be identical in form. Therefore, the surfaces of constant phase are spherical and of radius R when $z \gg z_0$. Furthermore, Eq. (3.3.71) shows that the profile of the $1/e$ points of the beam

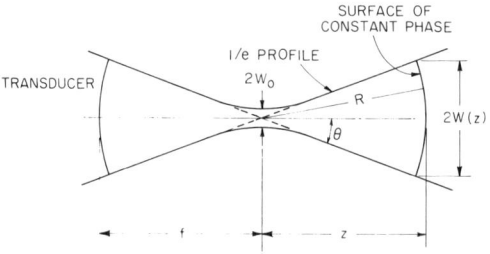

Figure 3.3.16 Profile of a Gaussian beam emitted from a transducer at $z = -f$.

$r = w$ is hyperbolic in shape, becoming a straight line in the regions where $z \gg z_0$. Thus the half-angle θ of the beam at the $1/e$ points is

$$\theta = \tan^{-1} \frac{w_0}{z_0} \approx \frac{\lambda}{\pi w_0} \tag{3.3.77}$$

We conclude that a Gaussian beam with an input half-angle θ will have a $1/e$ width at the focus (i.e., minimum beam diameter) of

$$d_r\left(\frac{1}{e}\right) = 2w_0 \approx \frac{2\lambda}{\pi \theta} \tag{3.3.78}$$

If the exciting transducer intercepts the axis at $z = -f$, the relation between f and $w(f)$ is

$$f = z_0 \left(\frac{w^2}{w_0^2} - 1\right)^{1/2} = \frac{\lambda}{\pi \theta^2} \left(\frac{w^2}{w_0^2} - 1\right)^{1/2} \tag{3.3.79}$$

Thus we can regard f as the focal length of a Gaussian transducer whose coordinates are defined by Eq. (3.3.79). The axial field drops by 3 dB from the maximum where $w = w_0\sqrt{2}$ or at $z = \pm z_0$, that is, the depth of focus is

$$d_z(3 \text{ dB}) = 2z_0 = \frac{2\lambda}{\pi \theta^2} \tag{3.3.80}$$

Apodization. A Gaussian beam would be very suitable for probing complicated structures because of its relatively simple form; the problem is how to produce it. In practice, the transducer cannot have an infinite width and a Gaussian taper to infinity; its width must be finite. In optical systems, this finite width is many wavelengths in extent, so the Gaussian taper is a very good approximation to the truth. In acoustic systems, however, where the wavelength is larger, this is not always the case.

We can use Eq. (3.3.64) to show that if the phase variation at a spherical acoustic transducer located at $z = -f$ is taken to be of the form $\exp[jk(x'^2 + y'^2)]/2z$, there will be no phase variation at the focal plane $z = 0$. Furthermore, the amplitude variation at this focal plane [see Prob. 5 and Eq. (3.3.40)] is the two-dimensional Fourier transform of the amplitude variation at the plane $z = -f$. This argument leads to the conclusion that a Gaussian amplitude profile at the transducer yields a Gaussian beam.

More generally, there are various types of aperture weightings that can form a beam at the focal plane that is very much like a Gaussian beam. The excitation obtained tends to be very close to that required for a Gaussian beam, but with a finite step in $u_z(x', y', 0)$ or $\phi(x', y', 0)$ to zero at the edges of the transducer. One example of a commonly used taper in signal processing that is appropriate for focusing, as discussed in Sec. 4.5, is *Hamming* weighting; this is also the one-dimensional form $u_z(x')$, which would have an amplitude at a cylindrical transducer of the form

$$A(x') = 0.08 + 0.92 \cos^2 \frac{\pi X'}{D} \tag{3.3.81}$$

A detailed study of the Fourier transform of $A(x)$, which is also given in Sec. 4.5, leads to the conclusion that the Hamming taper is similar to the Gaussian taper, but not as extreme. It yields a focal width 1.3 times the focal width for a uniform amplitude excitation at the transducer, with a maximum first sidelobe level of -4 dB, as compared to -13 dB for the sinc X response obtained with a focused, uniformly excited strip beam.

It is not easy to construct a Gaussian beam because it is not easy to taper the amplitude of the beam. It is far simpler to make such transducers when an array of separately excited elements is employed, for this allows us to excite each element with the correct amplitude (see Sec. 3.5.2) [21]. Thus the technique for making single transducers involves depositing electrodes on the piezoelectric substrate to simulate such an array, and either exciting the electrodes with suitable amplitude weighting or varying the area of the electrodes to simulate this weighting.

PROBLEM SET 3.3

1. In this problem we use some simple ways to estimate the 3-dB focal diameter and depth of focus.
   (a) Consider a spherical lens of radius a. Estimate the depth of focus for a point on the axis by assuming that the 3-dB point is where the ray lengths R from a point on the axis to the edge and from the same point on the axis to the center of the lens differ by $\lambda/2$. Compare your result with Eq. (3.3.27).
   (b) Estimate the diameter of the beam at the 3-dB points $d_t(3 \text{ dB})$ by finding the condition at the focal plane for which $R(\max) - R(\min) = \lambda/2$. In both cases you will find it convenient to use Eq. (3.3.5).
   (c) Consider a hollow beam from an annular ring focused transducer that extends from $r = a$ to $r = b$. Using the methods of parts (a) and (b), estimate the depth of focus and the 3-dB width of the beam when $b - a \ll a$. Compare your estimate of the 3-dB width of the beam with the exact solution for $(b - a) \to 0$.

2. (a) Consider an annular ring focused transducer (i.e., part of a sphere of radius z_0 extending from $r = a$ to $r = b$). Discuss, physically, why the depth of focus becomes infinite as $b \to a$. To do this, consider the fields at any point on the axis due to rays arriving from the transducer. By tracing these rays to the focal plane $z = z_0$, show, physically, that there will be considerable energy on the sidelobes at the plane $z = z_0$.
   (b) If $b - a \ll a$, use Eq. (3.3.7) to work out the total power in the main lobe of the beam at its focal plane, where the main lobe is defined as being the region between the first zero-amplitude points. Compare this with the total power $P(\text{total})$ supplied to the beam at $z = 0$ (assume that the acoustic impedance is the plane wave impedance). Find
   $$S = \frac{P(\text{total}) - P(\text{main lobe})}{P(\text{main lobe})} = \frac{P(\text{sidelobe})}{P(\text{main lobe})}$$
   and show that the parameter $S \to \infty$ as $(b - a)/a \to 0$. You will need the relation
   $$\int_0^{X_m} X J_0^2(\alpha X) = \frac{X_m^2}{2} J_1^2(\alpha X_m)$$
   where $J_0(\alpha X_m) = 0$, $\alpha X_1 = 2.405$, and $J_1(\alpha X_1) = 0.520$.

3. (a) Generalize Eq. (3.3.6) to determine the signal received at r, z, ψ due to the excitation $u_z(r', \psi') [\exp(-jkR_1]$ at r', ψ', z_1.
   (b) Use the method of stationary phase, as described in Appendix G, to show that the main contribution of the signal arriving at r, ψ comes from r'_0, ψ'_0, where the phase Φ is such that $(\partial \Phi/\partial r') = 0$ and $(\partial \Phi/\partial \psi') = 0$. Show that r, ψ lies on the optical ray path from r'_0, ψ'_0. This result is, in fact, the mathematical justification for ray tracing.
   (c) Following through with the method of stationary phase, carry out a Taylor expansion in terms up to $(\psi' - \psi'_0)^2$ and $(r' - r'_0)^2$ to find the amplitude of the beam at r, ψ, and compare your result with what you would expect from geometrical optics. This result gives a complete justification of ray tracing theory.
   (d) Discuss why this type of theory breaks down near the focus or when ka is small.
4. Prove Eq. (3.3.69) from Eq. (3.3.67) by carrying through the algebra.
5. An acoustic beam is excited at $z = 0$ with an amplitude

$$u_z(r', 0) = A(r') e^{j\alpha r'^2/2z_0}$$

   (a) Find the focal point of this beam using the paraxial theory of Sec. 3.3.3.
   (b) Assuming that the beam is excited with a Gaussian amplitude profile $A(r') = \exp[-(r'/w)^2]$, find the profile of the beam at the focal plane. You will find it convenient to write

$$r' \cos \psi' = x'$$
$$r' \sin \psi' = y'$$
$$r \cos \psi = x$$
$$r \sin \psi = y$$

   and to work out your results in Cartesian coordinates using Eq. (3.3.64).
6. Suppose that we construct a reflection acoustic microscope with a sapphire lens, using a liquid indium–gallium mixture as the operating medium, to observe internal cracks in Pyrex glass. The incident wave in the liquid excites a shear wave in the glass. For an operating frequency of 500 MHz, a spherical lens with an acceptance angle of $\theta_0 = 45°$ is employed. Estimate the Sparrow definition of the microscope and the 3-dB depth of focus of the scanning system.
   Note: For the depth of focus, you will need to modify the usual depth of focus formula to account for the fact that the lenses are used twice [see Eq. (3.3.52)].

   Indium–gallium: $V_l = 2.8$ km/s
   Pyrex glass: $V_s = 3.28$ km/s
   Sapphire: $V_l = 11$ km/s

7. Consider the solution of Eq. (3.3.61) for the beam from a flat piston transducer. Use the paraxial forms for the expression of Eq. (3.2.62) [see Eq. (3.2.5)] for the potential on axis for a flat piston transducer to find the potential at a point r, z by writing

$$f(r, z) = f_0(z) + r^2 f_2(z) + r^4 f_4(z) + \ldots$$

and equating terms of equal power in r in Eq. (3.3.61). You will find it convenient to write Eq. (3.3.61) in the form

$$\frac{1}{r} \frac{\partial}{\partial r} \left(r \frac{\partial f}{\partial r} \right) - 2jk \frac{\partial f}{\partial z} = 0$$

(a) By expanding $f(r, z)$ to fourth order in r, find the value of r for the 3-dB point at $z = a^2/\lambda$. How does this compare with the exact result ($r = 0.35a$) and with the result of an expansion to second order?

(b) By writing f_4 in terms of f_0, show that the potential initially increases with r at the planes where $\partial f_0/\partial(z) = 0$ and $\partial^2 f_0/\partial z^2$ is positive (i.e., where there is a minimum potential on the axis).

8. Use the method of Prob. 7 to calculate the potential in the neighborhood of the focus of a spherical lens. It is convenient to employ Eq. (3.3.25) in the approximate form of Eq. (3.3.29). Keep only second-order terms in r and z by using the approximation that near the focus, sinc $X \approx 1 - \pi^2 X^2/6$. Show that to this degree of approximation, the contours of constant amplitude are ellipsoids.

(a) Find the ratio of the major to minor axes of the ellipsoids when the amplitude of the potential on the surface of each ellipsoid is reduced to κ times its maximum value at the focus.

(b) Find the phase variation of the potential along the surface of an ellipsoid of constant κ.

9. Consider the situation shown in Fig. 3.3.13, where rays normal to the surface of the substrate and at the Rayleigh angle to the substrate excite return echoes.

(a) Assuming that the wave velocity in the lens material is infinite, work out, by ray tracing, the time-delay difference T between the two sets of echoes. Give your result in terms of the Rayleigh wave velocity V_R, the velocity in water V_w, and the radius a of the lens. Take the focal point to be at a depth h below the surface of the substrate.

Answer:

$$T = \frac{2h}{V_w}(1 - \cos\theta_R)$$

(b) Assuming that a CW signal of frequency $\omega/2\pi$ is being used to excite the lens, work out expressions for the values of h at the points where the output signal is maximum or minimum. Suggest how this information could be used to determine V_R.

10. (a) By expanding sinc X to second order in X, use Eq. (3.3.25) to find the form of the potential on the axis of a spherical lens near the geometrical focus $z = z_0$. Writing $z/z_0 = p$, differentiate with respect to p and find where the potential on axis is maximum. This is the true geometrical focus $z = f$ for the lens.

(b) Find an approximate expression for $Z(f)$ from your result.

(c) Compare your results with the exact solution of Fig. 3.3.8 for $S = 0.2$, $S = 0.5$, and $S = 1$.

3.4 PULSED EXCITATION OF TRANSDUCERS

In Sec. 3.2.1 we showed that there are relatively rapid variations of the fields in the Fresnel region and slower variations, with distance and angle, in the Fraunhofer region. These field variations are associated with phase differences of the rays that reach a given point from the transducer. When a transducer is excited with a short pulse that is only one or two RF cycles long, some of these phase cancellations and additions disappear and the response becomes smoother. For instance, the nulls found on the axis in the CW case disappear, to be replaced by minima in the pressure profile, and the variation of the fields across the beam are smoother.

In this section we carry out the analysis for pulsed excitation in a liquid. For simplicity and for consistency with the literature, we shall derive our results in terms of the initial velocity excitation of the transducer face, and determine the pressure at a point z, r, θ in space. We shall show that when a transducer is excited uniformly across its face, the signal received at the point z, r, θ is strictly dependent on the distance of this point from the nearest and farthest points on the transducer. The length of the signal received at z, r, θ depends on the time difference between the rays from these two points.

The particle velocity can be written in the form

$$\mathbf{v} = -\nabla \psi \qquad (3.4.1)$$

with $\psi = -\partial \phi/\partial t$ defined as the *Rayleigh velocity potential*.† Then, from the Green's function theory [Eq. (3.1.36)], it follows that

$$\psi = \frac{1}{2\pi} \int_s \frac{v(t - R/V)}{R} ds \qquad (3.4.2)$$

where s is defined as the area of the transducer and $v(t - R/V)$ is the velocity of the transducer face at a time $t - R/V$.

The equation of motion for a liquid can be written in the form

$$\nabla p = -\rho_{m0} \frac{\partial v}{\partial t} \qquad (3.4.3)$$

where $p = -T_1 = -T_2 = -T_3$ is the pressure. It follows from Eqs. (3.4.1) and (3.4.3) that

$$p = \rho_{m0} \frac{\partial \psi}{\partial t} \qquad (3.4.4)$$

where p is the pressure in the liquid.

Transient response on- and off-axis of a piston resonator. We now consider the transient response on the axis of a cylindrical piston transducer. We assume that the velocity $V(t)$ at the transducer is symmetric about the axis. Equation (3.4.2) is written in the form

$$\psi = \int_{r'=0}^{a} \frac{v(t - R/V)}{R} r' \, dr' \qquad (3.4.5)$$

Putting $R^2 = r'^2 + z^2$ and $R \, dR = r' \, dr'$, it follows that

$$\psi = \int_{z}^{\sqrt{z^2 + a^2}} v\left(t - \frac{R}{V}\right) dR \qquad (3.4.6)$$

---

†Note that the use of a minus sign with $v = -\nabla \psi$ agrees with the definition for the velocity potential used by Lord Rayleigh.

Sec. 3.4   Pulsed Excitation of Transducers

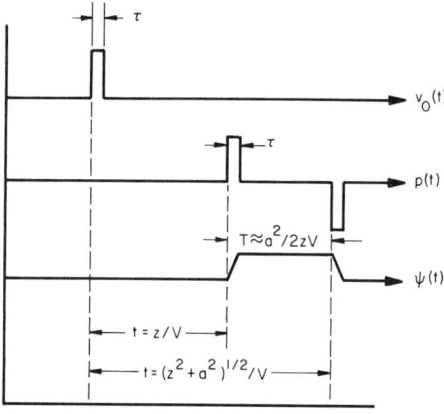

Figure 3.4.1 Exciting and received signals on-axis in the Fresnel region when a transducer is excited with a short pulse of length τ.

If the transducer is excited by an extremely short velocity pulse of amplitude v_0 and length τ, then Eq. (3.4.6) implies that at the plane z, the potential on-axis $\psi(0, z)$ will be of constant amplitude $\tau v_0 V$ and finite in the time range $z/V < t < (z^2 + a^2)^{1/2}/V$, that is, from the time the initial ray along the axis reaches point $0, z$ to the time the last ray from the edge of the transducer reaches the same point.

The pressure p at the point $0, z$ is the differential with respect to time of the velocity potential. Thus we can write

$$p = \rho_{m0} \int_z^{\sqrt{z^2 + a^2}} \frac{\partial}{\partial t}\left[v\left(t - \frac{R}{V}\right)\right] dR \qquad (3.4.7)$$

or differentiate ψ directly to find p. Writing $\partial v/\partial t = -V(\partial v/\partial R)$, it follows that

$$p = \rho_{m0} V\left\{v\left(t - \frac{z}{V}\right) - v\left[t - \frac{(a^2 + z^2)^{1/2}}{V}\right]\right\} \qquad (3.4.8)$$

Fresnel region. Equation (3.4.8) shows that the pressure at the point z due to an initial short velocity pulse v_0 of length τ consists of a positive pulse followed by a negative pulse, with the pulses delayed from each other by a time T, where

$$T \approx \frac{(z^2 + a^2)^{1/2} - z}{V} \approx \frac{a^2}{2zV} \qquad (3.4.9)$$

We have assumed in this derivation that $T \gg \tau$ or $a^2/2z \gg V\tau$, which corresponds to the observer being in the Fresnel, or near-field, region. These results are illustrated in Fig. 3.4.1. In this case, it follows from Eq. (3.4.8) that the pressure pulses have equal and opposite amplitudes of value

$$|p| = \rho_{m0} v_0 V \qquad (3.4.10)$$

where v_0 is the initial pulse amplitude. Hence, on-axis, *the pressure in the Fresnel region does not vary with z.*

Another way of looking at this result, which gives a great deal of physical insight into the behavior of pulsed transducers, has been given by Weight and

Hayman [24]. They suggest that there are two contributions to the signal, a plane wave and an edge wave. Consider a point r, z, which is off-axis with $r < a$. When the transducer is excited by a short pulse, there is a contribution to p from the nearest point to the receiver delayed by a time $T = z/v$. There will be contributions of opposite sign from points on the edge of the transducer, as we have described. Thus it is as though two types of waves are excited: (1) a quasi-plane wave, excited by the main surface of the transducer, whose amplitude is independent of r when $r < a$; and (2) following signals from the edge of the transducer, known as *edge waves*, which are of opposite sign from the quasi-plane wave pulse and delayed by a time $t = R/v$, where R is the distance between the receiving point and a point on the edge of the transducer. We observe that the quasi-plane wave contribution does not exist at points r, z where $r > a$.

This concept has been generalized to deal with waves excited by a transducer in contact with the surface of a solid. In this case, a longitudinal wave transducer, for example, can excite longitudinal and shear edge waves, as well as surface waves. When a finite-length pulse is used, the output can be calculated by taking the convolution of the exciting signal with the calculated response to a very short pulse.

Fraunhofer region. In the Fraunhofer region, where $a^2/2z \ll V\tau$, the pulses overlap and tend to cancel each other out. These results are illustrated in Fig. 3.4.2. From Eq. (3.4.8) it follows that with $a^2 \ll z^2$ and $a^2/2z \ll V\tau$, the pressure consists of two short pulses of opposite sign of length $a^2/2zV$ and amplitude $\rho_{m0}Vv_0$, separated by a time τ, as illustrated in Fig. 3.4.2. The velocity potential is the negative integral of the pressure pulse, and corresponds to a positive pulse of length τ and amplitude ψ, defined as

$$\psi = \frac{v_0 a^2}{2z} \qquad (3.4.11)$$

We see that the potential falls off as $1/z$, as we might expect.

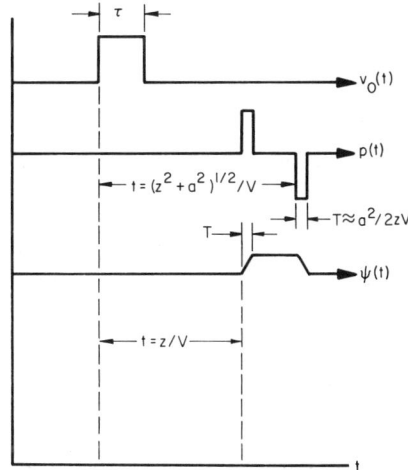

Figure 3.4.2 Exciting and received signals on-axis in the Fraunhofer region when a transducer is excited with a short pulse of length τ.

Excitation with a tone burst. The picture becomes clearer when the transducer is excited by an RF tone burst of constant amplitude, perhaps several RF cycles long. On the axis in the Fresnel, or near-field, region (defined at the frequency of the tone burst), the arriving signal corresponds to two tone bursts of opposite sign, delayed from each other by the time difference $T = a^2/2zV$. In the Fresnel zone, at a point where the pressure is normally at a maximum, these two tone bursts catch up to each other and add where they overlap, thus doubling the amplitude of the signal. At a CW pressure minimum, the overlapping tone bursts tend to cancel each other out. In the Fraunhofer, or far-field, region, the overlapping tone bursts are out of phase, and the signal amplitude is reduced.

Numerical calculations of this type have been made by Robinson et al. [25], Beaver [26], and Tancrell et al. [27]. Beaver has carried out the calculations for an exciting waveform one RF cycle long, as shown in Fig. 3.4.3. He assumed in his calculations a piston of radius 5λ.

Beaver has also carried out the calculations at different axial distances to account for the pressure variation over the radial cross section. As we might expect, the plane wave signals arriving from the transducer surface are essentially undistorted, while those from its edge add out of phase because of their different

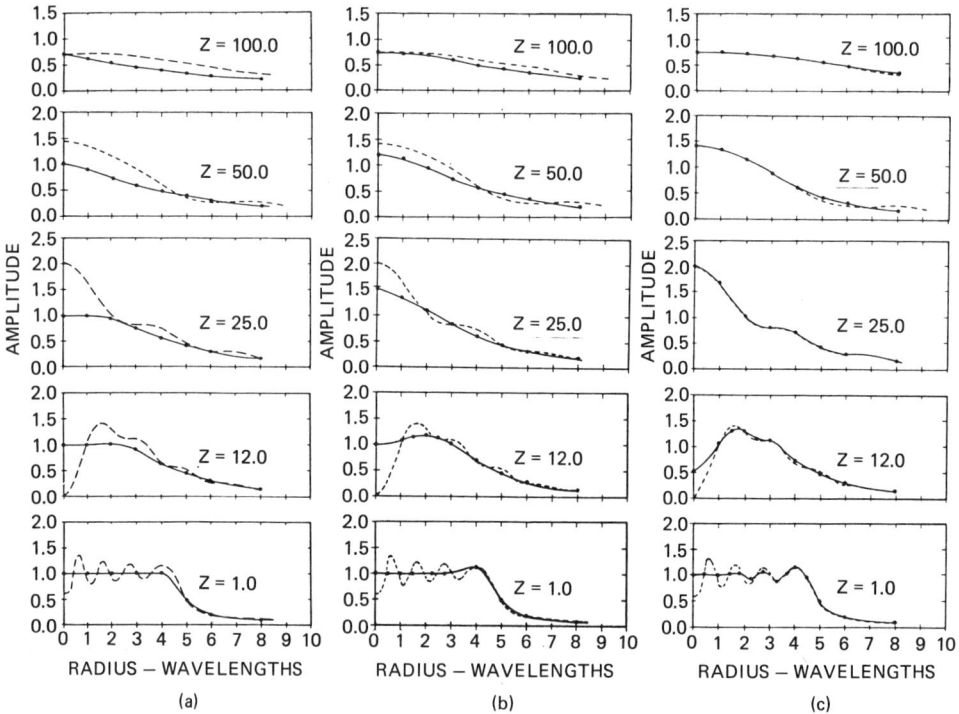

Figure 3.4.3 Sonic-field amplitude verus radius for various normalized axial positions Z expressed in wavelengths. $Z = 25$ corresponds to the Fresnel length $S = 1$. The CW field profile is shown by the dashed-line plots. (a) A 5.0λ radius piston, type I pulse (half an RF cycle); (b) a 5.0λ radius piston, type II (one full RF cycle); (c) a 5.0λ radius piston, type III pulse (four RF cycles). (After Beaver [26].)

time delays. At points off-axis, there is always a strong signal from the plane wave component, with a relatively weak signal arriving from the edge of the transducer.

Beaver calculated how the pressure varies with axial position for signals, or tone bursts, that he calls types I, II, and III, which correspond to half an RF cycle, one full RF cycle, and four RF cycles long, respectively. These are compared to the CW results in Fig. 3.4.3. As we might expect, the pressure variation with radius tends to be smoothed out by the use of short pulses. This is because there is very little interference between the RF signals that arrive from different parts of the transducer when short exciting pulses are used. As a result, when short-pulse excitation is employed, the field variations are smoother and the pronounced maxima and minima observed in the CW case tend to disappear. Thus experiments carried out with short RF pulses are in some ways not as difficult to interpret as those that employ long RF pulses.

Tancrell et al. obtained a very similar set of results for a rectangular transducer

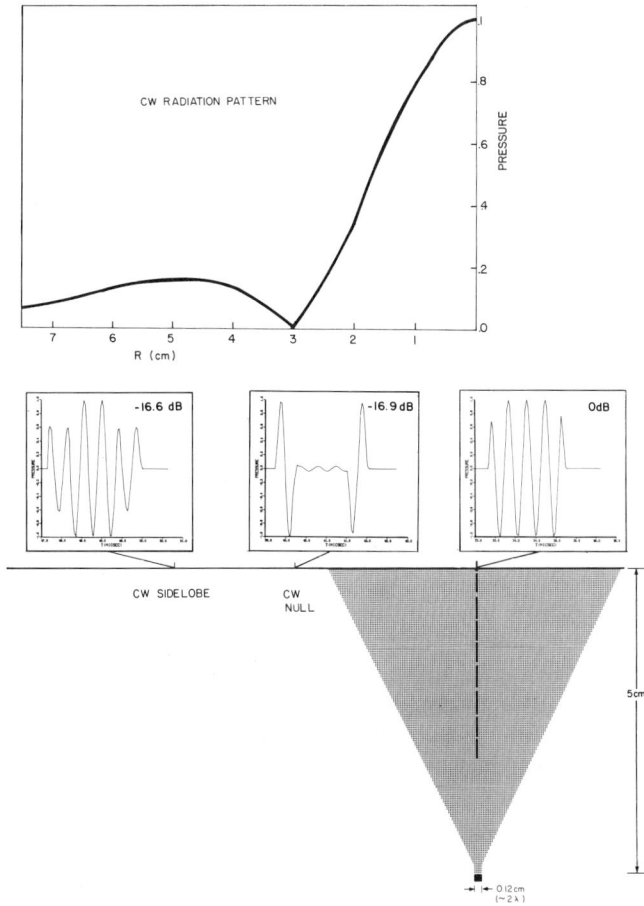

Figure 3.4.4 Far-field pattern for CW radiation and for a four-cycle exciting pulse, showing the time waveform at locations corresponding to the CW null and the CW sidelobe. (After Tancrell et al. [27].)

Sec. 3.4 Pulsed Excitation of Transducers

[27]. In Fig. 3.4.4, the pulse shapes obtained confirm the general picture given by Figs. 3.4.1 and 3.4.2. For several cycle-exciting pulses, the signals on axis in the far field tend to add, thus making the center three cycles larger than the ends of the pulse. At a CW null in the far field, however, the signals tend to cancel, except where they fail to overlap. Once more at a sidelobe maximum, they tend to add where they overlap, but are smaller where they do not. The maximum time-delay differences of the signals arriving at a particular point are determined by the distances of the nearest and farthest rays to the transducer from that point. The extremal rays arriving at a point on axis correspond to a ray from the center and another from the edge, but for a point far off the axis, the extremal rays correspond to those from each edge of the transducer.

PROBLEM SET 3.4

1. Consider a spherical transducer excited by a short unipolar positive velocity pulse. Work out the form of the pulses to be expected on axis, both in front of the optical focal point, at the optical focal point, and beyond the focal point. Show that the sign of the pulse changes from one side of the focal plane to the other. You will need to generalize the results of Eq. (3.4.8) using Eqs. (3.3.7) and (3.3.24).

3.5 LENSLESS ACOUSTIC IMAGING

3.5.1 Introduction

A. Applications of Acoustic Imaging

In Sec. 3.3.2 we described the scanned acoustic microscope, which employs a physical lens and mechanical scanning to produce high-quality acoustic images. Here we will emphasize an alternative approach to the problem, by reviewing other methods of acoustic imaging that use acoustic waves to probe a material and produce a visual image of its internal structure. We will describe how electronically scanned arrays and other synthetic imaging techniques, such as acoustic holography, can replace the physical lenses normally employed in optics. Such techniques give far more flexibility and speed, and tend to eliminate some of the difficulties due to internal reflections that are associated with physical lenses, but with the penalty of considerable complexity. They are used to detect flaws in materials and probing the human body in medical diagnostics; they are also employed in sonar systems for visualizing objects in the sea.

Because of their complexity, most of the work on electronically scanned imaging has been limited to two-dimensional systems. These systems typically have good definition in one transverse dimension, the x direction, and good range resolution in the z direction, but definition in the other transverse dimension, the *y* direction, is relatively poor. For this reason, much of our analysis will be limited to one-dimensional array systems.

In the nondestructive testing (NDT) field, the standard techniques employed for inspection of solid parts have included x-rays, radiographics, and acoustics, eddy current testing in metals, and dye penetrants. Noninvasive techniques currently employed in clinical testing include radiographic and acoustic methods and, to a more limited extent, infrared detection. Radiological techniques have, of course, been used in both medical and NDT applications for many years, with the major advantage that the resultant pictures are in a familar form. However, they also have several disadvantages. The most serious, of course, is the potential danger to human beings associated with ionizing radiation: for example, X-rays cannot safely be used to examine the fetus of a pregnant woman. In addition, radiological techniques are difficult to apply to moving objects, such as the valves in the heart. Neither are they suited to large metallic structures because of the highly penetrating radiation they require, which, as well as being dangerous, is also difficult to use, because it takes very large apparatus and requires the area where the structure is examined to be cleared of all personnel.

Acoustic waves can penetrate both the body and large metal structures without difficulty. They measure mechanical or elastic properties, which, in the NDT case, are directly associated with the strength and life of the structure. In the medical case, these properties are related to the elastic and mechanical properties of the body, which are of direct clinical interest.

We can gain important insights into the nature of the images obtained with acoustic waves from an understanding of acoustic wave propagation. An acoustic image will not have the familiar form of an optical image, which is a serious problem. As an example, we expect the optically displayed acoustic image of a sphere to look more complicated and less familiar than the optical image, for several reasons. First, most materials are at least partially transparent to acoustic waves, whereas an optical image normally corresponds to the visualization of the surface of an object. Visually, we can see only the surface of the body. An acoustic wave, however, sees its entire interior, including all the regions through which it passes, as though the object were semitransparent. This phenomenon complicates the nature of the image, as well as giving more information than is optically observable.

A second problem associated with acoustic wave imaging is that acoustic waves are usually excited by signals with a relatively narrow bandwidth. Thus, as discussed in Sec. 3.3.1, the same problems that occur when observing objects with laser illumination—namely, interference rings and speckle—also are observed in acoustic images. This again makes an acoustic image more difficult to recognize than the optical image we are used to.

A third important phenomenon in NDT is that both longitudinal and shear waves can propagate in the same medium. When, for instance, a longitudinal acoustic wave is reflected from an arbitrary object, it can give rise to both longitudinal and shear reflected waves. Suppose, for instance, that we want to obtain an image of a simple object such as a spherical defect in a solid. As discussed in Sec. 3.6, we expect to see signals reflected from its front face, as well as both shear and longitudinal waves, which pass through its interior, reflected from its back face. Thus the images obtained from even as simple an object as a sphere may be considerably more complicated than the equivalent optical images, and less easy to recognize.

One advantage of acoustical techniques is the ease with which extremely short acoustic pulses, one or two RF cycles long, can be employed. Consequently, good range definition can be obtained relatively easily with acoustics, whereas the range definition in optics is dictated essentially by the depth of focus of a lens. Thus an important advantage of acoustic imaging is that we can differentiate image planes at different distances from the transmitter by using short pulses. In Sec. 3.4 we reexamined some of the concepts of acoustic wave propagation to take account of transient phenomena that are not normally of importance in optics.

Because this new acoustic technology produces images that are not always easily recognizable to an untrained observer, it has taken time to be adopted in practice. As an example, a large background of clinical experience has long existed for the use of x-ray techniques in medicine; a similar body of experience with acoustic imaging had to be developed before acoustic methods could be put into widespread use in medical diagnosis.

B. A-Scan, B-Scan, and C-Scan Imaging

Before discussing imaging systems, it is worth reviewing the standard acoustic techniques employed in NDT and medical diagnosis. In the simplest and most common NDT applications, a simple, single piezoelectric transducer is used to excite an acoustic wave in the object being examined. Commonly, the transducer is placed in a water bath and the acoustic wave excited by the transducer propagates through the water and into the object. The advantage of this technique is that the transducer is readily movable. However, because there is a large impedance mismatch between the water and the metal, the system may also be relatively inefficient in NDT applications.

A-scan. A second technique, illustrated schematically in Fig. 3.5.1, is to place the transducer directly against the solid material to be examined, making contact between the transducer and the sample with grease or a thin layer of rubber. This technique is particularly convenient in medical diagnostic applications because of the flexibility of the human body and the consequent ease with which such contact can be made with it.

Now suppose that the acoustic transducer is excited by a short electrical pulse. If the transducer is correctly designed, it will emit an acoustic pulse of length $\tau_p \approx 1/\Delta f$ determined by the bandwidth Δf of the transducer. The generated acoustic pulse passes into the object and is reflected by the acoustic impedance discontinuities caused by the presence of flaws or internal structures of the body. The return echo signal is received at the transducer and is amplified and displayed as a function of time on an oscilloscope. The time delay of the echo is $T = 2z/V$, where z is the distance of the flaw from the surface and V is the acoustic wave velocity in the material being examined. Thus the distance of the flaw from the

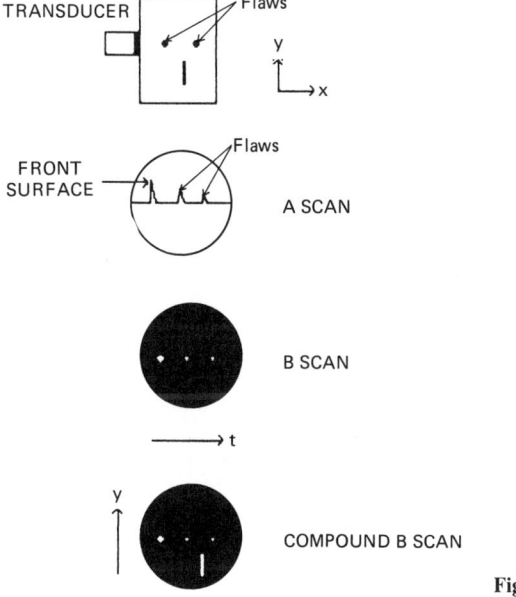

Figure 3.5.1 A-scan and B-scan operations. (After Kino [28].)

surface or its range can be determined from the time delay of the pulse observed on the oscillocope. The accuracy of the measurement of the range definition is determined by the pulse length τ_p and is $\Delta z_p \approx V\tau_p/2$. Furthermore, the amplitude of the return echo depends on the size and shape of the flaw, so that a rough estimate of the size of the flaw can be obtained by measuring the amplitude of the return echo. This technique is known as the amplitude scan, or the *A-scan*, technique.

By moving the transducer along the surface of the object being examined, various flaws can be detected while the A-scan information is being obtained, as shown in Fig. 3.5.1. Thus we can detect the transverse position of flaws, although the definition of the system in the transverse direction will be dictated basically by the size of the transducer. If the distance z of the flaw from the transducer is such that $z \gg a^2/\lambda$, where a is the radius of the transducer and λ is the wavelength of the center frequency of the pulse, then, because of diffraction, the transducer behaves almost like a point source, with the beam diameter at the flaw tending to be much larger than that of the transducer. If, on the other hand, the flaw is in the near field of the transducer ($z \ll a^2/\lambda$), the transverse definition is comparable to the radius of the transducer. Thus we obtain the best definition when the transducer diameter is chosen so that the flaw is located roughly at the boundary between the near and far fields of the transducer. This typically means that the definition in the transverse direction will be relatively crude, of the order of 1 cm at operating frequencies of a few megahertz. The definition can be improved by using a focused transducer; however, this will give good definition over only a limited range, the extent of which is determined by the depth of focus of the lens, as discussed in Sec. 3.2.

Although the transverse definition is limited, the range definition of such an A-scan system can be relatively accurate, as we have seen, because it is dictated by the length of the pulse. Such a device is normally used with a baseband pulse only two or three cycles long, with a transducer whose bandwidth is comparable to its center frequency; this is different from the conventional radar systems, which use a relatively long tone burst. At low frequencies, the definition is poor. As the frequency is increased to improve the definition, however, the attenuation of the signal in most solids typically increases as the square of the frequency, so that there is a limit to the upper frequency that can be used. Thus the larger the structures, the lower the frequency and hence the poorer the definition that can be obtained.

For nondestructive testing of materials such as nuclear reactor steel, where the walls of the reactor may be as much as 25 cm thick, the operating frequencies employed are of the order of 2.25 MHz. As the acoustic velocity of longitudinal waves in such materials is approximately 6 km/s, the best range definition that can be obtained is of the order of 3 to 5 mm, and the transverse definition will be several times worse. In aircraft materials, such as titanium or aluminum, frequencies as high as 20 MHz may be used, with a correspondingly better definition. Frequencies as high as 400 MHz have been employed for examining structural ceramics, yielding a definition of the order of 25 μm ($V = 10^6$ cm/s). Still higher frequencies, in the range 2 to 3 GHz, have been used to image integrated circuits with the acoustic microscope.

Similarly, in medical diagnostics, the maximum usable frequency for acoustic waves to penetrate as much as 20 cm into the human body is of the order of 2 to 5 MHz, because the attenuation in body tissue is approximately 0.8 dB/cm MHz and varies linearly with frequency. Higher frequencies are used to observe shallower objects, such as the internal organs of children, for instance, or the carotid artery near the body surface. Similarly, for observations of the human eye, which is less than 3 cm in extent, relatively high frequencies, in the range 10 to 20 MHz, can be employed. The scale is reduced still further for observing body cells and thin layers of tissue, where, as we have discussed in Sec. 3.3.2, very high frequencies, in the range 1 to 8 GHz, can be employed.

B-scan. A disadvantage of the A-scan method is that it is slow and tedious to use. Only one line of amplitude information can be observed at a time, and although mechanical means can be used to move a transducer relatively rapidly, this results in large amounts of information that must still be interpreted by a human operator.

An alternative technique is the brightness scan, or the *B-scan*, method, in which the return echo signal is used to modulate the intensity of the spot on an oscilloscope, while the time delay is represented by the horizontal position of the shot and the mechanical position along the surface of the object is represented by the vertical position of the spot on the oscilloscope. By this means, a crude picture of the structure within the material can be presented, as illustrated in Fig. 3.5.1.

The problem here is that few structures are completely flat, and it is difficult to make contact over very large regions with those that are not flat. Thus B-scan

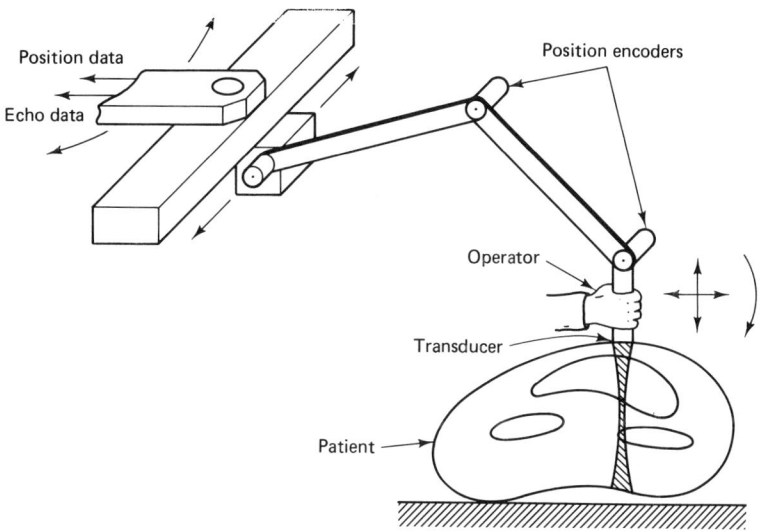

Figure 3.5.2 Scanning arrangement for a static medical B-scan system. (After Maginness [29].)

imaging is not often employed in NDT applications. On the other hand, the technique has been useful in the medical field because a greased transducer can make contact with the body very easily. Because the body itself is flexible, the transducer can be moved around it and tilted to direct the beam in an arbitrary direction.

The B-scan static medical imaging system illustrated in Fig. 3.5.2 employs a single fixed-focus transducer, typically focused to a point in the middle of the view region. The transducer is supported from a mechanical arm and can be moved along the body surface and tilted at will. Position data are encoded by the mechanical arm system, which makes it possible to display the reflected echo as an intensity display; a single point on the body always appears in the same position on the cathode ray tube. An image of a cross section perpendicular to the body surface is obtained. The transverse and range definition of the system are determined in part by the size of the transducer and in part by the play in the mechanical system, which can often be as poor as 1 cm. Although the mechanical system could be improved with enough care in design, the static B-scan system has been replaced by real-time systems in which the image is obtained at frame rates of at least 15 Hz.

C-scan. A third technique, illustrated in Fig. 3.5.3(a), is to employ transmission imaging in the *C-scan* mode to form an image in a plane that is perpendicular to the direction of propagation of the acoustic beam. For example, a focused transducer can be used to transmit an acoustic beam through the object of interest. A thin object is placed at the focus of the acoustic beam, which is received by a second confocal focused receiving transducer. The object is then mechanically scanned across the beam, while the beam itself is moved back and forth, to create a raster scan. The amplitude of the received signal is used to vary the intensity

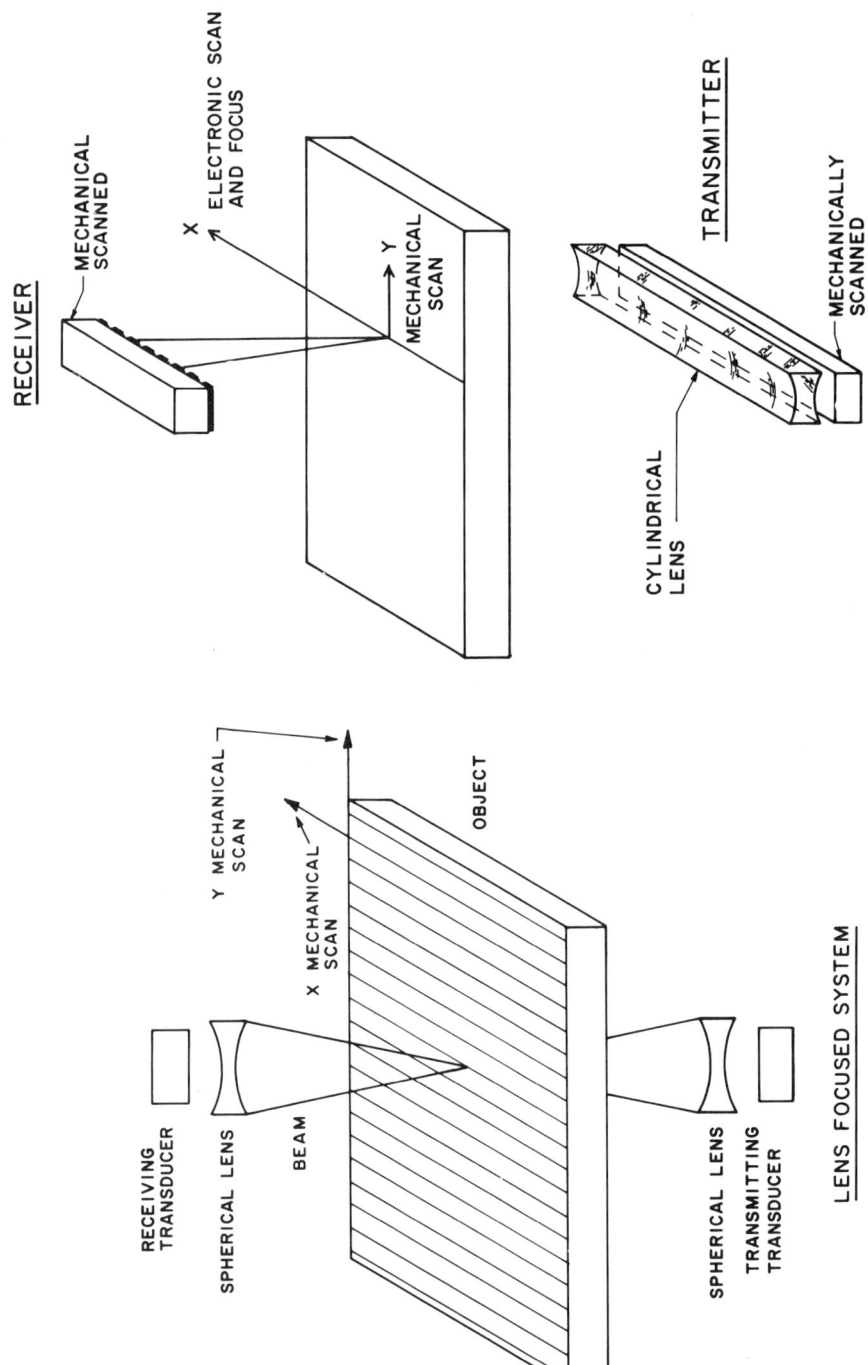

Figure 3.5.3 (a) C-scan transmission system: (a) mechanically scanned system focused with physical lenses; (b) mixed electronically and mechanically scanned system with electronic focusing, with electronic scanning and focusing in one direction and mechanical scanning and focusing with a lens in the other.

of a light spot and is displayed and recorded on either a TV screen or paper. The advantage of this method is that it gives good definition and a high-quality transmission image of sheet metal and other thin objects. An acoustic transmission microscope of the type described in Sec. 3.3.2 is a good example of the use of this technique.

The same C-scan technique is also employed for reflection imaging (as in the acoustic microscope), using a single transducer as transmitter and receiver. Figure 3.5.4 shows an example of such a reflection mode scan taken of an impact damaged fiber epoxy composite by Khuri-Yakub and Reinholdtsen with an F 0.9 transducer operating at 3 MHz [30]. The depth of focus of this transducer is small, so that damage to layers at different depths can be found by moving the transducer up or down to focus on the different layers.

C. Focusing Systems and High-Speed Scanning

By now it is apparent that two improvements in acoustic imaging techniques are needed. First, the process must be speeded up by using high-speed mechanical scanning or electronic scanning. Second, the acoustic beam must be focused so that both good transverse definition and range resolution through the depth of a thick sample can be obtained. The problem with focusing an acoustic beam is that, to do this, a physical lens must be immersed in a medium that can propagate acoustic waves. Typically, this medium is water, which means that for low-frequency imaging systems operating in the megahertz range, where the propagation path may be 10 cm or more, the imaging system tends to be very bulky and heavy when a very tight focus is required. However, a weakly focused lens with a large depth of focus may be made by using a small spherically shaped transducer. On the other hand, for very high frequencies in the gigahertz range, the lenses and water path are very small. Thus physical lenses have been used very successfully to obtain good definition in the acoustic microscope.

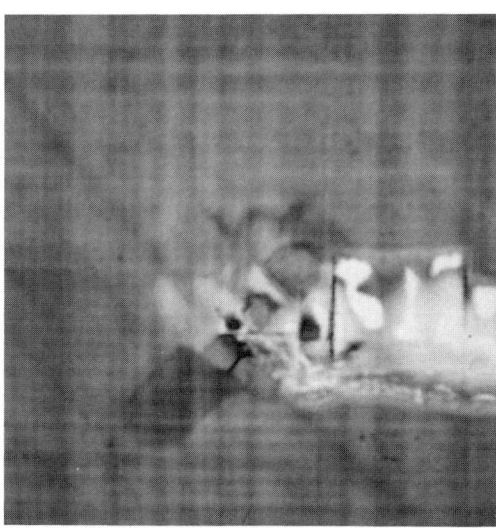

Figure 3.5.4 Impact-damaged fiber epoxy composite material. (Courtesy of B. T. Khuri-Yakub.)

A major difficulty with static B-scan imaging is the slow speed at which the image is formed. This requires use of a storage oscilloscope display with a very poor gray scale (the number of shades of contrast between black and white) or the use of an electronic scan-converter and storage system. In either case, slowly scanned B-scan devices are not suitable for observing rapidly moving targets, such as a moving heart valve, and may even have problems observing a patient who is breathing normally rather than holding his or her breath.

Figure 3.5.5 shows a very successful approach to high-speed imaging, which combines B-scan imaging with a high-speed mechanically scanned system that has a radial sector scan format. This technique is now being used by several manufacturers. A small transducer 1 to 2 cm in diameter is mounted in a small liquid bath in a plastic enclosure, one surface of which is placed against the body. The transducer itself is vibrated back and forth over an angular range of the order of $\pm 30°$ or more. This provides an excellent picture with a frame rate of the order of 15 to 30 Hz. By using more than one transducer and switching between them, the effective frame rate and the angle of the scan can be increased. The transverse definition over the region of most interest can be improved by using focused transducers, focused at the center of the range. At the present time, mechanical scan systems of this type are relatively low in cost compared to the highly complex, electronically focused systems that provide good focusing at all ranges but in only one transverse direction.

Elimination of physical lenses. It is desirable to eliminate physical lens systems because of the problem of varying their focal length and because of their size. Two techniques have been employed for this purpose. The first is *holography*, a lensless system that uses optical or computer reconstruction techniques

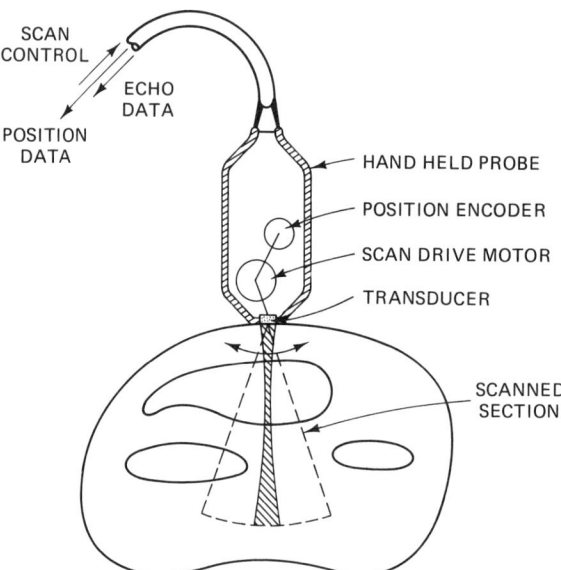

Figure 3.5.5 Mechanically scanned B-scan system with a radial sector scan. (After Maginness [29].)

to synthesize the phase change made by a lens. The second uses an array of transducers and processes the signals arriving at them, to make the array behave like an electronically focused lens.

The holographic system typically provides good transverse definition in both directions, but its range definition is limited to essentially the depth of focus of the equivalent lens. Furthermore, holographic systems tend to be complicated, costly, and inconvenient. The use of a water bath in some versions makes them insensitive and unwieldy. Other mechanically scanned versions are slow and inconvenient because optical photos of the holograms must be exposed and developed.

An electronically scanned system can provide good range resolution as well as good transverse resolution. Its disadvantages, however, are its great complexity and the large number of transducer elements and electronic components it requires; the quantity must often be limited by economic considerations, which means that the system usually gives good definition in only two dimensions: normally, range and one transverse dimension. One advantage, however, of using an array to provide electronic scanning over the face of the object, rather than mechanical scanning, is that it speeds up the process of forming an image, since a single mechanically scanned transducer is now replaced with an array of small transducers in which the signal can be switched electronically from one element to another, with the speed controlled by the electronic switching rather than by a relatively slow mechanical scan.

As an example, consider the transmission system illustrated in Fig. 3.5.3(b), using an N-element array, which transmits a cylindrically focused beam through a thin object to a similarly focused receiver array or the other side of the object. If the system is mechanically scanned at right angles to the electronic scan, we expect the speed of the scan to be increased by a factor N. The electronic scan rate is extremely fast, and mechanical scanning is carried out in only one direction, not two. We shall describe later how such improvements in scan speed have, in fact, been obtained.

Physical lens. To understand which components are needed for an electronically focused and scanned array or for a holographic system, let us first consider the action of a physical lens, which focuses the signal received from one point on an object onto the plane of a single large-area transducer, as shown in Fig. 3.5.6(a). The physical lens delays the rays passing through it so that all rays reaching the transducer from the focal point suffer the same phase and time delays. Thus if an object at this point is illuminated by a short pulse, all signals arrive at the transducer at the same time and with the same phase.

Time-delay focusing. To carry out this process electronically, all the signals received by the individual array elements must be delayed in such a way that they can be added to each other. The simplest way to do this, conceptually, is to connect electrical delay lines to each element of the array so that a pulsed RF signal emitted from a point on the object will arrive at the receiver, with all pulses passing through each element of the array arriving at the same time at a common sum line, as illustrated in Fig. 3.5.6(b). There will therefore be a strong response

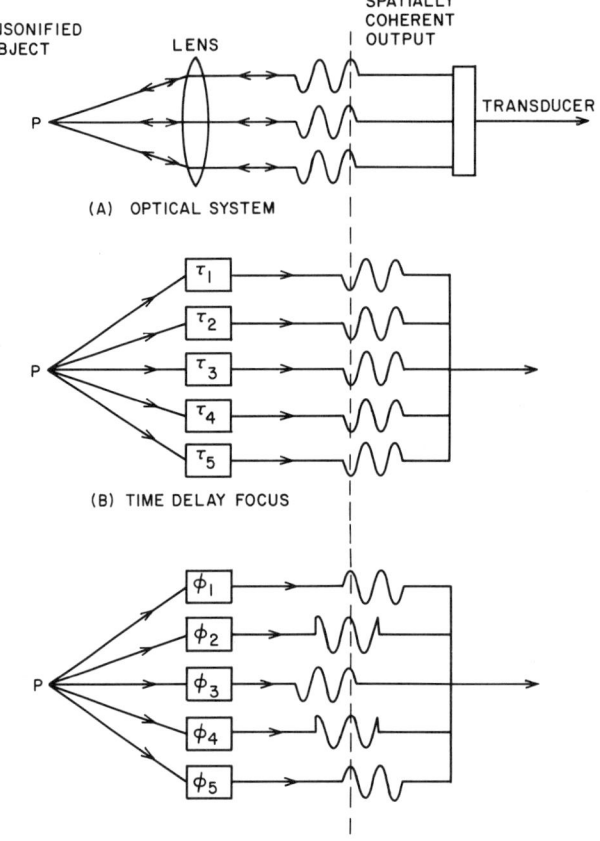

Figure 3.5.6 (a) Rays passing through a physical lens to a plane transducer; (b) time-delay system for focusing; (c) phase-delay system for focusing.

from the point P on the object, but the signals from some other point will arrive at different times and be out of phase. We call this a *time-delay imaging system*.

Phase-delay focusing. The *phase-delay system*, illustrated in Fig. 3.5.6(c), is often simpler to implement. An RF pulse several cycles long, arriving from a point in the object, passes through the individual transducer elements and into a phase-delay instead of a time-delay system. In this case, the nth RF cycle from one element can be added to the $(n + 1)$th RF cycle from another array element. Thus all signals from the array elements can be added to give a strong output corresponding to the point of interest. The rays emitted from any other point will give signals that will arrive out of phase, so the sum of the signals from the receiver elements will be relatively small. The disadvantage of such a phase-delay system is that it requires an RF pulse that is several RF cycles long. The range definition of the system therefore tends to be worse than that for a system in which the time delays of the signals arriving from all elements are equal; in the latter case, the RF pulse can be only one or two cycles long.

The change in phase required in a phase imaging system can be provided by phase changes in the transducer itself, for example, the Fresnel transducer, described in Sec. 3.5.3.A, or by mixing a reference signal with the signal arriving at the transducer. These two signals can be multiplied together in a mixer to produce an output at the sum or difference frequencies. The phases of the two input signals are similarly added or subtracted. It follows that if signals with the correct reference phases can be inserted into the mixers, they can compensate for the phase differences of the different signals arriving at the transducers from a point on the object. Therefore, the outputs from the mixers will all be in phase.

Alternatively, in one form of a holographic system, the mixing can be accomplished in two steps. Two signals of the same frequency, one scattered from an object in water and one a reference acoustic beam, are incident on the surface of a water bath. The static sound pressure, which is proportional to the square of the total incident signal amplitude, modulates the height of the surface of the water bath. Thus there is a spatial variation in the height of the water that is proportional to the product signal from the reference and the object and depends on the phase difference between the two waves. To reconstruct an image, a laser beam is reflected from the surface of the water bath and is deflected by the surface ripple. When a point in the water bath is to be observed, the laser beam deflected from the water surface is passed through an optical lens, which focuses it to an equivalent image point. More generally, the object illuminated by an acoustic beam in the water tank is reproduced as an optical image. This system will be described in more detail in Sec. 3.5.6.

3.5.2 Basic Imaging Theory

A. Matched Filter Concepts

The focusing systems discussed in Sec. 3.3, which use physical lenses or spherically shaped transducers, and those in Sec. 3.5.1, which use electronic focusing or holographic imaging, are all examples of matched filters. Here we will discuss how physical, electronic, or holographic lenses can be regarded as spatial matched filters for a spatially varying input signal.

For simplicity, we consider initially a two-dimensional receiver system with a line source of radian frequency ω at the point x, z, as illustrated in Fig. 3.5.7. Ideally, the acoustic imaging system must be able to reconstruct this line source as a δ function (line) in space.

A common way to study a signal processing system is to consider the response of the system to an impulse or mathematical delta function. As we show in Sec. 4.3, if the system has a response $f(t)$, the optimum signal-to-noise ratio or maximum peak signal (i.e., the best approximation to a δ function) can be obtained by passing the output of the system through a *matched filter* with a response $f^*(-t)$.

By analogy, the signal arriving at the plane $x', 0$ from a line source at x, z is of the form $f(x - x', z)$. The appropriate spatial response to this signal to obtain the maximum output is at matched filter with the response $f^*(x' - x, z)$.

Suppose that a one-dimensional receiving system is composed of continuous array of infinitesimally wide transducers at the plane $z = 0$. Let the coordinates of any point on the transducer array be $x', 0$. After the signal from the point x,

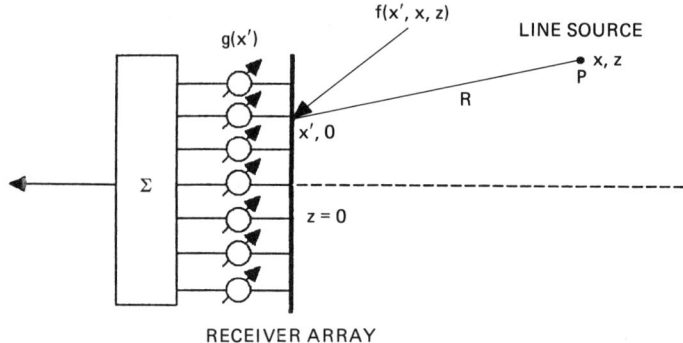

Figure 3.5.7 Two-dimensional receiver system with a line source at x, z.

z has passed through the medium, which in general may be nonuniform, the signal arriving at the plane $z = 0$ is of the form $f(x', x, z)$. We regard the receiving system as a continuous structure with a response $g(x')$. We will show that the maximum output is obtained from the point $0, z_i$ when we choose $g(x')$ so that it is a matched spatial filter for the spatially varying signal $f(x', 0, z_i)$.

The distance from $x', 0$ to x, z is

$$R = \sqrt{(x - x')^2 + z^2} \tag{3.5.1}$$

Thus the signal from a line source at x, z, which varies as $\exp(j\omega t)$, is of the form†

$$f(t, R) = \frac{e^{j\omega(t - R/V)}}{R^{1/2}} \tag{3.5.2}$$

As the spatial response of the system depends only on R, we can write $f(x', x, z)$ in the form $f(x - x', z)$.

We suppose that the electrical response of the receiver system in the region between x' and $x' + dx'$ of the transducer array is of the form $g(x')\, dx'$, and that all signals from the receiver elements are added and weighed by the response $g(x')$. The output from the receiver will therefore be

$$y(x, z) = \int f(x', z)\, g(x - x')\, dx' \tag{3.5.3}$$

where we assume that all signals vary as $\exp(j\omega t)$. The output $y(x, z)$ is the convolution of $f(x', z)$ and $g(x')$. We know from signal processing theory, as shown in Sec. 4.3, that if $g(x')$ is chosen so that $g(x') = \alpha f^*(-x', z)$, then $g(x')$ is the matched filter for $f(x', z)$ (i.e., for the signal emitted from the point $0, z$). Here α is a constant and f^* denotes the complex conjugate of f.

It is convenient to write $f(x', z)$ in the form

$$f(x', z) = a(x', z) e^{-j\phi(x', z)} \tag{3.5.4}$$

---

†We have used the asymptotic form for the potential from a line source, corresponding to the assumption that $(\omega R/V) \gg 1$ [see Probs. 3.1.3 and 3.2.2 and Eq. (3.2.59)].

where the amplitude term $a(x', z)$ is a real function. It follows that the matched filter response becomes

$$g(x', z) = \alpha f^*(-x', z) = \alpha a(-x', z)e^{j\phi(-x', z)} \quad (3.5.5)$$

With the matched filter, the output $y(0, z)$ from the point 0, z becomes

$$\begin{aligned} h(0, z) &= \alpha \int f(x', z) f^*(-x', z) \, dx' \\ &= \alpha \int [a(x', z)]^2 \, dx' \end{aligned} \quad (3.5.6)$$

More generally, $h(x, z)$ is the line spread function (LSF) of the focused system to the point 0, z. We derive the form of $h(x, z)$ in the paraxial approximation in Sec. 3.5.1.B.

We see that with this choice of $g(x')$ as a matched filter, all phase errors are removed by the matched filter and the integrand is always positive. When noise is present, it is implied by signal processing theory (see Sec. 4.3) that by using this matched filter we obtain the maximum possible signal-to-noise ratio when observing the point (0, z) (see Prob. 2).

In the simplest case, the signal emitted from a point source at x, z varies as $\exp(j\omega t)$. It follows from Eq. (3.5.2) that the signal arrives at the receiver in the form

$$f(x' - x, z) = \frac{e^{-j\omega[(x-x')^2 + z^2]^{1/2}/V}}{[(x - x')^2 + (z^2)]^{1/4}} \quad (3.5.7)$$

where we have omitted the $\exp(j\omega t)$ term for simplicity.

B. Paraxial Approximation

To keep the analysis simple, we shall use the paraxial approximation or Fresnel approximation of optics, and assume that $(x - x')^2 \ll z^2$, keeping only up to second-order terms in $(x - x')^2$ in the phase-varying term and zeroth-order terms in the amplitude variation. Thus we find that

$$f(x' - x, z) \approx \frac{1}{\sqrt{z}} e^{-j\omega z/V} e^{-j\omega(x-x')^2/2zV} \quad (3.5.8)$$

The resultant signal arriving at the plane $z = 0$ therefore has a square-law spatial phase variation with the coordinate x'; that is, in signal processing terms it is a spatial frequency-modulated (FM) chirp. The spatial filter response to an object point 0, z should therefore be of the form

$$\begin{aligned} f^*(x', z) &= e^{j\omega z/V} e^{j\omega(x')^2/2zV} \\ &= e^{2j\pi z/\lambda} e^{j\pi(x')^2/z\lambda} \end{aligned} \quad (3.5.9)$$

where we have taken the matched filter constant (see Sec. 4.3) to be $\alpha = \sqrt{z}$, and where $\lambda = 2\pi V/\omega$ is the wavelength of the acoustic wave. Using $g(x')$ for the matched filter to the point 0, z, we find that the output from a source at x, z, using a receiver system of finite length D, is

$$h(x, z) = \int_{-D/2}^{D/2} e^{-j\omega(x-x')^2/2zV} e^{j\omega x'^2/2zV} \, dx' \tag{3.5.10}$$

or

$$h(x, z) = e^{-j\omega x^2/2zV} \int_{-D/2}^{D/2} e^{j\omega xx'/zV} \, dx' \tag{3.5.11}$$

Line spread function. Integration of Eq. (3.5.11) yields

$$\begin{aligned} h(x, z) &= De^{-j\omega x^2/2zV} \frac{\sin(\omega xD/2zV)}{\omega xD/2zV} \\ &= De^{-j\pi x^2/z\lambda} \operatorname{sinc} \frac{xD}{z\lambda} \end{aligned} \tag{3.5.12}$$

where sinc $x = (\sin \pi x)/\pi x$, the wavelength $\lambda = 2\pi V/\omega$, and $h(x, z)$ is known as either the *point spread function* (PSF) in the x direction or, more correctly in this two-dimensional system, the *line spread function* (LSF) in the x direction for the point $0, z$. More generally, if the matched filter is designed to focus on an image point x_i, z_i, we choose $f(x') = h^*(x_i - x', z_i)$ and find that

$$h(x - x_i, z) = De^{-j\pi(x^2 - x^{2/i})/z_i\lambda} \operatorname{sinc} \frac{(x_i - x)D}{z_i\lambda} \tag{3.5.13}$$

The response is a sinc function, which, as $D \to \infty$, approaches a δ function at $x = x_i$. Thus, by using a matched filter to compensate for the phase differences of the rays arriving at the receiver, we can construct a system with a spatial response narrowly centered about the point of interest. It follows that the construction of a matched spatial filter for the point x_i, z_i is equivalent to designing a lens focused on this point. Thus *all lenses, whether electronic, physical or holographic, give basically the same transverse definition and spatial response if their apertures or F numbers, z_i/D, are the same.* The analysis we have given is valid only for the paraxial approximation, but it serves to give a very reasonable estimate of the basic results, even with wide-aperture lenses.

3 dB definition of a paraxial rectilinear system. From Eq. (3.5.13), the 3-dB points of the response are where $x - x_i = 0.45\lambda z_i/D$ (the 4-dB points are at $x - x_i = \pm 0.5\lambda z_i/D$). Thus the 3-dB points are a distance $d_x(3 \text{ dB})$ apart, where

$$d_x(3 \text{ dB}) = \frac{0.89\lambda z_i}{D} \tag{3.5.14}$$

We regard this parameter $d_x(3 \text{ dB})$ as the definition of the lens.

C. Radial Sector Scan System

It is also useful to consider the 3-dB definition in a cylindrical coordinate rather than a Cartesian coordinate system. We do this because many medical systems use a relatively small transducer array and carry out an azimuthal scan.

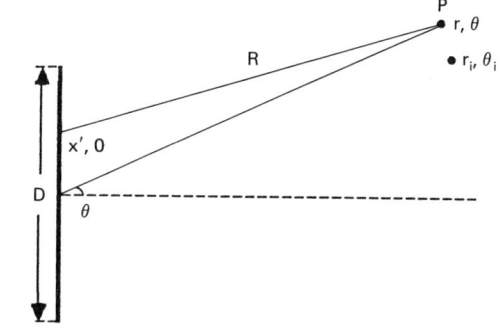

Figure 3.5.8 Imaging system in a cylindrical format.

As illustrated in Fig. 3.5.8, this is a *radial sector scan* in an r, θ coordinate frame, based on an axis normal to the page at the center of the receiver transducer array.

We consider this imaging system to be focused on the point r_i, θ_i with a flat transducer array of length D, with the center of the coordinate system at the center of the array. The distance to the point r, θ from a point x' on the array is

$$R = (r^2 + x'^2 - 2x'r \sin \theta)^{1/2} \tag{3.5.15}$$

We expand Eq. (3.5.15) to second order in x'/r and find that

$$R \approx r - x' \sin \theta + \frac{x'^2}{2r} \cos^2 \theta \tag{3.5.16}$$

Our previous treatment, in Cartesian coordinates, was for the case $x_i = 0$. More generally, if x_i is finite, we reverse the sign of x_i and $f(x'-x_i, z_i)$ to find the matched filter and write $g(x' - x_i) = \alpha f^*[-(x' - x_i), z_i]$. The equivalent operation in cylindrical coordinates is to change the signs of both x' and $\sin \theta_i$. The output obtained, when the system is focused on the point r_i, θ_i, is of the form

$$h(r, \theta, r_i, \theta_i) = Ae^{j\omega(r_i - r)/V} \int_{-D/2}^{D/2} e^{j\omega x'(\sin\theta - \sin\theta_i)/V}$$

$$\times e^{j\omega x'^2/2V} [(\cos^2 \theta_i)/r_i - (\cos^2 \theta)/r] \, dx' \tag{3.5.17}$$

where A is a constant.

The maximum output is obtained when $r = r_i$ and $\theta = \theta_i$; it is

$$h(\theta_i, r_i) = AD \tag{3.5.18}$$

When $\theta \neq \theta_i$, the square-law term in x' is zero, provided that

$$r = \frac{r_i \cos^2 \theta}{\cos^2 \theta_i} \tag{3.5.19}$$

Thus if $\theta_i = 0$, r decreases slightly as θ increases. In this case,

$$|h(r, \theta, r_i, \theta_i)| = AD \left| \text{sinc} \left[\frac{D}{\lambda} (\sin \theta - \sin \theta_i) \right] \right| \tag{3.5.20}$$

Taking $\delta\theta = \theta - \theta_i$ to be small, we see that to first order in $\delta\theta$,

$$|h(r, \theta, r_i, \theta_i)| = AD \left| \text{sinc}\left(\frac{D \cos \theta_i}{\lambda} \delta\theta\right) \right| \quad (3.5.21)$$

Therefore, the effective width of the transducer is $D' = D \cos \theta_i$ (i.e., the definition decreases as the angle θ_i is increased). In this case the 3-dB azimuthal definition is

$$d_\theta(3 \text{ dB}) = r_i \Delta\theta(3 \text{ dB}) \quad (3.5.22)$$

where $\Delta\theta$ (3 dB) is the total angular spread between the 3-dB points. Thus

$$d_\theta(3 \text{ dB}) = \frac{0.89 r_i \lambda}{\pi D \cos \theta_i} \quad (3.5.23)$$

It is apparent that the definition deteriorates as θ_i is increased from zero.

An important feature of this kind of focusing is that the phase change required for focusing on the point θ_i may be divided into two parts, a linear term, which varies as x', and a square-law term, which varies as x'^2. The system can be focused on the point 0, r_0 if the square-law term is chosen to vary as $\exp(j\omega x'^2/2Vr_0)$, where r_0 is the focal length at the center of the lens. In this case the focusing term is independent of the azimuthal angle θ_i. We then find that

$$h(r, \theta, r_i, \theta_i) = A e^{j\omega(r_i - r)/V} \int_{-D/2}^{D/2} e^{j\omega x'(\sin\theta - \sin\theta_i)/V} \\ \times e^{(j\omega x'^2/2V)(1/r_0 - \cos^2\theta/r)} \, dx' \quad (3.5.24)$$

We note that such a focusing system focuses on the point r_i, θ_i, where

$$r_i = r_0 \cos^2 \theta_i \quad (3.5.25)$$

and has the transverse definition given by Eq. (3.5.23).

The advantage of this system, in practice, is that it is far easier to construct than other scanning systems. The square-law phase variation can be programmed independently of the azimuthal linear phase variation, and the two phase variations can be added. Only the linear phase variation term must be changed with angle. The two operations are similar to what happens when a parallel beam passes through a lens and the beam is tilted to vary the angular position of the focal spot. We discuss further implications of this principle in Sec. 3.5.5, when we deal with time-delay focusing.

Rayleigh criterion. As we discussed in Sec. 3.3.1, a standard criterion for the transverse definition of an imaging system is to consider when two point sources of equal amplitude can be distinguished from each other. The *Rayleigh criterion* is based on the idea that this can be done when one point is placed at the position where the response to the other point is zero. This criterion yields a definition d_x (Rayleigh) that is equal to that for the 4-dB points, or

$$d_x(\text{Rayleigh}) = \frac{\lambda z_i}{D} \quad (3.5.26)$$

This formula, like the one for d_x (3 dB), can easily be generalized for a radial sector scan system by replacing D with $D' = D \cos \theta_i$, z_i with r, and d_x with d_θ.

Sparrow criterion. The Rayleigh result is inadequate when the signals from two neighboring points are in phase. In general, the total output from two equal amplitude point sources with a phase difference ϕ is of the form

$$|y(x, z)| = \left| \text{sinc} \frac{(x_i - a/2)D}{z_i \lambda} + e^{j\phi} \text{sinc} \frac{(x_i + a/2)D}{z_i \lambda} \right| \quad (3.5.27)$$

where the points are at $x = \pm a/2$, z_i, respectively, and the system is focused on the point x_i, z_i. For $\phi = 0$, the response at $x_i = 0$, with $a = \lambda z_i/D$, is $4/\pi$. This is larger than the response at $x_i = \pm \lambda z_i/2D$, where it is 1. Thus the two points cannot be distinguished from each other.

A similar problem occurs with signal processing and is discussed in Sec. 4.5.4; the same situation for spherical lenses is covered in Sec. 3.3. When $\phi = 0$, it follows from Eq. (3.5.27) that the signal at the midpoint $x = 0$ is just equal in amplitude to the signals at $x = \pm a/2$, if the points are a distance $d_x(\text{Sparrow})$ apart. Thus, for a coherent system,

$$d_{xi}(\text{Sparrow}) = \frac{1.33 \lambda z_i}{D} \quad (3.5.28)$$

This is known as the *Sparrow criterion* (see Sec. 3.3.1).

Range definition. By using a filter based on the point $x_i = 0$, z_i, we can also make a similar estimate for the definition in the z direction. We consider a correlation filter or lens with the characteristic $f^*(-x', z_i)$. We can write $h(x, 0; z, z_i)$ in a normalized form, as follows:

$$h(x, 0; z, z_i) = \frac{1}{D} \left(\frac{z_i}{z} \right)^{1/2} e^{j(2\pi/\lambda)(z_i - z)} \int_{-D/2}^{D/2} e^{-j\omega(x-x')^2/2zV} e^{j\omega x'^2/2z_i V} \, dx' \quad (3.5.29)$$

or

$$h(x, 0; z, z_i) = \frac{1}{D} \left(\frac{z_i}{z} \right)^{1/2} e^{j(2\pi/\lambda)(z_i - z)} e^{-j\omega x^2 2zV}$$
$$\times \int_{-D/2}^{D/2} e^{jx'^2/\lambda)(1/z_i - 1/z)} e^{2j\pi x x'/\lambda z} \, dx' \quad (3.5.30)$$

Equation (3.5.30) is, in general, a Fresnel integral. For $4 z_i \lambda / D^2 \ll 1$ and $|z - z_i| \ll z_i$, we can write the on-axis PSF in the form

$$|h(0, 0; z - z_i)| \approx \frac{1}{D} \left| \int_{-D/2}^{D/2} e^{j\pi x'^2 (z - z_i)/\lambda z_i^2} \, dx' \right| \quad (3.5.31)$$

It is convenient to normalize Eq. (3.5.31) by putting $Z = D^2(z - z_i)/4\lambda z_i$

and $u = x'/D$. Equation (3.5.31) can then be written in the form

$$|h(Z)| = \left| \int_{-1/2}^{1/2} e^{j\pi Z u^2} \, du \right| \tag{3.5.32}$$

This function has a response 3 dB below its maximum value, where the normalized distance $\delta z(3 \text{ dB})$ is defined as

$$\delta z(3 \text{ dB}) = \frac{D^2}{\lambda} \frac{z - z_i}{z_i^2} = 3.8 \tag{3.5.33}$$

This corresponds to the argument of the exponential (the phase error) being approximately π at its limits. The appropriate 3-dB range definition for a CW signal is therefore

$$d_z(3 \text{ dB}) \approx 7.6 \frac{\lambda z_i^2}{D^2} = 1.9 z_i S \tag{3.5.34}$$

where we define the *Fresnel parameter S* as

$$S = \frac{4\lambda z_i}{D^2} \tag{3.5.35}$$

We observe that, in terms of S, the tranverse Rayleigh definition is

$$d_x(\text{Rayleigh}) = \frac{DS}{4} \tag{3.5.36}$$

Thus, just as with a spherical lens, the transverse definition is much smaller than the aperture width D if the parameter S is such that $S \ll 4$.

Typically, for a small aperture system with $z_i/D \gg 1$, the definition in the range direction d_z is far poorer than in the transverse direction. Later we discuss how to avoid this difficulty by using short pulses to obtain good range definition without deteriorating the transverse definition.

More exactly, if we use the more general form of Eq. (3.5.30) for the on-axis ($x = 0$) LSF, then

$$|h(Z)| = \frac{1}{\sqrt{1 + ZS}} \left| \int_{-1/2}^{1/2} e^{j\pi Z u^2/(1 + ZS)} \, du \right| \tag{3.5.37}$$

The value of $|h(Z)|$ is plotted in Fig. 3.5.9 for several values of S. The curves are cut off for the region near $z = 0$ ($Z = -1/S$), where the variation of $|h(Z)|$ with Z becomes very rapid and the paraxial theory is no longer valid. The result obtained in Eq. (3.5.36) corresponds to the $S = 0$ solution. In this case, the first sidelobe is reduced in amplitude by 8.4 dB from the main lobe. In practice, with, for instance, $z_i = 10$ cm, $D = 4$ cm, and $\lambda = 0.5$ mm, $S = 0.125$. So, as we can see from Fig. (3.5.9), S may be large enough to make the use of normalized curves, at best, a rough approximation to the truth. However, provided that $S < 0.25$, this result is still adequate for estimating the range definition of a cylindrical lens.

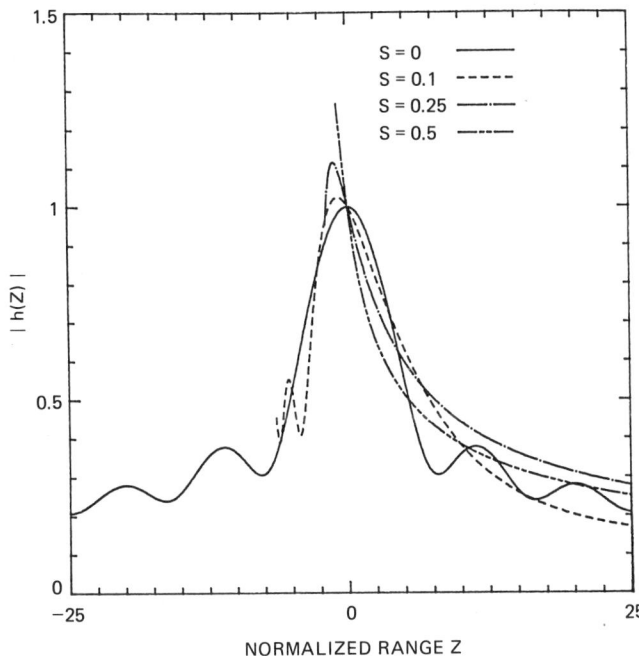

Figure 3.5.9 Normalized line spread function in the z direction. The normalized range Z is defined as $Z = D^2(z - z_i)/\lambda z_i^2$.

D. Generalization of the Matched Filter Theory

The matched filter theory is obviously much more general than the special case given above. Returning to Eq. (3.5.2), we see that if the rays pass through a nonuniform medium, the signals arriving at the receiver will be distorted. However, provided that $f(x - x', z)$ is known, we can still construct a matched filter that will image a point in a nonuniform medium. As with a uniform medium, this matched filter must be changed for each point in the image, but with a far more complicated algorithm. The process is still equivalent to signal processing with matched filters; that is, our focused system essentially operates as a matched filter to compensate for phase differences of the rays arriving at the receiver. If the process is generalized to a two- or three-dimensional system, two- or three-dimensional matched filters must be constructed to give good range definition in the z direction and good transverse definition in the x and y directions.

As an example for a three-dimensional system, we write

$$f(x', y', x, y, z) = \frac{e^{-j\omega R/V}}{R} \tag{3.5.38}$$

where

$$R = [(x - x')^2 + (y - y')^2 + z^2]^{1/2} \tag{3.5.39}$$

The matched filter for the point $0, 0, z_i$ is $f^*(-x', -y', 0, 0, z_i)$.

When the system is cylindrically symmetric and the transducer is of radius a, we can work in cylindrical coordinates by writing $x = r \cos \phi$, $y = r \sin \phi$, $x' =$

$r' \cos \phi'$, and $y' = r' \sin \phi'$. At the plane $z = z_i$, these substitutions yield $h(x, y, z_i) = h(r, \phi, z_i)$ in the form

$$h(r, \phi, z_i) = \int_0^{2\pi} \int_0^a [f(r' \cos \phi' - r \cos \phi, r' \sin \phi' - r \sin \phi, z_i)]$$

$$\times [f^*(-r' \cos \phi' - r \cos \phi, -r' \sin \phi' - r \sin \phi, z_i)] r' \, dr' \, d\phi' \qquad (3.5.40)$$

To the paraxial approximation, this implies that the PSF is

$$h(r, z_i) = C \int_{\phi'=0}^{2\pi} \int_{r'=0}^{a} e^{-j\omega r r' \cos(\phi' - \phi)/z_i V} r' \, dr' \, d\phi' \qquad (3.5.41)$$

where C is a constant.

Following the analysis of Sec. 3.3.1, it can be shown that

$$h(r, z_i) = K e^{-j\omega r^2/2 z_i V} \text{jinc} \frac{ra}{\lambda z_i} \qquad (3.5.42)$$

where jinc $(X) = J_1(2\pi X)/\pi X$, and the parameter $h(r, z_i)$ is known as the *radial PSF*. Here K is a constant and $J_1(x)$ is a Bessel function of the first kind and first order. This result, of course, is of the same form as the one we obtained in Sec. 3.3.1 for the transverse response in the geometrical focal plane of a spherical lens, which is plotted in Fig. 3.2.6.

E. Sidelobes and Grating Lobes

In the foregoing analysis, if the receiver is a spatial matched filter of infinite width, the response to the line source it is focused on is a delta function. On the other hand, if the receiver system has a finite width D, it is no longer a perfectly matched filter and the transverse LSF is of the form

$$f(x, z) = D \text{ sinc} \frac{xD}{z\lambda} \qquad (3.5.43)$$

In this case, sidelobes or subsidiary maxima in amplitude occur at the points x_{sn}, where

$$x_{sn} = \frac{(2n + 1)z\lambda}{2D} \qquad (3.5.44)$$

The amplitude of the first sidelobe is -13 dB relative to the main lobe at $x = 0$. The second sidelobe is -18 dB relative to the main lobe at $x = 0$ and so on.

As we discussed in Sec. 3.3.1, such effects can be a serious disadvantage for imaging. Suppose that we want to image two points, A and B, where B is located on a sidelobe of A. If the amplitude of B is less than the sidelobe amplitude of A, then B will be essentially undetectable. Thus the lower the sidelobe level of the system, the more easily we can detect a weakly reflecting or emitting object in the presence of another, much larger reflector or more powerful emitter.

Such problems are typical, for instance, in NDT, where the aim is to detect a small flaw near one surface of a metal object. In medical imaging, the problems are even more severe: there are so many scatterers present in the body that the sidelobes from them may add up to obscure completely the image of a small object of interest. As a result, much effort has been made to reduce the sidelobe levels in imaging systems; this criterion is often the most important in determining the practical quality of an imaging system.

Many of the same kinds of problems arise in signal processing systems when a sharp pulse is needed; for example, for use in a radar system. Sidelobes of a compressed pulse give rise to very similar difficulties. In Sec. 4.3 we show that the solution is to apodize the response of the filter.

The basic reason for sidelobes is because the matched filter is of finite spatial width. Thus, if we consider the amplitude response of an infinite width matched filter to be artificially weighted by a function $w(x')$, it follows from Eq. (3.5.11) that the output of the system at the plane $z = z_i$ is

$$h(x, z_i) = e^{-j\omega x^2/2z_i V} \int_{-\infty}^{\infty} w(x') e^{j\omega x x'/z_i V} \, dx' \quad (3.5.45)$$

The basic response of the system at the focal plane is the Fourier transform of the weighting function $w(x')$. The implications of using this result for apodization of the acoustic beam are discussed in Secs. 4.5.1 and 3.3.3.

It is apparent that if

$$w(x') = \begin{cases} 1 & |x'| < \dfrac{D}{2} \\ 0 & |x'| > \dfrac{D}{2} \end{cases} \quad (3.5.46)$$

the response is the sinc function of Eq. (3.5.12).

Grating lobes. All electronically scanned systems use an array of separate transducers to simulate the effect of the continuous matched filter or artificial lens described earlier in this chapter. Thus the individual transducer elements and the associated electronic circuitry are designed so that the response of the array, at the center point of each array element, matches that of the continuous system. It is therefore as if the array elements sample the received signal at points spaced a distance l apart. If we use a finite number of receiving elements (i.e., sampling), we introduce grating lobes, or aliasing. This is because the phases of signals arriving at the individual elements can change by 2π and still give rise to the same output acoustic beam.

Suppose, for simplicity, that a receiving array is focused on a point at infinity, as illustrated in Fig. 3.5.10. Individual rays entering the array will be in phase when they arrive normal to the surface of the array, as shown in Fig. 3.5.10(a). However, if the rays arrive at an angle θ_{Gp} to the normal to the array surface, as illustrated in Fig. 3.5.10(b), the difference in length of the rays arriving at neighboring elements, spaced a distance l apart, is $l \sin \theta_{Gp}$. If this distance is such that $l \sin \theta_{Gp} = p\lambda$, all the rays will arrive at the individual elements of the array in

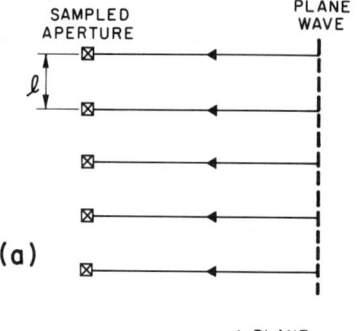

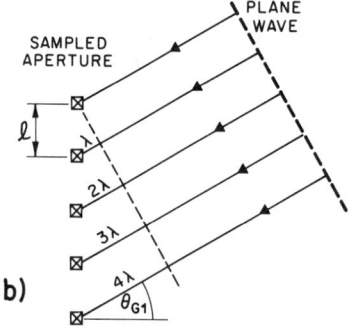

Figure 3.5.10 The $p = 1$ grating lobe for a system focused at infinity: (a) plane wave normally incident on a periodically sampled aperture; (b) plane wave obliquely incident on the same sampled aperture at angle $\theta_{G1} = \sin^{-1}(\lambda/l)$.

phase. Generally, a focused image will repeat itself, that is, aliasing or a repeated lobe will occur at a point that is d_{Gp} distant from the original main lobe, where

$$d_{Gp} = z\theta_{Gp} = \frac{zp\lambda}{l} \qquad (3.5.47)$$

We can obtain this result more rigorously using Eq. (3.5.45). For simplicity, we assume that each element of the receiving transducer is infinitesimally thin. We suppose that there are N elements in the array a distance l apart, as illustrated in Fig. 3.5.11; for simplicity, we assume N to be an even number. It follows that the weighting function $x(x')$ becomes

$$w(x') = \sum_{n=-N/2}^{n=N/2-1} w(x_n)\lambda(x' - x_n) \qquad (3.5.48)$$

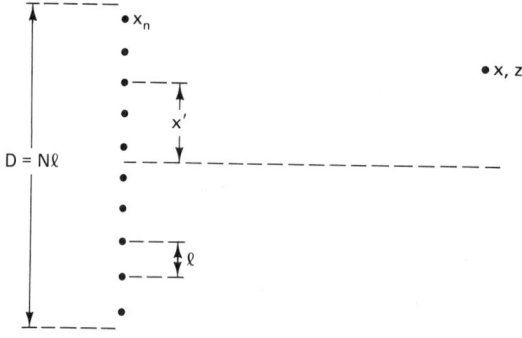

Figure 3.5.11 An N-element array with an even number of elements.

where $x_n = (2n + 1)l/2$ and $\delta(x)$ is the Dirac delta function. Inserting this expression in Eq. (3.5.45) leads to the following result at the focal plane:

$$h(x, z_i) = e^{-j\omega x^2/2z_iV} \sum_{n=-N/2}^{n=N/2-1} w(x_n)e^{j\omega x_n x/z_iV} \qquad (3.5.49)$$

where $w(x_n)$ is the weighting of the nth element. Note that this result is valid only for an even number of elements.

Now consider a uniform array $[w(x_n) = 1]$ with N elements a distance l apart, as illustrated in Fig. 3.5.11. Summing Eq. (3.5.49) results in

$$h(x, z_i) = \frac{e^{-j\pi x^2 \lambda z_i} \sin(N\pi lx/\lambda z_i)}{\sin(\pi lx/\lambda z_i)} \qquad (3.5.50)$$

Thus the maximum output amplitude is at $x = 0$ and is of amplitude N, or N times the output from a single element. When N is large and $\pi lx/\lambda z_i$ is small, the output is $h(x, z) \rightarrow N \, \text{sinc}\,(Dx/\lambda z_i)$, where the length of the array of infinitesimal elements is regarded as being $D = Nl$, just as in the uniform continuous receiving system. For larger values of x, the denominator of Eq. (3.5.50) becomes zero where $\pi lx/\lambda z_i = p\pi$ (i.e., where $x = d_{Gp}$), as given by Eq. (3.5.47).

We note that for the first sidelobe,

$$d_{G1} = Nd_x(\text{Rayleigh}) \qquad (3.5.51)$$

Therefore, as the image and sidelobe pattern repeats itself around each grating lobe, the number of resolvable points in the image that are free from aliasing (i.e., free of regions where the image repeats itself) is N, the number of elements in the array.

Examples of apodization and grating lobes. The sampled Fourier transform relation (3.5.49) yields the form of the LSF for any arrangement of tap weighting. The result for a 32-element unapodized system [31] is shown in Fig. 3.5.12(a). The 13-dB sidelobe level and the grating lobes can be seen clearly.

As a second example, we consider the use of Hamming weighting (see Secs. 3.3 and 4.3). In this case, taking $x = 0$ to be at the center of the array, $w(x_n)$ is chosen so that

$$w(x_n) = 0.08 + 0.92 \cos \frac{\pi x_n}{D} \qquad (3.5.52)$$

The maximum output or the sum of the signals from N elements is, for $N \gg 1$,

$$h(0) \approx 0.54N \qquad (3.5.53)$$

An illustration of Hamming weighting for a 32-element system, with the maximum amplitude normalized to unity, is shown in Fig. 3.5.12(b). At the expense of a slight loss in definition, the use of apodization clearly improves the sidelobe level.

We now consider the effect of errors on the sidelobe levels.

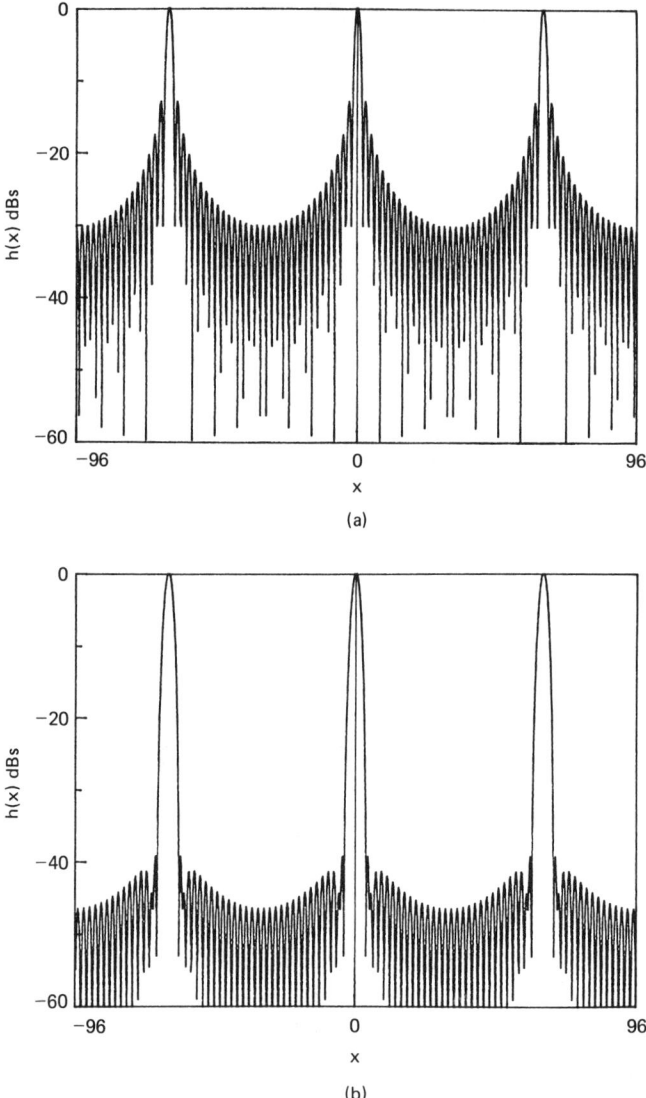

Figure 3.5.12 (a) LSF for a 32-element uniform array. Note the −13-dB sidelobes near the central lobe and the grating sidelobes. (b) LSF for a 32-element Hamming weighted array. Note the −43-dB near-in sidelobes. (c) LSF for a 32-element Hamming weighted array with one central element missing (31 elements present). The absence of this single element raises the sidelobe level from −43 to −23 dB. (d) LSF for a 32-element Hamming weighted array with two elements missing (30 elements present, central element and third from center element missing). The sidelobe level is −18 dB. (Two units of the abscissa correspond to the 4-dB size of one spot.)

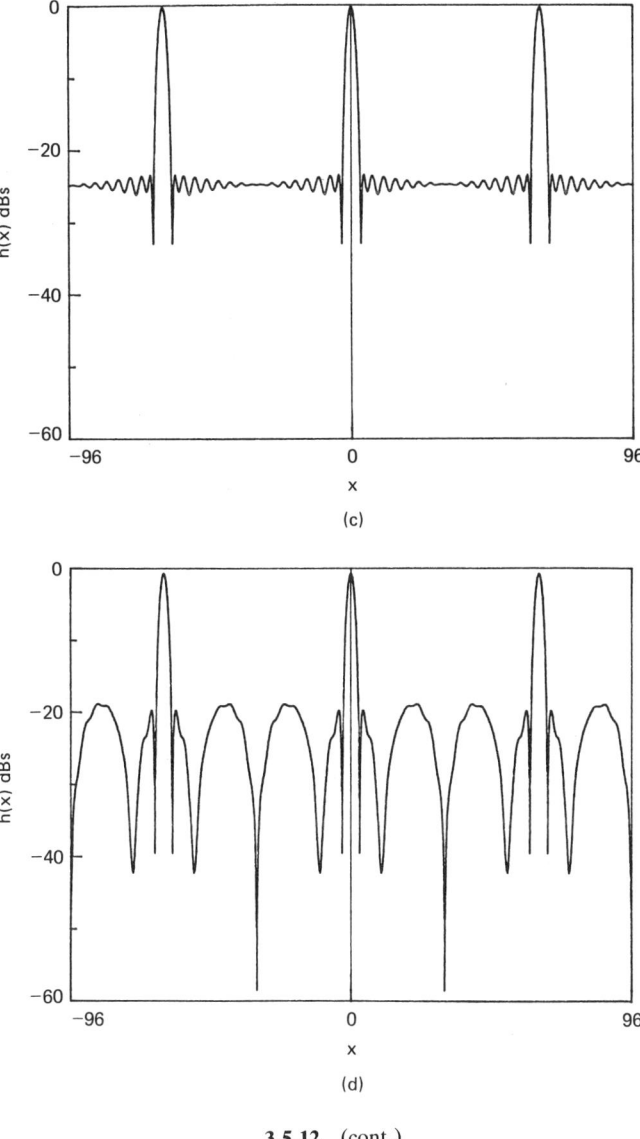

3.5.12 (cont.)

F. Effect of Missing Elements (Amplitude Errors)

The sidelobe level increases when transducer elements are missing or there are phase errors in the system. In this section, following the work of Fraser et al. [31], we analyze the response of an array to a point source and show how this is affected by missing elements. By using the formulas already given, we can calculate how both amplitude and phase errors affect the output. We can use a computer to do this, but it is helpful to try to obtain an analytic formula first, with which we can estimate errors and find how many elements we can afford to have missing.

Suppose that the qth element is missing. It follows from Eq. (3.5.49) that this is equivalent to subtracting an error term $e(x, x_q)$ from the output that is equal to the contribution of the qth element to the output. The error in the output is

$$e(x, x_q) = w(x_q)e^{2j\pi xx_q/\lambda z} \qquad (3.5.54)$$

where $w(x_q)$ is the amplitude of excitation of the correctly excited qth element. Thus the total output is

$$h(x) = h_0(x) - e(x, x_q) \qquad (3.5.55)$$

where $h_0(x)$ corresponds to the output when there are no missing elements.

The phase of the function $e(x)$ depends on the position x_q of the error. If it is in the center, $e(x)$ has virtually no phase change with position x. Assuming that the level of $e(x)$ is much larger than the sidelobe level in the error-free array, the magnitude of the ratio of the sidelobe to the main lobe amplitude, due to one missing element, will be $R(x_q)$, where

$$r(x_q) = \frac{w(x_q)}{\sum_n w(x_n) - w(x_q)} \qquad (3.5.56)$$

In a Hamming weighted system, this corresponds to

$$R(x_q) = \frac{w(x_q)}{0.54N - w(x_q)} \qquad (3.5.57)$$

Thus, for a missing element at the center of the array,

$$R(x_q) = \frac{1}{0.54N - 1} \qquad (3.5.58)$$

For $N = 32$ elements, this corresponds to $R(x_q) = 0.061$, or -24 dB.

A more exact calculation would take account of the sidelobe level that was present before the errors were introduced (i.e., 0.007 down from the main lobe). The worst possible sidelobe level in the presence of errors would then be 0.061 + 0.007, or -23 dB. This result agrees fairly well with the computer result shown in Fig. 3.5.12(c).

When there are several missing elements, the question is whether their effects are additive or tend to add only randomly; that is, if there are q missing elements, is the amplitude error proportional to q or $q^{1/2}$? Unfortunately, the effects are additive, for if there are several missing elements, the effect on the error signal is like that of an array made up of the missing elements. This array produces a signal with a main lobe in which the effect of all the elements is additive, and if the element spacing is periodic, the main lobe tends to repeat itself in a distance corresponding to that of the grating lobes of the error array. If most of the missing elements are near the center of the array, the main lobe of the error array will be wider than that of the full array. Thus there will be sidelobes near the main lobe of the original array that have the full amplitude of the main lobe of the error array. More than likely, there will also be other sidelobes of similar amplitude farther out from the main lobe, due to the quasi-spatial periodicity of the error signal. An illustration of what occurs when two elements are missing is shown in Fig. 3.5.12(d).

The results obtained from these simple concepts agree well with the computer results; they indicate that it is critically important to have all elements working correctly in the transducer array. Otherwise, the maximum sidelobe amplitude is of the order of q/N below that of the main lobe, where q is the number of missing elements. Because not all elements are fully excited in an apodized array, the maximum sidelobe level can actually be worse than the results given by this simple formula. This happens when elements are missing near the center of the array. If they are missing from one end of the array, however, the level is better than this estimate.

3.5.3 Fresnel Lenses and Digital Sampling

A. Basic System

As we discussed, a physical acoustic lens is often difficult to construct because it suffers from multiple reflections at each surface, must be immersed in an acoustic medium such as water, and has relatively large aberrations in comparison to optical lenses. These difficulties are caused in large part by the great difference in refractive index and impedance between water and the typical media available. In addition, mode conversion at the surface of the lens from longitudinal to shear waves, and vice versa, can give rise to unwanted signals that are extremely difficult to eliminate.

A reasonably good physical acoustic lens can be made using liquid, RTV rubber, urethane, or some other plastic. The refractive indices of these materials differ from that of water, but their impedances can be chosen to match that of water very closely. Furthermore, they all have high shear wave attenuation. The best alternative to using these materials, if it is convenient, is to construct a properly shaped acoustic transducer.

Another approach is to use the Fresnel lens concept of optics. In this case, the basic principle is to synthesize the spherical or cylindrical phase front produced by a physical lens.

Consider a plane disk transducer at $z = 0$ in a rigid baffle. Suppose that the wave excited at $z = 0$ varies as $\exp[j\Phi(r')]$ and has a displacement $u_z(0)$ of unit negative amplitude. Then the wave excited at $(0, z_i)$ has a potential of the form

$$\phi(0, z_i) = \frac{1}{2\pi} \int \frac{e^{j[\Phi(r') - kR]}}{R} r' \, dr' \qquad (3.5.59)$$

where

$$R = \sqrt{z_i^2 + r'^2} \qquad (3.5.60)$$

If $\Phi(r') = kR$, all signals arriving at $(0, z_i)$ will be excited in phase, which synthesizes the action of a perfect lens. The Fresnel lens approximates this desired

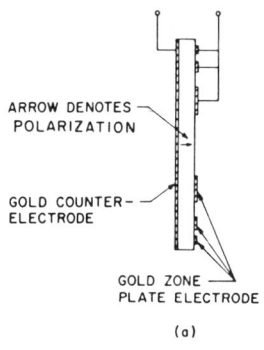

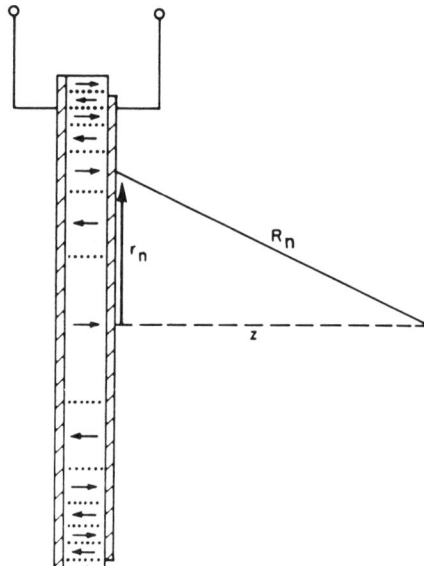

Figure 3.5.13 Fresnel lens imaging system: (a) the PZT polarization is determined by reverse poling an already poled transducer; (b) a simple disk electrode is placed over the phase plate pattern. (After Farnow and Auld [32].)

behavior by sampling the desired response, usually with two samples for every 2π phase change.

Consider the system due to Farnow and Auld [32], illustrated in Fig. 3.5.13(b), in which a disk-shaped PZT transducer is divided into rings of radius r_n. Suppose that a point on the object a distance z along the axis of the lens is a distance $R_n = \sqrt{r_n^2 + z^2}$ from the nth ring. Suppose also that the object is illuminated with a signal of frequency ω and wavelength λ. The phase delay of the ray reaching the nth ring will therefore be $2\pi R_n/\lambda$. If we choose the $n = 0$ element (i.e., the center element) as the reference, we choose the first element, $n = 1$, to have a radius at its center r_1 such that the arriving signal is π out of phase with the signal on the axis. We choose the $n = 2$ element so that the signal arriving at it is in phase, and so on in turn (i.e., all even elements are in phase and all odd elements are out of phase). This means that all the signals arriving at the rings can be added if an extra π phase shift is introduced into the electronic signals picked up by the odd elements.

Farnow and Auld built their device by using a PZT ceramic transducer with ring radii r_n chosen so that

$$\sqrt{r_n^2 + z^2} - z = \frac{n\lambda}{2} \qquad (3.5.61)$$

The poling of the ceramic (see Sec. 1.3.1) was reversed in sign at the appropriate positions, as shown in Fig. 3.5.13(a), to obtain the required π phase shift automatically. Operating this system at 10 MHz, in either reflection or transmission, with two Fresnel lenses placed opposite each other, they were able to mechanically scan the lenses across an object to form an image in the same way as with the acoustic microscope. By this means, they obtained good reflection and transmission pictures. A reflection image of a serrated metal sheet obtained with this system operating at a frequency of 10 MHz is shown in Fig. 3.5.14. The definition is excellent, and comparable to the theoretical prediction.

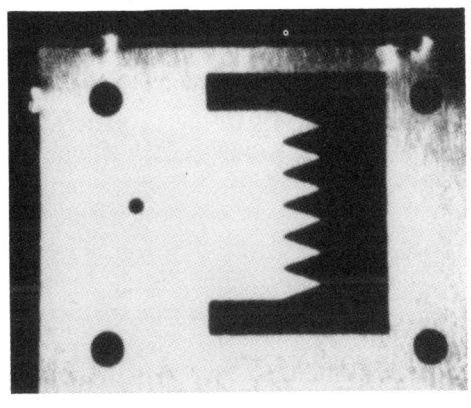

(a)

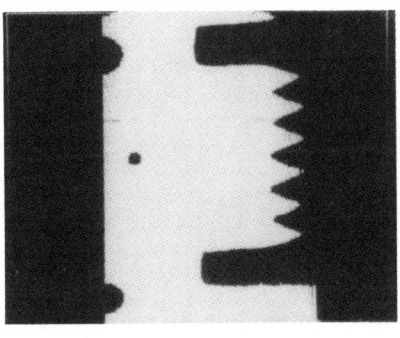

(b)

Figure 3.5.14 Comparison of optical (a) and acoustic reflection (b) images of a sawtooth pattern punched in a 3-mil nickel sheet. The results were taken with a Fresnel lens at a frequency of 10 MHz. (After Farnow and Auld [32].)

There are two problems with such a simple Fresnel imaging system. First, the only phases introduced are 0 and π; intermediate phases are required to obtain a perfect image with low sidelobes. As we shall see, because of the errors due to phase sampling, extra subsidiary foci occur, as well as a relatively high sidelobe level at the main focal plane. Second, if there are N rings, the total difference in length to the axial point between the outer and inner rings is $N\lambda/2$. Therefore, if a pulsed RF signal (i.e., a signal M cycles long) is employed, M must be chosen so that $M > N/2$ in order for signals from all the rings to arrive together at the focal point on the lens. This implies, in turn, that the range definition will be dictated mainly by the depth of focus of the lens, and that short pulses cannot be used unless the rings themselves are excited at different times so that all pulses arrive at the focus at the same time.

B. Fresnel Lens Sidelobes and Phase Sampling in Digital Systems

The ideal matched filter response to image the point 0, z_i is $2\pi z_i/\lambda\, g(r', z_i)$ = $\exp(j\pi r'^2/\lambda z_i)$, where we have omitted the term $\exp(2j\pi z_i/\lambda)$, which may for the present purposes be regarded as a constant. In general, with the development of digital processing and Fresnel lenses, it is important to consider the effect of sampling the phase $\phi = \pi r'^2/\lambda z_i$ at M points (i.e., dividing the possible phases into M steps). Such phase sampling leads to errors in the response of the receiving system, which in turn show up as a decrease in the main lobe amplitude and the production of additional sidelobes and spurious focal points.

To understand the principles involved, let us consider a one-dimensional rectilinear Fresnel lens. Suppose that the response of the transducer array is either $+1$ or -1 at its different elements. If we write $u = x'^2/\lambda z_i$, the optimum spatial response of the two-phase Fresnel transducer array can be written in the form

$$g(u) = \text{sgn}(\sin \pi u) \tag{3.5.62}$$

This expression has the form of a simple square wave in u, where $u = x'^2/\lambda z_i$ is defined only for u positive, with $\text{sgn}(s) = 1$ for $x > 0$ and $\text{sgn}(s) = -1$ for $s < 0$, where $s = \sin(\pi u)$.

We can carry out a Fourier expansion of $g(u)$ to obtain the results, stated as a function of u or x', respectively, in the form

$$g(u) = \frac{2}{\pi} \sum_{m=-\infty}^{\infty} \frac{e^{j(2m+1)\pi u}}{2m + 1} \tag{3.5.63}$$

or

$$g(x') = \frac{2}{\pi} \sum_{m=-\infty}^{m=\infty} \frac{e^{j(2m+1)\pi x'^2/\lambda z_i}}{2m + 1} \tag{3.5.64}$$

Note that in this case, as the response of the transducer array is $g(u) = \pm 1$, the transducer can consist of individual electrodes with infinitesimal spacing between them, connected to either the positive or negative terminals of an individual amplifier, or it can be made of a piezoelectric material in which individual elements are poled in the $+z$ or $-z$ directions, as we have already discussed. Each element of the transducer must decrease progressively in width from the one before it, for $x' \propto u^{1/2}$.

We observe that each harmonic term of the expansion in Eq. (3.5.64) corresponds to an individual matched filter or lens with a focal point at

$$z_m = \frac{z_i}{2m + 1} \qquad (3.5.65)$$

Thus the $m = 0$ term corresponds to a simple lens with a focal length z_i. In addition, however, there are subsidiary foci at $z_i/3$ ($m = 1$), $z_i/5$ ($m = 2$), and so on, as illustrated in Fig. 3.5.15, with virtual foci at $-z_i$ ($m = -1$), $-z_i/3$ ($m = -2$), and so on. As a receiver, the amplitudes of the transducer response to the individual harmonic terms are reduced by a factor of $2/(2m + 1)\pi$, and that of the main lobe ($m = 0$) by $2/\pi$, in comparison to a system with a continuous analog phase reference of unit amplitude. This result is identical to that for the cylindrical Fresnel lens.

Sidelobe level. It is interesting to consider, for a transmitting array, the sidelobes generated by the subsidiary beams passing through the subsidiary foci. The reciprocity theorem then yields results of the same form for a receiver system. For the mth focus, using Fig. 3.5.15, it follows from simple geometry that the beam passing through the focus at $z_m = z_i/(2m + 1)$ has a width $|2mD|$ at the focal plane $z = z_i$. It therefore produces a uniformly distributed field at the focal plane that is reduced in amplitude by $1/|2m|^{1/2}$ from its value at the array. The strongest subsidiary beams are the $m = -1$ and $m = +1$ beams. The interference between

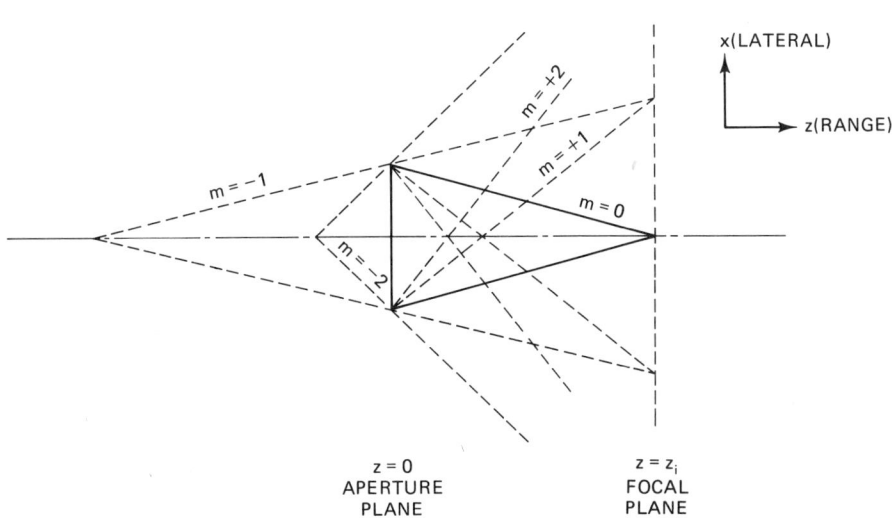

Figure 3.5.15 Subsidiary foci due to a Fresnel lens or to digital sampling of phase.

them gives rise to nonuniformly distributed sidelobes, which can be at a fairly high level in a two-phase system.

To treat this subject more quantitatively, we take the normal displacement at the transducer to be $u_z(x', 0) = Kg(x')$, where K is a constant. It then follows from the Rayleigh–Sommerfeld formula (see Prob. 3.1.3) that for a rectilinear system in the paraxial approximation, the LSF is of the form

$$h(x, z) = \frac{e^{-jkz}}{z^{1/2}} \int g(x') e^{-jk(x-x')^2/2z} \, dx' \tag{3.5.66}$$

where $k = 2\pi/\lambda$. Substituting Eq. (3.5.64) in Eq. (3.5.66), we obtain

$$h(x, z) = \frac{2e^{-jkz}}{\pi z^{1/2}} \int_{-D/2}^{D/2} \sum \frac{e^{j(2m+1)\pi x'^2/\lambda z_i} \, e^{-j\pi(x'-x)^2/\lambda z}}{2m+1} \, dx' \tag{3.5.67}$$

where D is the length of the array.

As we have already discussed, for the mth harmonic, the quadratic terms in x' cancel out where $z = z_i/(2m+1)$, and each harmonic term can be treated as if it were associated with a lens of focal length $z_i(2m+1)$.

At the focal point $(0, z_i)$, we find that the fundamental harmonic gives rise to a term

$$h_0(0, z_i) = \frac{2De^{-jkz_i}}{\pi z_i^{1/2}} \tag{3.5.68}$$

where the mth harmonic is denoted by the subscript m. Thus the strength of the main lobe is reduced by a factor of $2/\pi$, compared to a perfect continuously excited system.

Similarly, the amplitude of the mth harmonic at the focal plane is

$$h_m(x, z_i) = \frac{2e^{-jkz}}{\pi z_i^{1/2}} \int_{-D/2}^{D/2} \frac{e^{j(\pi/\lambda z_i)(2mx'^2 + 2xx' - x^2)}}{2m+1} \, dx' \tag{3.5.69}$$

The integral can be evaluated by the method of stationary phase (Appendix H) to yield the result

$$h_m(x, z_i) = \frac{2e^{-jkz_i}}{\pi(2m+1)} \left|\frac{\lambda}{2m}\right|^{1/2} e^{[j\pi\mathrm{sgn}(m)/4]} e^{-[j\pi(2m+1)x^2/2z_i\lambda m]} \tag{3.5.70}$$

It follows that

$$\left|\frac{h_m(x, z_i)}{h_0(0, z_i)}\right| = \left(\frac{\lambda z_i}{D^2}\right)^{1/2} \left|\frac{1}{2m}\right|^{1/2} \left|\frac{1}{2m+1}\right| \tag{3.5.71}$$

This agrees with our earlier physical derivation, which implied that the fields of the mth harmonic at the focal plane are reduced in amplitude by a factor $1/|2m^{1/2}|$. The $1/|2m+1|$ factor in Eq. (3.5.71) is proportional to the excitation of the mth harmonic.

We observe that the $m = -1$ harmonic is the one most strongly excited. It yields a background level reduced by $(\lambda z_i/2D^2)^{1/2}$ from the main lobe. As an example, with $z_i = 10$ cm, $\lambda = 0.5$ mm (3 MHz in water), and $D = 2$ cm, the background level is 12 dB below the main lobe level. The $m = +1$ term can add to the $m = -1$ term, bringing the peak sidelobe level to 9.5 dB below the main lobe. So the sidelobe problem of a Fresnel lens can be a serious one.

A similar analysis can be carried out for an M-phase system of the type used in a receiver. When lumped delay lines are employed to synthesize the phase, or when a signal is digitally sampled with a clock rate M times the signal frequency, there are M samples per cycle. In this case a similar analysis yields the result (see problem 4)

$$\left|\frac{h_m(x, z_i)}{h_0(0, z_i)}\right| = \left(\frac{\lambda z_i}{D^2}\right)^{1/2} \left|\frac{1}{mM}\right|^{1/2} \left|\frac{1}{mM + 1}\right| \qquad (3.5.72)$$

Note that the sidelobe level falls off with M. A five-phase system (e.g., an analog-to-digital converter operating at a clock frequency of five times the fundamental frequency) will therefore give a sidelobe level that is still by no means adequate for a high-quality medical imaging system. For example, a system with $M = 5$, $m = -1$, $z_0 = 10$ cm, $D = 2$ cm, and $\lambda = 0.5$ mm gives a background level of -28 dB. The inclusion of the $m = 1$ term raises the peak sidelobe level to approximately -22 dB below the main lobe. The reader is referred to more detailed numerical calculations for further information [33]. Note that such sidelobe levels are far higher than the -60-dB level required, which is obtained in some very high quality medical imaging systems.

3.5.4 Chirp-Focused Systems

A. Basic System

In Sec. 3.5.2 we showed that the use of a matched spatial filter makes it possible to obtain good definition of point or line sources. One way to realize such a filter is by employing SAW techniques of the type described in Chapter 4.

Here we will describe principles of the chirp-focused system because the concept is simple, the concepts used are related to holography (see Sec. 3.5.6), and experiments have been carried out, using the technique, that demonstrate many possible forms of electronically focused acoustic imaging.

Chirp-focused imaging systems have been constructed that operate as both receivers and transmitters of focused beams and, at the same time, provide automatic scanning along a line parallel to the acoustic array. The systems have been demonstrated mainly in NDT, rather than medical, applications, because they are basically phase-focused systems and thus suffer from relatively poor range definition and severe sidelobe problems. This tends to make them unsuitable for medical imaging.

Chirp-focused systems employing SAW delay lines are no longer used, because it is difficult and expensive to make the physically large, multiple tap delay lines they require. Since SAW delay lines are analog devices with fixed taps, they tend to be somewhat inflexible in this acoustic imaging application and may not, in practice, give the necessary accuracy. However, the same concept can also be implemented with more flexible digital systems to provide the required sampled phase references for focusing and scanning [31, 34–36].

Consider the FM chirp-focused receiver system using an array of transducer elements, which is illustrated schematically in Fig. 3.5.16. Suppose that the object is illuminated by a signal of frequency ω_s and wavelength λ in the medium being examined. The signal arriving at each array element is mixed with a signal from a corresponding tap on an SAW line.

The signal propagating along this delay line is chosen to have a parabolic variation of phase on the same magnitude but opposite sign as the signals arriving from the point x, z. When the two signals, one from the tap and one from the array element, are mixed, their frequencies add, as do their phases. Therefore, if the phase delay of the wave propagating along the delay line is chosen correctly, it is possible to cancel out the phase differences between the different rays arriving from the point x, z. Using this technique, we can construct an electronically focused lens focused on the point x, z; this lens is a matched spatial filter for the point x, z, as described in Sec. 3.5.2.

The correct signal to inject into the delay line is a *linear FM chirp*, one whose frequency varies linearly with time as $\omega = \omega_1 + \mu t$. This signal has its frequency at the nth tap given by the relation

$$\omega = \omega_1 + \mu\left(t - \frac{x_n}{V_R}\right) \qquad (3.5.73)$$

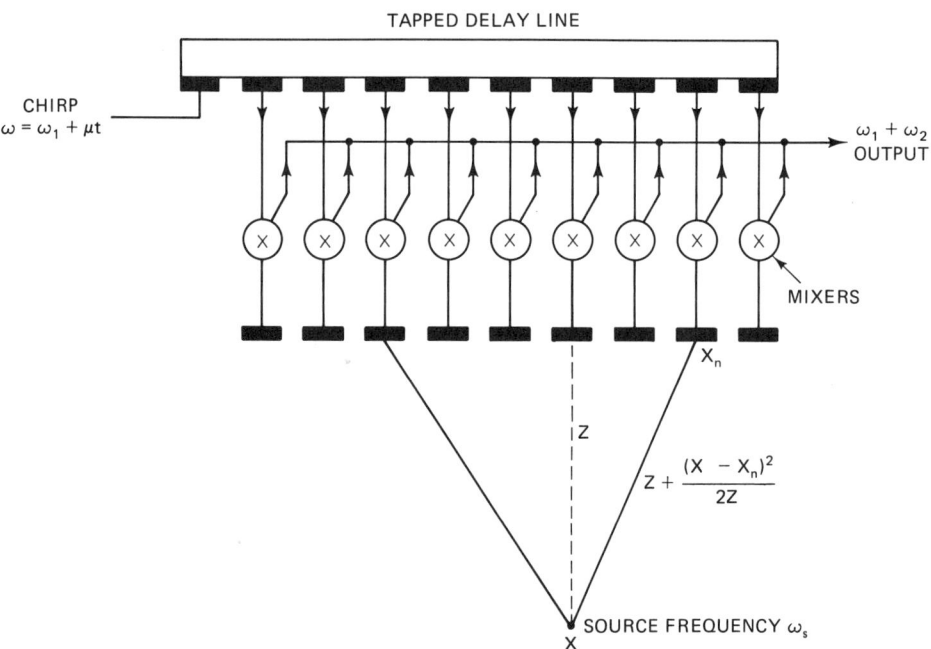

Figure 3.5.16 Arrangement used for mixing the signals from the array elements with a signal from the delay line.

Because the phase is $\phi = \int \omega \, dt$, it follows that the phase at the nth tap is

$$\phi_{Rn} = \omega_1\left(t - \frac{x_n}{V_R}\right) + \frac{\mu}{2}\left(t - \frac{x_n}{V_R}\right)^2 \tag{3.5.74}$$

that is, the filter response is of the form

$$g(x_n, t) = e^{j\omega_1(t - x_n/V_R) + (\mu/2)(t - x_n/V_R)^2} \tag{3.5.75}$$

where $V_R = l/\tau$ is the effective velocity of the surface acoustic wave (or reference wave) along the array, $x_n = nl$, l is the tap spacing, and τ is the time delay of the acoustic wave from tap to tap of the delay line. The FM chirp signal generates a spatial chirp, which moves at a velocity V_R along the delay line. The equivalent matched filter generated by this process therefore moves in the x direction at a velocity V_R, thus providing automatic scanning along one line of the image.

The summed output from the mixers is of the form

$$y(x, t, z) = \sum_{x_n} w(x_n) g(x_n, t) f(x_n - x, t) \tag{3.5.76}$$

where $w(x_n)$ is the amplitude response or weighting of the transducers and $f(x_n - x, t)$ is the signal arriving from a line source at the nth transducer.

It follows that for $w(x_n) = 1$, the output from the array is of the form

$$h(x, t, z) = \frac{1}{z^{1/2}} \sum_{x_n} e^{j[\omega_1(t - x_n/V_R) + \mu/2(t - x_n/V_R)^2]}$$
$$\times e^{j\omega_s[t - z/V - (x_n - x)^2/2zV]} cb \tag{3.5.77}$$

The first exponential term in Eq. (3.5.77) is the filter response $g(x_n, t)$. For $g(x_n, t)$ to be a sampled version of a matched spatial filter for the point x, z [i.e., equivalent to $f^*(-x', z)$ in Eq. (3.5.6)], the chirp rate μ must be chosen so that the square-law spatial phase variation terms cancel out, or for $z = z_i$

$$\mu = \frac{\omega_s V_R^2}{V z_i} = \frac{2\pi V_R^2}{\lambda z_i} \tag{3.5.78}$$

It is also convenient, but not necessary, to have $\omega_1 l / V_R = 2m\pi$, where $x_n = (2n + 1)l/2$ and m is an integer. This choice of ω_1 implies that at $t = 0$, all elements are at the same phase and the system is focused on $x = 0$. The choice of μ is equivalent to choosing the focal length of the lens so that it focuses on the plane $z = z_i$.

With these assumptions, Eq. (3.5.77) can be summed to give the result

$$h(x, t, z_i) = \frac{\sin N\pi l(x - V_R t)/\lambda z_i}{\sin \pi l(x - V_R t)/\lambda z_i} e^{j[(\omega_1 + \omega_s)t + \mu(t^2 - x^2/V_R^2)/2]} \tag{3.5.79}$$

Thus the output of the system is in the form of a modulated FM chirp with a frequency $\omega_1 + \omega_s + \mu t$, with the modulation corresponding to the LSF of the system.

The system behaves like a moving lens, focused on the plane $z = z_i$, that moves at a velocity V_R. By displaying the detected output of the system as a

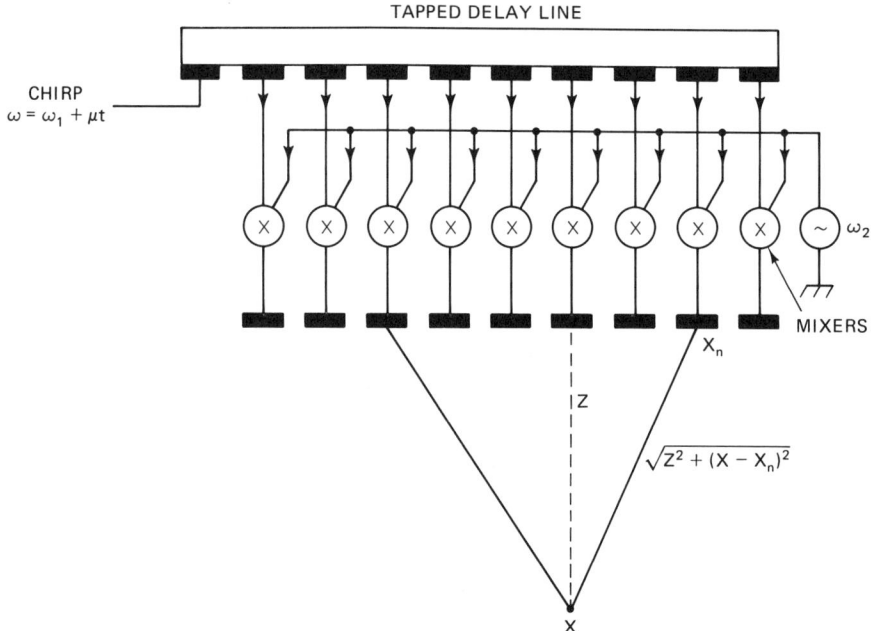

Figure 3.5.17 Chirp-focused transmitter system.

function of time, as an intensity modulated line on a cathode ray tube, we automatically obtain a display of one line of the image. The line x, z_i in the image is displayed at the time $x = V_R t$.

This system behaves in exactly the same way as the matched filter discussed in Sec. 3.5.2. The sidelobe level will be of the same form, and grating lobes will occur in the same relative positions as given by the earlier derivations. The only difference is that the use of the chirp obtains the very useful feature of automatic scanning, with the main lobe focused on the point $x_i = V_R t$, $z_i = \mu V/\omega_s V_R^2$.

This basic system is simple, in principle. In practice, however, it typically requires mixers and amplifiers on each element, and suitable summing networks. It also requires an SAW delay line several inches long, with perhaps as many as 100 taps, and a total delay comparable to a TV line scan (64 μs). It is not easy to construct such delay lines free of defects. This is a serious problem because, as was shown in Sec. 3.5.2.F, if q elements are missing from an N-element array, the maximum sidelobe amplitude due to the missing elements will be approximately q/N lower than the main lobe.

The same system can be used as a transmitter as well as a receiver, by exciting the mixers from the delay line and a separate oscillator, as shown in Fig. 3.5.17, to give an output signal centered about a frequency ω_s, which can be used to excite an element of the array. In this case, of course, separate transmitter amplifiers are needed for each element.

When this device is used in the transmit mode, it behaves like a moving lens traveling at a velocity V_R parallel to the array. The time for which a single spot in the image is illuminated is approximately

$$\tau_x(4 \text{ dB}) = \frac{d_x(4 \text{ dB})}{V_R} = \frac{\lambda z_i}{D V_R} \qquad (3.5.80)$$

where $d_x(4 \text{ dB})$ is the spot size.† The time difference between the ray paths from the center and the outside of the beam is

$$\tau_R = \frac{D^2}{8 z_i V} \qquad (3.5.81)$$

In order for the spot to be illuminated by all the rays, and hence for optimum definition to be obtained, the requirement $\tau_R < \tau_x$ is apparent. This, in turn, implies that when $\tau_R = \tau_x$, the minimum spot size to the 4-dB points is

$$d_x(4 \text{ dB}) \approx \left(\frac{z \lambda^2 V_R}{8 z_i V}\right)^{1/3} \qquad (3.5.82)$$

with

$$D = \left(\frac{(8 z_i^2 V \lambda)}{V_R}\right)^{1/3} \qquad (3.5.83)$$

provided that the central axis of the beam is opposite the point of interest.

The differences in transit time from the edge and the center of the array, respectively, to the focal point, give problems with such a phase focused system, but would not in a time-delay focused system. Similar difficulties occur with Fresnel lenses and with holographic imaging systems. In each case, the length of the excitation signal has a minimum value τ_R fixed by the requirement for phase focusing. The minimum range resolution of a phase focusing system in which the scan is very slow is of the order of τ_R. As we shall show below, the fast chirp scan gives better range resolution, albeit with other disadvantages.

B. Various Forms of Chirp-Focused Systems

Chirp-focused systems have been made in several versions, some with as many as 128 elements. Several examples are illustrated in Fig. 3.5.18. The left-hand figure of Fig. 3.5.18(a) shows a system employing transmitting and receiving transducer arrays placed opposite each other with the object placed between them and mechanically scanned up and down in the y direction. Such a system has the advantage, already discussed in Sec. 3.5.1 and illustrated in Fig. 3.5.3(b), of giving the speed and focusing in the x direction of electronic scanning, which speeds up the scan by approximately a factor of N. Also, because the receiver and the transmitter are both focused on the same point, the response of the system will vary as $[\text{sinc}(x/d_x)]^2$, giving the same advantages as those we have already discussed for the acoustic microscope in Sec. 3.3.3.

An example of an NDT application is shown in Fig. 3.5.18(a). The system

---

† It is convenient here to work with the 4-dB resolution, $d_x(4 \text{ dB}) = \lambda z/D$, thus omitting the 0.89 factor used in the definition of the 3-dB resolution.

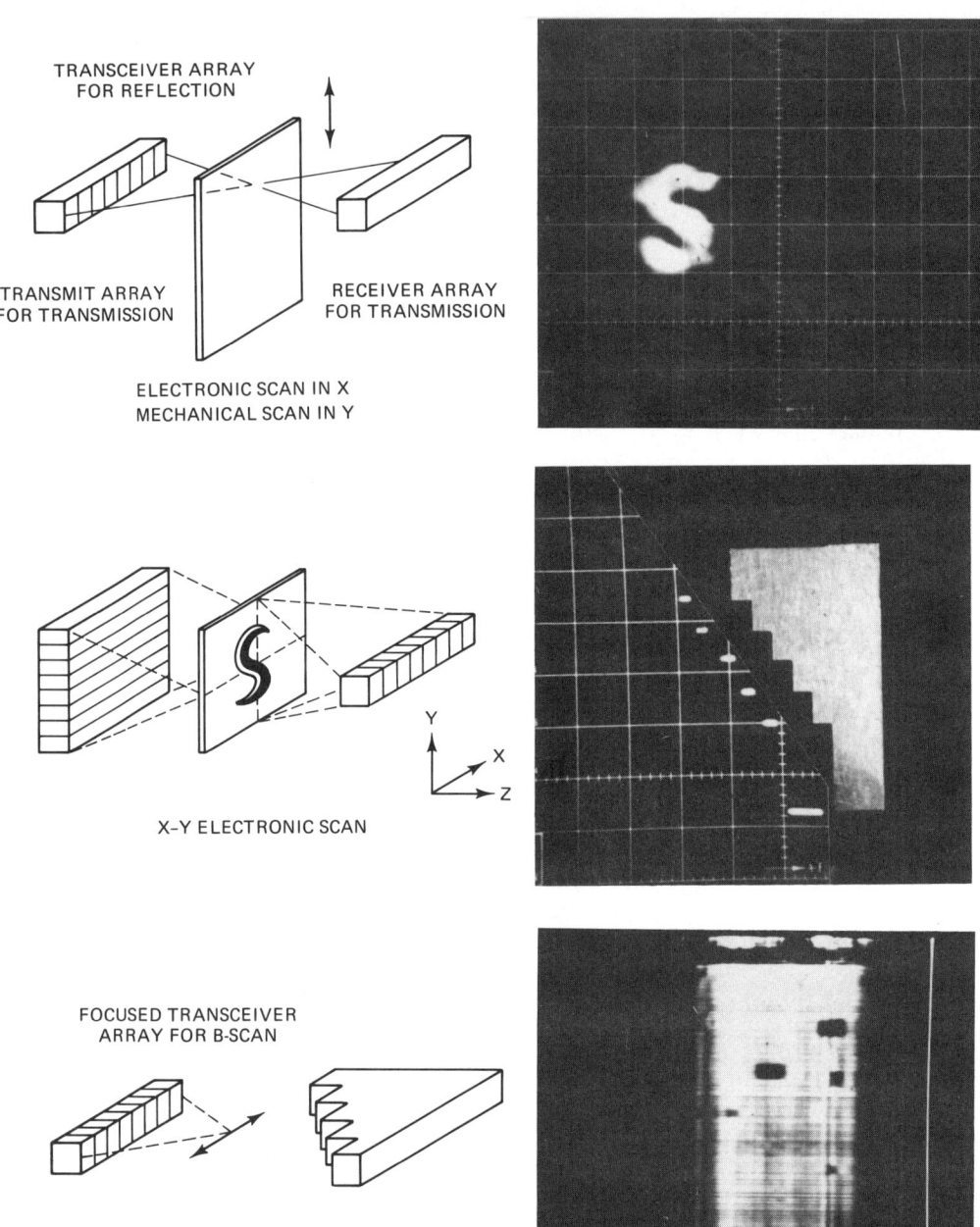

Figure 3.5.18 Chirp-focusing schemes. (a) Transmission image of a boron fiber epoxy laminate carried out with transmitting and receiving arrays placed opposite each other. (After Waugh et al. [36].) (b) Two-dimensional transmission image with the two arrays placed at right angles to each other. (After Fraser et al [31].) (c) Simple B-scan image, using the same array as transmitter and receiver. (After Waugh et al. [36].)

was used to scan a boron fiber–reinforced epoxy laminate laid down on titanium. The sample measured approximately 22 × 7.5 cm and defects were deliberately introduced into it, as can clearly be seen. The total scan took only a few seconds because the mechanical scanning was only in one direction.

A second example of the use of two arrays, illustrated in Fig. 3.5.18(b), is to work in a transmission mode with the transmitting and receiver arrays arranged at right angles to each other. As in the previous example, the transmitter and the receiver are both focused on the same plane, but the transmitter array scans in the x direction and the receiver array scans in the y direction. This makes it possible to obtain N^2 resolvable spots with only $2N$ array elements. The major disadvantage of such a system, though, is the very high sidelobe level. We have to pay for the saving in electronic complexity!

Systems of this type have also been used in a reflection mode. In this case, the same SAW delay line and array are used for both transmission and reception, with appropriate switching to transmit or receive. In this case, the basic transverse definition is comparable to that of the two-lens transmission system shown in Fig. 3.5.18(a) or that of the acoustic microscope; this implies that there is a $(\sin x/x)^2$ response function in the focal plane. The receiver is operated at a time $T_z = 2z/V$ later than the transmitter. By varying the time delay T_z, along with the chirp rate μ, a B-scan in the form of a series of lines or a raster in a plane perpendicular to the array can be scanned, as illustrated in Fig. 3.5.19. In principle, the range definition should be fairly good, because the object point is illuminated for a time $\tau_x = d_x/V_R$, so that the range definition becomes

$$d_z \approx V\tau_x \approx \frac{Vd_x}{V_R} \tag{3.5.84}$$

As the scan velocity V_R in such systems is comparable to the velocity V in the medium, the implication is that the range definition is comparable to the transverse definition. We use the relation that the total frequency excursion of the chirp is

$$\Delta f_c = \frac{\mu T}{2\pi} = \frac{\mu D}{2\pi V_R} \tag{3.5.85}$$

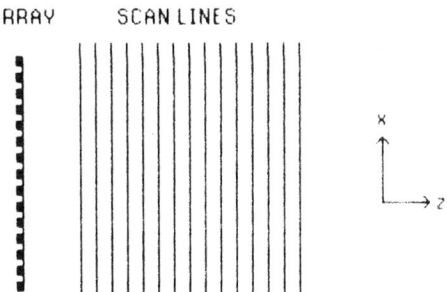

Figure 3.5.19 Scan format for a B-scan FM chirp system.

Sec. 3.5 Lensless Acoustic Imaging

where D is the length of the array. It then follows from Eqs. (3.5.80) and (3.5.85) that

$$\tau_x \approx \frac{\lambda z}{DV_R} = \frac{2\pi V_R}{\mu D} = \frac{1}{\Delta f_c} \tag{3.5.86}$$

Thus the range definition turns out to be just what we would expect from a pulse with approximately the same bandwidth as the chirp.

The FM chirp-focused system has provided fairly good images in NDT applications; an example is illustrated in Fig. 3.5.18(c). But work on systems of this type was abandoned because of their poor sidelobe response. In this system, a point in the image can be insonified by the main lobe of the transmitter and received on a sidelobe of the receiver, and vice versa. A further difficulty is that the sidelobes can be due to reflectors that are closer to the array than the focal point of, for example, the transmitter. This makes the device particularly difficult to use in medical applications, where the attenuation in body tissue is very high. In such cases, the effective sidelobe level is higher than that caused by a reflection at the focal plane by an amount corresponding to the two-way attenuation between the plane of the unwanted reflector and the required image plane.

These examples serve to point out the difficulties that may occur in the application of practical imaging systems. In principle, we expect the reflection imaging system to give a sinc^2 (x/d_x) response and a fairly good range definition. In practice, the system responds just this way for the transverse definition in the focal plane, but sidelobes occur in regions we might not have expected to see them in initially.

Systems of this type are at their best when used in a reflection or transmission mode to look at a thin object for which range resolution is not of great importance. In such cases, the sidelobe level is low. The phase reference of the chirp delay line can be synthesized digitally, and the scan in the x direction can be made at any velocity, from zero to a value larger than that for a typical SAW system.

Focused optical beams can also be used to give the required phase reference. This is the basic principle of holography, which will be described in Sec. 3.5.6.

Reflection mode imaging is best carried out with a system that defines at least two of the dimensions of the focal spot very precisely and has very low sidelobe levels. One example is time-delay focusing, discussed in Secs. 3.5.1 and 3.5.5, in which the region to be imaged is confined by the use of a short pulse to a short range (z direction). A second example is the acoustic microscope, in which both the transverse dimensions (x and y) of the focal spot are well defined, while the range resolution in the z direction can be very good because of the short range definition of a wide aperture confocal imaging system (see Fig. 3.5.4).

3.5.5 Time-Delay and Tomographic Systems

A. Introduction

In Sec. 3.5.1 we discussed the various types of scanned systems that can be employed for imaging. As we saw, time-delay systems are very attractive because their range definition is not necessarily limited by the aperture size, as it is in a

Figure 3.5.20 Standard B-scan pulse array system illustrating scan lines.

phase-focused system. However, a variable time-delay system is more difficult to make than a variable phase-delay system.

Simple implementations of unfocused array systems are commonly employed for medical imaging. As an example, several commercial systems use an array of transducer elements with a total length of the order of 10 cm in the x direction (center frequency 2.25 MHz). Typically, 64 or 128 elements, each 1 cm wide (y direction), are employed.

In one example of a 64-element system, these elements are excited sequentially with a short pulse in groups of four, to emit a parallel beam in the z direction, approximately 1 cm square. The beam is received by the same four elements that emitted it, and displayed in the form of one line of a B-scan image. The device then moves one element on, to the next group of four elements, and the process is repeated. Thus a 64-line image is obtained, with a scan form like that shown in Fig. 3.5.20. This image has accurate range definition, because of the short pulse employed, but relatively poor transverse definition, because the device behaves as if it were a 1-cm-square transducer being moved along the x direction in 2.5-mm steps.

This type of system has been developed by a number of laboratories; it has been applied mainly in medical and, to a limited extent, NDT applications. Because no delay lines are employed, it is possible to use the same technique at relatively high frequencies.

Other alternatives have been considered. One, employed by Becker et al. at Battelle [37], is to use 128 elements in a row, exciting them sequentially in much the same manner already described. However, by exciting the array elements with a programmed time delay between them, it is possible to excite a wave at an arbitrary angle to the array. One application of this technique is to look for faults in nuclear reactor walls. The standards that have been set up for pressure vessels require such observations to be made with several incident beam angles, to pick up specular reflectors. By using time-delay techniques, the system is scanned like the system described earlier, except that the scan is now carried out at several different angles. By mechanically scanning in the other direction, a large volume of material can be tested in a relatively short time. Using sophisticated display techniques, three-dimensional information can be obtained in the form of an isometric projection display that makes the results relatively easy to interpret. A sample of an image obtained with this system is shown in Fig. 3.5.21.

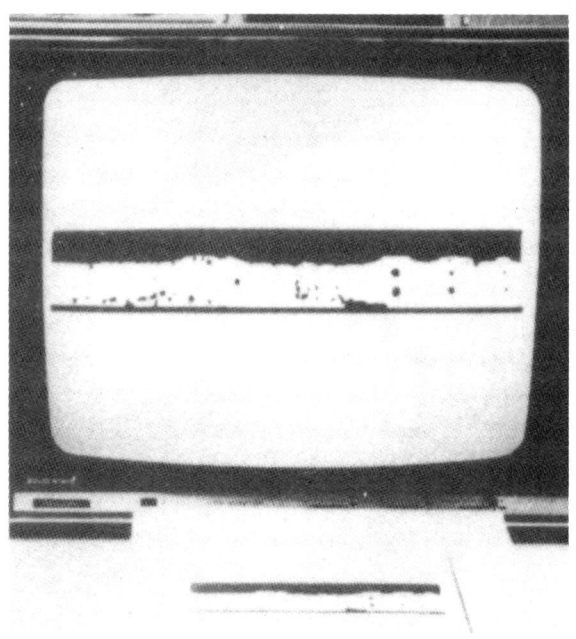

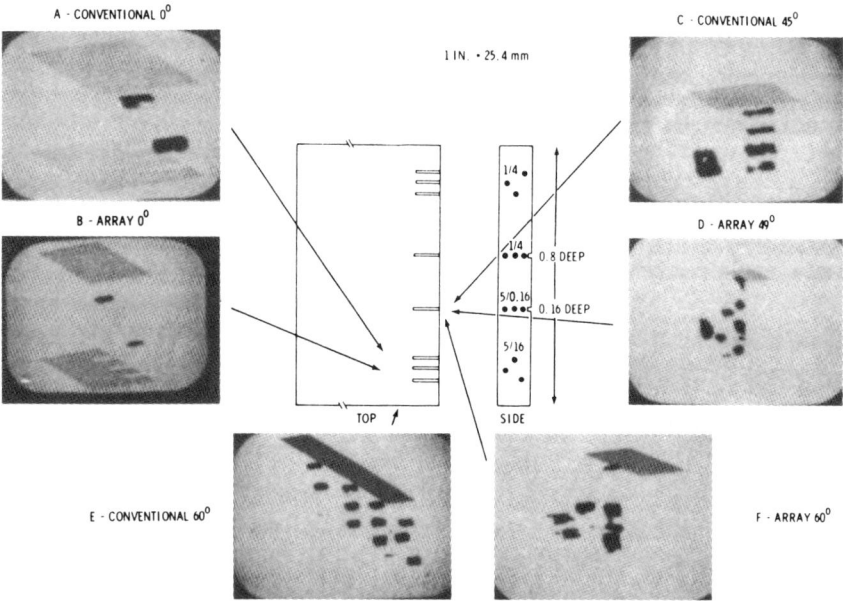

Figure 3.5.21 Isometric images of side-drilled holes taken with the Battelle imaging system. (After Becker et al. [37].)

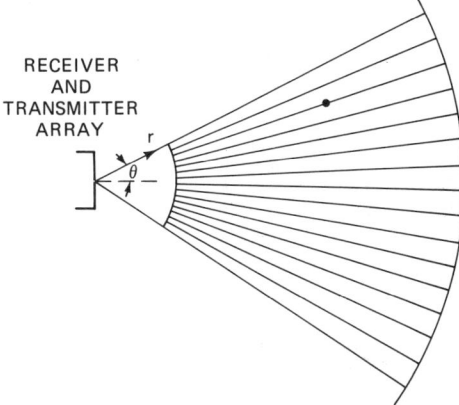

Figure 3.5.22 B-scan radial sector scan format. Scan lines are radial.

The design problem for a full-time-delay system is difficult because adequate delay lines must be provided for the receiver, with delays that can be varied with time to focus on the return echo signal for the transmitted pulse as it travels out from the array.

The simplest solution is to put a fixed-focus physical lens in front of the array to give optimum focusing at the center of the field. However, although a cylindrical lens can help focus a beam in the y direction, a lens to focus a long array in the x direction cannot be made easily.

An alternative technique is to excite four or more elements of the array from the transmitter, with suitable time delays, to provide an optimally focused beam at a fixed distance from the array. With a finite depth of focus, this is equivalent to using a focused transducer for the transmitter, which can be stepped along the array just as it is with the simpler system already described. At other ranges, the focusing will be worse than the optimum value. This stratagem is now employed in several commercial medical imaging systems to provide focusing in the x direction, in combination with a fixed cylindrical lens to provide better definition in the y direction.

In cardiac imaging systems, a dynamically focused receiver array is used with a radial sector scan format, with scans along radial lines extending from the center of the transducer array, as illustrated in Fig. 3.5.22. In this system, a parallel beam is transmitted at an angle θ to the axis by appropriately delaying the signals that excite the elements. The signals received on the elements are passed through delay lines so that the receiver is aimed in the θ direction. As the transmitted pulse travels out from the array, the focusing of the receiver must be changed. To do this, the original system, by Thurstone and Von Ramm [38], changed the delay time from each element to the detector by switching the taps on electromagnetic (EM) delay lines, with this switching controlled from a minicomputer. This system, which changes focusing as signals arrive from different depths, is known as a *dynamically focused system*.

The disadvantage of using lumped EM delay lines is that only a limited number can be used, and they tend to be bulky, which implies that the phase sampling of

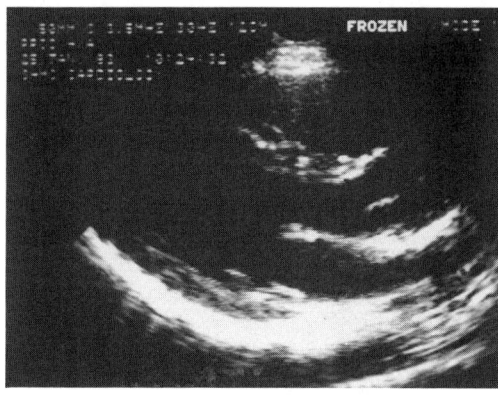

Figure 3.5.23 Image of the heart taken with a Hewlett-Packard radial sector scan real-time ultrasonic imaging device. (Courtesy of J. Larson.)

this system may be coarser than would be ideal. As shown in Sec. 3.5.3.B, this implies that the sidelobe level is higher than acceptable in a high-quality medical imaging system; often in such systems, it is desirable to have a sidelobe level no higher than 60 dB below the main lobe level.

Several stratagems have been adopted to circumvent these difficulties. One, for a relatively narrow aperture system, is to use lumped EM delay lines to provide the basic time-delay increments required for steering. Additional phase adjustments are made by mixing the received signals with phase references in a manner similar to that described in Sec. 3.5.3.B. An image of the heart taken at 3.5 MHz with a 64-element array, 1.5 cm wide, is shown in Fig. 3.5.23. This system has been adopted by Hewlett-Packard for cardiac imaging [39], where the access region to the heart is limited.

Another technique is to use CCD delay lines (see Sec. 4.3) to provide the necessary time delays; these are changed by varying the clock rates of the lines [40]. This approach has proven less simple than it appeared initially, basically because of interference from the clock signals.

Finally, there are several examples of such digital systems that provide the necessary delays. One approach is to use an analog-to-digital (A/D) converter to convert the received analog signal to, for example, an 8-bit digital signal, and then using variable digital delays. The digital signals from the different transducer elements are summed and then reconverted to analog signals with a D/A converter. In some cases, signal detection is carried out before A/D conversion, which makes it possible to design linear detectors that are almost perfect.

Another such system, which has been demonstrated with *synthetic aperture imaging* [28, 33, 41–45], is illustrated in Fig. 3.5.24. This system is based on the idea that the required information does not have to be gathered simultaneously. Since the processing operations required are all linear, we can obtain signals from individual elements at different times, store and delay them, and add them later, or put the delays in later, after storing them; it does not matter.

In one example of this system, a short pulse is emitted from a single transducer and enters the object of interest. Reflected echoes are received on the same transducer, then passed through an amplifier and an A/D converter into an 8 ×

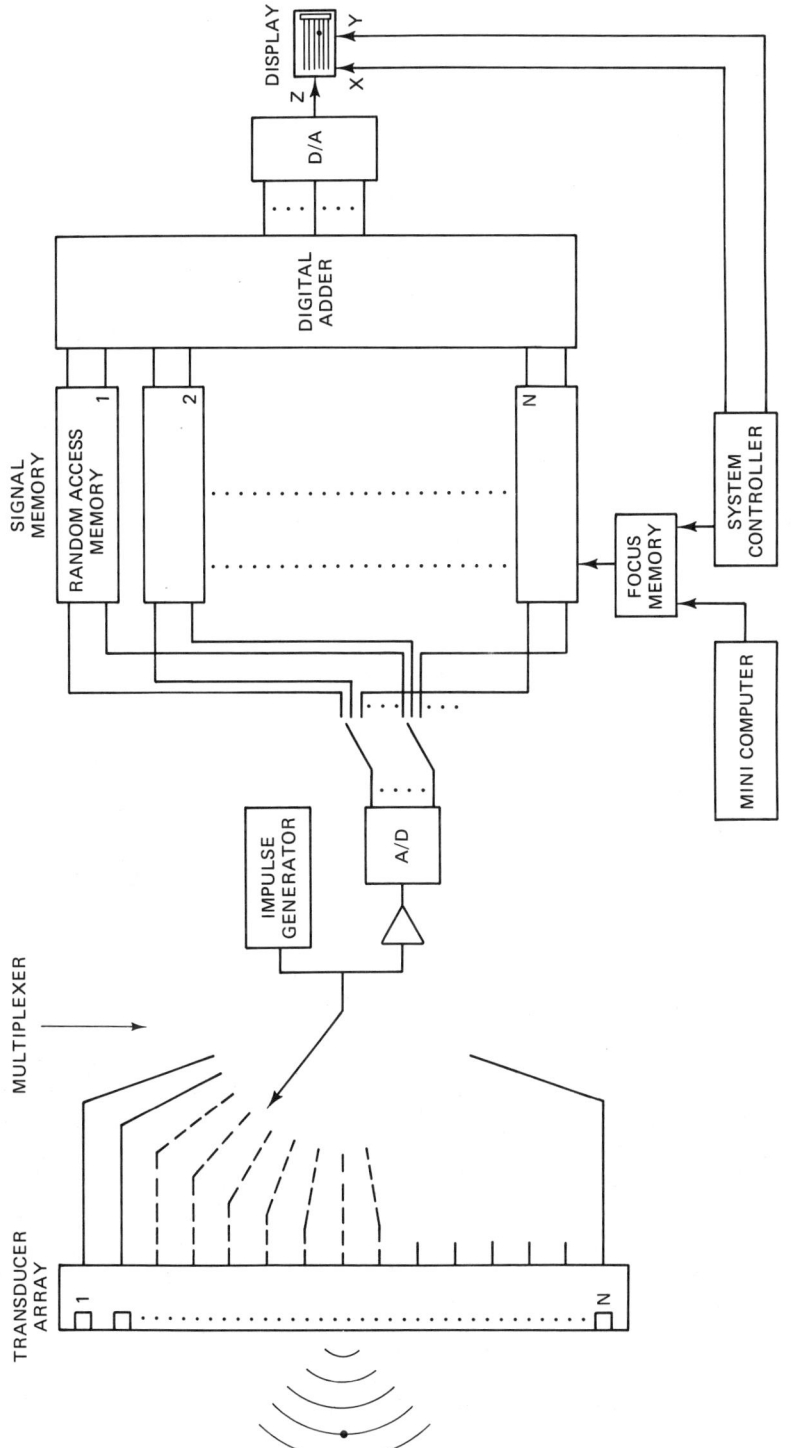

Figure 3.5.24 Schematic of a synthetic aperture digital imaging system. (After Corl et al. [41].)

1000-bit random-access memory (RAM). The process is then repeated for the next element in the array, and the signals from that element are stored in a separate RAM. Using a clock rate several times the signal frequency, several samples per RF cycle are obtained. Thus a relatively high frequency A/D converter must be used. Once the process is completed, information from the entire field of view is stored in the RAM memories and is available for reconstruction of an image. To image a particular point in the field, signals are taken from the appropriate points in the RAM, corresponding to the correct time delays. A basic system of this type was first demonstrated by Johnson et al. [42] at the Mayo Clinic, using a computer for storage and reconstruction of the image.

In the system designed by Corl et al. [41, 45], high-speed RAMs are employed with a separate RAM focus memory, to program the registers in the RAM from which the signals are read out. The digital signals are added in a digital adder, passed to a D/A converter, and then used to intensity modulate the signal on the screen of a cathode ray tube. By carrying out these processes in turn, any line in the image can be scanned in any direction to construct a complete raster image.

The same basic technique can be used to provide time delays in any focusing system. Signals can be emitted from all the elements of the transducer at once. As described above, separate A/D converters can be used on each element of the receiving transducer array, with the signals from all these elements processed simultaneously instead of separately, as in the synthetic aperture system. The advantage of simultaneous reception is that fast motion, like the flutter of a heart valve, can be reproduced in real time. Although a real-time synthetic aperture system takes only of the order of $\frac{1}{60}$ of a second to gather in one frame of information and another $\frac{1}{60}$ of a second to process it, the time delay from transducer to transducer in gathering information is of the order of 100 μs; this is obviously a disadvantage. Another difficulty with synthetic aperture imaging is the small size of an individual transmitting element and the wide area that must be insonified by this element. Consequently, the beam intensity at the object to be examined tends to be low, and hence the signal-to-noise ratio poorer, than with a more conventional time-delay system.

On the other hand, an important advantage of the synthetic aperture system is that it requires only a single front-end amplifier, regardless of the number of elements in the transducer array. This means that a great deal of effort can be put into the design of the front-end amplifier with little regard for its complexity, number of adjustments, expense, and so on, all of which are important considerations in a system where an amplifier is required for each element of the array. Furthermore, as the signal emitted from an array element must travel to a point in the field and back, its effective length of travel is double that of an equivalent receiver system, with a definition d_x, with the object illuminated by an unfocused transmitter. Thus the transverse definition in this system is $d_x/2$ and the sidelobe amplitude near the focal point varies as sinc $(2xD/\lambda z)$. On the other hand, as we have seen in Sec. 3.3, a system with a focused transmitter and receiver has a sinc^2 $(xD/\lambda z)$ response and hence a much lower sidelobe level, but not such good 3-dB definition, as the synthetic aperture system. In comparison to a system with an unfocused transmitter, such as the radial sector scan system described above, the

3-dB definition is better by a factor of 2, but the far-out sidelobes of the radial sector scan system are much weaker.

The range resolution is determined essentially by the pulse length (bandwidth), as it is with other imaging techniques. The system provides the same improvement in transverse resolution as a scanned holographic imaging system (see Sec. 3.5.6.D). But because time-delay rather than pulse-delay techniques are being used to reconstruct the image, excellent range resolution is also obtained.

B. TV Display

A major difficulty with real-time acoustic imaging systems is the format and nature of the display. The format of a radial sector scan, for instance, is quite different from that of a TV display, and the frame rates of each system may also be different. This problem is normally circumvented by using a digital scan converter, which can also provide additional sorts of image processing, such as averaging over several frames, contrast enhancement, artificial color displays, and quantitative measurements on the display. The synthetic aperture system, on the other hand, can use a rectilinear scan format like that of a TV image, which makes it possible, with the correct choice of clock rate, to display the image in real time on a TV screen. In all cases, it helps to use a magnetic deflection cathode ray tube of the type used in normal TV systems to obtain a good gray-scale display.

C. Sidelobes, Grating Lobes, and Sampling Lobes in Time-Delay Systems

We have already discussed how the use of short pulses can improve the range resolution of an acoustic imaging system. Using short pulses or tone bursts, instead of quasi-CW signals, also lowers the far-out sidelobe levels, and decreases the grating lobe and sampling lobe levels. We shall treat these effects here; most of the examples will refer to synthetic aperture imaging, since numerical calculations for this case are easily available to the author. However, the conclusions reached are pertinent to any kind of short-pulse focused imaging system.

We assume that the system is excited by a pulse of the form $f(t) \exp(j\omega t)$, where $\omega = 2\pi f_0$. The system is assumed to be focused on the point x_i, z_i. We shall consider both a synthetic aperture system in which the same transducer is used for transmission and reception, and a conventional pulsed system in which insonification of all points in the field is carried out by an unfocused beam and signals are received on separate transducers. The radial sector scan system is an example of the latter configuration. We will find it convenient, initially, to consider a continuous array system with points on the array located at $x', 0$. Later we consider a finite number of elements with the array elements located at the points $x_n, 0$, a distance l apart.

A time-delay focusing system focused on the point x_i, z_i introduces a time delay $T_0 - T_i$, where T_0 is a constant and

$$T_i = \frac{[z_i^2 + (x' - x_i)^2]^{1/2}}{V} \tag{3.5.87}$$

After suitable time delays have been introduced, the sum of the delayed signals returning to the transducers at x', 0 is of the form

$$h(t) = \int f\left(t - \frac{\gamma \Delta R}{V}\right) e^{j\omega(t - \gamma \Delta R/V)} w(x') \, dx' \qquad (3.5.88)$$

where $\gamma = 1$ for a standard pulsed imaging system and $\gamma = 2$ for a synthetic aperture system, and where $w(x') \, dx'$ is the amplitude response of the transducer in the receive-only or in the transmit–receive mode in the region between x' and $x' + dx'$. The $1/R^{1/2}$ or $1/R$ amplitude variation has been included in the definition of $f(t)$, and we define ΔR by the relation

$$\Delta R = \sqrt{z^2 + (x' - x)^2} - \sqrt{z_i^2 + (x' - x_i)^2} \qquad (3.5.89)$$

as the difference in range from x', 0 to x, z and x_i, z_i.

Range resolution. By making the paraxial approximation $(x' - x)^2 \ll z^2$, and taking $x_i = 0$, for simplicity we can write

$$\Delta R \approx \Delta z - \frac{x' \Delta x}{z_i} \qquad (3.5.90)$$

where, in general, $\Delta z = z - z_i$ and $\Delta x = x - x_i$. It follows that the range resolution (i.e., the result with $\Delta x = 0$) is determined by the function $f(t - \gamma \Delta z/V)$ for all transducers. So the range resolution is determined by the pulse shape and length.

Transverse definition. On the other hand, the transverse definition of a line reflector and sidelobe levels at the focal plane ($\Delta z = 0$) are determined by the integral

$$h(t, \Delta x, z_i) = \int f\left(t + \frac{\gamma x' \Delta x}{z_i v}\right) e^{j\omega(t + \gamma x' \Delta x/z_i V)} w(x') \, dx' \qquad (3.5.91)$$

with, if $f(t)$ is a sufficiently long pulse,

$$d_x(3 \text{ dB}) = \frac{0.89 \lambda z_i}{\gamma D} \qquad (3.5.92)$$

It is important to realize that the 3-dB width of the main lobe in a synthetic aperture imaging system ($\gamma = 2$) is half that of a conventional imaging system.

At $t = 0$, $\Delta x = 0$, and $\Delta z = 0$, the output is

$$h(0, 0, 0) = f(0) \int w(x') \, dx \qquad (3.5.93)$$

So the magnitude of $h(0, 0, 0)$ is determined by the maximum amplitude of the pulse and the spatial integral of the transducer response.

Grating lobes. We shall take the signal $f(t)$ to have the form of a Gaussian pulse:†

$$f(t) = e^{-(t/T)^2} \tag{3.5.94}$$

We represent the weighting function of the array by its Fourier series,

$$w(x') = \Pi\left(\frac{x'}{D}\right) \sum A_p e^{-j2\pi px'/l} \tag{3.5.95}$$

where $\Pi(x'/D) = 1$ when $|x'| \leq D/2$, $\Pi(x'/D) = 0$ when $|x'| > D/2$, and

$$A_p = -\int w(x') e^{j2p\pi x'/l} \, dx' \tag{3.5.96}$$

The pth grating lobe is associated with the pth harmonic of the aperture weighting. It follows from Eqs. (3.5.91) and (3.5.95) that if $f(t)$ is a very long pulse, the response is centered at the point

$$x_p = x_i + \frac{p\lambda z_i}{\gamma l} \tag{3.5.97}$$

By using the pth term of Eq. (3.5.95), $w_p(x') = A_p \exp(-j2p\pi x'/l)$, in Eq. (3.5.91), with $x_p = \Delta x$ ($x_i = 0$), we can determine the effect of the pulse envelope on the pth grating lobe amplitude, by writing

$$h_p(0, x_p, z_i) = A_p \int_{-D/2}^{D/2} f\left(\frac{px'}{f_0 l}\right) dx' \tag{3.5.98}$$

Therefore, when the duration of the pulse envelope is shorter than $D/\gamma f_0 l$ or $N/\gamma f_0$, corresponding to N RF cycles for conventional system ($\gamma = 1$), there will be a significant reduction in the grating lobe amplitude. In a system using Gaussian envelope pulses, the grating lobe will be diminished by the factor

$$\begin{aligned} h_p(0, x_p, z_i) &= \frac{A_p \int_{-D/2}^{D/2} e^{-(x'/f_0 Tl)^2} \, dx'}{A_0 \int_{-D/2}^{D/2} dx'} \\ &= \frac{A_p}{A_0} \frac{\sqrt{\pi} f_0 T}{|p|N} \operatorname{erf}\left(\frac{N}{|p|f_0 T}\right) \approx \frac{A_p}{A_0} \frac{\sqrt{\pi} f_0 T}{|p|N} \end{aligned} \tag{3.5.99}$$

where we have assumed in the last term that $pf_0 T \ll N$, so that we can use the asymptotic approximation to the error function erf (x).

For a 32-element array of infinitesimally wide transducers ($A_0 = A_p$) trans-

---

†The shape of a Gaussian pulse can be specified in many different ways. For instance, with $f_0 T = 1$, the 3-dB duration of the pulse is $1.2T$, or 1.2 RF cycles, and the 20-dB duration is $3T$, or 3 RF cycles. For the same pulse, the 3-dB bandwidth is 37% and the 20-dB bandwidth is 97%.

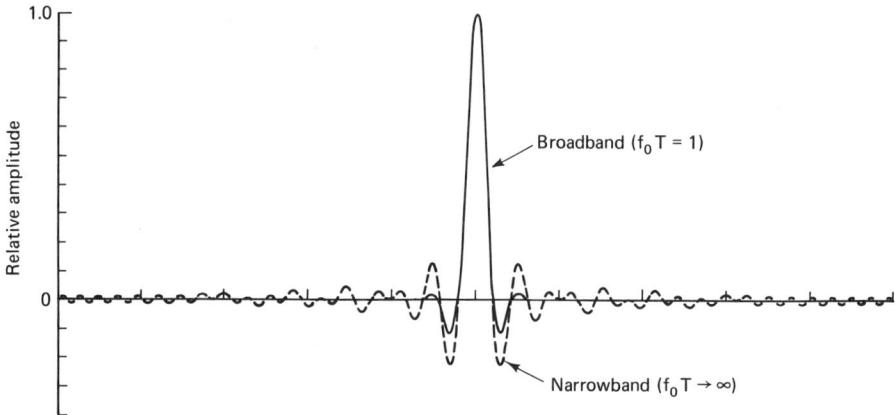

Figure 3.5.25 Comparison of narrowband (dashed line) and broadband (solid line) LSFs of continuous delay imaging system. The sidelobes of the broadband LSF diminish very quickly away from the main lobe. (From Peterson and Kino [33].)

mitting a broadband pulse with $f_0T = 1$, the grating lobe is reduced in amplitude by a factor of $0.06 = -25$ dB.

We conclude that for imaging systems using broadband pulses, the CW and broadband pulse LSFs resemble each other very closely near in to the main lobe. Features that are well removed from the main lobe (e.g., sidelobes and grating lobes) are "washed out," the effect being proportional to the distance from the main lobe.

To illustrate these effects, we first compare, in Fig. 3.5.25, numerical calculations for the LSFs of two imaging systems with continuous delays, one using narrowband CW imaging signals (dashed line) and the other using broadband ($f_0T = 1$) Gaussian-envelope pulses (solid line). Both have the same main lobe shape, but the sidelobes of the broadband LSF diminish very quickly away from the main lobe.

Broadband imaging systems have lower sidelobe levels, because the sidelobe patterns from different frequency components of the imaging signal shift with frequency, but the main lobe remains in the same location regardless of frequency. This means that the various frequency components add constructively at the main lobe but destructively elsewhere. Delay quantization sidelobes are reduced in a broadband system for exactly the same reason [33, 46].

In Fig. 3.5.26, we demonstrate this point by comparing the LSFs of two numerically calculated, synthetic aperture imaging systems, both of which suffer from delay quantization errors ($M = 4$). The narrowband CW LSF ($f_0T \to \infty$) is plotted with a dashed line and the broadband ($f_0T = 1$) LSF is plotted with a solid line. Again, the CW and broadband LSFs agree near the main lobe, but the sidelobes of the broadband LSF fall off more rapidly.

The analytical theory we have given here is a paraxial one, which we have evaluated only at the focal plane. More generally, suppose that the imaging system is focused on a point P distance R_n from a transducer at the point x_n, 0, as illustrated

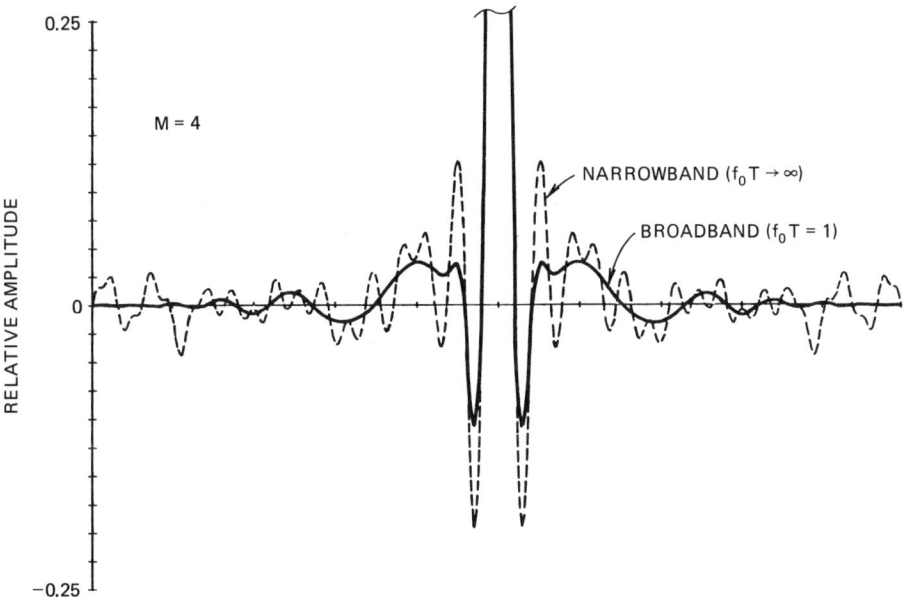

Figure 3.5.26 Comparison of narrowband (dashed line) and broadband (solid line) LSFs of quantized time-delay imaging system. (From Peterson and Kino [33].)

in Fig. 3.5.27. All signals received at the transducers from the point P suffer a total time delay T after processing. The signals from all the transducers are added so that the total amplitude is N times the amplitude of the signal arriving at one transducer. However, it is possible for a point Q, located anywhere on a circle of radius R_n and center x_n, 0, to produce a signal with the same delay time T. Hence sources located anywhere on circles of radius R_n, whose centers are at the transducer elements, can give rise to spurious out-of-focus signals or sidelobes. In

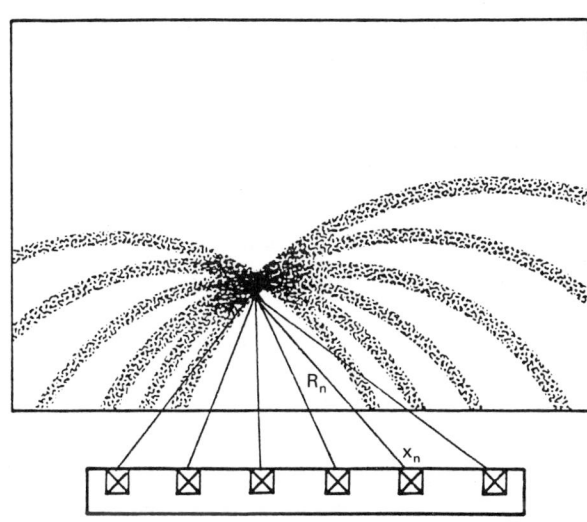

Figure 3.5.27 Causes of near-in and far-out sidelobes.

Sec. 3.5 Lensless Acoustic Imaging

practice, the pulse length is finite, so that the circles have a finite thickness. Near the main lobe, these "thick" circles overlap and there are phase additions and subtractions, yielding the typical $(\sin x)/x$ response. Farther out from the main lobe, the circles only partially overlap or do not overlap at all. Thus the sidelobe response tends to be washed out, until far out there are only isolated responses along the individual circles, whose amplitudes are reduced by $1/N$ from that of a beam focused on the same point.

D. Examples of Synthetic Aperture Imaging

We now consider some examples of synthetic aperture imaging. These will illustrate some of the theoretical results we have already obtained in Sec. 3.5.5.C, and also demonstrate how synthetic aperture imaging can be applied to NDT. We will draw our examples from work carried out in the author's own laboratory, where the concentration has been mainly on real-time imaging systems. The reader is also referred to work in other laboratories, where computer processing of synthetic aperture imaging has been used to make two- and three-dimensional images of internal features of large structures [43, 44]. In such cases, it is, in principle, possible to work with structures that are not flat, using the signals reflected from the surface as a time reference for the reconstruction process. This procedure makes it possible to correct for the shape of arbitrary structures. In practice, only relatively simple, large radius surfaces have been dealt with, so far, in this manner.

We first illustrate the results for an imaging system with a 32-element array, 1.6 cm long, with transducer element spacing of 0.5 mm, operating at a frequency of 3.5 MHz in water [45]. The acoustic wavelength was 0.42 mm, so we might have expected to see some grating lobes at an angle of the order of 30° to the normal of the array. In this case, however, the array response fell off fairly rapidly in this angular range; thus, due to this effect and because short pulses were used, the grating lobes did not show up very strongly. An image of three wires in front of a metal plate is shown in Fig. 3.5.28. The image in Fig. 3.5.28(a) was taken with an 8-element system that was 1.6 cm long. A similar image, taken with a 32-element system, is shown in Fig. 3.5.28(c). The incoherent sidelobes of the wires and the plate can easily be seen in the 8-element array image, while in the 32-element array image, they are less apparent. Mirror images of the wires in the plate can also be observed. Two additional images using nonlinear processing for the 8- and 32-element systems are shown in Fig. 3.5.28(b) and (d), respectively. Here the image quality was improved by nonlinear processing. But such techniques can often produce artifacts and false images due to the nonlinear interactions; thus their improvement in image quality is not consistent.

Imaging modes. Imaging systems of this type can be used for observing surface and internal features in a solid by either employing an array in water, with water between the array and the solid, or placing the array in direct contact with the solid. In the first case, aberrations occur in the image because angles of the rays entering the solid from water are changed in accordance with Snell's law. Thus we must take account of changes in time delay and phase in the synthetic

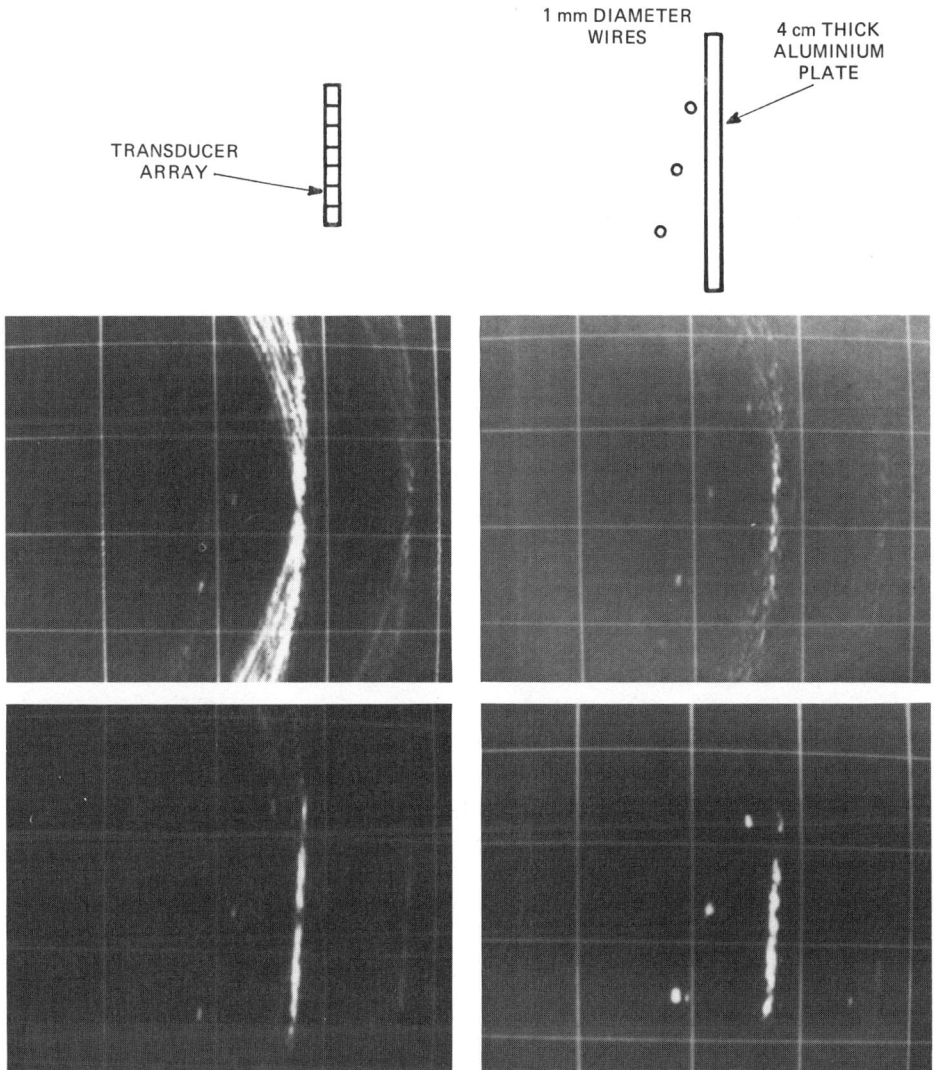

Figure 3.5.28 Schematic and computer-processed images for test targets in water tank with superimposed 1-cm reference grid. Upper photographs show images on left without and right with input gain compression for the 8-channel synthetic focus system. Lower photographs show similar images with 32 active channels. (After Corl and Kino [45].)

aperture reconstruction program and make different corrections every time the array is moved. Therefore, if it is possible, it is far easier to use a contacting array system. If the system is operating in real time (30 frames/s), the array and its contact to the surface can be conveniently adjusted while observing the image; this makes it possible to adjust for the best contact and spot artifacts that may occur in the image.

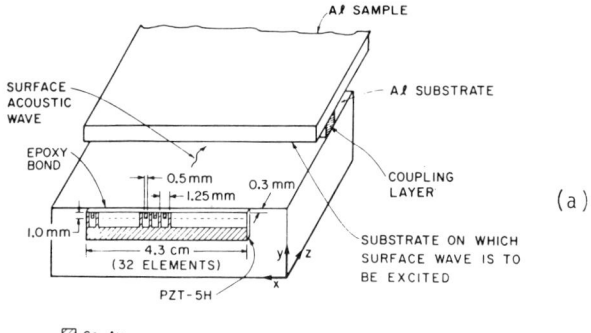

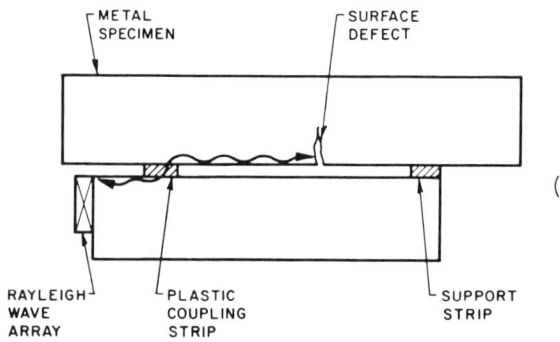

Figure 3.5.29 Edge-bonded transducer (EBT) surface wave array: (a) schematic of array construction; (b) diagram of a sample in contact with the array. (After Tuan et al. [47].)

Surface acoustic wave imaging. As our first example of a contacting array, we consider the Rayleigh wave array using an edge-bonded transducer, illustrated in Fig. 3.5.29. This transducer, which is approximately one wavelength deep, excites surface waves on a substrate of the same material as that being examined. Each element is a point source of surface acoustic waves, which are transferred from the array substrate to the specimens through a plastic strip approximately 10λ long (Sec. 2.5.3). The coupling process is relatively efficient, with a one-way loss of only −2 dB, and introduces no aberrations because the array substrate and specimen are made of the same material and thus have equal surface wave velocities.

Images of an 11-mm spark-machined (EDM) slot taken with this array are shown in Fig. 3.5.30. When the crack is rotated with respect to the array, the image changes. When the crack is parallel to the array, it behaves like a good specular reflector and signals transmitted to the crack are received on the same element of the array. Consequently, the expected bright line image of the crack is obtained. When the crack is rotated through a sufficiently large angle, however, specularly reflected rays are not returned within the array aperture.

Another example utilizes a shear wave array, as illustrated in Fig. 3.5.31, that is constructed on a metal buffer rod, again of the same material as the sample being examined. By using an incident wave at an angle of 45° to the normal, energy is easily transferred between the two substrates by the normal component

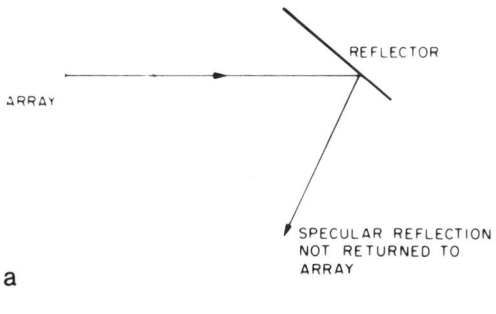

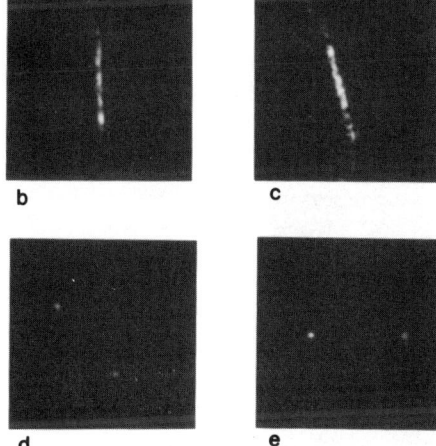

Figure 3.5.30 (a) When imaging specular reflectors at a large angle of incidence, only the two ends are seen. Images of an 11-mm EDM slot at: (b) 0° between normal to reflector and normal ray to array; (c) 15° between normal to reflector and normal ray to array; (d) 45° between normal to reflector and normal ray to array; (e) 90° between normal to reflector and normal ray to array. (After Kino et al. [49].)

of particle velocity at the surface [48, 49]. Typically, a thin layer of grease, whose thickness is kept constant by two Mylar pads, one at each end of the array, is used to keep a uniform contact. With arrays of this kind, contacting is extremely easy and images with a high degree of repeatability can be obtained without effort; good repeatability is most important in a practical system.

A shear wave image of a fatigue crack, approximately 4 mm deep and 10 mm long, obtained with this array is shown in Fig. 3.5.32(d), and compared to a similar image of the same crack taken with a surface wave array on the top surface [part (b)], and with an optical picture of the crack [part (a)]. In both acoustic images, the crack is seen with good resolution; in the shear wave image, a jog in the crack is clearly resolved. Two additional images of this crack are shown, one with a contacting longitudinal wave array [part (c)] and another in a pitch/catch mode [part (e)]. In the pitch/catch mode, the crack was insonified with longitudinal waves from a transducer on the back of the sample, and the image formed with energy scattered by the crack into other modes. The return signals were collected with a Rayleigh wave array.

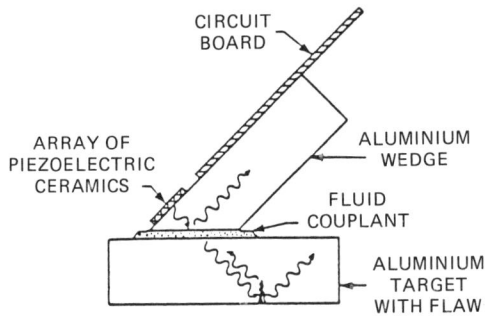

Figure 3.5.31 Shear wave array in contact with an aluminum block. (After Kino et al. [49].)

Figure 3.5.33 illustrates the pitch/catch mode in more detail. The results shown here provide multiple images of a crack corresponding to conversion of the incident longitudinal wave to other modes. The different time delays observed are related to: (1) conversion to a shear wave at the tip of the crack; (2) conversion to a Rayleigh wave at its tip; and (3) conversion to a Rayleigh wave at its root. This technique makes it possible to measure the crack depth accurately.

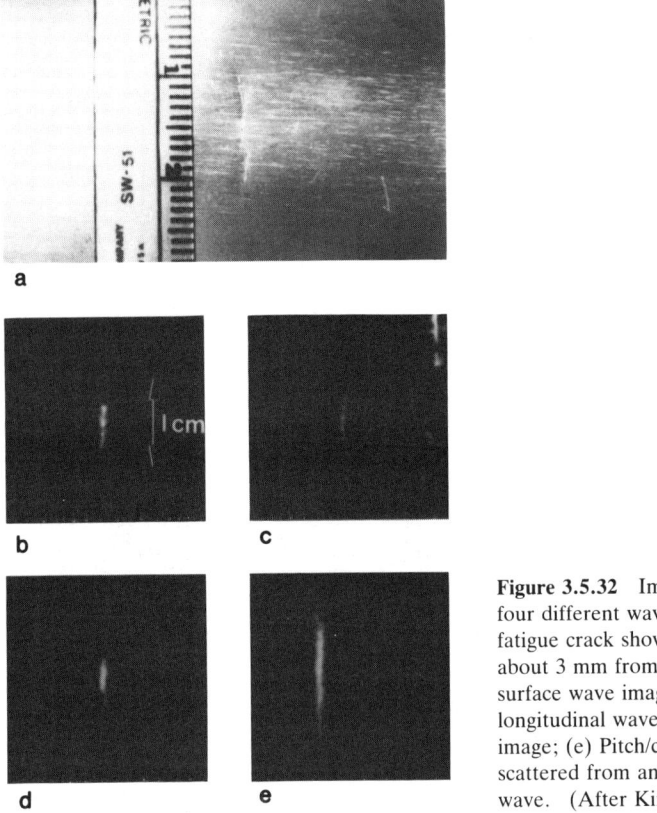

Figure 3.5.32 Imaging a crack with four different wave modes: (a) 10-mm fatigue crack showing the prominent jog about 3 mm from the bottom tip; (b) surface wave image of fatigue crack; (c) longitudinal wave image; (d) shear wave image; (e) Pitch/catch image of waves scattered from an incoming longitudinal wave. (After Kino et al. [49].)

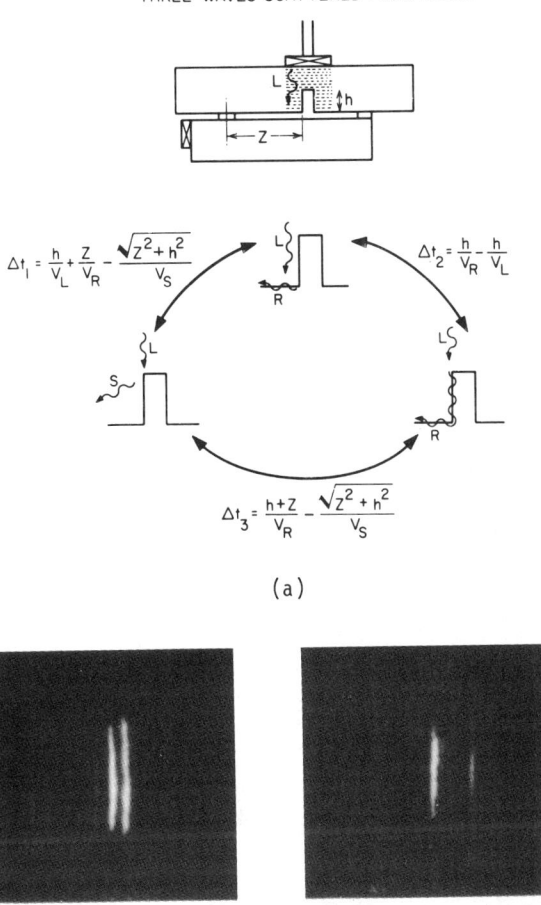

Figure 3.5.33 Pitch/catch imaging: (a) transmitter–receiver geometry and diagram of the three waves scattered from the crack; (b) pitch/catch image of a long 7-mm-deep slot; (c) pitch/catch image of the 10-mm-long fatigue crack. (After Kino et al. [49].)

E. Tomographic Imagining Systems

Synthetic aperture imaging is closely related to tomographic imaging, a type of imaging modality that has become very important since the development of the x-ray *computerized axial tomography scanner* (CAT scanner) [50, 51]. This system is based on the old idea of motion tomography, illustrated in Fig. 3.5.34, in which a film is moved in synchronism with an x-ray source, but in the opposite direction, so that one plane of the object remains in focus while all others are blurred. Since there is no phase information in this image, it can be shown that the PSF of a point in the image plane varies as $1/r$, where r is the distance from the point.

The CAT scanner uses a more complicated scan to construct an image of all points in a thin slice. In one version, illustrated in Fig. 3.5.35, an x-ray source is moved to N points, perhaps 32 along a row, and the collimated x-ray beams emitted are detected at points opposite the source positions. The whole assembly is then

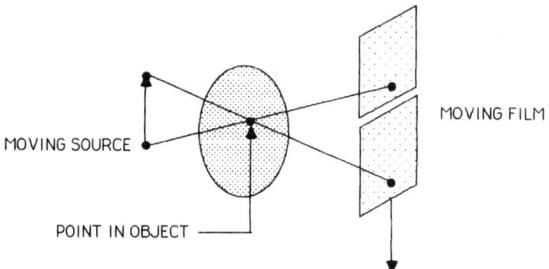

Figure 3.5.34 Motion tomography.

rotated by an angle $\Delta\theta$, perhaps 1°, and this process repeated until the full 360° is covered. Then, by suitable processing, excellent images can be obtained.

Similar techniques have been investigated for acoustic imaging. Either changes in amplitude, associated with the attenuation of the medium, or changes in phase, associated with changes in the refractive index of the medium, are measured. In some respects, acoustic tomography is more complicated than x-ray tomography because, due to diffraction, the beam cannot remain well collimated, and the perturbations in refractive index or attenuation may be so large that a ray path will not be the same as in a uniform medium. Consequently, progress in this field has been slow.

The technique can also be used for reflection-mode, rather than transmission-mode, imaging. For instance, a synthetic aperture quasi-CW system, using a short tone burst, can be constructed as a circular array enclosing the object; each array element is used in turn as a transmitter and receiver before passing on to the next element. It can be shown that the PSF of such a system, anywhere inside the array, with infinitesimally spaced array elements, is $J_0(2kr)$ where $J_0(x)$ is a zeroth order Bessel function of the first kind [52]. A similar analysis can be made for a transmission system in which an element at an angular position θ transmits to a receiver at a position $\pi + \theta$, with θ changed by $\Delta\theta$ until the full 360° circle is completed. Using processing similar to that employed for a synthetic aperture reflection mode system yields an LSF of the form $J_0(kr)$. Such a system can be employed to produce images of small perturbations in refractive index or attenuation. The theoretical definition is excellent because phase information with a full 360° scan is used.

The equivalent x-ray tomographic system uses only intensity information, as if the width of the beam had the form of a Dirac δ function, while in the acoustic

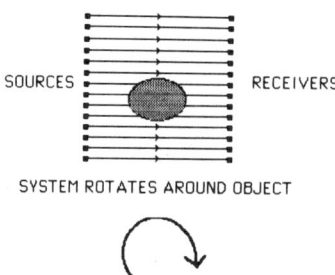

Figure 3.5.35 CAT scan tomographic imaging system.

system, there is phase variation across and along the beam due to the use of short tone bursts. The x-ray tomographic system introduces the phase variation in the computer, rather than in the transmitter, after the signal has been received. The processes are equivalent in a linear system. In a tomographic system, the transmitter is located at one point with the receivers located, ideally, at points around a 180° arc; the system is then rotated until the full 360° circle has been completed. Processing is carried out by convolving the outputs from receiving transducers at different angular coordinates, with a function like a sinc function in angular coordinates. This synthesizes the effect of a spatial phase variation; and also provides more information if a limited number of points are used: if there are N transmitters and M receivers, there will be MN pieces of information available and MN resolvable spots in the image.

A simple N-element synthetic aperture system will tend to have a far-out sidelobe amplitude of the order of $1/N$ below that of the main lobe, while the tomographic system will have much reduced levels of interference from artifacts. Thus the simple synthetic aperture system is not good enough for imaging a continuous medium where there are many equivalent sources present, and the more sophisticated algorithms developed for tomography are required. The reader is referred to the literature on CAT scanners and acoustic tomography for more information [51].

3.5.6 Acoustic Holography

A. Introduction

It was first shown by Gabor that it is possible to construct a system that can form an image without using lenses [53]. His immediate purpose was to eliminate aberrations in electron optical systems, but this basic idea has since been generalized to record and display three-dimensional optical images, and to display acoustical images in visual form. As used in acoustics, the method and its variations are suitable not only for image display, but also for such applications as measuring vibration amplitudes in crystal resonators and aircraft wings. Some closely related alternative techniques are the Schlieren optical system (see Secs. 4.9.4 and 4.9.5) and Bragg scattering of light by acoustic waves. These are suitable for measuring the amplitude distribution over the cross section of an acoustic beam; as a byproduct, the latter technique is also important for electronically deflecting and modulating a laser beam.

To understand the basic features of holography, we first discuss the formation of an optical hologram on a photographic film and how images are reconstructed from it. We then discuss general holographic techniques, in particular, the reconstruction of images using waves of a different wavelength from those used to illuminate the original object, and how these techniques are employed in acoustic applications.

We now consider the technique for lensless imaging. A signal arriving from a point x_O, y_O, z_O on an object at the point $x, y, 0$ on the plane $z = 0$ has a phase variation $\phi_O(x, y) = -k\sqrt{(x - x_O)^2 + (y - y_O)^2 + z_O^2}$. When a lens is employed

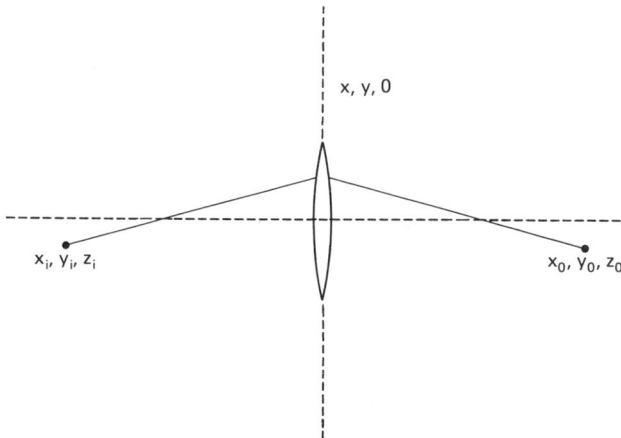

Figure 3.5.36 Imaging with a lens.

to reconstruct the image of this point x_O, y_O, z_O, as shown in Fig. 3.5.36, the lens changes the path lengths of the rays passing through it, and hence their phases, to produce an image at the point x_i, y_i, z_i. Thus the lens introduces a phase change $\Delta\phi$, where

$$\Delta\phi = k[\sqrt{(x - x_O)^2 + (y - y_O)^2 + z_O^2}$$
$$+ \sqrt{(x - x_i)^2 + (y - y_i)^2 + z_i^2} \qquad (3.5.100)$$

The phase change from a point $x, y, 0$ on a thin lens to the image point x_i, y_i, z_i is $\phi_i = -k\sqrt{(x - x_i)^2 + (y - y_i)^2 + z_i}$. Therefore, the phase of the rays at the image point x_i, y_i, z_i is $\phi_i + \phi_O + \Delta\phi$ and is independent of x and y.

Holography was originally developed as a technique that uses a laser beam for the phase reference and reconstruction needed to form an image. It can be employed for lensless imaging of light or acoustic waves. With light, the phase information is recorded on film in the form of a "hologram" and the necessary reconstruction or imaging is carried out by passing a second light beam through the film. With acoustic waves, the acoustic beam is scattered from the object and a reference acoustic beam ripples the surface of water or of a solid; the image is then reconstructed with a light beam, which introduces the necessary phase change, incident on this surface. In another method, an acoustic transducer is mechanically scanned over the plane $z = 0$ and used to modulate a light source, from which is recorded, on film, an optical hologram. This hologram is reconstructed by the normal optical techniques. A third technique is to synthesize the necessary phase changes in a computer and then reconstruct the image.

The holographic reconstruction of images yields a definition determined by the aperture of the original receiving system. This definition will be entirely equivalent to that obtained by a physical or electronic lens of the same aperture, for the holographic method is essentially an alternative technique for constructing the matched spatial filter described in Sec. 3.5.2.

The use of the direct optical reconstruction technique in acoustic holography has an advantage over electronically scanned systems or computer reconstruction in that it provides rapid parallel processing and a detector that is essentially continuous. Thus grating lobes do not occur. On the other hand, such systems, which record the hologram on photographic film, are slow in operation because of the time required to develop the film. They also require a laser optical system, which needs a large, stable, mechanical table as a platform on which to reconstruct the optical image. Some types of acoustic holography systems use the ripple of a water surface due to an acoustic wave to reflect a laser beam. Such systems need a high-power argon laser for reconstruction; in addition, they are not very sensitive and are mechanically quite restricted, because the water surface must be kept level. Thus they are unsuitable for medical imaging and have been employed only in NDT testing applications.

As we have already discussed, some of these difficulties can be eliminated by using computer rather than optical reconstruction techniques. Either a movable acoustic transducer, or the deflection of an optical beam reflected from the rippling surface of a solid, is used to detect the acoustic beam. If the number of detection points is limited, however, the problem of grating lobes may then reoccur.

Acoustic holography excited a great deal of interest when it was first introduced, and topical research conferences were devoted to the subject for several years. However, it is now rarely used because in addition to the unwieldy optics required, it employs a phase compensation process to reconstruct the image. This means only a very narrow band of frequencies can be used for insonification (illumination) of the object and, as a result, the problems of speckle and interference fringes that occur with coherent light illumination still occur in holography. Furthermore, using a narrowband signal means that short pulses cannot be used for insonification. Thus the range definition of the system will be the same as the depth of focus of the equivalent lens with the same aperture, and only very poor images of partially transparent acoustic objects, such as inclusions of one solid in another, or of small objects near large reflectors, can be obtained.

With these limitations in mind, it is of interest to study holography as a useful technique for reflection and transmission imaging of isolated flaws, and for its other applications to the measurement of vibrating structures.

B. Holographic Reconstruction with Spherical Reference Waves

We first consider the basic technique for recording an optical or acoustic hologram on film. In principle, the technique requires recording both amplitude and phase information. We shall illustrate some of the basic ideas here by considering optical holography.

A plane wave is incident on a film with an amplitude variation $A \exp(j\omega t)$ along the film surface. The film records only the intensity $I(x)$, where

$$I(x) = AA^* \qquad (3.5.101)$$

Thus all phase information is lost.

We consider, initially, the configuration for making a Leith–Upatnieks hologram using a plane wave source and reference [54, 55]. Two plane waves of frequency ω are incident on the film, as illustrated in Fig. 3.5.37. The first one

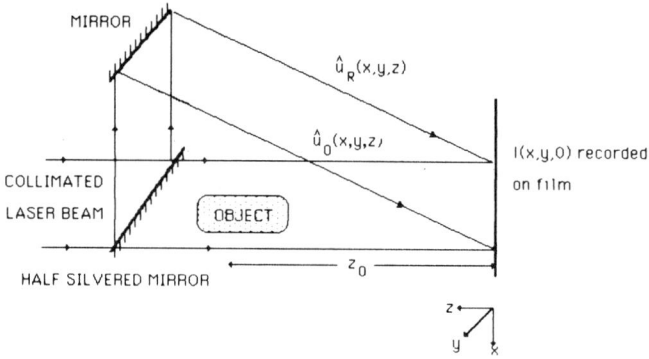

Figure 3.5.37 Recording of a Leith–Upatnieks hologram with a plane wave reference and source.

has the form

$$\hat{u}_O(x, y, 0) = u_O(x, y)e^{j\omega t}e^{-j\phi(x,y)} \quad (3.5.102)$$

The second is a reference plane wave of the same frequency and has fields of the form

$$\hat{u}_R(x, y, 0) = u_R(x, y)e^{j\omega t}e^{-j(k_x x + k_y y)} \quad (3.5.103)$$

along the photographic film.

The intensity of the illumination on the film is therefore

$$I(x, y) = [\hat{u}_O \hat{u}_R^* + \hat{u}_O \hat{u}_O^* + \hat{u}_O \hat{u}_O^* + \hat{u}_R \hat{u}_R^*] \quad (3.5.104)$$

or

$$I(x, y) = 2u_O u_R \cos[k_x x + k_y y - \phi(x, y)] + (u_O u_O^* + u_R u_R^*) \quad (3.5.105)$$

The first term in Eq. (3.5.105) can be written in the form

$$\{i_{RO}(x, y) = u_O u_R e^{j[k_x x + k_y y - \phi(x,y)]} + e^{-j[k_x x + k_y y - \phi(x,y)]}\} \quad (3.5.106)$$

We see from Eq. (3.5.105) that the use of a reference wave yields two terms in the intensity variation that retain both the phase and amplitude information in the wave $u_O(x, y)$. This recording of $I(x, y)$ made on the photographic film with plane waves is the Leith–Upatnieks hologram [54, 55]. Note that additional information associated with the incident waves (the $u_O u_O^*$ and $u_R u_R^*$ terms) is also recorded on the film. We shall find it convenient to ignore these terms, as well as one of the exponential terms in Eq. (3.5.106), in our initial treatment of holography. In practice, the terms $(u_O u_O^* + u_R u_R^*)$ can give rise to glare (i.e., a background light corresponding to the sum of the incident wave intensities), and one of the unwanted exponential terms gives rise to an unfocused virtual image in the reconstructed image. These terms can be eliminated by various stratagems, which we discuss later.

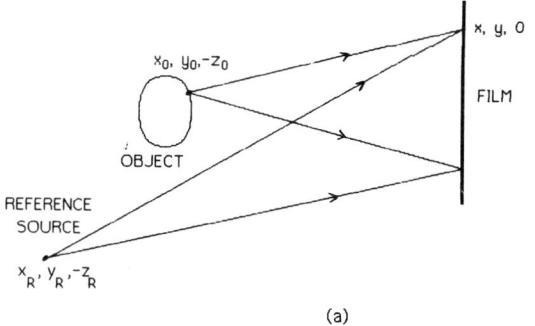

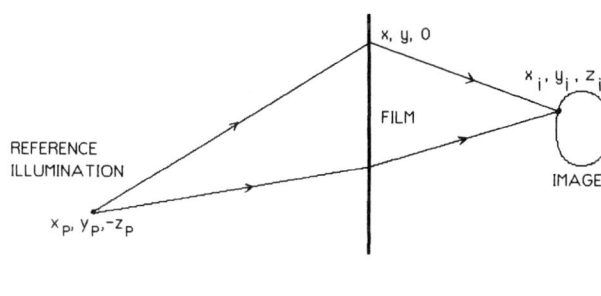

Figure 3.5.38 Spherical reference wave: (a) object and reference source; (b) reconstruction system.

Spherical wave sources. Before we discuss how this hologram is reconstructed, let us consider the more general case, where a spherical reference wave (a point source) is used to illuminate the film originally, and another spherical reference wave of different wavelength is used to form the holographic image.

We illuminate the film with a reference signal from a point source at x_R, y_R, $-z_R$, as shown in Fig. 3.5.38(a). A point on the object illuminated from the same coherent light source is taken to be at x_O, y_O, $-z_O$, and the wavelength to be $\lambda_R = 2\pi/k_R$. We use the paraxial approximation, so that the distances from x_O, y_O, $-z_O$ and x_R, y_R, $-z_R$ to a point x, y, 0 on the film are, respectively.

$$R_O = \sqrt{(x - x_O)^2 + (y - y_O)^2 + z_O^2}$$
$$\approx z_O + \frac{(x - x_O)^2 + (y - y_O)^2}{2z_O} \quad (3.5.107)$$

and

$$R_R = \sqrt{(x - x_R)^2 + (y - y_R)^2 + z_R^2}$$
$$\approx z_R + \frac{(x - x_R)^2 + (y - y_R)^2}{2z_R} \quad (3.5.108)$$

It is convenient to write

$$\mathbf{r} = \mathbf{a}_x x + \mathbf{a}_y y$$
$$\mathbf{r}_O = \mathbf{a}_x x_O + \mathbf{a}_y y_O \qquad (3.5.109)$$
$$\mathbf{r}_R = \mathbf{a}_x x_R + \mathbf{a}_y y_R$$

where $\mathbf{a}_x$ and $\mathbf{a}_y$ are unit vectors in the x and y directions, respectively. Equations (3.5.107) and (3.5.108) can then be written in the simple form

$$R_O = z_O + \frac{(\mathbf{r} - \mathbf{r}_O)^2}{2z_O} \qquad (3.5.110)$$

$$R_R = z_R + \frac{(\mathbf{r} - \mathbf{r}_R)^2}{2z_O} \qquad (3.5.111)$$

respectively. It follows, from an analysis similar to that used to derive Eqs. (3.5.105) and (3.5.106), that the intensity of the light on the film contains a product term of the form

$$I_{RO} = A \cos k_R \left[\frac{(\mathbf{r} - \mathbf{r}_O)^2}{2z_O} - \frac{(\mathbf{r} - \mathbf{r}_R)^2}{2z_R} + z_O - z_R \right] \qquad (3.5.112)$$

where A is a constant.

Reconstruction of the image. Suppose that we now illuminate the exposed film with a wave of wavelength $\lambda_p = 2\pi/k_p$, which is emitted from a point source at $x_p, y_p, -z_p$. This will give rise to a wave with a complex amplitude

$$A_i(x, y) = e^{-jk_p[z_p + (\mathbf{r}_p - \mathbf{r})^2/2z_p]} \qquad (3.5.113)$$

along the film surface. This situation is shown in Fig. 3.5.38(b).

When this wave passes through the developed film, which has a density variation proportional to the intensity variation given in Eq. (3.5.112), the amplitude variation of the light beam leaving the film is proportional to $A_i(x, y) I_{RO}$. So we can write that the amplitude of the wave passing through the film is of the form

$$F(x, y) = K \cos k_R \left[\frac{(\mathbf{r} - \mathbf{r}_O)^2}{2z_O} - \frac{(\mathbf{r} - \mathbf{r}_R)^2}{2z_R} + z_O - z_R \right] e^{-jk_p[z_p + (\mathbf{r}_p - \mathbf{r})^2/2z_p]z_p}$$

$$(3.5.114)$$

where K is a constant. The phase delay from a point $x, y, 0$ on the film to a point x'_O, y'_O, z'_O ($\mathbf{r}'_O, z'_O$) is

$$\phi(x'_O, y'_O, z'_O) - \phi(x, y, 0) = k_p \left[z'_O + \frac{(\mathbf{r}'_O - \mathbf{r})^2}{2z'_O} \right] \qquad (3.5.115)$$

We now split the cosine term in Eq. (3.5.114) into two exponential components with phases $\pm \phi(x, y, 0)$. It follows that the phase of each signal component reaching the point x'_O, y'_O, z'_O is $\phi'(x'_O, y'_O, z'_O)$, where

$$\phi(x'_O, y'_O, z'_O) = \phi_0 \mp k_R \left[\frac{(\mathbf{r} - \mathbf{r}_O)^2}{2z_O} - \frac{(\mathbf{r} - \mathbf{r}_R)^2}{2z_R} \right] \\ - k_p \left[\frac{(\mathbf{r}_p - \mathbf{r})^2}{2z_p} + \frac{(\mathbf{r}'_O - \mathbf{r})^2}{2z'_O} \right] \quad (3.5.116)$$

where $\phi_0 = \mp k_R(z_O - z_R) - k_p(z_p + z'_O)$ and is independent of $\mathbf{r}, \mathbf{r}_O, \mathbf{r}_p$ and $\mathbf{r}'_O$. Thus ϕ_O is a constant.

The r^2 terms in the phase cancel out at a plane $z'_O = z_i$, satisfying the relation

$$\pm k_R \left(\frac{1}{z_O} - \frac{1}{z_R} \right) - k_p \left(\frac{1}{z_p} + \frac{1}{z_i} \right) = 0 \quad (3.5.117)$$

It follows that the image lies at the point z_i, where

$$z_i = \left[\mp \frac{\lambda_p}{\lambda_R} \left(\frac{1}{z_O} - \frac{1}{z_R} \right) - \frac{1}{z_p} \right]^{-1} \quad (3.5.118)$$

Two images are formed: the one with z_i positive, on the right side of the hologram film, is real; the one with z_i negative, on the left side, is a virtual one. Thus the two exponential terms forming the cosine in Eq. (3.5.112) give rise to separate images at points determined, respectively, by the $+$ and $-$ signs in Eq. (3.5.118).

The quadratic terms have determined the z location of the image. However, we have not yet determined the x and y locations of the image. We can do this by using the condition that the linear terms in $\mathbf{r}$, like $k_R \mathbf{r} \cdot \mathbf{r}_O / z_O$ in Eq. (3.5.116), cancel out. The condition is

$$\mp k_R \left(\frac{\mathbf{r}_O}{z_O} - \frac{\mathbf{r}_R}{z_R} \right) + k_p \left(\frac{\mathbf{r}_p}{z_O} + \frac{\mathbf{r}_i}{z_i} \right) = 0 \quad (3.5.119)$$

Thus the transverse coordinates x_i and y_i of the reconstructed image are where

$$\frac{x_i}{z_i} = -\frac{x_p}{z_O} \pm \frac{\lambda_p}{\lambda_R} \left(\frac{x_O}{z_O} - \frac{x_R}{z_R} \right) \\ \frac{y_i}{z_i} = -\frac{y_p}{z_O} \pm \frac{\lambda_p}{\lambda_R} \left(\frac{y_O}{z_O} - \frac{y_R}{z_R} \right) \quad (3.5.120)$$

respectively. Note that a small change in object coordinates, $\Delta \mathbf{r}_O$, leads to a change in the image coordinates, $\Delta \mathbf{r}_i$, where

$$\Delta \mathbf{r}_i = \pm \frac{\lambda_p}{\lambda_R} \frac{z_i}{z_O} \Delta \mathbf{r}_O \quad (3.5.121)$$

Thus the transverse magnification M_T associated with the wavefront reconstruction process is defined as

$$M_T = \pm \frac{\lambda_p}{\lambda_R} \frac{z_i}{z_O} \qquad (3.5.122)$$

or, equivalently, as

$$M_T = \frac{1}{\left| \frac{z_O}{z_p} \frac{\lambda_R}{\lambda_p} \mp \left(1 - \frac{z_O}{z_R}\right) \right|} \qquad (3.5.123)$$

Similarly, the change Δz_i in the z coordinate of the image for a small change in the position of the object Δz_O gives a longitudinal magnification M_L, where

$$M_L = \frac{(\lambda_R/\lambda_p)^2}{\left| \frac{z_O}{z_p} \frac{\lambda_R}{\lambda_p} \pm \left(1 - \frac{z_O}{z_R}\right) \right|^2} = \left(\frac{\lambda_R}{\lambda_p}\right)^2 M_T^2 \qquad (3.5.124)$$

If the reference and reconstruction sources have the same wavelength ($\lambda_R = \lambda_p$), then $M_L = M_T^2$. This is identical to the similar magnification formulas for a lens:

$$M_T = \left|\frac{z_i}{z_O}\right|$$

$$M_L = \left|\frac{z_i}{z_O}\right|^2 \qquad (3.5.125)$$

where z_o is the position of the object and z_i is that of the image. If a different wavelength is used for reconstruction and illumination in a holographic system (i.e., if $\lambda_R \neq \lambda_p$), the results and the perspective to an observer will be quite different than they are for a normal imaging system.

The definition of the image may be determined directly by carrying out the integral

$$s(x'_O, y'_O, z_i) = \int e^{-j\phi(x,y,0,x'_o,y'_o,z_i)} \, dx \, dy \qquad (3.5.126)$$

over the area of the aperture, where x'_o, y'_o, and z_i are the coordinates in the image plane. This yields exactly the result we would expect; for a square aperture with sides D_x and D_y, for example, the 3-dB definitions in the x and y directions are

$$d_x = \frac{0.89 z_i \lambda_R}{D_x}$$

$$d_y = \frac{0.89 z_i \lambda_R}{D_y} \qquad (3.5.127)$$

with a sinc $(x'_o D_x/z_i\lambda_R)$ sinc $(y'_o D_y/z_i\lambda_R)$ variation for the point spread function. With a circular aperture, the point spread function is a jinc function, as we would expect. These definitions correspond, of course, to their scaled versions, using the magnification factor M_T in the object system. The range definitions also correspond to what we would expect for a lens of the same aperture.

C. Water–Air Surface as an Acoustic Imaging Intensity Detector

When sound waves impinge on the water–air interface in a water bath, there is a static radiation pressure associated with the sound wave of value $p_0 = 2I/V$, where I is the acoustic intensity of the wave approaching the surface; this causes the water surface to lift.† The movement of the water surface at a distance R is proportional to the sound wave intensity and hence, if it can be detected optically, this phenomenon can be used for holographic detection of sound. At the same time, the water is undergoing RF motion due to the presence of the sound waves, with a particle displacement $u = \lambda_a S$, where λ_a is the wavelength of the sound and S is the strain. With typical strains of the order of $S = 10^{-6}$, and wavelengths of the order of 1 mm, the RF displacement of the water is of the order of 10 Å, a very small value. In practice, the water displacement due to radiation pressure is normally comparable to this value and often much larger than it.

We can estimate the water displacement by considering the restoring force due to surface tension, and carrying out a one-dimensional analysis pertaining to displacement of the water surface by a sound wave whose intensity varies in the x direction along the water surface [55]. If σ is the surface tension, the restoring force due to surface tension, after the liquid is displaced a distance h, is $\sigma(d^2h/dx^2)\,dx$ in an elemental length dx. However, the downward force due to gravity is $g\rho_{m0}h\,dx$, and the upward force due to the radiation pressure of the sound wave is $p_0\,dx$. Hence we can write

$$\sigma\frac{d^2h}{dx^2} + g\rho_{m0}h = p_0(x) \qquad (3.5.128)$$

If we take a spatial Fourier transform of Eq. (3.5.128), we can write

$$H(v_x) = \int_{-\infty}^{\infty} h(x)e^{-jv_x x}\,dx$$
$$P(v_x) = \int_{-\infty}^{\infty} p_0(x)e^{-jv_x x}\,dx \qquad (3.5.129)$$

It follows from Eq. (3.5.128) that

$$H(v_x) = \frac{P(v_x)}{(v_x^2\gamma^2 + 1)\rho_{m0}g} \qquad (3.5.130)$$

where $\gamma^2 = \sigma/\rho_{m0}$.

If the second term in the denominator of Eq. (3.5.130) can be neglected

---

†The easiest way to understand this effect is to use quantum mechanics. We regard the wave as composed of acoustic phonons. If there are N acoustic phonons per unit area per second reaching the surface, the rate of change of momentum due to the phonons reflected at the surface is $p_0 = 2N\hbar k$, where $\hbar$ is Planck's constant and k is the propagation constant. Therefore, $p_0 = 2N\hbar\omega/V$. However, $N\hbar\omega = I$. Thus $p_0 = 2I/V$.

compared to the first term, we find that

$$H(v_x) = \frac{P(v_x)}{v_x^2 \sigma g} \qquad (3.5.131)$$

or, more generally, in a two-dimensional system,

$$H(v_x, v_y) = \frac{P(v_x, v_y)}{v^2 \sigma g} \qquad (3.5.132)$$

where $v^2 = v_x^2 + v_y^2$.

Equations (3.5.131) and (3.5.132) are valid if the spatial period of the hologram has a wavelength $\lambda = 1/v$ that is much smaller than the parameter γ, where γ is typically of the order of 2 mm. In this case, the second spatial derivative of the height h is proportional to the acoustic intensity at the surface. In a holographic system, this implies that the cross product term of the wave passing through the object and of the reference wave is the important term, and that d^2h/dx^2 is proportional to the amplitude of the wave passing through the object. More generally,

$$p_0(x, y) \propto \frac{\partial^2 f}{\partial y^2} + \frac{\partial^2 h}{\partial y^2} \qquad (3.5.133)$$

Smith and Brenden holographic technique. Many different types of acoustic holographic imaging methods that use the ripple of a water surface for reconstruction of the image have been described in the literature [55]. First we will describe the Holosonics industrial system of Brenden, which has been fairly widely used [56]. If a spherical reference source is employed, the water surface must, in principle, be lifted into a spherical shape, and distortion inevitably occurs. The Brenden system eliminates this problem by using a plane wave acoustic reference beam ($z_R = \infty$), a small separate tank for the region where the reference beam impinges on the water surface, and a lens between the object and the water surface to bring the image plane $z'_O = z_i$ of the optical system to a finite distance. A schematic of the system is shown in Fig. 3.5.39.

As illustrated, a lens is placed in front of the optical reference source to produce a collimated or parallel optical beam ($z_p = \infty$) at the water surface.

It follows from Eq. (3.5.118) that

$$z_i = \pm \frac{\lambda_R}{\lambda_p} z_O = \pm \frac{\lambda_a}{\lambda_L} z_O \qquad (3.5.134)$$

where $\lambda_R = \lambda_a$ is the acoustic wavelength and $\lambda_p = \lambda_L$ is the wavelength of light. By using a lens in front of the object, z_O has been made very small. However, it follows from Eq. (3.5.134) that because $\lambda_a \gg \lambda_L$, the image plane can be at a finite distance from the water surface.

Note that two images are formed, a virtual one, indicated by the negative sign, and a real one, indicated by the positive sign. An optical lens of focal length f, a distance d from the water surface, will focus the image to a plane a distance

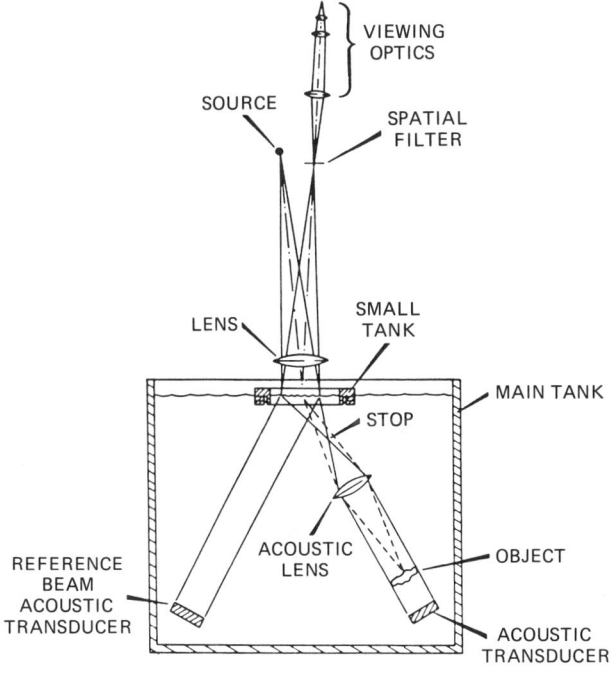

Figure 3.5.39 Liquid surface imaging system due to Brenden. This is the system employed by Holosonics. (After Brenden [56].)

z_i from the lens, where

$$\frac{1}{z_i'} = \frac{1}{f} - \frac{1}{d - z_i} \qquad (3.5.135)$$

The true and conjugate images will therefore be focused at different planes.

If the acoustic image is focused on the water surface, then $z_O = z_i = 0$, with the real and virtual images focused at the same plane, although they will be displaced in the x, y plane from each other. If, on the other hand, z_O is chosen to be finite, as above, the lens can be used to focus on one image or the other. Thus, by a combination of transverse and longitudinal displacements, the two images can be separated from each other.

We now consider the magnification of the image. At the plane $z = z_O$, the magnification M_T of the image is

$$M_T = \pm \left(\frac{\lambda_L}{\lambda_a}\right) \frac{z_i}{z_O} = 1 \qquad (3.5.136)$$

When light is reflected from the vibrating water surface, there is very little change in amplitude, as there is when light passes through a hologram on photographic film. Instead, the light is phase modulated by the change in the length of the light path. This phase modulation must be converted to amplitude modulation. We will show below how this is done in the Brenden imaging system [55, 56].

When a plane optical wave of amplitude A_0 is reflected from the water surface, the reflected wave is phase modulated by the change in its path, $2h$, and a component is obtained with an amplitude variation $a(x, y)$, defined as

$$a(x, y) = A_0 e^{-4j\pi h(x,y)/\lambda_L} \approx A_0 \left[1 - \frac{4j\pi h(x, y)}{\lambda_L} \right] \qquad (3.5.137)$$

where it has been assumed that $h \ll \lambda_L$.

Suppose that we observe the light at the back focal plane of the optical lens. At this plane, the amplitude of the light can be shown to be proportional to the Fourier spatial transform $A(v_x, v_y)$ of the function $a(x, y)$. This has the form

$$A(v_x, v_y) = \delta(v_x, v_y) - \frac{4j\pi H(v_x, v_y)}{\lambda_L} \qquad (3.5.138)$$

where $H(v_x, v_y)$ is the two-dimensional Fourier transform of $h(x, y)$. The spatial frequencies v_x and v_y are proportional to x'' and y'', respectively, where x'' and y'' are the coordinates at the back focal plane $z = z_i''$ of the lens.

The Dirac δ function represents the bright central spot corresponding to the light reflected directly from the surface of the water. It can be eliminated with a simple opaque dot located on the axis at the back focal plane of the lens. In addition, a neutral density filter with a density variation proportional to r''^2, where r'' is the distance from the axis, can be used along with the small stop. This filter will lower the response to the higher spatial frequencies, yielding an image with a light-amplitude variation of the form $v^2 H(v_x, v_y)$. However, we have shown [Eq. (3.5.133)] that $P(v_x, v_y) \propto v^2 H(v_x, v_y)$; hence the amplitude of the light passing through the filter is proportional to $P(v_x, v_y)$. Because the viewing optics can display a Fourier transform of the light amplitude at the back focal plane of the first lens, the image amplitude obtained can be made proportional to $p_0(x, y)$. However, by focusing the viewing optics on the plane $z = z_i$, a focused image whose amplitude is proportional to the acoustic amplitude of the wave passing through the object can be obtained.

An image of a deliberately disbonded aircraft honeycomb structure, taken some years ago with the Holosonics imaging system, is shown in Fig. 3.5.40. In this example, the honeycomb structure was immersed in the water tank; the operating frequency was 2.25 MHz. Note that the definition of the image is fairly good, but there is considerable speckle. This is to be expected with a coherent, phase-imaging, single-frequency system.

D. Scanned Holographic Imaging

There have been many ideas for producing holograms and forming images from them during the last few years. One of them, *scanned holography*, is interesting because of its close relation to synthetic aperture imaging [57, 58]. This system employs a focused transducer mechanically scanned in the x and y directions, as illustrated in Fig. 3.5.41. The transducer is immersed in water to form holograms of either objects in water or objects inside a solid.

In the latter case, the acoustic transducer is focused on the surface of the solid and forms a moving point source or receiver at that plane, which is the plane of the hologram, $z = 0$. The output signal from the receiving transducer is mixed

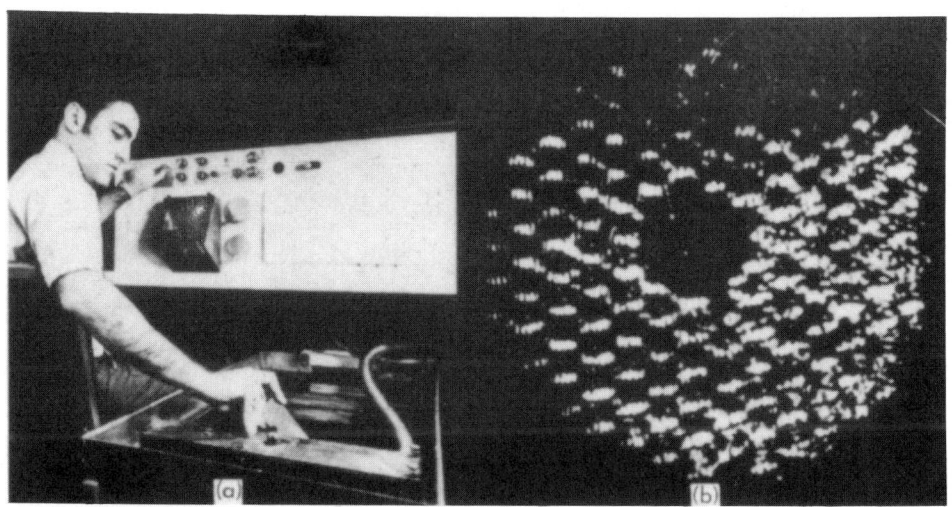

Figure 3.4.40 (a) Holosonics holographic ultrasound imaging system; (b) honeycomb structure ($\frac{1}{2}$-in. period) with intentionally introduced nonbonding areas as imaged by the Holosonics system at 5 MHz. (After Mueller [55].)

with a signal of the same frequency, which is equivalent to using a plane wave reference normal to the surface of the solid. However, as the phase of the reference signal can be changed with an adjustable phase shifter, the phase can be varied linearly with time, while the transducer is scanned in the x direction, to produce the effect of a tilted plane wave reference source.

The dc output from the mixer has an amplitude that varies with time in the same way as the light intensity passing through a hologram. Therefore, if a light source is modulated by this output signal and scanned in synchronism with the

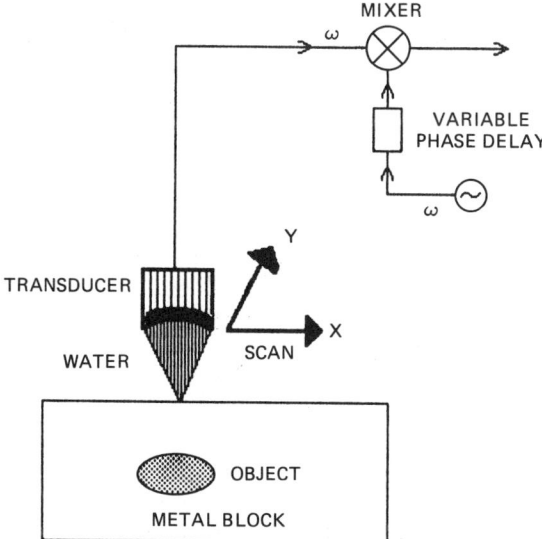

Figure 3.5.41 Simplified scanned holography system.

Sec. 3.5 Lensless Acoustic Imaging

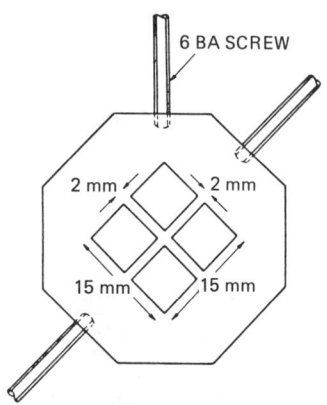

 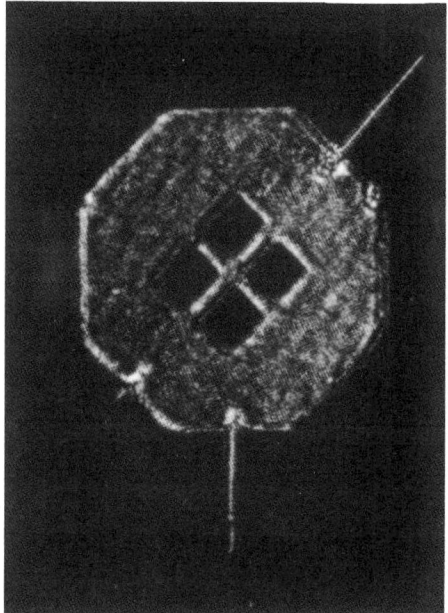

Figure 3.5.42 Reflection mode imaging of a simple object: (a) drawing of the object; (b) space frequency spectrum as seen optically; (c) holographic image. (After Aldridge et al. [57].)

movement of the acoustic transducer, it can be used to expose a photographic film to form a hologram. The hologram can then be reconstructed by the standard techniques already described.

Because of Snell's law, a transducer with a relatively narrow aperture produces a wide angle beam in a metal. Since the transducer can be scanned in the x and y directions over an aperture that is large in comparison to its size, it is possible by holography to synthesize the effect of a lens with a very large aperture.

The system can be used with a plane wave insonifying transducer, either at the top surface or incident on the object from another direction. Alternatively,

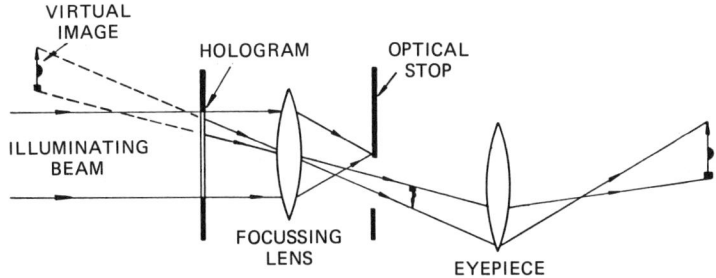

Figure 3.5.43 System used by Aldridge for reconstruction of a scanned holographic image. (After Aldridge et al. [57].)

the same focused transducer can be used as transmitter and receiver. In this case, just as with synthetic aperture imaging, the effective ray lengths are doubled, and the definition of the image is improved by a factor of 2 [57, 58].

A flat structure immersed in water, as illustrated in Fig. 3.5.42(a), was used for a reflection mode imaging experiment by Aldridge et al. [57]. With plane wave illumination at 10 MHz, they obtained the hologram shown in Fig. 3.5.42(b). The hologram was reconstructed with a helium–neon laser source, using the system illustrated in Fig. 3.5.43; the image obtained is shown in Fig. 3.5.42(c). Although there is some evidence of speckle, we see that this early system produced an image of very high quality.

The results obtained for objects inside a solid are less impressive, although they demonstrate that changing the focus of the optical reconstruction system makes it possible to image different planes within the solid. One reason they are less impressive is that there are often interfering signals present, due to reflections from the surface of the solid. Gating the transmitter signal helps, but a minimum pulse length is required to accommodate the time differences between the acoustic ray paths reaching the object at different angles to the axis of the transducer.

A very similar system has been used by a number of authors for short-pulse synthetic aperture imaging. Computers can be used either for holography or for short-pulse synthetic aperture time-delay reconstruction of images. If the phase and amplitude of the acoustic signal are recorded, it can be shown that a fast Fourier transform technique can be used to reconstruct a holographic image (see Prob. 8). But since it does not take much longer to form a true synthetic aperture time-delay image instead, which has the advantage of good range resolution, development of such computer techniques has been centered on synthetic aperture imaging or tomography.

E. The Scanning Laser Acoustic Microscope

Before leaving the subject of holography, we will describe a microscope that uses a laser readout; this is known as the *scanning laser acoustic microscope* (SLAM). We discuss it here because although it is not a strictly holographic system, it can be used in this mode, and it employs very similar techniques to those used in acoustic holography.

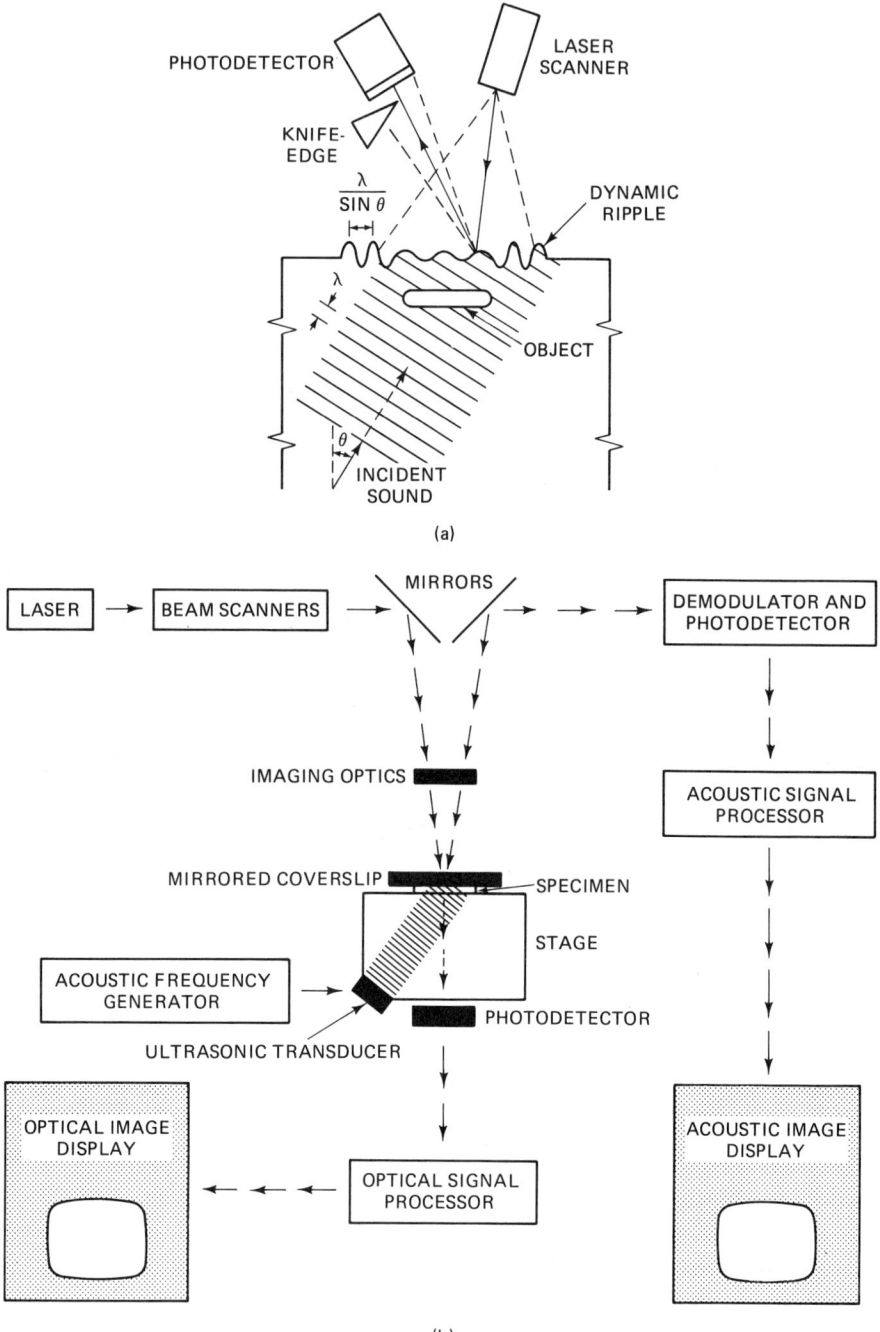

Figure 3.5.44 (a) Laser detection of acoustic energy at an interface. (From Kessler and Yuhas [59], and Korpel et al. [60].) (b) Schematic diagram of the scanning laser acoustic microscope. (After Kessler and Yuhas, as noted in Kessler and Yuhas [59, 61].)

Schematics of the device developed by Kessler and Yuhas are shown in Fig. 3.5.44. The sample to be examined is placed on the microscope stage, which consists of a solid block of quartz or other material, and is insonified by a longitudinal or shear acoustic plane wave directed at an angle to the normal to the surface. If the sample has a polished surface, the surface displacements due to the acoustic wave passing through it can be measured directly with the laser beam; alternatively, a plastic mirror in the form of a coverslip can be placed in contact with the sample and used to relay the sonic information to the laser.

An incident laser beam is reflected to a photodetector from the surface to be examined. When there is a surface ripple present due to an acoustic wave, the laser beam will be tilted by an angle $2\Delta\theta$ when the surface tilts by $\Delta\theta$. This small tilt in the laser beam would not normally affect the output from the photodetector. However, if a knife edge stop is placed in front of the beam, the slight tilt in angle will change the output amplitude. Thus the surface displacement due to the acoustic wave of frequency ω is changed to an amplitude modulation of the same frequency at the photodetector. A Bragg cell (see Sec. 4.9.5) is used to scan the laser beam rapidly over the surface in the x direction. A servo-controlled moving mirror (not shown) is used for the scan in the y direction. A lens (not shown) is placed in front of the knife edge to focus its image on the mirror image of the center of the Bragg cell in the reflecting surface. In this way, the scanned laser beam is always incident on the knife edge.

The operating frequency of the SLAM is usually between 100 and 500 MHz and the frame time is $\frac{1}{30}$ s, identical to the standard TV frame rate. The definition of the SLAM is not controlled, principally, by the acoustic wavelength, since the sample is insonifed by a plane wave. The system is basically a near-field shadow imaging system, with the definition controlled by the size of the laser beam at the reflecting surface. Since the beam must be focused on the image of the knife edge, its size is larger than it is at the focus. The definition for features below the surface rapidly deteriorates at a rate depending on the wavelength of the acoustic waves.

One major advantage of the SLAM, in addition to its real-time imaging capability, is the fact that the object can be viewed optically while the acoustic image is being obtained. An illustration of the configuration used for this purpose is shown in Fig. 3.5.44(b). When the sample is transparent, a half-silvered cover slip can be used and a photodetector placed below the transparent stage to detect the transmitted light. In this way, scanned optical and acoustic images can be obtained simultaneously.

Images of a live mouse embryo heart, taken at 100 MHz, are shown in Fig. 3.5.45. An optical image is shown in Fig. 3.5.45(a), with the acoustic image in Fig. 3.5.45(b). By mixing the output signal with the reference input signal to the insonifying transducer, an interferogram can be generated, as in Fig. 3.5.45(c). If the phase delay through the sample is uniform, the spacing of the fringes in this interferogram is constant and corresponds to $\lambda/\sin\theta$, where θ is the angle between the direction of propagation of the insonifying plane wave and the normal to the surface. When the thickness of the sample is nonuniform, as in the example shown

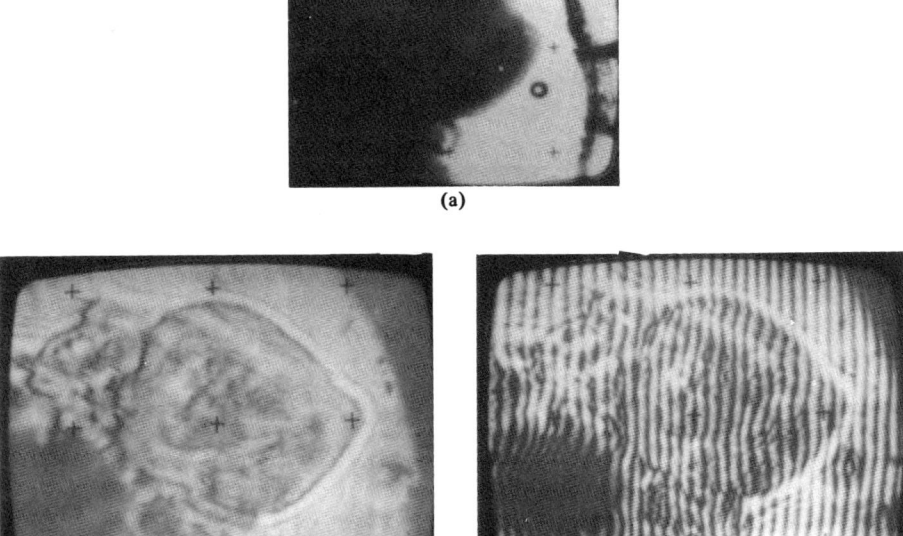

Figure 3.5.45 Live mouse embryo heart: (a) optical image; (b) SLAM acoustic micrograph at 100 MHz (cross-marks are placed 1 mm apart); (c) acoustic interferogram from which variations in elastic properties of the muscles are determined. (After Eggleton and Vinson, as noted in Kessler and Yuhas [59, 62].)

here, the finger spacing varies and can be used to measure the variation in thickness of the sample.

When the sample is far from the surface, the interferogram obtained is, in fact, a hologram. Some theoretical analyses involving digital techniques to reconstruct images from these holograms have been made [63, 64], but at the time of writing, there are few, if any, experimental results.

F. Holographic Imaging of Vibrating Objects

One of the most useful applications of holographic techniques is to the imaging of vibrating objects and the determination of the amplitude of vibration of any point on the object [2].

We suppose that any point x_O, y_O, z_O on the object is vibrating with an amplitude

$$\Delta z = m(m_O, y_O) \sin \Omega t \qquad (3.5.139)$$

When a point x_O, y_O, z_O on a stationary object, as illustrated in Fig. 3.5.46, is illuminated by a spherical source with its center at x_p, y_p, z_p, and the reference

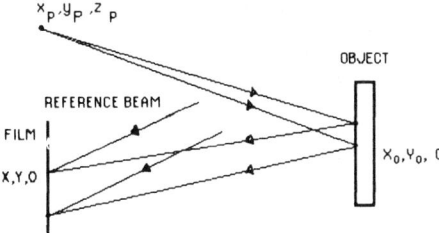

Figure 3.5.46 Holographic system for observing vibrating objects.

wave is a plane wave ($z_R = 0$), the image intensity at a point $x, y, 0$ on the film is

$$I = A \cos k \left[\frac{(x - x_O)^2 + (y - y_O)^2}{2z_O} \right.$$
$$\left. + \frac{(x - x_P)^2 + (y - y_P)^2}{2(z_O - z_P)} + 2z_O + \text{constant} \right] \quad (3.5.140)$$

$$= A \cos(\theta_O + \theta_P)$$

where $k = 2\pi/\lambda$ and λ is the optical wavelength.

If the object is vibrating, $I_{RO} = A \cos(\theta_O + \theta_P + \theta_1)$, where $\theta_1 \propto \Delta z$. If the angles of incidence of the reference beam and of the rays from the object at the film are small, it follows from the second-to-last term in Eq. (3.5.140) that $\theta_1 \approx 2k \Delta z$. Hence

$$I = A \cos(\theta_O + \theta_P + 2k_P m \sin \Omega t) \quad (3.5.141)$$

We can write I in the form

$$I = \frac{A}{2} [e^{-jkA(\theta_O + \theta_P + 2k_P m \sin \Omega t)} + \text{c.c.}] \quad (3.5.142)$$

where c.c. stands for complex conjugate. However, we can write

$$e^{-2jkm \sin \Omega t} = \int_{-\infty}^{\infty} c_n e^{-jn\Omega t} dt \quad (3.5.143)$$

with

$$c_n = \frac{1}{2\pi} \int_0^{2\pi} e^{j(n\Omega t - 2km \sin \Omega t)} dt = J_n(2km) \quad (3.5.144)$$

where $J_n(z)$ is an nth-order Bessel function of the first kind.

It follows that the static value of the intensity varies as

$$I_0(x_O, y_O) = A J_0[2km(x_O, y_O)] \quad (3.5.145)$$

due to the vibration. Thus the intensity of the holographic image of a plane object varies with $m(x_O, y_O)$ in the manner given by Eq. (3.5.145).

This result for the intensity variation gives us a way to measure the vibration

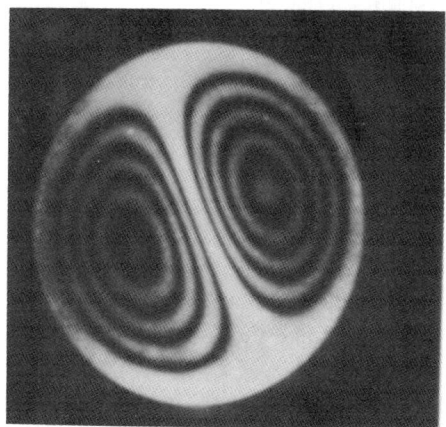

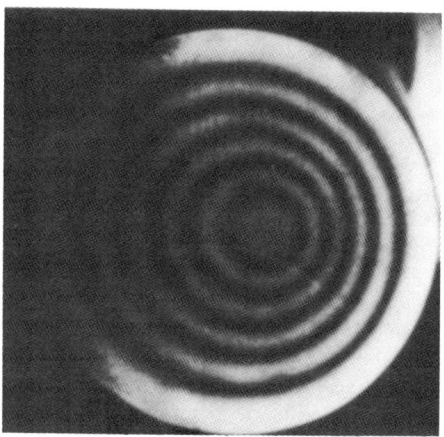

Figure 3.5.47 Holographic images of a diaphragm vibrating in (a) a circularly symmetric mode, and (b) a mode without circular symmetry. (After Powell and Stetson, as noted in Goodman [2, 65].)

amplitude. We see that $I_0(x_O, y_O) = 0$, where $J_o(2km) = 0$ (i.e., at $2km = 2.4$). With illumination from a helium–neon laser, this requires only a vibration amplitude of the order of 1250 Å. With small vibration amplitudes, the intensity I_0 varies as

$$I_o \approx A(1 - k^2 m^2) \qquad (3.5.146)$$

Thus, when m is small, the reduction in intensity is proportional to the square of the amplitude of vibration. A holographic image of a vibrating object is shown in Fig. 3.5.47.

PROBLEM SET 3.5

1. Use diffraction theory to treat the reflection of a finite beam of width w, incident at an angle θ to the normal on a perfect plane reflector (a specular reflector), as illustrated in the figure. Suppose that the incident beam can be regarded as a section of a plane wave

(i.e., to have uniform fields over its width w and a propagation constant k in the θ direction). Find the value of the normal particle displacement u_z^i of the incident wave at the plane $z = 0$ of the reflector.

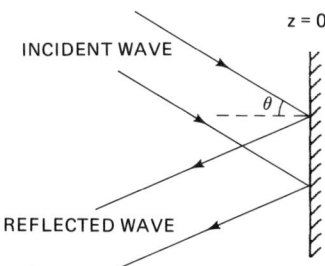

(a) Regard the specular reflector as a rigid object, so that the normal displacement of the reflected wave at the plane $z = 0$ is $u_z^R = -u_z^i$. Use the Rayleigh–Sommerfeld theory to determine the form of the reflected beam in the far field, and work out its 3-dB width and the angular position of its central axis. You will need the Rayleigh–Sommerfeld formula for a strip beam (i.e., a beam that is infinite in length in the y direction). This is given as the solution of Prob. 3.1.3. The result justifies ray optic theory in the limit $w/\lambda \to \infty$.

(b) Now consider sampling the reflected wave source by regarding it as a set of N point sources, with a spacing $w \cos \theta/(N - 1)$ between them. Work out the form of the far field, regarding each source as a δ function multiplied by the correct phase term. Show that as $N \to \infty$, your result converges to one of the same form as the solution of part (a). Show that when N is finite there are subsidiary maxima associated with the finite number of sampling points. These are called grating lobes. Find a formula for the positions of these grating lobes.

2. Consider an N-element transducer array at the plane $z = 0$. Suppose that the field due to a point source at x, z is $f(x - x_n, z)$ at the nth element of the array, and that the response of each element of the array is $g(x_n)$. If the random noise power at each element of the array is N_0, and the noise powers received at each element add linearly, the total noise power P_N into a 1-Ω load at the receiver will be

$$P_N = N_0 \sum_n |g_n^2(x)|$$

Work out the signal power S into a 1-Ω load at the receiver, and show that the signal-to-noise ratio S/P_N is maximum when $g(x_n) = f^*(x - x_n, z)$. Show also that when $|g(x_n)| = A$, where A is a constant, the signal-to-noise ratio is increased by a factor N by using N transducer elements instead of only one.

Note: You may find it helpful to use Schwarz's inequality in a similar manner to the derivation that leads to Eq. (4.4.6).

3. Prove Eq. (3.5.70) using the method of stationary phase in Appendix G.

4. Consider an M-phase digital imaging system with a continuous linear array. As discussed in Secs. 3.5.2.B and 3.5.3.B, the ideal matched filter response for a continuous array is

$$g(u) = e^{j\pi u}$$

where $u = x'^2/\lambda z_i$ and the phase is $\phi = \pi u$.

Assume that a digitally sampled system is designed to match this requirement as closely as possible, with M samples per cycle. Take the phase of the digital system to be φ with an error ε, defined as

$$\varepsilon = \frac{2n\pi}{M} - \pi u \qquad \left[\frac{(2n-1)\pi}{M} < u < \frac{(2n+1)\pi}{M}\right]$$

where n is an integer. The phase φ and error ε as functions of u are illustrated, respectively, in Figures (a) and (b) below.

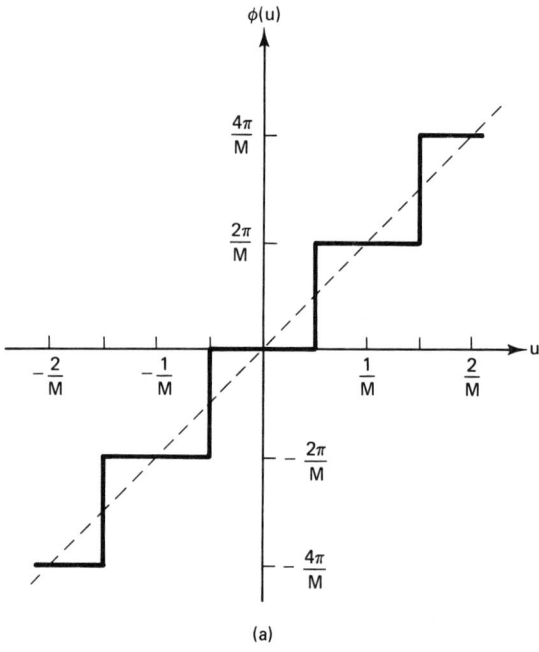

(a)

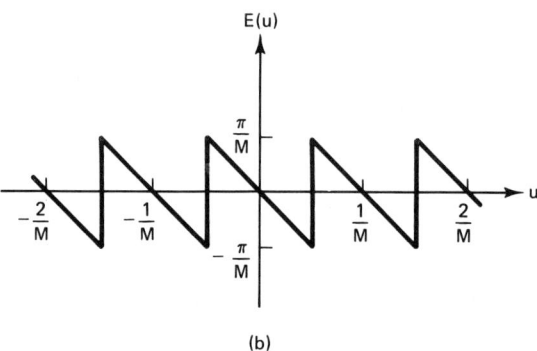

(b)

(a) Expand $\exp(j\varepsilon)$ in the form

$$e^{j\varepsilon} = \sum_m A_m e^{2jm\pi u}$$

and show that

$$A_m = \operatorname{sinc}\left(m + \frac{1}{M}\right) = \frac{(-1)^m \sin(\pi/M)}{\pi(m + 1/M)}$$

(b) Use your result and the method of stationary phase (Appendix G) to prove Eq. (3.5.72).

(c) Determine the positions of the subsidiary foci associated with the mth harmonics.

5. (a) Consider a nonparaxial chirp-focused system, used as a receiver, with transducer elements a distance l apart. Find the form of the chirp required to focus on a point x, z.

(b) Now consider the system as a transmitter. What is the form of the chirp required for nonparaxial focusing on the point x, z?

Note: You may find it helpful to consider the signal at each element of the transducer and allowing the phase to change by an extra 2π from element to element.*

6. (a) Consider a radial sector scan imaging system in which focusing and steering are handled separately, as described in Sec. 3.5.2.C. Suppose that we want to limit the number of lumped delay lines required for focusing in the receiver. Examine one strategy: Suppose that the receiver is focused on the point $0, z$, and that the maximum phase error allowed is $\pi/4$. Then, if the receiver is focused on $z = z_1 = \infty$, find the value of z for which the maximum phase error is $\pi/4$. Now find the value of $z = z_2$ for which the error is $\pi/2$ and suppose that the receiver is refocused at this point. Repeat the process to find how z_n depends on n, the nth refocusing point.

(b) Consider a system operating in body tissue (acoustic velocity 1.5 kM/s) at a frequency of 3.5 MHz. Suppose that we require the system to have a range of focus from 2 cm to ∞. How many focal points will be needed, and hence how many lumped delay lines per element will be required? Assume, in this case, that the error for an individual element may be less than the maximum error allowed for the outside elements.

7. Consider the Leith–Upatnieks hologram, which uses a plane wave reference, shown in Fig. 3.5.37.

(a) Show that an image may be reconstructed by using a plane wave source of the same wavelength to illuminate the film from the same angle as the original reference. You may regard any image as being composed of plane wave components incident at all angles to the film. If these plane wave components can be reconstructed, so can the original image.

(b) Consider the situation when reconstruction is attempted with a plane wave source of different wavelength. Will it work or not? Do not use the paraxial approximation.

8. Consider the problem of forming an image from a hologram by using fast Fourier transform (FFT) techniques on the computer. As an example, the hologram of a point x, y, z is formed with a SLAM. For simplicity, assume that a plane wave reference source is directed normal to the surface $z = 0$. This plane wave source insonifies the point x, y, z, so that the total signal arriving at the plane $z = 0$ may be regarded as the sum of a plane wave and a source at the point x, y, z.

(a) Using the paraxial approximation, work out the intensity of the image at the plane $z = 0$. This is the recorded signal.

(b) Now, in analogy to the FM chirp imaging method, multiply the recorded signal by a term of the form $\exp(j\{[(x - x')^2 + (y - y')^2]/2z\})$, where x', y', and 0 are the coordinates at the plane $z = 0$ being scanned by the laser beam. Now show how a Fourier transform technique can be used to reconstruct the image of the point x, y,

---

*Reference: W. H. Chen, F. C. Fu, and W. L. Lu, "Scanning Acoustic Microscope Utilizing SAW-BAW Conversion," *IEEE Trans. Sonics Ultrason.*, SU-32, No. 2 (Mar. 1985), 181–88.

z and, in general, the image of a thin layer at the plane z, whose attenuation varies as $f(x, y, z)$.

(c) Determine from your analysis the 3-dB definitions of the image in the x and y directions, for total laser scan distances of D_x and D_y in the x and y directions, respectively.

3.6 REFLECTION AND SCATTERING BY SMALL AND LARGE OBJECTS

3.6.1 Introduction

There are many applications in nondestructive testing (NDT), medical acoustics, and sonar for which we must be able to determine how different types of objects scatter acoustic waves. An object that is large in cross section compared to the wavelength, and whose surface roughness is small in scale compared to the wavelength, like a bone in the human body or a large hole in a metal or a ceramic, will tend to behave like specular reflectors. On the other hand, objects whose dimensions are small compared to the wavelength, such as grains in a metal, sand in the sea, fatty globules in tissue, or blood cells, will tend to give rise, individually, to weak waves, which are scattered in all directions. Thus a large number of fine particles will tend to behave, collectively, like a diffuse reflector. This scattering regime is known as *Rayleigh scattering*, named for Lord Rayleigh, who first investigated it in connection with the scattering of light by dust particles in the sky. He showed that the intensity of the scattered radiation varies as the fourth power of the frequency, which makes the sky appear blue.

The same law holds true for acoustic waves, so that Rayleigh scattering is one of the principal causes of attenuation in granular materials, although it is not present in a high-quality single crystal. The attenuation occurs because power is scattered from the incident acoustic beam by the grains. When the grain size is small compared to the wavelength, the implication is that the attenuation per unit length varies as the fourth power of the frequency [24, 66, 67].

Recently, considerable advances have been made in NDT by adapting the earlier electromagnetic (EM) theory and the theory used for scattering in liquids to the study of scattering in solids. This makes it possible to predict the frequency and angular variation of amplitude of the signals scattered from various types of objects, such as cracks, inclusions, and bounding surfaces, in terms of their size and material constants. Thus, by carrying out quantitative theory, it is becoming possible to establish the position, nature, and size of flaws in a solid body [1, 66, 67].

The NDT or sonar fields are convenient vehicles for such studies because for the simplest cases, scatterers of simple shape and composition are treated. Simple scatterers, like spheres and cylinders, can be treated theoretically and investigated in the laboratory. In the medical field, the problem is more difficult because of the complicated and less controllable nature of biological materials. A study of scattering from simple objects, however, is fundamental to understanding scattering from more complicated structures, and leads to very useful insights.

In Sec. 3.6.2 we will examine an easy way to treat specular scattering from spherical objects. Then, in Sec. 3.6.3 we will describe a more sophisticated theory to deal with scattering from different types of scatterers, applying it to scattering from both small and large objects.

3.6.2 Scattering by Large Objects (Physical Concepts)

Acoustic scattering of an incident wave, by objects in the sea, parts of the body, and flaws in materials, tends to be specular. This is because, unlike the case for optical waves, the surface roughness of these objects is typically small compared to the wavelength. The scattering from biological and solid objects at frequencies in the 1-GHz range is not specular, because the acoustic wavelength is of the order of 1.5 μm, comparable to the scale of surface roughness or to the size of biological cells. At low frequencies, as we have seen, the scale of surface roughness is small compared to the wavelength, and most discontinuities in impedance give rise to specular reflections. Thus it is difficult to obtain diffuse reflecting surfaces at low frequencies, even when we want to do so.

We first consider a large, perfectly rigid sphere insonified by a plane wave. We assume, for simplicity, that if the sphere is much larger in diameter than the wavelength, it behaves like an isotropic reflector. The power intercepted by the sphere, of radius a, is $\pi a^2 I_i$, where I_i is the intensity of the wave and πa^2 is the cross-sectional area of the sphere exposed to the incident wave. The reflected power radiated by the sphere will be $P_R = \pi a^2 I_i$. Thus the power per unit area in the reflected wave at a radius r will be the intensity $I_R = P_R/4\pi r^2$. We write

$$I_R = \frac{\pi a^2}{4\pi r^2} \qquad I_i = \frac{a^2}{4r^2} I_i \qquad (3.6.1)$$

Thus for $a \gg \lambda$, the reflected power intensity is proportional to a^2 and inversely proportional to the distance squared from the sphere. We obtain this result another way in Sec. 3.6.3.

We can extend the same general idea to account for reflection at normal incidence from any convex surface where all radii of curvature are large compared to a wavelength. We consider a plane wave normally incident on the convex surface at the point P, as shown in Fig. 3.6.1(c). Let its principal radii be R_1 and R_2. Imagine a series of planes that intersect the surface of the object: The *principal normal sections* are defined as those having a maximum and minimum radius of curvature, called R_1 and R_2, respectively. Suppose that AA' and BB' are lines on the convex surface that lie in the two respective principal normal sections, as shown in Fig. 3.6.1(a) and (b).

Consider the plan view shown in Fig. 3.6.1(c). The two incident rays subtending an angle $2d\theta_1$ at the center of curvature intercept a segment of length dl_1, where

$$dl_1 = R_1 \, d\theta_1 \qquad (3.6.2)$$

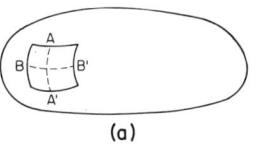

(a)

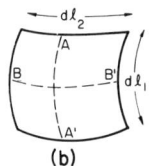

(b)

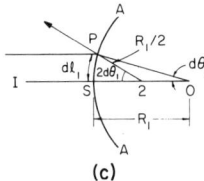

(c)

Figure 3.6.1 Reflection from convex surface: (a) cross-section through AA' with radius of curvature R_1; (b) reflection from a small element of area $dl_1\, dl_2$ with AA' and BB' as the principal normal sections; (c) ray incident from O at the point P on the convex surface.

If we consider the element dl_2 at right angles to this segment, we can write

$$dl_2 = R_2\, d\theta_2 \qquad (3.6.3)$$

The power incident on this element of area $dl_1\, dl_2$ is

$$dP = I_i\, dl_1\, dl_2 = R_1 R_2\, d\theta_1\, d\theta_2 \qquad (3.6.4)$$

The reflected wave in the plane view of Fig. 3.6.1(c) subtends an angle $2d\theta_1$ and appears to be emitted from a point $R_1/2$ from the center of curvature. Thus the area over which the reflected power is distributed at a radius r is

$$ds = r^2 (2\, d\theta_1)(2\, d\theta_2) \qquad (3.6.5)$$

Hence the reflected intensity I_R is

$$I_R = \frac{dP_R}{ds} = \frac{I_i R_1 R_2}{4r^2} \qquad (3.6.6)$$

For a sphere of radius a, Eq. (3.6.6) reduces to the relation

$$\frac{I_R}{I_i} = \frac{a^2}{4r^2} \qquad (3.6.7)$$

This result is the same as Eq. (3.6.1).

When the sphere is nonrigid and made of a material with a reflection coefficient Γ, we can write

$$\Gamma = \frac{Z_2 - Z_0}{Z_2 + Z_0} \qquad (3.6.8)$$

where Z_2 is the impedance of the sphere and Z_0 is the impedance of the surrounding material. We find, by the same arguments, that

$$\frac{I_R}{I_i} = |\Gamma|^2 \frac{a^2}{4r^2} \qquad (3.6.9)$$

In practice, a calculation for Γ in the simple form given in Eq. (3.6.8) is reasonable only for waves at normal incidence. As the main contribution to the backscattered wave is from this region, however, the use of Eq. (3.6.8) is valid. On the other hand, reflection of the incident plane wave to an off-axis transducer requires using a more general formula for Γ.

3.6.3 General Scattering Theory

Here, to acquaint the reader with the concepts used in the theory of scattering from small, arbitrarily shaped objects, we treat a simplified case of scattering by a hydrostatically compressible object in a liquid. We also discuss more general formulas for scattering in solids, and their implications.

Surface integral formulation. We use the Green's function theory given in Sec. 3.1.3 [see Eq. (3.1.27)]. There we showed that if the potential ϕ and its normal gradient $\nabla \phi \cdot \mathbf{n}$ are known on a surface s, the potential at any other point due to excitation at this surface is $\phi_s(x, y, z)$, where

$$\phi_s(x, y, z) = \int_s (\phi \nabla' G - G \nabla' \phi) \cdot \mathbf{n} \, ds' \qquad (3.6.10)$$

and the prime denotes source coordinates, in this case on the surface of the sphere. Here, to eliminate negative signs in Eq. (3.6.10), we have defined $\mathbf{n}$ as the *outward normal* from the sphere, and have redefined G as

$$G = \frac{e^{-jkR}}{4\pi R} \qquad (3.6.11)$$

with

$$R = \sqrt{(x - x')^2 + (y - y')^2 + (z - z')^2} \qquad (3.6.12)$$

Now suppose that we consider an object excited by an incident plane wave, as illustrated in Fig. 3.6.2. Suppose that the incident plane wave has a potential ϕ_i when there is no scattering object present. When a scattering object is present, the total potential is

$$\phi = \phi_i + \phi_s \qquad (3.6.13)$$

where ϕ_s is the potential of the scattered wave.

Therefore, in principle, if the total potential is known at the surface of the sphere, we can find $\phi^s(x, y, z)$ by writing the integral equation for ϕ, obtained using Eqs. (3.6.10) and (3.6.13), and solving to determine ϕ at any point. The

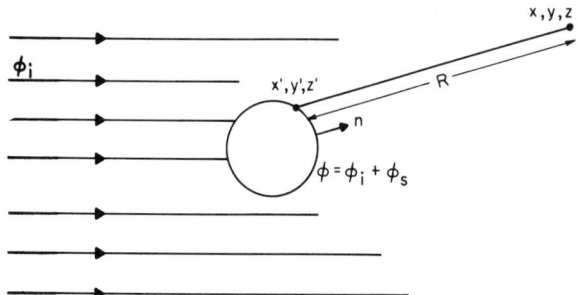

Figure 3.6.2 Object excited by an incident plane wave.

required integral equation is

$$\phi = \phi_i + \int (\phi \nabla' G - G \nabla' \phi) \cdot \mathbf{n} \, ds' \qquad (3.6.14)$$

Exact solutions to this problem can be obtained for spheres and cylinders, but not for scatterers of other shapes [66–69].

Volume integral formulation. It is convenient to write Eq. (3.6.10) in the form of a volume integral. Before doing this, however, note that the potentials defined in Eq. (3.6.10) are for a region just outside the perturbed volume V. At the surface of this volume the normal pressure p is continuous, as is the normal component of displacement $u_n = \partial \phi / \partial n$, but the potential is not. Using the relationship $p = \omega^2 \rho_{m0} \phi$ in Eq. (3.6.10), Eq. (3.6.10) can be written in terms of the pressure just inside the volume V in the form

$$p_s(x, y, z) = \int_s \left(p \nabla' G - \frac{\rho_{m0}}{\rho'_{m0}} G \nabla' p \right) \cdot \mathbf{n} \, ds' \qquad (3.6.15)$$

where ρ'_{m0} and κ' are the density and compressibility factors, respectively, inside the volume V (see Secs. 2.2 and 3.1). We have dropped the ' superscript for all other parameters at this point to keep the notation simple; here the ' superscript denotes differentiation with respect to coordinates inside the sphere.

We may now employ Gauss's integral theorem to write Eq. (3.6.15) in the form

$$p_s(x, y, z) = \int_V \left[\nabla' \cdot (p \nabla' G) - \frac{\rho_{m0}}{\rho'_{m0}} \nabla' \cdot (G \nabla' p) \right] dV' \qquad (3.6.16)$$

Using the vector relation $\nabla \cdot (\phi \nabla \psi) = \phi \nabla^2 \psi + \nabla \phi \cdot \nabla \psi$, it follows that

$$p_s(x, y, z) = \int_V \left[\left(1 - \frac{\rho_{m0}}{\rho'_{m0}} \right) \nabla' p \cdot \nabla' G + \left(\nabla'^2 G - \frac{\rho_{m0}}{\rho'_{m0}} G \nabla'^2 p \right) \right] dV'$$

$$(3.6.17)$$

where p and G obey the wave equations

$$\nabla^2 p + \frac{\omega^2 \rho'_{m0}}{3 \kappa'} p = 0 \qquad (3.6.18)$$

and

$$\nabla^2 G + \frac{\omega^2 \rho_{m0}}{3\kappa} G = 0 \qquad (3.6.19)$$

respectively, inside the volume V. Substituting Eqs. (3.6.18) and (3.6.19) in Eq. (3.6.17) and using the relation $k^2 = \omega^2 \rho_{m0}/3\kappa$ finally yields the scattered pressure in the form

$$p_s(x, y, z) = \int_V \left[\left(1 - \frac{\rho_{m0}}{\rho'_{m0}}\right) \nabla' p \cdot \nabla' G - k^2 p G \left(1 - \frac{\kappa}{\kappa'}\right) \right] dV' \qquad (3.6.20)$$

We can now use this scattering theorem to determine the scattered power or amplitude resulting from different types of scattering objects.

Born approximation. It is not easy to obtain an exact solution of Eq. (3.6.20) for the scattered pressure, because the total pressure p is not known within the perturbing object. A simple assumption, the *Born approximation*, which is used in many types of perturbation theories, takes the pressure inside the scatterer to be equal to the value of the unperturbed incident wave pressure p_i [1, 66–68] defined as

$$p_i = A_i e^{-jkz'} \qquad (3.6.21)$$

where the exciting wave propagates in the z direction. The Born approximation is equivalent to assuming that the perturbations in κ and ρ_{m0} are small. We can find the backscattered wave at the point $r, \theta, 0$, in spherical coordinates, for the simple situation $r \to \infty$ by using the expression for G of Eq. (3.6.11). We write, first,

$$R = \sqrt{(r \cos \theta - z')^2 + (r \sin \theta - x')^2 + y'^2} \qquad (3.6.22)$$

Because $r^2 \gg x'^2$, $r^2 \gg y'^2$, and $r^2 \gg z'^2$, it follows that

$$\begin{aligned} R &\approx r - z' \cos \theta - x' \sin \theta \\ z' &= r' \cos \theta' \\ x' &= r' \sin \theta' \sin \phi' \end{aligned} \qquad (3.6.23)$$

We then substitute for G in Eq. (3.6.20) and write

$$\frac{p_s(r, \theta)}{A_i} = -\frac{k^2 e^{-jkr}}{r} \int_{\text{object}} \left[\left(\frac{\rho_{m0}}{\rho'_{m0}} - 1\right) \cos \theta + 1 - \frac{\kappa}{\kappa'} \right] \\ \times \{e^{-jkr'[1 - \cos \theta) \cos \theta' - \sin \theta \sin \theta' \sin \phi']}\} r'^2 \sin \theta' \, d\theta' \, d\phi' \qquad (3.6.24)$$

In the simplest case, when the diameter of the object is very small compared to the wavelength, the integral becomes merely the volume of the object. Thus the scattered pressure varies with angle as

$$\frac{p_s(r, \theta)}{A_i} = -\frac{k^2 V e^{-jkr}}{r} \left[\left(\frac{\rho_{m0}}{\rho'_{m0}} - 1\right) \cos \theta + \left(1 - \frac{\kappa}{\kappa'}\right) \right] \qquad (3.6.25)$$

Observe that in the *Rayleigh scattering limit*, the scattered pressure varies as the square of the frequency and is proportional to the volume of the scattering object. This result is universal for small scattering objects. In addition, if Eq. (3.6.24) is written in Cartesian coordinates, the integral becomes a Fourier transform of the term in square brackets over the cross-sectional area of the object. Starting from the angular spectrum of the scattered wave, we can go through an inversion process, using Fourier transform techniques to determine the shape of the object. Thus, with only these simple assumptions, a great deal of information is available about the nature of the object.

Quasistatic approximation. Even when the diameter of the scattering object is very small, if there is a large change in its density or elastic constant, the fields inside it will be different from the exciting fields. When the object is very small in dimensions compared to the wavelength, we can assume that in the neighborhood of the object the field variations are quasistatic and can then use static solutions for the fields in this region. To put it another way, in this case the variations of the pressure p in the neighborhood of the object are very large. When the terms $\partial^2 p/\partial x^2$, $\partial^2 p/\partial y^2$, and $\partial^2 p/\partial z^2$ are large compared to $k^2 p$, the solution of the wave equation [Eq. (3.6.18)] will then depend on the solution of Laplace's equation, $\nabla^2 p = 0$.

The exciting pressure wave p_i [Eq. (3.6.21)] in the neighborhood of the object may be written in the approximate form

$$p_i \approx A_i(1 - jkz) = A_i(1 - jkr \cos\theta) \quad (3.6.26)$$

where we assume that the origin of the coordinates is at the center of the scattering object and that $|kz| \ll 1$. This pressure term p_i has associated with it a displacement u_{zi}, defined as

$$u_{zi} = \frac{\partial p_i/\partial z}{\omega^2 \rho_{m0}} = \frac{-jkA_i}{\omega^2 \rho_{m0}} \quad (3.6.27)$$

When an ellipsoid is placed in such an exciting field, the displacement fields within it will also be uniform [67, 68, 70]. The simple case of a sphere is derived in Appendix H, where it is shown that the internal displacement field u_z is

$$u_z = \frac{3u_{zi}}{1 + 2\rho_{m0}/\rho'_{m0}} \quad (3.6.28)$$

Substituting this result in Eq. (3.6.20) for a sphere of radius a yields the relation

$$\frac{p_s(r, \theta)}{A_i} = -\frac{e^{-jkr}}{r} \frac{k^2 a^3}{3} \left[\frac{3(1 - \rho'_{m0}/\rho_{m0})}{1 + 2\rho'_{m0}/\rho_{m0}} \cos\theta + \left(1 - \frac{\kappa}{\kappa'}\right) \right] \quad (3.6.29)$$

In the Rayleigh scattering regime, just as with the Born approximation, the scattered pressure varies as the square of the frequency and the cube of the radius of the scattering sphere (i.e., as the volume of the scattering sphere).

We can work out the scattered intensity per unit area at a radius r by taking the square of Eq. (3.6.29) and integrating over a unit solid angle. We find that as $I_s = (|p_s^2|r^2/2z_0) \, d\Omega$, where $d\Omega = \sin\theta \, d\theta \, d\phi$ is the differential solid angle and $Z_0 = \sqrt{\rho_{m0}c}$ is the impedance of the medium, the ratio of the scattered intensity I_s to the incident intensity I_i is

$$\frac{I_s(\theta)}{I_i} = \frac{k^4 a^6}{9r^2} \left[\frac{3(1 - \rho'_{m0}/\rho_{m0})}{1 + 2\rho'_{m0}/\rho_{m0}} \cos\theta + \left(1 - \frac{\kappa}{\kappa'}\right) \right]^2 \quad (3.6.30)$$

We now consider some examples of scattering of acoustic waves in a liquid.

Example: Scattering from a Rigid Sphere in a Liquid

Lord Rayleigh, by carrying out the exact analysis for a rigid sphere, showed that a small rigid sphere acts like a dipole source when illuminated by a plane wave [66, 69]. If θ is the angle between the scattering direction and the incident wave ($\theta = 0$ is forward-scattering and $\theta = \pi$ is backward-scattering), it follows from Eq. (3.6.30) that for a rigid sphere of radius $a \ll \lambda$, for which $\rho'_{m0}/\rho_{m0} \to \infty$ and $\kappa'/\kappa \to \infty$ (impedance large and velocity finite), the scattered intensity I_s at an angle θ is

$$\frac{I_s}{I_i} = \frac{k^4 a^6}{9r^2} \left(1 - \frac{3}{2}\cos\theta\right)^2 \quad (3.6.31)$$

Thus for backward scattering ($\theta = \pi$),

$$\frac{I_s(\text{backward})}{I_i} = \frac{25}{36} \frac{k^4 a^6}{r^2} \quad (3.6.32)$$

with

$$\left|\frac{p_s}{a_i}\right| = \frac{5}{6} \frac{k^2 a^3}{r} \quad (3.6.33)$$

The total scattered power $P_s(\text{tot})$ is

$$P_s(\text{tot}) = \int_{\phi=0}^{2\pi} \int_{\theta=0}^{\pi} r^2 I_s \sin\theta \, d\theta \, d\phi \quad (3.6.34)$$

or

$$P_s(\text{tot}) = \frac{7\pi}{9} k^4 a^6 I_i \quad (3.6.35)$$

Thus it is as if the incident beam were intercepted by an obstacle of *total cross section* $\sigma(\text{tot})$, where

$$\sigma(\text{tot}) = \frac{P_s(\text{tot})}{I_i} = \frac{7\pi}{9} k^4 a^6 \quad (3.6.36)$$

This parameter $\sigma(\text{tot}) = P_s(\text{tot})/I_i$ is known as the *total scattering cross section* of the scatterer. We note that when the scatterer behaves like a specular reflector, then $\sigma(\text{tot}) = 2\pi a^2$. This is twice the physical cross section of a sphere because as much power is radiated in the forward direction as in the backward direction. For a rigid sphere, it follows that

$$\frac{\sigma(\text{tot})}{2\pi a^2} = \frac{7}{18} k^4 a^4 \quad (3.6.37)$$

Example: Air Bubbles and Sand Grains in Water

Because air bubbles are easily compressed, I_R/I_i is very large for air bubbles in water. Thus, for air in water, $\kappa/\kappa' = 19{,}000$ and, from Eq. (3.6.30), the backscattering is 77 dB larger than for classical Rayleigh scattering from a rigid sphere. For sand in water, on the other hand, where the sand is regarded as equivalent to a compressible liquid, $\rho'_{m0}/\rho_{m0} = 2.6$, $\kappa/\kappa' = 0.1$, and the scattering from a spherical grain is 4 dB less than it is from a rigid sphere of the same volume.

Kirchoff approximation for scattering from a large sphere ($ka \gg 1$). It is interesting to consider the situation treated in Sec. 3.6.2 by simple specular reflection concepts, by using the field theory techniques developed in this section. When considering scattering from a large sphere, it is convenient to use the surface integral formula of Eq. (3.6.10). We assume that if $ka \gg 1$, any portion of the sphere around the point r, θ, ϕ acts like a planar reflector at an angle θ to the incident wave, and the reflection coefficient from that point is $\Gamma(\theta)$.

The pressure of the incident wave p_i is

$$p_i = A_i e^{-jka\cos\theta} \tag{3.6.38}$$

where we have taken the origin of the coordinates to be at the center of the sphere. The total pressure at the spherical surface, as with a planar reflector, is assumed to be doubled when $\Gamma = 1$. More generally, on the basis of the same types of assumptions,

$$p = -\omega^2 \rho_{m0} \phi = [1 + \Gamma(\theta')] A_i e^{-jka\cos\theta'} \tag{3.6.39}$$

We assume that $p = \phi = 0$ and $0 < \theta < \pi/2$ in the shadow region. For simplicity, we consider only backscattering in the $\theta = \pi$ direction. For $r \gg a$, we write $R \approx r + r' \cos\theta'$, where the coordinates just outside the sphere are r', θ'. Then the Green's function G at $r' = a$ can be written in the form

$$G\bigg|_{r'=a} \approx \frac{e^{-jka\cos\theta'}}{4\pi r} e^{-jkr} \tag{3.6.40}$$

with

$$\frac{\partial G}{\partial r'}\bigg|_{r'=a} \approx -\frac{jk\, e^{-jka\cos\theta'}}{4\pi r} e^{jkr} \cos\theta' \tag{3.6.41}$$

and

$$\frac{\partial p}{\partial r'}\bigg|_{r'=a} = -jk[1 - \Gamma(\theta')] A_i \cos\theta\, e^{-jka\cos\theta'} \tag{3.6.42}$$

It follows from Eq. (3.6.10), taken for a surface just outside the volume V, that

$$\frac{p_s(r, \theta)}{A_i} = \frac{jke^{-jkr}}{r} \int_{\theta=\pi/2}^{\pi} a^2 \cos\theta' \sin\theta'\, e^{-2jka\cos\theta'}\, \Gamma(\theta')\, d\theta' \tag{3.6.43}$$

Assuming that $\Gamma = $ constant, we find that when $ka \gg 1$, Eq. (3.6.43) can be integrated to yield

$$\frac{p_s(r, 0)}{A_i} \approx \frac{e^{-jk(r-a)}}{2r} \Gamma a \qquad (3.6.44)$$

Here we have ignored terms of the order of $1/ka$ down from the main term, because the theory should not be expected to be reliable for terms of this order. The backscattered intensity I_s is defined as

$$I_s = \frac{|p_s|^2}{2Z_0} \qquad (3.6.45)$$

Therefore, the ratio of the backscattered intensity to the incident intensity I_i is

$$\frac{I_s}{I_i} = \frac{\Gamma^2 a^2}{4r^2} \qquad (3.6.46)$$

As we might expect, this result agrees with Eq. (3.6.1). Note that we can estimate the backscattering cross section σ_{Rs} by assuming that the power is radiated uniformly in all directions, which implies, from the argument in Sec. 3.6.1, that

$$\sigma_{Rs} = \pi a^2 \qquad (3.6.47)$$

We note that this is just half the total scattering cross section, for as much power is radiated in the forward direction as in the backward direction.

Comparison with exact results and the concept of creeping waves.
The exact results for scattering from a rigid obstacle are plotted by the solid line in Fig. 3.6.3; for comparison, the dashed lines plot Rayleigh scattering theory, which is approximately true for $ka < 0.5$, and the asymptotic specular approximation for reflection for $ka \gg 1$. The exact results have an extra periodicity due to the presence of leaky waves, which propagate around the surface of the sphere with approximately the longitudinal wave velocity in the liquid. These leaky waves, known as *creeping waves*, are illustrated in Fig. 3.6.4. For this reason, it is rarely worthwhile to carry out the integral of Eq. (3.6.43) in more detail and keep terms that vary as $1/ka$. Such analyses can be improved to reproduce the theory of scattering from the front surface with accuracy. They cannot, however, reproduce the effect of creeping waves.

The fact that there are waves that can propagate around an obstacle is important. If the sphere of radius a is surrounded by another sphere of very large radius b, and if $ka \gg 1$, a series of longitudinal or shear waveguide modes propagates around the inner sphere. If the outer sphere were removed to infinity, a mode propagating around the inner sphere would tend to have infinite energy. Instead, as it radiates power radially, this creeping wave becomes a leaky wave, and its amplitude decreases along the circumference away from the point of excitation [66, 69, 71].

When the plane wave incident on the sphere is longitudinal, as it is in a liquid, it tends to excite a longitudinal creeping wave at the shadow boundary (i.e., where

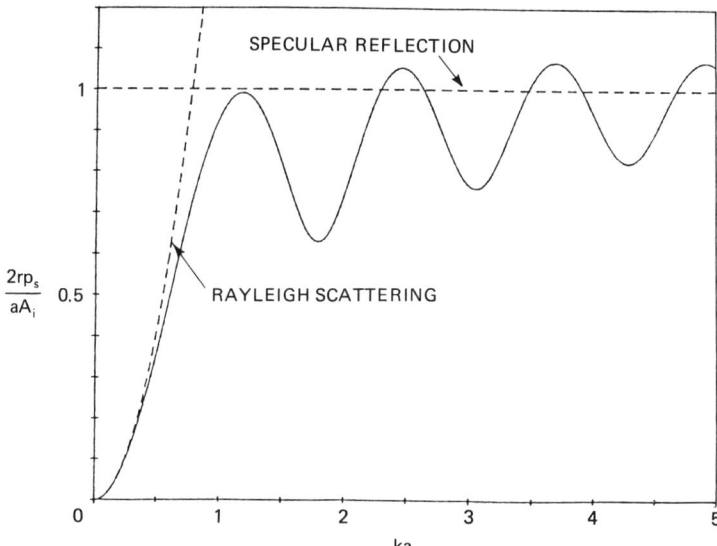

Figure 3.6.3 Solid line, plot of the exact result for backscattering from a rigid sphere in a liquid; Dashed lines, plots of the specular reflection theory [Eq. (3.6.1)] and the quasistatic Rayleigh scattering limit [Eq. (3.6.31)].

the phase velocity of the creeping wave is in the same direction as that of the incident wave). The creeping wave excites a reflected plane wave on the other side of the sphere, as illustrated in Fig. 3.6.4. The results of this concept are clear in Fig. 3.6.5, which shows the time-domain backscatter response to a short pulse. This has been found by taking the Fourier transform of Fig. 3.6.3.

Generalizations of the theory. We see that a pulse is reflected from the front surface of the sphere and that a further weak pulse, corresponding to the creeping wave, arrives at a time $t = a(2 + \pi)/V$ later. The response in the frequency domain thus corresponds to the interference between the specularly reflected wave and the later signal from the creeping wave, which travels around the circumference at a velocity close to that of a longitudinal wave [69].

The creeping wave concept can be generalized further for obstacles on which there are sharp points or corners. In this case, each sharp discontinuity behaves like a source for reflected waves [72]. Sharp discontinuities also generate creeping

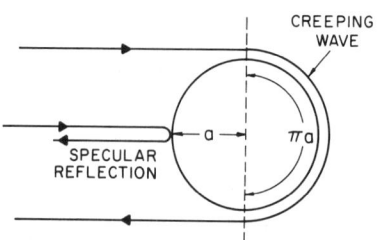

Figure 3.6.4 Specular reflection and creeping wave from a sphere.

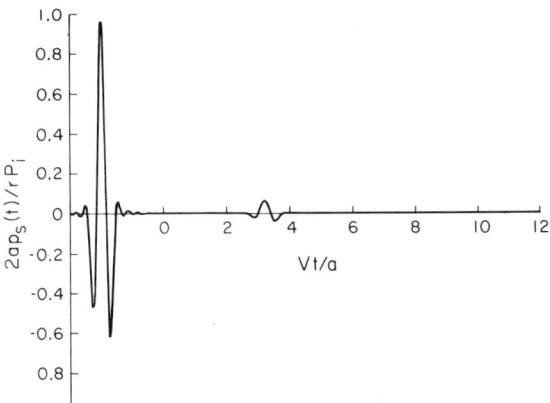

Figure 3.6.5 Plot of the backscatter reflected pressure from a rigid sphere in a liquid as a function of time when the excitation is a short pulse.

waves, which propagate along the surface of the object and are reemitted at another similar discontinuity. A good example is a smooth crack in a solid, which behaves like a specular reflector. The crack may not emit a reflected wave in the direction of the receiving transducer; the ends of the crack, however, scatter waves in all directions, to create a characteristic angular interference pattern between the two waves emitted, one from either end of the crack; this pattern varies with frequency [73, 74]. Creeping waves can also be excited at one end of the crack, propagate along the crack surface, and be emitted from the other end of the crack. The situation becomes even more complicated if the surface is rough. In this case, there may be a number of secondary sources along the surface of the object, as we discussed when considering the phenemonon of speckle in Sec. 3.3.2.

The results can also be generalized in other ways. For instance, the dependence of scattering with angle will change with the shape of the object, even when its size is small, and the intensity of the scattered wave can have a different frequency dependence in the Rayleigh limit. As an example, we consider scattering of a surface wave from a small cylinder of radius a, whose axis is normal to the direction of propagation of the incident wave. For $kR \gg 1$, the Green's function will vary as $[\exp(-jkR)]/\sqrt{kR}$, and the scattered wave intensity will vary as $f^3 a^4$ (see Prob. 3.1.3). This compares to scattering of a volume wave from a sphere of radius a, where the scattered intensity varies as $f^4 a^6$ [1, 66–69].

Similarly, if we consider scattering of a longitudinal wave in a solid from a small defect, we can find generalized forms of Eq. (3.6.20) [1, 68, 75]. In general, both shear and longitudinal waves are excited and the angular dependences of these scattered waves will be quite different and complicated, even in the Rayleigh limit, by their dependence on stress type and direction rather than pressure. Thus, even in the Rayleigh limit, we can obtain a great deal of information from the angular dependence of scattering on the elastic properties of the scatterer [1, 68, 70].

At higher frequencies, scattering behavior becomes even more complicated. Consider a void in a solid, which behaves partially like a specular reflector; the reflection coefficient of a short pulse reflected from the front surface is given by Eq. (3.6.1). In addition, as we have seen, creeping waves, which give rise to very weak reradiated signals, will propagate along the surface of the object. When excited by a pulse, such a void then gives a strong reflection from its front surface, followed by a short pulse corresponding to a creeping wave, as illustrated in Figs. 3.6.4 and 3.6.5 [71].

The behavior of inclusions is still more complicatted and is best understood on the basis of ray tracing theory. Once more there will be a reflection from the front surface, but there will also be waves that propagate through the middle of the inclusion and are reflected from its back surface. As mode conversions can take place from longitudinal to shear waves, and vice versa, quite complicated behavior occurs and there may be relatively strong reflections from the back surface of such an inclusion. Furthermore, waves can also propagate around the outside of the inclusion, which can give rise to still other pulses. Hence the behavior in the frequency domain is relatively complicated because of the interference pattern that occurs between these various, multiply reflected waves [71, 75, 76].

As with the simpler theories described in this section, the detailed behavior of scattering from spherical voids and inclusions in solids can be worked out with an exact theory, and approximate techniques, very similar in nature and form to those of Eqs. (3.6.15) and (3.6.20), can be derived. Both Born approximations and quasistatic approximations have been made in the manner described here. The quasistatic approximations are particularly interesting for dealing with scattering from small cracks, because they rest on a foundation of static fracture mechanics theory, which has been developed over the years [68, 70, 77–79]. The theories have been generalized to deal with excitation by Rayleigh surface waves as well as plane longitudinal and shear waves. Variational techniques and other methods have also been developed to deal with these problems. Thus a great deal of theoretical insight into scattering from flaws can be obtained. The inverse problem of determining the shape, size, and nature of a flaw from the scattered signals is also of great interest, and a great deal of theoretical work on the subject is being carried out.

PROBLEM SET 3.6

1. (a) Work out the scattering from a disk-shaped crack of radius a excited by a plane wave. Assume that $ka \gg 1$ and that the incident plane wave is incident on the disk at an angle θ to its axis. Assume that the reflection coefficient is zero at the crack, so that u_z is doubled at the crack from its incident value. Regarding the crack as a piston transducer surrounded by a rigid baffle, find the amplitude of the backscattered wave in the far field (the Fraunhofer region of the crack) as a function of angle ($\theta > \pi/2$). Thus consider only specular reflection and neglect creeping waves.
   (b) Compare the amplitude of your result for the backscattered signal ($\theta = \pi$) with that for a spherical void of the same radius. Again, ignore creeping waves.

2. Use a quasi-static theory to work out the scattering, as a function of angle, of a normally incident longitudinal plane wave from a penny-shaped (disk-shaped) crack of radius a, when $ka \ll 1$.

   Hint: Using quasistatic theory, we can show that the crack opening displacement Δu_z (the incremental distance between the two crack faces), when an RF field T_{3i}, or pressure, is applied, is of the form

$$\Delta u_z = \frac{4}{\pi} \frac{1-\sigma}{\mu} (a^2 - r^2)^{1/2} T_{3i}$$

Treat the crack like a small piston transducer, or use Eq. (3.6.10) with $\phi = 0$ (i.e., $T_3 = 0$ on the crack). Here σ is Poisson's ratio and μ is the second Lamé constant.

3. The attenuation per unit length of a plane wave is

$$\alpha = \frac{P_L}{2P_T}$$

where P_L is the power lost per unit length and P_T is the incident power transmitted.

(a) Consider the problem of observing objects in turbid water, in which there are N sand particles per unit volume. Suppose that the particles are spherical and 100 µm in diameter. Derive the total scattered power of a light wave from a particle, using the concepts of Sec. 3.6.1 and the discussion after Eq. (3.6.36). Then work out the attenuation per unit length for N sand particles per unit volume.

(b) Work out a formula for the attenuation in turbid water, due to sand, of a 0.5-MHz acoustic wave ($V = 1.5$ km/s). Suppose that there are 10^4 particles/cm^3. Find the attenuation per meter of light and sound.

Hint: See the example after Eq. (3.6.37).

4. Consider a spherical reflector of iron in a medium such as silicon nitride. In this case, the velocity of a wave inside the sphere is almost exactly half the acoustic wave velocity in the surrounding medium. Show by ray tracing in the paraxial limit that rays entering the sphere will return along a parallel path after being reflected from the back of the sphere. For this reason, reflections from the back of a sphere are often larger than those from the front surface. This phenomenon is known as "the scotchlight effect."

REFERENCES

1. G. S. Kino, "The application of Reciprocity Theory to Scattering of Acoustic Waves by Flaws," *J. Appl. Phys.*, 49, No. 6 (June 1978), 3190–99.
2. J. W. Goodman, *Introduction to Fourier Optics*. New York: McGraw-Hill Book Company, 1968.
3. B. Delannoy, H. Lasota, C. Bruneel, R. Torguet, and E. Bridoux, "The Infinite Planar Baffles Problem in Acoustic Radiation and Its Experimental Verification," *J. Appl. Phys.*, 50, No. 8 (Aug. 1979), 5189–95.
4. H. Seki, A. Granato, and R. Truell, "Diffraction Effects in the Ultrasonic Field of a Piston Source and Their Importance in the Accurate Measurement of Attenuation," *J. Acoust. Soc. Am.*, 28 No. 2 (Mar. 1956), 230–38.
5. R. A. Lemons and C. F. Quate, "Acoustic Microscopy," Chapter 1 in *Physical Acoustics: Principles and Methods*, Vol. XIV, W. P. Mason and R. N. Thurston, eds. New York: Academic Press, Inc., 1979, pp. 1–92.
6. A. Papoulis, *Systems and Transforms with Applications in Optics*. (orig.: New York: McGraw Hill, 1968; rpt.: Melbourne, Fla.: R. E. Krieger Publishing Co., Inc. 1981).

7. K. K. Liang, G. S. Kino, and B. T. Khuri-Yakub, "Material Characterization by the Inversion of $V(z)$," *IEEE Trans. Sonics Ultrason.*, SU-32, No. 2 (Mar. 1985), 213–24.
8. R. N. Bracewell, *The Fourier Transform and Its Applications*, 2nd ed. New York: McGraw-Hill Book Company, 1978.
9. A. Atalar, "An Angular-Spectrum Approach to Contrast in Reflection Acoustic Microscopy," *J. Appl. Phys.*, 49 No. 10 (Oct. 1978), 5130–39.
10. J. Heiserman, D. Rugar, and C. F. Quate, "Cryogenic Acoustic Microscopy," *J. Acoust. Soc. Am.*, 67, No. 5 (May 1980), 1629–37.
11. H. K. Wickramasinghe and C. R. Petts, "Gas Medium Acoustic Microscopy," *Scanned Image Microscopy*, E. A. Ash, ed. London: Academic Press, Inc. (London) Ltd., 1980, pp. 57–70.
12. W. Parmon and H. L. Bertoni, "Ray Interpretation of the Material Signature in the Acoustic Microscope," *Electron. Lett.*, 15, No. 21 (Oct. 11, 1979), 684–86.
13. R. D. Weglein, "Acoustic Microscopy Applied to SAW Dispersion and Film Thickness Measurement," *IEEE Trans. Sonics Ultrason.*, SU-27, No. 2 (Mar. 1980), 82–86.
14. R. C. Bray, C. F. Quate, J. Calhoun, and R. Koch, "Film Adhesion Studies with the Acoustic Microscope," *Thin Solid Films*, 74, No. 2 (Dec. 15, 1980), 295–302.
15. E. A. Ash, ed., *Scanned Image Microscopy*. New York: Academic Press, Inc., 1980.
16. C. F. Quate, "The Acoustic Microscope," *Sci. Am.*, 241, No. 4. (Oct. 1979), 62–70.
17. R. A. Lemons, *Acoustic Microscopy by Mechanical Scanning*, Ph.D. dissertation, Stanford University, Stanford, Calif., 1975.
18. R. A. Lemons and C. F. Quate, "A Scanning Acoustic Microscope," *1973 Ultrason. Symp. Proc.* (IEEE), 73 CHO 807-8 SU, 18–20.
19. B. Hadimioglu and C. F. Quate, "Water Acoustic Microscopy at Suboptical Wavelengths," *Appl. Phys. Lett.*, 43, No. 11 (Dec. 1, 1983), 1006–7.
20. A. Rosencwaig, *Photoacoustics and Photoacoustic Spectroscopy*, Vol. 57 of *Chemical Analysis: A Series of Monographs on Analytical Chemistry and Its Applications*, P. J. Elving and J. D. Winefordner eds., I. M. Kolthoff, ed. emeritus. New York: John Wiley & Sons, Inc., 1980.
21. R. A. Burrier, R. O. Claus, J. W. Gray, and W. T. O'Connor, "Circularly Symmetric Gaussian Field Transducer with Equal Impedance and Equal Voltage Electrode Design," *1983 Ultrason. Symp. Proc.* (IEEE), Vol. 1, 83CH1947-1, 570–72.
22. A. Yariv, *Quantum Electronics*, 2nd ed. New York: John Wiley & Sons, Inc., 1975.
23. H. Kogelnik and T. Li, "Laser Beams and Resonators," *Proc. IEEE*, 54, No. 10 (Oct. 1966), 1312–29.
24. J. P. Weight and A. J. Hayman, "Observations of the Propagation of Very Short Ultrasonic Pulses and Their Reflection by Small Targets," *J. Acoust. Soc. Am.*, 63, No. 2 (Feb. 1978), 396–404.
25. D. E. Robinson, S. Lees, and L. Bess, "Near Field Transient Radiation Patterns for Circular Pistons," *IEEE Trans. Acoust. Speech Signal Process.*, ASSP-22, No. 6 (Dec. 1974), 395–403.
26. W. L. Beaver, "Sonic Nearfields of a Pulsed Piston Radiator," *J. Acoust. Soc. Am.*, 56, No. 4 (Oct. 1974), 1043–48.
27. R. H. Tancrell, J. Callerame, and D. T. Wilson, "Near-Field, Transient Acoustic Beam-Forming with Arrays," *1978 Ultrason. Symp. Proc.* (IEEE), 78CH 1344-1SU, 339–43.

28. G. S. Kino, "Acoustic Imaging for Nondestructive Evaluation," *Proc. IEEE*, 67, No. 4 (Apr. 1979), 510–25.
29. M. G. Maginness, "Methods and Terminology for Diagnostic Ultrasound Imaging Systems," *Proc. IEEE*, 67, No. 4 (Apr. 1979), 641–53.
30. B. T. Khuri-Yakub, private communication.
31. J. Fraser, J. Havlice, G. Kino, W. Leung, H. Shaw, K. Toda, T. Waugh, D. Winslow, and L. Zitelli, "An Electronically Focused Two-Dimensional Acoustic Imaging System," in *Acoustical Holography*, Vol. 6, N. Booth, ed. New York: Plenum Press, 1975, pp. 275–304.
32. S. A. Farnow and B. A. Auld, "An Acoustic Phase Plate Imaging Device," in *Acoustical Holography*, Vol. 6, N. Booth, ed. New York: Plenum Press, 1975, pp. 259–73.
33. D. K. Peterson and G. S. Kino, "Real-Time Digital Image Reconstruction: A Description of Imaging Hardware and an Analysis of Quantization Errors," *IEEE Trans. Sonics Ultrason.*, SU-31, No. 4 (July 1984), 337–51.
34. J. F. Havlice, G. S. Kino, J. S. Kofol, and C. F. Quate, "An Electronically Focused Acoustic Imaging Device," in *Acoustical Holography*, Vol. 5, P. S. Green, ed. New York: Plenum Press, 1974, pp. 317–33.
35. K. N. Bates, E. Carome, K. Fesler, R. Y. Liu, and H. J. Shaw, "Digitally Controlled Electronically Scanned and Focused Ultrasonic Imaging System," *1979 Ultrason. Symp. Proc.* (IEEE), 79CH1482-9, 216–20.
36. T. M. Waugh, G. S. Kino, C. S. DeSilets, and J. D. Fraser, "Acoustic Imaging Techniques for Nondestructive Testing," *IEEE Trans. Sonics Ultrason.*, SU-23, No. 5 (Sept. 1976), 313–17.
37. F. L. Becker, J. C. Crowe, V. L. Crow, T. J. Davis, B. P. Hildebrand, and G. J. Posakony, "Development of an Ultrasonic Imaging System for the Inspection of Nuclear Reactor Pressure Vessels," *Electric Power Research Institute*, EPRI RP 606-1 (Sept. 1977).
38. F. L. Thurstone and O. T. von Ramm, "A New Ultrasound Imaging Technique employing Two-Dimensional Electronic Beam Steering," in *Acoustical Holography*, Vol. 5, P. S. Green, ed. New York: Plenum Press, 1974, pp. 249–59.
39. R. D. Gatzke, J. T. Fearnside, and S. M. Karp, "Electronic Scanner for a Phased-Array Ultrasound Transducer," *Hewlett-Packard J.*, 34, No. 12 (Dec. 1983), 13–20.
40. J. T. Walker and J. D. Meindl, "A Digitally Controlled CCD Dynamically Focussed Phase Array," *1975 Ultrason. Symp. Proc.* (IEEE), 75 CHO 994-4SU, 80–83.
41. P. D. Corl, P. M. Grant, and G. S. Kino, "A Digital Synthetic Focus Acoustic Imaging System for NDE," *1978 Ultrason. Symp. Proc.* (IEEE), 78CH 1344-1SU, 263–68.
42. S. A. Johnson, J. F. Greenleaf, F. A. Duck, A. Chu, W. R. Samayoa, and B. K. Gilbert, "Digital Computer Simulation Study of a Real-Time Collection, Post-Processing Synthetic Focusing Ultrasound Cardiac Camera," in *Acoustical Holography*, Vol. 6, N. Booth, ed. New York: Plenum Press, 1975, pp. 193–211.
43. V. Schmitz and P. Höller, "Reconstruction of Defects by Ultrasonic Testing Using Synthetic Aperture Procedures, in *Quantitative Nondestructive Evaluation*, Vol. 4A, D. O. Thompson and D. E. Chimenti, eds. New York: Plenum Press, 1979, pp. 297–307.
44. J. R. Frederick, C. Vanden Broek, S. Ganapathy, M. Elzinga, W. De Vries, D. Papworth, and N. Hamano, "Improved Ultrasonic Nondestructive Testing of Pressure Vessels," U.S. Nuclear Regulatory Commission Progress Report NUREG/CR-0909 R5

(September 1979), Dept. of Mechanical Engineering, The University of Michigan, Ann Arbor, Mich.

45. P. D. Corl and G. S. Kino, "A Real-Time Synthetic-Aperture Imaging System," in *Acoustical Imaging*, Vol. 9, K. Y. Wang, ed. New York: Plenum Press, 1980, 341–55.

46. A. Papoulis, *Probability, Random Variables, and Stochastic Processes*. New York: McGraw-Hill Book Company, 1965.

47. H. C. Tuan, A. R. Selfridge, J. E. Bowers, B. T. Khuri-Yakub, and G. S. Kino, "An Edge-Bonded Surface Acoustic Wave Transducer Array," *1979 Ultrason. Symp. Proc.* (IEEE), 79CH1482-9, 221–25.

48. R. L. Baer, A. R. Selfridge, B. T. Khuri-Yakub, and G. S. Kino, "Contacting Transducers and Transducer Arrays for NDE," *1981 Ultrason. Symp. Proc.* (IEEE), Vol. 2, 81CH1689-9, 969–73.

49. G. S. Kino, D. K. Peterson, and S. D. Bennett, "Acoustic Imaging," in *New Procedures in Nondestructive Testing*, P. Holler, ed. New York: Springer-Verlag, New York, Inc., pp. 113–25.

50. W. B. Meredith and J. D. Massey, *Fundamental Physics of Radiology*. Bristol, England: John Wight & Sons, Ltd., 1977.

51. A. Macovski, *Medical Imaging Systems*. Englewood Cliffs, N.J.: Prentice-Hall, Inc., 1983.

52. K. Liang, B. T. Khuri-Yakub, C-H. Chou, and G. S. Kino, "A Three-Dimensional Synthetic Focus System," in *Acoustical Imaging*, Vol. 10, P. Alais and A. F. Metherell, eds. New York: Plenum Press, 1981, pp. 643–68.

53. D. Gabor, "A New Microscopic Principle," *Nature*, 161, No. 4098 (May 15, 1948), 777–78.

54. E. N. Leith and J. Upatnieks, "Wavefront Reconstruction with Continuous-Tone Objects," *J. Opt. Soc. Am.*, 53, No. 12 (Dec. 1963), 1377–81.

55. R. K. Mueller, "Acoustic Holography," *Proc. IEEE*, 59, No. 9 (Sept. 1971), 1319–35.

56. B. B. Brenden, "Real Time Acoustical Imaging by Means of Liquid Surface Holography," in *Acoustical Holography*, Vol. 4, G. Wade, ed. New York: Plenum Press, 1972, pp. 1–9.

57. E. E. Aldridge, A. B. Clare, and D. A. Shepherd, "Ultrasonic Hollography in Nondestructive Testing," in *Acoustical Holography*, Vol. 3, A. F. Metherell, ed. New York: Plenum Press, 1971, pp. 129–45.

58. B. P. Hildebrand and H. D. Collins, "Evaluation of Acoustical Holography for the Inspection of Pressure Vessel Sections," *Mat. Res. Stand.*, 12, No. 12 (Dec. 1972), 23–31.

59. L. W. Kessler and D. E. Yuhas, "Acoustic Microscopy—1979," *Proc. IEEE*, 67, No. 4 (Apr. 1979), 526–36.

60. A. Korpel, L. W. Kessler, and P. R. Palermo, "Acoustic Microscope Operating at 100 MHz," *Nature*, 232, No. 5306 (July 9, 1971), 110–11.

61. L. W. Kessler and D. E. Yuhas, "Structural Perspective," *In. Rev.*, 20, No. 1 (Jan. 1978), 53–56.

62. R. C. Eggleton and F. S. Vinson, "Heart Model Supported in Organ Culture and Analyzed by Acoustic Microscopy," in *Acoustical Holography*, Vol. 7, L. W. Kessler, ed. New York: Plenum Press, 1977, pp. 21–35.

63. Z. C. Lin, H. Lee, and G. Wade, "Scanning Tomographic Acoustic Microscope: A Review," *IEEE Trans. Sonics Ultrason.*, SU-32, No. 2 (Mar. 1985), 168–80.

64. C. H. Chou, B. T. Khuri-Yakub, and G. S. Kino, "Transmission Imaging: Forward Scattering and Scatter Reconstruction," in *Acoustical Imaging*, Vol. 9, K. Y. Wang, ed. New York: Plenum Press, 1980, pp. 357–77.

65. R. L. Powell and K. A. Stetson, "Interferometric Vibration Analysis by Wavefront Reconstruction," *J. Opt. Soc. Am.*, 55, No. 12 (Dec. 1965), 1593–98.

66. P. M. Morse and H. Feshbach, *Methods of Theoretical Physics*. New York: McGraw-Hill Book Company, 1953, Part II, Chapters 9–13.

67. C. F. Ying and R. Truell, "Scattering of a Plane Longitudinal Wave by a Spherical Obstacle in an Isotropically Elastic Solid," *J. Appl. Phys.*, 27, No. 9 (Sept. 1956), 1086–97.

68. J. E. Gubernatis, E. Domany, and J. A. Krumhansl, "Formal Aspects of the Theory of the Scattering of Ultrasound by Flaws in Elastic Materials," *J. Appl. Phys.*, 48, No. 7 (July 1977), 2804–11.

69. M. C. Junger and D. Feit, *Sound, Structures, and Their Interaction*. Cambridge, Mass.: The MIT Press, 1972.

70. J. D. Eshelby, "The determination of the Elastic Field of an Ellipsoidal Inclusion, and related Problems," *Proc. Roy. Soc.*, A, 241, No. 1226 (Aug. 20, 1957), 376–96.

71. Y. H. Pao and W. Sachse, "Interpretation of Time Records and Power Spectra of Scattered Ultrasonic Pulses in Solids," *J. Acoust. Soc. Am.*, 56, No. 5 (Nov. 1974), 1478–86.

72. A. Freedman, "A Mechanism of Acoustic Echo Formation," *Acustica*, 12, No. 1 (1962), 10–21.

73. J. D. Achenbach, L. Adler, D. K. Lewis, and McMaken, "Diffraction of Ultrasonic Waves by Penny-Shaped Cracks in Metals: Theory and Experiment," *J. Acoust. Soc. Am.*, 66, No. 6 (Dec. 1979), 1848–56.

74. J. D. Achenbach and A. K. Gautesen, "Geometrical Theory of Diffraction for Three-D Elastodynamics," *J. Acoust. Soc. Am.*, 61, No. 2 (Feb. 1977), 413–21.

75. R. Hickling, "Analysis of Echoes from a Solid Elastic Sphere in Water," *J. Acoust. Soc. Am.*, 34, No. 10 (Oct. 1962), 1582–92.

76. C. H. Chou, B. T. Khuri-Yakub, G. S. Kino, and A. G. Evans, "Defect Characterization in the Short-Wavelength Regime," *J. Nondestructive Evaluation*, 1, No. 4 (Dec. 1980), 235–47.

77. B. Budiansky and J. R. Rice, "On the Estimation of a Crack Fracture Parameter by Long-Wavelength Scattering," J. Appl. Mech., 45, No. 2 (June 1978), 453–54.

78. M. T. Resch, B. T. Khuri-Yakub, G. S. Kino, and J. C. Shyne, "The Acoustic Measurement of Stress Intensity Factors," *Appl. Phys. Lett.*, 34, No. 3 (Feb. 1, 1979), 182–84.

79. J. J. Tien, B. T. Khuri-Yakub, and G. S. Kino, "Long Wavelength Measurements of Surface Cracks in Silicon Nitride, in *Review of Progress in Quantitative Nondestructive Evaluation*, Vol. 1, D. O. Thompson and D. E. Chimenti, eds. New York: Plenum Press, 1982, pp. 569–71.

Chapter 4

Transversal Filters

4.1 INTRODUCTION

We are all familiar with the simple low-pass, high-pass, and bandpass filters formed out of resistances, capacitances, and inductances. Such filters are convenient for eliminating signals at unwanted frequencies, and for choosing signals in a desired frequency range, such as those used for television or radio.

As radar and communication techniques have developed, far more sophisticated applications of filtering techniques have become of interest. These can now be realized with *transversal filters*, which are basically tapped delay lines with each tap connected to a common input or output line. If we want to detect a signal that is weaker or lower in amplitude than an interfering noise signal, it is a great advantage to know the nature of the signal we want to find beforehand. For instance, a narrowband filter selects a particular narrowband signal and increases its amplitude relative to noise while eliminating interfering signals, thus improving the signal-to-noise ratio. This filter is a familiar example of one that operates in the frequency domain. If, instead, the signal is a digital code, a time-domain filter matched to this code, known as a *correlation filter*, will improve the signal-to-noise ratio. In general, such filters enable us to insert a long code into a matched transversal filter and obtain a narrow output pulse. Since the output signal contains the same amount of energy as the input signal, the peak amplitude of the pulse will be much higher than that of the input signal. Noise is treated differently because the filter is not matched to it; thus the signal-to-noise ratio is improved.

Another type of transversal filter removes distortion. If a television signal, for instance, suffers from ghost echoes because of reflections from buildings, it is possible to construct an *inverse filter* to reverse the distortion process and remove the unwanted echo. In a linear system, if the filter removes an unwanted echo from a known signal, such as the synchronizing pulse of a television signal, it will also remove unwanted echoes from an unknown television signal that has been distorted in the same way.

There are many more interesting examples. We will devote much attention to *frequency-modulated chirps* (FM chirps), which are signals of approximately constant amplitude whose frequency varies linearly with time over a limited frequency range. We can construct a matched filter for such signals to produce a narrow output pulse in the same way as we described for digital codes. Thus a filter can be used to emphasize the wanted chirp signal relative to noise, thereby improving the signal-to-noise ratio.

FM chirp filters have far more general applications. We shall show, for instance, that when a modulated chirp signal is injected into such a filter, its output is an instantaneous Fourier transform of the modulation. Thus such filters are very powerful tools for carrying out real-time Fourier transforms. This process in turn leads to a wide range of devices that can carry out Fourier transforms for various types of filtering operations, such as eliminating unwanted frequency components while retaining desirable ones.

One reason FM chirps are widely used in radar systems is because it is easy to produce a chirp signal of relatively long duration with the same amount of energy as a short, intense pulse. Another reason is that Doppler shift changes the frequency of the chirp. This change in frequency, which can be measured instantaneously with a chirp filter, indicates the target's velocity [1].

Sophisticated pseudorandom digital codes are similarly used in spread-spectrum communication systems. The very long code of a matched filter can improve the signal-to-noise ratio by many orders of magnitude, and because so many long codes are possible, intercepting an unknown code becomes extremely difficult.

It is impractical, of course, to design such sophisticated filters using only lumped circuits, for a sophisticated code consisting of many chips could require almost as many lumped elements for its construction. Such filters require a simpler construction technique, preferably one employing the photolithographic methods now common in integrated circuits. Fortunately, filters suitable for both analog and digital signal processing have become available in the last few years. The original *transversal filters* of this type were first demonstrated in the form of *surface acoustic wave devices* (SAW devices), which typically operate in the frequency range 10 MHz to 1 GHz [2, 3].

More recently, the same principles have been applied to the *charge-coupled device* (CCD) [4–6], the *single transfer device* (STD) [7], a type of switched capacitor filter, and the *bucket brigade device* (BBD) [8]. Other types of transversal filters using acousto-optic interactions have also been demonstrated. Research is under way on transversal filters that employ tapped fiber-optic delay lines [9–11] and tapped superconducting lines [12]; these systems should lead to devices with bandwidths of several gigahertz.

The *switched capacitor filter* is a related device [7]. In its original form it is not a transversal filter, and so is outside the scope of this book; however, the STD, a development of the switched capacitor filter, can be regarded as a type of transversal filter and will be described together with the CCD, BBD, and SAW devices.

For completeness, we shall give here a short description of the operation of the conventional form of the switched capacitor filter, which works on the principle that a capacitor switched rapidly on and off, as shown in Fig. 4.1.1, can simulate a resistor. Suppose that the capacitor in Fig. 4.1.1(a) is rapidly switched to the right. The charge flowing from V_2 is $Q = C(V_2 - V_1)$. If the switch is switched back and forth at a clock rate f_c, the current flowing is $I = Qf_c$. Hence the switched capacitor circuit behaves like a series resistor of value $R = 1/Cf_c$. Provided that the clock frequency is much higher than the signal frequency, we can use a circuit with two MOSFETs (metal-oxide semiconductor field-effect transistors), shown in Fig. 4.1.1(b), to synthesize the resistor. We can easily construct such a circuit by using a silicon integrated-circuit technology, together with operational amplifiers which are used as fundamental components for analog low-pass, high-pass, and bandpass filters at frequencies usually below 50 kHz.

In this section we briefly review some of the operating principles of transversal filters. Later we describe the various types of devices and their applications in more detail. We concentrate our attention on transversal filter types that either are now or will ultimately be suitable for radar or communication applications at frequencies above 100 kHz. We will also cover examples of lower-frequency signal processing applications that illustrate the state of the art as it stands at the time of this writing. We start with SAW devices for our illustrations because of their basic simplicity and flexibility, and because many applications of transversal filters were first demonstrated with SAW devices. Later we show the connection between these devices and the wide range of analog devices based on silicon technology that are now beginning to be widely employed. We will also briefly discuss some concepts for tapped fiber-optic and superconducting delay lines, on which research is now being conducted [9–12].

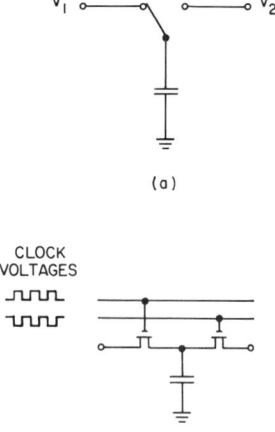

Figure 4.1.1 (a) Capacitor circuit switched to simulate a resistor. (b) MOS implementation of this circuit driven by a square-wave clock.

Surface acoustic waves provide a means for storage, delay, and complex parallel processing of long-duration wideband signals. It is possible to operate them at frequencies up to 500 MHz quite easily, and, with some care, up to 2 GHz, with bandwidths of as much as 30 to 40% of the center frequency. The corresponding data rates are as high as several hundred megabits per second. Certain types of these devices will store signals for as much as a second, and it is easy to construct tapped delay lines for surface acoustic waves. By arranging the spacing and the strength or weight of the taps correctly, it is possible to design broadband and narrowband filters, digital filters, and analog filters that recognize or generate particular signals. With further refinements, it is also possible to make programmable filters whose characteristics can be altered at will.

Analogous devices that can be constructed on silicon, such as CCDs, BBDs, and STDs, are at present more suitable for lower-frequency applications; with the development of gallium arsenide CCDs, however, the operating frequency is moving into the UHF range (100 MHz to 1 GHz). Silicon devices can carry out most of the signal processing functions provided by SAW devices, but in frequency ranges typically below 5 MHz, with correspondingly longer time delays. A major advantage of this technology is that the rate at which signals can be read into and out of these devices can be varied, so it is possible to expand or contract a given signal entering the device by reading it out at a different rate from that at which it was read in.

Interactions between two types of input signals are sometimes employed. For instance, because any semiconductor device can be made sensitive to light, it is possible to construct a light-sensitive CCD where a signal corresponding to the light intensity along the length of the device can be read out as one line of an optical image. By employing interactions of surface acoustic waves with semiconductors, it is possible to carry out Fourier transforms and other signal processing functions on the optical image.

Nonlinear interactions between two surface acoustic waves in SAW devices can lead to real-time processing functions in which the convolution or correlation of the two surface acoustic waves is obtained. Thus one signal can be used as a reference, instead of using a transversal filter with fixed properties. In this manner a transversal filter whose properties can be varied at will can be constructed merely by changing a reference code. A very similar device can also be constructed by using two tapped CCD delay lines connected to each other. One delay line controls the outputs from the taps of the other line; thus signals passed into one line act as a reference for signals passing through the other line [4].

Alternatively, we can utilize the interaction between a light beam and an acoustic wave to deflect the light beam, since the acoustic wave forms a type of optical diffraction grating. Such techniques have made it possible to devise devices that can carry out Fourier transforms or can correlate acoustic signals, either with each other or with a signal used to modulate a light beam. The basic principle of this device can also be used for deflecting a light beam to form a multiple-address system or an optical beam deflection system suitable for projection television systems.

The common feature of all these devices is the *transversal filter*, a delay line on which an acoustic or electrical signal propagates and then is accessed electrically, optically, or acoustically, at different points along it. These points form delay-line "taps." If we use a light beam or acoustic beam to tap the system, we can regard the taps as almost continuous; if we use electrical taps instead, we are essentially sampling the wave propagating along its surface. In either case, we can choose the amplitude and the phase weighting of the taps to perform a wide variety of signal processing functions.

4.2 LINEAR PASSIVE SURFACE WAVE DEVICES

4.2.1 Interdigital Transducer

The SAW filter has the advantage of being constructed by depositing thin films of metal on a piezoelectric substrate; thus the technology required is identical to that used for integrated circuits. This makes it possible to design filters with desirable characteristics that need none of the adjustments required by inductive or capacitive filters; these filters exhibit excellent characteristics over very long lifetimes and are also relatively simple and economical to manufacture.

These filters (whose basic structure is described here and in Sec. 4.2.2) employ a substrate of a piezoelectric material, such as quartz or lithium niobate, on which an interdigital transducer is deposited, as illustrated in Fig. 4.2.1. This transducer excites a surface acoustic wave whose fields fall off exponentially into the interior of the substrate, with a penetration depth on the order of one wavelength. At 40 MHz, with a wave velocity of approximately 3 km/s, the penetration depth of the waves is of the order of 75 μm. The interdigital transducer has neighboring fingers connected to the two bus bars (the input or output lines). Normally, the finger pair spacing (center-to-center spacing between pairs of fingers) is approximately one wavelength; for example, a 40-MHz television intermediate frequency (IF) filter has a finger pair spacing of 75 μm. Typical finger widths are approximately one-fourth of this value, or in this case 20 μm, so the devices can easily be manufactured by standard integrated-circuit technology.

As we shall see, devices of this kind can be tailored to yield a given amplitude and phase response over the frequency range of interest. They can also be designed to yield a given time-domain response, or even to respond to special analog or digital codes. Such devices can be made for operation anywhere in the range from 10 MHz to 2 GHz.

Digitally coded devices. It is very easy to forecast the time-domain response of a SAW transducer from simple geometrical considerations, and once the time-domain response of a filter is known, it is again easy to determine its frequency response by Fourier transform techniques. Thus we can construct transversal filters with a given frequency response quite simply, once we know the equivalent time-domain response.

We first consider physically the simple example of a digitally coded filter,

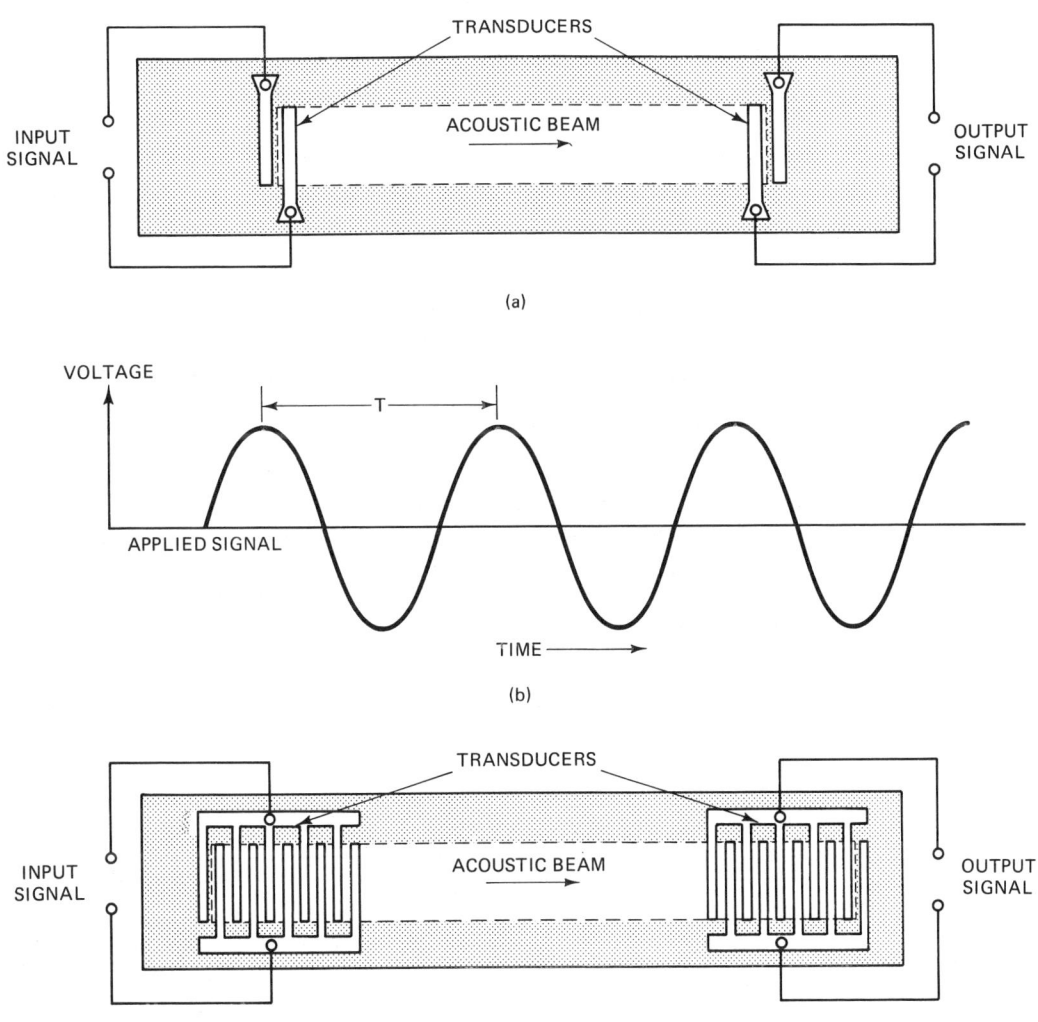

Figure 4.2.1 Excitation and detection of Rayleigh waves can be accomplished by using a simple transducer consisting of two metal electrodes deposited on a piezoelectric crystal (a). When a sinusoidal electric signal (b), which repeats itself in time T, is applied to the input electrodes, the alternating electric field sets up alternating vibrations in the piezoelectric material that give rise to Rayleigh waves. When the waves reach the output electrodes, they generate an alternating voltage between the two metal fingers. If the input electrodes have a uniform interdigital pattern (c), the separate waves excited by each pair of fingers will reinforce one another if the time required for the Rayleigh wave to travel between electrode pairs corresponds to the frequency of the electrical signal. If the output electrodes have the same spacing as the input electrodes, they will be "tuned" to receive the passing acoustic waves. (After Kino and Shaw [13].)

used both to produce given digital codes and as a matched receiver for a particular digital code. We then demonstrate mathematically that the geometry of the transducer is directly related to the time domain response of the filter.

Sec. 4.2   Linear Passive Surface Wave Devices

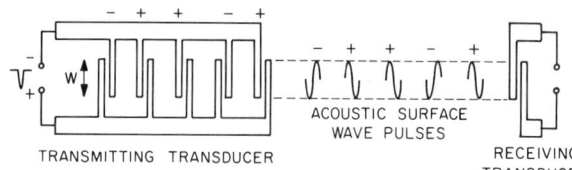

Figure 4.2.2 Digitally coded SAW filter.

Consider the interdigital transducer illustrated in Fig. 4.2.2. Metal fingers are laid down on a piezoelectric substrate and are connected to one of two input bus bars. A simple two-finger pair receiving transducer is laid down on the same substrate. When a voltage is applied across the two input bus bars, charges of opposite sign are induced on the two fingers. These charges, in turn, excite surface acoustic waves in the piezoelectric material because of the electric fields associated with them. We expect the amplitude of the surface acoustic wave signal excited by a metal finger to be proportional to the charge on the finger, or to the normal component of electric field at the surface of the finger.

In the configuration shown, the first pair of fingers on the right-hand side of the transmitting transducer may be regarded as positively polarized, and each finger pair may be regarded as a capacitor. Thus when a voltage is applied to the transducer, the charge on the finger will be proportional to the applied voltage. The charge on the first finger excites a positive acoustic pulse, followed by a negative pulse excited by the charge of opposite sign on the second finger. We note that the second pair of fingers is connected to the bus bars of opposite sign from the first pair. The first RF acoustic pulse travels down the delay line in the form of a single-cycle tone burst; it is followed by a second similar pulse of opposite sign, excited by the second pair of fingers, whose polarity is reversed from the first pair. In turn, each pair of fingers excites an RF pulse, so that a coded series of pulses, determined by the geometry and connections of the fingers, is excited on the SAW delay line. This digitally coded series of RF pulses may be detected at the receiving transducer, and the transversal filter yields a particular coded output, $+ - + + -$, determined by its geometrical configuration. This is known as a *biphase digital code*. Each element of the code, an RF pulse or tone burst, is called a *chip*.

The same device can be used as a coded receiver. In this case the charge excited on a finger of the output transducer is proportional to the amplitude of the acoustic signal passing underneath it, and the voltage output from the transducer is proportional to this charge. When an electrical signal $- + + - +$, corresponding to the time-reversed version of this digital code, is inserted into the left-hand transducer, as illustrated in Fig. 4.2.2, the first element of the code emitted from the left-hand pair of fingers will be a negative RF electrical pulse injected into a transducer of negative sign; thus it will produce a positive RF acoustic pulse. It reaches the receiving transducer at the same time the second element of the code is received from the second finger pair. As the input code matches the transducer code, all elements of the code reach the corresponding parts of the receiving transducer simultaneously and yield signals of the same sign. These signals add to give a pulse five times the amplitude of an individual RF pulse from a single-finger pair.

If a different signal, having a different sequence of positive and negative RF pulses, is introduced into the delay line, there will be no instant at which it will match the polarities of all the tapping transducers. In other words, there is very little correlation between the signal and the filter response; therefore, the peak output signal will be reduced. This is a simple example of the ability of transversal filters to perform pattern recognition and to select a signal with a given code from all other signals. A so-called orthogonal set of codes is defined as one for which the output is of unit amplitude when the codes are not matched, and of N units for matched codes N chips long.

More generally, it will be seen that the minimum output pulse width τ is determined by the bandwidth B of the system, and is approximately $\tau = 1/B$. If T is the length of the coded pulse train, τ will be the length of one chip, or of the pulse of minimum length (in our example, this is the pulse corresponding to the signal from one finger pair or one chip). The ratio of the length of the input pulse to that of the output pulse is then T/τ, or approximately TB. This is the product of the bandwidth and the time delay of the filter, the *time–bandwidth product*, or, equivalently, a parameter known as the *pulse compression ratio*. The magnitude of the output pulse is increased over that of the input pulse by the same ratio (see Prob. 5); this is referred to as the *processing gain*. We note that for a digital code, the pulse compression ratio is N times the number of chips in the code, and $N \approx TB$.

A useful way of looking at this idea of pulse compression is to consider a digitally coded input signal N chips long, each of value ± 1, to a matched filter in the presence of a random code (noise). The output from the matched filter will only be one chip of time τ long. Thus its amplitude will be increased by a factor N and the peak output power by a factor of N^2. A very long random noise signal—an uncorrelated digital code, for instance—will not be correlated with the matched filter response, and the amplitudes of the individual chips will not necessarily add. On the other hand, the noise power output will still be N times as large as the power output from a single chip of a purely random code or noise signal, because it is being received by an N-chip-long filter and the powers are added at the individual elements of the filter. Thus the average noise power output from the random code will be increased by a factor N over that from a single chip, which will increase the peak signal-to-noise power ratio at the output by a factor N^2/N, or by the compression ratio $N = T/\tau$. More rigorous proofs of this property of matched filters are given in Secs. 4.4.2 and 4.4.3.

Experimental input and output signals from a tapped delay line of this type are shown in Fig. 4.2.3. This delay line has 127 taps, spanning a time delay of 25.4 μs, and the pulse compression ratio is approximately 127. In this system we produce a digital code by sending a short pulse into a coded delay line. The receiver delay line must have a time-reversed response to that of the input delay line. Therefore, it is usually called the *conjugate filter*.

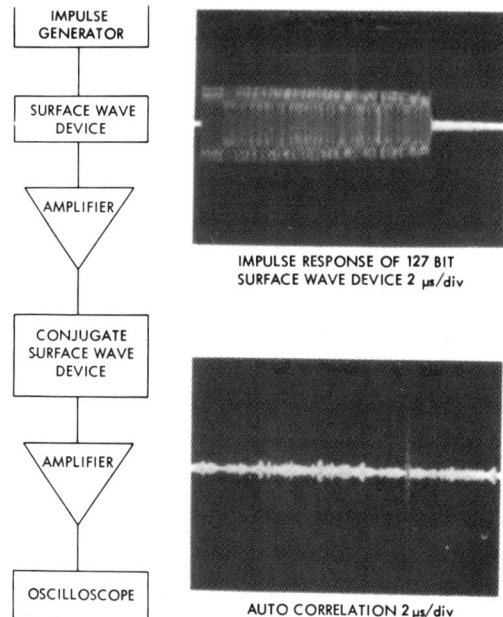

Figure 4.2.3 Impulse response and autocorrelation response obtained with a 127-chip PN-coded SAW device, using the system shown.

Mathematical model of the transducer. We now consider the response of an interdigital transducer using a simple mathematical formulation that is often called the *impulse model* [15]. As we shall see, this model applies in an almost unchanged form to CCD analog delay line filters. We shall use the symbols t' and z' to apply to the source, and t and z to apply to the observation point.

Suppose that at time t', a voltage $V(t')$ is applied to an electroded region between z' and $z' + dz'$ on the surface of the piezoelectric substrate, as shown in Fig. 4.2.4. Because of the capacity of this element to other electrodes or ground, the input voltage will induce a charge $\sigma(t', z') \, dz'$ proportional to the voltage $V(t')$ applied to this element. We might also expect the charge $\sigma(t', z')$ to be proportional to the width $w(z')$ in the x direction of the electroded region. The acoustic wave signal at the plane z at time t is defined as having an amplitude $a(t, z)$, where the power of the acoustic wave is $P = |a|^2$. This signal reaches the plane z from the element between z' and $z' + dz'$ after a delay time $(z - z')/V_R$, where V_R is the Rayleigh wave velocity. In a linear medium, the acoustic amplitude $a(t, z)$ must be proportional to the excitation charge $\sigma(t', z')$. The total acoustic signal at the plane z, $a(t, z)$, must therefore be of the form

$$a(t, z) = \alpha \int_{-\infty}^{z} \sigma\left(t - \frac{z - z'}{V_R}, z'\right) dz'$$
$$= \alpha \int_{-\infty}^{z} \sigma(t', z') \, dz' \quad (4.2.1)$$

where α is a coupling constant independent of the value of $\sigma(t', z')$, $t' = t - (z - z')/V_R$, and $\sigma(t', z')$ is the charge density induced at the plane z' at a time t'.

If we take $\sigma(t', z') = \sigma(z')\sigma(t' - t_0)$, which corresponds to a very short input voltage pulse at $t' = t_0$, we find that

$$a(t, z) = \alpha \int_{-\infty}^{\infty} \sigma(z')\delta\left(t - t_0 - \frac{z - z'}{V_R}\right) dz' \qquad (4.2.2)$$

and

$$a(t, z) = \alpha\sigma[z - V_R(t - t_0)] \qquad (4.2.3)$$

where we have assumed that $\sigma(z') = 0$ for $z' > z$, and $\delta(t)$ is the Dirac delta function. Thus the impulse response $a(t, z)$ of the transducer is proportional to the charge distribution $\sigma(-V_R t + \text{constant})$ along it, where $z' = z - V_R(t - t_0)$. Note that as t is increased, the value of z' decreases; in fact, the impulse response $a(t, z)$ depends directly on $\sigma(-z')$.

This property is very useful because it enables us to design a transducer for a given frequency response merely by spacing or shaping its fingers to give a certain impulse response. We now have only a problem of simple geometrical design.

We shall assume, for simplicity, that the charge distribution is uniform over each finger, although it is fairly easy to modify this assumption to account for a charge distribution closer to reality, like the electrostatic charge distribution of a periodic array of strips. We shall define $\sigma(z')$ to be finite only where there are metal fingers.

More generally, if the finger width or position is altered, and the input voltage is $V(t')$, we can write

$$\sigma(t', z') = \mu V(t')w(z') \qquad (4.2.4)$$

where μ is a constant.

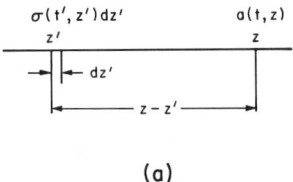

(a)

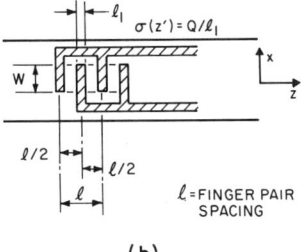

(b)

Figure 4.2.4 Notation used (a) to estimate the response of a transducer, and (b) for the finger pair spacing and width.

We shall call $w(z')$ the *finger weighting*. Typically, $w(z')$ is proportional to the finger width w, but changes sign depending on how the finger is connected to the external source.

It follows from Eqs. (4.2.1) and (4.2.4) that

$$a(t, z) = \kappa \int_{-\infty}^{z} V\left(t - \frac{z - z'}{V_R}\right) w(z') \, dz' \tag{4.2.5}$$

where $\kappa = \alpha\mu$ is a constant. The response to a short pulse at time $t' = 0$, which may be represented as $V(t') = \delta(t')$, is

$$h(t) = \kappa \int_{-\infty}^{z} \delta\left(t - \frac{z - z'}{V_R}\right) w(z') \, dz' \tag{4.2.6}$$

or

$$h(t) = \kappa w(z - V_R t) \tag{4.2.7}$$

If we put $t' = t - (z - z')/V_R$ in Eqs. (4.2.5) and (4.2.7), it follows that

$$a(t, z) = \kappa V_R \int V(t') w[z - V_R(t - t')] \, dt' \tag{4.2.8}$$
$$= V_R \int V(t') h(t - t') \, dt'$$

We may compare Eq. (4.2.5) or (4.2.8) with the usual forms for correlation and convolution:

$$\text{Correlation:} \quad s(t) = \int g(t + \tau) f(\tau) \, d\tau = g(t) * f(t) \tag{4.2.9}$$
$$\text{Convolution:} \quad s(t) = \int s(t - \tau) f(\tau) \, d\tau = g(t) * f(t)$$

It will be seen that the signal induced in the acoustic delay line is either the *correlation* of the transducer weighting in the $+z$ direction with the input signal, or the *convolution* of the transducer weighting in the $-z$ direction with the input signal. This property makes the SAW device a powerful tool for signal processing. To understand it in more physical terms, the reader is referred to the example of digitally encoded transducers discussed earlier in this section.

The frequency response of the transducer is the Fourier transform of its time-domain response. Thus we can write

$$A(\omega, z) = \mathcal{F}[a(t, z)] = \int_{-\infty}^{\infty} a(t, z) e^{-2j\pi ft} \, dt \tag{4.2.10}$$

where the symbol $\mathcal{F}$ denotes the Fourier transform, and $f = \omega/2\pi$ is the frequency.

The frequency response $A(\omega, z)$ of the excited acoustic signal will be the product of the frequency response of the input signal $V(\omega)$ and the frequency response of the transducer $H(\omega)$:

$$A(\omega, z) = V(\omega) H(\omega) \tag{4.2.11}$$

It will be seen from Eq. (4.2.7) that $H(\omega)$ is given by the relation

$$H(\omega) = \kappa \int_{-\infty}^{\infty} e^{-j\omega t} w(z - V_R t) \, dt \qquad (4.2.12)$$

Putting $t = (z - z')/V_R$, it follows that

$$H(\omega, z) = \gamma e^{-j\omega z/V_R} \int_{-\infty}^{\infty} e^{j\omega z'/V_R} w(z') \, dz' \qquad (4.2.13)$$

where $\gamma = -\kappa/V_R$. We have now obtained the important relation that *the frequency response of the transducer is the Fourier transform of the finger weighting*.† For a more direct derivation of this relation, see Sec. 2.5.

4.2.2 Bandpass Filter

Uniform transducer. We shall consider the example of a standard uniform transducer (Fig. 4.2.1) with constant finger length and uniform finger pair spacing, and derive how its response varies with frequency. Suppose that the fingers have a pair spacing of l and a finger width l_1, and that the charge on every other finger is Q and $-Q$, respectively. We shall assume that l_1 is very small (i.e., $kl_1 \ll 1$), where $k = \omega/V_R = 2\pi/\lambda$ and λ is the wavelength of the acoustic wave. Thus each finger is taken to be very narrow in comparison to a wavelength. We shall also assume, for simplicity, that σ is uniform over a finger, so that $\sigma l_1 = \pm Q$. With the correct normalization, since we are only interested in the relative amplitude of the response as a function of frequency, we can put $w(z') = \pm 1/l_1$. It follows from Eq. (4.2.13) that the contribution to the response $H(\omega)$ by a pair of fingers spaced $l/2$ apart, with the center of the pair at $z' = nl$, and the centers of the fingers at $z' = nl + l/4$ and $z' = nl - l/4$, respectively, is

$$H(\omega) = \frac{\gamma}{l_1} \left(-\int_{nl-l/4-l_1/2}^{nl-l/4+l_1/2} e^{jkz'} \, dz' + \int_{nl+l/4-l_1/2}^{nl+l/4+l_1/2} e^{jkz'} \, dz' \right) \qquad (4.2.14)$$

or

$$H(\omega) = \gamma(e^{jkl/4} - e^{-jkl/4}) \frac{\sin(kl_1/2)}{kl_1/2} e^{jknl} \qquad (4.2.15)$$

where we have omitted the $\exp(-j\omega z/V_R)$ term. This is equivalent to defining the response at the plane $z = 0$. When $kl_1 \ll 1$, it follows that

$$H(\omega) = 2j\gamma \sin \frac{kl}{4} e^{jknl} \qquad (4.2.16)$$

---

†This is actually the inverse Fourier transform of $w(z)$. However, it is the Fourier transform of $w(-z)$, or the finger weighting in the direction of decreasing z.

The contribution from the N-finger pair of the transducer is therefore the sum of the contributions from the individual finger pairs, or

$$H(\omega) = 2j\gamma \sin \frac{kl}{4} \sum_0^{N-1} e^{jknl}$$

$$= 2j\gamma \sin \left(\frac{kl}{4}\right) \frac{\sin(kNl/2)}{\sin(kl/2)} e^{jk(N-1)l/2} \quad (4.2.17)$$

$$= j\gamma \frac{\sin(kNl/2)}{\cos(kl/4)} e^{jk(N-1)l/2}$$

The parameter $H(\omega)$ is the Fourier transform of the charge distribution on the fingers.

The center frequency of the transducer $\omega = \omega_0$, $k = k_0$ is where $kl = 2\pi$ and the wavelength is $\lambda = l$. Using l'Hospital's rule on Eq. (4.2.17), it follows that $|H(\omega_0)| \to 2N\gamma$. The response is zero where $kl = 2\pi[1 \pm (1/N)]$; thus the bandwidth is narrowed as N increases. The bandwidth between the zeros in the response is given by the relation

$$\frac{\Delta\omega(\text{zero})}{\omega_0} = \frac{2}{N} \quad (4.2.18)$$

It is convenient to normalize Eq. (4.2.17) in terms of kNl, so that in the normalized form all transducers have the same response and have a maximum response when the normalized parameter is zero. We write

$$x = \frac{N\pi(k - k_0)}{k_0} = \frac{N\pi(\omega - \omega_0)}{\omega_0} \quad (4.2.19)$$

In this case, if $|\omega - \omega_0| \ll \omega_0$, we can see that when $x/N \ll 1$, $\cos kl/4 \to -x/2N$ and

$$|H(x)| \approx 2\gamma N \left|\frac{\sin x}{x}\right| \quad (4.2.20)$$

Thus all transducers have the same normalized response. It follows that the total bandwidth to the 4-dB points ($x = \pm\pi/2$) is

$$\frac{\Delta\omega(4\text{ dB})}{\omega_0} = \frac{1}{N} \quad (4.2.21)$$

The 3-dB points occur where $\Delta\omega(3\text{ dB})/\omega_0 = 0.89/N$. This behavior is illustrated in Fig. 4.2.5.

Apodized transducer. It is apparent that if we taper the length of the fingers and their spacing, we can tailor the response to any given characteristic. Such variation of the finger length is called *apodization*. For example, with a uniform finger spacing, but the finger lengths varying as $|\sin \mu z'/\mu z'|$ and the finger polarity depending on the sign of $\sin \mu z'$, the total charge on the finger will vary

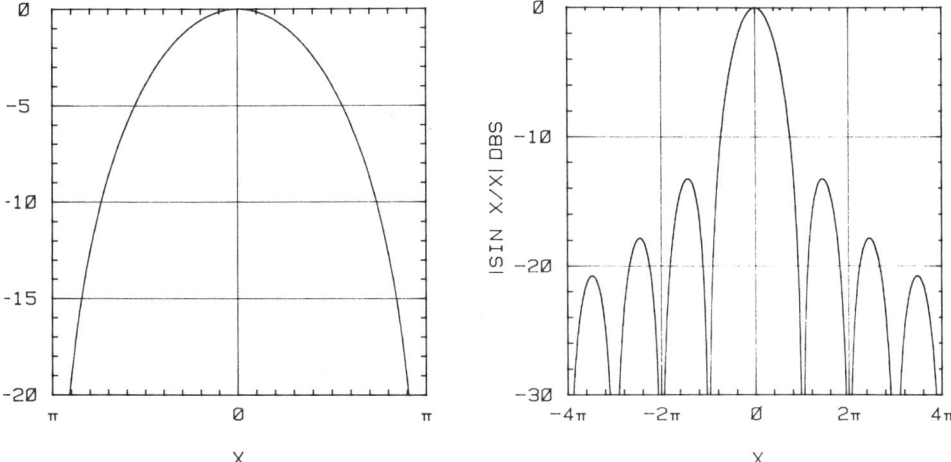

Figure 4.2.5 Plots of $|(\sin x)/x|$ in decibels.

as $\sin \mu z'/\mu z'$. The transducer output will be the Fourier transform of the finger excitation, and the result will be a square-topped frequency response. The difficulty, of course, is that the transducer length L must theoretically be infinite. Furthermore, the fingers at each end of the transducer must be progressively shorter in length, and acoustic wave diffraction effects then become dominant.

An illustration of how filters with such a configuration are constructed is given in Fig. 4.2.6(a) and (b). The simple apodized transducer shown in Fig. 4.2.6(a) can yield a distorted response, for the velocity of a surface acoustic wave passing under a metal finger will be slightly different from the velocity in an unmetalized region of the piezoelectric delay line; this causes the acoustic beam to have phase distortion across its width. One way to eliminate this distortion is to place dummy fingers in the gap that are connected to the bus of the same polarity as the neighboring fingers. Then all parts of the acoustic beam suffer the same phase delay,

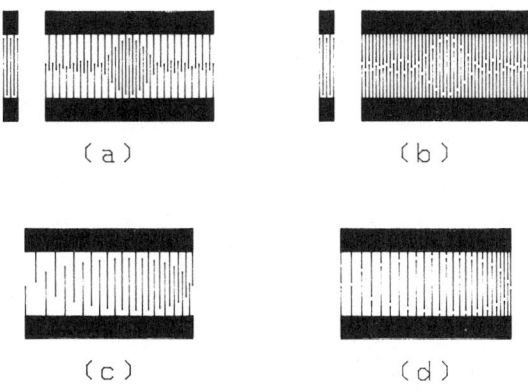

Figure 4.2.6 (a) Nondispersive bandpass filter configuration: one transducer is unapodized and one transducer is apodized with a $\sin x/x$ variation; (b) similar filter corrected with "dummy" electrodes to produce straight-crested surface waves (i.e., waves with no phase distortion across their width due to the presence of fingers); (c) electrode configuration of apodized dispersive transducer with uncorrected comb; (d) electrode configuration of apodized dispersive transducer with corrected comb.

and we obtain a beam with straight wavefronts, a *straight-crested wave*. A dispersive filter with variable finger spacing is shown in Fig. 4.2.6(c) and (d). This approach, taking into account such modifications as the charge distribution over a finger, diffraction, and weak reflections from the fingers, is the basis of design for interdigital transducer filters.

4.2.3 FM Chirp Analog Filter

Another important example of signal processing within an SAW delay line is one that uses an analog signal of the type illustrated in Fig. 4.2.7. The signal shown is an FM chirp signal whose amplitude is constant, but whose instantaneous frequency varies linearly with time. The finger spacing of the transducer array is varied along its length to match the frequency variation across the chirp (i.e., to correlate with it). The left-hand end of the array responds to the higher frequencies and the right-hand end to the lower frequencies. At the instant shown, the chirp signal, which is traveling to the right, registers exactly with the array, much as we have observed for the digital filter. An intense output pulse is obtained at the right-hand terminals. This is a dispersive filter in which the low-frequency end of the signal is delayed more than the high-frequency end, allowing the trailing edge of the long input pulse to catch up with the leading edge, thus collapsing the pulse.

Pulse compression techniques of this type are of great importance in a variety of systems. Perhaps the best known examples are radar systems using pulse compression in which the signal transmitted from the radar is an FM chirp; after returning from a target, it is passed through a pulse compression filter which compresses it into a short pulse [1]. In this way, it is possible to use a long pulse containing large amounts of energy for long-distance ranging, and compress it in a receiver into a short intense pulse for accurate timing and range resolution.

In secure radar and communications systems, an FM chirp represents one way of coding a signal. The listener needs to know the code (i.e., the chirp rate) and needs to have a chirp compression filter which operates at this rate in order

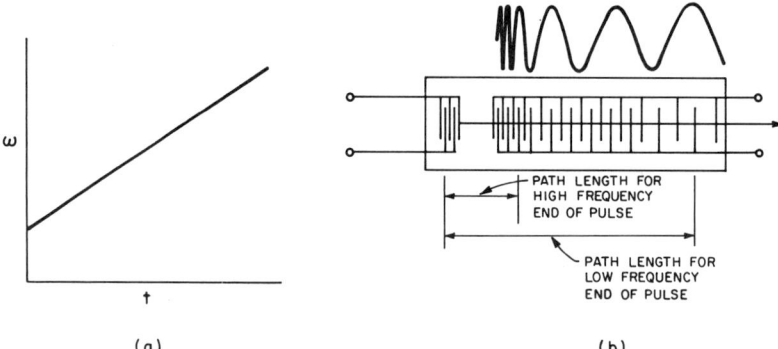

Figure 4.2.7 FM chirp pulse compression filter: (a) chirp frequency variation as a function of time; (b) schematic representation of the SAW filter.

to receive the signal. Indeed, the signal can be below the thermal or background noise level and, when received by a compressive receiver containing a filter matched to its chirp rate, can still be extracted from the background noise level because of the improvement in signal-to-noise ratio resulting from pulse compression. On the other hand, an ordinary receiver would not be able to detect the presence of the signal.

Surface acoustic waves fit naturally into this picture in applications requiring high chirp rates (i.e., a high rate of change of frequency versus time across the chirp), together with a large total frequency excursion (bandwidth). Compression ratios in the range 1000 to 10,000 can be reached with SAW systems [2]. SAW pulse compression filters have been built with bandwidths exceeding 500 MHz and with time delays (chirp lengths) of the order of a microsecond. The range resolution of a radar system is the inverse of its bandwidth, and this bandwidth (500 MHz) corresponds to a target range resolution capability of 1 ft (0.31 m). Far larger time delays have been achieved (more than 100 μs) at smaller bandwidths.

Fourier transform chirp filters are also applicable to nonscanning real-time spectrum analyzers. In fact, they are capable of calculating the complete complex Fourier transform of an arbitrary incoming analog signal. Consider the chirp signal illustrated in Fig. 4.2.8, whose frequency is $\omega = \omega_0 + \mu t$, and thus varies linearly with time. Suppose now that the chirp is modulated by a sinusoidal signal $\sin \Omega t$. This will introduce two sidebands of frequencies $\omega = \omega_0 + \mu t \pm \Omega$, in addition to the carrier of frequency $\omega = \omega_0 + \mu t$. Thus the system behaves as if there are three identical chirps delayed from each other by times Ω/μ. When these chirps are inserted into the correct matched dispersive delay line, three output pulses will be obtained. The differences in their time delays will indicate their frequency differences. More generally, an arbitrarily modulated chirp, when inserted into a chirp filter matched to the carrier, gives an output that is the Fourier transform of the modulated signal.

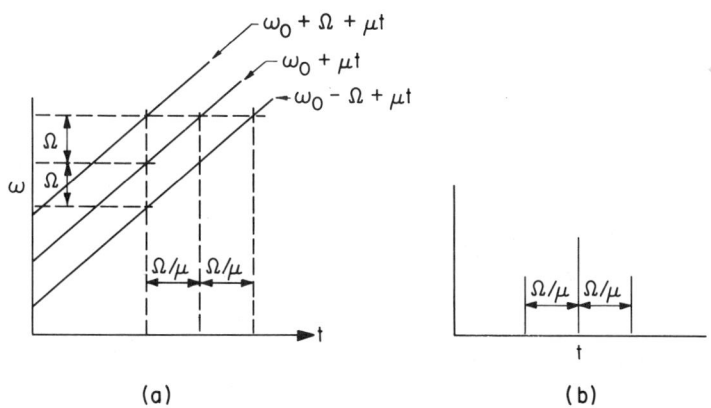

Figure 4.2.8 (a) Chirp modulated by a sinusoidal signal; (b) output from the matched filter.

4.2.4 Resonators

The conducting electrodes used to form interdigital arrays cause small reflections of surface waves which can lead to spurious low-level signals. In designing long filters, these reflections must be limited to specified levels. On the other hand, it is possible to use the reflection phenomenon constructively, by forming an SAW resonator from an array of reflectors [16, 17]. These are useful for electronic circuits, where it is advantageous to have very small, inexpensive, accurate, and pretuned resonators that can be mass-produced by photolithographic methods. Arrays of isolated parallel conducting strips deposited on the surface, as shown in Fig. 4.2.9, can behave as efficient reflectors for surface waves if a large number of strips (typically hundreds) are used, and if they are spaced so that the reflected waves from individual strips reinforce each other. As shown in Fig. 4.2.9(b), this occurs when the strips are $\lambda/2$ apart, for the waves reflected from two neighboring strips will trace distances differing by λ, and will therefore be in phase. Thus all the waves reflected from the strips will be in phase when the strips are spaced $\lambda/2$ apart.

In a surface wave resonator, the surface waves are trapped between the two reflectors of Fig. 4.2.9(a), making multiple transits between them and creating a standing wave, like electromagnetic waves in a cavity resonator or optical waves in a Fabry–Perot interferometer. Interdigital transducers are used to couple this resonant standing wave to an external electrical circuit. Grooves etched into the substrate can also be used effectively to form the reflective gratings.

Values of resonant Q ($Q = f_0/\Delta f$, where f_0 is the resonant frequency and Δf is the 3-dB bandwidth) as high as 20,000 have been achieved, with such resonators operating at frequencies in the range from 50 MHz to several gigahertz. The maximum Q is limited by the finite reflectivity of the arrays, for the propagation loss on quartz substrates (at least in the lower-frequency part of the range) is small enough to allow approximately one more order-of-magnitude increase in Q. As we might expect, these resonators can also be used as circuit elements to build ladder networks and other types of classical filter networks.

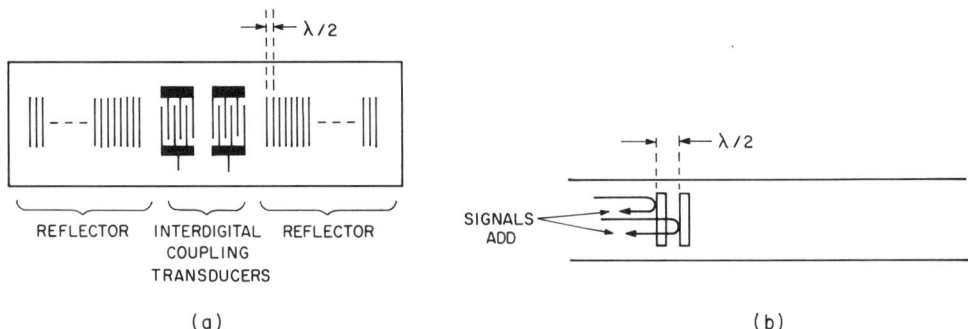

Figure 4.2.9 (a) SAW resonator; (b) waves reflected from neighboring fingers.

4.2.5 RAC Filters

The *reflective array compressor* (RAC) is closely related in its principle of operation to the surface wave resonator [18–20]. This device is an alternative form of transversal filter in which arrays of reflecting parallel grooves or metal strips on the delay-line surface perform the function of tapping, which is normally performed by interdigital electrode arrays. Just as with the resonator, we take advantage of the reflections resulting from the grooves or strips, instead of regarding them as a difficulty to be eliminated. The key idea is that a groove acts as a tap for the surface acoustic wave, because the portion of the surface acoustic wave reflected from the groove can be collected elsewhere by an interdigital transducer. The basic system is shown in Fig. 4.2.10: a surface acoustic wave is excited by transducer A and reflected by the two sets of grooves back to transducer D.

We can understand the operation of such devices by considering two extreme surface acoustic wave paths. A signal from the interdigital transducer A will be partially reflected into a surface acoustic wave traveling from B to C, where it will again be partially reflected and travel to interdigital transducer D. If rays reflected from the two sets of neighboring grooves in the two reflective arrays have a total difference in path length of λ, the acoustic wavelength, the effect will be strong. In this case, the rays traveling from B to C are in phase. The total time delay for this signal will be proportional to the path length $ABCD$. Thus if the spacing between the grooves is tapered, high-frequency signals will be reflected strongly from the front region of the grating array. However, a signal of lower frequency will follow the path $AEFD$ and experience a longer time delay. Thus this device has the same type of dispersion characteristic, with time delay as a function of frequency, as the FM chirp interdigital structure of Fig. 4.2.7, and can be used to perform the same functions. The folded paths in Fig. 4.2.10, however, give twice the time delay for a given substrate length, and chirp pulse compression filters with time delays exceeding 100 μs can be achieved.

Reflective arrays are less defect sensitive than interdigital arrays. They have an additional advantage for very high frequency operation, because the grooves in reflective arrays are generally spaced by the order of a half wavelength, compared to the quarter-wavelength spacing, which is more characteristic of split-finger, nonreflecting, interdigital arrays (see Prob. 3), the dimensional requirements are less stringent; furthermore, internal reflections no longer present a problem, so very long arrays may be made. This system has a major advantage over one using a very long interdigital transducer, because the many small reflections inherent in such transducers tend to build up and make it impossible to obtain the characteristics

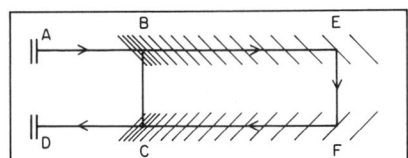

Figure 4.2.10 Reflective array compressor.

predicted by theory. RAC delay lines have been made with compression ratios of as much as 10,000 to 1 [18–20]; these results are probably a factor of 10 better than those obtained by any interdigital array system.

4.2.6 Stabilized SAW Oscillators

Another device of great importance is the SAW oscillator, which consists of a standard amplifier whose output is fed back to its input through a surface wave delay line, as illustrated in Fig. 4.2.11. This forms an oscillator operating in the UHF range (100 to 1000 MHz), whose frequency and frequency stability are determined by the properties of the SAW delay line crystal. This oscillator is simpler and cheaper to make than alternative approaches producing highly stable signals in this frequency range, such as relatively low frequency quartz crystal–controlled oscillators followed by multiplier chains.

The key point is that the delay line can be thousands of wavelengths long between the input and output transducers because of the very low propagation velocity of acoustic waves. As is well known, the frequency of an oscillator with an external feedback loop adjusts itself so that the phase shift around the loop is $2N\pi$, where N is an integral number. The larger the value of N, the better the short-term stability of the oscillation frequency. Both the basic frequency selectivity associated with the length of the delay-line path and the frequency-filtering characteristics of interdigital transducers can be brought into play. The stability of the delay-line crystal gives high stability for any of a number of different longitudinal modes (different values of N) with different center frequencies, and the interdigital transducers are used to select one oscillation mode from the entire set possible.

Methods of compensating for those phase shifts within the oscillator that result from voltage and temperature variations have been demonstrated. These methods make it possible to come very close to the ultimate frequency stability of the delay line itself. Techniques have also been devised to tune the voltage of the oscillator frequency over a small range by changing the phase in the external loop, for use in tracking or frequency modulation applications. Frequency synthesizers, which can be programmed to operate at any of a number of equally spaced frequencies with high short-term stability, have been demonstrated as well [22, 23].

At the time of writing, stable oscillators operating in the frequency range 100 MHz to 4 GHz have been demonstrated. Another class of SAW oscillators employs the SAW resonator; this form of oscillator, with an internal feedback loop, can be as useful as one with an external feedback loop.

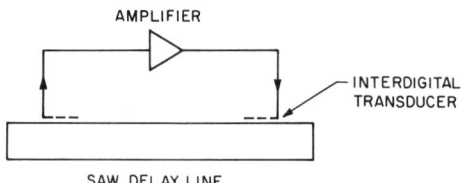

Figure 4.2.11 SAW oscillator with an external feedback loop. (From Crabb et al. [21].)

PROBLEM SET 4.2

1. (a) Work out the response of an N-finger-pair interdigital transducer with fingers of finite width l_1 (kl_1 finite) and finger pair spacing l. Assume that the charge distribution on the fingers is uniform. The result should be a slightly more general version of Eq. (4.2.17).
   (b) Show that if $l_1 = l/4$, the response at the third harmonic $\omega = 3\omega_0$ is zero.

   Note: In practice the charge density on the fingers is not uniform, so the third harmonic response is finite.

2. An acoustic resonator uses N equally spaced strips as reflectors. Let each strip be of width d and their spacing be l. Suppose that we take the reflected wave from the nth strip to be of amplitude b and the incident wave, of amplitude a. Suppose also that each strip is of negligible width d ($kd \ll 1$), and that the reflection coefficient of each strip is $\Gamma = b/a$. Assuming, for simplicity, that $a_{n+1} = a_n \exp(-jkl)$, where $k = \omega/V_R$ (i.e., the forward wave is unperturbed, with a similar phase delay for the reflected waves). Find the total reflection coefficient from N strips. At what freqeuncy is this $N\Gamma$? Your result should be similar in form to Eq. (4.2.17). By using the same arguments as given after Eq. (4.2.17), find the 4-dB bandwidth of the reflected wave.

3. (a) Following the analysis of Prob. 2, modify the theory to take account of fingers of finite width d. Assume that k is unchanged, both under the fingers and in the region in between. Assume that the reflection coefficient of the wave from the back edge of the finger is Γ and from the front edge, $-\Gamma$ (this assumption can be justified from transmission-line analogs, where the region under the strip is taken to have an impedance different from that between the strips).
   (b) At what frequency is the total reflected wave amplitude identically zero?
   (c) Note that an interdigital transducer is normally operated with a finger spacing of approximately $\lambda/2$, where the reflection from the fingers is maximum. Suggest a configuration in which each finger is split down its middle with a gap between its two halves (both connected to the same bus bar) and which eliminates the reflection problem.

4. An SAW oscillator is made with the configuration shown below. Two transducers with M and N finger pairs, respectively, each with a finger pair spacing l, are used. The centers of the two transducers are a distance L apart, as shown in the figure. The amplifier should be taken to be an inverter that gives a π phase shift.

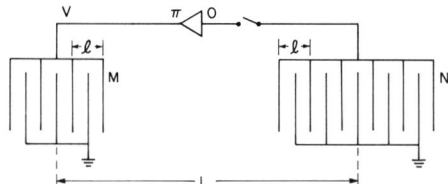

(a) If the M-finger transducer is excited with a voltage $V \exp(j\omega t)$, find an expression for the frequency response of the output signal from the N-finger transducer when the switch to the amplifier is open. Assume that the signal induced on a finger whose center is at the plane z is proportional to $a(z, t)$, and that the wave velocity on the substrate is V_R.

Hint: By reciprocity, the response of a receiving transducer is the same as a transmitting transducer.

(b) Suppose that $M = N$. Show how to choose the length L so that the phase delay from the input to the output of the delay line at the center frequency is $(2R + 1)\pi$, where R is an integer. This causes the device to oscillate when the switch is closed. If the response of the delay line is known at the center frequency, work out an expression for the response when the phase delay changes to $(2R - 1)\pi$.

(c) With $M = N$, suggest a choice for M that will make the loss very large at all other possible oscillation frequencies so as to avoid oscillations at these spurious frequencies.

5. A 7-bit Barker code has the form $+ + + - - + -$. A matched SAW receiving filter is made up for this code. Show that the correlation output of this filter to the code consists of a single peak of amplitude 7 and subsidiary pulses of amplitudes -1 or 0. How many subsidiary pulses are there?

6. A radar system uses an FM chirp signal of length T and frequency $\omega = \omega_0 + \mu t$. It is used to detect a target traveling with a velocity V away from the radar. The velocity of electromagnetic waves in air is c, and $c \gg V$. Find the delay time to the target which is at a distance $R = R_0 + Vt$. Then find the delay time of the return echo from the transmitter back to the receiver.

(a) If the signal emitted by the radar is $f(t)$, find the form of the return echo signal. Assume that $V \ll c$ and the bandwidth of the signal is much smaller than its center frequency. Use the formula for a Fourier transform to show that there is a simple linear Doppler shift of frequency proportional to V/c (keep only terms linear in V/c).

Hint: a signal $f(t)$ after travelling a distance $2R$ has the form $f(t - 2R/c)$. Consider the effect of R varying with time.

(b) Now suppose that a chirp signal of length T is inserted into a matched filter of length T. Estimate how the output falls off with target velocity. Show that for $T = 2$ ms, $f_0 = 10$ GHz, a bandwidth $B = 2$ MHz, and a velocity V of 100 m/s, there is essentially no change in the output peak pulse amplitude. Thus we cannot differentiate between a change in range and a change in velocity of the target.

Hint: The output is proportional to the number of active taps in the filter (i.e., the taps that respond to their corresponding frequencies). If there is a large frequency shift in the chirp, some taps in the filter are not excited.

(c) Consider a V-FM chirp waveform of the type shown in the figure below.

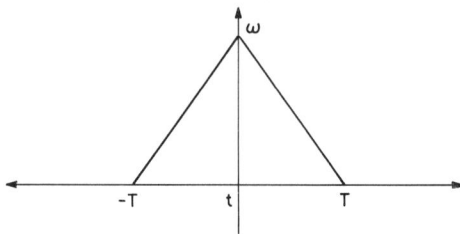

If this signal of length $2T$ is inserted into a matched filter of length $2T$, it will give a simple pulse output. Show that by using this waveform, it should be relatively easy to distinguish a target with a small velocity $V \ll c$, as well as its range (the time width of a peak of a correlated FM chirp signal is approximately $\tau = 1/B$, corresponding to a frequency change in the input signal of B, where $B = \mu T/2\pi$. Take $T = 1$ ms, $V = 200$ ms, $B = 500$ kHz, and $f_0 = 10$ GHz. This example is a simple illustration of the use of the ambiguity function of radar processing. The problem is to obtain good range definition as well as good velocity or Doppler resolution.

Note: The numbers given here are more appropriate for a CCD filter than for an SAW filter.

4.3 CHARGE-TRANSFER DEVICES

4.3.1 Introduction

Here we describe the principles of three types of *charge-transfer devices* (CTDs), which can be employed to process analog signals in much the same manner as the SAW devices discussed in Secs. 4.1 and 4.2. These devices employ the principle that a free electron or charge can be stored in a *metal-oxide semiconductor* (MOS) capacitor or in a PN diode capacitor. In the *charge-coupled device* (CCD) [4, 5] and the *bucket brigade device* (BBD) [8], the charge is transferred along a chain of capacitors; in the *single transfer device* (STD) [24], which is basically a switched capacitor system [7], the charge is read into a capacitor and then read out from the same capacitor at a later time.

In all these devices, the charge read into the system can be made to be linearly proportional to an input signal voltage, and the output signal can similarly be made proportional to the charge arriving at the output register. Because only a finite number of registers exists in such systems, all these devices must basically be sampled data signal processing devices (i.e., they store several samples during one RF cycle of the analog signal). Sampled data signal processing devices have been developed for use with digital signal processing. Here, however, because the charge can be made proportional to the input and output voltages, these devices can be used as analog sampled data systems. Thus CCDs, BBDs, and STDs offer the possibility of performing many sampled data filtering functions in the analog domain. While these devices combine some of the best features of digital and analog techniques, they also encounter some of the same difficulties. Like digital filters, CTDs are controlled by a master oscillator, but the need for analog-to-digital conversion is eliminated and all functions are performed in the analog domain. Furthermore, analog devices are usually much smaller and consume less power than those required for the equivalent digital signal processing systems. For instance, a digital device must use eight separate stages for an 8-bit system, while the equivalent sampled analog system requires only one stage. One advantage of analog devices is that they can be used as image sensors for television line scans, because light can generate charge within the registers or in neighboring registers. On the other hand, digital devices can be made with greater accuracy, and far more flexibility and programmability, than analog devices.

The analog signal processing functions that can be implemented with these devices are very similar to those made available by SAW devices. The nature of silicon technology gives us much more flexibility, albeit at a somewhat lower frequency range. A disadvantage, however, is the need to provide power, for these are "active" rather than "passive" devices. A further disadvantage is that the "clock" signal used to control the transfer of charge from one register to another, or from input to output, can give rise to interference with the signals passing through the device. This is not a problem in digital devices, although the clock signals used in them can interfere with analog signals.

There are five broad applications of interest:

1. Analog delay, where the input waveform is sampled at a sampling frequency f_s and delayed by a time T_d.
2. Multiplexing, where a number of inputs are loaded in parallel into a register and clocked out serially. The complementary function of demultiplexing is also important, as is a combined multiplexer and demultiplexer, as used in the STD.
3. Transversal filtering, where the CTD device is nondestructively tapped at a number of points and the outputs are weighted and summed together.
4. Recursive filtering, where the CTD is tapped as in a transversal filter and the weighted outputs are fed back to the CTD input.
5. Correlation or convolution, in which two analog waveforms are multiplied and integrated together to form an output signal that is the correlation or convolution of these two signals.

Because these analog devices can easily be operated at lower frequencies than SAW devices, they are in a class by themselves for use at frequencies below 10 MHz. Furthermore, because the time delay through these devices can be varied at will, they can perform functions that would be impossible with surface acoustic waves. For many possible applications, there is an overlap between the two types of delay devices, so that either could be used. Relative cost of technological development and production determines which one shall be used; so do trade-offs in detailed performance characteristics important to the designer.

Some examples of simple delay-line applications are:

1. Line delays for the PAL and SECAM TV systems. In these systems the signals from neighboring lines are compared to assure perfect color reproduction. Fused quartz bulk wave acoustic wave delay lines have been employed for this purpose because of their low cost, but CCD and BBD delay lines are now being used in the same application.
2. Ghost suppression in television. Signals result from multipath interference (i.e., from signals reflected from buildings which arrive with a time delay T). The signal passing through this delay line is reversed in sign, and its amplitude is changed appropriately and added to the original signal. Thus it will cancel out the ghost, leaving a still weaker signal at a time $2T$. The signal produced by this process will normally be weak enough not to be noticed. Thus the use of a variable time-delay line with appropriate circuitry and possibly more complex algorithms with microprocessor control can, in principle, lead to a simple ghost eliminator. SAW delay lines have also been used for this purpose.

3. Time-delay beam-forming in sonar and acoustic imaging applications. By combining the appropriately delayed return signals from an array of acoustic transducers, a high-resolution beam can be formed in any direction [25]. Sonar and radar applications have been described by Jack et al. [26].
4. Audio delay lines for artificial reverberation and for changing the speed of taped speech without changing its frequency.
5. Devices for time stretching or compression. If a signal at one frequency is clocked into the delay line and then clocked out at a different rate, an input signal can be changed in frequency and total time length so as to reproduce it faithfully at either a higher or lower frequency. This allows us to inject high-frequency signals or fast transients into low-frequency devices, such as a computer. Alternatively, it enables us to take very low frequency signals and display them rapidly.
6. Time axis equalization in a videotape playback unit. If we monitor a constant-frequency tone recorded on tape, we can correct variation (resulting from stretching of the tape or of motor speed) by passing the video through a delay line whose clock frequency is controlled to keep a pilot tone at a constant frequency.
7. Analog memories for electronic counter measures. There are also closely related delay-line functions, such as rejection of *clutter* (i.e., interference from stationary ground targets) in moving target indicator (MTI) radar. The required operations can be performed with recursive filtering or transversal filters and will be discussed more fully when we deal with applications.

4.3.2 Bucket Brigade Devices

A schematic of the BBD configuration is shown in Fig. 4.3.1. In the integrated-circuit version illustrated in Fig. 4.3.1(b), it consists of a chain of *field-effect transistors* (FETs). Charge is transferred through this chain of FETs, which are connected source to drain–drain to source, by applying a clock signal to the gates. The system is arranged so that when an FET is cut off by a voltage on its gate electrode, the charge is stored at the drain in the so-called overlap capacity between the drain and the gate. As illustrated in Fig. 4.3.1(a), the input signal is applied at the first source and taken out at a last drain. In the configuration shown in Fig. 4.3.1(b), a layer of silicon dioxide is laid down over the silicon and extends over the $n+$ islands and the p-type channel regions that separate them. The only contacts that need to be made are to the first and last $n+$ islands; generally, these islands act as both source and drain.

We may understand the operation of the device by referring again to Fig. 4.3.1(b). We suppose that the clock signals V_1 and V_2 are square waves of opposite phase. Thus V_1 is positive when V_2 is negative, and vice versa. We take $|V_1| = |V_2|$. The gates themselves are supplied with a dc potential V_0 in addition to the clock signal.

We suppose that the input and output are both held positive. If V_1 is sufficiently positive, an n-type channel (inversion layer) under the odd numbered

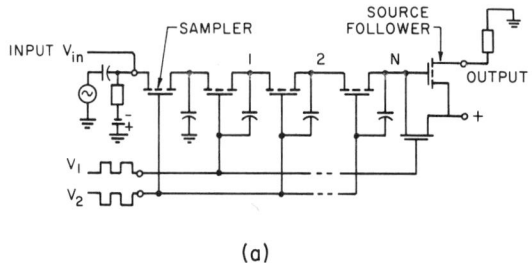

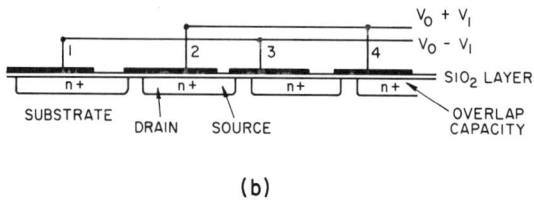

Figure 4.3.1 (a) Circuit configuration of an MOS delay line; (b) Sectional view of the integrated circuit.

gates will be induced (i.e., the odd channels will be conducting). At the same time, the *n*-type islands to the right of the odd-numbered gates will be driven positive because of the overlap capacity C, while the even-numbered gates are driven negative. Thus their overlap capacity with the *n* islands on the right will drive the islands more negative than those on the right-hand side of the odd-numbered gates. The more negative *n*-type islands therefore act as sources, and the more positive islands, as drains.

Referring now to Fig. 4.3.1(a), it will be seen that when the first FET is switched on, current can flow from the drain; thus the charge stored in its source capacitor moves from the source to the drain capacitor. When the next FET is turned on, the first is turned off, and the charge flows through it from its source (the drain of the first FET) to the capacitor at its drain.

We call a device of this kind a *bucket brigade device* (BBD) because the individual capacitors can be regarded as buckets that store the charge and the device transfers charge from bucket to bucket.

Suppose that V_T is the threshold voltage between the gate and the source for a conducting channel to form. Charge is injected into the conducting channel when the input is pulsed more negative than $V_0 + V_1 - V_T$ during the time that V_1 is positive. Thus the clock pulses sample the input waveform and the charge transfers along the system. The quantity of charge induced at the input FET is proportional to the potential $V_1 - V_T - V_{\text{in}}$. At the output the charge is detected as a momentary voltage drop across the output load. Provided that there are at least two output current pulses per RF input cycle, Nyquist's sampling theorem indicates that the input signal can be reproduced. This implies the use of a clock frequency at least twice that of the maximum input frequency.

Performance limitation of the BBD. We now consider the performance limitations of the BBD. We consider the BBD to be operated at a clock frequency

f_c. The maximum time for each transfer is therefore

$$\tau = \frac{1}{2f_c} \tag{4.3.1}$$

There will always be a finite amount of charge left behind after a time τ, and the charge left behind at each register will be cumulative. Ultimately, the signal passing through the registers becomes unacceptably degraded, and a limitation on either the number of registers or the clock frequency will result. It is apparent that the higher the clock frequency, the worse the degradation of the signal.

To provide an understanding of some of the basic concepts of CTDs, let us carry out a mathematical treatment of the performance of this device. We first consider the situation when no charge is transferred along the system. If there is no charge on a drain just before the clock voltages switch from one polarity to the other, the potential on the new source V_s just after switching is

$$V_s = V_0 - V_1 \tag{4.3.2}$$

However, assuming current saturation, the current that flows in an FET channel (the inversion layer) can be shown to be of the form

$$\begin{aligned} I &= \beta(V_G - V_s - V_T)^2 \\ &= \beta(V_0 + V_1 - V_T - V_s)^2 \end{aligned} \tag{4.3.3}$$

where $V_G = V_0 + V_1$ is the gate voltage, $\beta = \mu k_1 Z C_{Ox}/L$, μ is the *mobility* of the charge carriers, k_1 is a constant ~ 0.5, Z is the width of the channel (inversion layer), C_{Ox} is the capacity per unit area of the oxide, and L is the length of the channel. It will be apparent that the current is zero if

$$V_s = V_0 + V_1 - V_T \tag{4.3.4}$$

It therefore follows from Eqs. (4.3.2) and (4.3.4) that there is no charge on the source initially, and no charge flows from source to drain if the clock voltage is such that

$$V_1 = \frac{V_T}{2} \tag{4.3.5}$$

We now consider what occurs when charge is inserted at the input of this FET, and we keep $V_1 = V_T/2$. The charge on a source whose voltage is V_s is

$$\begin{aligned} Q &= -C(V_0 - V_1 - V_s) \\ &= -C(V_0 + V_1 - V_T - V_s) \end{aligned} \tag{4.3.6}$$

where C is the capacity between the gate and the source. It follows that the current flowing from source to drain is

$$I = \frac{-dQ}{dt} \tag{4.3.7}$$

Equations (4.3.3), (4.3.6), and (4.3.7) yield the following relation for Q:

$$\frac{dQ}{dt} = \frac{-\beta Q^2}{C^2} \quad (4.3.8)$$

This expression can be integrated to obtain the following formula for the charge as a function of time:

$$Q = \frac{Q_0}{1 + \beta Q_0 t/C^2} \quad (4.3.9)$$

Thus at a time $\tau = 1/2f_c$, the charge remaining on the source is

$$Q(\tau) = Q_0\left(1 + \frac{\beta Q_0}{2f_c C^2}\right)^{-1} \quad (4.3.10)$$

where Q_0 is the initial charge on the source.

We assume that $(\beta Q_0/2f_c C^2) \gg 1$ [i.e., $Q(\tau) \ll Q_0$]. Thus most of the charge is transferred from source to drain in time τ. Using a Taylor expansion in the parameter $2f_c C^2/\beta Q_0$, it follows that

$$\begin{aligned} Q(\tau) &= \frac{2f_c C^2}{\beta}\left(1 + \frac{2f_c C^2}{\beta Q_0}\right)^{-1} \\ &= \frac{2f_c C^2}{\beta} - 4\left(\frac{f_c C^2}{\beta}\right)^2 \frac{1}{Q_0} \end{aligned} \quad (4.3.11)$$

Remember that $Q(\tau)$ is the charge left behind on the source. The first term is a constant charge that depends on the frequency (i.e., the average charge left behind because a finite time is used for transfer of charge). The second term is dependent on the initial charge on the source Q_0.

Note that if we use a different clock voltage, $V_1 > V_T/2$, the charge transferred along the system will still be given by Eq. (4.3.9). The only difference will be that the source will assume an average dc level corresponding to a nontransferable charge Q_F, for which the current in the channel is zero, and V_s is given by Eq. (4.3.4).

We consider the results of Eq. (4.3.11). When we take representative values of $\beta = 10^{-5} A/V^2$, $Q_0 = 1$ pC, $C = 0.5$ pF, and $f_c = 1$ MHz, the remaining charge $Q(\tau)$ will be approximately 0.05 pC; this is mainly due to the first term. Thus 5% of the charge remains. On the other hand, with $Q_0 = 2.5$ pC and $Q(\tau) = 0.05$ pC, as before, only 2% of the charge remains because there is an approximate linear relation between $Q(\tau)$ and f_c. Thus after 20 transitions, 36% of the charge would be transmitted in the first case and 67% in the second, and the BBD would not be useful.

The basic problem is that the source charge cannot easily be reduced to zero; this would take a very long time. Instead, it is better to work with what is called a "fat zero." Suppose that we do not require the charge to be reduced to zero, and consider instead what occurs if we work with a finite minimum level of charge.

If the injected charge is Q_1, the corresponding remaining charge will be

$$Q_1(\tau) = \frac{2f_c C^2}{\beta} - 4\left(\frac{f_c C^2}{\beta}\right)^2 \frac{1}{Q_1} \qquad (4.3.12)$$

If the injected charge is changed to Q_2, it follows that

$$Q_2(\tau) = \frac{2f_c C^2}{\beta} - 4\left(\frac{f_c C^2}{\beta}\right)^2 \frac{1}{Q_2} \qquad (4.3.13)$$

If we regard the charge Q_1 as the minimum charge, the dc level, or the fat zero, we now need only consider $\Delta Q = Q_2(\tau) - Q_1(\tau)$ as the remaining difference charge. Thus our criteria for efficiency will now be based on a parameter ε, defined by the relation

$$\varepsilon = \frac{Q_2(\tau) - Q_1(\tau)}{Q_2 - Q_1} \qquad (4.3.14)$$

We wish to have ε as small as possible. We see that

$$Q_2(\tau) - Q_1(\tau) = 4\left(\frac{f_c C^2}{\beta}\right)\left(\frac{1}{Q_1} - \frac{1}{Q_2}\right) \qquad (4.3.15)$$

Hence

$$\begin{aligned}\varepsilon &= 4\left(\frac{f_c C^2}{\beta}\right)^2 \frac{1}{Q_2 - Q_1}\left(\frac{1}{Q_1} - \frac{1}{Q_2}\right) \\ &= 4\left(\frac{f_c C^2}{\beta}\right)^2 \frac{1}{Q_1 Q_2}\end{aligned} \qquad (4.3.16)$$

Originally, a similar criterion was

$$\varepsilon_0 = \frac{2f_c C^2}{Q_0 \beta} \qquad (4.3.17)$$

Now, with the use of a fat zero, $\varepsilon \sim \varepsilon_0^2$. For example, with $Q_2 = 2.5$ pC, $Q_1 = 1$ pC, and the same parameters as before, $\varepsilon = 0.094\%$, where $\varepsilon_0 = 5\%$ and 2%, for $Q_0 = 1$ pC and 2.5 pC, respectively. Thus the difference charge passed through the register is 0.99906 of the difference charge entering it, and the signal is transmitted faithfully. After N transfers, the signal charge transferred through the system will be

$$(\Delta Q)_N = \Delta Q(1 - \varepsilon)^N = \Delta Q e^{-N\varepsilon} \qquad (4.3.18)$$

Thus ΔQ has dropped to 98% of its original value after 20 transfers. Note, however, that ε now varies as f_c^2. Thus if we increase the frequency to 2 MHz, $\varepsilon = 0.37\%$, we can only allow five transfers for a 2% degradation. We can make further improvements by using shaped clock waveforms and far shorter gates.

In real BBD devices, the configuration is somewhat more complicated, as shown in Fig. 4.3.2. In the device shown in Fig. 4.3.2(a), the input signal usually

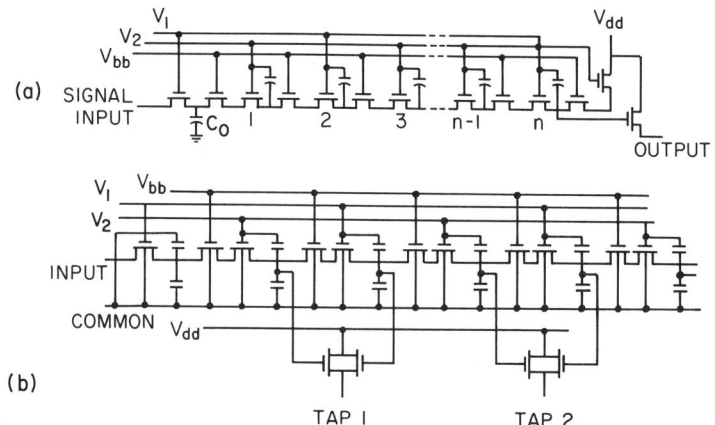

Figure 4.3.2 BBD structure used by Reticon: (a) input circuit; (b) method of tapping. (From Reticon Corporation [24].)

consists of a band-limited analog signal which varies around a dc bias level. This automatically introduces a fat zero and gives better incremental transfer efficiency. A further change in the configuration in Fig. 4.3.2(a) can be made by using tetrode rather than triode type FETs (i.e., intermediate FETs with fixed voltages on the gates). This isolates the drain capacitor from the source capacitor, eliminating the Miller effect (i.e., the feedback between output and input).

As we shall see, the CCD may be simpler to make in principle but a great deal more difficult in practice. In certain forms, however, it does lend itself to higher-speed operation. BBD devices have an advantage over CCDs because they are easily employed for making externally tapped devices. The configuration used by Reticon in their tapped device is shown in Fig. 4.3.2(b). The output is taken from a pair of balanced source followers at the tap, giving an output that is the difference between the charges on two neighboring registers (i.e., between the source and the drain). Thus the fat zero is removed from the output on these taps. This means it is possible to construct a tapped delay line with individual taps that have adjustable weights. Typical tapped delay lines have 32 or more taps.

A similar system can be constructed in the BBD configuration to give an input, instead of an output, at each tap, so that the output can be taken out of the delay line with parallel inputs. This configuration was used by Reticon and is shown in Fig. 4.3.3. In this device, two parallel BBD lines are used to provide isolation between adjacent even and odd inputs. In both delay lines, each signal sample is isolated on either side by a reference sample. The two parallel bucket brigade delay lines are driven from two-phase complementary clocks. The delay line receives the input signal charges in parallel, from corresponding even or odd input storage capacitors that have previously stored samples of the input signal. These new input signal charges are transferred into the bucket brigade capacitor storage sites under alternate V_1 and V_2 gates. The even-numbered input taps transfer signal samples to nodes under alternate even-numbered V_2 gates of the

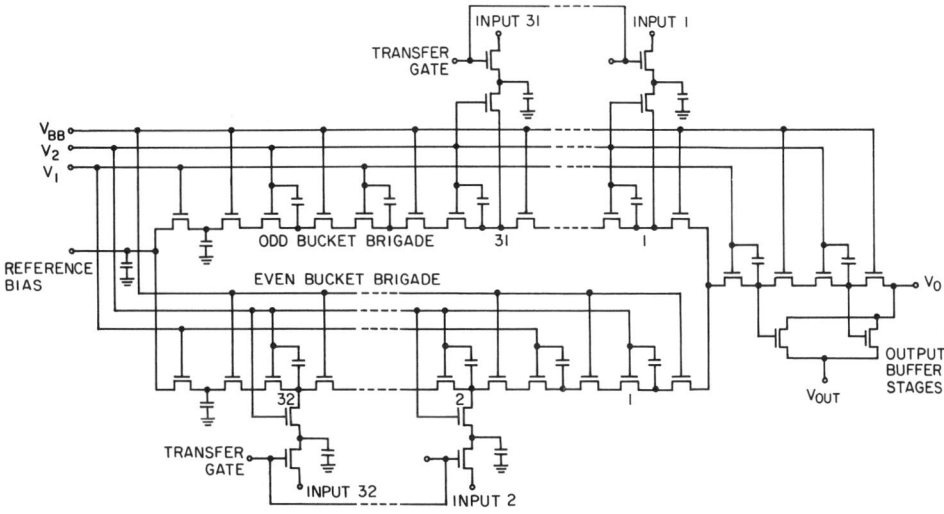

Figure 4.3.3 Equivalent circuit of Reticon tapped BBD. (From Reticon Corporation [27].)

even bucket brigade, while the odd-numbered taps transfer this sample to nodes of the odd-numbered gates of the odd bucket brigade.

The sampling to odd and even stages permits the outputs of the two bucket brigades to be multiplexed (i.e., switched alternately into a single channel) and yields a serial stream of pulses at the output. The combined signal channel output appears at two successive storage sites. It has the advantage of giving a balanced output, and multiplexing makes it possible to use fewer stages in a series for a given number of registers. This implies that the clock frequency can be reduced by a factor of 2, so that the requirements of high-frequency operation are reduced by such multiplexing schemes. Thus the multiplexing arrangement is useful not only for tapped systems, but for any system where high-frequency operation is a problem.

4.3.3 Charge-Coupled Devices

The CCD is even simpler in concept than the BBD. The essence of its operation is to store minority carriers in a spatially defined depletion region (defined by the clock potentials) at the surface of a semiconductor, and to move this charge about by moving the potential well in which the charges are stored. The charge can be injected at one end of the device, moved along it, and detected at the output. Alternatively, charge can be injected electrically through taps at points along the device, or by incident light, which generates charge. This makes it possible to construct a camera that can detect one line of an image and then read out the information sequentially from the output of the storage device [4, 5, 28].

To understand the storage mechanism, let us first consider a single *metal insulator semiconductor* (MIS) structure on an *p*-type semiconductor. We could, for instance, consider one element of a system like that shown in Fig. 4.3.4, in

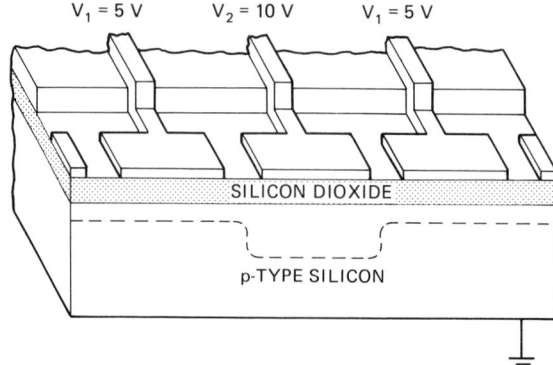

Figure 4.3.4 Cutaway of a CCD in the storage condition. The dashed line represents both the edge of the depletion region and the potential distribution. Voltages are typical. (From Sequin and Tompsett [5].)

which the center electrode is kept at a potential more positive than that of the two surrounding electrodes, thus forming a potential well. Normally, in thermal equilibrium, an inversion layer or surface charge of electrons is generated and forms just under the silicon dioxide. Such a layer may take seconds to form. Under nonequilibrium conditions, when there is no charge present in this inversion layer, the surface is depleted of minority charge, a potential well that is available to accept electrons. If we now suppose that charge (in this case, electrons) is introduced into the depletion region, the presence of electrons will cause the surface to assume a more negative potential and the depletion layer will reduce in width. At a time $\tau = \infty$, after thermal equilibrium has been reached by generating electrons from surface taps, the interference potential reaches its equilibrium value. Any further introduction of electrons, which will make the interface potential still more negative, will eventually lead to injection of electrons into the bulk; in other words, we cannot store more charges than would possibly be formed in the inversion layer in thermal equilibrium.

We now consider the three-phase charge-coupled device shown in Figs. 4.3.4, 4.3.5, and 4.3.6. In this type of device, every third electrode is connected to a common conductor. Each of these common conductors is a separate clock line and forms a three-phase clocking system. Initially, a voltage V_2 is applied to electrodes 1, 4, 7, and so on, and a voltage V_1, $|V_2| > |V_1|$ is applied to the other electrodes. The semiconductor is held at zero potential.

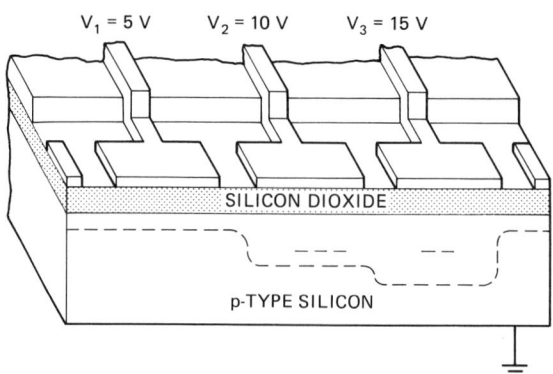

Figure 4.3.5 Cutaway of a CCD in the transfer condition. (From Sequin and Tompsett [5].)

It is assumed that $|V_1| > |V_T|$, where V_T is the threshold voltage for inversion under thermal equilibrium conditions. Under conditions shown in Figs. 4.3.4 and 4.3.6(a), and those that we have described, charge can be stored in the potential well so that electrons which have been introduced into this region will be stored at the oxide interface. We now suppose that a voltage V_3 is applied to electrodes 2, 5, 8, and so on. As shown in Figs. 4.3.5 and 4.3.6(b), $V_3 > V_2$. The charge will now transfer from electrode 1 to the potential minimum under electrode 2, and so on. The voltages are changed to those shown in Fig. 4.3.6(c). Charge is stored under electrodes 2, 5, 8, and so on, and the process is repeated once more to transfer charge along the system.

In this discussion, it is of course assumed that charge is transferred along the system in times much shorter than the storage time or generation time τ_g, where $\tau_g = Q/I_g$ is the time for generation of the charge Q by the thermally generated current I_g in the depletion layer and the surface states. In carefully constructed devices with high-quality bulk material and a small number of surface states, such times can be of the order of several seconds, for the main source of generated current is due to surface states.

So far, we have discussed devices with a three-phase clock. The disadvantages of such a system are apparent. For one thing, the minimum clock frequency must be three times the maximum frequency of the signal passed through the device, which puts more stringent criteria on the storage time in the device and the highest frequencies used. When the devices were first developed, another difficulty resulted from bringing three clock lines to the electrodes: one of the clock lines had to cross over another, which meant that a more complicated multiple-layer technology was needed. However, this technology is now routinely accomplished.

The problem with a two-phase system is that when the system is symmetrical, charge can pass equally well in either direction. Therefore, if a two-phase system is used, an asymmetry must be built into it. One method of doing this is illustrated in Fig. 4.3.7. Here, by depositing the electrodes on top of layers of silicon dioxide

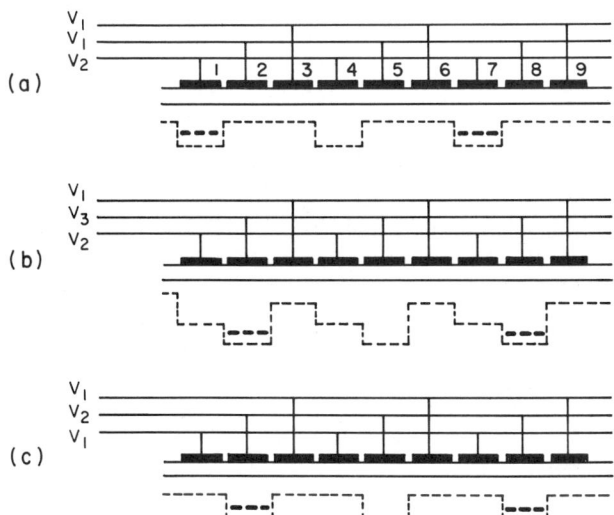

Figure 4.3.6 Three-phase CCD; illustration of its operation.

with appropriate windows, an asymmetry is built into the electrodes themselves, and a two-phase system moves carriers to the right. Other techniques involve ion implantation at one end of the electrode, so as to produce a built-in surface charge and to provide a bias that is built in at one end of the electrode. All in all, the two-phase system still tends to be as complex to manufacture as three-phase systems.

The signal carriers basically move between electrodes by diffusion, somewhat aided by their own space-charge fields and the fields between the electrodes, as discussed in Appendix J. If there were a large gap between the electrodes, the transfer would be increased. Thus it is often convenient to overlap the electrodes, as shown in Fig. 4.3.7, and so introduce transverse field components. Such a system is the natural one for a two-phase clock.

The charge transfer rate is determined by the mobility of the carriers; the higher the mobility, the better. For this reason it is preferable to use n-type carriers. Sometimes we can remove electrons from the surface region by trapping them in a potential maximum at a buried p-type layer formed by ion implantation; this is called a *buried channel*. This makes it far easier to provide a large transverse component of field for high charge-transfer rates. A further development is to use a material such as gallium arsenide, where the mobility of the electrons is of an order of magnitude greater than for silicon. This means that a buried channel must be used, together with so-called Schottky gates, to eliminate problems with surface states in gallium arsenide. However, this method has resulted in CCDs capable of handling signal frequencies of several hundred megahertz [29].

In all cases, the transfer efficiency is never perfect, so the fat zero concept applies equally well to CCD and BBD operation. A short discussion of the mathematical theory of operation of these devices is given in Appendix I. The detailed theory tends to be more complicated than that of the BBD and is outside the scope of this book.

A tapped CCD delay line can be implemented by taking advantage of the fact that when charge passes under an electrode, it induces an equal and opposite charge in the electrode. Thus by reducing the current flowing into the clock lines, and integrating it, the charge on a possible electrode can be determined. A technique is also needed for implementing a weighted transversal filter in which the weights can be positive or negative.

The technique employed is known as the *split-electrode method*. In one phase, the electrodes are split into two sections of different length, as shown in Fig. 4.3.8, with a schematic of the circuit used shown in Fig. 4.3.9. Two clock lines are employed instead of the V_3 clock line. These are supplied through the capacitors C_1^+ and C_1^-. Transistors T_1 and T_2 are used to discharge the clock lines after the clock is turned off. We shall call these the V_3^- and V_3^+ clock lines. These clock lines are individually connected to each side of a differential current integrator, so that if the strips on each side are of equal length, the output from the

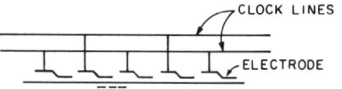

Figure 4.3.7 Two-phase CCD structure.

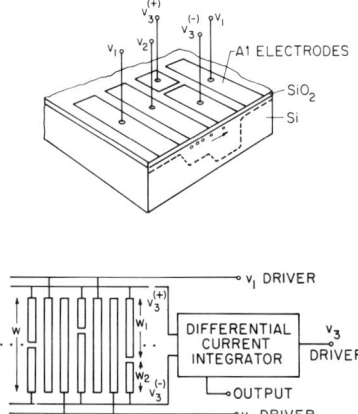
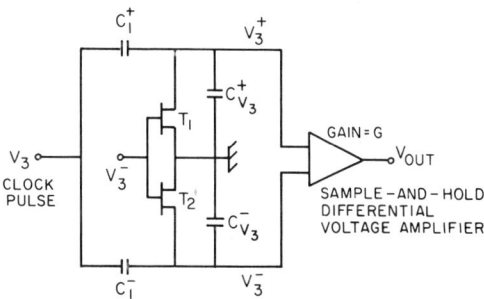

Figure 4.3.8 Split-electrode technique for implementing weighting coefficients of a transversal filter. (From Buss et al.[4].)

Figure 4.3.9 Differential current integrator (DCI) output amplifier. (From Buss et al. [4].)

differential current integrator will be zero when the charge passes underneath a particular strip. Now suppose that the total length of a strip is w, and the individual parts of a strip are of lengths w_1 and w_2, such that $w_1 + w_2 = w$. The total output from the integrator would then be proportional to $w_1 - w_2$, and both positive and negative weightings could be easily obtained.

By using a sample-and-hold circuit in the output (essentially, a capacitor connected through the FET switches T_1 and T_2 to the output circuit only when a signal is present), we can differentiate against higher-frequency clock noise, and the device will act faithfully as a transversal filter in exactly the manner already described for the SAW delay line. The charge induced on the individual electrodes is proportional to the weighting and is obtained at a time corresponding to the time delay through the delay line. The system may thus be used to obtain the convolution of the input signal with the weighting of the CCD. The design problem is *exactly* like that of SAW devices, except that the device is now operated at baseband. It follows that the device lends itself well to the design of a wide range of digital and analog transversal filters.

This method differs in one respect from the SAW delay line—it is not normally convenient to vary the distance between electrodes. Because of this, these devices are usually operated with a fixed spacing between electrodes, and the signals are sampled at these fixed points. We shall discuss the difference presented by this design problem later, when we discuss the chirp z transform implementation in both CCD and SAW delay lines.

It is apparent, however, that it is possible to construct low-frequency CCD and BBD equivalents of SAW delay lines for use as digital filters and bandpass filters, as described above. Figure 4.3.10 shows an example of a bandpass filter using the $(\sin x)/x$ response already described for SAW delay lines. Figure 4.3.10(a) shows the pattern cut into the electrodes, while Fig. 4.3.10(b) shows the time

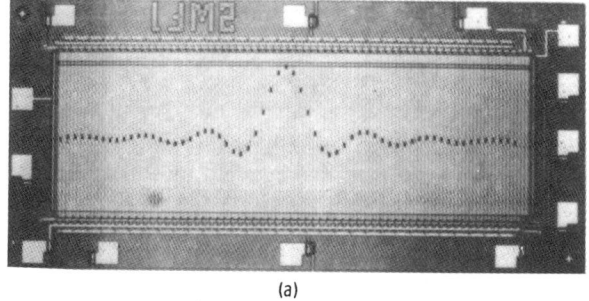

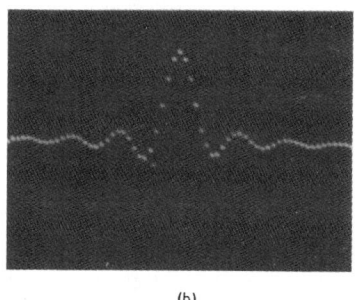

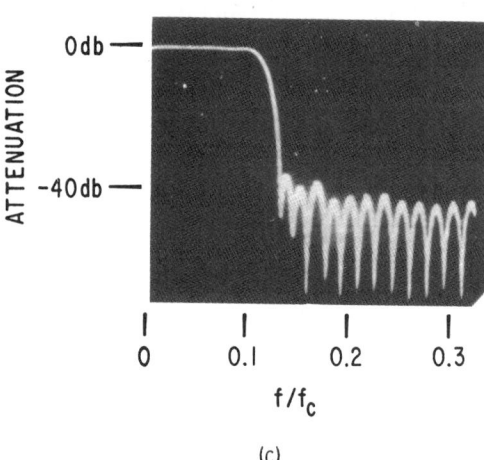

Figure 4.3.10 (a) Photomicrograph of a 63-stage CCD low-pass filter; (b) measured impulse response of the filter; (c) measured frequency response of the filter. (After Baertsch et al., as noted in Buss et al. [4, 30, 31].)

response of the clock delay line. Note that the frequency response of the filter in Fig. 4.3.10(c) is of excellent quality.

4.3.4 Effect of Imperfect Charge-Transfer Efficiency

When charge is transferred through a CCD or BBD delay line, a fraction ε of the charge is left behind after each clock pulse. Consequently, we might expect that after transfer through N stages, the output amplitude will be diminished from that of an ideal delay line. In addition, because charge from earlier transfers is still present in the registers, weak interfering signals will be present and will cause phase distortion of the input signal. The fractional losses of commercial CCDs and BBDs in their operating frequency ranges are usually of the order of $\varepsilon \sim 10^{-4}$, and at worst, $\varepsilon \sim 10^{-3}$. Although such fractional losses appear to be small, with a large number of registers with transfers of as many as $N = 1000$, even a small loss per stage can be vitally important. Examples of typical transfer efficiencies of Reticon commercial devices as a function of frequency are shown in Fig. 4.3.11.

In this section we derive an expression for the response of a charge-transfer device, assuming that ε is invariant with the frequency of the signal being transferred, although it will normally be a function of the transfer time (i.e., the clock period $T_c = 1/f_c$, where f_c is the clock frequency). We consider the properties of a device N registers long.

We consider the amplitude of a signal of the form $\exp(j\omega t)$. We call the time at which charge is injected into the input of the CTD, t'. After passing through N perfect registers, the signal at time $t = t' + NT_c$ is still exactly $\exp(j\omega t')$, or $\exp[j\omega(t - NT_c)]$. Thus the original signal input is reproduced at the output, and is of the form

$$H_N(\omega) = e^{-j\omega NT_c} \qquad (4.3.19)$$

We note that with sample time T_c apart, there is aliasing; that is, the response at frequency ω is the same as it is at frequency $\omega + 2m\pi/T_c$, where m is an integer. It is convenient to use a notation suitable for such sampled systems; as we shall see, this will simplify the algebra considerably. Adopting the z transform notation, we write [4, 32, 33]

$$z = e^{j\omega T_c} \qquad (4.3.20)$$

Here we have essentially expressed the frequency ω in terms of z. Thus for a perfect system, in the z-transform notation,

$$H_N(z) = z^{-N} \qquad (4.3.21)$$

In this notation, the ideal response after a delay T_c (one register) is $1/z$, and through N registers, z^{-N}.

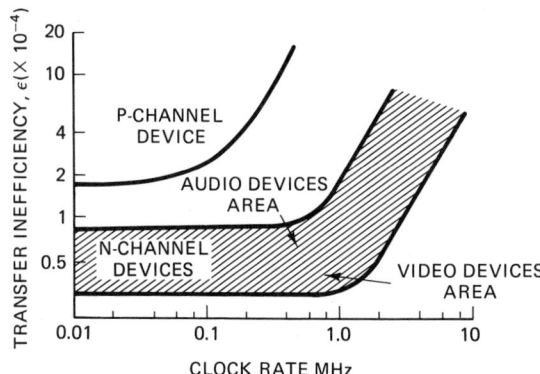

Figure 4.3.11 Charge-transfer inefficiency versus clock frequency. (After Reticon Corporation [24].)

Consider now the transfer function of the nonideal charge-transfer device. Suppose that the voltage at the kth register at a sampling time nT_c is $v_k(nT_c)$. This voltage consists of a term due to the incomplete transfer of charge from the previous register of the form $(1 - \varepsilon)v_{k-1}[(n - 1)T_c]$, and the charge remaining in the register from the previous transfer is $\varepsilon v_k[(n - 1)T_c]$. Thus we may now write

$$v_k(nT_c) = (1 - \varepsilon)v_{k-1}[(n - 1)T_c] + \varepsilon v_k[(n - 1)T_c] \quad (4.3.22)$$

Assuming that all quantities vary with time as $\exp(j\omega t)$, we could write that at a time $t = nT_c$, a signal component of frequency ω at the kth register would be of the form $V_k(\omega) \exp(j\omega nT_c)$. It follows that Eq. (4.3.22) can be written in the form

$$V_k(\omega)e^{j\omega nT_c} = (1 - \varepsilon)V_{k-1}(\omega)e^{j\omega(n-1)T_c} + \varepsilon V_k(\omega)e^{j\omega(n-1)T_c} \quad (4.3.23)$$

or

$$z^n V_k(z) = (1 - \varepsilon)V_{k-1}(z)z^{n-1} + \varepsilon V_k(z)z^{n-1} \quad (4.3.24)$$

Equation (4.3.24) reduces to the expression

$$zV_k(z) = (1 - \varepsilon)V_{k-1}(z) + \varepsilon V_k(z) \quad (4.3.25)$$

It follows from Eq. (4.3.25) that

$$V_k(z) = \left(\frac{1 - \varepsilon}{1 - \varepsilon/z}\right)\frac{1}{z}V_{k-1}(z) \quad (4.3.26)$$

Thus the response of the line after the signal has passed through N registers is

$$H_N(z) = \left(\frac{1 - \varepsilon}{1 - \varepsilon/z}\right)^N z^{-N} \quad (4.3.27)$$

We can simplify Eq. (4.3.27) by using the relation $\ln(1 + \alpha)^N \approx N\alpha$ when $\alpha \ll 1$, so that $(1 + \alpha)^N \approx \exp(N\alpha)$. Assuming that $\varepsilon \ll 1$, Eq. (4.3.27) can be written in the form

$$H_N(z) = e^{-N\varepsilon(1 - 1/z)}z^{-N} \quad (4.3.28)$$

or

$$H_N(\omega) = e^{-N\varepsilon(1 - e^{-j\omega T_c})}e^{-j\omega NT_c} \quad (4.3.29)$$

In turn, we can write $H_N(\omega)$ in terms of its amplitude and phase variations:

$$H_N(\omega) = A_N(\omega)e^{-j\phi_N(\omega)} \quad (4.3.30)$$

where

$$A_N(\omega) = e^{-N\varepsilon(1 - \cos\omega T_c)} \quad (4.3.31)$$

and

$$\phi_N(\omega) = N(\varepsilon \sin \omega T_c + \omega T_c) \quad (4.3.32)$$

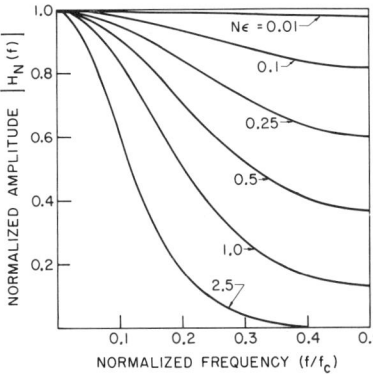

Figure 4.3.12 Magnitude of the frequency response of a CCD delay line. Eq. (4.3.29) is plotted for various values of $N\varepsilon$. This curve assumes that impulse sampling at the output must, in general, be multiplied by the frequency characteristic of the output amplifier stage. (From Buss et al. [4].)

It is convenient to write $\omega T_c = 2\pi f/f_c$ and then to plot Eq. (4.3.30) in the form shown in Fig. 4.3.12. In order to satisfy the sampling theory, the clock frequency must be at least $f_c = 2f$. But these results imply that the amplitude loss is maximum at this frequency. Thus it is usually wiser to operate a charge-transfer device with a clock frequency considerably larger than the maximum signal frequency.

It is also of interest to determine the time response or impulse response of the system after N transfers. We do this by calculating the time response from the inverse Fourier transform of the frequency response, and write

$$h_N(t) = \frac{1}{2\pi} \int_{-\infty}^{\infty} e^{j\omega t} H_N(\omega)\, d\omega \qquad (4.3.33)$$

Although it is difficult to evaluate Eq. (4.3.33) exactly, it is convenient to expand it in a series form as an expansion in powers of $1/z$. Thus, using Eq. (4.3.27), we write

$$H_N(z) = e^{-N\varepsilon(1-1/z)z - N} \qquad (4.3.34)$$

$$= z^{-N} e^{-N\varepsilon} \left[1 + \frac{N\varepsilon}{z} + \frac{1}{2}\left(\frac{N\varepsilon}{z}\right)^2 + \cdots + \frac{1}{n!}\left(\frac{N\varepsilon}{z}\right)^n + \cdots \right]$$

By writing $z^n = \exp(j\omega n T_c)$ and substituting Eq. (4.3.34) in Eq. (4.3.33), it follows that the output can be written in the form

$$h_N(t) = e^{-N\varepsilon} \{\delta(t - NT_c) + N\varepsilon \delta[t - (N+1)T_c] \cdots \} \qquad (4.3.35)$$

where $\delta(t - nT_c)$ is a δ function, the inverse transform of $\exp(j\omega n T_c)$. Thus we might expect a series of output pulses reduced in amplitude from the expected pulse at the time $t = NT_c$. More exactly, we could have used the formula of Eq. (4.3.27) and expanded it to obtain an nth term in the form $(1 - \varepsilon)^N \varepsilon^n z^{-(N+n)}$ $(N + n)!/n!N!$. The corresponding expansion for the time response is similar to that of Eq. (4.3.35), and is plotted in normalized form in Fig. 4.3.13 for an input function consisting of a short pulse. As we might expect, when $N\varepsilon$ is small, the initial response will be very close to that of a single pulse, but followed by a weak

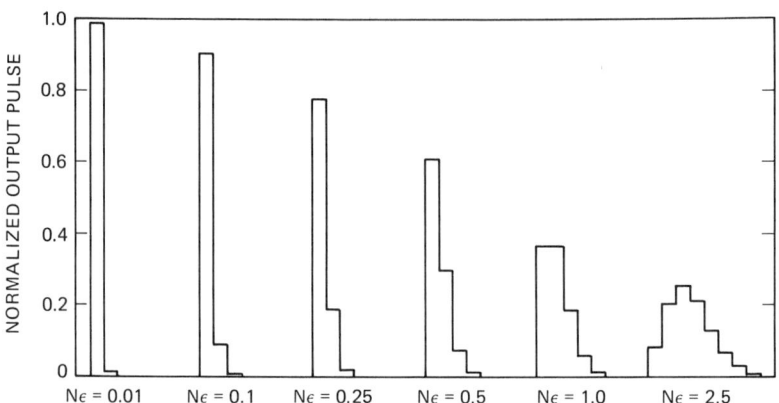

Figure 4.3.13 Impulse response of an N-stage CTD delay line having loss per stage ε. The output given by using Eq. (4.3.27) to obtain a more exact form of Eq. (4.3.35) is plotted for various values of $N\varepsilon$. (After Buss et al. [4].)

second pulse reduced in amplitude by $N\varepsilon$. As $N\varepsilon$ is increased so that $N\varepsilon \rightarrow 1$, the initial pulse may be weaker than the later pulses coming out of the device, and the signal is radically distorted (see Prob. 2).

4.3.5 Single Transfer Devices (Switched Capacitors)

One way to eliminate the difficulty caused by loss of charge in CTD devices when the charge passes through a very large number of registers is to employ M delay lines, each with N registers, and to inject the signal in turn into each delay line by multiplexing. Thus the system can be made to behave like a CTD with NM registers, with the corresponding loss through only the N registers.

An extreme example of this uses only one register, injects a signal into this register, and stores it, taking it out when needed. Such devices are known as *single transfer devices* (STDs). STDs essentially integrate sets of multiplexed sample-and-holds. Each successive sample is stored in a separate, discrete memory cell, where it stays until read out, as illustrated in the simplified diagram of Fig. 4.3.14. It is now only necessary to use digital shift registers to switch from one sample-and-hold to the next, as illustrated in Fig. 4.3.15. In the device illustrated, which is manufactured by Reticon, the input signal is read into a series of sample-and-hold capacitors, by means of FET switches controlled from the read-in shift register. The signal can then be read out in turn from the capacitors by means of the read-out shift register. This technique allows us to construct a simple delay line in which the clock rate of the output signal may be different from the input signal. Thus the device lends itself to time-base modification or correction of analog signals.

Devices of this kind are useful in the audio–video range and provide simple and inexpensive analog memory delays. Because of the single transfer, they do not have the same problem of transfer inefficiency as the CTDs, but they do have performance limitations associated with clocking noise and *fixed pattern noise*; that

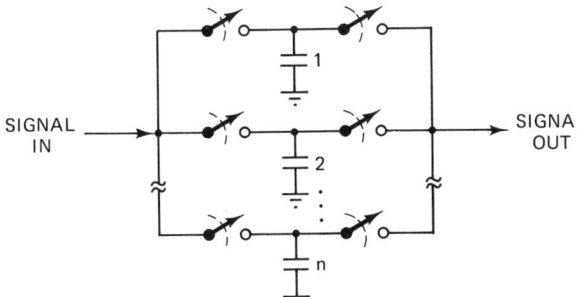

Figure 4.3.14 Sample-and-hold model of an STD.

is, there is a fixed pattern in the variation of the output signal when the memory cells or switches are not uniform, which implies some limitation on the dynamic range of the device.

The problem of clock noise can be radically decreased by using balanced circuits for readout. As an example, consider the self-scanning photodiode array system, developed by Reticon and illustrated in Fig. 4.3.16. In this device, the sample-and-hold diode is photosensitive, so that when it is illumunated with light, charge is generated within it; thus the potential across the diode is altered by the illumination during the given integration time. The charge read into a diode can then be read out by switching the output line to the diode.

In a manner similar to the CCDs and BBDs already discussed, both odd and even shift registers can be employed to lower the clock frequency. A further addition is to use a set of dummy photodiodes which are not exposed to light. Each photosensitive photodiode has a corresponding dummy diode, and the outputs from the two diodes are read out at the same time into opposite sides of a balanced circuit. The output obtained from the balanced circuit is therefore the difference in the charge on the two diodes and is proportional to the illumination. However, as the same clock pulses are used for both diodes, the clock output is eliminated, or at least differentiated against.

Note that a similar balanced system can be used with tapped BBD or CCD delay lines. Now, however, the charge is read from the diode into a tap on a CCD or BBD delay line and the output is read out serially through the delay line, rather

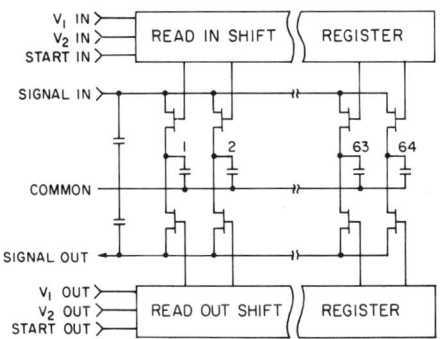

Figure 4.3.15 Simple equivalent circuit of the Reticon SAM-64. (From Reticon Corporation [24].)

Sec. 4.3   Charge-Transfer Devices

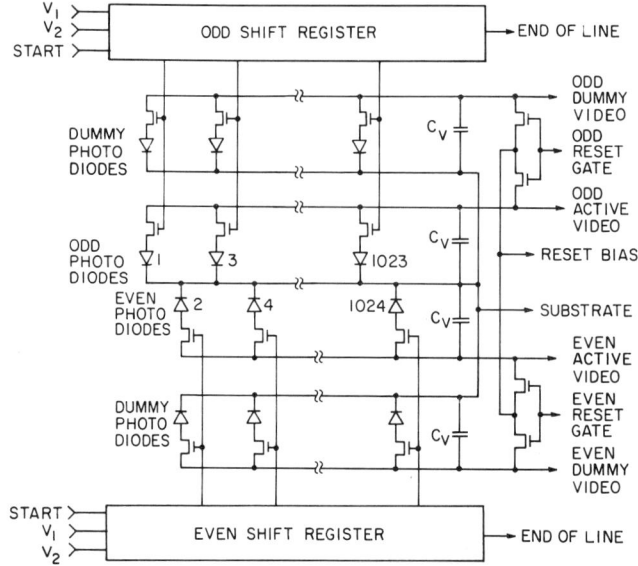

Figure 4.3.16 Equivalent circuit of Reticon self-scanning photodiode system. (From Reticon Corporation [28].)

than in the manner shown for the STD device. Such a system lends itself better to a two-dimensional configuration, although combinations of both systems are sometimes used in two-dimensional imaging devices.

PROBLEM SET 4.3

1. A unit step function $u(t) = 0, t < 0; u(t) = 1, t \geq 0$ is injected into an imperfect charge transfer delay line. Assuming ε is small use Eq.(4.3.34) to find the form of the output signal from the Nth register. Carry out the problems in two parts
   (a) With a short pulse input find the output, as in the text. Plot the output for a short pulse and $N\varepsilon = 2.5$ for the three cycles before and after the largest pulse. Note that the largest pulse is not the first output pulse.
   (b) Plot the amplitudes for the step function input for $t = (N-1)T_c$ to $t = (N+8)T_c$ with $N\varepsilon = 0.01, 2.5$. You can do this problem by convolution.

2. Derive the result of Fig. 4.3.13 for $N\varepsilon = 2.5$ by using an expansion of Eq. (4.3.27) without the approximation of Eq. (4.3.28) [see the discussion after Eq. (4.3.35)].

3. Compare the use of one CTD, N registers long, operating at a clock frequency f_c and signal frequency f, with the use of two multiplexed CTDs, $N/2$ registers long, operating at a clock frequency of $f_c/2$. The signal of frequency f is switched alternately from one CTD to the other on every half-cycle of the signal, and the output signals are added. Determine the amplitude and phase response of the second system, and compare its performance with that of the first system when $N\varepsilon = 0.1$. Plot your results as in Fig. 4.3.12.

4. Consider the use of a sample-and-hold circuit to eliminate clock signals and aliasing in the output waveform. Suppose that a signal $f(t)$ is sampled at times $t - nT_c$ and injected in a sample-and-hold circuit. Between the times $t = nT_c$ and $t = (n+1)T_c$, the sample-

and-hold circuit maintains the output at a constant level $f(nT_c)$. So we can write the sampled signal in the form

$$f_s(t) = \sum_n f(nT_c) \Pi\left(\frac{t - nT_c - T_c/2}{T_c}\right)$$

where

$$\Pi(x) = \begin{cases} 1 & 0 < |x| < 0.5 \\ 0 & |x| > 0.5 \end{cases}$$

When the input signal is $f(t) = \exp(j\omega t)$, find the amplitudes of the frequency components $\omega + 2\pi m/T_c$ as a function of ωT_c. How much are the $m = 1$ and $m = 2$ components reduced from that of the $m = 0$ component when $fT_c = 0.5, 0.25,$ and 0.1, where $f = \omega/2\pi$?

One way to carry out the analysis is to regard the output as of the form

$$f(t) = e^{j\omega t} \sum_m A_m e^{2j\pi mt/T_c}$$

and to find the amplitude of the mth harmonic component. Another way to solve the problem is to carry out the Fourier transform directly.

5. The outputs from an array of sonar receivers used to detect submarines are inserted into a tapped BBD delay line, as shown. Suppose that the signal frequency is f and the clock frequency, f_c. For simplicity, regard the taps as connected to neighboring registers. Assume that the parallel rays coming from a distant source travel at an angle θ to the normal to the plane of the array. Let there be N elements in the array spaced a distance L apart, and let the wave velocity in the medium be c.
   (a) Find how the output varies as a function of the angle θ and frequency f_c when there are N transducers.
   (b) Show how to choose f_c ($f_c \neq 0$) to be such that the signal from this electronically steerable array is maximum at $\theta = 0$. What are the possible values of f_c for $L = 1$ m, $c = 1.5$ km/s, and $f = 300$ Hz, for $\theta = 0$ and $\theta = 10°$?

4.4 MATCHED FILTERS

4.4.1 Introduction

SAW devices and charge-transfer devices can be designed as filters for a particular passband, as already described. We have seen that they can also be designed (by changing the spacing between the taps in addition to the polarity of the taps) as *transversal filters*, which can respond to or generate a particular digital code. The merit of such devices is that they are particularly simple in form, and so can be made by the standard photolithographic integrated-circuit technology.

Another broad application of SAW devices is to recognize analog codes rather than digital codes. A good example of such an analog code is the so-called linear FM chirp, a signal of constant amplitude with a frequency that varies linearly with time. The reason we are interested in the use of FM chirps and other analog codes is best illustrated by the example of radar, for which this kind of coding was initially developed. As we shall see later, there are also other broad applications of FM chirps, the prime example being the *chirp z transform*, a method of obtaining a fast Fourier transform.

The initial form of radar consists of a transmitter emitting a short pulse, which is then reflected from an object and detected at the receiver. The range of the object is determined by measuring the transit time of the pulse to the object and back. For good range resolution, we must obviously use as short a pulse as possible; for the best signal-to-noise ratio, we must use as high a peak power as possible. Thus the necessity for a transmitter with a very large peak power can limit the system.

The development of radar techniques was stimulated by an important monograph by Woodward [34], which proved that it is not necessary to work with high peak powers. Instead, we can use complex waveforms with much less peak power and longer time duration. A linear filter that maximizes the return signal-to-noise power ratio is called a *matched filter*. For noise with a uniform frequency spectrum, the matched filter has a time response that is the time reverse of the input signal. The output signal is a narrow pulse that is the autocorrelation of the input signal. The maximum peak power output from such a matched filter, and hence the sensitivity and signal-to-noise ratio of the radar, is determined by the total energy in the input signal [1, 34–36].

The range accuracy of a radar is inversely proportional to the bandwidth, while the Doppler shift velocity accuracy is inversely proportional to the time length of the signal. In both cases, the accuracy improves with signal-to-noise ratio. Therefore, another important advantage of a matched filter is that it yields the optimum range and velocity accuracy; this accuracy depends on the total energy in the input signal.

The choice of input signal waveform and an appropriately matched filter is an extremely important parameter to be chosen by the radar designer. On one hand, a large bandwidth provides good range resolution; on the other, a large signal length provides good frequency resolution. As we shall see in Sec. 4.4.5, a Gaussian pulse is therefore the worst possible choice, for it makes the product of these two resolutions minimum.

The key to obtaining optimum performance is to use a signal with as large a time–bandwidth product as possible. An important advantage of a system using a matched filter is that it has a built-in preference for its own signals rather than for other, competing signals. Thus the use of a matched filter gives optimum performance in the presence of coherent interference as well as random noise.

Other filter functions are also possible. For instance, if the signal-to-noise ratio is high, the criterion may be to obtain as narrow a pulse as possible, and hence the best range definition. The ideal filter for this purpose is known as an *inverse filter* or *equalizer*. Its basic purpose is to compensate for any distortions in the signal from the ideal delta function (δ function) response [36].

As the ideal inverse filter requires infinite bandwidth, a practical system must be designed on the basis of different criteria. One possible criterion is to design a filter that gives an output signal with the best mean-square fit to a desired signal (this may be a δ-function pulse, i.e., an infinitesimally narrow pulse) over a given time range or frequency range. Such a filter is known as a *Wiener filter* [36].

In this section we derive the theoretical properties of the matched filter. We then give examples of how such properties are realized in practice with analog tranversal filters. In Sec. 4.8 we follow the same procedures for inverse and Wiener filters.

4.4.2 Matched Filter Theory

The maximum peak signal-to-noise ratio for a given input signal in the presence of random noise is obtained by the use of a matched filter, that is, one for which the time response of the filter is the time-reversed version of the input signal. Here we demonstrate this general property mathematically.

We consider a signal $X(\omega)$ in the frequency domain passed into a filter with a response $H(\omega)$. The output from the device will be

$$Y(\omega) = X(\omega)H(\omega) \quad (4.4.1)$$

We may define the following Fourier transform pairs for an input signal $x(t)$ using a notation similar to Bracewell's [32]:

$$\begin{aligned} X(\omega) = \mathcal{F}[x(t)] &= \int_{-\infty}^{\infty} x(t)e^{-j\omega t}\, dt \\ &= \int_{-\infty}^{\infty} x(t)e^{-2j\pi ft}\, dt \\ x(t) = \mathcal{F}^{-1}[X(\omega)] &= \frac{1}{2\pi}\int_{-\infty}^{\infty} X(\omega)e^{j\omega t}\, d\omega \\ &= \int_{-\infty}^{\infty} X(f)e^{2j\pi ft}\, df \end{aligned} \quad (4.4.2)$$

with similar relations for the output signal $y(t)$. We shall define the transforms in terms of the radian frequency $\omega = 2\pi f$, and write, for example,

$$\begin{aligned} Y(\omega) = \mathcal{F}[y(t)] &= \int_{-\infty}^{\infty} y(t)e^{-j\omega t}\, dt \\ y(t) = \mathcal{F}^{-1}[Y(\omega)] &= \frac{1}{2\pi}\int_{-\infty}^{\infty} Y(\omega)e^{j\omega t}\, d\omega \end{aligned} \quad (4.4.3)$$

For a causal system, with a signal $x(t)$ and filter response $h(t)$ starting at a time $t = 0$, the response of the system in the time domain, equivalent to Eq. (4.4.1), is given by the convolution integral

$$\begin{aligned} y(t) = x(t) * h(t) &= \int_0^t x(\tau)h(t - \tau)\, dt \\ &= \int_0^t x(t - \tau)h(\tau)\, d\tau \end{aligned} \quad (4.4.4)$$

where $h(t)$ is the response of the filter to an impulse $\delta(t)$. The upper limit of the integrals is t so that $x(t - \tau) = 0$ and $h(t - \tau)$ if $t - \tau < 0$. We define the signal-to-noise ratio of the output from the filter as

$$\frac{S}{N} = \frac{\text{peak instantaneous output signal power}}{\text{output noise power}}$$

Our objective will be to determine the optimum form of $h(t)$ or $H(\omega)$, which yields the maximum value of S/N.

Let us suppose that the signal output is maximum at the time $t = T_d$, where T_d is the delay time through the filter. At this time

$$y(T_d) = \frac{1}{2\pi} \int_{-\infty}^{\infty} X(\omega)H(\omega)e^{j\omega T_d} \, d\omega \qquad (4.4.5)$$

It follows from Eq. (4.4.4) that we could equally well write

$$y(T_d) = \int_0^{T_d} x(\tau)h(T_d - \tau) \, d\tau$$
$$= \int_0^{T_d} x(\tau)h(T_d - \tau) \, d\tau \qquad (4.4.6)$$

The real instantaneous output power into a 1-Ω load is

$$S = y^2(t) \qquad (4.4.7)$$

Suppose that the amplitude of the noise input to the filter between the frequencies f and $f + df$ is $X_N(f) \, df$. Then the noise output from the filter at frequency f is

$$Y_N(f) = X_N(f)H(f) \qquad (4.4.8)$$

The total normalized mean-square noise power σ^2 at the filter output is the expectation value of the total noise output power, or

$$\sigma^2 = \left\langle \int_{-\infty}^{\infty} |Y_N(f)|^2 \, df \right\rangle = \left\langle \int_{-\infty}^{\infty} |X_N(f)H(f)|^2 \, df \right\rangle \qquad (4.4.9)$$

where $X_N(f)$ is defined for both positive and negative frequencies. Since the noise is assumed to be uncorrelated, or random, we can write

$$\sigma^2 = \frac{1}{2} \int_{-\infty}^{\infty} N_0(f)|H(f)|^2 \, df = \int_0^{\infty} N_0(f)|H(f)|^2 \, df \qquad (4.4.10)$$
$$\sigma^2 = \frac{1}{4\pi} \int_{-\infty}^{\infty} N_0(\omega)|H(\omega)|^2 \, d\omega$$

where $N_0(f) = N_0(-f)$ is the noise power spectral density in W/Hz, defined for positive frequencies. It follows that

$$\tfrac{1}{2} N_0(f) = \langle |X_N(f)|^2 \rangle \qquad (4.4.11)$$

We observe that $N_0(f)$ has the dimensions of energy, the product of power and time. As an example for Gaussian noise from a blackbody at temperature T, $N_0(f) = kT$, where k is Boltzmann's constant. It is usually assumed that filters have equal responses at positive and negative frequencies. It also follows from Parseval's theorem that when $N_0(f) = N_0$ is constant,

$$\sigma^2 = \frac{N_0}{2} \int_0^\infty h^2(t)\, dt \qquad (4.4.12)$$

Our aim is to maximize $y^2(T_d)/\sigma^2$. From now on we shall assume that $N_0(f) = N_0 =$ constant (see Prob. 1 for a more general relation), and write

$$\frac{S}{N} = \frac{\left| \int_0^{T_d} x(T_d - \tau) h(\tau)\, d\tau \right|^2}{\tfrac{1}{2} N_0 \int_0^\infty h^2(t)\, dt} \qquad (4.4.13)$$

We have taken the upper limit of the integral in the numerator as T_d in order to make $x(T_d - \tau) = 0$ if $T_d - \tau < 0$. It follows from Schwarz's inequality† that if $x(t)$ and $h(t)$ are real, then

$$\frac{S}{N} \leq \frac{\int_0^{T_d} x^2(T_d - t)\, dt \int_0^{T_d} h^2(t)\, dt}{\tfrac{1}{2} N_0 \int_0^\infty h^2(t)\, dt} \qquad (4.4.14)$$

The maximum value of S/N is given when the right-hand side of Eq. (4.4.14) is equal to the right-hand side of Eq. (4.4.13). Therefore, the maximum value of S/N is obtained at a time T_d if

$$\begin{aligned} h(t) &= \alpha x(T_d - t) & t < T_d \\ &= 0 & t > T_d \end{aligned} \qquad (4.4.15)$$

where α is a constant. Eqs. (4.4.13) and (4.4.14) become identical and the upper limit of the integral in the denominator is T_d.

The filter given by Eq. (4.4.15) is known as a *matched filter*. Its response is a time-delayed version of the time-reversed signal, and the output is the autocorrelation function of the signal to which it is matched [Eq. (4.4.19)]. Equations (4.4.13) and (4.4.14) show that the matched filter gives a maximum signal-to-noise ratio of

$$\left(\frac{S}{N} \right)_{max} = \frac{\int_0^{T_d} x^2(T_d - t)\, dt)}{N_0/2} = \frac{\int_0^{T_d} x^2(t)\, dt}{N_0/2} \qquad (4.4.16)$$

†One form of Schwarz's inequality is

$$\left| \int f(x) g(x)\, dx \right|^2 \leq \int |f(x)|^2\, dx \int |g(x)|^2\, dx$$

The left- and right-hand sides of this equation are equal only if $|g(x)| = \alpha |f(x)|$, where α is a constant. [If $f(x)$ and $g(x)$ are real, the modulus symbols may be omitted.]

We conclude that as the numerator in Eq. (4.4.16) is the total energy in the signal, the maximum signal-to-noise ratio of a signal, after passing through a matched filter, is determined only by its *total energy* E, not by the detailed structure of the signal. Thus we can write

$$\left(\frac{S}{N}\right)_{max} = \frac{2E}{N_0} \tag{4.4.17}$$

where for a signal of length T, it follows from Rayleigh's theorem that the energy in the waveform is

$$E = \int_0^T x^2(t)\, dt = \int_{-\infty}^{\infty} |X^2(f)|\, df \tag{4.4.18}$$

In practice, as the numerator of Eq. (4.4.16) must have the maximum possible value, the time T_d should be chosen to equal the length of the pulse T. This also implies that the delay time through the filter must be the length of the pulse for optimum signal-to-noise ratio. An infinitely long signal would ideally need an infinitely long delay line as a matched filter. In all cases, if the length of the signal is of a finite length T, the delay line must have a delay time $T_d \geq T$, in order to obtain the convolution of the signal within the time response of the delay line. When the tranversal filter is longer than necessary for a matched filter, it picks up more noise than the minimum possible value, and its response is decreased. We have already seen examples of devices using such principles in Sec. 4.2, where we discussed FM chirp filters and time-delay filters.

The response of a matched filter, with a delay time T to a pulse of length T, is the autocorrelation function of $x(t)$. We can write that the output from a matched filter is

$$y(t) = \int_{t_0}^{t} x(\tau)x(T - t + \tau)\, d\tau \tag{4.4.19}$$

The upper limit of the integral is determined by the condition that $h(t - \tau) = 0$ for $\tau > t$. The lower limit of the integral is determined by one of two conditions: When $t < T$, then $x(\tau) = 0$ if $\tau < 0$; thus $t_0 = 0$ for $t < T$. Alternatively, when $T < t < 2T$, then $x(T - t + \tau) = 0$ if $\tau < t - T$; thus $t_0 = t - T$ for $T < t < 2T$. Finally, if $t > 2T$, then $x(T - t + \tau) = 0$ for $\tau < t$; thus the output is zero.

Example: Matched Filter for a Square Pulse

We can put this concept on a more physical basis by considering a square pulse of length T, injected into a tapped delay line with a delay time $T_d = T$, as illustrated in Fig. 4.4.1. We assume that there is a large number of closely spaced taps which sample a small portion of the signal passing through the line. The output from these taps is then summed. The line is a matched filter because its response to a δ function pulse is a square pulse of length T. When the pulse is injected into the delay line, the output builds linearly with time, as the signal from an increasing number of taps is summed when the pulse reaches them. This increase in output signal continues until a time $t = T$. Suppose that z is the distance along the delay line, and the pulse moves into the delay line with a velocity V. Then the output from the line is from

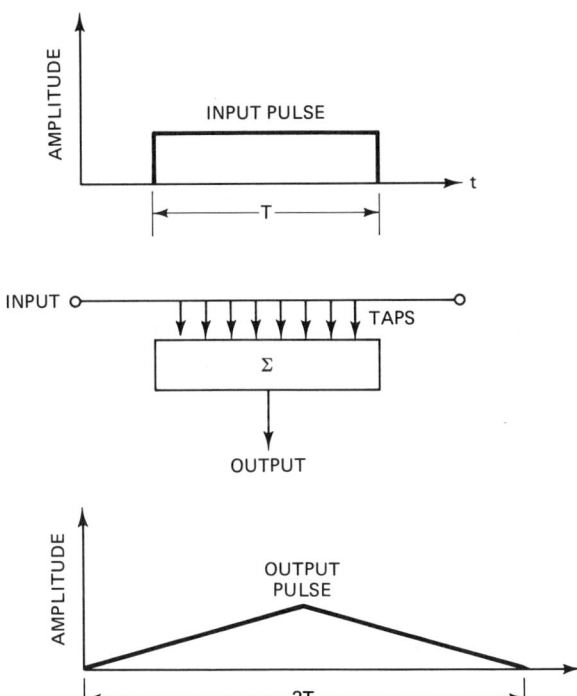

Figure 4.4.1 Tapped delay line of delay length T excited by a square pulse of length T.

a point $z = 0$ to $z = Vt$, so the limits of the integral in Eq. (4.4.19) are at 0 and t. Beyond this point, the taps at the front end of the line are no longer excited by the pulse of length T, and the output decreases linearly with time. Correspondingly, the lower limit of the integral corresponds in time to the point $z = V(t - T)$, where T is the pulse length, or in Eq. (4.4.19), $t_0 = t - T$. Finally, after a time $2T$, the pulse has passed through the line and there is no output.

We note that if the tapped delay line has a longer delay T_d than T, the output will increase to its maximum value, stay constant for a time $T_d - T$, and then decrease linearly with time to zero at a time $T_d + T$. As more taps are present, the noise output will be larger. Furthermore, the unused taps will tend to place a load across the output and decrease it.

If, on the other hand, the length of the pulse is $T > T_d$, the delay time through the line, the output signal cannot increase to as large a value as it would if $T_d = T$. Thus, once more, the line is not the ideal matched filter.

Matched filter in the frequency domain. We could just as well have carried out this derivation in the frequency domain, using Eqs. (4.4.5) and (4.4.9). It then follows that

$$\frac{S}{N} = \frac{1/(2\pi)\left|\int_{-\infty}^{\infty} X(\omega)H(\omega)e^{j\omega T_d}\,d\omega\right|^2}{(N_0/2)\int_{-\infty}^{\infty} |H(\omega)|^2\,d\omega} \qquad (4.4.20)$$

By using Schwarz's inequality, we find that

$$\left| \int_{-\infty}^{\infty} X(\omega)H(\omega)e^{j\omega T_d} \, d\omega \right|^2 \leq \int_{-\infty}^{\infty} XX^* \, d\omega \int_{-\infty}^{\infty} HH^* \, d\omega \quad (4.4.21)$$

where $X^*(\omega)$ is the complex conjugate of $X(\omega)$. Thus S/N is maximum when

$$H(\omega) = \alpha X^*(\omega)e^{-j\omega T_d} \quad (4.4.22)$$

where α is a constant. The matched filter has a frequency response that is the complex conjugate of the signal, and an additional phase delay associated with its length ωT_d. Note that a matched filter has an output in the frequency domain $Y(\omega) = \alpha XX^* \exp(-j\omega T_d)$; the maximum signal-to-noise ratio obtained with such a filter is

$$\left(\frac{S}{N}\right)_{max} = \frac{2 \int_{-\infty}^{\infty} X(f)X^*(f) \, df}{N_0} = \frac{2E}{N_0} \quad (4.4.23)$$

Thus the signal-to-noise ratio at the output of a matched filter is determined by the total energy in the input signal.

4.4.3 Pulse Compression

Now let us consider how the signal-to-noise ratio is improved by using a matched filter. We have shown, by a heuristic argument in Sec. 4.2, that if the input signal consists of a code of equal amplitude, positive or negative tone bursts, or an analog signal (such as an FM chirp of uniform peak amplitude), the output pulse will be compressed and will have a higher peak value than any part of the input signal. The improvement in signal-to-noise ratio is essentially determined by the time–bandwidth product of the input signal.

We may use the formulas for a matched filter to prove this concept directly. Suppose that the peak value of $x(t)$ is x_M. If the detector is band-limited, the input signal-to-noise ratio can be determined by taking $H(f) = 1$ over a bandwidth B. From Eq. (4.4.10), with $S(\max) = x_M^2$, we get the result

$$\left(\frac{S}{N}\right)_{in} = \frac{x_M^2}{N_0 B} \quad (4.4.24)$$

It follows, however, from Eq. (4.4.16), that at the output of a matched correlation filter, the signal-to-noise ratio is

$$\left(\frac{S}{N}\right)_{out} = \frac{\int_0^T x^2(t) \, dt}{N_0/2} \quad (4.4.25)$$

The *correlation gain* G, or the improvement in signal-to-noise ratio due to the filter, is therefore

$$G = \frac{(S/N)_{\text{out}}}{(S/N)_{\text{in}}} = \frac{2B \int_0^T x^2(t)\, dt}{x_M^2} \qquad (4.4.26)$$

Thus it follows that the improvement in signal-to-noise ratio, obtained by using a matched filter with a code of uniformly weighted tone bursts $\langle x^2(t) \rangle = x_M^2/2$, is

$$G = BT \qquad (4.4.27)$$

Therefore, the ideal improvement in signal-to-noise ratio gained by using a matched filter is equal to the time–bandwidth product of the signal. We note that if the amplitude weighting of the signal over the time T is not uniform or the frequency response of the filter is not uniform over the bandwidth of the system, the improvement in maximum signal-to-noise ratio will be less than BT, the time–bandwidth product of the signal.

4.4.4 Matched Filters for Complex Signals

Before proceeding further, it is worthwhile considering how a correlation filter processes a complex signal of the form

$$u(t) = a(t)e^{j\phi(t)} \qquad (4.4.28)$$

where $a(t)$ is the amplitude of the signal and ϕ is its phase. We normally take the real signal as

$$\begin{aligned} x(t) &= \text{Re}\,[u(t)] \\ &= a(t)\cos[\phi(t)] \end{aligned} \qquad (4.4.29)$$

Following the derivation of Eq. (4.4.16), we consider the power in a complex signal $u(t)$, across a 1-Ω load, to be $S = u(t)u^*(t)/2$. Hence the signal-to-noise ratio of a signal $u(t)$, when it is inserted into a filter $h(t)$, is

$$\left(\frac{S}{N}\right)_{\max} = \frac{\tfrac{1}{2}\int_{-\infty}^{\infty} u(T_d - \tau)h(\tau) \int_{-\infty}^{\infty} u^*(T_d - \tau)h^*(\tau)\, d\tau}{(N_0/2)\int_{-\infty}^{\infty} h(t)h^*(t)\, dt} \qquad (4.4.30)$$

From Schwarz's inequality, the signal-to-noise ratio is maximum (i.e., the system is a matched filter) when a complex matched filter is used with the property

$$h(t) = \alpha u^*(T_d - t) \qquad (4.4.31)$$

or

$$h_R(t) = \alpha x(T_d - t) \qquad (4.4.32)$$

as before. In Sec. 4.5 we see how a real filter processes a complex signal, and how a true complex filter can be constructed by using two filters.

When the signal is of finite length T, it follows that the maximum signal-to-noise ratio is

$$\left(\frac{S}{N}\right)_{max} = \frac{\frac{1}{2}\int_0^T u(t)u^*(t)\, dt}{N_0/2} = \frac{\frac{1}{2}\int_0^T a^2(t)\, dt}{N_0/2} \quad (4.4.33)$$

If we substitute the form of $u(t)$ given in Eq. (4.4.28) in the numerator of Eq. (4.4.33), at a time $t = T$, a matched filter exactly compensates for phase errors in the input signal, and the real output signal is just

$$y(t' = 0) = \frac{\alpha}{2}\int_0^T a^2(t)\, dt \quad (4.4.34)$$

Since $a^2(t)$ is always positive, all components of the signal add together to give the maximum output. The energy in the waveform is

$$y(t - T) = \frac{1}{2}\int_0^T a^2(t)\, dt \quad (4.4.35)$$

The maximum signal-to-noise ratio is therefore

$$\left(\frac{S}{N}\right)_{max} = \frac{2E}{N_0} \quad (4.4.36)$$

Thus the signal-to-noise ratio at the output of a matched filter is determined by the total energy in the signal.

4.4.5 Range and Velocity Accuracy of a Radar System

We are not necessarily interested in measuring the amplitude of a radar or communication signal. For example, a radar is used to determine the time delay of the return echo from a target as precisely as possible. Radar systems are also used to determine the Doppler frequency shift of a return echo; the Doppler frequency shift Δf is

$$\Delta f = \frac{2V f_0}{c} \quad (4.4.37)$$

where V is the velocity of the target, c is the velocity of light, and f_0 is the carrier frequency of the transmitted signal (see Prob. 4.2.6).

The signal returning from the target is corrupted by noise; this decreases the accuracy of both the range and frequency resolutions. We shall show that the resolution depends directly on the signal-to-noise ratio, and hence on the energy in the signal, and that the optimum time resolution is obtained with a matched filter.

Simple physical treatment. This result can be illustrated by a simple physical argument. Consider a rectangular pulse of amplitude A and rise time t_r, as illustrated in Fig. 4.4.2. Suppose that there is a random noise signal $n(t)$ present, as illustrated by the dashed line.

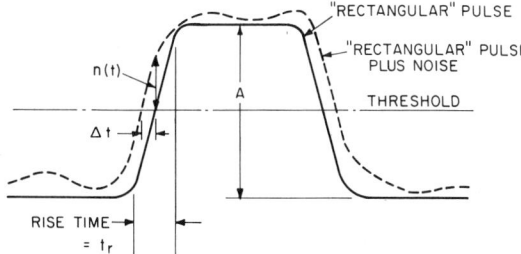

Figure 4.4.2 Measurement of time delay using the leading (or trailing) edge of the pulse. Solid curve, echo pulse uncorrupted by noise; dashed curve, effect of noise.

The slope of the leading edge of the pulse is approximately A/t_r. Thus at the point of measurement of the leading edge of the pulse, as shown in the figure, the error in rise time is

$$\Delta t = \frac{t_r n(t)}{A} \tag{4.4.38}$$

The mean-square error in Δt after many pulses is therefore

$$\langle (\Delta t)^2 \rangle = \frac{t_r^2 \langle n^2(t) \rangle}{A^2} \tag{4.4.39}$$

The maximum signal power into a 1-Ω load is $S = A^2$, and the average noise power is $N = \langle n^2(t) \rangle$. Hence we can write

$$[\langle (\Delta t)^2 \rangle]^{1/2} = [\langle N^2 \rangle]^{1/2} \frac{t_r}{A}$$
$$= \frac{t_r}{(S/N)^{1/2}} \tag{4.4.40}$$

If the pulse is the output from a matched filter, Eq. (4.4.36) yields the result

$$[\langle (\Delta t)^2 \rangle]^{1/2} = \left(\frac{N_0}{2E}\right)^{1/2} t_r \tag{4.4.41}$$

If the rise time is limited by the bandwidth B of the system, the output is of the form $A(\sin \pi Bt)/(\pi Bt)$.† The maximum slope is approximately $1.3 A/B$. The result is

$$[\langle (\Delta t)^2 \rangle]^{1/2} \approx \frac{1.3(N_0/2E)^{1/2}}{B} \tag{4.4.42}$$

Thus the error in measured range is least when the signal-to-noise ratio is as large as possible, and so is determined directly by the energy in the pulse. We therefore expect the optimum range resolution to be obtained with a matched filter.

More rigorous treatment. A more rigorous argument for the error in the measured time can be established. Let the total signal and noise-free signals

---

†We are concerned with radar signals, which are modulated carriers. Therefore, we have defined B as the total bandwidth of the radar signal and $B/2$ as the bandwidth of the video signal.

returning from the target be $x(t)$ and $x_0(t)$, respectively. Then
$$x(t) = x_0(t) + n(t) \tag{4.4.43}$$
where $n(t)$ is the noise signal. The output signal from a radar with a matched filter response $h(t) = x_0(-t)$, and maximum output at $t = 0$, is
$$y(t) = \int_{-\infty}^{\infty} x_0(\tau)x_0(\tau - t)\,d\tau + \int_{-\infty}^{\infty} n(\tau)x_0(\tau - t)\,d\tau \tag{4.4.44}$$
where the noise-free output signal, $y_0(t)$, is defined by the relation
$$y_0(t) = \int_{-\infty}^{\infty} x_0(\tau)(x_0(\tau - t)\,d\tau \tag{4.4.45}$$
Let the maximum value of $y_0(t)$ be $y_0(0)$. For short times Δt from the maximum, we use a Taylor expansion of $y_0(t)$ to obtain the result
$$y_0(\Delta t) = y_0(0) + \frac{(\Delta t)^2}{2}y_0''(0) + \cdots \tag{4.4.46}$$
where $y_0'(0) = 0$ when $y_0(t) = y_0(0)$, its maximum value. We will need $y_0^2(\Delta t)$ to second order in Δt. It follows from Eq. (4.4.46) that
$$y_0^2(\Delta t) = y_0^2(0) + (\Delta t)^2 y_0''(0)y_0(0) + \cdots \tag{4.4.47}$$
From Eq. (4.4.45),
$$y_0(0) = \int x_0^2(\tau)\,d\tau = E \tag{4.4.48}$$
and
$$y_0''(t) = \int x_0''(\tau - t)x_0(\tau)\,d\tau \tag{4.4.49}$$
Thus
$$y_0''(0) = \int_{-\infty}^{\infty} x_0''(\tau)x_0(\tau)\,d\tau = -4\pi^2 \int_{-\infty}^{\infty} f^2 X_0(f)X_0^*(f)\,df \tag{4.4.50}$$
It is convenient to define a quantity β, related to the bandwidth B as follows:
$$\beta^2 = \frac{4\pi^2 \int_{-\infty}^{\infty} f^2 X_0(f)X_0^*(f)\,df}{\int_{-\infty}^{\infty} X_0(f)X_0^*(f)\,df} \tag{4.4.51}$$
or
$$\beta^2 = \frac{4\pi^2}{E} \int_{-\infty}^{\infty} f^2 X_0(f)X_0^*(f)\,df \tag{4.4.52}$$
where E is the total energy in the pulse.

This definition of bandwidth was first introduced by Gabor and then used by Woodward in his treatment of range accuracy by inverse probability [34, 35]. As an example, a Gaussian pulse of the form $\exp[-(1.18t/\tau)^2]$ has its 3-dB points τ apart, a 3-dB bandwidth of $B = 0.44/\tau$, and a value of $\beta = 1.18/\tau$. We note that β^2 is the normalized second moment of the spectrum $|X(f)|^2$ and is expressed in rad/s. From its definition, for a fair comparison with B, we should use β/π. For a Gaussian pulse, $\beta/\pi = 0.38/\tau$.

It follows from Eqs. (4.4.50) and (4.4.52) that

$$y_0''(0) = -\beta^2 E \tag{4.4.53}$$

and

$$y_0^2(\Delta t) - y_0^2(0) = -\beta^2 E^2 (\Delta t)^2 \tag{4.4.54}$$

We wish to determine, as accurately as possible, the time when $y_0(t)$ is maximum. We therefore estimate the mean square error $\langle \Delta t^2 \rangle$. We define this time from the condition that the mean square error $\langle \varepsilon^2 \rangle$, in estimating $y_0(0)$, is equal to the noise, or

$$\langle \varepsilon^2 \rangle = \langle y_0^2(0) - y_0^2(\Delta t) \rangle = \beta^2 E^2 \langle (\Delta t)^2 \rangle \tag{4.4.55}$$

We can write from Eqs. (4.4.43), (4.4.44), and (4.4.55) that as the noise is uncorrelated with the signal,

$$\langle \varepsilon^2 \rangle = \left\langle \left| \int_{-\infty}^{\infty} n(\tau) x_0(\tau) d\tau \right|^2 \right\rangle = \beta^2 E^2 \langle (\Delta t)^2 \rangle \tag{4.4.56}$$

We use the following relation, obtained from Parseval's theorem [32] [Eqs. (4.4.11) and (4.4.48)]:

$$\left\langle \left| \int_{-\infty}^{\infty} n(\tau) x_0(\tau) d\tau \right|^2 \right\rangle = \left\langle \int_{-\infty}^{\infty} |N^2(f) X_0(f)|^2 df \right\rangle$$
$$= \tfrac{1}{2} N_0 \int_{-\infty}^{\infty} |X_0^2(f)| df = \frac{N_0 E}{2} \tag{4.4.57}$$

where N_0 is the noise power per Hz in the positive frequency range, and $N(f)df$, which is the same as $X_N(f)df$ in Eqs. (4.4.8) and (4.4.11), is the noise amplitude between frequencies f and $f + df$. It follows from Eqs. (4.4.56) and (4.4.57) that

$$\langle (\Delta t)^2 \rangle = \frac{N_0}{2\beta^2 E} \tag{4.4.58}$$

or

$$[\langle (\Delta t)^2 \rangle]^{1/2} = \frac{1}{\beta(2E/N_0)^{1/2}} \tag{4.4.59}$$

This result is therefore the more rigorously derived equivalent of Eq. (4.4.41). Note that the use of this more rigorous formula tends to lead to a slightly better time error than the approximate formula of Eq. (4.4.41).

We could have carried out a more complete derivation by taking the response of the radar receiver to be $h(t)$, and finding the optimum value of $h(t)$ for $\langle \Delta t^2 \rangle^{1/2}$

to be minimum. With the proper choice of normalization for the frequency response, it can be shown that the optimum filter is indeed a matched filter.

Accuracy of frequency measurement. The accuracy with which we can determine the Doppler shift of an echo depends on the signal duration T, rather than its bandwidth. By analogy to his normalized bandwidth function, Gabor has suggested a normalized signal duration defined by the relation [34, 35]

$$\alpha^2 = \frac{4\pi^2 \int_{-\infty}^{\infty} t^2 x_0^2(t)\, dt}{\int_{-\infty}^{\infty} x_0^2(t)\, dt} \qquad (4.4.60)$$

We note that $(\alpha/2\pi)^2$ is the normalized second moment of $x_0^2(t)$, and that the effective signal duration time is α/π. It can be shown by methods very similar to those we have used already (see Prob. 5) that the minimum rms error in the measurement of frequency is

$$[\langle(\Delta f)^2\rangle]^{1/2} = \frac{1}{\alpha(2E/N_0)^{1/2}} \qquad (4.4.61)$$

Thus the longer the time duration of the measuring signal, the better the accuracy in determining the Doppler shift in radar frequency; the larger the bandwidth, the better the range resolution.

Uncertainty relation. It is possible to obtain an uncertainty relation for radar systems [see the derivation following Eq. (4.4.65)], which states that the product of the effective bandwidth β and effective time duration α of a signal is such that

$$\beta\alpha > \pi \qquad (4.4.62)$$

This implies, from Eqs. (4.4.58) and (4.4.61), that as

$$[\langle(\Delta t)^2\rangle]^{1/2} [\langle(\Delta f^2)\rangle]^{1/2} = \frac{1}{\alpha\beta(2E/N_0)} \qquad (4.4.63)$$

then

$$[\langle(\Delta t)^2\rangle]^{1/2} [\langle(\Delta f^2)\rangle]^{1/2} < \frac{N_0}{2E\pi} \qquad (4.4.64)$$

We conclude that the time delay and frequency of a radar signal can be measured simultaneously to any desired accuracy by designing the system to yield a sufficiently large ratio of energy in the pulse-to-noise power per unit bandwidth. To measure both of these quantities accurately, the signal must have a long duration and large bandwidth. This conclusion, of course, is almost the opposite of the uncertainty principle in quantum mechanics, since in the latter case the $\beta\alpha$ product is fixed by the energy of one quantum rather than an energy quantity chosen by the observer.

Proof of the inequality. We can derive the inequality we have used, fairly simply, by noting that

$$\beta^2 = \frac{-\int_{-\infty}^{\infty} x_0''(t) x_0(t)\, dt}{\int_{-\infty}^{\infty} x_0^2(t)\, dt} \tag{4.4.65}$$

Integrating by parts, and assuming $x_0(t) \to 0$ as $t \to \pm\infty$, we find that

$$\beta^2 = \frac{\int_{-\infty}^{\infty} [x_0'(t)]^2\, dt}{\int_{-\infty}^{\infty} x_0^2(t)\, dt} \tag{4.4.66}$$

and

$$\beta\alpha = \frac{\left\{ 2\pi \int_{-\infty}^{\infty} [x_0'(t)]^2\, dt \int_{-\infty}^{\infty} t^2 x_0^2(t)\, dt \right\}^{1/2}}{\int_{-\infty}^{\infty} x_0^2(t)\, dt} \tag{4.4.67}$$

From Schwarz's inequality, it follows that

$$\int_{-\infty}^{\infty} x_0'(t)^2\, dt \int_{-\infty}^{\infty} t^2 x_0^2(t)\, dt \geq \left[\int_{-\infty}^{\infty} t x_0'(t) x_0(t)\, dt \right]^2 \tag{4.4.68}$$

Integrating by parts, and assuming that $x_0(t) \to 0$ as $t \to \pm\infty$, we can write

$$\int_{-\infty}^{\infty} t x_0'(t) x_0(t)\, dt = \tfrac{1}{2} \int t \frac{d}{dt}[x_0^2(t)]\, dt$$
$$= -\tfrac{1}{2} \int x_0^2(t)\, dt \tag{4.4.69}$$

Thus it follows from Eqs. (4.4.66)-(4.4.69) that

$$\beta\alpha \geq \pi \tag{4.4.70}$$

The equality sign in Eq. (4.4.68) holds only when $-Atx(t) = x'(t)$, where A is a constant. This relation yields a solution in the form of a Gaussian pulse, $\exp(-At^2/2)$. Therefore, the Gaussian pulse gives poorer simultaneous measurements of range and frequency than any other signal.

Narrowband waveforms. It is usually more convenient to measure the maximum amplitude or zero crossing of a waveform by working with the detected video signal rather than the carrier itself. This procedure avoids aliasing problems that result from the repetitive nature of the carrier. In this case, we may use the results we have already obtained with only a slight modification, provided that the signal has a "narrowband" waveform centered about a frequency ω_0. It is convenient to write

$$u(t) = a(t)e^{j\omega_0 t}$$

or (4.4.71)

$$u(t) = f(t)e^{j[\omega_0 t + \theta(t)]}$$

where the real part of $u(t)$ is $x(t)$, and we define $f(t)$ as a real quantity. For a *narrowband waveform*, $f(t)$ and $\theta(t)$ are defined as amplitude and phase functions that vary relatively slowly compared to $\omega_0 t$. We can write the real part of $u(t)$ in the form

$$\begin{aligned} x(t) &= f(t) \cos [\omega_0 t + \theta(t)] \\ &= \tfrac{1}{2}[u(t) + u^*(t)] \end{aligned} \quad (4.4.72)$$

The employment of the modulation function $a(t)$ in Eq. (4.4.71) is often no more convenient analytically than employing $u(t)$, unless $x(t)$ or $u(t)$ is a narrowband waveform. A narrowband waveform can be defined as one for which

$$\begin{aligned} U_+(\omega) &= 0 \quad \omega < 0 \\ U_-(\omega) &= 0 \quad \omega > 0 \end{aligned} \quad (4.4.73)$$

where

$$U_+(\omega) = \tfrac{1}{2}\int_{-\infty}^{\infty} u(t)e^{-j\omega t}\, dt \quad (4.4.74)$$

it follows from Eq. (4.4.71) that

$$U_+(\omega) = \tfrac{1}{2}\int_{-\infty}^{\infty} f(t)e^{-j(\omega - \omega_0)t}e^{j\theta(t)}\, dt \quad (4.4.75)$$

Similarly, we define $U_-(\omega)$ as follows:

$$U_-(\omega) = \tfrac{1}{2}\int_{-\infty}^{\infty} u^*(t)e^{-j\omega t}\, dt \quad (4.4.76)$$

or

$$U_-(\omega) = \tfrac{1}{2}\int_{-\infty}^{\infty} f(t)e^{-j(\omega + \omega_0)t}\, dt \quad (4.4.77)$$

where $u^*(t)$ is the complex conjugate of $u(t)$. Thus a narrowband waveform, as illustrated in Fig. 4.4.3(a), is one for which there is no spillover of $U_+(\omega)$ into the negative frequency region, due to a modulation signal $a(t)$. This can be seen from Eq. (4.4.75) if we assume that the only contributions to $U_+(\omega)$ are from the region near $\omega = \omega_0$, where the phase of the integrand varies only slowly. The narrowband assumption will be satisfied for most radar signals, as illustrated in Fig. 4.4.3(a), but not always for sonar or acoustic imaging devices that are operated with extremely short baseband pulses; this condition is illustrated in Fig. 4.4.3(b).

A linear detector, or envelope detector, can be regarded as obtaining $a(t)$ directly, that is, because of the narrowband nature of $u(t)$ or, more correctly, $x(t)$,

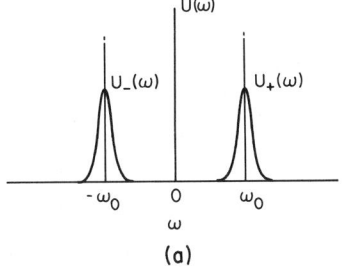

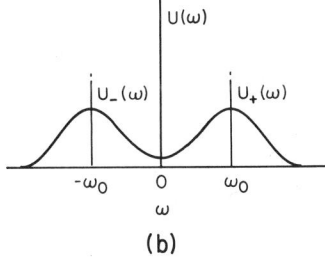

Figure 4.4.3 (a) Narrowband waveform with no "spillover" from positive frequencies to negative frequencies; (b) wideband waveform.

we can carry out sufficient filtering. After passing $u(t)$ through a matched filter, we obtain an output

$$y(t) = e^{j\omega_0 t} \int_{-\infty}^{\infty} a(\tau)a(\tau - t)\, d\tau \qquad (4.4.78)$$

$$= \frac{1}{2\pi} e^{j\omega_0 t} \int_{-\infty}^{\infty} |A(\Omega)|^2\, e^{j\Omega t}\, d\Omega$$

where $A(\Omega)$ is the Fourier transform of $a(t)$ and Ω is the radian modulation frequency. Thus we can write

$$A(\Omega) = \int_{-\infty}^{\infty} a(t)e^{-j\Omega t}\, dt \qquad (4.4.79)$$

$$= \int_{-\infty}^{\infty} f(t)e^{j\theta(t)}e^{-j\Omega t}\, dt$$

The stored energy in the real signal is

$$E = \int_{-\infty}^{\infty} x^2(t)\, dt \qquad (4.4.80)$$

$$= \tfrac{1}{4} \int_{-\infty}^{\infty} \{u^2(t) + [u^*(t)]^2 + 2u(t)u^*(t)\}\, dt$$

It follows from Parseval's theorem [32] that

$$E = \tfrac{1}{2} \int_{-\infty}^{\infty} [U(f)U(-f) + U(f)U^*(f)]\, df \qquad (4.4.81)$$

For a narrowband signal, $U(f) = 0$ when $U(-f)$ is finite, and vice versa. Hence

$$E = \tfrac{1}{2} \int_{-\infty}^{\infty} U(f)U^*(f)\, df$$
$$= \tfrac{1}{2} \int_{-\infty}^{\infty} A(f)A^*(f)\, df \quad (4.4.82)$$

We define a bandwidth parameter β_M for the modulated signal as follows:

$$\beta_M^2 = \frac{4\pi^2 \int_{-\infty}^{\infty} (f - f_0)^2 U(f)U^*(f)\, df}{\int_{-\infty}^{\infty} U(f)U^*(f)\, df} \quad (4.4.83)$$

When the waveform is narrowband, there is no spillover from the negative frequency terms near $f = -f_0$. Thus we can write

$$\beta_M^2 = \frac{4\pi^2 \int_{-\infty}^{\infty} (f - f_0)^2 |A(f)|^2\, df}{\int_{-\infty}^{\infty} |A(f)|^2\, df} \quad (4.4.84)$$

4.4.6 Ambiguity Function

In Sec. 4.4.5 we showed that the use of a matched filter provides optimum range and Doppler shift resolution. The choice of waveform to optimize both these parameters is a principal task of the radar designer. The designer's aim is to choose a waveform that will leave the least amount of ambiguity after the matched filtering process. To do this, it is convenient to define a function, the *ambiguity function* $|\chi(\Delta t, \Delta f)|^2$, where

$$\chi(\Delta t, \Delta f) = \int_{-\infty}^{\infty} u(s)u^*(s - \Delta t)e^{2j\pi s \Delta f}\, ds \quad (4.4.85)$$

Here we have used s as a floating variable instead of τ, which we shall reserve for other purposes. The ambiguity function is the basis of modern radar technology in the systematic search for the optimum waveform. As we shall show, it is proportional to the peak power output from the matched filter when the input signal time changes by Δt, and its frequency changes by Δf from the optimum values for peak power output from the matched filter.

We generalize Eq. (4.4.45) for a complex signal and write that the output at a time Δt due to an input signal $u(t)$ is

$$\chi(\Delta t, 0) = \int_{-\infty}^{\infty} u(s)u^*(s - \Delta t)\, ds \quad (4.4.86)$$

The input signal can be written in the form

$$u(t) = a(t)e^{j\omega_0 t} \quad (4.4.87)$$

If the carrier frequency is changed to a frequency ω so that

$$\omega = \omega_0 + 2\pi \, \Delta f \tag{4.4.88}$$

then when the input is changed to a radian carrier frequency ω, the output from a matched filter designed for the original signal of radian carrier frequency ω_0 is the parameter $\chi(\Delta t, \Delta f)$ defined in Eq. (4.4.85).

In accordance with the conventional notation for the ambiguity function, we put $\phi = -\Delta f$ and $\Delta t = -\tau$, and write

$$\chi(\tau, \phi) = \int_{-\infty}^{\infty} u(s)u^*(s + \tau)e^{-2j\pi s\phi} \, ds \tag{4.4.89}$$

By using Parseval's theorem [32], it follows that

$$\chi(\tau,\phi) = \int_{-\infty}^{\infty} U^*(f)U(f + \phi)e^{-2j\pi f\tau} \, df \tag{4.4.90}$$

From Eqs. (4.4.89) and (4.4.90) it is easy to show that

$$|\chi(\tau, \phi)|^2 = |\chi(-\tau, -\phi)|^2 \tag{4.4.91}$$

Thus the ambiguity function is symmetric around the origin, as might be expected, because the output power obtains its peak value when $\tau = 0$ and $\phi = 0$. We can also show that

$$\int_{-\infty}^{\infty}\int_{-\infty}^{\infty} |\chi(\tau, \phi)|^2 \, d\tau \, d\phi = |\chi(0, 0)|^2 \tag{4.4.92}$$

This property is usually referred to as the radar uncertainty principle, which is closely related to the uncertainty principles already derived in Sec. 4.4.5; it shows that the total potential ambiguity is the same for all signals that possess the same energy. The radar designer's aim is to distribute the ambiguity in the optimum way for a particular system.

Some examples of the distribution of ambiguity, plotted as functions of τ and φ, are given in Fig. 4.4.4. The first example is a short monotone pulse, shown in Fig. 4.4.4(a), for which a contour of constant amplitude is plotted. This takes the form of an ellipse with major and minor axes of lengths $\Delta\tau = T/2$ and $\Delta\phi = 1/2T$, respectively. Multiple peaks occur when a periodic pulse train is used, as shown in Fig. 4.4.4(c). These correspond to aliasing, or sidelobes, which must be considered in the design of the optimum radar system. Similar sets of ambiguity functions for FM pulses are shown in Fig. 4.4.4(b) and (d). The uniform pulse train shown in Fig. 4.4.4(c) degenerates to a distribution of points when the pulse train takes the form

$$u(t) = \sum_{n=-\infty}^{\infty} \delta(t - nT) \tag{4.4.93}$$

The corresponding ambiguity function is known as a "bed of nails." It is illustrated in Fig. 4.4.5.

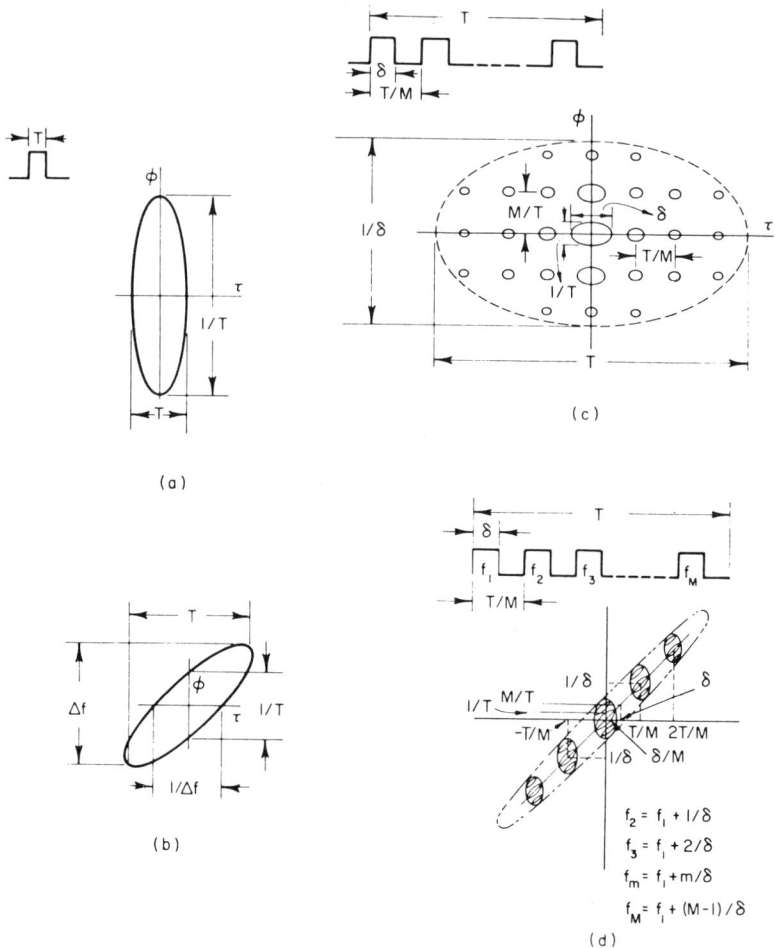

Figure 4.4.4 Examples of ambiguity function distributions, with contours at a constant amplitude level relative to $|\chi(0, 0)|$: (a) short monotone pulse; (b) linear FM pulse; (c) uniform pulse train; (d) pulse-to-pulse stepped FM pulse train. (After Bernfeld as noted in Cook and Bernfeld [1, 37].)

We can generalize the analysis for the range and velocity accuracy of a radar system given in Sec. 4.4.5, and write

$$\frac{|\chi(\tau, \phi)|^2}{|\chi(0, 0)|^2} = 1$$

$$+ \frac{1}{|\chi(0,0)|^2}\left[\tau^2\left(\frac{\partial^2\chi}{\partial\tau^2}\right)_{\tau=\phi=0} + 2\tau\phi\left(\frac{\partial^2\chi}{\partial\tau\partial\phi}\right)_{\tau=\phi=0} + \phi^2\left(\frac{\partial^2\chi}{\partial\phi^2}\right)_{\tau=\phi=0}\right] \quad (4.4.94)$$

Following the same kind of analysis as in Sec. 4.4.5, it can be shown that

$$\frac{|\chi(\tau, \phi)|^2}{|\chi(0, 0)|^2} = 1 - \beta^2\tau^2 + 2\alpha\beta\rho\tau\phi + \alpha^2\phi^2 \quad (4.4.95)$$

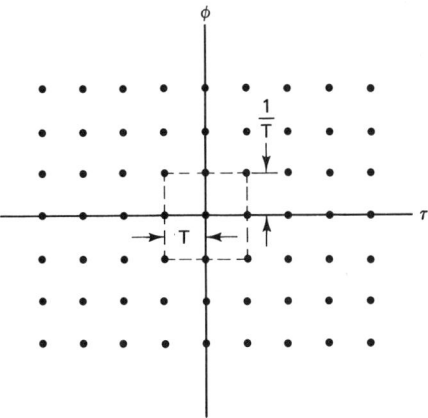

Figure 4.4.5 "Bed of nails" ambiguity function as an infinite train of impulses.

where

$$2E\alpha^2 = 4\pi^2 \int_{-\infty}^{\infty} t^2 |u(t)|^2 \, dt \tag{4.4.96}$$

$$2E\beta^2 = 4\pi^2 \int_{-\infty}^{\infty} f^2 |U(f)|^2 \, df \tag{4.4.97}$$

and

$$2E\rho = -\frac{2\pi}{\alpha\beta} \, \text{Im}\left\{\int_{-\infty}^{\infty} tu(t)[u^*(t)]' \, dt\right\} \tag{4.4.98}$$

The curve formed by the intersection of $|\chi(\tau, \phi)|^2$ with a level plane close to the maximum of $|\chi(0, 0)|^2$ is

$$\beta^2 \tau^2 + 2\alpha\beta\rho\tau\phi + \alpha^2 \phi^2 = \gamma^2 \tag{4.4.99}$$

This curve is an ellipse. By putting $\gamma^2 = N_0/2E$, Eq. (4.4.99) describes the *uncertainty ellipse*, corresponding to a generalized form of the mean-square error in range and frequency described in Sec. 4.4.5. A plot of this uncertainty ellipse is given in Fig. 4.4.6. It will be seen that, in general, the maximum errors in time and frequency are

$$\Delta\tau_{max} = \pm \frac{1}{2\beta\sqrt{E/N_0}} \frac{1}{\sqrt{1-\rho^2}} \tag{4.4.100}$$

and

$$\Delta\phi_{max} = \pm \frac{1}{2\alpha\sqrt{2E/N_0}} \frac{1}{\sqrt{1-\rho^2}} \tag{4.4.101}$$

These results correspond to generalizations of our earlier results for time and frequency error.

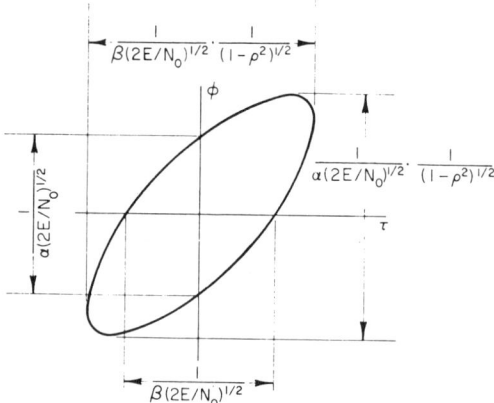

Figure 4.4.6 Uncertainty ellipse. (After Cook and Bernfeld [1].)

We have discussed general properties of the ambiguity function. The idealized ambiguity function would have the form of a spike at $\tau = 0$, $\phi = 0$, with zero amplitude elsewhere. Obviously, this is not possible.

We have discussed many times the linear FM pulse waveform, whose properties are illustrated in Fig. 4.4.7 together with a plot of $|\chi(\tau, \phi)|$. In this case it is very difficult to distinguish between changes in τ and ϕ. On the other hand, the V-FM pulse waveform consisting of a down-chirp followed by an up-chirp, as illustrated in Fig. 4.4.8, makes it relatively easy to distinguish between time and

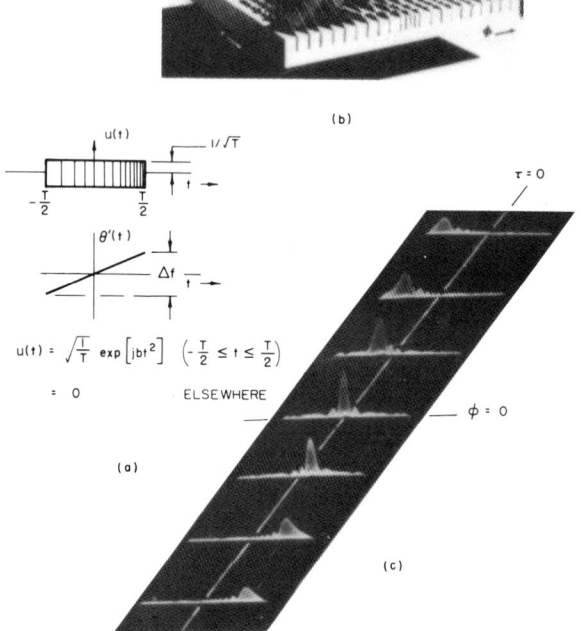

Figure 4.4.7 Linear FM pulse waveform and response function properties $T\Delta f = 10$: (a) waveform; (b) calculated surface; (c) experimental composite surface. (After Cook and Bernfeld [1].)

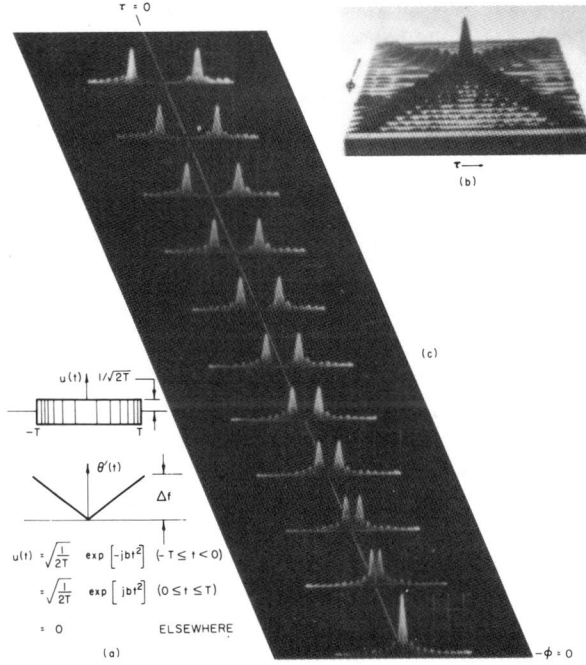

Figure 4.4.8 V-FM pulse waveform and response function properties $2T\Delta f = 200$: (a) waveform; (b) calculated composite response surface (note that pedestal is not included); (c) experimental composite response surface. (After Cook and Bernfeld [1].)

frequency changes. Time differences correspond to the difference in the τ direction between two peaks, while frequency differences correspond to movements along the ϕ axis. Parabolic FM pulse waveforms and other kinds of waveforms can provide still better improvements in the form of the ambiguity function (see Prob. 9).

A possible set of idealized waveforms are ones that are rotationally invariant in the ambiguity plane. The *Hermite polynomials* can be used for such a set. For the nth-order Hermite polynomial $H_n(x)$, the required signal has the form

$$u_n(t) = \frac{2^{1/4}}{\sqrt{n!}} e^{-\pi t^2} H_n(2\sqrt{\pi} t) \qquad (4.4.102)$$

where $H_0(x) = 1$; $H_1(x) = x$; $H_2(x) = x^2 - 1$; $H_3(x) = x^3 - 3x$; .... A plot of the composite response surface for the Hermite waveform, with $n = 10$, is shown in Fig. 4.4.9.

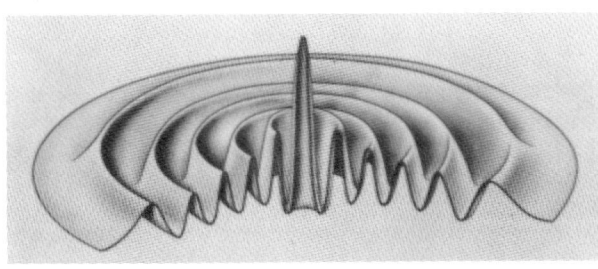

Figure 4.4.9 Composite response surface $\chi(\tau, \phi)$ for Hermite waveform, $n = 10$. (After Klauder, as noted in Cook and Bernfeld [1, 38].)

Obviously, there is no ideal ambiguity function. The ambiguity function is a measure of how the radar system distinguishes between similarly coded waveforms at the receiver that differ in time and frequency. The reader is referred to the standard references for more information on ambiguity functions [1].

PROBLEM SET 4.4

1. Determine the optimum matched filter and maximum signal-to-noise ratio when the noise power spectral density $N_0(f)$ in W/Hz is nonuniform over the frequency range.

    Hint: You will find it convenient to define a new function $M(f) = H(f)[N_0(f)]^{1/2}$.

2. The example in Sec. 4.4.2 of a matched filter for a square pulse of length T and frequency response $\sin(\omega T/2)/(\omega T/2)$ implied the use of a filter with an identical frequency response $\sin(\omega T/2)/(\omega T/2)$. Find an expression to determine how much an ideal low-pass filter, with a response

$$H(\omega) = \begin{cases} 1 & 0 < |\omega| < \omega_c \\ 0 & |\omega| > \omega_c \end{cases}$$

will reduce the signal-to-noise ratio from that of the ideal matched filter, when used with the same input pulse of length T. Plot the reduction factor R as a function of $\omega_c T$. From your numerical calculations, show that the signal-to-noise ratio is reduced by 0.84 dB, when $\omega_c T = 4.4$, where R is maximum. You can carry out a computation or use tables for

$$\text{Si}(a) = \int_0^a \frac{\sin x}{x} \, dx$$

You will also need the result

$$\int_{-\infty}^{\infty} \left(\frac{\sin x}{x}\right)^2 dx = \pi$$

3. Repeat Prob. 2 for a Gaussian filter with a response characteristic $\exp(-\omega/\omega_c)^2$. Find $\omega_c T$ for the maximum signal-to-noise ratio.

4. Consider the value of the root-mean-square (RMS) time error $[\langle \Delta t^2 \rangle]^{1/2}$ for an input signal $x(t) = x_0(t) + n(t)$, received by a radar with a response $h(t)$. Define a generalized bandwidth parameter β' as follows:

$$(\beta')^2 = \frac{4\pi^2 \int_{-\infty}^{\infty} f^2 X_0(f) H(f) \, df}{\int_{-\infty}^{\infty} X_0(f) H(f) \, df}$$

Show that with the effective bandwidth β' kept constant, the optimum response for minimum time error [i.e., a minimum value of $[\langle (\Delta t)^2 \rangle]^{1/2}$ near a time $t = 0$] will be minimum when

$$H(f) = K X_0^*(f)$$

where K is a constant.

5. Consider a narrowband radar signal of the form $u(t) = a(t)e^{j2\pi f_0 t}$. Write down the form of the matched filter for this waveform. Find $y_0(0)$, the maximum output at $t = 0$ when there is no noise present. Now suppose that we change the carrier frequency of the input signal to this matched filter, from f_0 to $f_0 + \Delta f$. Using the methods leading to Eq. (4.4.59), write down an expansion for $|Y_0(0, \Delta f)|^2$ (i.e., the change in output power when the input frequency is changed by Δf). Use this result to show that in the presence of noise, the mean-square error in determining f by measuring the point where the output amplitude is maximum is

$$\langle (\Delta f)^2 \rangle^{1/2} = \frac{1}{\alpha \sqrt{2E/N_0}}$$

where

$$\alpha^2 = \frac{4\pi^2}{2E} \int_0^T t^2 u(t) u^*(t) \, dt$$

and

$$E = \tfrac{1}{2} \int_0^T u(t) u^*(t) \, dt = \tfrac{1}{2} \int_0^T a^2(t) \, dt$$

6. Prove Eq. (4.4.92).
7. Prove Eq. (4.4.99).
8. (a) Show that for a Gaussian envelope $u(t) = (2k^2/\pi)^{1/4} \exp(-k^2 t^2)$, the function $\chi(\tau, \phi)$ has the form

$$\chi(\tau, \phi) = \exp\left[-\tfrac{1}{2}\left(k^2 \tau^2 + \frac{\pi^2 \phi^2}{k^2}\right)\right]$$

   (b) What are α, β, and ρ for this signal when τ and ϕ are small?

9. (a) Consider a radar waveform in the form of a parabolic FM chirp of the form

$$u(t) = \sqrt{\frac{1}{T}} \cos(ct^3) = \sqrt{\frac{1}{T}} \operatorname{Re}\left(e^{jct^3}\right)$$

   By using the method of stationary phase, show that

$$|\chi(\tau, \phi)| = \frac{1}{T}\sqrt{\frac{\pi}{3c\tau}} \cos\left(\frac{\pi^2 \phi^3}{3c\tau} - c\tau^3\right)$$

   Note that the result is similar to that for the V-FM waveform.

   (b) Find an expression for the contours of maximum $|\chi(\tau, \phi)|$, that is, the peaks of $|\chi(\tau, \phi)|$.

4.5 FM CHIRP FILTERS AND CHIRP TRANSFORM PROCESSORS

4.5.1 Introduction

In Sec. 4.2.3 we discussed the basic concept of an FM chirp filter. As we saw, if a dispersive delay line (i.e., delay line whose time delay varies with frequency) is used as a matched filter for a linear FM chirp of length T, an output pulse of length

τ_p will be obtained. The pulse width τ_p is determined by the bandwidth B of the delay line, and $\tau_p \approx 1/B$. The signal-to-noise ratio is therefore improved by a factor of $T/\tau_p \approx BT$, the compression ratio.

Furthermore, as we have seen, the system can be used to obtain real-time Fourier transforms of modulated chirp signals. In this section we first consider a mathematical treatment of these concepts; we then describe various SAW and CCD realizations for chirp filters, and their applications to filtering complex signals.

4.5.2 Mathematical Treatment of FM Chirp Filter

We now examine in more detail what occurs when a linear FM chirp signal is inserted into a matched filter, as well as the problems associated with sidelobes and the techniques used to decrease them. We consider more general filters derived from the basic matched filter, which have time delays not necessarily equal to the length of the input pulse. We also consider how real physical filters can be used for complex signals.

A linear FM chirp signal, with a radian frequency that varies with time as $\omega = \omega_0 + \mu t$, has a phase

$$\phi(t) = \int \omega \, dt = \omega_0 t + \frac{\mu t^2}{2} \tag{4.5.1}$$

where the slope of the chip μ has the dimensions of rad/s^2. We can therefore write a modulated chirp signal in the form

$$u(t) = f(t) e^{j(\omega_0 t + \mu t^2/2)} \tag{4.5.2}$$

where $f(t)$ is, in general, the complex amplitude modulation of the FM chirp. If we insert this signal in a matched filter, it follows from Eq. (4.4.31) that the matched filter characteristic is of the form

$$h(t) = \alpha u^*(T_0 - t) = f^*(T_0 - t) e^{j(\omega_0 t - \mu t^2/2)} \tag{4.5.3}$$

where α is a constant of the filter and T_0 is the time at which the output signal will reach its peak value. To simplify the algebra, we shall assume from now on that $T_0 = 0$, and that the FM chirp can start at a time earlier than $t = 0$, as dictated by the value of $f(t)$. We will generalize the form of the filter from that of a simple matched filter and use it as a matched filter only for the carrier $\exp[j(\omega_0 t + \mu t^2/2)]$. Therefore, we write

$$h(t) = \alpha w_0(t) e^{j(\omega_0 t - \mu t^2/2)} \tag{4.5.4}$$

The parameter $w_0(t)$ is called the *weighting* of the filter, while $f(t)$ is called the *modulation of the chirp*. In the simplest case, $f(t) = 1$ and $w_0(t) = 1$ for $-T/2 < t < T/2$ with $f(t) = 0$, and $w_0(t) = 0$ for $|t| > T/2$. This is an exact matched filter.

It is convenient to use one of the convolution formulae for the output signal

$$y(t) = \int_{-\infty}^{\infty} u(\tau) h(t - \tau) \, d\tau \tag{4.5.5}$$

or

$$y(t) = \int_{-\infty}^{\infty} u(t-\tau)h(\tau)\,d\tau \qquad (4.5.6)$$

When the filter is of finite length T and the chirp length is much longer than T, we take $f(t) = 1$ for $-\infty < t < \infty$, with $w_0(t) = 1$ for $-T/2 < t < T/2$, and $w_0(t) = 0$ for $|t| > T/2$. In this case, it is convenient to use Eq. (4.5.6). If the output is maximum at $t = 0$, and $h(t) = \alpha \exp[j(\omega_0 t - \mu t^2/2)]$, then

$$y(t) = \alpha \int_{-T/2}^{T/2} e^{j[\omega_0(t-\tau) + \mu(t-\tau)^2/2]} e^{j(\omega_0\tau - \mu\tau^2/2)}\,d\tau \qquad (4.5.7)$$

It follows that

$$y(t) = \alpha e^{j(\omega_0 t + \mu t^2/2)} \int_{-T/2}^{T/2} e^{-j\mu t\tau}\,d\tau \qquad (4.5.8)$$

This expression can be integrated to give the result

$$y(t) = \alpha T \frac{\sin(\mu T t/2)}{\mu T t/2} e^{j(\omega_0 t + \mu t^2/2)} \qquad (4.5.9)$$

Width of main lobe. The output is now sharply peaked around the time $t = 0$, with a maximum amplitude proportional to the time length of the delay line or filter T, as we might expect. Furthermore, the amplitude of the signal drops 4 dB at the points where $\mu T t = \pi$. Thus we can define the 4-dB width of the pulse in seconds as

$$\tau_p(4\text{ dB}) = \frac{2\pi}{\mu T} = \frac{1}{B} \qquad (4.5.10)$$

and the 3-dB width as

$$\tau_p(3\text{ dB}) = \frac{0.89}{B} \qquad (4.5.11)$$

where $B = \mu T/2\pi$ is the bandwidth of the chirp in hertz and μT is its bandwidth in rad/s. Thus the pulse compression ratio (to the 3-dB points) is $\tau_p/T = 0.89/BT$, close to what we would have expected from general considerations.

Sidelobes. We observe that the first sidelobe is at $t = \pm 3/2B$, with an amplitude $2/3\pi$ smaller than the main lobe. Thus the first sidelobe has a level only 13 dB down from the main lobe. In a radar system (or any other signal processing system) this can cause serious problems: If a later weak signal (from a weak reflector) enters the filter and has an amplitude more than 13 dB below that of the main signal, it may be masked by a sidelobe from the main signal, or else a sidelobe might be mistaken for an echo from a weaker target. For this reason it is usually necessary to design radar systems with a modified filter or chirp signal, which is used to decrease the sidelobe level at the expense of slightly increasing the width of the main lobe and slightly worsening the optimum signal-to-noise ratio. The procedures used are similar to those employed in designing antennas or interdigital transducers to obtain low sidelobe levels.

Physics of FM chirp filter. To put this derivation on a more physical basis, we can consider the output obtained directly from an SAW chirp filter of the type illustrated in Fig. 4.2.7. Suppose that the input to the filter were of the form given in Eq. (4.5.2). After the signal has traveled along the delay line a distance z from the input toward the matching dispersive transducer, as shown in Fig. 4.2.7, it will have an amplitude

$$u\left(t - \frac{z}{V}\right) = f\left(t - \frac{z}{V}\right) e^{j\omega_0(t - z/V)} e^{j\mu(t - z/V)^2/2} \qquad (4.5.12)$$

where V is the velocity of the wave along the delay line, and we have included a general modulation function $f(t)$. Suppose that the length of the finger at the plane z is $w(z)$, and that the weighting $w(z)$ may be positive or negative. Using the apodizing technique illustrated in Fig. 4.2.6, with closely spaced fingers of infinitesimal width, we would expect the current induced on a finger to be proportional to $w(z)$. Thus the total output is of the form

$$y(t) = \gamma \int_{-L/2}^{L/2} w(z) u\left(t - \frac{z}{V}\right) dz \qquad (4.5.13)$$

where L is the spatial length of the delay line, taken to start at $z = -L/2$, and γ is a constant.

We choose $w(z)$ to have the physically realizable form

$$w(z) = w_0(z) \cos\left(\frac{\omega_0 z}{V} - \frac{\mu z^2}{2V^2}\right) \qquad (4.5.14)$$

so that $w(z)$ is positive, negative, or zero. We can split the expression for $w(z)$ into two parts

$$w(z) = \frac{w_0(z)}{2} \left[e^{j(\omega_0 z/V - \mu z^2/2V^2)} + e^{-j(\omega_0 z/V - \mu z^2/2V^2)} \right] \qquad (4.5.15)$$

and we now take $w_0(z) = w_0 = $ constant for $|z| < L/2$. Then the contribution of the dominant cumulative term from Eqs. (4.5.14) and (4.5.15) [i.e., the first term in Eq. (4.5.15)] to the integral in Eq. (4.5.13) is

$$y(t) \approx \frac{\gamma w_0}{2} \int_{-L/2}^{L/2} f\left(t - \frac{z}{V}\right) e^{j[\omega_0(t - z/V) + \mu(t - z/V)^2/2]} e^{j(\omega_0 z/V - \mu z^2/2V^2)} dz \qquad (4.5.16)$$

where, in analogy with our previous example, we have taken the chirp to be very long, and the delay line to have its input at $-L/2$ and to be of length L. Setting $L = VT$ and $z = V\tau$, Eq. (4.5.16) reduces to the form

$$y(t) = \frac{\gamma V w_0}{2} e^{j(\omega_0 t + \mu t^2/2)} \int_{-T/2}^{T/2} f(t - \tau) e^{-j\mu t \tau} d\tau \qquad (4.5.17)$$

If $f(t) = 1$, Eq. (4.5.17) will be identical in form to Eq. (4.5.8). In this case, the maximum amplitude of the output is at $t = 0$ and of value $y(0) = \lambda V w_0 T/2$. But note that if we had chosen the delay line to have a time delay much longer than the length of the chirp, our results would have been different (see Prob. 1).

In using a physical delay line, we have chosen the elements to have a cosine weighting. Therefore, in deriving the filter response, we had to neglect the contribution of the second term in Eq. (4.5.15). It can be shown that in a SAW device, the output signal due to this effect is negligible (see Prob. 2). A device that operates at baseband, however, like a CCD filter, can give rise to spurious signals associated with the interaction of an unwanted cosine or sine component of the chirp signal with the delay line. In Sec. 4.5.5 we describe how a pair of sine and cosine weighted delay lines can be used to synthesize a complex filter to eliminate this problem.

We have described here a system with fingers that are infinitesimally spaced (i.e., that continuously sample the signal). A real SAW or CCD delay line would employ sampling elements a finite length apart. One possibility would be to choose the elements to be equally spaced and amplitude-weighted in the manner prescribed by Eq. (4.5.14), with the elements at positions $z_n = z = nl$, where l is the spacing of the elements. Alternatively, as shown in Fig. 4.2.7, the elements would be placed at positions $z = z_n$ corresponding to the maxima of the cosine function, such that

$$\frac{\omega_0 z_n}{V} - \frac{\mu z_n^2}{2V^2} = n\pi \qquad (4.5.18)$$

In both cases, there are usually at least two elements per RF wavelength to satisfy the Nyquist sampling criterion.

4.5.3 Filter Weighting for Sidelobe Reduction

Before we proceed, let us consider how we can weight the filter to give a response with low sidelobe levels [1, 39]. Otherwise, if the sidelobe levels are high, as shown in Eq. (4.5.9), two chirp signals arriving at different times from two different targets will both give $\sin(\mu tT/2)/(\mu tT/2)$ responses after passing through the matched filter. If the sidelobe level of this response is too high, the main lobe of the weak signal may be obscured by the sidelobe from the strong signal. The basic reason for this difficulty comes from the fact that the FM chirp signal and/or the matched filter may be finite in length.

It is obviously desirable to smooth out the response of the matched filter; therefore, we consider weighting the filter by the weighting function $w_0(t)$ given in Eq. (4.5.4). Similarly, if we were using a physical transversal filter, we would have chosen an equivalent filter of the form given by Eq. (4.5.14). But in this case, it follows from Eq. (4.5.6) that a similar derivation to that of Eq. (4.5.17) yields the output from the matched filter in the form

$$y(t) = \alpha e^{j(\omega_0 t + \mu t^2/2)} \int_{-\infty}^{\infty} e^{-j\mu t\tau} w_0(\tau)\, d\tau \qquad (4.5.19)$$

Here we have used infinite limits for the integral by assuming that $w_0(\tau)$ has a finite length, and we have taken $\tau = z/V$ in the physical filter as before. We call the weighting of a finite-length filter *apodization*.

Note that the output signal is the *Fourier transform* of the weighting of the filter. To obtain optimum weighting, we want to choose $w_0(\tau)$ such that $S(\mu t)$, defined by the integral

$$S(\mu t) = \int_{-T/2}^{T/2} e^{-j\mu t \tau} w_0(\tau) \, d\tau \qquad (4.5.20)$$

with a main lobe centered at $t = 0$, has a minimum width and minimum sidelobe levels.

As a simple example, suppose that we take $w_0(\tau)$ to be a Gaussian function with the filter infinitely long, such that

$$w_0(\tau) = e^{-8(\tau/T)^2} \qquad (4.5.21)$$

The effective length of this filter between the $1/e^2$ amplitude points is T, and the bandwidth between these points is $B = \mu T/2\pi$. The transformed output then takes the form

$$U(\mu t) = \frac{T\sqrt{\pi}}{2\sqrt{2}} e^{j(\omega t + \mu t^2/2)} e^{-(\mu t T)^2/32} \qquad (4.5.22)$$

The output is a Gaussian pulse with no sidelobes, and with an effective pulse length $\tau_p(3 \text{ dB})$ between 3-dB points of $\tau_p(3 \text{ dB}) = 6.66/\mu T$, or $\tau_p(3 \text{ dB}) = 1.06/B(1/e^2)$. Thus the use of an infinitely long Gaussian waveform, of approximately the same bandwidth or time length between $1/e^2$ points as that of the rectangularly weighted chirp, increases the width of the main lobe by a factor of 1.06/0.89, or 1.19.

In practice, of course, it is not possible to use a *Gaussian* weighting, for this would imply a filter of infinite length. We might approximate the filter by cutting it off where the amplitude drops to $1/e^2$; to do so, however, would increase the sidelobe level considerably, and this design method is still inefficient because it requires a much longer filter length than necessary. So techniques identical to those used for antenna array design and digital filter design have been developed. The aim is to design a finite-length weighted filter with the lowest possible sidelobe level and the narrowest possible main lobe. Such a filter normally has a fairly close approximation to a Gaussian weighting.

We conclude that the basic aim of apodization is to use extremely smooth functions for the windowing. The difficulty, of course, is that the filter must be finite in length. Some examples of apodization functions are given in Table 4.5.1. Plots of these functions, and of the main lobes and sidelobes arising from them, are given in Fig. 4.5.1; Table 4.5.2 summarizes the results obtained with them.

As an example, observe that *Hanning* weighting with $w_0(\tau) = \cos^2 \pi\tau/T$ is smooth and fairly close in form to a Gaussian function. It yields a maximum sidelobe level of -32 dB, with an increase in compressed 3-dB pulse width of 1.62 times that for *rectangular* weighting. Its advantage compared to rectangular weighting is that the far-out sidelobe amplitudes fall off as $1/t^3$. On the other hand, *Hamming*

Table 4.5.1 SOME EXAMPLES OF APODIZATION FUNCTIONS

Dirichlet (rectangular)	$w_0(\tau) = 1$	
Bartlett (triangular)	$w_0(\tau) = 1 + 2\tau/T$	$-T/2 < \tau < 0$
	$w_0(\tau) = 1 - 2\tau/T$	$0 < \tau < T/2$
Hanning	$w_0(\tau) = \cos^2(\pi\tau/T)$	$-T/2 < \tau < T/2$
	$= 0.5[1 + \cos(2\pi\tau/T)]$	
Hamming	$w_0(\tau) = 0.08 + 0.92 \cos^2(\pi\tau/T)$	$-T/2 < \tau < T/2$
	$= 0.54 + 0.46 \cos(2\pi\tau/T)$	
Blackman	$w_0(\tau) = 0.42 + 0.5 \cos(2\pi\tau/T) + 0.08 \cos(4\pi\tau/T)$	$-T/2 < \tau < T/2$
Finite Gaussian	$w_0(\tau) = \exp[-12.5(\tau/T)^2]$	$-T/2 < \tau < T/2$
Infinite Gaussian	$w_0(\tau) = \exp[-12.5(\tau/T)^2]$	

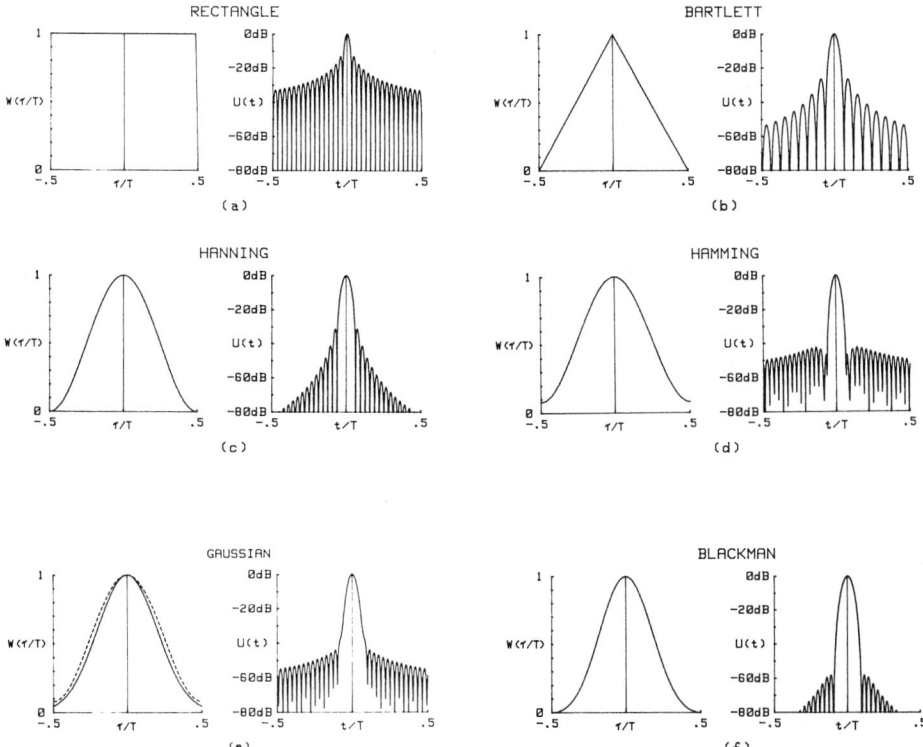

Figure 4.5.1 Some examples of windows and their Fourier transforms. All examples are for a time–bandwidth product of $BT = 50$. (a) Rectangular window and log magnitude of transform; (b) Bartlett or triangular window and log magnitude of transform; (c) Hanning window and log magnitude of transform; (d) Hamming window and log magnitude of Fourier transform; (e) finite Gaussian window $\exp[-12.5(\tau/T)^2]$ (solid line) and log magnitude of its transform compared to the Hamming window (dashed line); (f) Blackman window and log magnitude of transform.

Sec. 4.5   FM Chirps Filters and Chirp Transform Processors

Table 4.5.2 PROPERTIES OF WEIGHTING FUNCTIONS

Weighting function	Peak sidelobe level (dB)	$B\tau_p$(3 dB)[a]	$B\tau_p$(6 dB)[a]	R (dB) [Eq. (4.5.26)]	Far sidelobe falloff rate
Rectangular	−13	0.89	1.21	0	$1/t$
Bartlett (triangle)	−27	1.28	1.78	−1.23	$1/t^2$
Hanning $\cos^\alpha(\pi\tau/T)$					
$\alpha = 1$	−23	1.20	1.65	−0.90	$1/t^2$
$\alpha = 2$	−32	1.44	2.00	−1.76	$1/t^3$
$\alpha = 3$	−39	1.66	2.32	−2.20	$1/t^4$
$\alpha = 4$	−47	1.86	2.59	−2.88	$1/t^5$
Hamming	−43	1.3	1.81	−1.34	$1/t$
Finite Gaussian $\exp[-12.5(\tau/T)^2]$	−42	1.33	1.86	−1.43	$1/t$
Infinite Gaussian $\exp[-12.5(\tau/T)^2]$	—	1.32	1.87	—	$\exp\left[\dfrac{-(\pi Bt)^2}{12.5}\right]$
Blackman	−58	1.68	2.35	−2.38	$1/t^3$

[a] The bandwidth in this table is $B = \mu T/2\pi$.

weighting, a commonly used weighting function, yields a somewhat narrower pulse with lower sidelobe levels, with the disadvantage of a lower rate of amplitude falloff for the far-out sidelobes. We shall see later that improvement of this latter characteristic can be useful for designing high-quality bandpass filters with very low spurious levels outside the passband. Note that Hamming weighting is very close in form and gives similar results to the cutoff Gaussian $\exp[-12.5(\tau/T)^2]$.

Weighting functions other than the ones given here are sometimes used to obtain an optimum response. The reader is referred to the literature for discussion of *Dolph–Tchebysheff* weighting, to which Hamming, Hanning, and *Blackmann* weightings form approximations [1, 39]. A very flexible form of weighting is *Kaiser* weighting, defined by the relation

$$w(\tau) = \frac{I_0[(\omega_a T/2)(1 - 4\tau^2/T^2)^{1/2}]}{I_0(\omega_a T/2)} \quad (4.5.23)$$

where I_0 is a modified Bessel function of zero order, and ω_a is a parameter that can be adjusted to trade the main lobe width against the sidelobe amplitude. Kuo and Kaiser have shown that these windows tend to yield the largest energy in the main lobe for a given peak sidelobe amplitude [40].

Effect of apodization on signal-to-noise ratio. Now that the filter is no longer perfectly matched, we might expect nonuniform weighting to decrease its signal-to-noise ratio. The signal-to-noise ratio is

$$\frac{S}{N} = \frac{\left|\int_0^\infty u(T-t)h(t)\,dt\right|^2}{\frac{1}{2}N_0 \int_0^\infty |h^2(t)|\,dt} \tag{4.5.24}$$

With $t = 0$, we take

$$h(t) = \alpha w_0(t)u^*(-t) \tag{4.5.25}$$

and assume that the input signal has a rectangular envelope. The reduction in signal-to-noise ratio, due to weighting, will be

$$R = \frac{(S/N)_{\text{weighted}}}{(S/N)_{\text{rectangular}}} = \frac{\langle w_0 \rangle^2}{\langle w_0^2 \rangle} \tag{4.5.26}$$

In most cases, as we can see from Table 4.5.2, the penalty paid in signal-to-noise ratio with the use of common apodizations is small.

4.5.4 Fourier Transform Operations with FM Chirp Filters

In Sec. 4.5.2 we showed that the output from a weighted chirp filter is the Fourier transform of the weighting. We have also shown heuristically in Sec. 4.2.3 that if an FM chirp signal is modulated by a signal $f(t)$ injected into a matched filter, we might expect to obtain an output that is the Fourier transform of $f(t)$. We shall now consider this operation mathematically and give some examples of its use.

We shall take the input signal to be a modulated "down-chirp," that is, a chirp whose frequency is decreasing with time

$$u(t) = f(t)e^{j(\omega_0 t - \mu t^2/2)} \tag{4.5.27}$$

We assume that the matched filter is unapodized [$w(\tau) = 1$] and very long. Thus we use Eq. (4.5.5) [i.e., $y(t) = \int u(\tau)h(t-\tau)\,d\tau$] to show that the output from the chirp filter, with a modulated input chirp of length T, is

$$Y(\mu t) = \alpha \int_{-T/2}^{T/2} f(\tau)e^{j(\omega_0\tau - \mu\tau^2/2)}e^{j[\omega_0(t-\tau)+\mu(t-\tau)^2/2]}\,d\tau$$

$$= \alpha e^{j(\omega_0 t + \mu t^2/2)} \int_{-T/2}^{T/2} f(\tau)e^{-j\mu t\tau}\,d\tau \tag{4.5.28}$$

We put

$$F(\mu t) = \alpha \int_{-T/2}^{T/2} f(\tau)e^{-j\mu t\tau}\,d\tau \tag{4.5.29}$$

Then in this case, with the chirp length much shorter than the filter length, the output from the chirp filter is just the Fourier transform $F(\mu t)$ of the input signal,

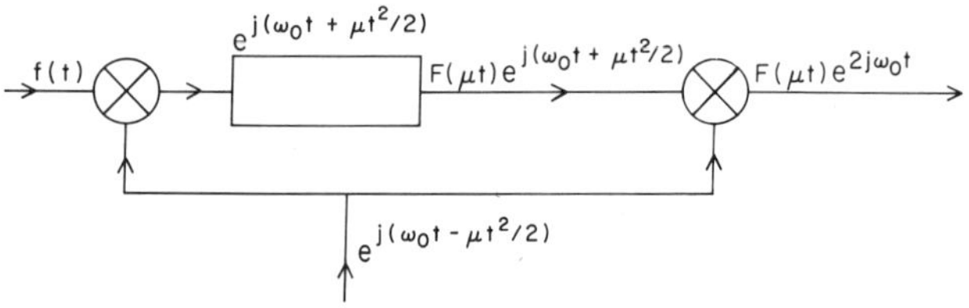

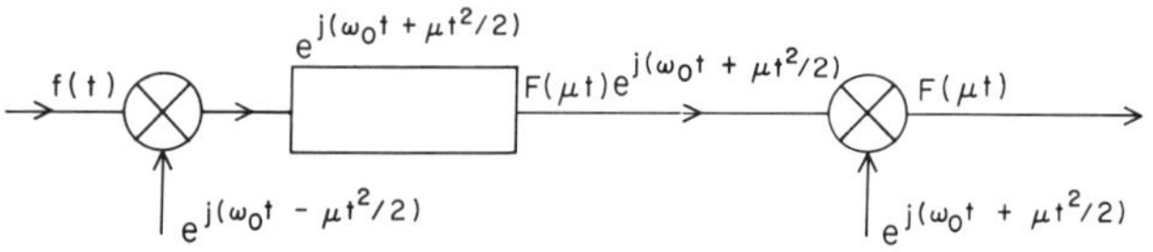

Figure 4.5.2 Two alternative schemes for obtaining a complex Fourier transform: (a) frequency transform obtained as the modulation of a carrier of frequency $2\omega_0$. (b) Fourier transform obtained as a baseband signal.

multiplied by an "up-chirp." The output signal is now a function of time, but μt represents the radian frequency in the Fourier transforms. If we require only $|F(\mu t)|$, it is sufficient to determine $|Y(\mu t)|$. However, if we need both the phase and amplitude of $F(\mu t)$, we can post-multiply the output signal $Y(\mu t)$ by a further down-chirp, $\exp[j(\omega_0 t - \mu t^2/2)]$, and obtain $F(\mu t) \exp(2j\omega_0 t)$. In this case, $F(\mu t)$ is the modulation of a carrier of frequency $2\omega_0$. Alternatively, as a mixer multiplies two signals together, signals of the forms $\cos(\omega_1 t + \phi_1)$ and $\cos(\omega_2 t + \phi_2)$ yields outputs at their sum and difference frequencies and phases, that is, with form $\cos[(\omega_1 + \omega_2)t + \phi_1 \pm \phi_2]$. We could therefore multiply by an up-chirp and obtain the output $F(\mu t)$ at baseband.

We summarize the required system in Fig. 4.5.2. In the first case, the signal is multiplied by a chirp $\exp[j(\omega_0 t - \mu t^2/2)]$, inserted into the matched filter for this chirp, and then post-multiplied by the same chirp. Observe that if the input is at baseband ($\omega_0 = 0$), the output obtained will be the direct Fourier transform of $f(t)$. Alternatively, with ω_0 finite, the modulation of the carrier is $F(\mu t)$, provided that the mixing chirp is delayed by a time T. Otherwise, we obtain a different modulated carrier as the output Fourier transform.

To determine the resolution of this Fourier transform device, let us consider what occurs when we insert a signal $f(t) = \exp(j\Omega t)$ into the Fourier transform system. The output we obtain is

$$Y(\mu t) = \alpha e^{j(\omega_0 t + \mu t^2/2)} \int_{-T/2}^{T/2} e^{j(\Omega - \mu t)\tau} \, d\tau \qquad (4.5.30)$$

This formula may be integrated to give the result

$$|Y(\mu t)| = \alpha T \left| \frac{\sin(\Omega - \mu t)T/2}{(\Omega - \mu t)T/2} \right| \qquad (4.5.31)$$

As we predicted from the simple physical model discussed in Sec. 4.2.3, the output will be maximum at a time $t = \Omega/\mu$. Thus the time at which the maximum output occurs depends on the input frequency.

An important critical parameter is the number of separate frequency components that can be resolved. The problem is closely analogous to the definition of resolution in an optical or acoustic image, discussed in Chapter 3. The simplest way to define resolution is to take it as the spacing between the 3-dB points of the response [i.e., $\tau_p(3\text{ dB}) = 0.89/B$], where $B = \mu T/2\pi$ is the bandwidth. In this case, from Eq. (4.5.31), the frequency variation $\Delta f = \Delta\Omega/2\pi$ between 3-dB points is

$$\Delta f(3\text{ dB}) = \frac{0.89}{T} \qquad (4.5.32)$$

Thus the number of resolvable frequencies in the spectrum is $N = 1.12BT$. However, this definition of resolution is not entirely adequate. Consider, for instance, what occurs when the two equal amplitude signals of frequencies $f_1 = \Omega_1/2\pi$ and $f_2 = \Omega_2/2\pi$ are present. Let $\Delta f = f_2 - f_1$. Then the magnitude of the Fourier transformed signal is

$$|Y(t)| = \alpha T \left| \left[\frac{\sin \pi(Bt - \Delta f T/2)}{\pi(Bt - \Delta f T/2)} \right] + e^{j\phi} \left[\frac{\sin \pi(Bt + \Delta f T/2)}{\pi(Bt + \Delta f T/2)} \right] \right| \qquad (4.5.33)$$

where $t = 0$ is now defined to correspond to the frequency $f_0 = (f_1 + f_2)/2$ and ϕ is the phase difference between the two signals. This function is plotted for $\phi = 0, \pi/2$, and π in Fig. 4.5.3, with $\Delta f = 0.89/T$. As we might expect, peak signals corresponding to the two frequencies are easily resolved when $\phi = \pi$; for $\phi = \pi/2$, they are barely resolved; and when $\phi = 0$, there is actually a peak in amplitude at the midway frequency (corresponding to $t = 0$) of value 2. To ensure good resolution of equal-amplitude in-phase signals, we must increase Δf. Calculation shows that in this case, a value of $\Delta f T = 1.33$ is just adequate for resolution of the two frequencies.

A simple way to estimate the optimum value of Δf, required to resolve two in-phase equal-amplitude components, is to have each component be half its maximum amplitude point at the midway (6-dB point). These values are tabulated in Table 4.5.2 for various apodizing functions. They tend to be slightly optimistic,

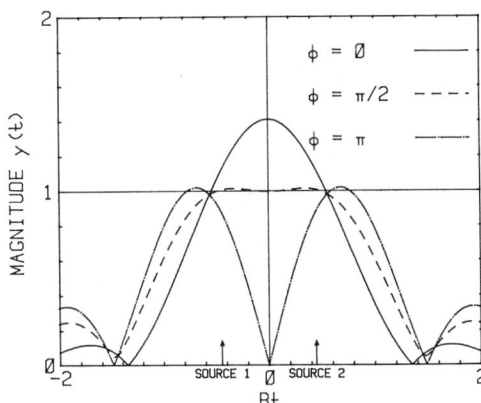

Figure 4.5.3 Plot of the magnitude of the Fourier transform of two equal-amplitude signals separated by a frequency $\Delta f = 0.89/T$, with phase differences $\phi = 0$, $\phi = \pi/2$, and $\phi = \pi$, respectively.

because the observable maximum amplitude of each component tends to be slightly lowered by the out-of-phase term contributed by the other component.

It follows from this treatment that the number of resolvable points in the spectrum is given by the approximate relation

$$N \approx BT \tag{4.5.34}$$

This is an important general result. The number of resolution points in any transversal filter system basically depends on the time–bandwidth product of the system. The exact figure will depend on the relative amplitudes and phases of the frequency components of interest, on the shape of the main lobe, and on the amplitudes of the sidelobes.

We must approach this result with caution when referring to the limitations of a real system, because we are concerned with the time length and bandwidth of both the chirp and the filter. Up to now, we have assumed that the filter is much longer than the chirp. The frequencies present in a single sideband chirp waveform, partially modulated by a frequency Ω, as in Eq. (4.5.30), are $\omega_0 - \mu t$ and $\omega_0 + \Omega - \mu t$. Thus the total bandwidth of the filter is required to be $f + B$, where $f = \Omega/2\pi$ is the signal bandwidth or bandwidth of the modulation and $B = \mu T/2\pi$ is the bandwidth of the chirp. This implies that if f can vary from 0 to B, the entire Fourier transform can only be produced if the filter has a bandwidth $B_d = 2B$ and a time length of $T_d = 2T$. In this case the maximum frequency excursion of the modulation can be B. Therefore, the limitation on the number of resolvable frequencies becomes, in practice,

$$N \approx \frac{B_d T_d}{4} \tag{4.5.35}$$

Unless we make other modifications to the system, such as with the sliding transform, where the input chirp has twice the bandwidth and time length of the filter, as discussed in Sec. 4.5.5, the system must have a much larger bandwidth than that required by the signal alone. As we will see, such stratagems destroy the phase information in the transform, but are useful when only the magnitude of the transform is required.

We first consider applications of this Fourier transform technique to situations where we require only the magnitude or real part of the Fourier transform $|Y(t')|$.

In practice, the Fourier transform of a complex signal requires that the signal be divided into its real and imaginary parts. Real [$\cos(\omega_0 t + \mu t^2/2)$] and imaginary [$\sin(\omega_0 t + \mu t^2/2)$] filters must be used, each having a response for each real and imaginary input signal component. This means that we need four filters in all. To put it another way, extra filters are required to reproduce both the amplitude and phase of the Fourier transform. We shall describe such processes in Secs. 4.5.5 and 4.5.6. If the signal is real and its Fourier transform is also real (a symmetric input waveform), only one real transform is required. Thus all the necessary operations can be obtained by using a premultiplied chirp of the form $\cos(\omega_0 t - \mu t^2/2)$ and a filter of the form $\cos(\omega_0 t + \mu t^2/2)$.

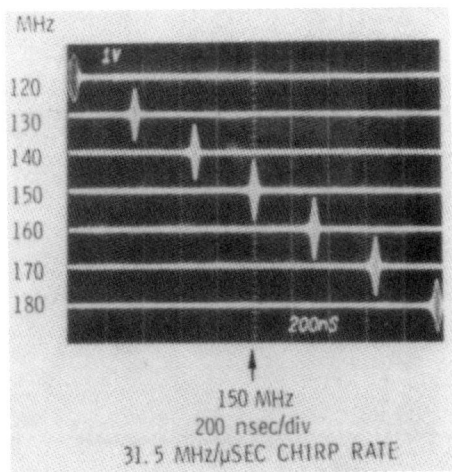

Figure 4.5.4 Prototype chirp transform results for seven successive CW input signals, stepped from 120 to 180 MHz in 10-MHz steps. (After Hays et al. [41].)

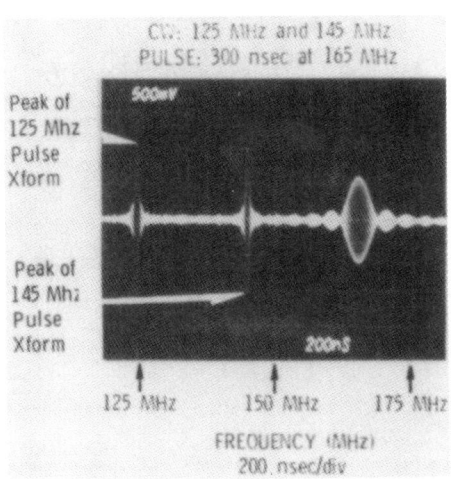

Figure 4.5.5 Chirp transform of three simultaneous input signals, including CW and a tone burst. (After Hays et al. [41].)

Compressive receiver. Perhaps the simplest application of a real transform is to the so-called *compressive receiver*. This device instantaneously detects and determines the frequencies of unknown signals. There are obvious applications to radar detection systems, where only a few RF pulses might be available for this purpose. In this case, an input signal is mixed with a chirp; the chirp can be generated by injecting a pulse into a dispersive delay line. The modulated chirp signal is then passed through the matched filter and into a detector. Thus we obtain an output that is the spectrum of the input modulating signal. This output will occur at a time that is independent of the time of the input signal, provided that the input occurs during the time the chirp is entering the matched filter. Thus the device is capable of interrogating several signals at once and displaying the frequencies of the separate input signals. Results obtained with a system of this nature are shown in Figs. 4.5.4 and 4.5.5.

Variable bandpass–bandstop filters and bandshape filters. A second application of this chirp transform technique is to use the device for variable bandpass–bandstop filtering, as illustrated in Figs. 4.5.6 and 4.5.7. In this case, because a particular frequency in the waveform corresponds to an output signal at a certain time, it is possible to gate the output signal to eliminate a particular frequency. Then by taking the inverse Fourier transform, the original signal may be obtained with the interfering signal filtered out.

More generally still, if we multiply the transformed signal by a pulse that is the Fourier transform of the filter function required, we can design a general filter function system. An illustration of such a filter realization for differentiation of the input signal [multiplying by ω (or by t in the transform domain) corresponds to differentiation] is shown in Fig. 4.5.8.

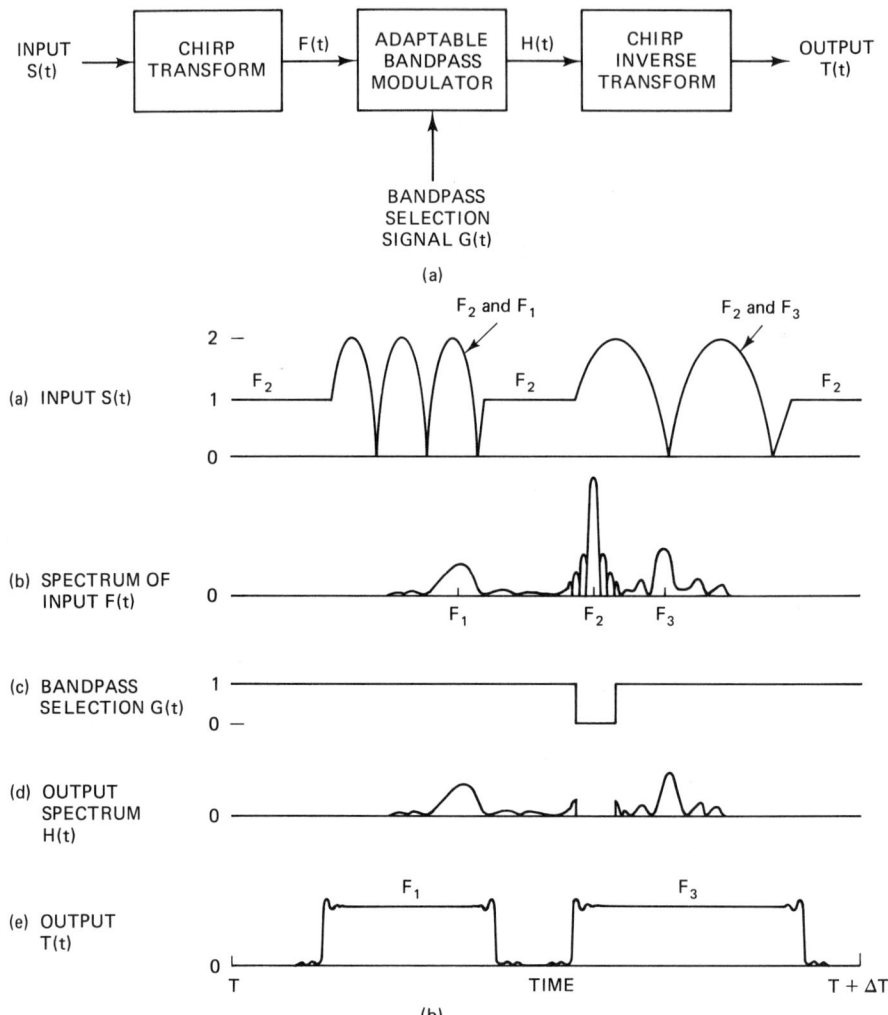

Figure 4.5.6 (a) Bandpass–bandstop filter; (b) chirp transform use for variable bandpass/bandstop filtering. (After Hays et al. [41].)

Variable-time-delay filters. The Fourier transform filter can give a variable time delay. Translation by T_0 of a waveform $f(t)$ results in the multiplication of its Fourier transform by a phase shift term $\exp(j\omega T_0)$ or, in our case, $\exp(j\mu\tau T_0)$. To put it more simply, if we translate the frequency of the transform by $\omega_T = \mu T_0$, the time of the output signal will change by T_0. Thus all that is needed to vary the delay is to multiply the Fourier transform by a signal of frequency ω_T, and then to find the inverse transform. Results of this kind are illustarted in Fig. 4.5.9.

Correlation or convolution of two signals. A final application is the correlation or convolution of signals. Suppose the two signals are $f(t)$ and $g(t)$.

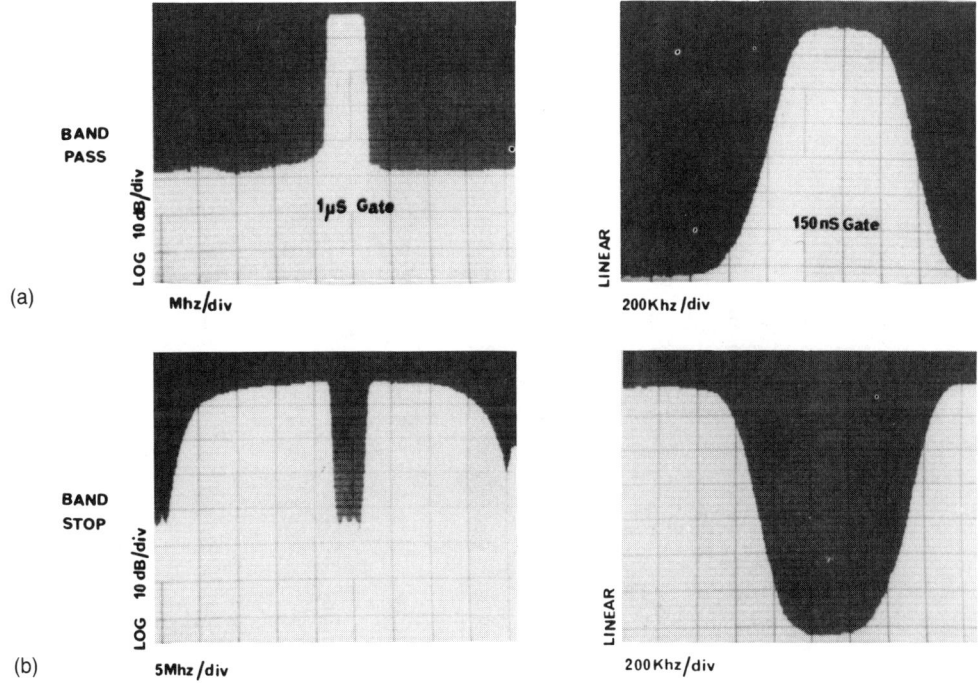

Figure 4.5.7 (a) Wide-gate and narrow-gate passband and stopband results for a time-domain filter transmission as a function of input frequency, with gate fixed (experiment). (After Maines et al. [42].) (b) Lowpass and bandpass filtering of the chirp transform of two different-frequency (225 kHz and 600 kHz) time-overlapped signals. The outputs are still modulated by a chirp factor. (After Atzeni et al. [43].)

Their convolutions are

$$h(t) = \int f(t - \tau)g(\tau)\,d\tau \qquad (4.5.36)$$

The Fourier transform of this result is

$$H(\omega) = F(\omega)G(\omega) \qquad (4.5.37)$$

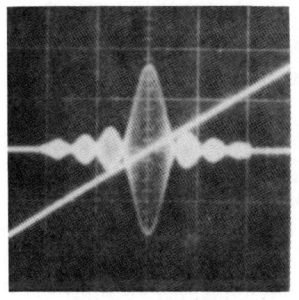 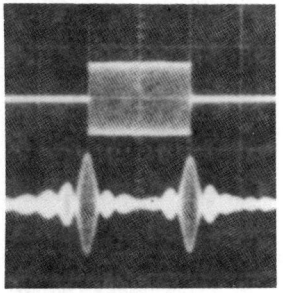

Figure 4.5.8 Linear ramp synthesizing the differentiator transfer function, superimposed to the CT of a rectangular pulse (left) and derivative of the pulse (right). Chirp modulation is not removed. (After Atzeni et al. [43].)

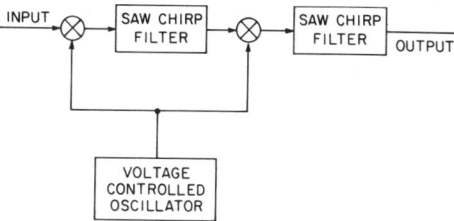

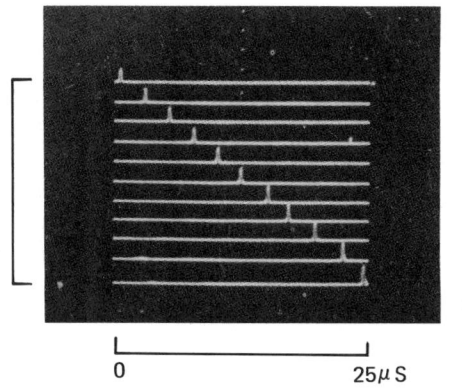

Input pulse 200nS ; 50MHz

Output

0   25μS

Bmax=36MHz
Tmax=24μS

Figure 4.5.9 (a) Schematic of SAW variable delay line using chirp filters. (b) SAW variable-delay-line operation. The input pulse is 200 ns at 50 MHz. The remaining traces show the delayed long 50-MHz outputs. (After Maines and Paige [44].)

Thus if we take the Fourier transforms of the two signals, multiply them together, and then find the inverse Fourier transform of the result, the output will be the convolution of the two signals. Similarly, if we multiply $F(\omega)$ by $G^*(\omega)$, the complex conjugate of $G(\omega)$, we will obtain the correlation of the two signals after finding the inverse transform of the product.

To carry out convolution with chirp transform devices, we simply carry out the Fourier transforms of $f(t)$ and $g(t)$, multiply the Fourier transforms $F(\mu t)$ and $G(\mu t)$ together, and find the inverse Fourier transform of the resultant signal.

Alternatively, as shown in Fig. 4.5.10, if correlation is required, the Fourier transform $G(\mu t)$ can be mixed with a signal $\exp[j(\omega_1 t + \mu t^2/2)]$ + c.c., where c.c. is shorthand for the complex conjugate. The output signal from the mixer has a form

$$S(\mu t) = G(\mu t)e^{j[(\omega_0 + \omega_1)t + \mu t^2]} + G^*(\mu t)e^{j(\omega_1 - \omega_0)t} + \text{other terms} \quad (4.5.38)$$

If this signal is passed through a bandpass filter to keep only the latter term, we can obtain the necessary conjugate signal. Results obtained with such a pair of Hamming-weighted chirps and a 127-chirp biphase code are shown in Fig. 4.5.11.

We can also employ the technique to measure the difference in arrival time of two arbitrary signals from the same source. Suppose that we find the autocorrelation of the signals $f(t)$ and $g(t - T)$, from two antennas, for example. The

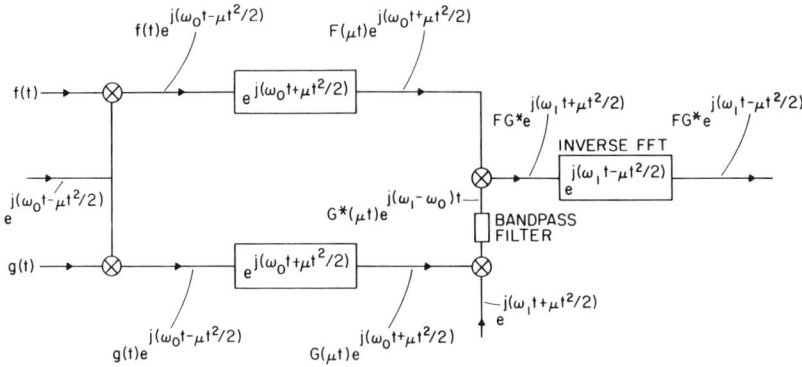

Figure 4.5.10 Circuit required for cross-correlations.

time delay T from a noisy source such as a radio star will vary with the position of the source. When the source is located on a line midway between the antennae, $T = 0$. The correlation will be $\int f(\tau)f(\tau + t - T) \, d\tau$. The output correlation peak occurs at a time $t = T$. Thus we can measure the time delay and the inclination of the source. Figure 4.5.12 shows a result of this kind.

Spectral whitening and nonlinear processing. A third application of these techniques is to eliminate an interfering CW signal, as illustrated in Fig. 4.5.13. A CW signal will give rise to a large peak in the Fourier transform. Therefore, by clipping the output of the Fourier transform processor to limit the amplitude of this peak and then taking the inverse transform of the resultant signal, we can, to a large extent, eliminate the interfering signal. In practice, this can give a reduction of as much as 40 dB in an interfering signal.

4.5.5 Implementation of Chirp z Transforms with SAW Devices

Here we describe the implementation of complex Fourier transform processes. We call these transforms *chirp z transforms* because of their close relation to digitally implemented sampled data transforms. The relation to z transform theory will be discussed in Sec. 4.5.6. Before doing this, however, let us discuss why it is necessary to employ separate transform processors to process the real and imaginary parts of an input waveform, and the advantages of such processors, even for processing real input waveforms.

A complex function $f(t)$ can be represented by two real components $f_R(t)$ and $f_I(t)$, with

$$f(t) = f_R(t) + jf_I(t) = a(t)e^{j\phi(t)} \quad (4.5.39)$$

where

$$f_R(t) = a(t) \cos [\phi(t)]$$
$$f_I(t) = a(t) \sin [\phi(t)] \quad (4.5.40)$$

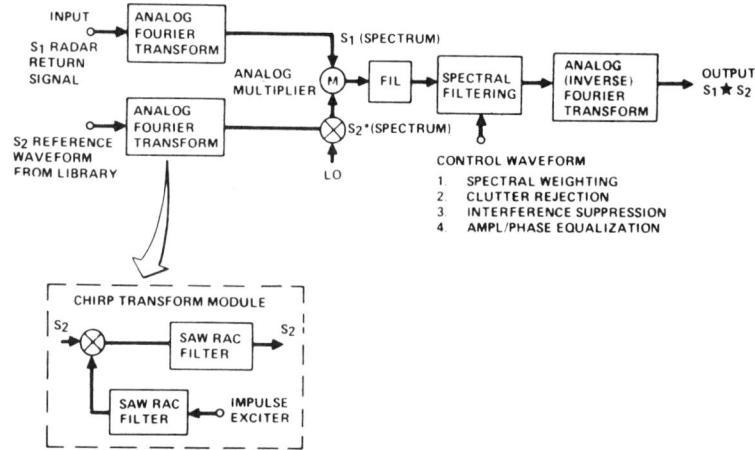

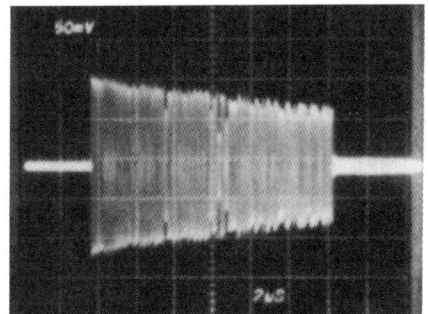

127 CHIP BIPHASE CODED
WAVEFORM (0.1 μSEC CHIPS)

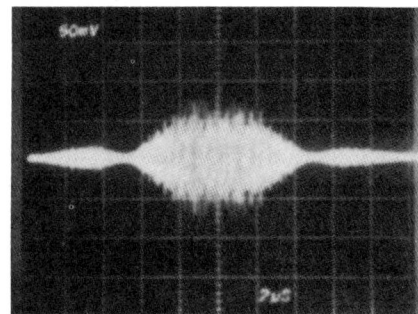

CHIRP TRANSFORM
OF BIPHASE CODE
HORIZ SCALE → 2.0 MHz/μSEC

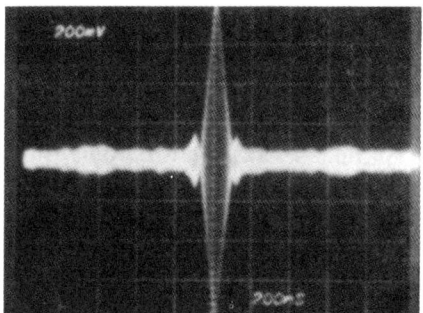

CORRELATION PULSE
SIDELOBE LEVEL = -23 dB

Figure 4.5.11 Autocorrelation of two signals: (a) block diagram of the system used. (b) correlation of a 127-chip biphase-coded waveform. (After Gerard et al. [45].)

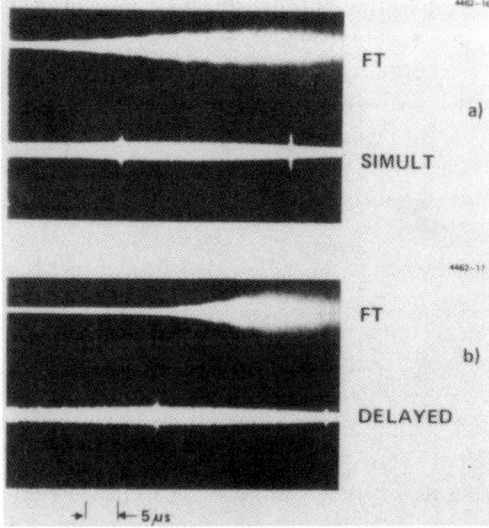

Figure 4.5.12 Relative-time-of-arrival determination. (After Nudd and Otto [46].)

Both the real and imaginary parts of $f(t)$ are needed if we wish to obtain the amplitude and phase of its Fourier transform.

Conversely, if we wish to take the inverse Fourier transform of a function and reconstruct the original signal, we must know both the amplitude and phase of the function.

Consider, as an example, a rectangular pulse with a time duration T. This has a real Fourier transform $\sin(\omega T/2)/(\omega T/2)$. However, if the pulse were passed through a delay line with a time delay T_D, its Fourier transform would be $\exp(-j\omega T_D)\sin(\omega T/2)/(\omega T/2)$, with

$$F_R = \cos \omega T_D \sin \frac{(\omega T/2)}{(\omega T/2)}$$
$$F_I = -\sin \omega T_D \sin \frac{(\omega T/2)}{(\omega T/2)} \quad (4.5.41)$$

It is apparent that in order to reconstruct the delayed pulse using an inverse Fourier transform, we would have to know both F_R and F_I, although in this case $f(t)$ is real. In general, we must be able to deal with complex signals or the amplitude

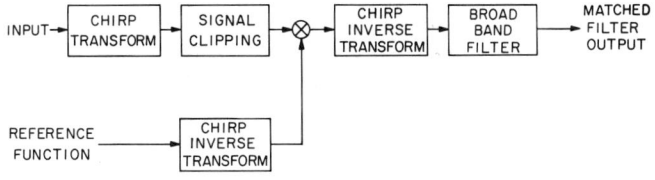

Figure 4.5.13 Chirp transform programmable matched filter, with suppression of narrowband interference by transform clipping to approximate spectral pre-whitening.

Sec. 4.5   FM Chirps Filters and Chirp Transform Processors   401 and phase of real signals, because the output of a real-time Fourier transform device itself is one or more time-varying signals.

Now let us consider how to process an input signal $f(t)$ modulated by a down-chirp $\exp[j(\omega t - \mu t^2/2)]$, which we write in the form

$$u(t) = f(t)e^{j(\omega t - \mu t^2/2)} \tag{4.5.42}$$

where, from Eqs. (4.5.39) and (4.5.42), the real physical signal is

$$s_1(t) = f_R(t) \cos\left(\omega t - \frac{\mu t^2}{2}\right) - f_I(t) \sin\left(\omega t - \frac{\mu t^2}{2}\right) \tag{4.5.43}$$

Thus we can obtain the physical signal required by multiplying $f_R(t)$ and $f_I(t)$ by a cosine and sine chirp, respectively, and adding the resultant signals, as shown in Fig. 4.5.14.

The matched filter for the complex chirp is of the form $\exp[j(\omega t + \mu t^2/2)]$. Here we use a physical matched filter of the form $\cos(\omega t + \mu t^2/2)$. The output from the system, when a signal $u(t)$ is inserted into it, can be written in the form

$$\begin{aligned} S_2(t) &= \int f(\tau)e^{j(\omega\tau - \mu\tau^2/2)} \cos\left[\omega(t-\tau) + \frac{\mu(t-\tau)^2}{2}\right] d\tau \\ &= \tfrac{1}{2} \int f(\tau)\left[e^{j(\omega t + \mu t^2/2)}e^{-j\mu t\tau} + e^{-j(\omega t + \mu t^2/2)}e^{j\mu(t-\tau)\tau}e^{2j\omega\tau}\right] d\tau \end{aligned} \tag{4.5.44}$$

The second term in the integrand contains a rapidly varying term that varies as $\exp(-j\mu\tau^2)$. This term will yield very little contribution to the integrand and may

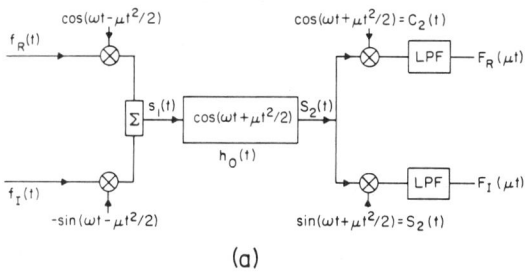

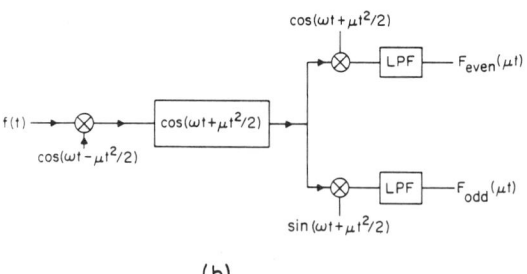

Figure 4.5.14 (a) Fourier transform processor for a complex function. The real input is represented by $f_R(t)$, the imaginary input is represented by $f_I(t)$, and F_R and F_I are the real and imaginary parts of the Fourier transform. (b) Fourier transform processor for odd and even real functions.

be neglected. Therefore, we write

$$S_2(t) \approx \tfrac{1}{2} e^{j(\omega t + \mu t^2/2)} \int f(\tau) e^{-j\mu t \tau} \, d\tau \qquad (4.5.45)$$

Thus the output from this signal processing operation is the Fourier transform of $f(\tau)$ multiplied by an up-chirp.

We are interested in the real and imaginary parts of this Fourier transform, which are

$$\begin{aligned} F_R(\mu t) &= \operatorname{Re} \int f(\tau) e^{-j\mu t \tau} \, d\tau \\ &= \int f_R(\tau) \cos \mu t \tau \, d\tau + \int f_I(\tau) \sin \mu t \tau \, d\tau \qquad (4.5.46) \\ F_I(\mu t) &= \operatorname{Im} \int f(\tau) e^{-j\mu t \tau} \, d\tau \\ &= \int f_I(\tau) \cos \mu t \tau \, d\tau - \int f_R(\tau) \sin \mu t \tau \, d\tau \end{aligned}$$

respectively, where

$$\operatorname{Re}[S_2(t)] = F_R(\mu t) \cos \left(\omega t + \frac{\mu t^2}{2} \right) - F_I(t) \sin \left(\omega t + \frac{\mu t^2}{2} \right) \qquad (4.5.47)$$

If $f(t)$ is real, we need only multiply the original signal by a cosine chirp, as shown in Fig. 4.5.14(a), to obtain the result

$$\begin{aligned} F_R(\mu t) &= \int f_R(\tau) \cos \mu t \tau \, d\tau \\ F_I(\mu t) &= \int f_R(\tau) \sin \mu t \tau \, d\tau \end{aligned} \qquad (4.5.48)$$

In this case, $F_R(\mu t) = F_{\text{even}}$ is the Fourier transform of the even part of $f_R(t)$, and $F_I(\mu t)$ is the Fourier transform of the odd part of $f_R(t)$.

Suppose now that we multiply the output $S_2(t)$ of the filter by a chirp signal of the form $\cos(\omega t + \mu t^2/2 + \phi)$, and pass the resultant output signal of the mixer through a low-pass filter (i.e., look at the output at baseband). The term of interest, then, is

$$\operatorname{Re}(Y_{\text{out}}) = \cos \phi \, F_R(\mu t) + \sin \phi \, F_I(\mu t) \qquad (4.5.49)$$

By choosing $\phi = 0$ or $\pi/2$ [i.e., by multiplying by $\cos(\omega t + \mu t^2/2)$ or $\sin(\omega t + \mu t^2/2)$], we can separate the real and imaginary parts of $F(\mu t)$, as shown in Fig. 4.5.14(a).

Thus the output from the top filter in Fig. 4.5.14(a) is the real part of the complex Fourier transform $F_R(\mu t)$, and the output from the bottom filter is the imaginary part $F_I(\mu t)$ of the complex Fourier transform of $f(t)$. When $f(t)$ is real, $F_R(\mu t)$ is the transform of the even part of $f(t)$ and $F_I(\mu t)$ is the transform of the odd part of $f(t)$, as shown in Fig. 4.5.14(b). When $f(t)$ is complex, it follows from

Eq. (4.5.46) that $F_R(\mu t)$ can be stated in terms of Fourier transforms of $f_R(t)$ and $f_I(t)$.

An example of how this type of SAW processor is operated has been given by Jack and Collins. The following quote (slightly modified) is from their paper "Fast Fourier Transform Processor Based on the SAW Chirp Transform Algorithm" (slight modifications have been made to correspond with our earlier examples) [47].

Experimental SAW FFT Processor Realization. The SAW FFT processor shown in schematic form in Figs. 4.5.14(a) and 4.5.15 was configured to demonstrate computation of the Fourier transform in accordance with the theory of the previous section. The processor employs three commercially available SAW chirp filters (MESL types WB041 B-2 and WB045B-2). Here the input chirp C_1 is centered at 22 MHz and sweeps over 4 MHz bandwidth in 25 μs ($TB = 100$, dispersive slope $\mu = 2\pi \times 160$ kHz/μs). The chirp filter C_2 and the post-multiplier chirp C_3 are centered at 32 MHz and sweep over 8 MHz in 50 μs ($TB = 400$, $\mu = 2\pi \times 160$ kHz/μs). In view of the offset center frequencies of C_1, C_2 a 54-MHz reference carrier must be inserted at the input to permit acceptance of baseband input signals. However, since the center frequencies of C_2, C_3 are identical, the output—after the required low-pass filtering discussed previously—is automatically at baseband. The timing electronics associated with the SAW FFT processor are arranged to operate in synchronism with the inserted 54-MHz reference carrier using divider circuits. Pre- and post-multiplier chirps in phase quadrature

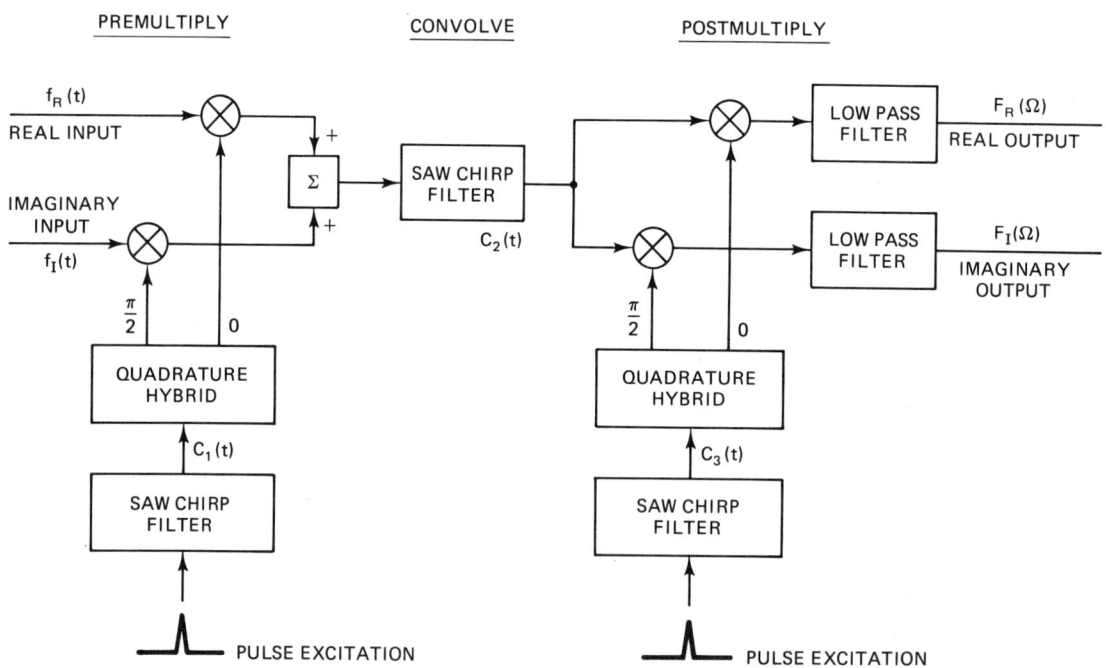

Figure 4.5.15 SAW chirp transform arrangement for baseband operation.

were derived from C_1 and C_3, respectively, by means of a commercial quadrature hybrid circuit (Merrimac type QHS-3-30). The input waveforms used in this experiment were purely real, analog and continuous. These waveforms were derived from a function generator (Tektronix type FG504), which permits synchronous operation (i.e., stationary relative to the premultiplier chirp) with variable input signal phase.

The results shown in Fig. 4.5.16(a) relate to a *real* and *even* (cosine) input, with period 2.08 μs (480 kHz). Choice of such an input period (a basic vector) means that an integer number of cycles of the input waveform, fit within the time window defined by the premultiplier chirp. The center and bottom traces represent the real, $F_R(\Omega)$, and imaginary, $F_I(\Omega)$, components, respectively, of the Fourier transform as measured at the outputs of the SAW FFT processor. $F_R(\Omega)$ and $F_I(\Omega)$ represent frequency-domain functions over both positive and negative regions of the frequency domain relative to zero frequency (dc). The horizontal scale for these traces can be calculated as 160 kHz/μs × 1 μs/div = 160 kHz/div. The center and bottom traces share a common oscilloscope delaying time base. The delay has been adjusted such that the center of the display corresponds to zero frequency (dc). As stated earlier, the upper trace in Fig. 4.5.16(a) represents the input time-domain signal. Here in view of the propagation delay through the SAW processor, it is not possible to show input and outputs simultaneously. How-

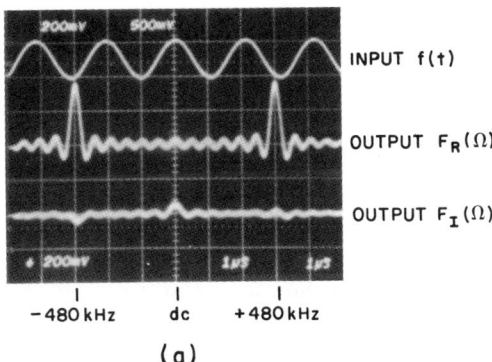

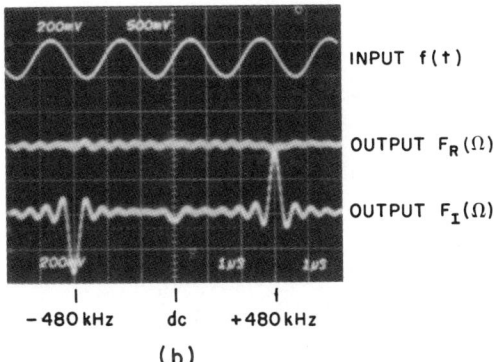

Figure 4.5.16 SAW FFT performance with (a) 480-kHz cosine input, and (b) 480-kHz sine input. (After Jack and Collins [47].)

ever, the upper trace uses a separate oscilloscope delaying time base where the (arbitrary but fixed) delay has been set such that the time-domain signal, $f(t)$, in Fig. 4.5.16(a) is *even* relative to the center of the trace.

The upper trace of Fig. 4.5.16(b) shows a *real* and *odd* (sine) input, with period 2.08 μs (480 kHz). It is important to note that the photographs of Fig. 4.5.16 were taken with the two time-base delays unchanged. The photographs confirm the presence of cosine and sine inputs, respectively, these being produced by variation of the start phase of the function generator output.

The output Fourier transform of the (truncated) cosine input (480 kHz), $f(t)$, shown in Fig. 4.5.16(a) displays *real* and *even* responses as can be predicted theoretically. These responses peak at ± 3 μs × 160 kHz/μs = ± 480 kHz and consist of $(\sin x)/x$ responses with nulls at spacings of 250 ns × 160 kHz/μs. This structure can be directly related to signal truncation by the input window of 25-μs duration. In comparison, the output Fourier transform of the (truncated) sine input (480 kHz) as shown in Fig. 4.5.16(b), displays *imaginary* and *odd* responses at ± 3 μs × 160 kHz/μs = ± 480 kHz. The unwanted signals visible in the respective "zero" output components of Fig. 4.5.16 have been investigated, and these appear to be due predominantly to the practical difficulty of ensuring the existence of a purely even or purely odd input.

A very similar set of results obtained with a larger bandwidth system have been given by Jack and Paige [48], who demonstrated the Fourier transform of 7.5-MHz signals.

4.5.6 Chirp z Transform with a CCD

A baseband system cannot be used in the same manner as an SAW chirp z-transform system. In the latter case, we could use low-pass filters and a high carrier frequency to eliminate unwanted components of the output. However, a baseband system must employ separate chirp filters to obtain the complex Fourier transform. Consider, for example, the implementation of the same process in a CCD that operates at baseband. The system may now be regarded as a sampled system where we can write the discrete Fourier transform in the form

$$F_k = \sum_{n=0}^{n=N-1} f_n e^{-2jn\pi k/N} \qquad (4.5.50)$$

$$= \sum_{n=0}^{n=N-1} f_n z^{2n}$$

where $z = \exp[-(j\pi k/N)]$.
Using the substitution

$$2nk = n^2 + k^2 - (n - k)^2 \qquad (4.5.51)$$

we find that

$$F_k = e^{-j\pi k^2/N} \sum_{n=0}^{n=N-1} f_n e^{-j\pi n^2/N} e^{j\pi(k-n)^2/N} \quad (4.5.52)$$

$$= z^k \sum_{0}^{n=N-1} f_n z^{n^2/k} z^{-(k-n)^2/k}$$

This operation of the chirp z transform is identical to that described earlier for a CW system in Eqs. (4.5.7) and (4.5.16). In this case, the input signal is multiplied by a sampled baseband chirp $\exp(-j\mu t^2/2)$, where for the nth sample, $\mu t_n^2 = 2\pi n^2/N$. The resultant signal is then inserted into a chirp filter with the opposite slope and finally the output is multiplied by a chirp. As with SAW devices, final multiplication by a chirp can be eliminated only if the power density spectrum is required. To process this signal, we require four filters, as illustrated in the block diagram in Fig. 4.5.17. In this case, the input data would be premultiplied by a chirp of a frequency varying from $-f_1/2$ to $f_1/2$, with a duration T.

The expression in Eq. (4.5.52) shows that the filter, which has an $\exp[j\pi(k-n)^2/N]$ response, must have $2N$ stages, that is, the maximum value of $k+n$, corresponding to a delay length $2T$ and a frequency variation from $-f_1$ to f_1. This ensures that all frequency components of the input, multiplied by the chirp, can be processed by the filter, and that the chirp is present in the filter at all times while the processing is being carried out. This requirement, of course, is identical to that of the SAW device, the only difference being that the chirp frequency varies from $f_0 - f_1/2$ to $f_0 + f_1/2$, where f_0 is the carrier frequency. As the chirp length is shorter than that of the filter, we must weight the chirp rather than the filter.

Provided that these processes are carried out, we can obtain, with a post-multiplier, the correct component of the transform (i.e., real or imaginary, even or odd). In principle, it is not always necessary to use four filters, because if the input signal were purely real, we could use only two filters. The advantage of a

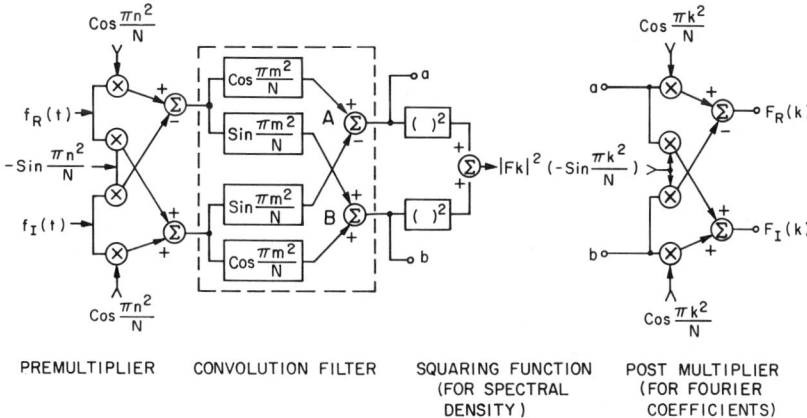

Figure 4.5.17 Implementation of FFT chirp transform or for obtaining spectral-density.

four-filter system, however, is that it can eliminate the sidelobes that arise from unwanted signal components, which correspond to the second term in the second equation of Eq. (4.5.44), and which we normally ignore.

Another way to look at this result might be to consider what happens when the input is a single-frequency exp $(j\Omega t)$, so that $f_R(t) = \cos \Omega t$ and $f_I(t) = \sin \Omega t$. When these signals are premultiplied by the chirps $\cos (\mu t^2/2)$ and $-\sin (\mu t^2/2)$, respectively, and the resultants subtracted, we obtain a single-sideband chirp $\cos (\Omega t - \mu t^2/2)$ at the input to the upper set of filters in Fig. 4.5.17. Similarly, the lower set of filters has an input $\sin (\Omega t - \mu t^2/2)$. Therefore, the output from the summer A in the diagram is

$$Y_U(t) = \int \left\{ \cos\left(\Omega \tau - \frac{\mu \tau^2}{2}\right) \cos\left[\Omega(t - \tau) + \frac{\mu(t - \tau)^2}{2}\right] \right.$$

$$\left. - \sin\left(\Omega T - \frac{\mu \tau^2}{2}\right) \sin\left[\Omega(t - \tau) + \frac{\mu(t - \tau)^2}{2}\right] \right\} d\tau \quad (4.5.53)$$

$$= \int \cos\left(\Omega t + \frac{\mu t^2}{2} - \mu t \tau\right) d\tau$$

We note that there are no terms in the argument of the cosine function that vary as τ^2. Thus the equivalent of the second term of the integrand of the second equation of Eq. (4.5.44) has been eliminated in the output of the filter. A similar argument applies to the further processing through the post-multiplier section. The use of four multipliers yields the one required sideband without further filtering.

Furthermore, as real input components are composed of two complex components, one of positive frequency and one of negative frequency, the use of real filters alone, even for a real input signal, can give rise to aliasing. Each real input component, such as the signal $\cos \Omega t$, is composed of complex components exp $(j\Omega t)$ and exp $(-j\Omega t)$, whose imaginary parts cancel. If these signals are sampled over a range 0 to $f_s/2$, where f_s is the sampling frequency, negative frequency components occur in the band f_s to $f_s/2$ as an alias, thus restricting the useful input bands to components lying below the Nyquist limit $f_s/2$. However, a complex four-channel system will eliminate these alias terms and make it possible to work over a band to f_s without aliasing, for $2f_s$ samples are effectively employed per second. Another way to look at this is as a consequence of the fact that we are using a single-sideband chirp modulator, and are therefore eliminating the aliasing frequencies.

The CCD system normally operates with split-gate sampled weighting of the type already described [4]. For instance, the weighting amplitude would be $\cos \pi k^2/N$ for the cosine chirp, as illustrated in Fig. 4.5.18.

Sliding transform. When only the power spectrum is required, it is often more convenient to employ longer chirps relative to the length of the filter. If the chirp length is longer than that of the filter, only a portion of the chirp waveform

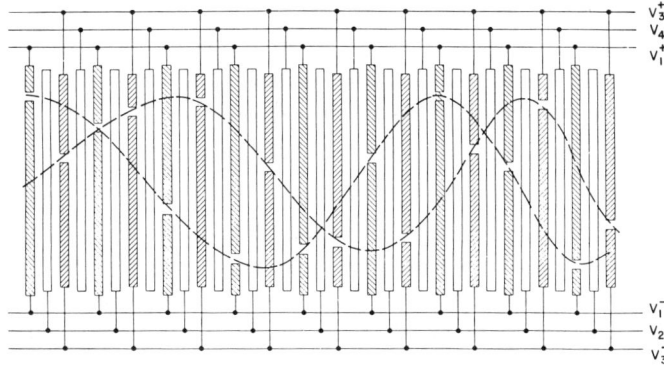

Figure 4.5.18 Layout of the CCD gates showing the split-gate weighting. (From Reticon Coroporation [49].)

is being sampled at any one time. The output from the filter then takes the form

$$F_k = \sum_{n=k}^{k+N-1} f_n e^{-2j\pi nk/N} = e^{-2j\pi k^2/N} \sum_{n=0}^{N-1} f_{n+k} e^{-2j\pi nk/N} \qquad (4.5.54)$$

This is essentially equivalent to the continuous form of Eq. (4.5.17) for a matched filter response when the chirp is longer than the filter. In this case, the input chirp is modulated by $f(t)$, a down-chirp is used, and increasing n in this case is equivalent to decreasing z. The difference between this transform and the usual DFT is that the input data are shifted by one sample each time the spectral component is calculated. For a general waveform, this procedure destroys the phase information. However, because the power spectrum does not contain any phase information, this is unimportant. Now we can use a filter N registers long, with a bandwidth f_1, to process signals with the same bandwidth. The system is thus more efficient than one employing a chirp only half the length of the total filter. Furthermore, as we saw in our treatment of weighting (Sec. 4.5.3), its major advantage is that in order to decrease the sidelobe level, the output filter can be weighted instead of the chirp.

4.5.7 Chirp Transforms with Superconductive Delay Lines

Research has been and is being carried out on other technologies for transversal filters. Chirp transforms were demonstrated many years ago using dispersive wire or tape delay lines [1]. (Applications of fiber-optic delay lines are described in Sec. 4.6.) Here we discuss early research on superconductive delay lines, and demonstrate their use for carrying out chirp transforms [12].

So far, the application of superconductive delay lines to carry out chirp transforms has extended the operating frequency range of transversal filters from the UHF up into the range 2 to 20 GHz. The device concepts are very similar to those of SAW delay lines, but the implementation must, of course, be changed to suit the particular attributes of superconductive delay lines.

The basic idea is to employ an electromagnetic delay line, instead of an SAW delay line, as a transversal filter. Since the velocity of an electromagnetic wave is approximately 10^5 times larger than that of an acoustic wave, the obtainable delays of an electromagnetic delay line will be far less than for an acoustic device. Thus if we want to obtain large delays of the order of 100 ns or more, we require an electromagnetic delay line at least 30 m long. But such delays lines are extremely bulky and their attenuation becomes very large as their diameter is reduced. To circumvent these difficulties, we may use superconductive materials, such as niobium, laid down on a high-quality, low-loss insulating substrate, such as sapphire or silicon, which will be an insulator at temperatures of the order of 4°K. By reducing the width of a strip of superconductive material to dimensions of the order of 25 μm, a relatively small line can be laid down in the form of a spiral, as illustrated in Fig. 4.5.19(a), or a meander line, as illustrated in Fig. 4.5.19(b). The attenuation as a function of distance along the line l is $\exp(-\alpha l)$, where

$$\alpha = \alpha_d + \alpha_c + \alpha_r \qquad (4.5.55)$$

The parameters α_d, α_c, and α_r are the contributions of dielectric loss, conductor loss, and radiation losses, respectively. The dielectric loss for sapphire is very small and can be written in the form

$$\alpha_d = 27.3 \sqrt{\varepsilon} \, (\tan \delta)/\lambda_0 \qquad \text{(dB/unit length)} \qquad (4.5.56)$$

Reported values for the loss tangent $\tan \delta$ of crystalline sapphire range from 10^{-5} to 10^{-8}. The surface resistance of niobium at 4.2°K equals 2.6×10^{-5} Ω/square

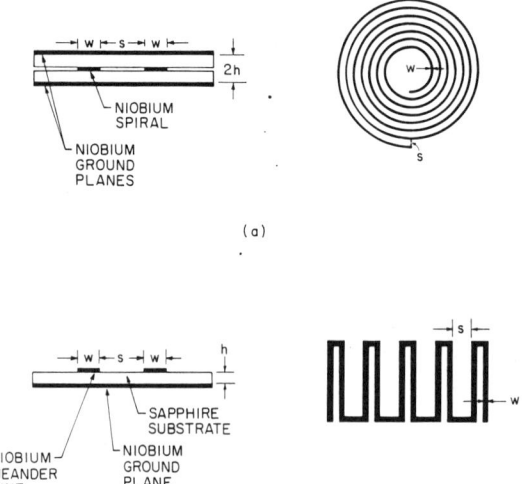

Figure 4.5.19 Superconductive delay lines: (a) spiral microstrip; (b) meander line.

at 10 GHz. The conductor loss α_c for strip line is

$$\alpha_c = \frac{8.68(R_{s1} + R_{s2})}{2wz_0} \quad \text{(dB/unit length)} \quad (4.5.57)$$

where R_{s1} and R_{s2} are the surface resistances of the strip and the neighboring ground planes, respectively. The loss per wavelength for a microstrip design with a strip width w and sapphire thickness h, with $w/h = 0.85$ and $h = 25$ μm, is plotted as a function of frequency, as shown in Fig. 4.5.20. As we might expect, this loss figure is considerably lower than for copper, and in fact the attenuation per wavelength is much lower than for an SAW delay line.

A simple delay line can be made out of a single spiral. To construct a tapped delay line, two interleaved spirals are laid down on the same substrate to make a double spiral dispersive delay line, as shown in Figs. 4.5.21 and 4.5.22. In Fig. 4.5.21, the input signal on electrode 1 to the first spiral is coupled out to excite a wave propagating in the opposite direction on the second. The coupling is between the wider regions, where the two spirals are close to each other.

The action of this coupler can be described by coupled wave theory (see Prob. 6). The coupling of two neighboring TEM modes propagating in the same direction

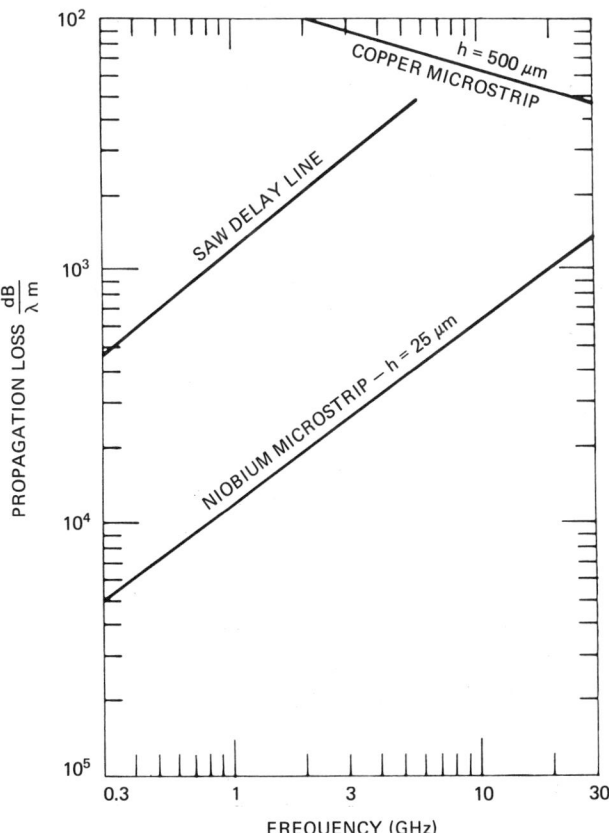

Figure 4.5.20 Propagation loss per wavelength for an SAW delay line and a superconductive delay line. (After Reible [12].)

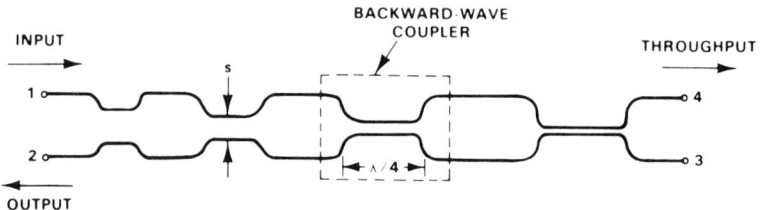

Figure 4.5.21 Proximity-tapped delay line. (After Reible [12].)

tends to be small because the inductive and capacitive coupling terms cancel each other out. The inductive and capacitive coupling from one line to another, of two waves propagating in opposite directions, adds. But the total excitation would also be very small if the coupling were over a long length. If the coupling is regarded as uniform over a length L, it can be shown that optimum coupling is obtained from one line to another when $L = \lambda/4$, where λ is the wavelength of the waves in both spirals. In early work, Reible made a delay line of this kind that was highly dispersive. It gave 27 ns of dispersion (i.e., of delay change) over a 2-GHz bandwidth centered at 4 GHz. The device had 101 proximity taps, with the potential for a time–bandwidth product of 50. The strip-line structure consisted of a 0.2-μm-thick niobium film sandwiched between two 5-cm-diameter, 125-μm-thick sapphire wafers. The pattern had two parallel lines, with a total length of 25 m, wound in a spiral pattern. Figure 4.5.23 shows the results for one of these devices used as an expander, when a 200-mV dc step with a 25-ps rise time is applied to its input. The resulting 27-ns chirp was amplified and time-gated to produce the pulse shown in the lower figure of Fig. 4.5.23(a). This was supplied to the input of the compressor, the up-chirp device. The resulting compressed pulse is displayed in the upper trace of Fig. 4.5.23(a) and a version expanded in time is shown in Fig. 4.5.23(b). The width of the main lobe of the compressed pulse is 2.5 ns, exactly what would be expected from an input with an 800-MHz bandwidth. The first three sidelobes levels were in good agreement with what would be expected from the theoretical sinc response.

This demonstration of a superconductive delay line is of great interest because it can obtain such large bandwidths. Its disadvantage is the low temperature required for its operation.

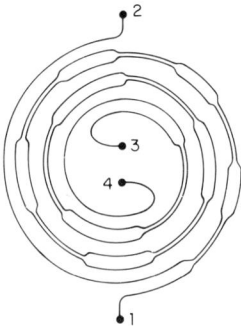

Figure 4.5.22 Double-spiral dispersive delay line. (After Reible [12].)

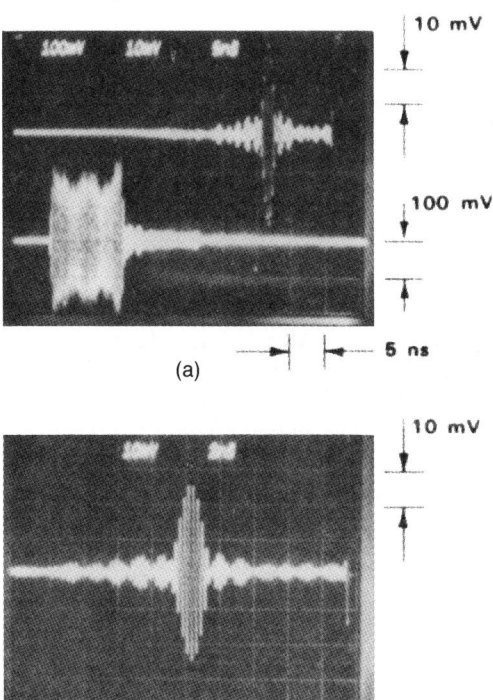

Figure 4.5.23 Outputs of superconductive dispersive filters: (a) top trace shows compressed pulse output, while bottom trace shows expanded waveform; (b) compressed pulse with expanded scale. (After Reible [12].)

PROBLEM SET 4.5

1. Consider Eq. (4.5.12) for a signal entering a physical filter. Following the derivation of Eq. (4.5.17), consider the FM chirp signal to be of finite length T, rather than infinitely long, as has been implicitly assumed in this derivation, and the length L of the delay line to be finite, with $L = VT$. Point z on the delay line can be excited by a signal only if it reaches that point at a time t. Thus if the signal has either not reached the plane z or already passed by it at a time t, there is no contribution to the output from the plane z. This implies that for a finite-length chirp signal, the limits on Eq. (4.5.13) must be reconsidered.

   (a) Determine the form of the output when the chirp is of uniform amplitude between times $-T/2 < t < T/2$ and zero, for $t < -T/2$ and $t > T/2$. Find the form of the output signal for $t < -T$, $-T < t < 0$, $0 < t < T$, and $t > T$.

   (b) Show that if $\mu T^2 \gg 1$, the main and first few subsidiary lobes (peaks and valleys) of the output signal are not much affected by changing the length of the chirp, as compared to keeping the length of the delay line L constant and using an infinitely long chirp.

2. When an FM chirp signal is inserted into a matched dispersive delay line whose characteristics are given by Eqs. (4.5.14) and (4.5.15), the output consists of a correlation peak at $t = 0$ with some sidelobes. In the text, we neglected the noncumulative second

term in Eq. (4.5.15). However, this gives rise to sidelobes or a background signal level that may be of some importance. Consider the neglected term, and find its amplitude as a function of time when the delay line is of length L and extends from $-L/2$ to $L/2$, and the chirp length can be regarded as being much longer than L/V. It is of particular interest to find the amplitude of this background signal at times near $t = 0$. To carry out the problem, it will be convenient to make use of the method of stationary phase given in Appendix G. This is equivalent to completing the square in the argument of the exponential, as in Eqs. (4.5.51) and (4.5.52).

(a) Show that the position z, which makes the main contribution to the integral, moves along the delay line at a velocity $V/2$. Show also that at time $t = 0$, this point is outside the delay line if the bandwidth B is such that $B < \omega_0/2\pi$, so that the contribution of this term is negligible in a SAW chirp transform device.

(b) Consider what occurs for a baseband chirp in a CCD filter ($\omega_0 = 0$). Find the amplitude of the unwanted signal relative to the amplitude of the main lobe. (Regard the dispersive CCD filter as a simple continuous filter of time delay T or length L.)

3. Consider a sampled system with N elements spaced a time t_0 apart. Some examples are a CCD chirp transform system or an amplitude-weighted chirp transform system of the type described in Sec. 4.5.3. By taking the weighting of a chirp filter $w(t)$ in Eq. (4.5.4) to be of the form

$$w(t) = \sum_{0}^{N-1} \delta(t - nt_0)$$

find the output of this system with the appropriate input chirp. Show that aliasing can occur, that is, that the output waveform will repeat itself at times $t = M/B$ apart, where B is the bandwidth and M is an integer. Explain physically, in terms of a spatial harmonic expansion representing the response of a quasiuniform region of tapped Fourier delay line, why this effect occurs.

4. One way to look at the sidelobe design problem is to cancel out the sidelobes. Consider Hamming weighting in a general form such that the weighting function is

$$w_0(\tau) = (1 - A) + A \cos \frac{2\pi\tau}{T}$$

Show that this weighting function gives rise to a Fourier transform, that is, to a time response with three main lobes arising from the first term and the cosine term, respectively. Choose the amplitude of the main lobes resulting from the cosine term so that it will cancel out the first sidelobe at $t = 3\pi/\mu T$ that resulted from the contribution of the constant term. With this choice, what is the ratio of the sidelobe to main lobe response at $t = 5\pi/\mu T$? Compare this to the unweighted system with $A = 0$ and to the result at $t = 3\pi/\mu T$ and $t = 5\pi/\mu T$ for the following Hamming weighting:

$$w_0(\tau) = 0.54 + 0.46 \cos \frac{2\pi\tau}{T}$$

5. Consider that the scales of a Fourier transform and an inverse Fourier transform need not necessarily be the same. On this basis, carrying out the two operations in turn can give rise to a change in the scaling, or time length, of an input signal. With the circuit shown in the figure, note that the input signal $f(t)$ is multiplied by a chirp exp $[j(\omega_1 t - \mu_1 t^2/2)]$, and inserted into the matched filter, which has a response exp $[j(\omega_1 t + \mu_1 t^2/2)]$. The output signal is then mixed with (multiplied by) a chirp of different slope exp

$[j(\omega_2 t + \mu_2 t^2/2)]$, and the resultant signal is passed through a low-pass filter to keep only the difference frequency term. This signal is inserted into the appropriate matched filter. Show that the output chirp is modulated by a scaled form of $f(t)$, namely $f[t(\mu_2 - \mu_1)/\mu_1]$.

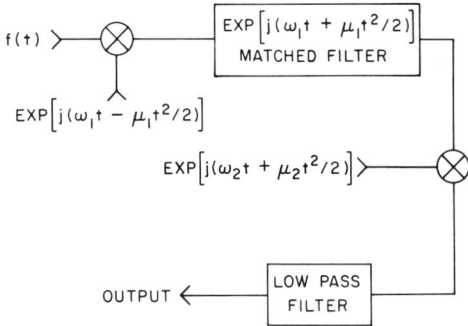

Hint: You will need to write the result in terms of a double integral (i.e., the two Fourier transforms). Invert the order of integration and use the relation

$$\int_{-\infty}^{\infty} e^{j\omega(t-t_0)}\, d\omega = \delta(t - t_0)$$

where $\delta(t)$ is a Dirac delta function.

6. Consider two coupled transmission lines coupled over a length L. Suppose that a wave of amplitude $a_1(z)$, propagating in the forward direction on line 1, is coupled to a wave of amplitude $a_2(z)$, propagating in the backward direction on line 2. We can write the following coupled mode equation for the excitation of a_2:

$$\frac{da_2}{dz} - jk_0 a_2 = j\kappa a_1$$

Assume that the coupling is weak ($\kappa \ll k_0$) so that we can write

$$a_1(z) = A_1 e^{-jk_0 z}$$

Show that the maximum value of $a_2(0)$ is obtained when the length L of the coupling region is $\lambda/4$. Assume that $a_2(L) = 0$.

4.6 BANDPASS FILTERS

4.6.1 Introduction

The basic method of designing bandpass filters was described in Sec. 4.2. It involved using the Fourier transform technique to predict the time response of the required transducer. In this section we discuss the use of this technique in more detail. In addition, we discuss other techniques, such as phase weighting and the design of unidirectional transducers.

The techniques employed in the earlier chapter change little when applied to CCD delay lines, except that CCDs operate at baseband in a lower-frequency range than is typical of SAW devices. They do not, however, suffer from reflections of signals at the taps; thus the design problem for CCDs is somewhat simpler than for SAW devices.

4.6.2 Strip Coupler

Before describing the amplitude-weighted design, let us discuss the strip coupler [50, 51], a device used to eliminate reflections, excitation of bulk waves, and diffraction problems in SAW filters. Although the strip coupler is an important component in the design of SAW devices, it is also interesting for its own sake, because it is an excellent illustration of the coupled mode principle. We have already seen how this technique can yield information on the excitation of waves by a metal structure without requiring a detailed knowledge of the field distribution of the waves of interest. Here we show how this formalism can be applied to describe the coupling between two waves.

The basic strip-coupled device transfers a surface acoustic wave beam from track 1 to a neighboring track 2, as illustrated in Fig. 4.6.1. The device requires a series of closely spaced metal strips laid down across the two acoustic beam paths. Beam 1 excites a signal on these strips; the strips in turn reexcite the second beam on track 2. This device eliminates problems with unwanted volume wave modes that can give rise to interfering signals; the strip coupler does not couple strongly to these volume waves. It also lets us employ very short fingers for apodized transducers, as illustrated in Fig. 4.6.2(a). If these short fingers are placed very near the multistrip coupler, then diffraction loss, due to the beam expanding sideways, is no longer a problem, as can be seen from the experimental results shown in Fig. 4.6.2(b) and (c).

It is fairly simple to demonstrate the principle of these strip couplers and of other coupled-mode devices. Consider the expression for excitation of a surface acoustic wave at the plane z, propagating in the forward direction [Eq. (4.2.1)]. This is

$$a(t, z) = \alpha \int_{-\infty}^{z} \sigma\left[t - \frac{(z - z')}{V_R}, z'\right] dz' \qquad (4.6.1)$$

Here the upper limit of the integral is taken to be z, to account for the fact that although the strip coupler extends beyond the point z, a forward wave is excited only by strips for which $z' < z$.

Figure 4.6.1 Transfer of an acoustic wave in a multistrip coupler from track 1 to track 2 for equal trace widths.

The Fourier transform of this expression is

$$A(\omega, z) = \alpha \int_{t=-\infty}^{\infty} e^{-j\omega t} \int_{z'=-\infty}^{z} \sigma\left[t - \frac{(z-z')}{V_R}, z'\right] dz' \, dt$$

$$= \alpha \int_{-\infty}^{z} e^{-j\omega(z-z')/V_R} \int_{t=-\infty}^{\infty} \sigma(t', z') e^{-j\omega t'} dt' \, dz' \quad (4.6.2)$$

$$= \alpha \int_{-\infty}^{z} \sigma(\omega, z') e^{-jk(z-z')} dz'$$

where $k = \omega/V_R$ and $t' = t - (z - z')/V_R$. By differentiating this equation with respect to z,† it follows that

$$\frac{dA}{dz} = \alpha\sigma(z) - jk\alpha \int_{-\infty}^{\infty} \sigma(z') e^{-jk(z-z')} dz' \quad (4.6.3)$$

$$= \alpha\sigma(z) - jkA(z)$$

or

$$\frac{dA}{dz} + jkA = \alpha\sigma(z) \quad (4.6.4)$$

where, for convenience, we have dropped the ω in $\sigma(\omega, z)$, $A(\omega, z)$.

Now consider the beams traveling along two neighboring paths, as illustrated in Fig. 4.6.1, which we shall denote by subscripts 1 and 2, respectively. Either beam can be excited by charge on the strips. For simplicity, we suppose that the strips have infinitesimal spacing and are infinitesimally narrow. If the voltage at the surfaces of the beams on both paths are equal, and we take a to be proportional to these voltages, then $A_1 = A_2$, where the subscripts 1 and 2 correspond to the parameters of the waves on each beam path, respectively. In this case we expect the charge on the strips, $\sigma(z)$, to be zero, for no current flows along the strips or is supplied externally to them. More generally, we expect that if the beams are of equal width, the charge on the strips must be proportional to $A_1 - A_2$, as discussed in Prob. 2. Thus we can write a *coupled mode equation* to express the excitation of beam 1, in the form

$$\frac{dA_1}{dz} + jk_0 A_1 = j\kappa(A_2 - A_1) \quad (4.6.5)$$

where κ is an arbitrarily chosen real coupling constant. The j is used on the right-hand side so that the eventual solution for the propagation constant of the waves will be real. For instance, if $A_2 = 0$, then A_1 must vary as $\exp(-jkz)$, where k

†If

$$G(z) = \int_{-\infty}^{z} F(z, u) \, du$$

then it follows that

$$\frac{\partial G}{\partial z} = F(z, z) + \int_{-\infty}^{z} \frac{\partial F(z, u)}{\partial z} du$$

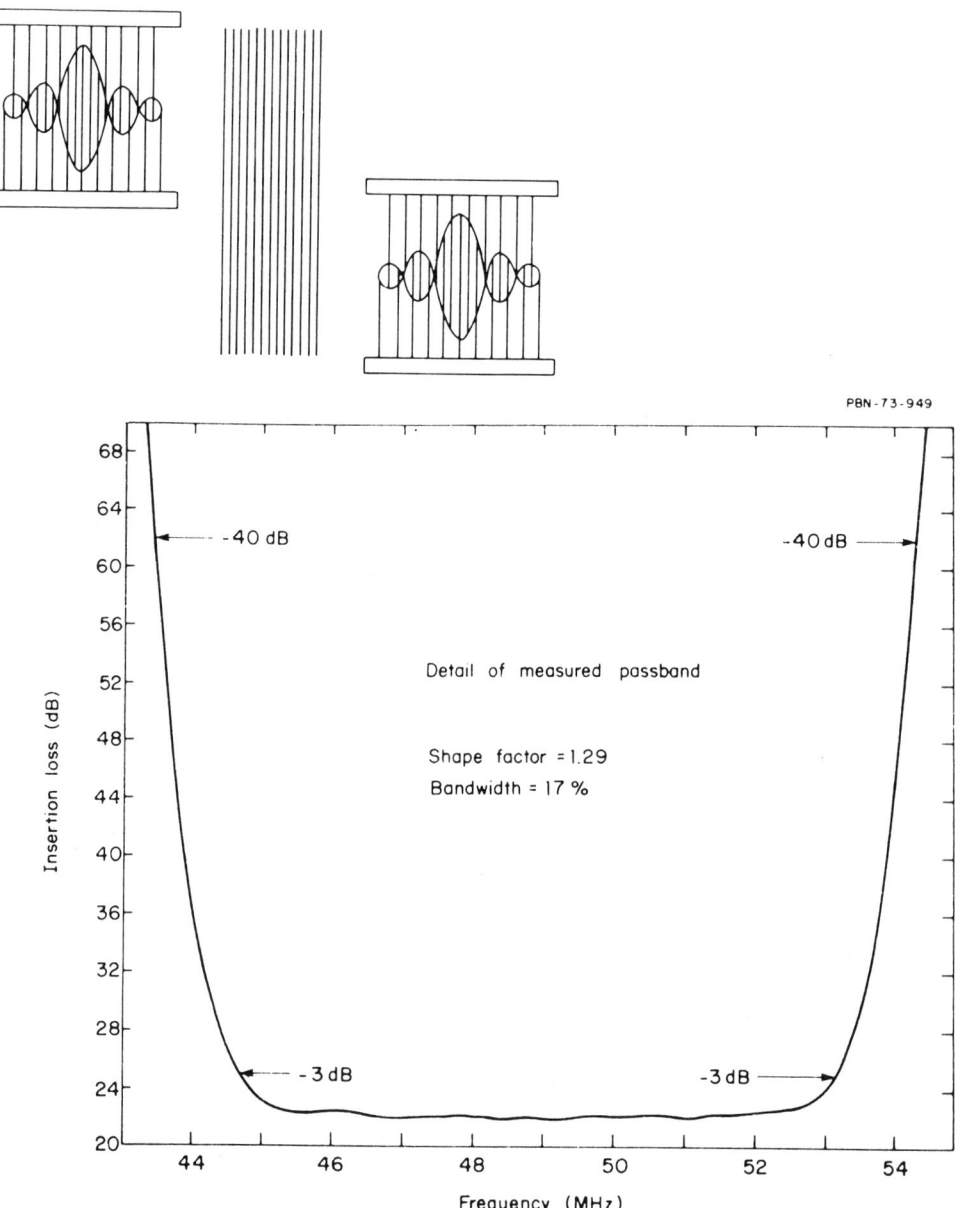

Figure 4.6.2 Filter incorporating a multistrip coupler, and the measured amplitude and phase responses for a filter with sharp skirts. The abrupt change in impedance at the band edges causes most of the phase error. (After Tancrell and Engan [52].)

is real. Additional justification for this choice is given in Prob. 2. By symmetry, we can also write

$$\frac{dA_2}{dz} + jk_0 A_2 = j\kappa(A_1 - A_2) \qquad (4.6.6)$$

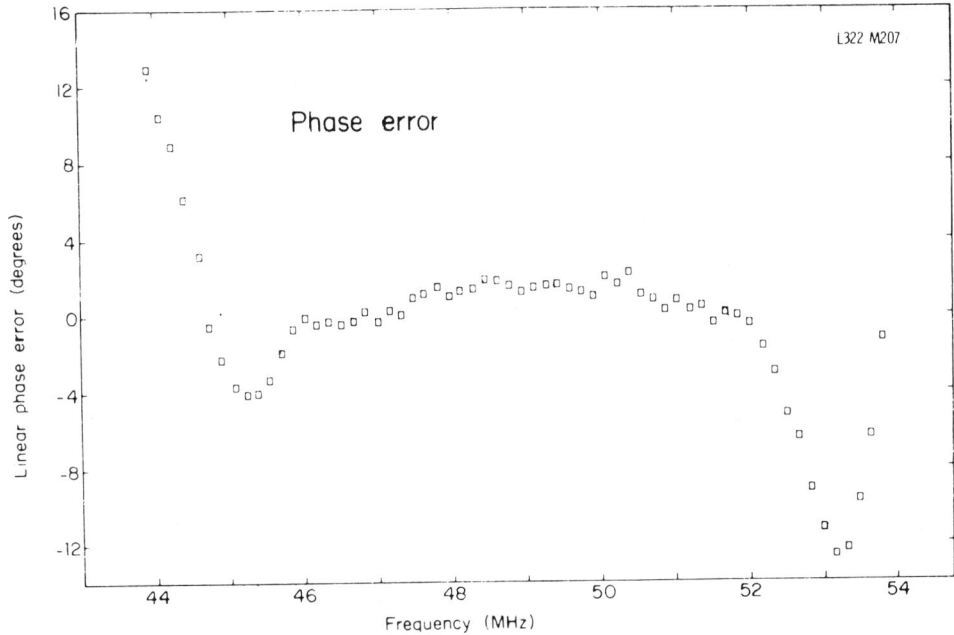

Figure 4.6.2 (continued)

Note that we have assumed that the propagation constants of the waves on both beam paths are equal and of a value k_0 when there are no strips present.

These equations can be solved by assuming a solution in the form $A_1 = A_{10} \exp(-jkz)$, $A_2 = A_{20} \exp(-jkz)$. This procedure leads to the results

$$A_{10}(k_0 - k + \kappa) = \kappa A_{20} \tag{4.6.7}$$

and

$$A_{20}(k_0 - k + \kappa) = \kappa A_{10} \tag{4.6.8}$$

Multiplying Eq. (4.6.7) by Eq. (4.6.8), we find that

$$(k_0 - k + \kappa)^2 = \kappa^2 \tag{4.6.9}$$

By taking the square root of Eq. (4.6.9), we find two solutions:

$$k = k_0 \tag{4.6.10}$$

and

$$k = k_0 + 2\kappa \tag{4.6.11}$$

In turn, this implies that for the two solutions, we can write

$$A_{10} = A_{20} \quad (k = k_0) \tag{4.6.12}$$

Sec. 4.6 Bandpass Filters

and

$$B_{20} = -B_{10} \quad (k = k_0 + 2\kappa) \tag{4.6.13}$$

respectively, where we have replaced A_{10} by B_{10} and A_{20} by B_{20} for the second solution.

When the amplitudes of the two waves are equal and of the same sign, the propagation constant of the system is unperturbed, as we might expect. When the amplitudes of the two waves are equal and of opposite sign, the propagation constants of the waves are increased (i.e., the wave velocity is slowed down).

We may now solve for the situation when a signal is excited on track 1, and is initially of zero amplitude at $z = 0$ on track 2. The signal on track 1 generally excites both waves of the system; these waves are called the *normal modes* of the system. The normal-mode solutions, as we have seen, have propagation constants $k = k_0$ and $k = k_0 + 2K$, respectively. Thus the initial boundary condition is $A_2(0) = 0$. We can write the solution on track 2 in the form

$$A_2(z) = e^{-jk_0 z}\left(A_{20} + B_{20} e^{-2j\kappa z}\right) \tag{4.6.14}$$

We need amplitude coefficients to solve for the total signals on tracks 1 and 2. We will call these A_{20} and B_{20}, respectively. To satisfy the boundary condition $A_2(0) = 0$, it follows that $A_{20} + B_{20} = 0$, or

$$A_2(z) = 2jA_{20} e^{-j(k_0 + \kappa)z} \sin \kappa z \tag{4.6.15}$$

It also follows from Eq. (4.6.13) that

$$\begin{aligned} A_1(z) &= 2A_{20} e^{-j(k_0 + \kappa)z} \cos \kappa z \\ &= A_1(0) e^{-j(k_0 + \kappa)z} \cos \kappa z \end{aligned} \tag{4.6.16}$$

At the point where $z = 0$, $A_2 = 0$ and $A_1(0) = 2A_{20}$. Farther along the line, however, where $\kappa z = \pi/2$, $A_1 = 0$ and $|A_2| = 2A_{20} = A_1(0)$. Thus all the power is transferred from one beam path to the other in a transfer distance $z = L_T = \pi/2\kappa$.

Furthermore, if the strip coupler is made longer, such that $\kappa z = \pi$, all the power is transferred back again to track 1. Thus there is a periodic transfer of energy back and forth between the two beam paths.

We can obtain a measure of the coupling parameter κ by noting that in the situation where the signals on each track are of opposite sign, that is, the solution for which $A_1(z) = -A_2(z)$ or $A_{10} = -A_{20}$ [Eq. (4.6.13)], one beam tends to excite the strip so it is positive, while the other tends to excite the strip so that it is negative. Under these conditions, current flows along the strips but the voltage on them must, by symmetry, be zero; therefore, the strips behave as if they are shorted to each other. Thus if we know the relative perturbation in surface acoustic wave velocity $\Delta V/V$, resulting from short-circuiting the surface of the device with a thin metal film, we can write

$$\frac{2\kappa}{k_0} = \frac{-\Delta V}{V} \tag{4.6.17}$$

As this quantity $\Delta V/V$ is easily calculable and measurable for most piezoelectric substrates and it is possible to determine κ, and hence the length required to transfer energy from one beam to the other.

For a material such as lithium niobate, with $\Delta V/V = 0.023$, Eq. (4.6.17) indicates that the transfer length is $L_T \approx \lambda/(2\Delta V/V)$, which is approximately 22λ, where $\lambda = V/f$ is the acoustic wavelength. This corresponds to a transfer length of approximately 0.7 mm at 100 MHz. In practice, the transfer length is approximately a factor of 2 larger than this figure because the strips have a finite gap between them; consequently, the entire region occupied by the strip coupler is not employed in coupling. Again in practice, the maximum transfer efficiency is usually better than -0.5 dB, with the minimum signal left in track 1 corresponding to a level less than -30 dB below the input signal. The directivity of the signal transfer with the coupler is usually better than 30 dB.

The basic principles described here apply to a wide range of systems. For instance, two optical beams in a fiber-optic waveguide system can be coupled to each other by lapping the two optical waveguides down to their cores, and arranging them so that they have a close spacing over the coupling length required (see Sec. 4.6.6). Microwave waveguides can be coupled to each other in a similar manner, by arranging them so that they have a common wall, with either a slot or closely spaced holes cut in it. In addition, acoustic beams may be coupled to optical beams by arranging for a suitable interaction region. In all cases, we can describe the qualitative behavior of the coupling mechanism by the simple coupled mode equations we have derived. Estimating the coupling coefficient itself is normally more difficult.

4.6.3 Amplitude-Weighted Bandpass Filters

The basic steps in designing an amplitude-weighted baseband bandpass filter involve deciding on the frequency response required, taking its Fourier transform, and then arranging the apodization of the filter to give the time response needed. A problem arises because such filters should, in principle, be of infinite length. Suppose, for instance, that we require a filter with a flat frequency response from $\omega_0 - \Omega/2$ to $\omega_0 + \Omega/2$ or $\Pi(\omega - \omega_0)/\Omega$, where $\Pi(x)$ is the rectangle function defined as

$$\Pi(x) = \begin{cases} 1 & |x| < \frac{1}{2} \\ 0 & |x| > \frac{1}{2} \end{cases}$$

Thus the bandpass in the positive frequency domain extends from $\omega_0 - \Omega/2 < \omega < \omega_0 + \Omega/2$. In this case, the time response of the filter will be

$$h(t) = \frac{\Omega}{2\pi} \frac{\sin \Omega t/2}{\Omega t/2} e^{j\omega_0 t} \qquad (4.6.18)$$

The time response required is therefore infinite in length.

Suppose that it is decided to cut off the time response of the filter at times $t = \pm T/2$, that is, to use a filter with a finite length T, as illustrated in Fig. 4.6.3(a).

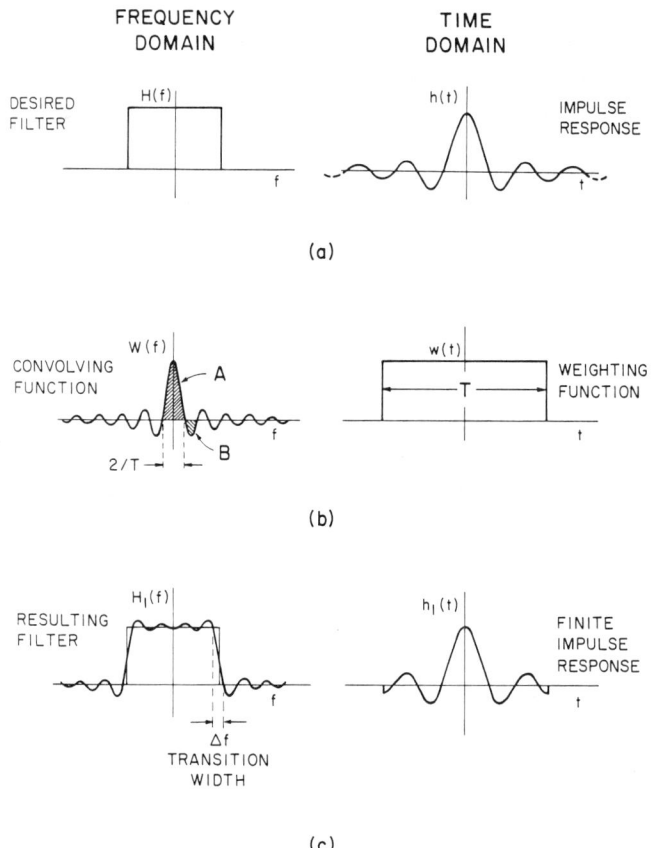

Figure 4.6.3 Convolution of the desired function $H(f)$ with the frequency spectrum of a finite impulse of length T: (a) ideal frequency and time responses; (b) weighting functions in the time and frequency domains for a finite-length filter; (c) behavior of a finite-length filter.

For an SAW device, this would imply using a filter of physical length $L = VT$, where V is the surface wave velocity. In a CCD, T would correspond to the total time delay in the device itself. This corresponds to a weighting function $w(t) = \Pi(t/T)$, whose Fourier transform is

$$W(\omega) = T \frac{\sin \omega T/2}{\omega T/2} \qquad (4.6.19)$$

The output from the device in the frequency domain will therefore be the convolution of the frequency responses $W(\omega)$ and $H(\omega) = \pi(\omega - \omega_0)/\Omega$, that is, the frequency response of the device will be

$$H_1(\omega) = \int H(\omega - s)W(s)\,ds \qquad (4.6.20)$$

We can illustrate this convolution process graphically by the diagram shown in Fig. 4.6.3. The length of the weighting function $W(\omega)$, shown in Fig. 4.6.3(b), between its two zeros is $\Delta\omega = 4\pi/T$, or

$$\Delta f = \frac{2}{T} \qquad (4.6.21)$$

Therefore, by limiting the length of the impulse response to a time T, we might expect the transition width of the frequency response to be approximately $2/T$, corresponding to the width between the zero of the main lobe of $W(\omega)$, as illustrated in Fig. 4.6.3. Furthermore, we can estimate the minimum attenuation outside the passband, from the area A of integration under the first sidelobe of $W(\omega)$, as compared to the area B of integration under the main lobe of $W(\omega)$, as indicated in Fig. 4.6.3(b). By taking the regions A and B to be triangular, we can show that the ratio of these two areas is approximately $(2/3\pi)/2 = 1/3\pi$, or 19.5 dB. The accurate result, given in Table 4.6.1, indicates -21 dB as the minimum stopband attenuation. On the same basis, the ripple at the edge of the passband would depend on the area under the first and second sidelobes, and would vary approximately from $1 - 1/3\pi$ to $1 - 1/3\pi + 1/5\pi$, or 0.6 dB.

If we could use a smoother weighting function, $W(\omega)$, the ripple within the passband would obviously be radically decreased. In addition, the attenuation outside the passband would be increased. Thus it is useful to employ the kinds of windowing functions described in Sec. 4.5.3. By doing this, we obtain the results given in Table 4.6.1. Typically, the stopband attenuation is improved at the expense of a wider transition width. The longer the filter length, the better the transition width of the filter.

An example of such a filter configuration is shown in Figs. 4.6.4 and 4.6.5. The results obtained with a low-frequency experimental filter terminated at the third lobe on either side of the main lobe (i.e., at $T = 8/B$) are shown in Fig. 4.6.5. In this case, with a Bartlett (triangular) weighting, the response outside the passband is improved considerably. Before weighting, the level outside the passband was -14 dB, worse than the -21 dB predicted by theory, while after weighting, the level outside the passband was -35 dB, better than the value of -25 dB predicted by theory. Since these early results were obtained, more carefully made devices have been constructed, and the results have usually been in very good agreement with theory.

TABLE 4.6.1 SIDELOBE LEVELS AND TRANSITION WIDTHS FOR PASSBAND FILTERS

Window	Transition width of main lobe	Minimum stopband attenuation (dB)
Rectangular	$2/T$	-21
Bartlett	$4/T$	-25
Hanning	$4/T$	-44
Hamming	$4/T$	-53
Blackman	$12/T$	-74

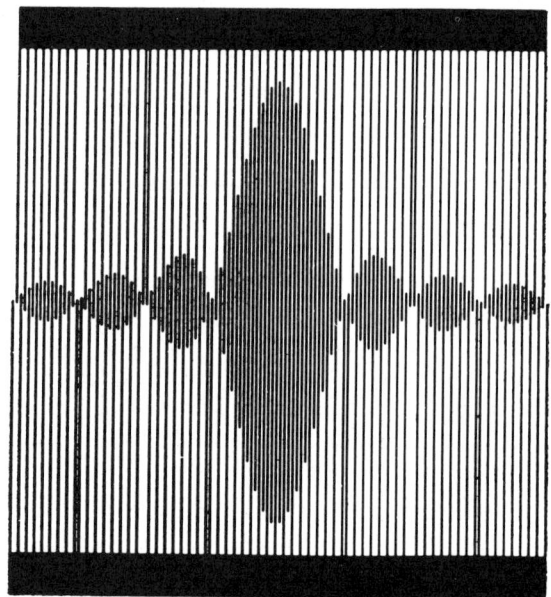

Figure 4.6.4 Interdigital electrode configuration in an SAW filter. (After Atzeni and Masotti [53].)

Other methods of weighting may also be used. For instance, the Dolph–Tchebysheff weighting technique allows us to maximize the attenuation outside the passband, and also to obtain the narrowest transition width possible for this attenuation. Because this particular technique is complicated to design and tends to lead to a weighting function with sharp maxima at the ends of the transducer, it is normally approximated by other methods, Hamming weighting being one of them. Thus the basic design approach is normally similar to that already described, with the added proviso that an attempt be made to optimize the transition width for a given attenuation outside the passband. Using these techniques, successful transducer designs for a wide range of bandwidths have been formulated with attenuations outside the passband as great as 60 dB.

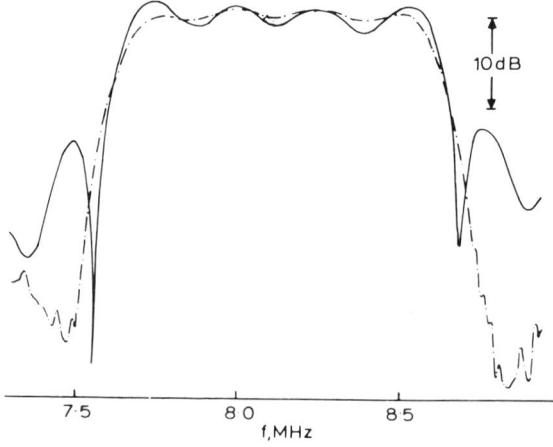

Figure 4.6.5 Amplitude against frequency response of the triangular weighted SAW filter (dashed line), compared with that of the unweighted filter (solid line). (After Atzeni and Masotti [53].)

Usually, the problem of obtaining sufficient attenuation outside the passband is associated with spurious effects. Simple designs, which might theoretically give rise to as much as a 70-dB attenuation outside the passband, often lead to attenuations of only 30 dB. This is due to unwanted excitation of volume waves, reflections from the fingers and the edges of the substrate, and undesirable diffraction phenomena.

The bulk wave excitation problem and the diffraction problem can, to a large extent, be eliminated by using strip couplers, because the strip coupler does not couple very strongly to bulk waves. Strip-coupled designs are thus often used for these purposes. Another approach, favored in Japan, is to use special cuts of lithium niobate and other materials for which the excitation of bulk waves is very weak. Successful intermediate-frequency (IF) filters for television receivers have been made with this method. It is normally fairly simple to eliminate reflections from the edges of the substrate. When a soft material, such as wax or a plastic resin, is deposited on the edges of the substrate, the surface waves are rapidly attenuated. Reflections from the fingers can usually be eliminated with the split-finger technique. Each finger is split into two parts, so that the periodic spacing of fingers is approximately $\lambda/4$, as illustrated in Fig. 4.6.6 (see Prob. 4.2.3). Neighboring fingers now give reflections π out of phase that cancel each other out. The excellent characteristic of a television IF filter, designed by Kodama using a combination of these techniques, is illustrated in Fig. 4.6.7.

The equivalent CCD devices do not suffer from most of these problems, but they can have problems with spurious effects resulting from charge-transfer efficiency. This tends to limit their design to a relatively low frequency, below 1 MHz, in addition to limiting the total length of the device. As the transition width depends on the total length of the device, this limits the transition times obtainable at a given frequency. As we have seen in Sec. 4.3 and Fig. 4.3.10, very successful low-frequency CCD filters have been made.

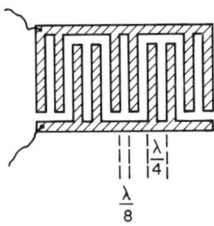

Figure 4.6.6 Transducer with split fingers to eliminate reflections.

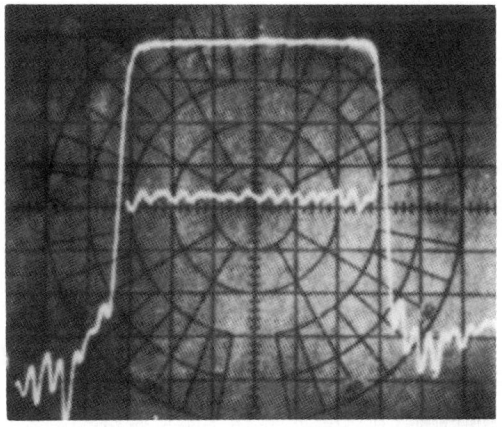

Figure 4.6.7 Measured vestigial sideband (VSB) filter amplitude and group-delay response of television IF filter designed by Kodama. (After Kodama [54].)

One problem with the standard interdigital transducer is that its symmetry excites equal-amplitude signals in both directions. Consequently, the transducer has a minimum loss of 3 dB; a device using two interdigital transducers has a minimum loss of 6 dB. A further consequence of this fact is that the interdigital transducer is a three-port device (one electrical port and two acoustic ports), and all three ports cannot be matched in a three-port network. For instance, suppose that a receiving transducer is perfectly matched at both acoustic ports (1 and 2), and an attempt is made to obtain maximum power into the electrical port (3). Then, by reciprocity, we know that if the input signal into port 1 is a_1 and the input power is a_1^2, the maximum power across the load at port 3 will be $a_1^2/2$ (a 3-dB loss). By reciprocity and symmetry, this signal across the load excites signals of value $a_1^2/4$ at ports 1 and 2. Thus a reflected signal is emitted from port 1 that is 6 dB lower in amplitude than the incident signal. The signal leaving port 3 is also 6 dB less than the input signal. Consequently, if both transducers of a delay line are matched as well as possible, there is a *triple transit echo*, that is, a signal reflected from the output transducer and then re-reflected from the input transducer, which is 12 dB down at the output electrical port 3 from the directly transmitted signal.

Reflections can be a serious problem when an interdigital transducer is matched for best efficiency. Thus the transducers often are not loaded for optimum efficiency. To improve the efficiency, a central input transducer is sometimes used to supply two output transducers, connected in parallel and placed on each side of the input transducer.

Another stratagem is to employ a three-phase transducer, illustrated in Fig. 4.6.8. In this transducer there are three fingers per wavelength, excited with phases 0, $2\pi/3$, and $4\pi/3$. This set of three fingers excites only a wave in the forward direction, since the wave excited by the three fingers in the backward direction has an amplitude $1 + \exp(-4j\pi/3) + \exp(-8j\pi/3) = 0$. Thus the device is a one-way transducer. The disadvantage is that a special electrical network must be used to excite this transducer; in addition, the geometry of the device becomes more difficult because it requires three separate connections. However, bandpass filters with excellent characteristics, high directivity, and only a 2-dB total loss from terminal to terminal have been made by this technique [55].

Finally, there are various permutations of the strip coupler configuration that can be used to give one-sided excitation. Such techniques, although useful, tend to introduce losses, so the improved performance desired is not always obtained (see Prob. 1) [50, 51].

4.6.4 Phase-Weighted Transducers

The treatment already carried out for chirp delay lines shows that such delay lines exhibit good frequency response over a finite bandwidth. However, because the input-output response of these delay lines is dispersive, their phase response or group delay may not be what is normally required in a signal processing system. A simple way to eliminate this difficulty is shown in Fig. 4.6.9(c). The dispersion

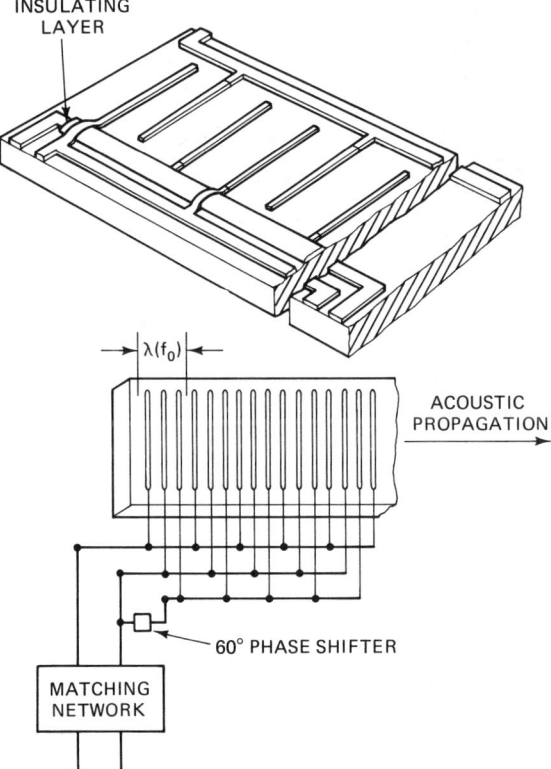

Figure 4.6.8 Geometry of a multilayer three-phase transducer. (After Hartmann et al. [55].)

is eliminated with two phase-weighted chirp transducers aligned in the same direction; one is used as a transmitting transducer and the other, as a receiving transducer. This implies that the delay time of any particular frequency component through the filter will be constant. The results obtained with such a filter are shown in Fig. 4.6.9(a) and (b), where the time response is of the form sin t/t, as we might expect. Thus this provides another simple way of designing filters (see Prob. 3).

4.6.5 Building-Block Design Technique

Many filters require not only a good response within the passband, but also extremely high attenuations near specific frequencies outside the passband; that is, they require the provision of "traps." A technique that has been utilized in the design of television IF filters, where deep traps are required for the neighboring channel and the sound channel, is illustrated in Fig. 4.6.10. Several transducers are placed in parallel, and the energy excited by them is transferred by a strip coupler to a single output transducer. Thus the individual transducers are designed for a given response and the responses are added.

Sec. 4.6   Bandpass Filters

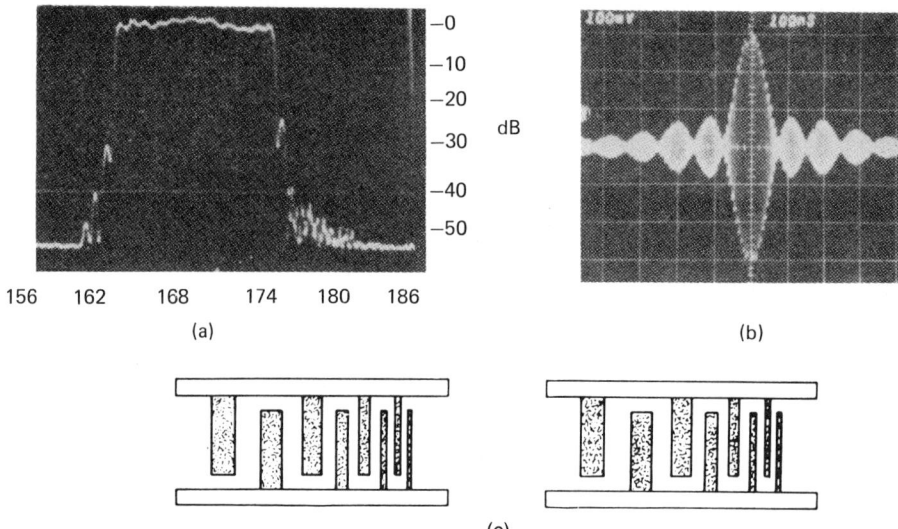

Figure 4.6.9 Low-shape-factor surface wave filter module: (a) 12 MHz at 168-MHz frequency response; (b) (sin x)/x impulse response indicates linear phase response and rectangular bandpass; (c) two phase-weighted transducers with opposite dispersion for canceling phase nonlinearity. (After Hartmann et al. [15].)

Suppose that we have two transducers, of center frequencies ω_1 and ω_2, with N_1 and N_2 fingers, respectively. Then the total response is

$$H(\omega) = W_1 \operatorname{sinc} \frac{N_1(\omega - \omega_1)}{\omega_1} + W_2 \operatorname{sinc} \frac{N_2(\omega - \omega_2)}{\omega_2} \quad (4.6.22)$$

where sinc $x = (\sin \pi x)/\pi x$, and W_1 and W_2 are the weights of the two transducers; that is, W_1 and W_2 are proportional to the finger lengths of the two transducers, respectively.

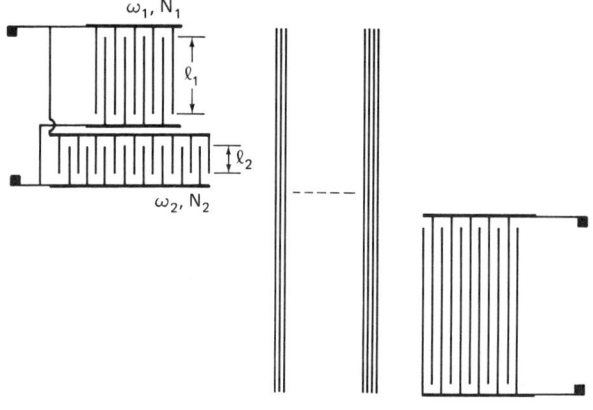

Figure 4.6.10 System for utilizing multiple transducers to affect frequency response. (After DeVries [56].)

Suppose, for instance, that we wished to compensate for the first upper sidelobe of the transducer with a center frequency ω_1. If we choose the second transducer to have a finger number N_2, such that the zero of the response of the second transducer is at the same point as the zero of the response of the first transducer, we must have $\omega_2 = \omega_1(1 + 3/2N_1)$, $N_2 = 2N_1 + 3$, and $W_2 = 2W_1/3\pi$ or $W_2 = 0.21W_1$. This will give a calculated response for the first upper sidelobe like that of Fig. 4.6.11, with a zero response or trap in its original position.

A similar technique can be used to broaden the frequency response of the main lobe. We may now choose $\omega_2 = \omega_1$, $N_2 = 2N_1$, and $W_2/W_1 < 0$. We will obtain a response of the type shown in Fig. 4.6.12.

Thus, in general, we design a system out of building blocks, so that the total response is of the form

$$H(\omega) = \sum_n W_n \operatorname{sinc} N_n \frac{(\omega - \omega_n)}{\omega_n} \qquad (4.6.23)$$

By taking the Fourier transform of this response, we note that each term consists of a finite length RF burst, as illustrated in Fig. 4.6.13. By summing these responses, we can make a simple filter corresponding to this total time response. A configuration of a TV IF filter designed by this technique is shown in Fig. 4.6.14, where the zeros of the response have been chosen as the trap frequencies. The results obtained are shown in Fig. 4.6.15.

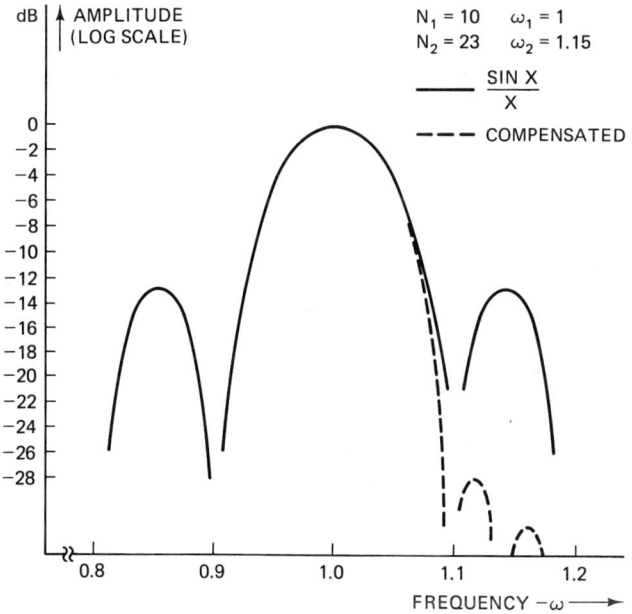

Figure 4.6.11 Calculated response curve for first upper sidelobe compensation. (After DeVries [56].)

Sec. 4.6   Bandpass Filters

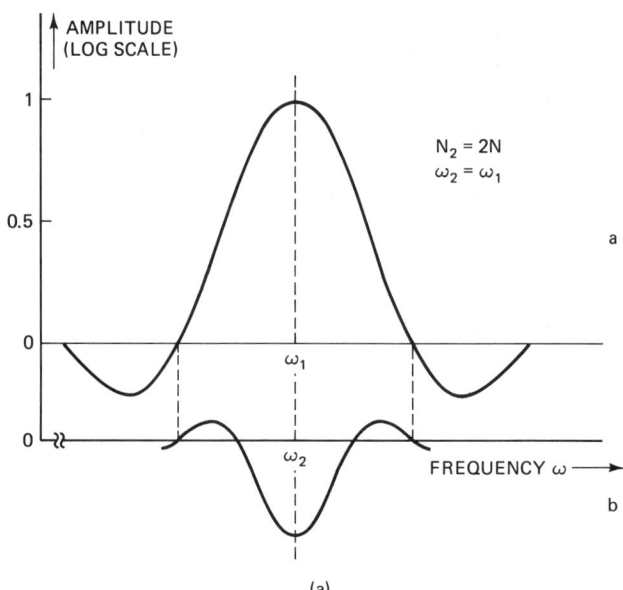

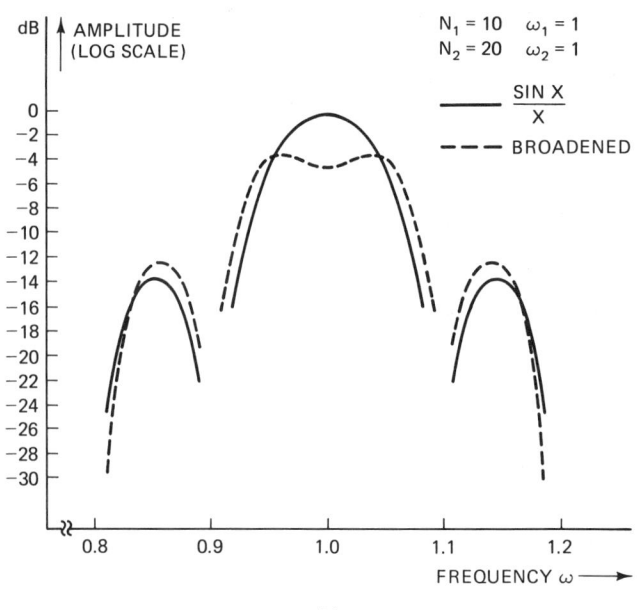

Figure 4.6.12 (a) System for broadening of frequency response; (b) calculated broadened frequency response. (After DeVries [56].)

Note that in this design both the finger spacing and the finger length vary. A multistrip transducer of a hundred strips was employed to eliminate bulk wave interactions. The output transducer did not use split fingers because the design was simplified by using fingers with a width of approximately $3\lambda/8$, which generally tends to eliminate spurious reflections.

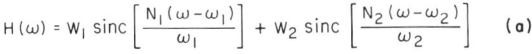

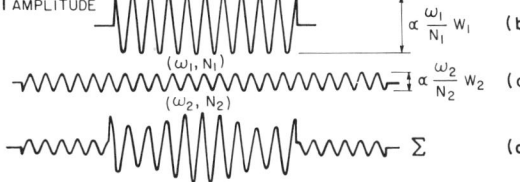

Figure 4.6.13 Method for synthesizing transducers. The desired frequency response is given by part (a). The respective Fourier transforms of the first and second terms are shown in parts (b) and (c). In part (d), the composite waveform from which the transducer configuration can be derived is shown. (From DeVries [56].)

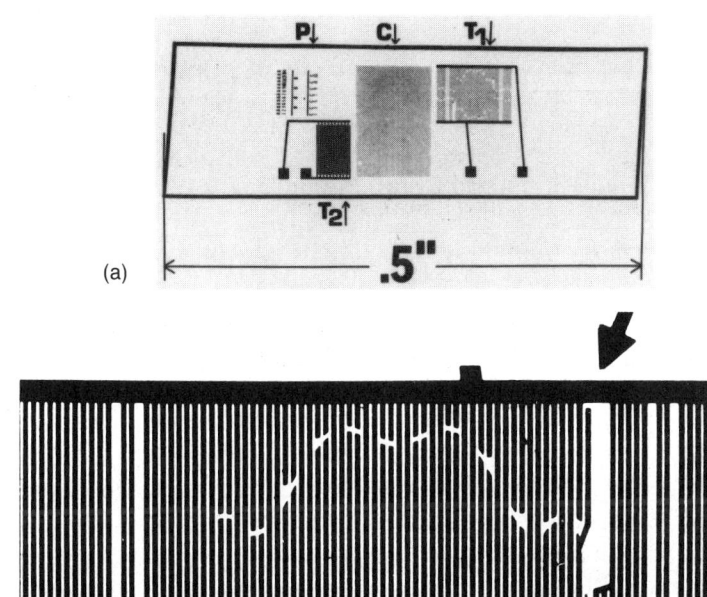

Figure 4.6.14 (a) Filter element showing transducer structure on lithium niofate substrate. The structure above the left transducer identifies the position of the element on the plate. (After DeVries and Adler [57].) (b) Configuration of weighted input transducer. An added transducer section that corrects for a finite trap depth is indicated by an arrow. (After DeVries et al. [58].)

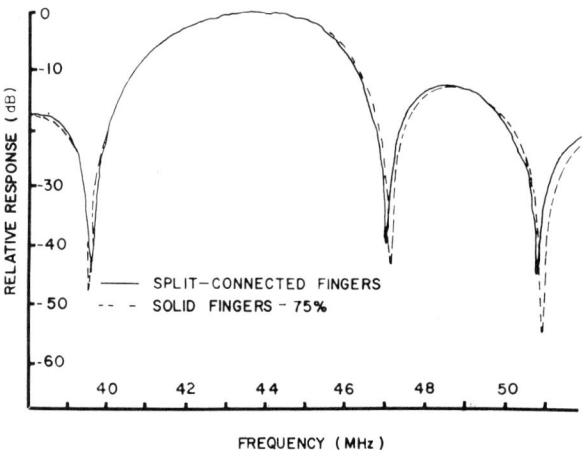

Figure 4.6.15 Comparison of the frequency response of a transducer with split connected fingers with a duty factor of 75%. (After DeVries et al. [58].)

4.6.6 Recursive Filters and Comb Filters

A. Introduction

Most transversal filters, because of their limited length, do not have as sharp a transition bandwidth as is sometimes desired. Better transition bandwidths can often be obtained in standard filters, because in z transform terms there are poles as well as zeros in the characteristic, which can give rise to very sharp changes in the transfer function. Such a pole in the characteristic can be generated in a transversal filter by using feedback between the output and the input. Zeros in the characteristic can be obtained using feedforward from input to output. Filters based on this principle are called *recursive filters*.

Suppose that the delay time in a simple delay line is T, as illustrated in Fig. 4.6.16; we feed the signal forward as shown, and add the input and output signals. We can write

$$y(t) = x(t) + \alpha_1 x(t - T) \tag{4.6.24}$$

If we choose $\alpha_1 = -1$, it follows that

$$y(t) = x(t) - x(t - T) \tag{4.6.25}$$

For an input signal that varies as $x(t) = \exp(j\omega t)$,

$$y(t) = e^{j\omega t}(1 - e^{-j\omega T}) \tag{4.6.26}$$

$$y(t) = 2je^{j\omega(t - T/2)} \sin \frac{\omega T}{2}$$

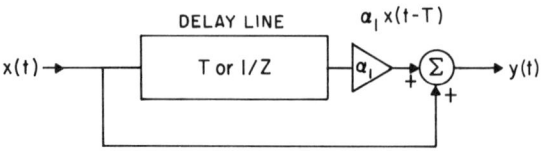

Figure 4.6.16 Recursive filter.

In z transform terms, we would write $z = e^{j\omega T}$, so that

$$H(z) = H(\omega) = (1 - e^{-j\omega T}) \qquad (4.6.27)$$
$$= 1 - \frac{1}{z}$$

This filter has a zero at $z = 1$, corresponding to notches in the output at $f_s = 1/T$ and all integral multiples of this frequency. Such a filter is called a *comb filter*.

This is the basic delay-line canceler element used for clutter rejection in moving target indicator (MTI) radar. The purpose is to eliminate *clutter*, that is, signals from stationary targets such as the ground or the sea. The frequency reflected from stationary targets is the transmitted frequency. A moving target, however, will give a Doppler shift of the return signal frequency and therefore can be observed while the clutter from the stationary targets is removed in the ideal system. As an example, a BBD delay line has been constructed for this purpose with 400 stages. This is clocked at a frequency $f_c = 1.6$ MHz, making the time delay through the device $T = 250$ μs. The effective sample rate for the data is $f_s = 1/T = 4$ kHz, and the delay-line canceler has notches at $f = 0, 4$ kHz, 8 kHz, and so on. Thus the radar is operated with a repetition frequency for the transmitted pulse of $f_s = 4$ kHz.

Note that signals arriving from different ranges arrive at different times. The 400-stage delay line described here has 400 possible "buckets," each spaced by 1/1.6 MHz (i.e., 0.67 μs), to store the information from the different ranges approximately 100 m apart. Thus we regard this device as having 400 "range bins." The effective sampling rate for each range bin is, in fact, 4 kHz (i.e., each range bin is filled every 250 μs). We can arrange to multiplex or switch the output $y(t)$ to 400 different storage units, where the information (time averaged over many samples, for example) can be processed.

In operation, the Nth pulse from the transmitter is received as a series of radar echoes spaced over a time of 250 μs. These echo signals pass into the delay line and fill individual buckets. As they leave the delay line, after being present in it for a time of 250 μs, they are added to the equivalent echo from the $(N + 1)$th pulse. Each echo can be switched or multiplexed from the delay line into a separate storage bin; echoes processed later can be added to the earlier ones to average the time of the signal.

We note that if we add M pulses of amplitude V, we obtain a total voltage output MV. However, as noise is incoherent, the total noise signal only adds as $\sqrt{M}$. To put it another way, the signal power into a 1-Ω load is

$$P_S = M^2 V^2 \qquad (4.6.28)$$

The noise power out is

$$P_N = M \langle N^2 \rangle \qquad (4.6.29)$$

where $\langle N^2 \rangle$ is the expectation value of the noise power. Thus the signal-to-noise ratio is improved by a factor M after averaging M times.

The signal entering the BBD is, in fact, a short pulse. Thus at best it is a sampled version of the RF signal. Suppose, for simplicity, that the return echo signal is an RF pulse or tone burst of length T_p and Doppler shifted carrier frequency ω, and that the repetition time is T. Then the nth RF pulse has the form $\cos \omega(t - nT)$. If the signal is synchronously detected by mixing it with the carrier signal $\cos[\omega_0(t - nT)]$, a baseband product signal $y_{\text{out}}(t)$ will be obtained at a frequency $\omega - \omega_0$, where

$$y_{\text{out}}(t) = \int_{-T_p/2}^{T_p/2} \cos[\omega(\tau - nT)] \cos[\omega_0(\tau - nT)] \, d\tau$$

$$\approx \frac{\cos(\omega - \omega_0)nT \sin[(\omega - \omega_0)T_p/2]}{\omega - \omega_0} \quad (4.6.30)$$

Thus the output from the radar detector has an amplitude that varies as $\cos(\omega - \omega_0)nT$. We now observe that each pulse entering a bucket is a sampled version of the detected signal $\cos(\omega - \omega_0)t$. Thus by averaging over many repetition cycles, the recursive filter treats the signal as if it were a continuous wave of frequency $\omega - \omega_0$, and the cancellation of signals for which $\omega = \omega_0$ can be obtained from stationary targets.

B. Recursive Filter with Feedback

The basic problem with the simple delay-line canceler described above is, of course, that it lacks an extremely sharp cutoff, which means that the signals from slowly moving targets will also tend to be attenuated. A recursive filter with feedback as well as feedforward has a characteristic in z transform terms as follows:

$$H(z) = \frac{1 + \alpha_1/z}{1 + \beta_1/z} \quad (4.6.31)$$

In our first example, we took $\beta_1 = 0$, $\alpha_1 = -1$. Suppose that we had used the circuit shown in Fig. 4.6.17. It is fairly easy to show that this circuit has the characteristic given in Eq. (4.6.31), with a pole at

$$z = e^{j\omega T} = -\beta_1 \quad (4.6.32)$$

Therefore, for $\beta_1 \to 1$, it has sharply peaked resonances at multiples of $f_s = 1/T$. By varying β_1, we can change the position of these resonances and their Q. At the same time, the recursive filter has zeros at

$$z = e^{j\omega T} = \alpha_1 \quad (4.6.33)$$

Thus it is possible to obtain a sharply peaked resonance very near zero frequency.

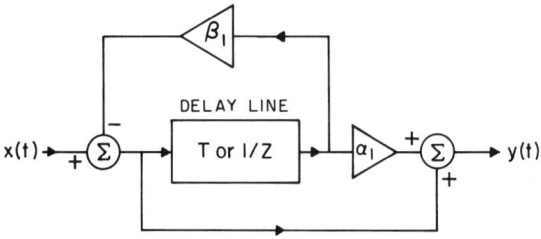

Figure 4.6.17 Single-pole recursive filter.

By cascading delay lines of this kind, it is possible to make far more sophisticated filters with well-understood characteristics using relatively short delay lines. We can write a general characteristic of such a recursive filter in the form

$$H(z) = \frac{\sum_0^N \alpha_n z^{-n}}{\sum_0^N \beta_n z^{-n}} \quad (4.6.34)$$

Consider the filter shown in Fig. 4.6.18, which consists of a first-order filter with $\alpha_1 = -1$, $\beta_1 = -K_1$, followed by a second-order filter with $\alpha_1 = -2$, $\alpha_2 = 1$, $\beta_1 = -K_2$, and $\beta_2 = K_3$. The operation of this filter can best be understood by setting the feedback equal to zero; then $K_1 = K_2 = K_3 = 0$. In this case

$$H(z) = (1 - z^{-1})^3 \quad (4.6.35)$$

It follows that the frequency characteristic is

$$|H(\omega)| = 8 \sin^3 \frac{\omega T}{2} \quad (4.6.36)$$

Thus this filter rejects zero frequency and all multiples of the pulse repetition frequency (the reciprocal of the transit time). However, it severely attenuates nonzero frequencies near zero that correspond to slow-moving targets, although it has a better characteristic than the first-order filter. More generally, when feedback is present, the filter has the characteristic

$$H(z) = \frac{(1 - z^{-1})^3}{(1 - K_1 z^{-1})(1 - K_2 z^{-1} + K_3 z^{-2})} \quad (4.6.37)$$

By selecting the amplifier gains $K_1 = 0.0881$, $K_2 = 1.2001$, and $K_3 = 0.7012$, we achieve the frequency response shown by the solid curve in Fig. 4.6.19. The incorporation of the feedback results in very little attenuation for $f > 0.1f_s$ and in a uniform amplitude response over a wide range of frequency between the notches.

The MTI canceler of Fig. 4.6.17 has been implemented by using eight-stage CCDs for the delay line. In this configuration, eight range bins can be filtered by clocking the CCD eight times each pulse repetition frequency. Feedback amplifiers are implemented with resistor networks around operational amplifiers. When the CCD is clocked to simulate a 38.4-kHz pulse repetition frequency, the filter rejection at $f = 0$ and $f = 38.4$ kHz is -32 dB, relative to the maximum output. The great advantage of this CCD over the equivalent acoustic delay-line cancelers is that its time delay can be controlled at will: therefore, the timing for the CCD

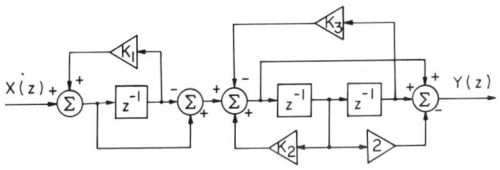

Figure 4.6.18 Block diagram of the three-delay canceler for MTI. It consists of a first-order recursive filter cascaded with a second-order recursive filter. (From Buss et al. [4].)

Sec. 4.6   Bandpass Filters

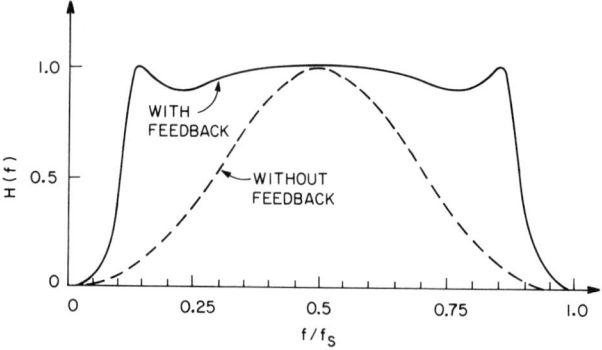

Figure 4.6.19 Calculated frequency response of the three-delay canceler of Fig. 4.6.18. The solid line shows the result for $K_1 = 0.0881$, $K_2 = 1.2001$, and $K_3 = 0.7012$. The dashed line shows the result with $K_1 = K_2 = K_3 = 0$. (From Buss et al. [4].)

can be synchronized with the timing of the transmitter to achieve a delay of exactly one pulse repetition frequency. On the other hand, the CCD has problems because of its imperfect charge-transfer efficiency, which does alter the filter characteristics. SAW delay lines have also been used in this application, with the clocks of the radar essentially controlled by the delay time through the delay line.

Example of an SAW recursive comb filter. Recursive filters may be constructed with SAW devices in the manner illustrated in Fig. 4.6.20. A relatively short input transducer is employed with two output transducers that have a center-to-center time separation T, and hence a separation between response maxima of $\Delta f = 1/T$. The passband will have the form of a "comb response," as illustrated in Fig. 4.6.21(b). Because of the characteristics of both the input and the separate output transducers, this comb response will be multiplied by their bandpass responses, as shown in Fig. 4.6.21(a).

One application of this comb response characteristic, which has been employed commercially, is for a channel indicator on a TV set. In a TV set, the input signal is mixed with the output of a varactor tuned oscillator to produce the IF frequency. When this oscillator frequency is swept through the TV channels in turn, the response of the comb filter passes through a series of peaks corresponding to the different channels, which are equally spaced in frequency. The detected output is connected to a counter which indicates the channel. An additional grating reflector trap is used to indicate the starting frequency for counting, as shown in Fig. 4.6.21(c) and (d). Four sets of filters each are required for the

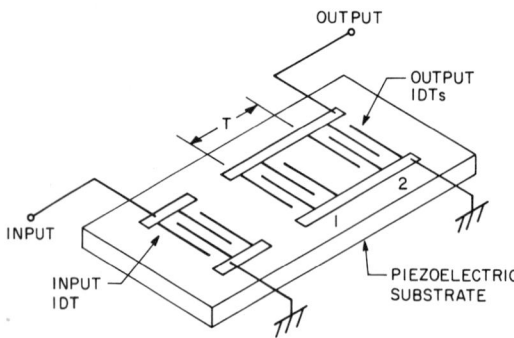

Figure 4.6.20 Fundamental structure of SAW comb filter. (From Kishimoto et al. [59].)

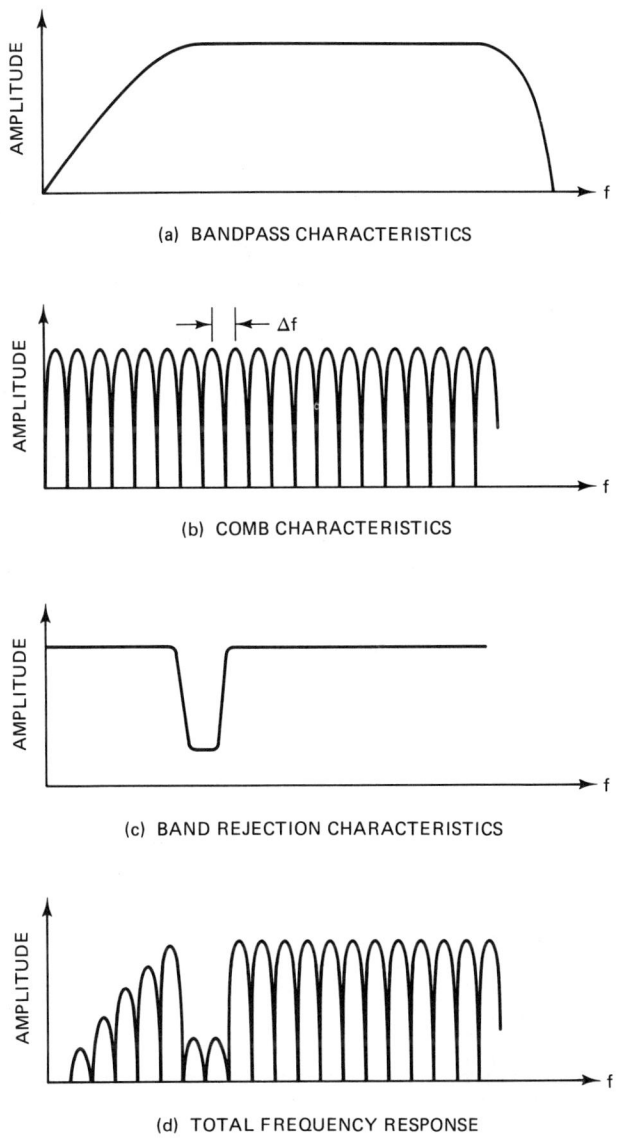

Figure 4.6.21 Frequency response of comb filter: (a) bandpass characteristics; (b) comb characteristics; (c) band rejection characteristics; (d) total frequency response. (From Kishimoto et al. [59].)

VHF low band (150 to 162 MHz), the VHF high band (230 to 276 MHz), and the UHF band (530 to 824 MHz) in current use in Japan. A photograph of the system and its characteristics is shown in Fig. 4.6.22.

A later version of this system has also been used in a feedback circuit, controlling the varactor-tuned oscillator frequency to provide automatic tuning of the oscillator to designated channels.

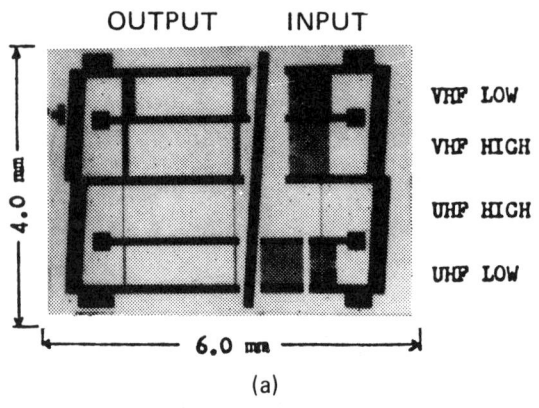

		INPUT IDT	GRATING REFLECTOR	OUTPUT IDT
VHF BAND	LOW	APODIZED SPLIT N = 30.5 PAIRS W = 2.0 μm		UNAPODIZED N = 10 PAIRS × 2 W = 5.4 μm
VHF BAND	HIGH	APODIZED SPLIT N = 46.5 PAIRS W = 1.2 μm		UNAPODIZED N = 8 PAIRS × 2 W = 3.4 μm
UHF BAND	LOW	APODIZED N = 61.5 PAIRS W = 1.2 μm	N = 200 W = 1.7 μm	UNAPODIZED N = 7 PAIRS × 2 W = 1.4 μm
UHF BAND	HIGH	UNAPODIZED N = 4 PAIRS W = 1.1 μm		UNAPODIZED N = 4 PAIRS × 2 W = 1.1 μm

N: NUMBER OF ELECTRODES
W: MINIMUM ELECTRODE WIDTH

(b)

Figure 4.6.22 Frequency characteristics and design features of comb filter, with a photograph of the filter. (After Matsu-ura et al. [60].)

C. Fiber-Optic Recursive Filters

Fiber-optic delay lines also lend themselves well to use as recursive filters for very high frequency signals in the range from several megahertz to several gigahertz. In such cases, we are normally interested in carrying out signal processing on the

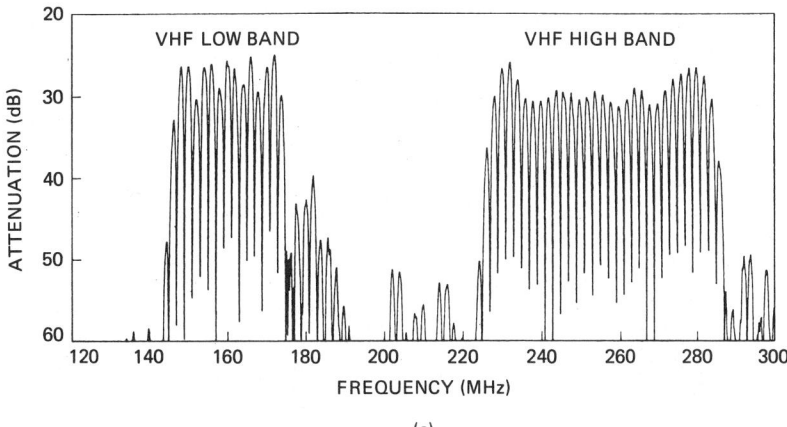

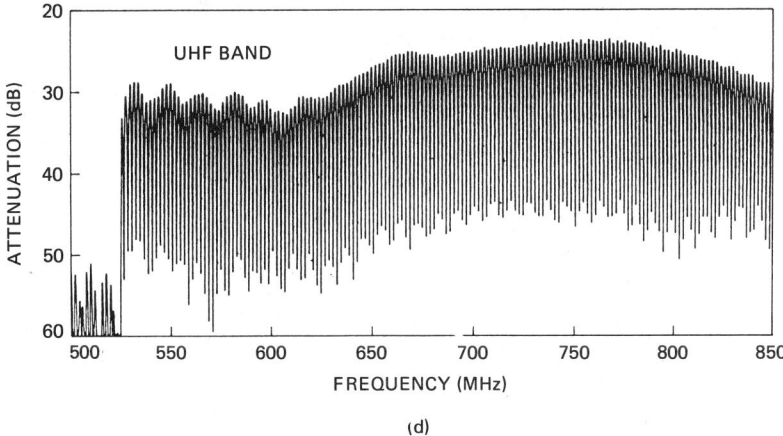

Figure 4.6.22 Continued

modulation instead of the carrier, which is at optical frequencies. Thus it is important to eliminate interference effects between carriers. For this reason, a relatively incoherent laser beam is used that has considerable phase fluctuation within the shortest modulation time required. This is equivalent to requiring that the laser should have an output bandwidth larger than that of the highest frequency in the modulation.

One basic component used in fiber-optic signal processing devices is a directional coupler. Normally, the light beam is confined to a glass core that, in the case of a *single-mode fiber*, is of the order of 4 to 10 μm in diameter. This glass core is embedded in a glass cladding, with the glass chosen so that the velocity of the wave in the core of the material is slightly slower than that in the cladding. The outer diameter of the cladding is typically of the order of 100 μm. This implies that a wave approaching the boundary of the core material at a small angle to the axis will be totally reflected at the edge of the core, and that any fields associated with this wave will fall off exponentially into the outer glass cladding. Consequently, the light traveling down the core is inaccessible from outside the fiber.

There are two possible ways to tap a fiber-optic delay line. If, as illustrated in Fig. 4.6.23, the delay line is coiled in a tight coil, some light will leak from the core to the outer radius of the coil because the effective velocity in the cladding will be larger than that in the core, which means that a small amount of light will "leak" outside the fiber. A second technique, illustrated in Fig. 4.6.24, is to lap off the cladding and place the flat surface so formed in contact with or very close to a similarly lapped fiber. This forms a directional coupler, which operates on the same principle as the strip coupler already described, and makes it possible to transfer energy from one fiber to another. By adjusting the length or width of the overlap region, or the spacing between the two fibers, the coupling from one fiber to the other can be changed; thus almost all the power can be transferred from one fiber to the other, or only a small proportion can be transferred. In all cases, the directivity of such couplers is usually better than 60 dB.

A fiber-optic recursive filter can be made using two directional couplers in the configurations shown in Figs. 4.6.25 and 4.6.26, then coupling the signal out of the main fiber into a coiled fiber with a delay time T, and coupling the signal from the delay fiber back into the original fiber. In the simplest case, we can use two 3-dB couplers where half the power is coupled into the delay line and half of that power is coupled back into the original line. At the same time, half the power in the original line is coupled through the second directional coupler into the fiber following the directional coupler, and is not used. In the case illustrated in Figs. 4.6.25 and 4.6.26, a variable delay line is employed so that the output fiber can be translated from one fiber tap to another, to vary the delay T in controllable increments of time.

We consider the operation of this device with a modulated incoherent optical signal. It is not desirable to use purely coherent optical signals, because then there would be interference of the carriers in the two paths, and the interference would

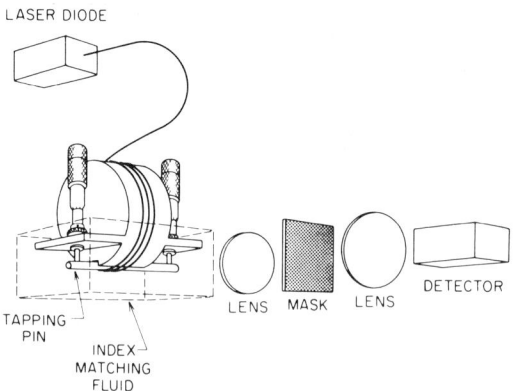

Figure 4.6.23 Macrobend tapped delay line. (After Jackson et al. [10].)

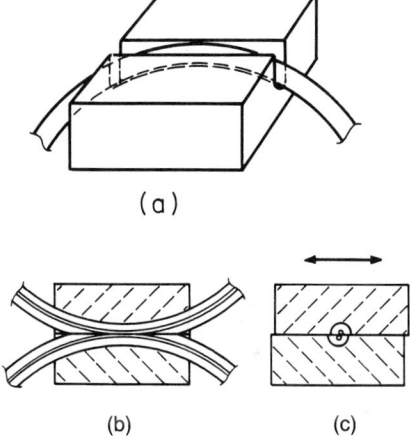

Figure 4.6.24 Construction of a fiber-optic coupler: (a) the fiber is placed in a channel of a solid material, such as silicon or quartz, and bonded in place with epoxy; (b) the two fibers in the coupler are lapped to the core and placed in contact; (c) the coupling may be varied by changing the amount of overlap of the cores. (After Digonnet and Shaw [62].)

depend critically on small delay changes through the fibers due to bending or slight flexing of the fiber. We take the input signal to be of the form

$$u(t) = (1 + A \cos \Omega t)e^{j[\omega t + \phi(t)]} \quad (4.6.38)$$

where ω is the carrier frequency and Ω is the modulation frequency, with A the amplitude of the modulation. The phase of the laser $\phi(t)$ is regarded as fluctuating with time in a random fashion, so that $\langle \phi(t) \rangle = 0$.

After passing through the delay-line path, the signal obtained from the delay line will be of the form

$$u(t - T) = [1 + A \cos \Omega(t - T)]e^{j[\omega(t - T) + \phi(t - T)]} \quad (4.6.39)$$

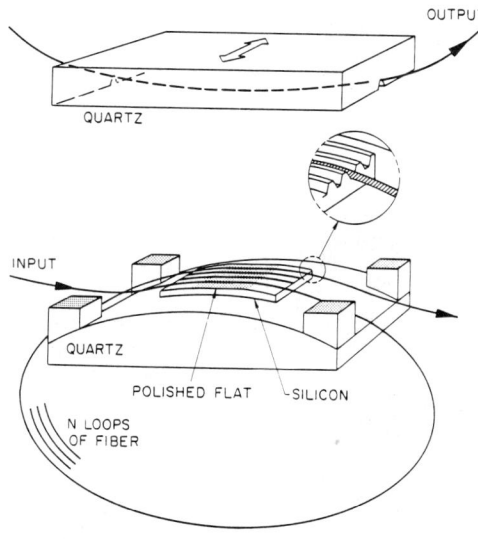

Figure 4.6.25 A variable delay line, in which the output fiber is translated from one input fiber to another. (After Bowers et al. [11].)

Sec. 4.6 Bandpass Filters

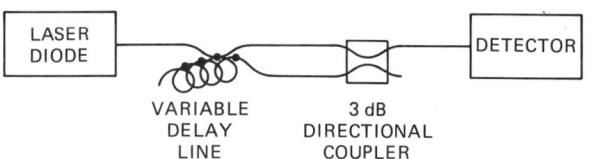

(a)

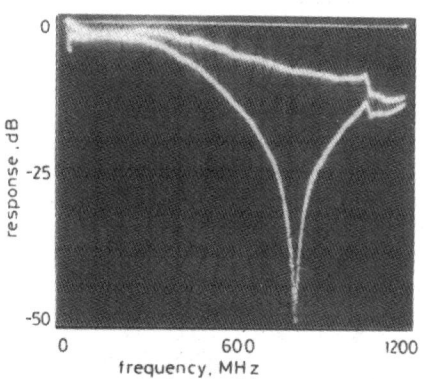

(b)

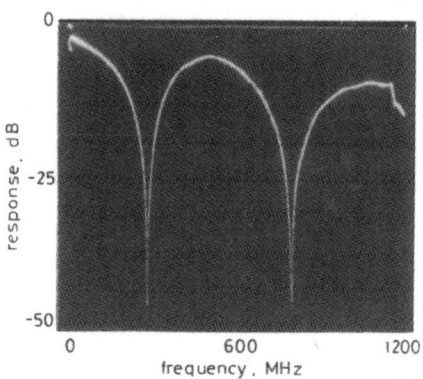

(c)

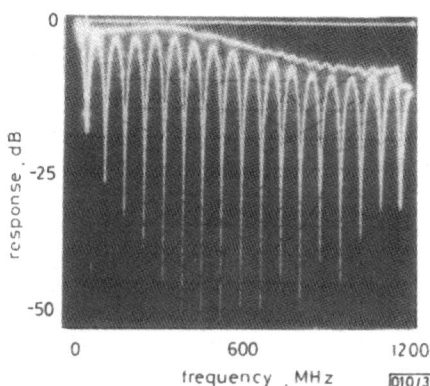

(d)

Figure 4.6.26 Use of variable delay line in a fiber-optic programmable filter: (a) schematic diagram showing source, programmable filter, and detector; (b) frequency response for relative delay of 0.6 ns; (c) frequency response for relative delay of 1.9 ns; (d) frequency response for relative delay of 14.4 ns. Upper trace in parts (b) and (d) shows frequency response of laser diode, amplifier, and detector. (After Bowers et al. [11].)

The sum of two signals $u(t)$ and $u(t - T)$ is inserted into a photodetector that gives an output proportional to the intensity of the light incident on it.

We are concerned with a signal that is the expectation value of the output from the photodetector in a time scale comparable to that of the modulation. This signal is of the form

$$V(t) = \langle [u(t) + u(t - T)]^2 \rangle \qquad (4.6.40)$$

Because the laser phase is incoherent, cross-products of the two terms in Eq. (4.6.40) do not give any output. Consequently, the output signal will be of the form

$$V(t) = (1 + A \cos \Omega t)^2 + [1 + A \cos \Omega(t - T)]^2 \qquad (4.6.41)$$

If this output signal is passed into a filter so that only signals of frequency Ω can pass through it, we will obtain the result

$$V_\Omega(t) = 4A \cos \Omega \left(t - \frac{T}{2} \right) \cos \frac{\Omega T}{2} \qquad (4.6.42)$$

We conclude that the fiber-optic system behaves like a standard recursive delay line, with sharp notches at frequencies corresponding to

$$f = \frac{2m + 1}{T} \qquad (4.6.43)$$

where m is an integer and $f = \Omega/2\pi$. Results obtained by Bowers et al. [11] with a device of this kind are shown in Fig. 4.6.26. Notches approximately 50 dB deep are easily observed, so that the device operates as a recursive filter for the modulation frequency. Coherence of the laser can sometimes be a problem. One way to eliminate this difficulty is to use only one directional coupler, and to take the outputs from the two fibers to separate photodetectors whose outputs are summed electrically. In this case there will be no cross-coupling between the two lasers.

In practice, this type of filter can be very convenient for very high frequency signals. A semiconductor laser source, which is normally fairly incoherent, can be employed, and the laser can be amplitude modulated by modulating the current passing through the laser. Modulation frequencies up to several gigahertz can be obtained, although a very high speed photodetector is required to detect them. Detectors are now being developed for use at modulation frequencies as high as 100 GHz.

PROBLEM SET 4.6

1. Consider a strip coupler with $k_1 \neq k_2$ and coupling coefficient κ. Work out the maximum transfer efficiency of the coupler as a function of $(k_1 - k_2)/\kappa$.
2. Consider a multistrip coupler with an optimum transfer length L_T from track A to track B (i.e., the length for complete power transfer is $L_T = \pi/2\kappa$), as shown [50, 51].

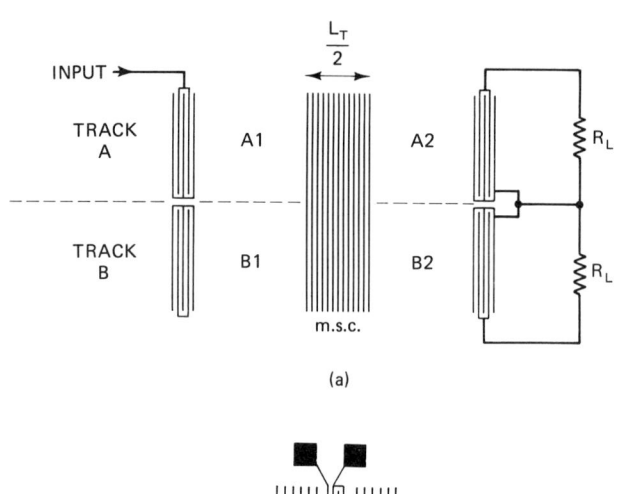

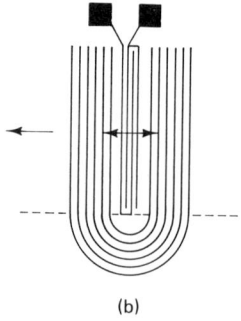

(a) Show that if a length $L_T/2$ is employed, the powers in tracks A and B are equal, and the signals are $\pi/2$ out of phase in each track.

(b) Consider the system shown in figure (a). Normally, there is a reflection from an output transducer, which gives rise to the so-called triple transit echo. This is caused by reflections from the output and input transducers in turn, which give rise to a later signal from that desired. In the schematic shown, a strip coupler of length $L_T/2$ is placed across the beam. Two identical output transducers 1 and 2 have identical loads R_L and are placed in tracks A and B, respectively. Show that the signal reflected back into track A is zero, so that the reflections from the output transducers are eliminated.

(c) Consider the system shown in figure (b), where the strip coupler is curved in shape, but still chosen to be of length $L_T/2$. If the input transducer radiates equally in both directions, as it must, show that when it is placed $\lambda/8$ from the centerline between the two strip couplers, all the power radiated from this system will be radaiated in one direction. Thus, by using a "multistrip mirror," we have constructed a unidirectional surface wave coupler with relatively broadband operation. In this problem, ignore the capacity of the curved strip region. Both these systems have been demonstrated as practical configurations.

3. Consider a phase-weighted transducer of the type illustrated in Fig. 4.6.9. Suppose that each transducer is of length $L = VT$, and each is weighted as $w(z) = \cos(\omega_0 z/V + \mu z^2/2V^2)$, where the origin of z is at the center of the transducer. Let the centers of the transducers be separated by a length $H > L$. Write down the time responses $h_1(t)$ and $h_2(t)$ of the two transducers. Then determine an approximate expression for the time response of the complete filter using only the term in the integrand for which square-

law terms cancel out, as in chirp transform theory. From this result, use the method of stationary phase to find the frequency response of the filter.

Answer: The frequency response $H(\omega)$ can be written in the following form:

$$H(\omega) = (1 + j)\sqrt{\frac{\pi}{\mu}} e^{j(\omega_0 - \omega)^2/2\mu} \frac{\sin(\omega_0 - \omega)T}{(\omega_0 - \omega)T} \Pi\left(\frac{\omega - \omega_0}{\mu T}\right)$$

4. Consider a strip coupler in which the two acoustic beam paths are of different widths, w_1 and w_2. Let $w_1 = Rw_2$. If a_1 and a_2 are proportional to the voltages at the surface of each beam path, we might expect the coupled mode equations to take the form

$$\frac{da_1}{dz} + jk_1 a_1 = j\kappa(a_2 - Ra_1)$$

$$\frac{da_2}{dz} + jk_2 a_2 = j\kappa(Ra_1 - a_2)$$

where k_1 and k_2 are the propagation constants in the two beam paths, respectively, and the coupling to the strips is proportional to the width of the beam path. Show that if

$$k_2 - k_1 = \kappa(R - 1)$$

complete power transfer from one beam to the other can be obtained. This process is carried out in practice by keeping $k_1 = k_2 = k_0$ and varying the strip spacing, as illustrated in Fig. 4.7.9(a).

Note: If a_1 and a_2 are taken to be proportional to the square root of the power in each beam, a_1 must be replaced by $a_1/\sqrt{R}$ in these equations. The results obtained will be identical, but the equations will display conservation of power.

5. Consider the problem of designing a recursive filter for a given passband characteristic. Such a filter samples the input signal with a sample spacing T. Using Laplace transform notation $s = j\omega$, a filter can be shown to have a response

$$H(s) = \sum_{p=1}^{P} \frac{A_p}{s - s_p}$$

The corresponding impulse response is of the form

$$h(t) = \sum_{p=1}^{P} A_p e^{s_p t} u(t)$$

where $u(t)$ is a unit step function.

Now suppose that we sample $h(t)$ at times $t = nT$. Then, with the proper normalization,

$$h(nT) = \sum_{p=1}^{P} A_p T e^{s_p n T} u(nT)$$

The Laplace transform of the sampled function is

$$H(s) = \sum_{p=1}^{P} A_p T \sum_{n=0}^{\infty} e^{(s_p - s)nT}$$

$$= \sum_{p=1}^{P} \frac{A_p T}{1 - e^{(s_p - s)T}}$$

Thus we can write $H(z)$ in normalized form as

$$H(z) = \sum_{p=1}^{P} \frac{A_p T}{1 - z^{-1} e^{s_p T}}$$

where $z = e^{sT}$.

Provided that $\omega T \ll 1$, we expect the sampled filter to give the same response as that of the equivalent analog filter. The result given here is equivalent to that of Eq. (4.6.34), although it is expressed as a sum, not as a product.*

(a) Consider the design of a sampled low-pass Butterworth recursive filter with a response

$$|H(s)|^2 = \frac{1}{1 + (\omega/\omega_c)^{2N}} = \frac{1}{1 + (s/s_c)^{2N}}$$

where $s_c = j\omega_c$. We choose $H(s)$ to have poles at

$$s_p = s_c e^{j\pi/2N} e^{jp\pi/N}$$

where we take only terms for a causal filter for which $\Re e(s_p) < 0$. The complex conjugate terms $H^*(s)$ account for the other poles of $|H(s)|^2$.

Taking $\omega_c T = 0.125\pi$ and $N = 2$, find the position of the poles in the s domain and the z domain. Note that the poles occur in complex conjugate pairs.

(b) Express your result in a form similar to Eq. (4.6.34), that is,

$$H(z) = \frac{\alpha_0 + \alpha_1/z}{\beta_0 + \beta_1/z + \beta_2/z^2}$$

and find the values of $\alpha_0, \alpha_1, \beta_0, \beta_1, \beta_2$.

(c) Following the derivation leading to Fig. 4.6.18, design a recursive Butterworth filter.

(d) Plot the response $|H(z)|^2$ in terms of frequency ω, and compare it with $|H(\omega)|^2$. Plot your results from $\omega = 0$ to $\omega = 16\omega_c$.

(e) Suggest, *without working out numerical values*, the form of the circuit for a recursive filter with $N = 4$. You may place recursive filters in parallel as well as in series.

6. A fiber-optic sensor is used to detect high-frequency vibrations of a surface at a frequency Ω, that is the displacement of the surface is

$$\Delta y = A \cos \Omega t$$

The measuring system illustrated in Fig. 1 is arranged so that light can travel the four possible paths there and back, as shown. In case 3 the light arrives at the substrate earlier than in case 2, but suffers the same total delay T in both cases. In cases 1 and 4 the total delay is either 0 or $2T$, respectively.

(a) Suppose that each directional coupler couples a proportion α of the light power into the neighboring path, leaving $1 - \alpha$ traveling along the original path. What should α be so that light arriving along each path A, B is maximum at the surface being tested?

(b) Suppose that the signal from the laser is incoherent, that is, is of the form $\exp[j(\omega t + \phi)]$, where ϕ changes randomly with time. Then a signal arriving at the substrate with no delay will be phase modulated as

$$e^{j(\omega t + \phi + 2kA\cos\Omega t)}$$

Reference: A. V. Oppenheim and R. W. Schafer, *Digital Signal Processing* (Englewood Cliffs, N.J.: Prentice-Hall, Inc., 1975).

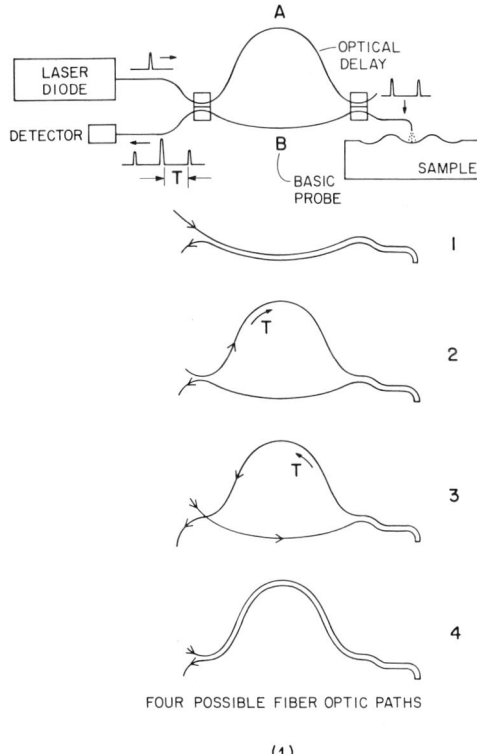

ALL FIBER PROBE

FOUR POSSIBLE FIBER OPTIC PATHS

(1)

where $k = 2\pi/\lambda$ is the optical wave number. Show that only for the light traveling on paths 2 and 3, will there be coherent modulation effects in the expected value of the output signal from the square-law detector.

(c) Assume that because light can be polarized, the effective path length changes slightly; thus signals traveling on path A suffer an extra phase shift $\pi/2$ greater than that of path B, in addition to the modulation effects at frequency Ω, as illustrated in Figs. 2 and 3. This can be accomplished by coiling the fiber a few turns around a disk. The stress induced in the fiber changes the velocity of a wave polarized in the plane parallel to the disk and hardly affects the perpendicularly polarized wave. Thus with the correct number of turns and disk radius, the fiber coil will be effectively a quarter-wave polarization plate [9]. The devices are used in the configuration shown as the fiber optic paths 2 and 3. Show that when $kA \ll 1$, the output from the photodetector at frequency Ω will vary as

$$A \sin \frac{\Omega T}{2}$$

where T is the delay time in the loop. What is the disadvantage of leaving out the extra $\pi/2$ phase shift when A is small?

Sec. 4.6   Bandpass Filters

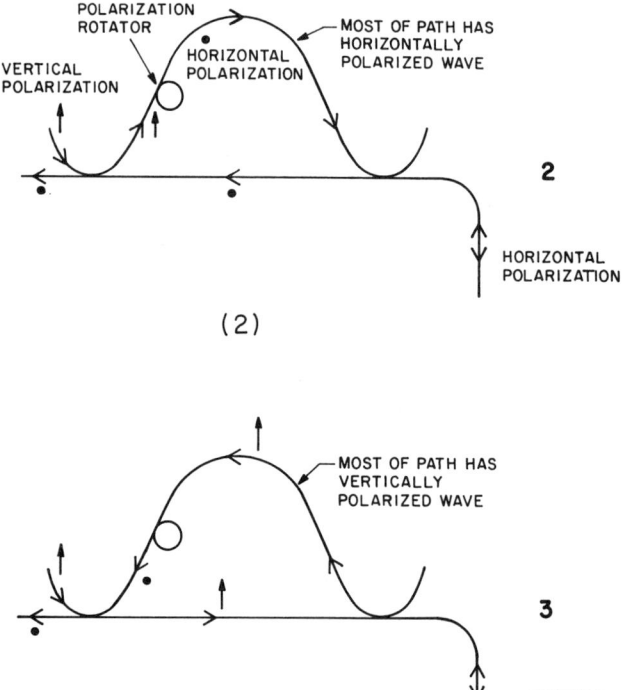

(2)

(3)

Note: Typically, we might expect A to be 1 Å or less, and the optical wavelength to be 6000 Å. So you can assume that $kA \ll 1$, and write $\exp(2jkA \cos \Omega t) = 1 + 2jkA \cos \Omega t$.

4.7 CONVOLVERS

4.7.1 Introduction

The transversal filters described so far use fixed tap weights, or tap weights that can be controlled externally at a relatively slow rate that is not comparable to the signal frequencies. It is also useful to have a real-time processing device that can compare an unknown signal to a reference signal and determine the correlation or convolution of the two. Such a device might even change the reference code from pulse to pulse. For example, a spread-spectrum system employing long codes to represent each bit of information could use different codes for each receiver, and these codes could be changed at will. By inserting a reference code, we could feed the tap weights serially into the system instead of adjusting them one by one as we do with an adjustable transversal filter.

One way to convolve two signals, already discussed in Sec. 4.5, is to take the inverse Fourier transform of the product of the Fourier transforms of two signals; this is mathematically their convolution. Similarly, correlation is obtained by taking the inverse transform of the product of the Fourier transform of one signal and the complex conjugate of the Fourier transform of the other. Such Fourier transform techniques are frequently used in digital processing systems, because they require basically three FFTs, each of order $n \log_2 n$ operations instead of at least n^2 operations, to form a correlation directly.

As we shall see, these considerations do not apply in analog processing because the necessary processes can be carried out in the time domain rather than the frequency domain. Performing the required multiplications and integrations with an analog device is much faster, and certainly requires a less complex system, than with the digital devices needed to take three sets of Fourier transforms. However, an analog system is typically less accurate than a digital system.

In this section we discuss SAW devices that can carry out the direct convolution of two signals by using nonlinear interactions between them; we also describe other kinds of convolvers and correlators based on the use of internal or external nonlinear semiconductor elements. In Sec. 4.8 we discuss closely related nonlinear processing devices in which two tapped CCDs are employed with external multipliers. In Sec. 4.9 we describe acousto-optic devices, which are also used for such processing operations.

4.7.2 Acoustic Convolver

Consider the SAW device illustrated in Fig. 4.7.1. Suppose that complex input signals $f(t) \exp(j\omega_1 t)$ and $g(t) \exp(j\omega_2 t)$ are supplied to the left- and right-hand sides of the device, respectively. At a position z along a device of length L, the individual signals will be of the forms

$$\hat{f}(t, z) = f\left(t - \frac{z}{V}\right) e^{j\omega_1(t - z/V)} \tag{4.7.1}$$

and

$$\hat{g}(t, z) = g\left(t - \frac{L - z}{V}\right) e^{j\omega_2[t - (L - z)/V]} \tag{4.7.2}$$

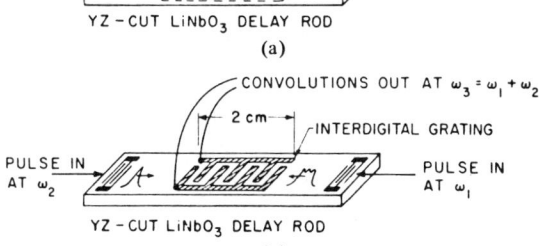

Figure 4.7.1 Transducer configurations used for harmonic generation, frequency mixing, and obtaining convolution between two signals: (a) pair of metal electrodes used as the output transducer for the second harmonic; (b) relatively coarse interdigital transducer used for the sum frequency output at $\omega = \omega_3$. (After Kino [64].)

respectively. We use the notation to indicate a modulated carrier and omit it when describing only the modulation. We assume the medium on which these two waves propagate to be slightly nonlinear. An acoustic wave of amplitude $a(t, z)$ is associated with an electric polarization field in a piezoelectric medium. In turn, a displacement field $\hat{D}(t, z)$ and a displacement current density $\hat{\imath}(t, z) = \partial \hat{D}(t, z)/\partial t$ will be generated by the applied signal, where $\hat{D}(t, z)$ is of the form

$$\hat{D}(t, z) = K_1 \hat{a}(t, z) + K_2 \hat{a}^2(t, z) + \cdots \qquad (4.7.3)$$

with K_1 and K_2 constants, and the higher-order terms in $\hat{a}(t, z)$, due to the nonlinearity. Because we are concerned with nonlinear interactions, we must take care to express product terms correctly. Thus we use the real part of the waves involved. The interaction between two real waves in a nonlinear medium can be expressed in the form

$$\hat{D}(t, z) = \frac{K_1}{2}[\hat{f} + \hat{f}^* + \hat{g} + \hat{g}^*] + \frac{K_2}{4}[\hat{f} + \hat{f}^* + \hat{g} + \hat{g}^*]^2 \qquad (4.7.4)$$

where $\hat{f}^*$ and $\hat{g}^*$ are the complex conjugates of $\hat{f}$ and $\hat{g}$, respectively, and K_1 and K_2 are arbitrary constants. Note that there are product terms due to the second-order nonlinear interaction at frequencies $2\omega_1$, $2\omega_2$, $\omega_1 + \omega_2$, and $\omega_1 - \omega_2$, and at the corresponding negative frequencies.

Suppose that we concentrate on the term of frequency $\omega_3 = \omega_1 + \omega_2$. We can write the displacement field at this frequency in the form

$$\hat{D}_3(t, z) = \alpha f\left(t - \frac{z}{V}\right) g\left(t - \frac{L-z}{V}\right) e^{j\omega_3 t} e^{j(\omega_1 - \omega_2)z/V} \qquad (4.7.5)$$

where α is a nonlinear coupling parameter that, for simplicity, includes the phase term $\exp(-j\omega_2 L/V)$.

We see that the phase variation with z of the product signal has the form $\exp[j(\omega_1 - \omega_2)z/V]$. We can therefore detect the product signal with an interdigital transducer of period l, such that

$$\frac{(\omega_1 - \omega_2)l}{V} = 2\pi \qquad (4.7.6)$$

When $|\omega_1 - \omega_2|$ is much less than ω_1 or ω_2, the period of output transducer becomes large and a relatively coarse interdigital transducer can be used to detect the output. In the degenerate case in which $\omega_1 = \omega_2 = \omega$, the output is obtained at a frequency $\omega_3 = 2\omega = \omega_1 + \omega_2$, and there is no phase variation with z. In this case, the generated signal in a piezoelectric material will consist of a uniform E field, which can be detected by two metal electrodes on the top and bottom surfaces of the SAW device, as illustrated in Fig. 4.7.1(a). Otherwise, the more general form of the device shown in Fig. 4.7.1(b) must be employed.

We can detect other frequency components by using the correct electrodes. For example, there is a dc term due to the input $\hat{f}$ proportional to $\hat{f}\hat{f}^*$; this has no phase variation with z and can therefore be detected on plate electrodes (see, e.g., Prob. 3). There is also a term of the form $f^2 \exp[2j(\omega t - kz)]$ that can be detected by an interdigital transducer with half the period of the input transducer. This f^2 term excites an acoustic wave at the second harmonic frequency whose amplitude grows linearly with distance; this second harmonic term can also give rise to saturation effects as it removes power from the fundamental (see Prob. 1).

We shall treat only the parallel-plate output electrode system here. In this case, the output obtained must be proportional to the total charge induced on the plates. If the charge is, in turn, proportional to the displacement field because of the nonlinear interaction, the total output will be of the form

$$y_3(t) = Ke^{2j\omega t} \int f\left(t - \frac{z}{V}\right) g\left(t - \frac{L}{V} + \frac{z}{V}\right) dz \qquad (4.7.7)$$

where K is a coupling parameter. By writing $z = V(t - \tau)$, Eq. (4.7.7) becomes

$$y_3(t) = -KVe^{2j\omega t} \int f(\tau) g(2t - \tau - T) \, d\tau \qquad (4.7.8)$$

where $T = L/V$ is the time delay between the transducers. Note that we can take the limits of the integral in Eq. (4.7.8) and $+\infty$ and $-\infty$ when each signal consists of a pulse whose length is less than the transit time under the detecting transducer, and when the signals overlap each other while passing under the transducer. The output obtained at a frequency 2ω is the convolution of the two input signals, but it is compressed in time by a factor of 2 because of the $2t$ term in the argument of G. The two signals are passing each other at a velocity $2V$, instead of one signal moving at a velocity V through a stationary filter; the output is therefore a scaled version of the convolution of the two signals. For the same reason, the output is at a carrier frequency of 2ω or, more generally, $\omega_1 + \omega_2$.

We can put this result on a more physical basis by considering what occurs when two square pulses of the type illustrated in Fig. 4.7.2 are injected into the device. As the two pulses pass by each other, we obtain an output that is the integral of the product of the two pulses over their overlap length. This integral increases linearly with time until the two pulses are coincident with each other. After they pass each other, the area of overlap begins to drop, and the amplitude decreases linearly with time until it becomes zero; thus we obtain the convolution of the two signals. If each pulse is of length T, the overlap between the zero output points will also be T, as we would expect from Eq. (4.7.8). Simple results obtained with a device of this type are shown in Fig. 4.7.3. In each case, a symmetric waveform has been convolved with itself.

We often require the correlation between two signals instead of their convolution; for this we need the function

$$y(t) = \int f(\tau) g(t + \tau) \, d\tau \qquad (4.7.9)$$

If we compare Eq. (4.7.9) with Eq. (4.7.8), we see that a convolver can be used to obtain correlation only if one of the signals, $f(\tau)$ or $g(\tau)$, is inverted in time. This can usually be done if one of the signals is used as a reference, but generally the device yields only the correlation of two symmetrical signals and the convolution of two arbitrary signals.

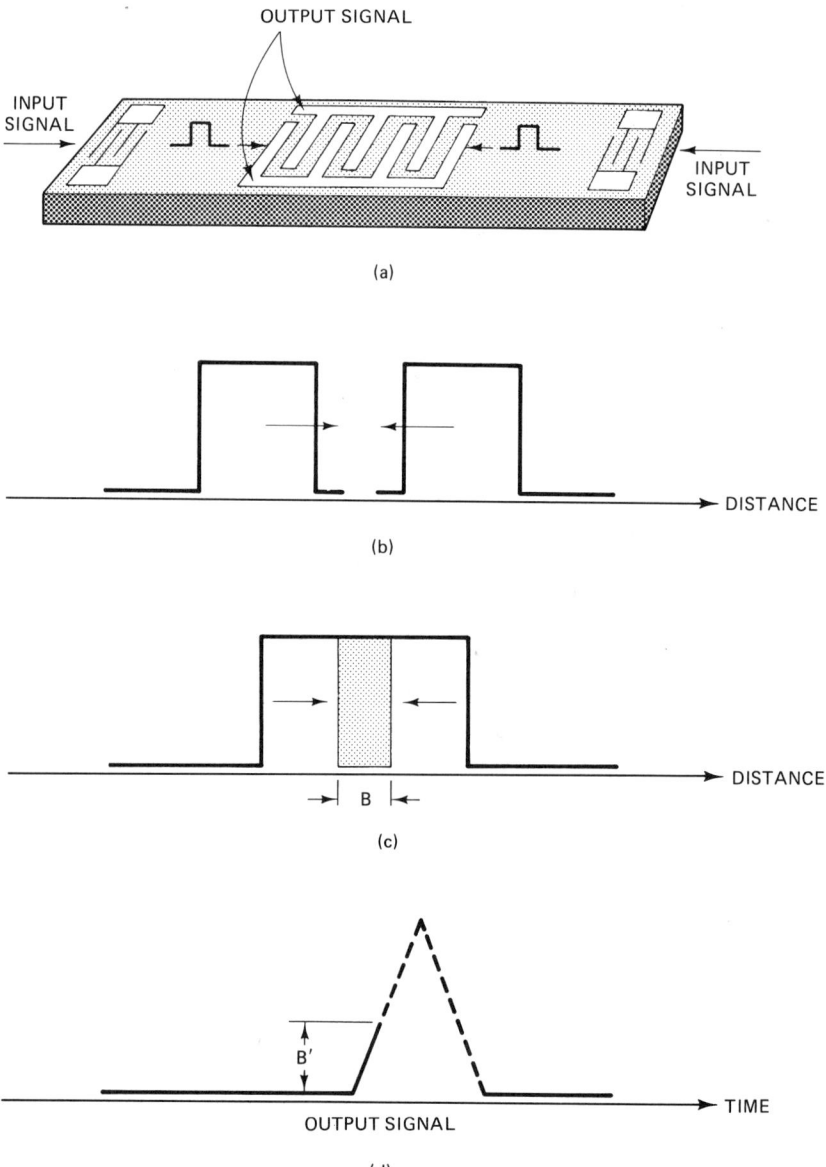

Figure 4.7.2 Nonlinear signal processing is accomplished by introducing two signals at opposite ends of an acoustic wave device (a), and extracting the mathematical result from transducers in the middle. The acoustic wave representations of two rectangular pulses approach each other (b) and begin to overlap (c). When the acoustic signals first touch, the output signal begins to rise (d). When the overlapping signals have proceeded to B, the output signal has risen to B'. The output signal reaches a peak when the two acoustic signals are exactly superposed, and then begins to fall again. Thus the output signal is proportional to the shaded area in part (c). The output is termed the convolution of the two input signals. (After Kino and Shaw [13].)

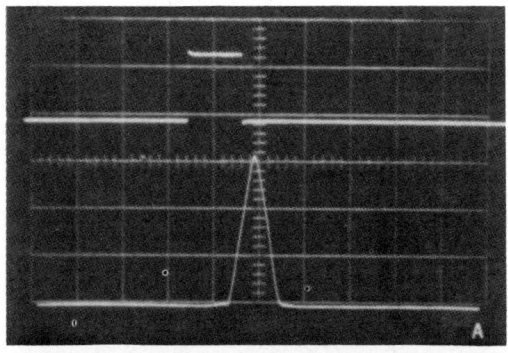

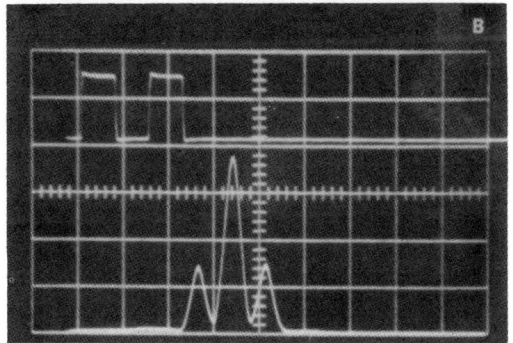

Figure 4.7.3 Autoconvolution of (a) a rectangular pulse observed using a surface wave convolver with an interdigital output transducer; and (b) a double pulse, with the same configuration and frequencies. (After Kino and Matthews [63].)

Devices of this kind can be used to reverse a signal in time to obtain the necessary reference for correlation by using the configuration shown in Fig. 4.7.4. A signal of frequency ω_1 is inserted into the left-hand transducer while a short-pulsed signal of frequency ω_3 is inserted into the center transducer. It can be shown by the process already illustrated that we will obtain an output signal of frequency $\omega_2 = \omega_3 - \omega_1$ traveling in the left-hand direction. When an asymmetrical pulse arrives under the center transducer, a signal will be generated that is the product of the two signals. However, since the regeneration occurs directly under the transducer, the tail end of the original pulse will arrive back at the

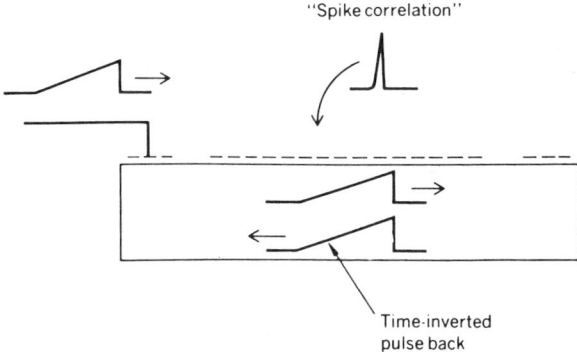

Figure 4.7.4 Configuration used to obtain time reversal of an input RF pulse.

original input transducer first and the front part of the pulse will arrive last; thus the output will be a time-inverted version of the input pulse.

Figure 4.7.5 shows an illustration of results obtained by this process. A short pulse or delta function is applied to the center transducer and a signal is introduced in the left transducer. The time-inverted signal then appears at the original input transducer. This process has been used with two convolvers to generate the correlation between two signals. However, this two-step process is not a practical way to obtain correlation, for it results in a poor signal-to-noise ratio.

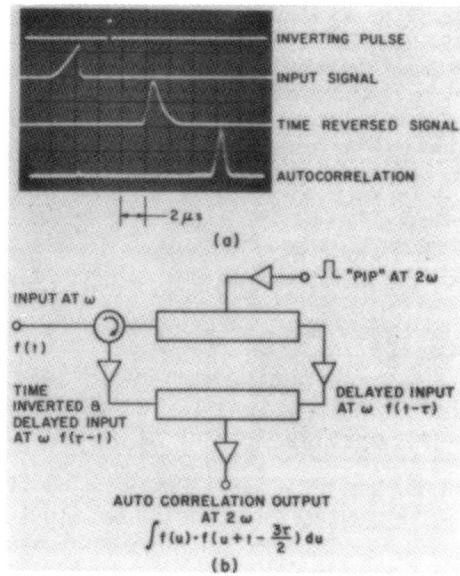

Figure 4.7.5 Inversion of electronic signal can be accomplished by nonlinear signal processing using, for example, the acoustic wave device illustrated in Fig. 4.7.1(a). (a) A signal (top oscilloscope trace) is introduced at the transducer at the left-hand side of the device. A short pip (middle trace) is inserted by way of the transducer in the center of the device. A time-inverted signal (bottom trace) then appears at the original input transducer. The autocorrelation of an asymmetrical triangle with two semiconductor SAW convolvers, is also shown. (b) A schematic of the circuit used. (After Kino [64].)

Nonetheless, the principle of time reversal or phase conjugation is still important (see Probs. 4.7.2 and 4.8.1). When a signal $\exp[j(\omega t - \phi)]$ is injected into a device excited by a signal $\exp(2j\omega t)$ on its plate, the signal generated by the nonlinearity is of the form $\exp[j(\omega t + \phi)]$; thus the phase is reversed. If the phase ϕ contains a term due to distortion in a transmission medium, the convolver will reverse the sign of the phase distortion; this process is called *phase conjugation*. If the phase-conjugated signal is retransmitted through the medium, the frequency component $\exp(j\omega t)$ will arrive without distortion. Thus a two-dimensional optical beam passing through a distorting medium, such as a turbulent atmosphere, and reflected from a target can in principle be regenerated in a crystal excited by a uniform field at a frequency 2ω. If the phase-conjugated beam is reemitted from the crystal and sent through the medium, it will produce an undistorted reflected image (see Prob. 4.8.1) [66].

4.7.3 Spread-Spectrum Communications

The convolver has a sophisticated application to *spread-spectrum communications*. If biphase (0 and π phase) or quadriphase (0, $\pi/2$, π, $3\pi/2$)-coded waveforms are used to represent one bit of a digital signal, then one bit of a digital signal can consist of several hundred "chips" of a spread-spectrum code. If the bit is 0, the code will not be transmitted; if the bit is 1, it will be. A coded waveform may consist of many "chips" per bit and typically employs a far larger bandwidth than the original digital signal. When the biphase-coded waveform is correlated with itself, it will produce a correlation peak corresponding to the bit of interest, but when the biphase-coded waveform is absent, there will be no correlation peak. A spread-spectrum system of this kind, because it employs a much larger bandwidth than the actual signal itself, is better able to reject unwanted signals and will improve signal-to-noise ratio considerably.

An important application of this system uses different codes for different receivers, so that several codes can use the same wideband communication channel, with the signals addressed to different receivers. In a conventional communication system, the receiver is tuned to the frequency of the signal; in a spread-spectrum system, the receiver code is changed to select the signal required. This gives us two advantages: Because relatively long codes and correlation techniques are employed, the signal-to-noise ratio of the received signal can be very small; and since the types of codes employed are pseudorandom codes, it is very difficult to intercept spread-spectrum signals without prior knowledge of the code.

A spread-spectrum system needs a device that can correlate with the coded waveform over relatively long integration times, of perhaps 1 ms or more. A *PN-coded waveform* (pseudonoise or pseudorandom code) is often used for this purpose. In one example, due to Morgan et al. [67, 68], an SAW convolver is employed to correlate segments of the waveform approximately 30 μs long. This produces correlation peaks, which are subsequently summed coherently in a recirculating loop of the kind illustrated in Fig. 4.7.6. The receiver signal input to the convolver is a biphase PN waveform normally buried in noise. The reference input is a biphase waveform, coded so that segments of length T_d match the time-reversed version of corresponding segments of the signal code. The convolver output is therefore a series of correlation peaks with a spacing T_d. To add these peaks coherently, the recirculation loop has a loop delay equal to T_d. The loop is cleared by opening a gate after a number of circulations designated N_c. The total signal-to-noise ratio ideally is increased by the same ratio as for a matched filter, correlating a length $N_c T_0$ of a waveform, where T_0 is the time delay in the convolver parametric region.

Perhaps the most critical part of this system is the recirculation loop, which consists of a wideband delay line. The gain in this loop must be equal to unity; if it is too large, the signals corresponding to early correlation peaks will continue to increase in amplitude, as will the noise associated with them; if it is too small, the signals from the earlier circulations will become negligible in amplitude. The loop is adding the $(N + 1)^{\text{th}}$ correlation peak to the sum of the previous N peaks. If the bandwidth of the earlier peaks decreases as they circulate, again there will be severe distortions and we will lose the peak amplitude of the correlation pulse. A simple delay line with a $(\sin f/f)^2$ response would have a response after N loops of $(\sin f/f)^{2N}$. Therefore, a delay line must be designed with a very flat response over the bandwidth of interest. In the examples illustrated in Figs. 4.7.6 and 4.7.7,

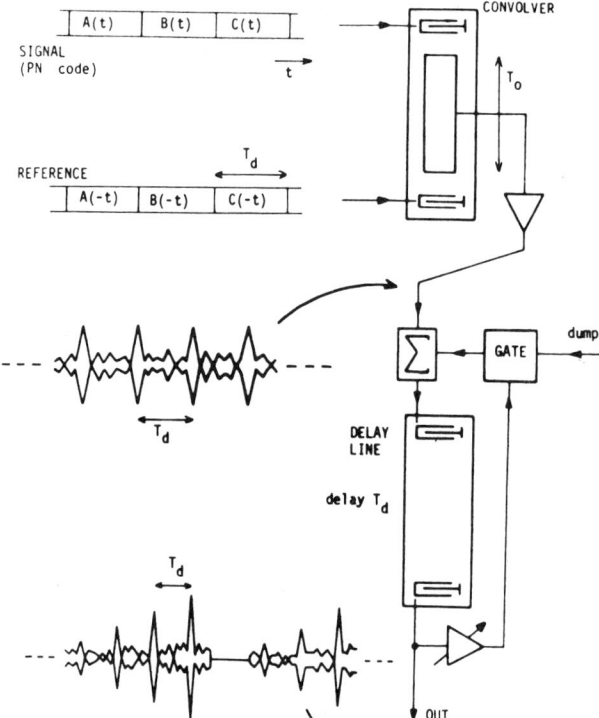

Figure 4.7.6 Principle of spread-spectrum correlator. (After Morgan and Hannah [67].)

two sets of transducers in parallel, with slightly different center frequencies, are used.

Small temperature drifts that, again, would distort the output signals must also be eliminated. This is done by using a quartz delay line with a crystal cut, which has a very small change in transit time with temperature.

Figure 4.7.7 illustrates the buildup of the output in the recirculation unit after 30 circulations with a 9.2-Mbit/s code applied to the input convolver. When the gain is adjusted correctly, the signal grows linearly with time. An automatic-gain-control (AGC) system, with a test pulse passed through the loop, will keep the gain exactly constant, but this precaution is not usually necessary.

Figure 4.7.8 shows results obtained by testing the system with an applied biphase PN signal waveform with a rate of 10 Mbits/s. The output for the seventieth circulation, when noise was added to the signal at the convolver input to give an input signal-to-noise ratio of -41 dB, is shown. The thermally generated noise was filtered to a bandwidth of 25 MHz, and the time delay T_D was 30 μs. The top trace is a signal without any input noise; the bottom trace is a signal with noise. The measured output signal-to-noise ratio was 4 dB, with an input signal-to-noise ratio of -41 dB. In theory there should have been an increase of $2BN_cT_D = 2 \times 25 \times 70 \times 30 \approx 100{,}000$ or 50 dB (in a convolver, the output bandwidth is

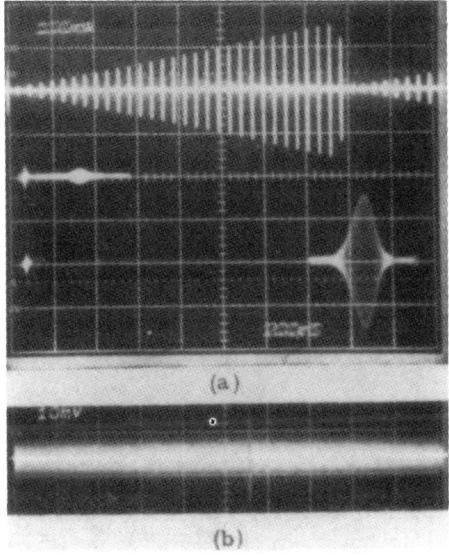

Figure 4.7.7 Loop output waveforms for input waveforms: (a) with no noise added at the convolver input (details of the output pulses are shown at 200 ns/div); (b) twenty-eighth output pulse, for a signal-to-noise ratio (SNR) of −30 dB at the convolver input, at 2 μs/div. (After Morgan et al. [68].)

twice the input bandwidth), which would correspond to a signal-to-noise ratio at the output of 9 dB. But in practice, due to the limited bandwidth of the recirculation loop and its distortions, the gain was 5 dB less than this value. Nevertheless, this technique can detect signals 41 dB below the noise level.

The same system was also demonstrated with a narrower-band input signal, using a 1-Mbit PN code and a noise bandwidth of 2MHz, with the input signal 34 dB below noise. After 70 circulations, the output signal-to-noise ratio was 4.5 dB, which corresponded very closely to the theoretical value of 5.2 dB.

This system demonstrates the great advantage of using spread-spectrum waveforms where a wide bandwidth input signal is employed with a matched filter to

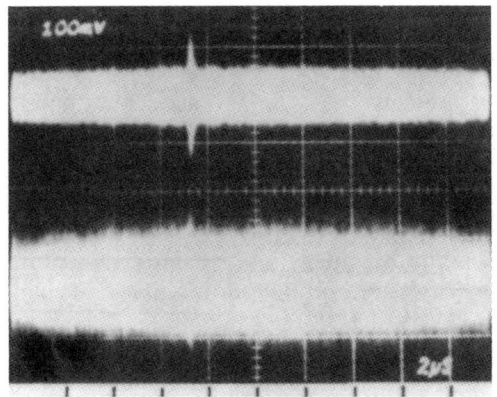

Figure 4.7.8 Output of recirculation unit for seventieth circulation, for 10-Mbit code. Lower trace, with input SNR = −41 dB; upper trace, no noise applied; horizontal, 2 μs/div. (After Morgan and Hannah [67].)

improve the signal-to-noise ratio. Analog systems of this kind are relatively simple to use for side bandwidth signals. The same sort of results have been obtained in digital systems with CCDs and very long codes, but as the need for wider bandwidths increases, SAW devices become particularly suitable.

Spread-spectrum systems for communications. One might think that a spread-spectrum system of this nature makes highly uneconomical use of the radio-frequency spectrum available. Its advantages, of course, are the great improvement it makes in signal-to-noise ratio and its ability to employ many different codes of signals within the same frequency band. This nullifies the disadvantage of using a wide bandwidth for an individual signal. The separation between the different signals is now carried out essentially in the time domain rather than the frequency domain.

The example of spread-spectrum operation given here uses a digital biphase code, but this is not the only possibility. Other techniques employ *frequency hopping modes*, where the frequency rather than the sign of the amplitude is changed from pulse to pulse. Such modes offer the advantages already discussed, and are also extremely difficult to jam or intercept unless the observer knows the exact code.

Systems of this kind require only the correlation peak, not the full correlation function. In principle, such a correlation peak can be obtained by inserting two identical signals into a mixer, taking their product, detecting the output, and integrating the resultant detector output for a time T. The output would then be proportional to $\int_0^T f^2(t)\,dt$, where $f(t)$ is the input signal. An analog system is useful for this purpose, because high-speed multiplication requires high-speed A/D converters and a large, more complex, digital system. But on the other hand, a digital system is well suited to very long narrowband spread-spectrum systems, such as those used for communicating with distant satellites in the neighborhood of the outer planets.

This still leaves us with the problem of finding where the two input codes, the reference and the received signal, are coincident in time, for without coincidence between the codes, no output is obtained. In practice, we find it by altering the time of the reference code several times until we obtain an output. Reference codes can be used at the start of a message for this purpose.

This locking-on process may be relatively slow, because the input signal-to-noise ratio (SNR) is typically -30 to -40 dB. Hence, for a bandwidth of 10 MHz, integration over 1 ms is necessary for each trial to obtain a satisfactory SNR to make a decision. The key advantage of a matched filter, such as a convolver, is that the input signal and the reference signal do not have to be coincident. Provided that the time difference between them is less than the time delay through the convolver, we will obtain an output, and the relative improvement in acquisition time will be equal to the time–bandwidth product of the filter. Thus the locking-on process is greatly simplified and sped up by a factor comparable to the time–bandwidth product of the convolver. Again, this system is far simpler and smaller than the equivalent digital system, which would have to be extremely large and complex to handle large-bandwidth short-pulse signals.

Range finding. We have described the use of spread-spectrum signals in communications. Another important application is *range finding*. Suppose that a fixed transmitter A emits a signal code with a carrier frequency f_1; the position of a receiver B is required. The received code is reemitted from B with a carrier frequency f_2 and received back at A. The signal received at A is correlated with the original transmitter code. By determining the time of the correlation peak, we can directly measure twice the delay time from A to B, and hence find the range from A to B. We can measure this time, under very noisy conditions, to an accuracy of one chip of the code. Digital versions of this system have been used to measure, with extreme accuracy, the distance of satellites approaching the planets and of course, to communicate with them as well.

This method is also used in the NAVSTAR global navigation system [69]. Each satellite in the system carries an atomic frequency standard and transmits two 1-GHz spread-spectrum signals whose codes are synchronized with the atomic clock. A user with an accurate knowledge of time (i.e., with an atomic clock in his or her receiver), can determine the arrival time of signals from as many as six satellites overhead, and can determine his or her position with an accuracy of 10 to 50 m.

4.7.4 Waveguide Convolver

The convolver we have described above has a major disadvantage: Its operation is based on the use of a weak nonlinearity in the medium. We shall give a heuristic argument to predict how the output signal depends on the input signal amplitudes and the dimensions of the device. The amplitudes of the input signal fields are proportional to the square root of the input signal power intensities of $(P_1/wd)^{1/2}$ and $(P_2/wd)^{1/2}$, respectively, where w is the acoustic beamwidth, the parameter d is proportional to the wavelength and approximately equal to the penetration depth of the surface acoustic wave, and P_1 and P_2 are the acoustic input powers. The dipole polarization field P_D generated by the two input signals can be shown to be proportional to the products of the input fields or power intensities. This polarization is developed over a region of effective thickness d, the generation layer. The voltage V_D across this dipole layer is $V_D = P_D d/E$, where E is the dielectric constant and we have taken P_D to be the average value of polarization in this region. Hence it follows that

$$V_D = \frac{\kappa}{\varepsilon}\left(\frac{P_1}{wd}\right)^{1/2}\left(\frac{P_2}{wd}\right)^{1/2} d \qquad (4.7.10)$$

where κ is a coupling constant.

Suppose that the length of the plates is L and that of the generating region is L'. The capacities of these regions are $C = \varepsilon Lw/h$ and $C' = \varepsilon L'w/h$, respectively, where h is the plate spacing. It follows that the open-circuit voltage across the plates, loaded by their self-capacity C, is

$$V_{op} = \frac{V_D C'}{C} = \frac{\kappa L'}{wL\varepsilon} (P_1 P_2)^{1/2} \qquad (4.7.11)$$

Thus the open-circuit voltage output of these devices is proportional to $1/w$. If the pulse length is less than that of the output plate, the signal will rise linearly with pulse length, as we would expect from our earlier arguments.

The output power P_3 into a resistive load is proportional to V_{op}^2. Thus P_3 will be proportional to the product of the input powers P_1 and P_2, and we can write the efficiency of the device in the form

$$F_T = \frac{P_3}{P_1 P_2} \qquad (4.7.12)$$

For a typical, practical device made of lithium niobate, working at a center frequency of 100 MHz with a beamwidth of 1.25 mm, this number is approximately -85 dBm. This implies that when $P_1 = 20$ dBm (0.1 W), the ratio P_3/P_2 equals -65 dB. The noise level of a receiver is a few decibels worse than kTB, where k is Boltzmann's constant, T is room temperature, and B is the bandwidth. A typical noise figure for a 30-MHz input bandwidth and a 60-MHz output bandwidth is approximately -95 dBm. The maximum input signal level allowed by the transducers before arcing is approximately 20 dBm, so the maximum output level is $(-65 + 20)$ dBm $= -45$ dBm. This result implies that the maximum dynamic range, from the largest input signal to the minimum detectable level, is 50 dB. This dynamic range can severely limit the use of such devices.

Two alternative strategies have been adopted to decrease the problem. One involves the use of a very narrow acoustic beam in an acoustic waveguide; the other makes use of the interaction of a surface acoustic wave with a semiconductor. The first technique uses a thin metal strip, perhaps 50 to 100 μm wide (two to three wavelengths), laid down on the piezoelectric substrate to form an acoustic waveguide. A wider interdigital transducer, perhaps 1 mm wide, is used with a strip coupler to reduce the beamwidth to 100 μm, as illustrated in Fig. 4.7.9(a). The wide input beam is needed so that the impedance of the input transducers will be low enough for matching and the voltage across the fingers for a given input power will be low enough to avoid arcing between the fingers. The metal strip is laid down on the substrate, as illustrated in Fig. 4.7.9(b). The metal short-circuits the electric fields of the wave, causing the acoustic wave propagating beneath it to have a slightly slower velocity than that of the wave in the material outside the metal strip. Consequently, by Snell's law, as illustrated in Fig. 4.7.9(b), a beam approaching the edge of the strip at an angle θ_i will be transmitted at an angle θ_T, such that

$$\frac{\sin \theta_i}{\sin \theta_T} = \frac{V_M}{V} \qquad (4.7.13)$$

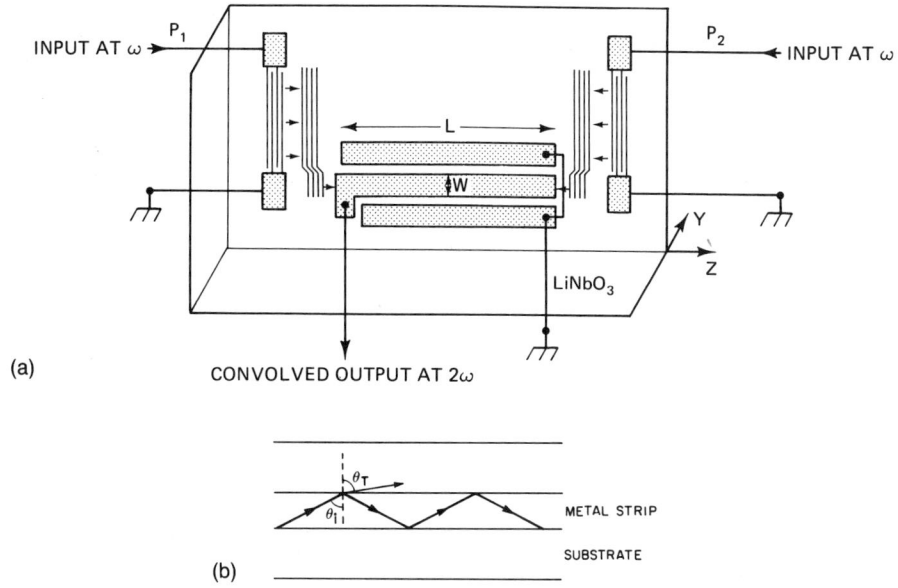

Figure 4.7.9 (a) Waveguide convolver configuration. (After Defranould and Maerfeld [70].) (b) Schematic of the interaction of the waves in the guide. Normally, θ_T is imaginary for the waveguide mode of interest. Unwanted modes may have θ_T real, causing them to radiate sideways and be rapidly attenuated.

where V_M is the wave velocity under the metal. Over a certain range of θ_i, if $V > V_M$, then θ_T will be imaginary and there will be total reflection of the waves at the two edges of the metal. If we arrange the input beam to the strip to be almost parallel to the edges of the strip, we can satisfy this requirement fairly easily, for the strip acts as an SAW waveguide to propagate the narrow beam.

Convolvers of this type have been constructed by Defranould and Maerfeld[70], with the metal strip itself used as one output electrode and two neighboring strips used as the ground electrodes, as illustrated in Fig. 4.7.9(a). By this means, they were able to obtain an improvement in efficiency of approximately 20 dB, an F_T of -65 dBm, and a dynamic range of 70 dB. Such devices are now being employed in military radar systems and developed for use in spread-spectrum communications. Other methods that have been used to compress the beam employ waveguide horns or focused transducers [71, 72].

4.7.5 Semiconductor Convolver

An alternative is to employ one of the two configurations shown in Fig. 4.7.10, in which the electric fields associated with the acoustic waves interact with a semiconductor [64, 73–75]. In the first case, a slab of silicon is separated from a piezoelectric substrate by rails, which eliminates the mechanical loading of the surface acoustic wave. The air gap required is typically a few thousand angstroms. In the second case, a layer of zinc oxide, a piezoelectric material, is laid down on the silicon; this gives piezoelectric coupling to a surface acoustic wave on the silicon substrate. The nonlinearity is now due to the interaction of the electric field with the semiconductor, with the semiconductor behaving like a type of varactor.

Suppose that the doping density of an *n*-type semiconductor is N_d per unit volume and the semiconductor is depleted at its surface, with the potential at its surface negative in respect to the bulk. If the normal component of the electric field is E_0 at the surface of the semiconductor, the surface will be depleted to a depth h. Integrating the one-dimensional form of Poisson's equation, it follows that

$$\frac{dE_0}{dy} = \frac{\rho}{\varepsilon} = \frac{qN_d}{\varepsilon} \qquad (4.7.14)$$

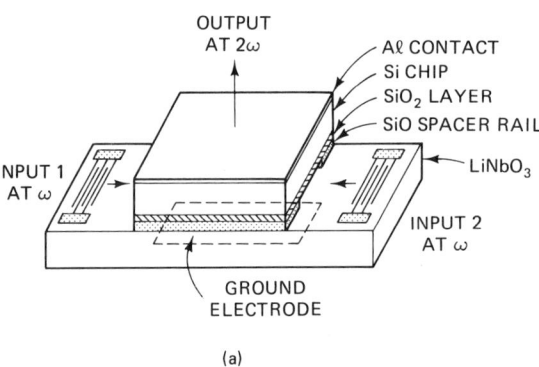

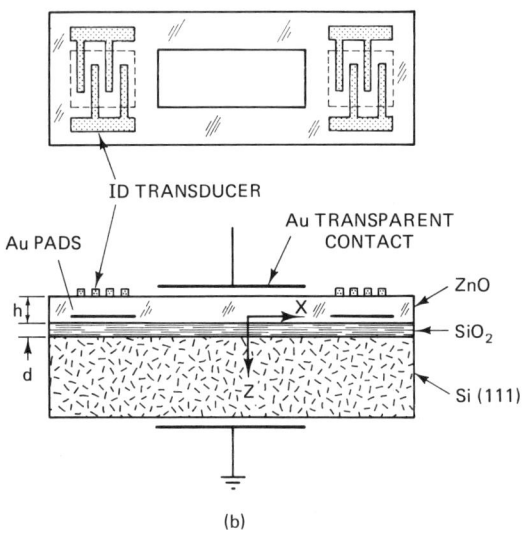

Figure 4.7.10 (a) Schematic of the air gap convolver configuration. (After Kino [64].) (b) Schematic of the monolithic zinc-oxide-on-silicon convolver. (After Khuri-Yakub and Kino [74].)

where q is the charge of a donor, y is a coordinate normal to the surface, and ρ is the charge density per unit volume. Therefore, the field at the surface is

$$E_0 = \frac{qN_d h}{\varepsilon} \tag{4.7.15}$$

where it has been assumed that $E_0 = 0$ at the edge of the neutral region. By a further integration, it follows that the potential at the surface is

$$\phi_0 = -\frac{qN_d h^2}{2\varepsilon} = -\frac{E_0^2 \varepsilon}{2qN_d} \tag{4.7.16}$$

We suppose that RF fields $E_1(z, t)$ and $E_2(z, t)$, excited by the input signals at each end, respectively, are applied at the surface in addition to the dc field already present. It is apparent that in this case a voltage proportional to $(E_0 + E_1 + E_2)^2$ is generated at the surface. This yields an output potential with a component $E_1 E_2$ that can be detected by the metal electrode illustrated in the two configurations of Fig. 4.7.10.

This system is far more sensitive, albeit more complex, than the simple lithium niobate convolver. The measured efficiencies F_T of devices of this kind are in the range -42 to -60 dBm. Although the devices saturate at lower power levels than the simple nonlinear acoustic convolvers, the dynamic range is approximately 60 to 70 dB, and devices of this kind have been used in radar systems with time–bandwidth products of up to 2000 (22 µs length, 96 MHz bandwidth) [75–77].

Most of the existing development has been made in the *air gap convolver*. The versions made at Lincoln Laboratories and Texas Instruments are extremely stable, with spacings between the semiconductor and the piezoelectric material of only 3000 Å [75–77]. However, the optical finishing required makes these devices expensive and difficult to reproduce on a large scale. The zinc oxide on silicon monolithic convolver, as developed at Stanford University and elsewhere, shows good efficiency but has a narrower bandwidth than the air gap convolver [74]. Because of its simplicity, the use of the lithium niobate waveguide convolver appears to be a better choice.

At the present time, perhaps the most important feature of the semiconductor convolver is that it has led to other types of related devices, such as the storage correlator, which will be discussed in Sec. 4.7.8. Furthermore, as surface acoustic waves can propagate directly on silicon, it follows that such devices can be incorporated with other circuit components on the same substrate. The system has even broader implications, because it follows that any type of SAW device could be made on silicon or, for that matter, gallium arsenide, as part of a total integrated circuit.

Where code processing is concerned, the results obtained with semiconductor convolvers of this type are similar to those obtained with the earlier lithium niobate convolvers. Lincoln Laboratories and Texas Instruments have demonstrated semiconductor convolvers with as much as a 240 MHz input bandwidth and a delay time of 10 µs, corresponding to a time-bandwidth product of 2400 [77]. This implies that the analog equivalent of $(2400)^2 = 5.8 \times 10^6$ bits of information can be processed in 10 μs when the convolution is taken between two signals. An example of using a convolver to process a pseudorandom code of 1024 chips at a 96-MHz rate is shown in Fig. 4.7.11.

4.7.6 Acoustic Convolver Using External Mixers

An alternative to the semiconductor convolver, pioneered by Montress and Reeder [78], conducts the mixing process in external nonlinear devices rather than in

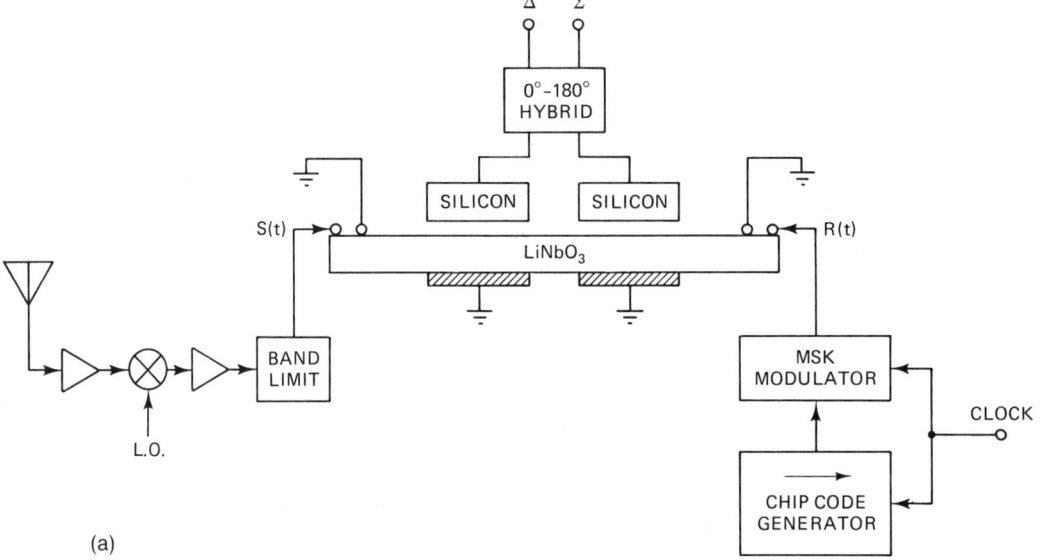

(a)

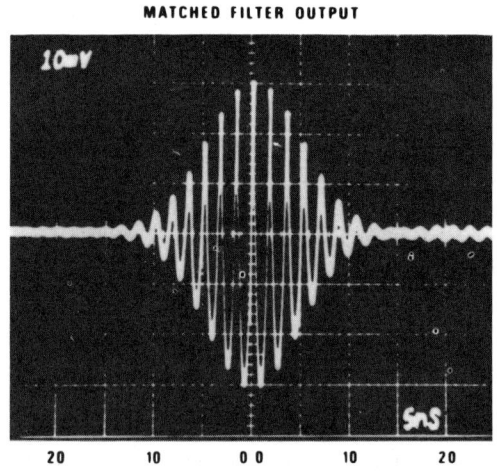

(b)

Figure 4.7.11 (a) Split-output electrode convolver used as part of a spread-spectrum receiver. Depending on the relative phases of the two parts of the reference code and signal code used, the output comes out of the Σ or Δ port. In this first case the bit code is 1; in the second case it is 0.
(b) Correlation output with a running pseudorandom code. Each segment contains 1024 chips at a 96-MHz rate. The ideal response would be a width of 22 ns at the nulls of the main response. (c) Response of the device to a bit code 1. The peak Σ output (top trace) is more than 20 dB larger than the Δ output (bottom trace). (d) The output with a bit code 0. The Δ output (lower trace) is more than 20 dB larger than the Σ output. (After Goll and Malocha [76].)

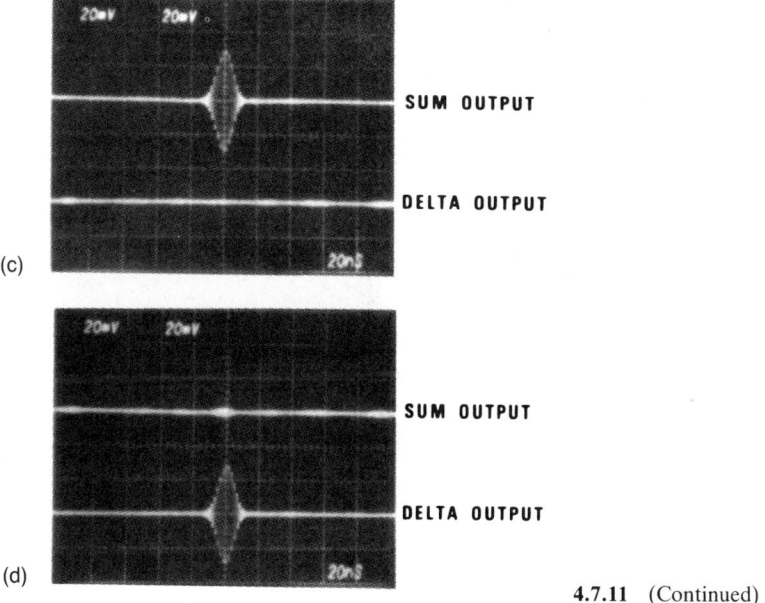

4.7.11 (Continued)

internal devices. They employ an SAW delay line with taps along it, each tap connected to an external mixer, as illustrated in Fig. 4.7.12. This device can be optimized for maximum efficiency because the nonlinear properties and impedances of the mixers can be chosen independently of the acoustic properties and piezoelectric coupling coefficient of the delay line material. The mixers themselves can use the nonlinear I–V characteristic of a diode or another semiconductor device, or they can use a nonlinear reactive effect, as does a normal convolver.

The optimum design of these arrays becomes a trade-off between choosing the diode impedance for maximum power sensitivity or for minimum reflection from the individual elements of the array of tapped transducers. The tapped transducers themselves, because there are normally a large number, should be made with split fingers of one-eighth of a wavelength to minimize reflections. The best sensitivity that Montress and Reeder obtained with diode mixers corresponds to $F_T \sim -22$ dBm, with saturation powers of the order of 0 dBm. By adjusting the current, they obtained a saturation power of 10 dBm, with $F_T \sim -33$ dBm. Obviously, a considerable improvement in convolver efficiency can be obtained by this technique, although as it reaches optimum efficiency, the saturation power is correspondingly decreased. External mixers do not necessarily increase the dynamic range of the system, but they can considerably improve the sensitivity of the device.

A disadvantage of tapped devices is that they use only a finite number of sampling points along the delay line, instead of interacting continuously, as does

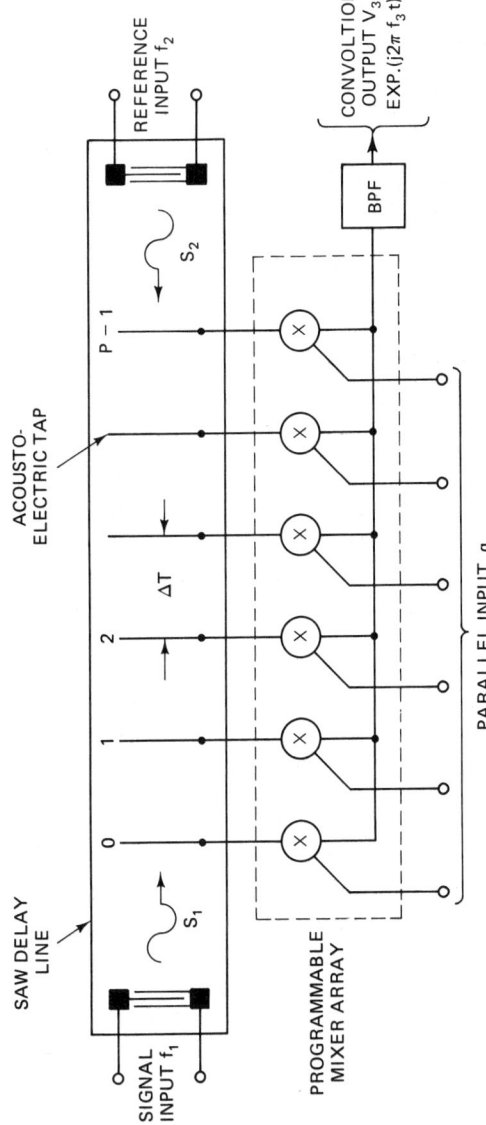

Figure 4.7.12 Fourier transform convolver. (After Montress and Reeder [78].)

a normal convolver. Thus the device can operate as a convolver with a number of possible input frequencies differing by integral multiples of a frequency that depends on the tap spacing L_T. Consider the situation where the two input frequencies are ω_1 and ω_2, respectively, so that the outputs obtained at the nth tap from the center of the device have the forms $\exp[j\omega_1(t - nL_T/V)]$ and $\exp[j\omega_2(t + nL_T/V)]$, respectively. After mixing, the output obtained from the system at the sum frequency will be of the form

$$y(t) = Ae^{j(\omega_1+\omega_2)t} \sum_n \exp\left(\left|\frac{-jnL_T(\omega_1 - \omega_2)}{V}\right|\right) \qquad (4.7.17)$$

where A is a constant. It is apparent that the output at the sum frequency, $\omega_3 = \omega_1 + \omega_2$, is strong when the signals from all taps are in phase. Therefore, the mixed output signals from the taps can be summed in phase if

$$\frac{(\omega_1 - \omega_2)L_T}{V} = 2M\pi \qquad (4.7.18)$$

where M is an integer. Because the taps are periodically spaced, the convolver can operate at several different input frequencies spaced by V/L_T. This result implies that the maximum bandwidth is limited by this frequency excursion; with larger bandwidths we would find aliasing of the input signals. Such a result is also implied by the sampling theorem, which indicates that at least two samples per RF modulation cycle are required. One advantage, however, is that if the bandwidth required is much less than either of the carrier frequencies, we need only a relatively coarse tap spacing.

The general form of this device is shown in Fig. 4.7.12. Using VHF bipolar transistors as the mixers, the taps themselves can be weighted over a wide dynamic range. We can obtain a significant reduction in spurious output signal levels because the high reverse isolation of the transistors, as compared to diodes, provides greater intertap isolation.

A device of this kind has been operated in a nondegenerate mode. As an example, with a 32-tap system, the two center input frequencies were taken as -83 MHz and 70 MHz. The results obtained, using Hamming weighting for a Fourier transform, are shown in Fig. 4.7.13.

4.7.7 CCD Convolver or Correlator

If the same principles are applied to CCDs, the outputs from two tapped CCDs can be mixed to provide convolution or correlation. A block diagram of a CCD convolver is given in Fig. 4.7.14. It operates by first loading CCD 1 with a reference signal and then clocking a second signal through CCD 2. Since the charge is stored dynamically in CCD 1, it decays because of thermal leakage and must be refreshed after 10 to 100 ms. Each stage of the CCD is tapped. The outputs of the corresponding taps on the two CCDs are taken to corresponding multipliers, using

4 MHz OUTPUT WORD RATE

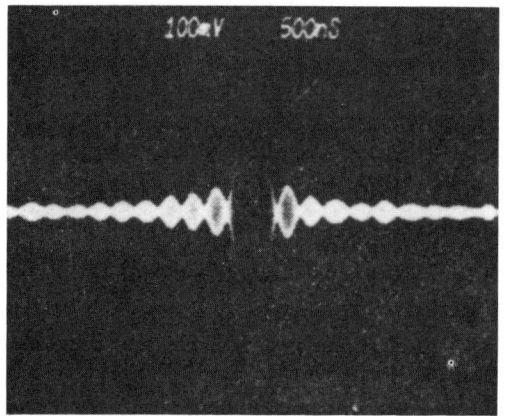

4MHz OUTPUT WORD RATE

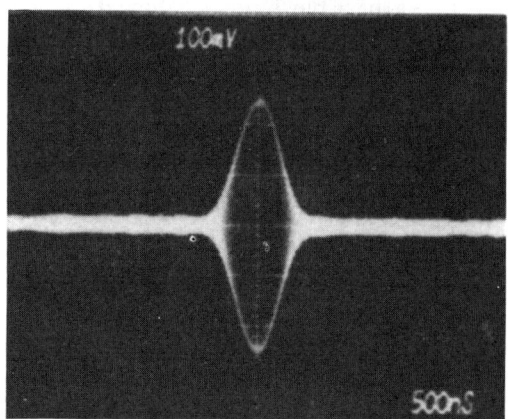

10 MHz OUTPUT WORD RATE

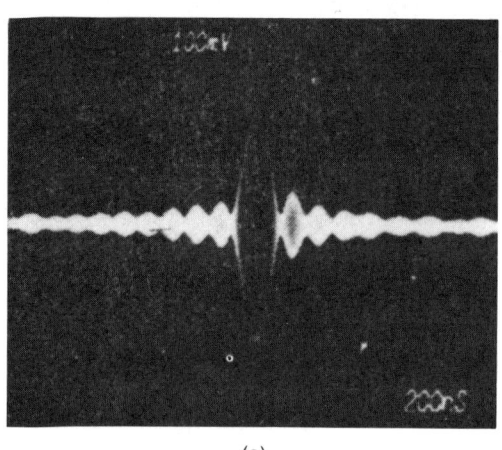

10 MHz OUTPUT WORD RATE

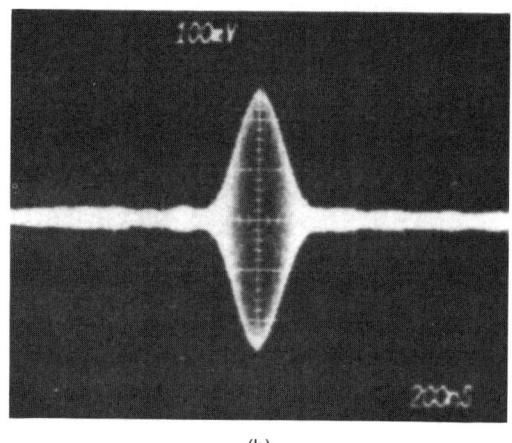

(a)   (b)

Figure 4.7.13 Fourier transform output: (a) uniform tap weighting; (b) Hamming tap weighting. (After Montress and Reeder [78].)

the circuit of Fig. 4.7.15. When the signal in CCD 1 is stationary in this configuration, it is possible, in principle, to feed the signal in either direction to CCD 2 and obtain either correlation or convolution. In the circuit shown, transistors Q_3 and Q_4 act as voltage-controlled resistors in a source-follower configuration. The balanced configuration overcomes the inherent nonlinearity of MOS transistors used as resistors.

A 100-stage correlator has been designed by Buss et al. [4] that uses floating-gate output taps for high charge-transfer efficiency. The device has been demonstrated to perform 100-point convolutions at clock rates up to 5 MHz. Because floating-gate outputs are used in the device, neither CCD can be stationary. Therefore, a device operated this way will be suitable only for convolution.

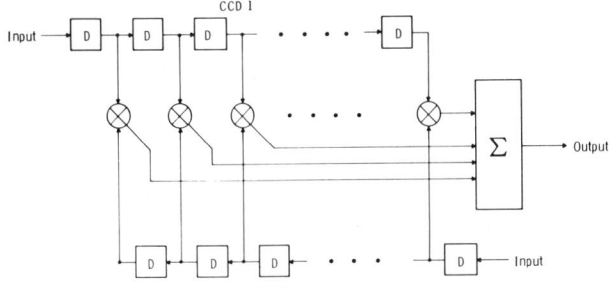

Figure 4.7.14 CCD correlator. (After Buss et al. [4].)

4.7.8 SAW Storage Correlator

The acoustic convolver has three major disadvantages that have limited its use to certain applications of the type already described:

1. The reference signal must be available at the time when the signal being interrogated is entering the device. This effect may be partially eliminated by using a periodically repeated reference and two convolvers, with one convolver in operation at all times. This does not necessarily cause problems in spread-spectrum operation.
2. The processing time in an acoustic convolver is at best only a few tens of a microsecond. In some cases, we may need much larger processing times of several milliseconds. As we have seen in the example of spread-spectrum communication, we can obtain this result to some extent by using an external integrator.
3. Because the device takes only the convolution between two signals, to obtain correlation we must time-reverse one of the signals. If the time-reversed version of one of the signals is not available in the form of a reference, a second convolver must be employed for the time reversal. Such a system has very limited dynamic range.

In this section, after describing a historically important concept, we shall describe a single device that can carry out both the required operations of long-term integration and internal storage of a reference signal. In this way we can

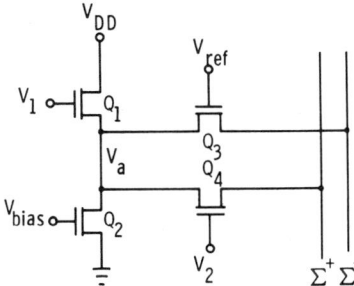

Figure 4.7.15 Circuit for four-quadrant multipliers. (After Buss et al. [4].)

Sec. 4.7 Convolvers

obtain correlation as easily as we can convolution, and we eliminate the need for both signals to be present in the device at the same time. Furthermore, by operating in a different mode, we can correlate two signals, each several milliseconds long, in a device that normally has an acoustic signal delay time of several microseconds.

Tapped-delay-line correlator. One way to eliminate some of the difficulties of the convolver is to adopt the configuration shown in Fig. 4.7.16. A tapped delay line is employed with external mixers using integrators on the output of each mixer. One of the input signals $f(t)$ is fed directly to the mixers, while the other input signal $g(t)$ is fed through the SAW delay line via the taps to the mixers. The output from the nth tap is therefore

$$y(\tau_n) = \int f(t)g(t - \tau_n)\, dt \qquad (4.7.19)$$

where $\tau_n = z_n/V$ is the time delay from the input to the nth tap at position z_n. The output is the correlation function at time τ_n. Thus by switching from one output tap to another, we can sample the correlation function over a limited range of time corresponding to the delay time of the delay line.

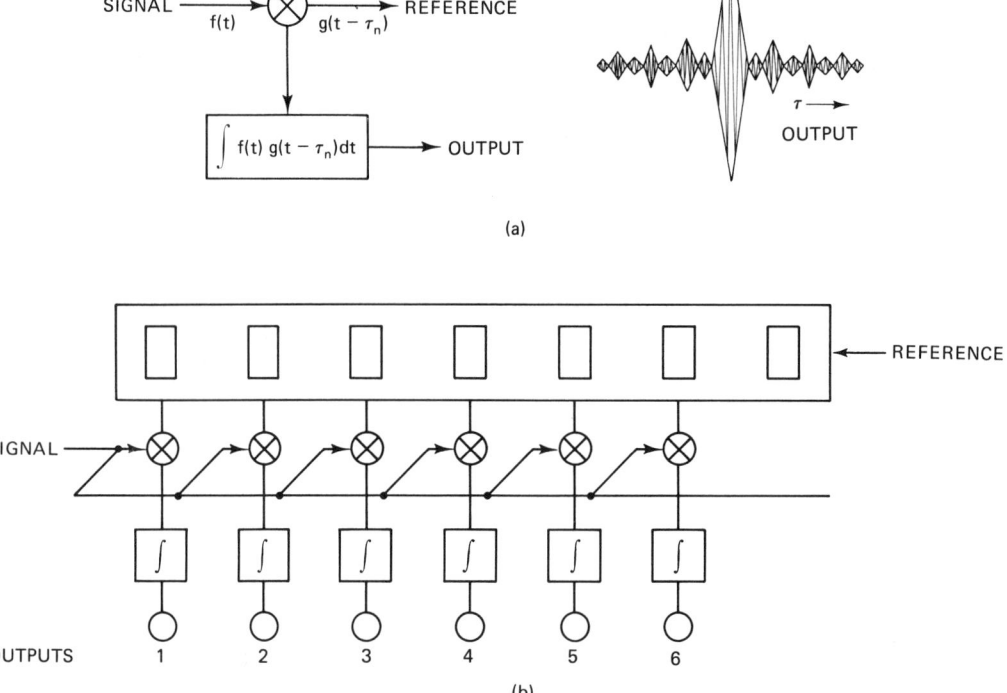

Figure 4.7.16 (a) Common form of active correlator; (b) tapped-delay-line correlator that is less sensitive to synchronization of signal and reference. (From Maines and Paige [44].)

We might imagine other versions of this device with internal mixers (i.e., strips of silicon placed near the substrate). But a technique is needed that will put all these components in one device, including the final multiplexing (i.e., switching to the output taps). The *storage correlator* satisfies these criteria.

Monolithic and air gap storage correlators. Consider the monolithic SAW storage correlator employing zinc oxide on silicon, illustrated in Fig. 4.7.17. In this device, a row of diodes with a spacing of the order of one-fourth of a wavelength of the acoustic wave is laid down in the silicon. SAW transducers are employed at each end of the device, and a top plate electrode is used as an input or output port in much the same manner as with the convolver.

Consider now what occurs when a signal is read into the device. Suppose that a short pulse less than half an RF cycle long, with voltage V_p, is applied to the top plate. This will turn on the individual diodes, as illustrated in Fig. 4.7.18. The capacity between an individual diode and a top plate is therefore charged up to a voltage approximately equal to V_p, with a corresponding charge in the capacitor of $Q_p = C_p V_p$, where C_p is the capacitance between the top plate and an individual diode. The diode itself is now reverse-biased, so that the only current flowing in the circuit comes from the leakage current I_s of the diode, which normally has a very low value. Thus the capacitance remains charged for a time $T_s \approx Q_p/I_s$, typically in the range 10 ms to 5 s, depending on the design of the diode and its temperature.

Now suppose that at the same time the pulse is applied to the top plate, an input signal $f(t) \cos \omega t$ is also applied to the device, exciting a surface acoustic wave so that the signal under a particular diode at position z becomes $f(t - z/V) \cos \omega(t - z/V)$. If the short pulse is applied at a time $t = 0$, we observe that the total charge stored in the diode is

$$Q(z) = C_p V_p + \alpha f\left(\frac{-z}{V}\right) \cos\left(\frac{\omega z}{V}\right) \qquad (4.7.20)$$

A spatial pattern of charge, representing the input signal, has been stored in the capacitors, and the device is thereby capable of storing a reference signal.

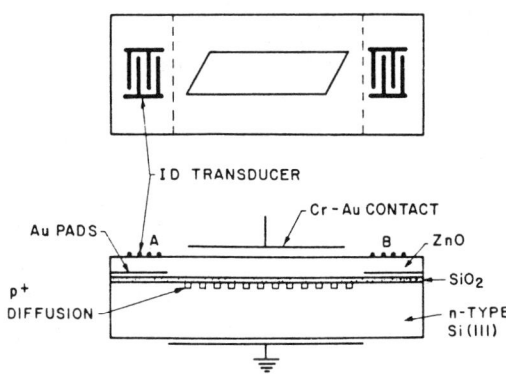

Figure 4.7.17 Surface acoustic wave storage correlator configuration. (After Tuan et al. [79].)

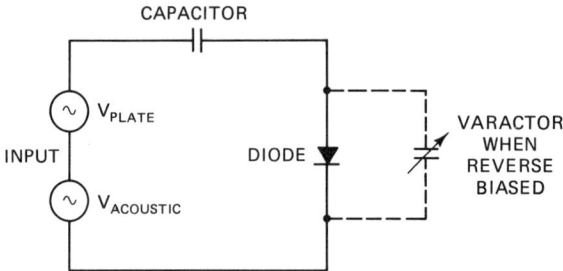

Figure 4.7.18 Equivalent circuit of correlator on read-in and read-out.

The reference charge is actually stored in the top plate, in a combination of the top plate capacitor and the capacitance of the diode itself. The total charge stored in the depletion layer of the diode with a depletion layer length h is

$$Q = \frac{qN_d hA}{\varepsilon} \tag{4.7.21}$$

where A is the area of the diode. We assume that the depleted region is in an n-type semiconductor of doping density N_d and dielectric constant ε. The voltage across the diode will therefore be

$$\phi = -\frac{qN_d h^2 A}{2\varepsilon} = -\frac{Q^2 \varepsilon}{2AN_d q} \tag{4.7.22}$$

and we can write that its effective capacitance is

$$C = \frac{\partial Q}{\partial \phi} = \frac{\varepsilon A}{h} = \frac{qN_d A}{Q\varepsilon} \tag{4.7.23}$$

The capacity of the diodes depends on the stored charge and hence on the initial reference signal. Suppose that the initial reference signal has a variation of the form $\cos \omega(t - z/V)$. There will then be a stored charge component $Q_1 \cos(\omega z/V)$, because of the exciting acoustic wave. The total charge Q will therefore be of the form

$$Q = Q_0 + Q_1 \cos \frac{\omega z}{V} \tag{4.7.24}$$

This implies that the capacity C has a spatially varying component $C_1(z)$ that is proportional in amplitude to $Q_1 \cos(\omega z/V)$, where it is assumed that $Q_1 \ll Q_0$.

Suppose that a "readout" signal $g(t)$ is now applied to the top plate. This readout signal will develop a voltage across the diode capacitance of the form

$$V(t, z) = g(t) \frac{C_p}{C_0 + C_1 + C_p} \cos \omega t \tag{4.7.25}$$

From Eqs. (4.7.24) and (4.7.25), it follows that if $C_p \ll C_0 + C_1$, as it often is, then $V(t, z)$ will have a component proportional to $Q_1 g(t) \cos(\omega z/V) \cos \omega t$. The individual diodes behave like fingers of an interdigital transducer and can excite acoustic waves propagating in both directions along the device, as though the device had been excited by a voltage with a spatially varying pattern $\cos \omega z/V$.

The input, as we have observed, is stored in the form $f(-z/V) \cos \omega z/V$. The output from a particular point is proportional to the product of the readout signal and the stored signal, so the output obtained from an output transducer at $z = 0$ will be of the form

$$y(t, 0) = \alpha \cos \omega t \int g\left(t - \frac{z}{V}\right) f\left(\frac{-z}{V}\right) dz \qquad (4.7.26)$$

This is the correlation of the stored signal and the readout signal. Similarly, the output obtained from an output transducer at $z = L$ will be

$$y(t, L) = \alpha \cos \omega t \int g\left(t - \frac{L}{V} + \frac{z}{V}\right) f\left(\frac{-z}{V}\right) dz \qquad (4.7.27)$$

This is the convolution of the two input signals.

To put it more simply, if we were to use a short pulse $g(t) = \delta(t)$ to read out from the device, we would obtain either the time-reversed waveform $f(-t)$ at $z = 0$, or the stored signal $f(t - T)$ itself at $z = L$, where $T = L/V$.

It follows that the device can be operated as a programmable matched filter for an arbitrary input waveform if a reference signal is read into it at an earlier time. An example of results taken by Ingebrigtsen, who constructed the first Schottky diode correlator in an air gap configuration, is shown in Fig. 4.7.19, with a schematic of his device in Fig. 4.7.20. In this device, a large number of Schottky diodes in the form of small square diodes were constructed in the silicon substrate, and the silicon was separated from a lithium niobate substrate by means of spacer rails. The results shown in Fig. 4.7.19 correspond to a 1.5-μs linear FM chirp, stored for 1 ms and correlated with a similar chirp in the readout. This example

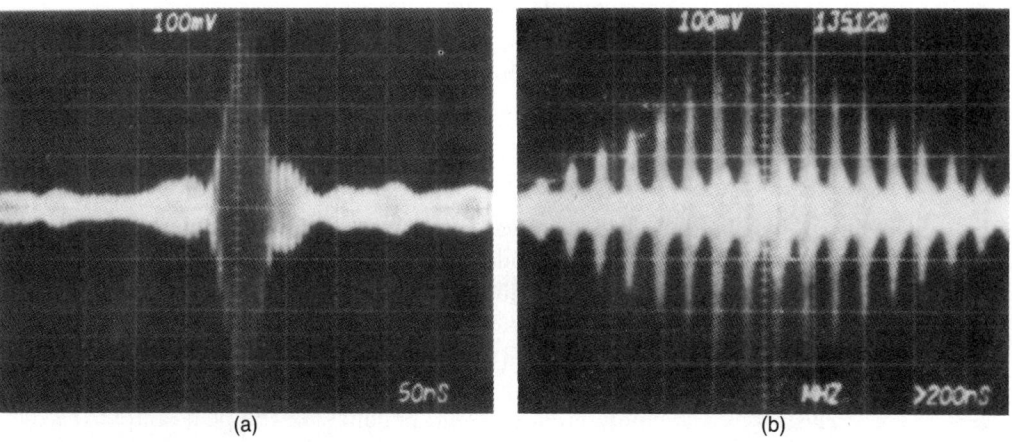

Figure 4.7.19 Correlation of 30-MHz-bandwidth, 1.5-μs linear FM chirps stored 1 ms: (a) details of single chirp correlation output; (b) nine stored chirps, each delayed by 180 ns, correlated with a similar sequence of nine chirps. Delay-line data: center frequency 145 MHz, 3 dB bandwidth, 30 MHz; Silicon data: 25 Ω-cm *n*-type platinum silicide diodes 4 μm in diameter on a 5.8-μm center distance. (After Ingebrigtsen [80].)

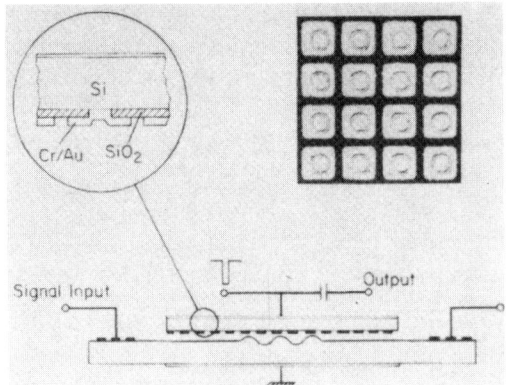

Figure 4.7.20 Diode memory correlator. (After Ingebrigtsen [80].)

corresponds, in fact, to nine stored chirps, each delayed by 180 ns, correlated with a similar sequence of nine chirps. The envelope of the function obtained, of course, is like that of two square pulses correlated with each other, and is hence triangular in form.

Input correlation mode. A second mode of operation is the *input correlation mode*. Suppose that two signals $f(t)$ and $g(t)$ are injected into a transducer and the top plate, respectively. If neither signal is sufficiently large in amplitude to switch on the diodes, no storage will take place unless the sum of the amplitudes of the two signals is high enough to do so. In this case, storage occurs only at the peak of the combined RF waveform. Furthermore, if the diodes are operated relatively slowly, as they would be with a *pn* diode, relatively little charge will be stored on each peak of the RF signal, because the diodes do not turn on completely and do not store all the available charge.

Suppose the current that flows during the peak of the RF cycle is $I(t)$; then the total charge injected into the diode and passing through the capacitor will be

$$\Delta Q = \int_0^\tau I(t)\, dt \tag{4.7.28}$$

where τ is the time for which forward current flows in the diode. If the recombination time for carriers in the diode is τ_p, we expect a proportion τ/τ_p of the total charge to recombine in the diode; therefore, the change in charge on the capacitor after one RF cycle will be approximately $\tau \Delta Q/\tau_p$. Because the effect is cumulative, after many RF cycles we would expect to see a gradual buildup of stored charge in the capacitor.

This process is highly nonlinear and depends on the peak amplitude of the two signals present; thus there will always be a stored charge component that depends on the product of the two signals. This stored charge component, in turn, depends on the integral of the product of the two signals over many RF cycles. Therefore, with two input signals $f(t)$ and $g(t)$ on the interdigital transducer and top plate, respectively, we would expect the stored charge, at any position z, to have a component

$$Q(z) = \alpha \int_0^t f\left(t - \frac{z}{V}\right) g(t) \, dt \qquad (4.7.29)$$

The spatial variation of the stored component of charge corresponds to the correlation of the two input signals.

In this system, the floating variable in the correlation function is time t; this can be relatively long compared to the delay time T of a surface acoustic wave in the device itself, and may be comparable to the storage time of the device. Thus correlation can take place over extremely long time periods, and the device can handle signals with a time–bandwidth product comparable to the product of the storage time and the bandwidth of the device. The storage times can be tens of milliseconds for any bandwidth that can be accepted by the input transducer; thus the possibility arises of correlating signals with time–bandwidth products as large as 10^6. In practice, correlation has been observed in these devices with time–bandwidth products up to 100,000 (in one case corresponding to a bandwidth of 20 MHz with an integration time of 5 ms) [80–82].

Note that the correlation peak occurs at $z = 0$; this may correspond to a spatial position outside the top plate itself and therefore may not be a readable point. To place the correlation peak at the center of the device, we must use a delay line to delay $g(t)$ by half the time delay in the device, as illustrated in Fig. 4.7.21. The correlation peak itself may be read out using a short interrogating pulse for the readout. If we use a more complicated waveform for the readout, of course, we can obtain correlation or convolution of the readout signal with the stored information. It is apparent that a wide range of signal processing techniques is possible with this configuration.

In one example, shown in Fig. 4.7.22, two chirps, each 5 ms long with a bandwidth of 6 MHz, were read into the device. One of the chirps was modulated with a signal of 186 Hz, using two sidebands on the carrier 372 Hz apart. The correlation of the function therefore consisted of two peaks. In this case, two peaks approximately 350 ns apart were observed, corresponding to a theoretical difference of 420 Hz in frequency. The agreement is reasonable, for it was not known how linear the chirps were, and there was some uncertainty as to their exact bandwidth. By varying the chirp lengths, components in different ranges could be observed, and thus the device could be used as a variable-bandwidth spectrum analyzer. The top trace shows the output signal and the short readout pulse alone as applied to the device. This represents the spurious signal levels of the device, which created considerable difficulties. These spurious signals occurred because there was some direct feedthrough of the signal injected on the top plate into the output transducers.

Spread-spectrum communications. The storage correlator is ideally suited to signal processing of spread-spectrum signals, for it carries out the required integration and correlation functions. Ideally, to correlate a PN code or another

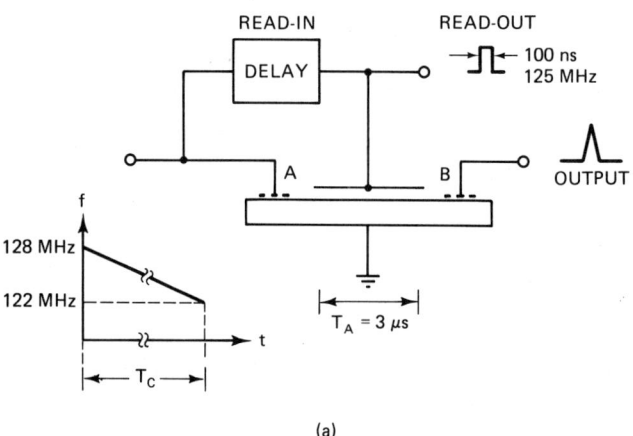

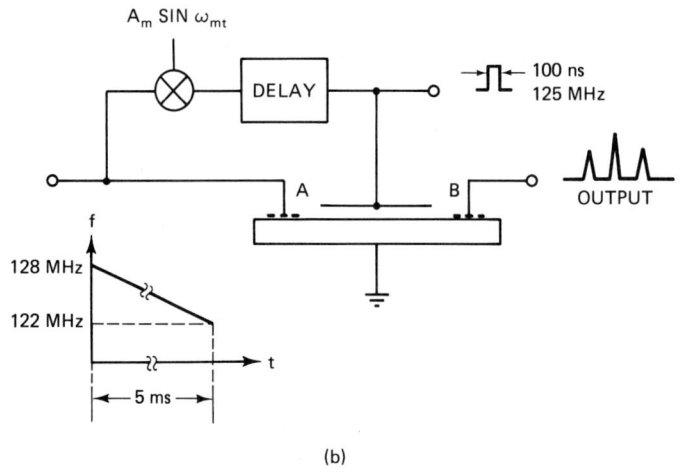

Figure 4.7.21 Modes of operation: (a) input correlation; (b) AM modulation of one of the input chirps. (After Tuan and Kino [81].)

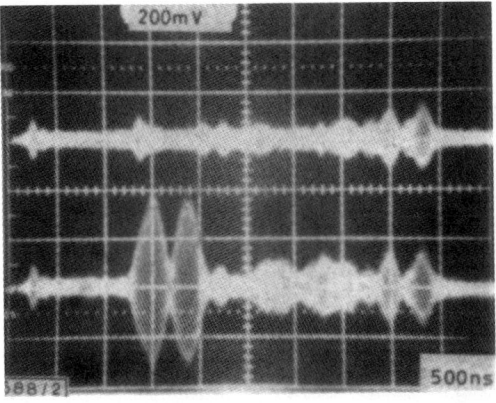

Figure 4.7.22 Correlation output of AM modulation experiment. Top trace, no signal present; bottom trace, a *modulated* chirp modulated at 186 Hz is used as the input signal. (After Tuan and Kino [81].)

type of code suitable for spread-spectrum communications, the device should build up the correlation peak continually with integration time. This experiment has been carried out with a reference code 1 ms long [83]. As shown in Fig. 4.7.23, we see that the correlation peak amplitude increases linearly with integration time, as it also does with the amplitude of the input signal. Figure 4.7.24 shows an example of the correlation peak obtained with a PN code 1 ms long and 5 MHz in bandwidth, corresponding to a time–bandwidth product of 5000. The three traces shown correspond to codes read in with slightly different delays between signal and reference ports. Similar results were obtained with codes as long as 10 ms, corresponding to a time–bandwidth product of 50,000.

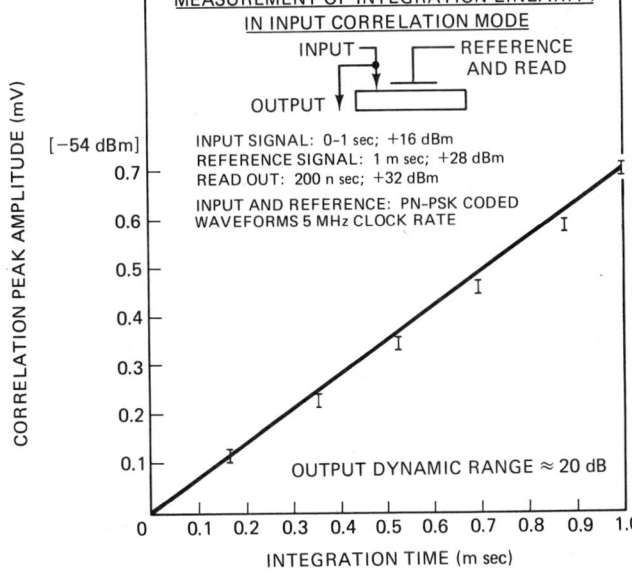

Figure 4.7.23 Measurement of device linearity with input signal duration. (After Tuan et al. [83].)

Sec. 4.7 Convolvers

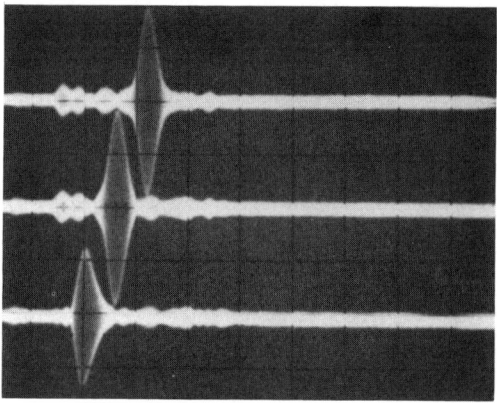

Figure 4.7.24 Output signal from large time–bandwidth product PN-PSK matched filter. (After Tuan et al. [83].)

PROBLEM SET 4.7

1. Consider nonlinear interactions in an SAW structure. Suppose that the propagation constants of the waves at the fundamental frequency are k_1, and at the second harmonic, $k_2 = 2k_1 + \Delta k$. We put

$$\hat{a}_1(z, t) = a_1(z)e^{j(\omega_1 t - k_1 z)}$$

$$\hat{a}_2(z, t) = a_2(z)e^{j(\omega_2 t - k_2 z)}$$

   (a) Assuming that the system possesses a second-order nonlinearity so that there is an excitation at the second harmonic frequency proportional to $\hat{a}_1(z, t)^2$, find a coupled-mode equation at the frequency ω_2 with an arbitrary coupling parameter to $(\hat{a}_1)^2$, which represents how $\hat{a}_2(z, t)$ changes with distance.
   (b) Now, assuming that the power in the second harmonic is small compared to that in the fundamental, we can write $a_1(z)$ = constant. In this case, find how $a_2(z)$ varies with distance from the input when $k_2 = 2k_1$ and when $k_2 \neq 2k_1$. The problem illustrates that slight dispersion will tend to make second harmonic generation very weak. Without dispersion, the second harmonic will build up indefinitely, and the fundamental amplitude will eventually decrease (i.e., there will be saturation with increase of the fundamental input signal).

2. (a) Consider mathematically what occurs when a signal $f(t)$ exp $(j\omega t)$ is injected in one port of a convolver, and a modulated RF signal $g(t)$ exp $(2j\omega t)$ is injected at the center plate such that $g(t)$ is a short pulse approaching a δ function. Show that a time-reversed signal of the form $f(T - t)$ exp $(j\omega t)$ is emitted from the input port.
   (b) What are the forms of the outputs from each end of the device when two signals $f(t)$ exp $(j\omega t)$ and $g(t)$ exp $(2j\omega t)$ are injected at one acoustic transducer and the center plate, respectively?

3. Find the dc potential generated on the surface of a semiconductor placed near a piezoelectric substrate by a surface acoustic wave RF signal field normal to the semiconductor surface of value E_1 exp $[j(\omega t - kz)]$. This phenomenon is known as the transverse acoustoelectric effect. Suggest how measurement of this potential might be used to determine the carrier density in the semiconductor.

4. A zinc-oxide-on-silicon convolver has a dispersive characteristic so that the phase velocity in the device varies with frequency. The phase velocity is defined as $V_p = \omega/k$ and the group velocity is $V_g = d\omega/dk$. The group velocity is the velocity at which the modulation travels through the system. Let us assume that the propagation constant is a parabolic function of frequency. Thus we can write

$$k \approx k_0 + (\omega - \omega_0) \left.\frac{\partial k}{\partial \omega}\right|_{\omega_0} + \tfrac{1}{2}(\omega - \omega_0)^2 \left.\frac{\partial^2 k}{\partial \omega^2}\right|_{\omega_0} \quad (1)$$

or

$$k = k_0 + \frac{\omega - \omega_0}{V_{g0}} - \frac{(\omega - \omega_0)^2}{2} \frac{\tau}{V_{g0}} \quad (2)$$

where

$$\frac{1}{V_{g0}} = \left.\frac{\partial k}{\partial \omega}\right|_{\omega_0} \quad (3)$$

and

$$\tau = \frac{1}{V_{g0}} \left.\frac{\partial V_g}{\partial \omega}\right|_{\omega_0} \quad (4)$$

(a) First, consider physically what occurs when two short pulses of frequency ω_0 are inserted into each end of a convolver of length L at a time $t = 0$. When does the output pulse occur? If the signal frequency is changed to ω_1, when does the output pulse occur? Finally, if a linear up-chirp varying in frequency from ω_0 to ω_1 is convolved with a linear down-chirp whose frequency varies from ω_1 to ω_0, and both chirps are of a time length T_1 less than the phase or group delay time through the convolver, what would be the approximate minimum length of output pulse that you would expect if: (i) there were no dispersion? (ii) there were dispersion? *Use only physical arguments and no detailed mathematical derivations* other than differentiating Eq. (2).

(b) Try the same argument on a dispersive storage correlator of length L, in which the same up-chirp is injected at an acoustic port and stored by putting a very short electrical pulse on the plate. The correlation is then read out with the same chirp as the stored chirp. Would dispersion now be expected to affect the output pulse width?

(c) From physical arguments, what effect would you expect a uniform attenuation per unit length to have on the form of the output of both devices (the convolver and correlator) when they are excited with a pulsed asymmetric signal with a carrier frequency ω_0, assuming that they are nondispersive?

(d) Now treat the dispersive convolver and correlator mathematically. Consider signals $f(t)$ and $g(t)$ inserted at each end of the convolver. Write

$$f(t) = \frac{1}{2\pi} \int_{-\infty}^{\infty} F(\omega) e^{j\omega t}\, d\omega \quad (5)$$

and

$$g(t) = \frac{1}{2\pi} \int_{-\infty}^{\infty} G(\omega) e^{j\omega t}\, d\omega \quad (6)$$

where

$$f(t) = a(t)e^{j\omega_0 t}$$
$$g(t) = b(t)e^{j\omega_0 t} \tag{7}$$

and the signals are narrowband, as defined at the end of Sec. 4.4.5.

Show that the convolver output with two input-modulated signals of carrier frequency ω_0 is like that of a theoretically perfect nondispersive convolver in which the output is passed into a chirp filter with a time-domain response

$$Ke^{j(2\omega_0 t - t^2/\tau'^2)} \tag{8}$$

where K is a constant and

$$\tau' = \sqrt{\frac{L\tau}{2v_{g0}}} \tag{9}$$

Taking $L = 2$ cm, $\tau = 5 \times 10^{-10}$ s, and $V_{g0} = 2.8 \times 10^5$ cm/s, estimate the maximum usable input bandwidth of the convolver. Suggest what filter could be used on the output terminal to eliminate the problem with dispersion.

Note: You will find it convenient to use the method of stationary phase in Appendix G to work out what happens to each Fourier component of the input signal, and to find the inverse Fourier transform of the output signal and how it is distorted.

4.8 ADAPTIVE FILTERS

4.8.1 Introduction

We have thus far described various types of fixed transversal filters, and convolvers and correlators whose parameters can be adjusted by using a reference signal as the input to match a required characteristic. In some cases, however, the required filter characteristic is unknown; we know only the output required from a test input signal. When a signal is distorted by transmission through a random medium, such as the atmosphere, or an imperfect communication system, we need a filter that can deconvolve or equalize the distorted signal to produce the original, undistorted signal. Such a filter is termed an *adaptive filter*: it must adapt itself to the distortion present in the system.

In some cases, such as propagation of waves through a turbulent atmosphere, the filter need only remove the phase distortion of a signal; in others, where we need to improve the response characteristics of a sonar transducer, for example, in order to obtain the shortest possible pulse, we may also want the filter to obtain a flatter amplitude response as a function of frequency. To remove unwanted echoes in a television ghost signal, for instance, we need a filter that can change both the amplitude and the phase characteristics of the received signal.

Another common application of an adaptive filter is to eliminate an interfering signal. Consider, for example, the difficulty of observing a fetal heartbeat in the presence of the mother's much stronger heartbeat; an adaptive filter is needed to eliminate the mother's heartbeat. Similarly, in a communication system, when the presence of an interfering signal masks the required signal, we can use an adaptive filter to eliminate the interfering signal.

We describe here a number of adaptive filter concepts that can be used for these purposes. We consider examples employing SAW filters, CCD filters, and acoustic storage correlators.

4.8.2 Removing Phase Errors with an Adaptive Matched Filter

An adaptive filter can eliminate phase errors that arise from distortion in a propagating medium or a signal processing system. Suppose that the distortion of a test impulse $x(t) = \delta(t)$, which has a uniform amplitude as a function of frequency when it passes through a transmission medium, is $H(\omega)$, where

$$H(\omega) = A(\omega)e^{-j\phi(\omega)} \tag{4.8.1}$$

and $A(\omega)$ is a real function.

If a compensating matched filter is constructed with a response $H^*(\omega)$, the resultant output, when a signal $X(\omega)$ is passed through the medium, will be

$$Y(\omega) = X(\omega)H(\omega)H^*(\omega) = A^2(\omega)X(\omega) \tag{4.8.2}$$

All phase errors due to the transmission medium are removed by this process, and the output of a test impulse, after passing through the distorting medium and the matched filter, has no phase variation with frequency. The transform of this output signal will be $y(t)$, and the combined response of the transmission system and the equalizer will be the autocorrelation function of $h(t)$ convolved with $x(t)$.

By removing the phase errors, an output at $t = 0$ that corresponds to the peak of the autocorrelation function of $h(t)$ can be obtained from an impulse input $[x(t) = \delta(t)]$. If $H(\omega)H^*(\omega)$ is constant in amplitude over a finite frequency range Ω, so that $H(\omega) = \Pi[(\omega - \omega_0)/\Omega]$, the output with an impulse input will be a sinc function of time. If there are major peaks in the frequency spectrum, the time response will exhibit severe ringing, but provided that the input spectrum is reasonably smooth (ideally, a Gaussian function), the time-domain response corresponding to $H(\omega)H^*(\omega)$ should be reasonably compact with no severe ringing.

Suppose that a reference impulse is sent through a distorting medium. If a distorted pulse $h(t)$ is stored in a storage correlator, phase distortions caused by the medium can be removed from it. Now any other signal $x(t)$, sent along the same path into the correlator where the reference pulse is stored, will be correlated with the stored reference $H^*(\omega)$ or $h(-t)$ and will have its phase distortions removed.

A practical application of this technique is to improve resolution when a transducer with a poor pulse response is used in a pulse echo NDT or sonar system [84]. To demonstrate this application, a 3.25-MHz-center-frequency, 2.5-MHz-bandwidth PZT transducer was placed 16 cm from a plastic block in a water tank, as illustrated in Fig. 4.8.1. The transducer was excited with a linear FM chirp of constant amplitude. A 6-μs-long, 2.5-MHz segment of the first reflected pulse from the plastic block was gated, mixed with 98 MHz, and stored in the correlator. The gate to the top plate was then reopened to allow a second echo pulse to correlate with the stored first reflected pulse. This second echo was the triple transit signal returning from the plastic block, reflected from the transducer face, reflected a second time from the plastic block, and received approximately 210 µs after the first one was used as the stored reference. Figure 4.8.2(a) and (b) show the correlation peak obtained and, for comparison, the impulse response of the transducer. The width of the correlation peak corresponded to a compression ratio of 9, compared with the time–bandwidth limit of 15.

The same experiment was tried a second time with a poor-quality transducer that exhibited severe ringing in its impulse response. Figure 4.8.2(c), (d), and (e) show the first reflected pulse, the correlation peak, and the transducer impulse response, respectively. The correlation peak width was essentially the same as that obtained with the original high-quality transducer. The results showed that by using a storage correlator with a poor-quality transducer, we could considerably improve the response of the pulse-echo system.

Similar techniques have been used in optics to remove phase errors. A distorted optical beam is used to store a holographic grating in a photosensitive material. This grating is then used to remove distortions in an image when the optical beam has passed through the same distorting medium [85].

4.8.3 Inverse Filter

The basic principle for constructing an equalizer or deconvolver is to send a reference pulse through the distorting medium and then use a transversal filter to remove the distortion. Consider a general input signal of the form $X(\omega)$. After passing through the distorting medium, whose response is $H(\omega)$, it will be of the

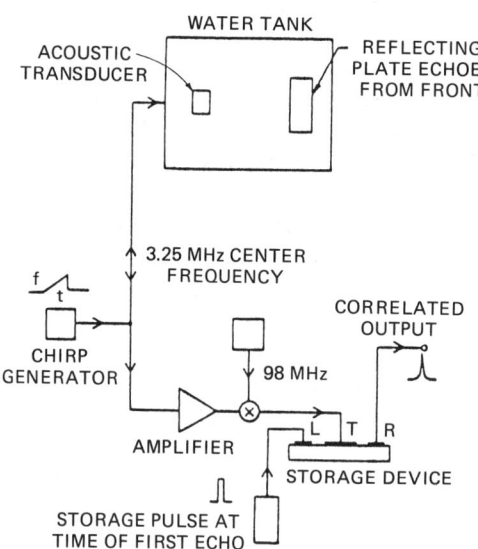

Figure 4.8.1 Acoustic pulse-echo system. (After Borden and Kino [84].)

GOOD TRANSDUCER

(a) CORRELATION PEAK

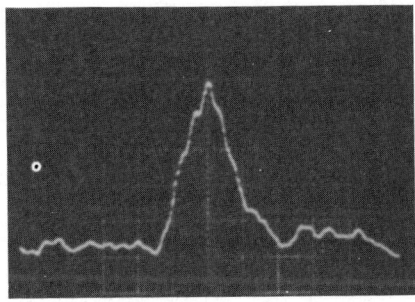

400 nsec/div

POOR TRANSDUCER

(c) FIRST REFLECTION

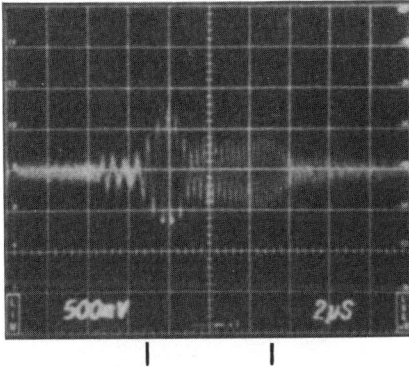

2 MHZ   4.5 MHZ

(b) IMPULSE RESPONSE

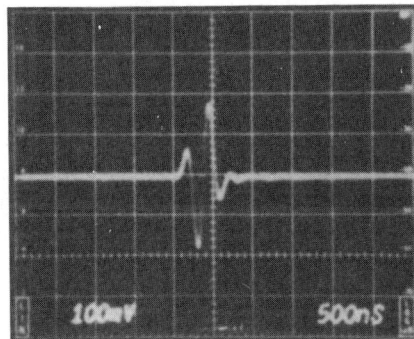

(d) CORRELATION PEAK

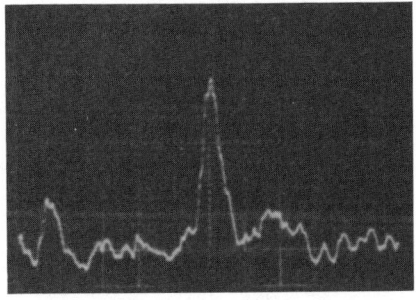

1 μsec/div

(e) IMPULSE RESPONSE

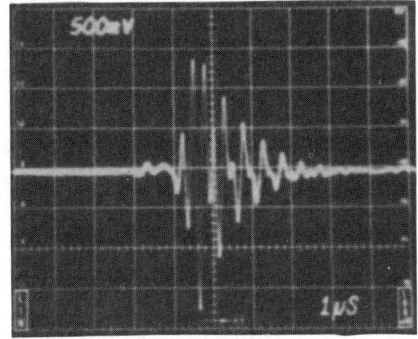

Figure 4.8.2 Pulse-echo experiment results carried out with both good and poor transducers. (After Borden and Kino [84].)

Sec. 4.8 Adaptive Filters

form

$$Y_1(\omega) = X(\omega)H(\omega) \tag{4.8.3}$$

We have already seen that a matched filter of the form $H^*(\omega)$ can remove the phase distortions due to $H(\omega)$. Such a filter will, however, accentuate amplitude variations over the frequency range of interest, for the response will now be $H(\omega)H^*(\omega)$, or essentially the square of the original amplitude response. To recover the original signal $X(\omega)$, we ideally want to construct an *inverse filter*, that is, a transversal filter with a characteristic frequency response $W(\omega) = 1/H(\omega)$. The output after passing the distorted signal through the inverse filter will be

$$Y_2(\omega) = \frac{X(\omega)H(\omega)}{H(\omega)} = X(\omega) \tag{4.8.4}$$

and the effective value of $H(\omega)$ of the combined system, which we shall call $H'(\omega)$, will be

$$H'(\omega) = \frac{H(\omega)}{H(\omega)} = 1 \tag{4.8.5}$$

A filter with a response $W(\omega) = 1/H(\omega)$ is known as an *inverse filter*. This filter is obviously impossible to realize because it needs an infinite response at the points where $H(\omega) = 0$. Thus only some approximate form of equalizer, at best, can be constructed. One way to do this, of course, is to measure the characteristic response $H(\omega)$ of the system. If $H(\omega)$ is finite over the frequency range $\omega_1 < \omega < \omega_2$, we can multiply the output by $W(\omega) = 1/H(\omega)$ over this frequency range and put $W(\omega) = 0$ at all other frequencies. This procedure yields a uniform response $H'(\omega) = 1$ in the frequency range $\omega_0 - \Omega/2 < \omega < \omega_0 + \Omega/2$ with zero output outside this frequency range, as illustrated in Fig. 4.8.3. The time response of this filtered system is then

$$h'(t) = \frac{\Omega}{2\pi} e^{j\omega_0 t} \frac{\sin(\Omega t/2)}{\Omega t/2} \tag{4.8.6}$$

Another possibility is to apodize the response of the inverse filter with a Hamming weighting to obtain a smooth frequency response or a smooth and compact time response. In all cases, unless some type of gating or apodizing function is used, the noise output will vary as $1/H(\omega)$ and will be very large in the regions where $H(\omega) \, \omega \to 0$ and $1/H(\omega) \, \omega \to \infty$.

A CCD or an acoustic transducer with the required inverse filter response over this frequency range can be constructed on this principle. This process has been carried out by Kerber et al. [86] using an SAW filter to compensate for the characteristics of an acoustic bulk wave transducer. There are also several examples in the literature that use a computer-based system for this process [87]. Computer-based systems, however, will not easily give large bandwidths and real-time processing.

Consider the system illustrated in Fig. 4.8.4. A reference pulse is sent into a tapped analog delay line. The output taps can be varied in amplitude by a weighting network that is controlled by computer or microprocessor. If the weight

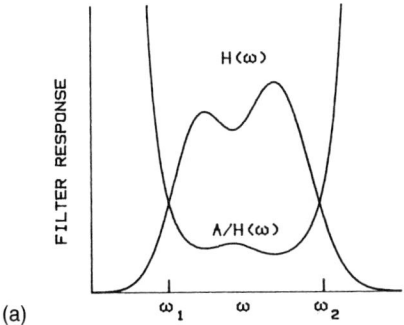

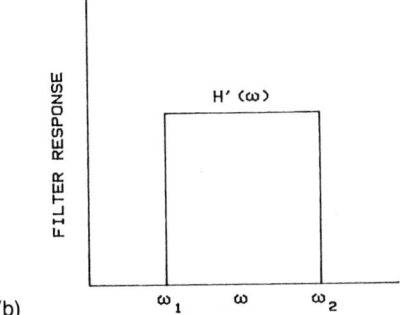

Figure 4.8.3 (a) Response $H(\omega)$ of the system and that of the inverse filter $A/H(\omega)$; (b) inverse filter for which the response of the system would be $H'(\omega)$, where $H'(\omega) = 1$, $\omega_1 < \omega < \omega_2$, $H'(\omega) = 0$, $\omega < \omega_1$, and $\omega > \omega_2$.

of the delay line can be adjusted to compensate for the distortion of the medium, the delay line becomes a real-time inverse filter, which can compensate for the distortion of any signal passing through the same distorting medium. The device is trained relatively slowly, using a repeated distorted input pulse that eventually produces an output pulse of the same form as the known undistorted reference pulse. The signal $x(t)$ entering the delay line is sampled at a rate $f_s = 1/\tau$, which corresponds to the delay time between taps. If the weighting of the nth tap is w_n, the output from the device at a time $t = m\tau$ will be

$$y(m\tau) = \sum_{n=1}^{N} w_n x[(m-n)\tau] \quad (4.8.7)$$

To train the device in the manner prescribed, we therefore require that

$$y(m\tau) = 1; \; m = k$$
$$y(m\tau) = 0; \; m \neq k \quad (4.8.8)$$

where the kth tap is called the reference tap.

The process was carried out in this example with a Reticon TAD-32 tapped analog delay line controlled by a microprocessor. The filter tap weights were adjusted by a process known as a *zero-forcing algorithm*, which is now commonly used for automatic equalization of digital communication systems employed in the

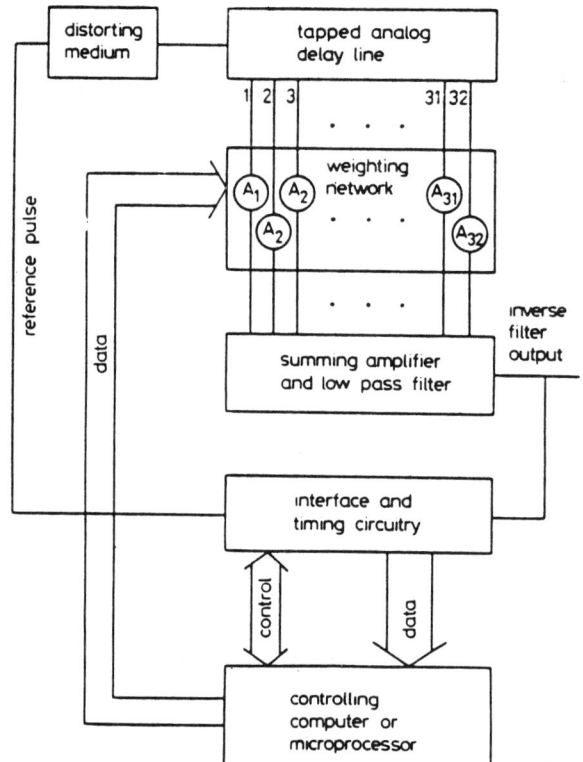

Figure 4.8.4 Adaptive inverse filter processor. (After Corl [88].)

telephone system. In the telephone system, for instance, by adjusting the weights of a tapped electromagnetic delay line, such methods have made it possible to upgrade standard telephone lines, which formerly operated in the frequency range below 3 kHz, to process digital codes with frequencies as high as 64 kHz. Such equalizers automatically come into operation to compensate for distortions over the changing paths used in a switched telephone network; this eliminates errors when telephone lines are used for transmitting digital codes.

The zero-forcing algorithm used in this example was first employed for telephone lines by Lucky [89]. It uses one tap, designated as the reference tap, whose weight is maintained at a constant nonzero value. The algorithm operates by increasing or decreasing each tap weight by one step, according to whether the filter output is positive or negative during the corresponding clock period. This forces the filter output to a maximum during the clock period corresponding to the reference tap, and to zero during the clock periods corresponding to the other taps of the delay line. The algorithm has the advantage of great simplicity and relatively high speed convergence. Its disadvantage is that it does not converge for any arbitrary distortion; however, it does converge for a large class of waveforms of the types associated with digital codes. As it was relatively easy to implement, it was the first one tried with CCD and BBD filters.

In the example shown, an *LC* resonator was constructed to simulate the response of a distorting medium [88]. This resonator was pulsed and gave a ringing signal, as shown in Fig. 4.8.5, and the CCD adjusted itself until it converged on an output pulse after 300 iterations. We expected the required filter to superimpose a second ringing waveform on the first one, delayed by one-half an RF cycle but reduced in amplitude, which would just cancel out the rest of the ringing waveform. Thus the inverse filter had to consist of a weighting corresponding to two impulses of opposite sign, one following the other. This was basically the form of the filter produced, as shown in Fig. 4.8.6. In another example, illustrated in Fig. 4.8.7., more complicated signals were injected into the device after the filter was adjusted and the distortion was then eliminated from them.

A completely monolithic LSI transversal filter has been constructed on the same principles [90]. This device uses BBD filters in the configuration shown in Fig. 4.8.8. To avoid making a tapped BBD, 16 separate BBDs, with time delays in steps from T to $16T$, are used with their outputs summed. The amplitude of the input signal to each BBD is varied by multiplying D/A converters. These devices behave like variable attenuators whose attenuations can be set by a digital input. In this case the digital inputs are set by a microprocessor. The 16 BBDs are therefore equivalent to a tapped delay line in which the signals are inserted

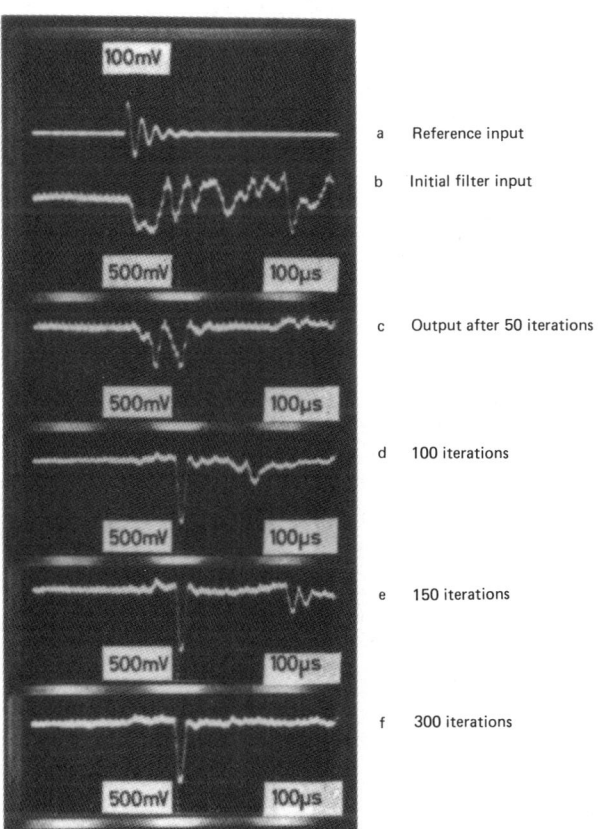

a   Reference input

b   Initial filter input

c   Output after 50 iterations

d   100 iterations

e   150 iterations

f   300 iterations

Figure 4.8.5 Convergence of the CCD filter. (After Corl [88].)

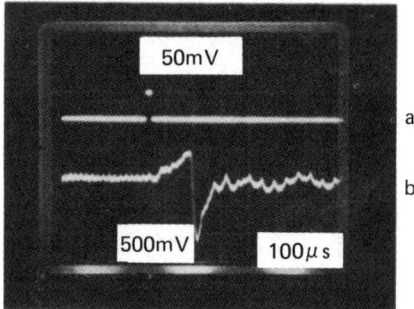

Figure 4.8.6 Inverse filter impulse response. (After Corl [88].)

into the taps. An adaptive inverse filter using the Lucky forcing algorithm was made with this system to obtain the results shown in Fig. 4.8.9.

4.8.4 Wiener Filter

A matched filter can remove phase errors but not amplitude errors. A more general filter, as we have seen, is the inverse filter. Unfortunately, this filter cannot be practically realized, because at certain frequencies it would require an infinite response.

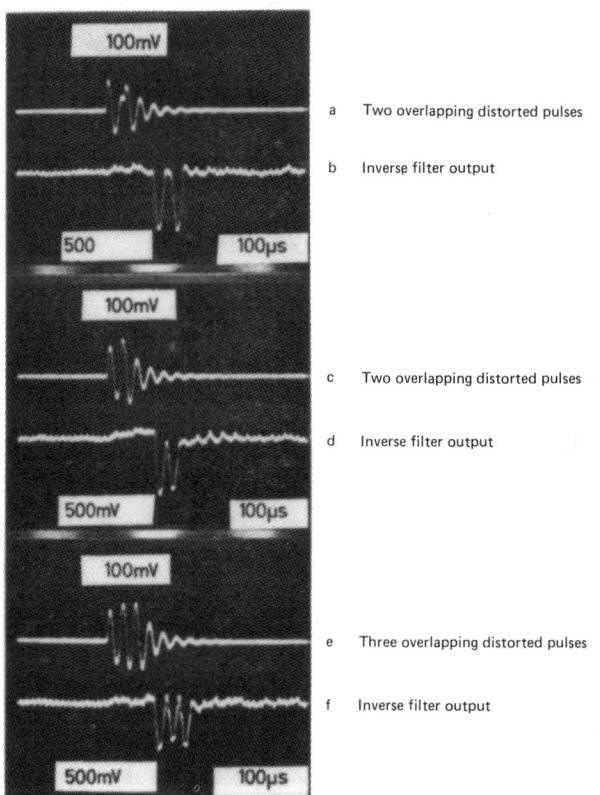

a Two overlapping distorted pulses
b Inverse filter output
c Two overlapping distorted pulses
d Inverse filter output
e Three overlapping distorted pulses
f Inverse filter output

Figure 4.8.7 Deconvolution of overlapping distorted pulses using the inverse filter. (After Corl [88].)

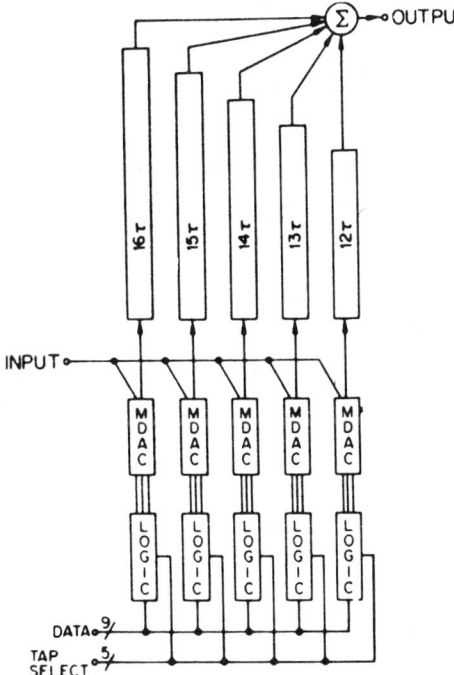

Figure 4.8.8 Tapped filter using separate BBD delay lines. Only five delay lines are shown. (After Tanaka et al. [90].)

The best approximation to such a filter is one with the best mean-square amplitude error over the passband; this is known as a *Wiener filter*. Suppose that the distorted signal is of the form $x(t)$ with a Fourier transform $X(\omega)$. After removing the distortion, the required signal is $d(t)$ and its Fourier transform is $D(\omega)$. Suppose that in addition to the input signal, there is noise present with a spectrum $N(\omega)$. When the input signal plus noise is inserted into a filter with a response $W(\omega)$, the output is

$$Y(\omega) = W(\omega)[X(\omega) + N(\omega)] \qquad (4.8.9)$$

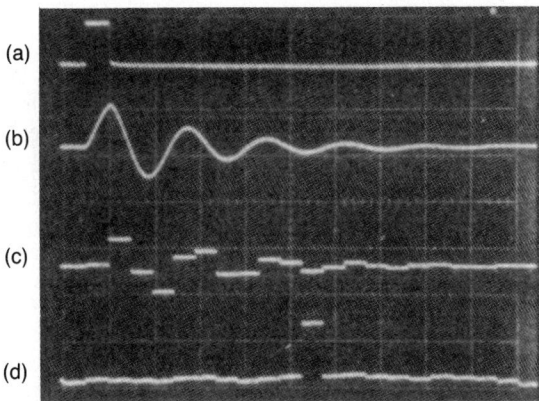

Figure 4.8.9 Adaptive inverse filter waveforms: (a) input impulse; (b) distorted input signal; (c) sampled input signal to the filter; (d) restored impulse output of the filter. (After Tanaka et al. [90].)

Sec. 4.8   Adaptive Filters

The error between the desired signal and the filter output is

$$e(t) = d(t) - y(t) \tag{4.8.10}$$

or

$$E(\omega) = D(\omega) - Y(\omega) \tag{4.8.11}$$

Our criterion in the time domain for an optimum filter will be to choose the parameter ε defined by the relation

$$\varepsilon = \int_0^T \langle e^2(t) \rangle \, dt \tag{4.8.12}$$

to be minimum in the time length T of the signal. This is equivalent to requiring that

$$\varepsilon = \frac{1}{2\pi} \int \langle |E^2(\omega)| \rangle \, d\omega \tag{4.8.13}$$

be minimum over the frequency range of interest. In each case, the symbol $\langle \; \rangle$ indicates the expectation value after a long period or many repeated pulses. Thus we choose the Wiener filter because it gives the least-mean-square error between a desired function and the output of the filter over the time or frequency range of interest.

It will be seen that

$$|E^2(\omega)| = E(\omega)E^*(\omega) = (D - Y)(D^* - Y^*) \tag{4.8.14}$$

so that

$$\begin{aligned}|E^2(\omega)| &= [D - W(X + N)][D^* - W^*(X^* + N^*)] \\ &= DD^* + WW^*(X + N)(X^* + N^*) - WD^*(X + N) \\ &\quad - W^*D(X^* + N^*)\end{aligned} \tag{4.8.15}$$

Because the noise signal can be assumed to be uncorrelated with x or d, there is no phase coherence between N or X and D. Hence, over a long period, $\langle ND^* \rangle = 0$, $\langle NX^* \rangle = 0$, and so on. Thus we can see that

$$\langle |E^2(\omega)| \rangle = DD^* + WW^*[XX^* + \langle NN^* \rangle] - DW^*X^* - D^*WX \tag{4.8.16}$$

The conditions for minimum error are $\partial P/\partial W^* = 0$ and $\partial P/\partial W = 0$. We differentiate inside the integral of Eq. (4.8.13) and equate the differential of the integrand to zero; this process yields the conditions that the differentials with respect to W and W^* of the integrand must be zero. In addition, we have regarded W and W^* as independent quantities, as are their real and imaginary parts. This condition yields the relation

$$\frac{\partial}{\partial W^*} \langle |E^2(\omega)| \rangle = W(XX^* + \langle NN^* \rangle) - DX^* = 0 \tag{4.8.17}$$

Thus the simplest condition for minimum mean-square error between the desired function and the filtered distorted function is that at any frequency within the range of interest

$$W = \frac{DX^*}{XX^* + \langle NN^* \rangle} \quad (4.8.18)$$

This equation defines the *Wiener filter*.

We note that if $XX^* \gg \langle NN^* \rangle$ and the signal-to-noise ratio is large, with D = constant, then $W \to 1/X$, and the filter behaves like an inverse filter. On the other hand, if $XX^* \ll \langle NN^* \rangle$, then $W \to DX^*/\langle NN^* \rangle$. If $D(\omega)$ and $\langle NN^* \rangle$ are constants, the latter response has the form of a matched filter.

It is possible to use Fourier transform techniques to program a computer to set the taps of a delay line transversal filter to the filter of Eq. (4.8.18). This filter will then give the lowest mean-square error between the response desired and the response obtained.

Pseudo-inverse filter. A Wiener filter, as defined in Eq. (4.8.18), is an ideal filter. It is difficult to realize such a filter in practice for two reasons: (1) we must know $N(\omega)$ precisely; and (2) a filter based directly on this principle would not be useful because slight changes in the noise level or the input distortion could change the output signal radically.

There are many approaches to make the design less critical. One method is to construct a filter with a characteristic

$$W(\omega) = \frac{DX^*}{XX^* + \Phi} \quad (4.8.19)$$

where Φ is chosen to be a constant considerably larger than the maximum value of $\langle NN^* \rangle$ in the passband of interest. Such a filter is known as a *pseudo-inverse filter*.

An illustration of the power of this type of filter has been made using a digital computer to process real signals; a schematic of the system is shown in Fig. 4.8.10. An ultrasonic transducer operating at a center frequency of 300 MHz was driven by a 2-ns electric pulse. The transducer excited an acoustic pulse in a solid; this pulse was reflected from the back surface of the solid and used to probe for flaws within it. For good range definition, the shortest possible acoustic pulse is desirable; thus, in this example, a test echo (the reutrn signal from the rear surface of the sample) was made as short as possible after filtering to improve the effective response of the transducer when used as both transmitter and receiver.

The signal received at the transducer was passed into a sampling oscilloscope; this yielded an output that was a slowed-down version of the pulse. An analog-to-digital converter sampled the output and digitized it for insertion into a PDP 11-10 minicomputer. The computer was used to take fast Fourier transforms and to correct the received signal by constructing a pseudo-inverse filter of the form given in Eq. (4.8.19). The value of Φ was fixed at a constant arbitrary value, typically 20 dB below the peak signal value at the center frequency of the transducer. This made the filter less sensitive to slight changes in input conditions. In practice, the smaller the effective value of $\langle |N^2| \rangle$ or Φ used in the design, the more critical the parameters required. We expect this because we need the pseudo-inverse filter to make far more severe corrections when Φ is effectively very small.

An example of the results obtained are shown in Fig. 4.8.10(a) and the frequency response in Fig. 4.8.10(b). We chose $\langle NN^* \rangle$ or Φ with a value 20 dB below the peak value of $X(\omega)X^*(\omega)$; thus the effective bandwidth of the system now became that of the transducer between its 20-dB points. The output of the Wiener filter is shown in Fig. 4.8.10(c). After gating this output over the frequency range between the original 20-dB points, and retransforming it, we obtained the much improved pulse response shown in Fig. 4.8.10(d). With the pseudo-inverse filter in place, the transducer was successfully used to probe for flaws in solid materials. The transducer could be moved over the surface of the object without readjusting the signal compensation in the computer.

4.8.5 Storage Correlator as an Adaptive Wiener Filter

The storage correlator lends itself well to such applications as an adaptive Wiener or pseudo-inverse filter. One way to construct this type of adaptive filter is to let a computer set the taps of a transversal CCD or SAW filter; this has been done with good results. Another possible application uses a storage correlator in a feedback loop in which the correlator sets itself for minimum error after several iterations. The iteration technique employed is closely related to the *LMS algorithm* (least-mean-square algorithm), due to Widrow and his co-workers [91].

We demonstrate this iterative mechanism with a mathematical derivation carried out in the frequency domain. Suppose that the filter is adjustable and that after the kth iteration, its response is $W_k(\omega)$. Then with a signal $X(\omega)$ inserted, the output is

$$Y_k = W_k(X + N) \qquad (4.8.20)$$

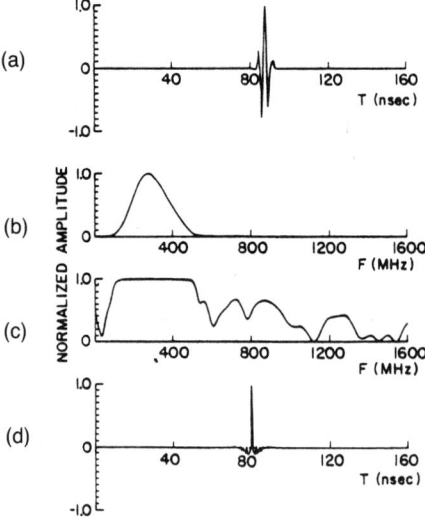

Figure 4.8.10 (a) Impulse response of zinc-oxide-on-sapphire transducer; (b) frequency response of zinc-oxide-on-sapphire transducer; (c) frequency response of the output signal through the filter (note that frequencies above 600 MHz are not used in the final transform; (d) impulse response of the output signal through the Wiener filter. (After Murakami et al. [87].)

where $N(\omega)$ is the noise spectrum. The error between the desired signal and the output is

$$E_k = D - (X + N_k)W_k \qquad (4.8.21)$$

Suppose that we try to minimize $|E_k(\omega)|^2$. If we do this, the final value of $W_k(\omega)$ should correspond to a Wiener filter. On the kth iteration $W_k(\omega)$ changes to $W_{k+1}(\omega)$, where

$$W_{k+1} - W_k = \Delta W_k \qquad (4.8.22)$$

If we write $S_k = \langle |E_k|^2 \rangle$, it follows that

$$\Delta S_k = S_{k+1} - S_k = \langle E_k^* \, \Delta E_k + E_k \, \Delta E_k^* \rangle \qquad (4.8.23)$$

Therefore, from Eqs. (4.8.21) through (4.8.23), we see that

$$\Delta S_k = \langle -E_k^*(X + N_k) \, \Delta W_k - E_k(X^* + N_k^*) \, \Delta W_k^* \rangle \qquad (4.8.24)$$

We wish to choose ΔW_k so that ΔS_k is always negative and $S_{k+1} < S_k$. This is the condition for the error to decrease on each iteration. A good choice for this purpose is to put

$$\Delta W_k = W_{k+1} - W_k = \mu E_k(X^* + N_k^*) \qquad (4.8.25)$$

where μ is a constant. In this case it follows that

$$\Delta S_k = S_{k+1} - S_k = -2\mu S_k \, (XX^* + \langle NN^* \rangle) \qquad (4.8.26)$$

For simplicity, we assume that each frequency component is independent and the noise is negligible. We can then write

$$S_{k+1} = S_k[1 - 2\mu(XX^* + \langle NN^* \rangle)] \qquad (4.8.27)$$

Thus after k iterations, it follows that

$$S_k = S_0[(1 - 2\mu(XX^* + \langle NN^* \rangle)] \qquad (4.8.28)$$

The error converges to zero as $K \to \infty$, provided that $1 - 2\mu(XX^* + \langle NN^* \rangle) < 1$ or $\mu XX^* < 1$ at all frequencies of interest. It also follows that the filter converges to a Wiener filter as $k \to \infty$.

We can explain the operation of the device equally well in the time domain. The output from the filter, when the weighting is $w_k(t)$ and the input is $x(t)$, is

$$y_k(t) = x(t) * w_k(t) \qquad (4.8.29)$$

This is the time domain form of Eq. (4.8.20) with the noise term omitted. We subtract the output from the desired signal $d(t)$, and it follows from Eq. (4.8.25) that the weighting must be changed by

$$\Delta w_k(t) = \mu e_k(t) \star x(t) \qquad (4.8.30)$$

or by correlating $e_k(t)$, the error in the time domain, with the input signal.

The storage correlator is well suited to carry out these operations. Such a procedure has been carried out by Bowers et al., and the two configurations used

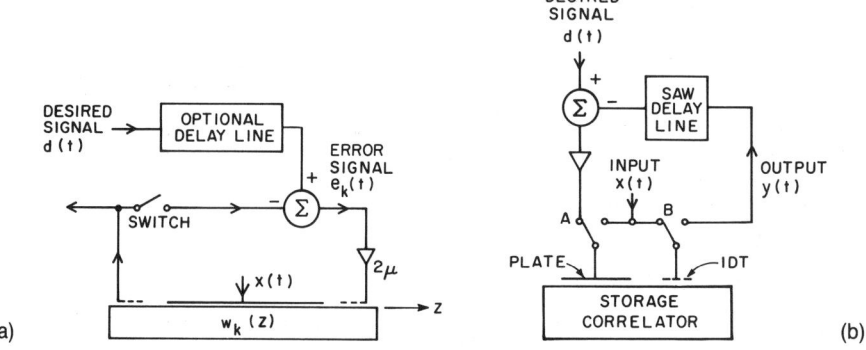

Figure 4.8.11 (a) Adaptive Wiener filter using a storage correlator; (b) second implementation of the adaptive Weiner filter. (From Bowers et al. [92].)

are illustrated in Fig. 4.8.11. The experiment consisted of applying a modulated RF signal $x(t)$ at the acoustic port of a monolithic storage correlator to represent, for instance, a television signal followed by a later echo or its ghost. Using a test signal, such as the line sync pulse, the correlator was required to give a desired output $d(t)$, which ideally would be the signal with the ghost echo removed. The feedback loop carried out the iterative process to make the output $y(t)$ approach the desired single $d(t)$ (a signal pulse) as closely as possible. The received signal $x(t)$ and desired signal $d(t)$ were entered repetitively into the storage correlator for several iteration cycles, where one cycle consisted of two samples of $x(t)$ and one sample of $d(t)$. The whole process involved two basic operations, illustrated in Fig. 4.8.11(a):

1. The convolution between $x(t)$ and the stored impulse response of the correlator $w_k(t)$ was obtained on the kth iteration. The output after the kth iteration was

$$y_k(t) = \int_0^L x\left(t - \frac{z}{V}\right) w_k(z)\, dz \qquad (4.8.31)$$

where z is the distance along the correlator, V is the acoustic wave velocity, and L is the length of the device, where $L = VT$. The error $e_k(t) = d(t) - y_k(t)$ was then obtained.

2. The cross-correlation between $x(t)$ and the error signal $e_k(t)$ was taken by inserting $e_{k(t)}$ and $x(t)$ into the device. Note that a delay line was used to give time to clear the device of the original input signal. This process changed the stored weights to $w_{k+1}(z)$, where

$$w_{k+1}(z) = w_k(z) + 2\mu \int_{T_0} x(t) e_k\left(t - \frac{L}{V} + \frac{z}{V}\right) dt \qquad (4.8.32)$$

and T_0 is the length of the error signal. The factor 2μ is now a multiplication factor proportional to the gain in the feedback loop and determines the convergence rate. The delays in the system are chosen so that $w_{k+1}(z)$ and

$w_k(z)$ are lined up with each other. This implies that the signals $d(t)$ and $x(t)$ are repeated periodically and that $e_k(t)$ starts at a time $t = T = L/V$. We recognize the integral term as the cross-correlation between $x(t)$ and $e_k(t)$. The process was then repeated. Finally, the output was taken from point A after opening the feedback loop with the switch.

If the $x(t)$ and $d(t)$ are modulated carriers, with the same carrier for both signals, a delay line with the same delay as the storage correlator can be used to delay $d(t)$. In this case the response will be independent of the carrier frequency, provided that this frequency is within the bandwidth of the device.

A somewhat more complicated circuit, like that of Fig. 4.8.11(b), has also been used. The actual configuration uses an RF switch A at the acoustic port, driven synchronously with switch B at the plate. This configuration requires less signal for error correction, using the acoustic port as the input, and does not suffer from reflections at a second interdigital transducer.

An experiment was carried out using this configuration with a zinc-oxide-on-silicon monolithic storage correlator with a bandwidth of 10 MHz, a center frequency of 124 MHz, and a time delay of 3 μs. The signal $x(t)$ consisted of a pulse followed by an echo 6 dB lower in amplitude than the main pulse; each pulse was 0.4 μs long and separated from the next by 0.8 μs. The aim was to remove the echo pulse and to obtain an output consisting of a single pulse.

Figure 4.8.12 shows the signal output before iteration, after one iteration (when the output is the correlation of the distorted signal with itself), and after 22 iterations. The device has been shown to decrease the echo level by approximately 10 dB. This type of filter performs far better even after two iterations than a simple correlation filter, for it is a better approximation of an inverse or a Wiener filter. Other experiments indicate that the time for final convergence is of the order of 200 μs and that it should be possible to improve the performance. The results clearly agree with computer simulation, so the operation of the device is well understood.

4.8.6 Elimination of an Interfering Signal with an SAW Correlator

One technique for eliminating a CW interfering signal has already been described, in Sec. 4.5.4 [46]. This employs a Fourier transform system with a limiter on the Fourier transform output to whiten the spectrum. Another technique uses a storage correlator as an adaptive filter; this forms a *notch filter*, which will eliminate a CW interfering signal. In this case, if we regard the interfering signal as an error signal, we program the filter to make the error zero. One method of doing this is similar to the system already described. Another, which has been demonstrated, is to use the system shown in Fig. 4.8.13. In this system, the device is used to eliminate CW interference from a wideband PN code. Both signals are inserted into the storage correlator at the same time, so that both switches shown are in position 1; the storage correlator then correlates the input signal with itself.

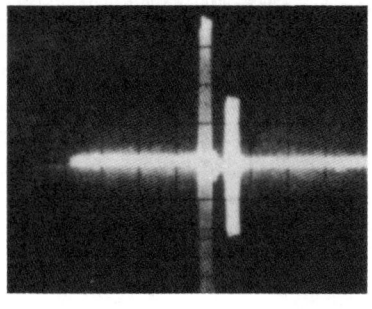

(a) INPUT SIGNAL

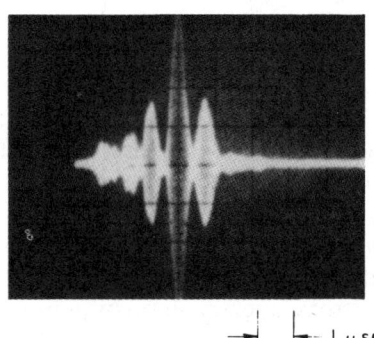

(b) CORRELATION SIGNAL OUTPUT
AT ZERO ITERATION

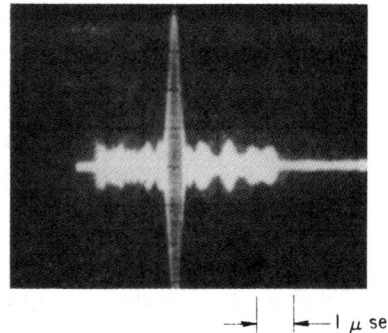

(c) OUTPUT SIGNAL AFTER
22 ITERATIONS

Figure 4.8.12 (a) Input signal. (After Behar et al. [93].) (b) Correlation signal output after one iteration. (c) Output signal after 22 iterations. (After Behar et al. [93].)

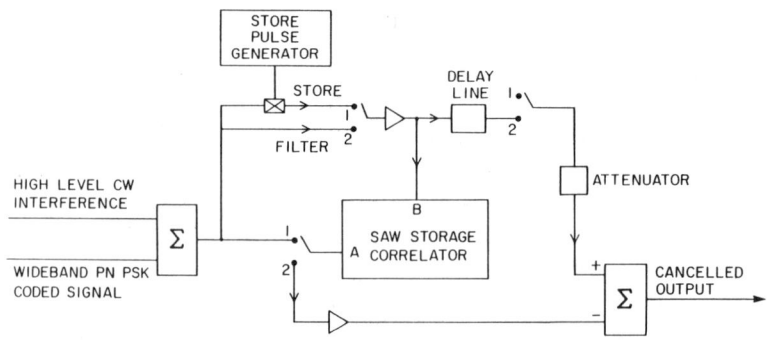

Figure 4.8.13 Surface acoustic wave implementation of narrowband interference canceler. (From Tuan et al. [83], and Grant and Kino [94].)

The PN code gives a correlation peak outside the device, so it has no effect here. With the CW interfering signal, however, the stored weight is of the form

$$w(z) = A \int_0^T \cos \omega_0 t \cos \omega_0 \left(t - \frac{z}{V} \right) dt \approx \tfrac{1}{2} A T \cos \frac{\omega_0 z}{V} \qquad (4.8.33)$$

where A is a constant. If a later signal is read into this device, it will form a narrowband filter. This filter will have a response of the form

$$W(\omega) = \frac{\sin (\omega - \omega_0) L/2V}{(\omega - \omega_0) L/2V} \qquad (4.8.34)$$

Thus its bandwidth will be approximately $\Delta f = 1/T$, where T is the time delay in the storage region of the device.

Returning to Fig. 4.8.13, the switches are switched to position 2 and the storage correlator now functions as a narrowband filter for the CW signal. By adjusting the attenuator, we can use this device to cancel out the direct CW signal passing into the summer. Thus only the PN code is observed at the output. In general, once the attenuator has been adjusted, and provided that the amplitude of response of the device is constant over a given band and the delay in the delay line is equal to that in the correlator, the device will act as a filter for any frequency within this band. Using this technique in the experiment, the 5-MHz rate PN code on a 120-MHz carrier was injected at the same time as an interfering CW signal. By this means, a 30-dB notch with a bandwidth of 330 kHz was introduced into the CW signal, and the CW interference was thus reduced by 30 dB. The correlation peak in the final matched filter for the PN code can be seen clearly in Fig. 4.8.14. The device is capable of simultaneously notching out one or more interfering frequencies anywhere within its 8-MHz bandwidth.

Other and better forms of an interference removing filter are based on the premise that as the error between the filter output and the desired signal decreases on each iteration, the system can be trained with the interfering signal entering each port of Fig. 4.8.11, or with $d(t) = x(t) = i(t)$, where $i(t)$ is the interfering signal. The system makes the error zero. If $d(t)$ is replaced by a signal corresponding to the required signal $s(t)$ and the interfering signal $i(t)$, the output from the operational amplifier will consist of only $s(t)$. Thus the system automatically nulls out interference [95]. If a radar system, for example, is trained on a jamming signal and its own transmitter is then turned on, the jammer will still be canceled out.

Other uses for adaptive filters. Several other types of adaptive filters that use the LMS algorithm to find the optimal set of tap weights have been developed. In one example, a CCD adaptive filter was made with analog tap weights held in sample-and-hold circuits. [96]. Another system uses an analog–digital hybrid approach with MOS-LSI technology. This has the advantage of lowering the power consumption and allowing 32 taps to be used without undue complexity of the external circuit. This technique yielded a large dynamic range, but the bandwidth was limited to less than 1 MHz [97].

An alternative approach for implementing filters with as much as a 500-MHz

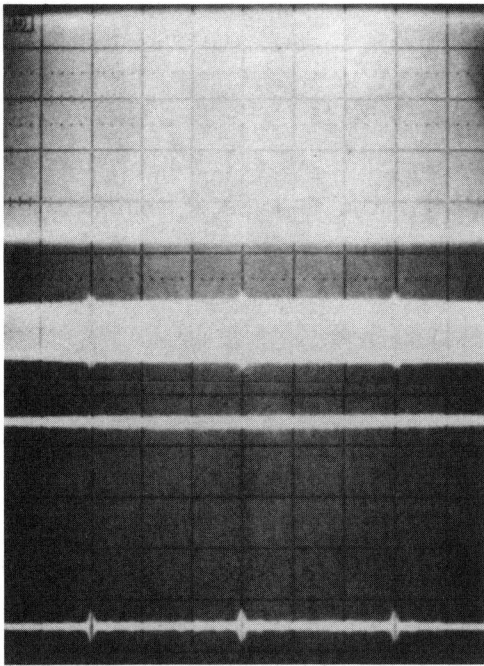

Figure 4.8.14 SAW adaptive filter demonstration. (After Tuan et al. [83].)

bandwidth has been developed by Masenten [98]. It uses tapped SAW filters with a large, complex, computer-controlled system to adjust the tap weights. This system has been used to cancel out interference and to put nulls into an antenna pattern to remove an interfering source.

PROBLEM SET 4.8

1. (a) A modulated RF acoustic pulse (a tone burst) of length τ is injected from a source of very small width D ($D \ll \lambda$) into a piezoelectric crystal. This pulse passes into the crystal and may be reflected several times from its various surfaces. A very short RF tone burst of frequency 2ω and length τ_0, where $\tau_0 \ll \tau$ excites an electric field between the two plates, is excited at a time $T > \tau$, as shown. Show that an undistorted but time-reversed pulse should arrive back at the original source after a further time T.

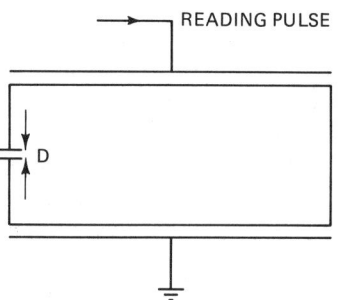

One way to deal with this problem is to take an excited signal of frequency $\omega_0 + \Omega$, amplitude $F(\Omega)$, and phase $\phi(x, y, z)$, together with a reading signal of frequency $2\omega_0$, and consider in general what occurs when a carrier of frequency ω_0 is amplitude-modulated by a signal $f(t)$. The other way is to deal with the problem directly in the time domain and consider the reading pulse to be very short.

(b) Show that if a spatially varying tone burst $f(x, y) \exp(j\omega t)$ is injected at the surface $z = 0$ (the front surface of the crystal) and interacts with a short RF E field pulse of frequency 2ω after a time T, there will be a signal received at the input surface at a time $2T$. Suppose that this received signal is retransmitted at this time and a further 2ω pulse excites the crystal at a time $t = 3T$. Show that the main component of the received signal returning to $z = 0$ at a time $t = 4T$ will be $f(x, y) \exp(j\omega t)$, so that all spatial and time distortions due to multiple reflections have been removed. You will need to consider carefully the phase of the signals involved, writing phase $\phi(x, y, z)$ along the beam as $\phi(x, y, z) = \omega t - k_x x - k_y y - k_z z$.

Hint: In part (a), the source on the crystal of width D can be regarded as a point source. The problem is set up to avoid difficulties, with the source being a finite number of wavelengths wide. Carry out a simple analysis for a ray propagating at an angle θ to the axis, and derive the rest from physical arguments in parts (a) and (b).

2. (a) Consider the system shown below. A sonar device emits a plane parallel beam from a pulsed source of frequency ω_0 through a plane parallel crystal which has nonlinear characteristics. The parallel beam passes through a lens and is focused in water at the point $0, z$. A signal is reflected from an object at this point, returns to the lens and the crystal, and is received on the same transducer, which can only respond to a parallel beam $[\phi(x, y, z) = \text{constant}]$.

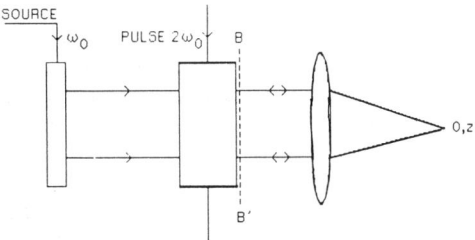

Now suppose that the water through which the focused beam passes is turbulent, so that the beam is distorted. The return echo signal from $0, z$ then arrives with distorted wave fronts after a time T. When the return echoes pass through the short crystal, it is excited with a frequency $2\omega_0$. Show that an acoustic beam will be emitted that will return from $0, z$ as a parallel beam, whatever the distortion of the medium. You can regard the lens and turbulent medium as a system that introduces a phase change $\phi(x, y, z)$. If there is no turbulence, $\phi(x, y, z) = $ constant at the plane BB'. You do not need to know $\phi(x, y, z)$.

Hint: For simplicity, ignore multiple reflections at the surface of the crystal.

(b) You may wonder if it is possible to obtain a signal large enough to be retransmitted out of the crystal. Consider an equivalent one-dimensional problem by coupled-mode theory. Let the signal entering the crystal be

$$a_1(z, t) = A_1(z) e^{j\omega t} \tag{1}$$

and the signal emitted from the crystal in the backward direction be

$$a_2(z, t) = A_2(z)e^{j\omega t} \tag{2}$$

and the signal (or pump) exciting the crystal between the plates be

$$a_3(t) = A_3 e^{2j\omega t} \tag{3}$$

where A_3 does not vary with z. As the generated signal must depend on the product of the other signals, we can write

$$\frac{dA_1}{dz} + jk_0 A_1 = j\alpha A_3 A_2^* \tag{4}$$

The second wave is traveling in the backward direction. So we can put

$$\frac{dA_2}{dz} - jk_0 A_2 = -j\alpha A_3 A_1^* \tag{5}$$

where α is a coupling constant. Note the changes in sign in Eq. (5) from the equivalent forms in Eqs. (4.6.5) and (4.6.6), or Eq. (4).

Beginning at the line before Eq. (6), the problem should read: Prove that it is still true, in this case, that

$$A_1 A_1^* + A_2 A_2^* = \text{constant} \tag{6}$$

Note that this result is a degenerate form of a more general formula. The power in a transmission line of impedance Z_0, in which there is a voltage $V_1 e^{-jk_1 z}$ propagating in the forward direction, is $P_1 = V_1 V_1^*/2Z_0$. The power propagating on a transmission line in the backward direction with a voltage $V_2 e^{jk_2 z}$ is $P_2 = V_2 V_2^*/2Z_0$. Thus, Eq. (6) does, in fact, correspond to $P_1 = P_2 = \text{constant}$ if P_2 is defined as being in the backward direction, and $V_1 \propto A_1$, $V_2 \propto A_2$.

Power P_3 is supplied to the system, so this is an "active" system, an amplifier which can supply more power output than the input power. In a passive system, in which a forward wave is coupled to a backward wave, conservation of power requires $P_1 + P_2 = \text{constant}$. Here, the situation is more complicated, and the Manley-Rowe relations (conservation of the number of quanta $N = P/\hbar\omega$, or $N_3 = N_1 + N_2$), and conservation of power P must be satisfied.

Now solve for A_1 and A_2^* using the appropriate boundary conditions *at each end* of a crystal of length L. Show that $A_1(z)$ has the form

$$A_1(z) = A_1(0)e^{-jk_0 z} \frac{\cos \kappa(z - L)}{\cos \kappa L} \tag{7}$$

where $\kappa = \alpha(A_3 A_3^*)^{1/2}$. You will find it convenient to assume that A_3 is real. Find $A_2(0)/A_1(0)$ and show that the system can self-oscillate if $\kappa L > \pi/2$ (i.e., there will be infinite gain). A device of this kind is known as a *parametric amplifier* or a *parametric oscillator*.

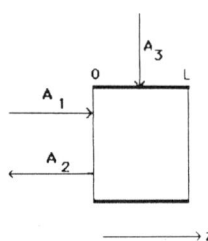

4.9 ACOUSTO-OPTIC FILTERS

4.9.1 Introduction

Light and sound waves interact when they are passed through the same medium. Light is scattered by sound in much the same way as by a diffraction grating, so we can regard the sound wave as a *moving* diffraction grating. Since the wavelength of a very high frequency acoustic wave can be comparable to the wavelength of light, the light beam can be scattered through relatively large angles, thus allowing us to observe a number of important effects. For example, this scattering phenomenon provides an important diagnostic tool for measuring the wavelength, type, and direction of propagation of sound waves and thermal phonons in solid and liquid media. It is also a very useful technique for deflecting a light beam: The amplitude of the deflected beam depends on the amplitude of the sound wave, and the angle of deflection of the beam depends on the acoustic frequency or its wavelength.

The deflection of the optical beam varies almost linearly with the frequency of the acoustic wave, which implies that the light pattern formed by the deflected beam is the Fourier transform of the acoustic beam modulation. If we utilize this type of interaction, a wide range of signal processing methods become available. If we also employ the ability of the optical lens to perform Fourier transforms or use interactions with two acoustic beams, we open many other possibilities for signal processing with acoustic waves in an acousto-optic system, including applications to real-time wideband convolution and correlation of electrical signals.

We can also use the light wave as a "pump" to generate very high frequency sound waves. This technique provides an important means of generating optical phonons in the far-infrared range. Its disadvantage, however, is that laser light of high intensity may unintentionally generate sound waves in certain materials, creating fine damage lines or fractures along its path.

The coupling between light and sound waves results from two effects:

1. The *photoelastic* or piezo-optic effect. Strain due to the presence of a sound wave causes a change in the atomic lattice spacing and hence to the dielectric constant. Thus the relative change in dielectric constant $\Delta\varepsilon/\varepsilon$ is generally proportional to the amplitude of an acoustic wave passing through it, a relation to be described quantitatively by Eq. (4.9.8).
2. The converse *electrostrictive* effect. The presence of an electric field causes a strain essentially proportional to the square of the electric field.

4.9.2 Photoelastic Effect

It is convenient to define the *dielectric impermeability* B as the reciprocal of the relative dielectric constant. In one-dimensional form, we define B by the relation

$$B = \varepsilon_0 \frac{\partial E}{\partial D} \qquad (4.9.1)$$

where E is the electric field and D the displacement density. For small signals, the dielectric constant ε would be defined in the same way as $\partial D/\partial E$.

We can write Eq. (4.9.1) in the more general tensor form

$$B_{ij} = \varepsilon_0 \frac{\partial E_i}{\partial D_j} \tag{4.9.2}$$

It follows that the one-dimensional relation between ε and B is

$$B = \frac{\varepsilon_0}{\varepsilon} \tag{4.9.3}$$

More generally, we write

$$\sum_j \frac{\varepsilon_{ij}}{\varepsilon_0} B_{jk} = I \tag{4.9.4}$$

or,

$$\sum_j \frac{\varepsilon_{ij}}{\varepsilon_0} B_{jk} = \delta_{ik} \tag{4.9.5}$$

where $\delta_{ik} = 1, i = k; \delta_{ik} = 0, i = k$.

The photoelastic effect relates the change in dielectric impermeability to the strain (the relative change in a length l of the material, defined as $S = \Delta l/l$), as follows:

$$\Delta B = pS \tag{4.9.6}$$

where p is the piezo-optic coefficient. More generally, we can write

$$\Delta B_{ij} = p_{ijkl} S_{kl} \tag{4.9.7}$$

where p_{ijkl} or p_{IJ} is the piezo-optic tensor.

For a cubic crystal, which has an isotropic permittivity, the effective photoelastic constant will depend on the polarization of light relative to the crystal axes and the sound waves. The one-dimensional form of B_{ij} or B_I, which we can use for plane waves, varies with the angle between the direction of sound wave propagation and the direction of the light waves. Different coefficients are required to describe the interaction of light with longitudinal or shear waves, but the relation retains the same form.

It is interesting to note that because of the symmetry relations of the tensors, light diffracted by a longitudinal wave normally gives rise to a diffracted light beam with the same plane of polarization. Light diffracted by a shear wave, however, results in a light wave with a plane of polarization perpendicular to the polarization of the incident light.

It follows that if we keep only first order terms in $\Delta B/B$, $\Delta \varepsilon/\varepsilon$ or $\Delta n/n_0$ in the one-dimensional relations, then

$$\frac{2\Delta n}{n_0} = \frac{\Delta \varepsilon}{\varepsilon} = -\frac{\varepsilon}{\varepsilon_0} pS \tag{4.9.8}$$

where n_0 is the refractive index of the medium and $n_0^2 = \varepsilon/\varepsilon_0$.

We expect the relative change in the dielectric constant to be of the same order as the relative change in density, so that

$$\frac{\Delta\varepsilon}{\varepsilon - \varepsilon_0} \sim -S \tag{4.9.9}$$

Thus we expect $p \sim \varepsilon_0(\varepsilon - \varepsilon_0)/\varepsilon^2$. This is a reasonable approximation to the truth for many materials.

We can carry out a simple analysis, based on the Lorentz–Lorenz law, for a gas or liquid. If the number of molecules per unit volume in the medium is N, the Lorentz–Lorenz law states that

$$\frac{\varepsilon - \varepsilon_0}{\varepsilon + 2\varepsilon_0} \frac{1}{N} = \text{constant} \tag{4.9.10}$$

For a plane wave, where the strain $[S = (\Delta l/l)]$ is assumed to be small, we may write

$$N = N_0(1 - S) \tag{4.9.11}$$

It follows that

$$\frac{\Delta\varepsilon}{\varepsilon} = -\frac{(\varepsilon + 2\varepsilon_0)(\varepsilon - \varepsilon_0)}{3\varepsilon_0\varepsilon} S \tag{4.9.12}$$

or

$$p = \frac{(\varepsilon + 2\varepsilon_0)(\varepsilon - \varepsilon_0)}{3\varepsilon^2} \tag{4.9.13}$$

Example: Piezo-Optic Constant of Water

For water with an index of refraction $n_0 = 1.33$ for light, $\varepsilon/\varepsilon_0 = 1.769$, $\Delta\varepsilon/\varepsilon = -0.546S$, and $p = 0.31$. This compares with a measured value of $p = 0.31$. Values of p for a number of solid materials are given in Appendix B.

4.9.3 Light Diffraction by Sound

An acoustic wave propagating through a solid or a liquid diffracts a light beam much as a diffraction grating does. The sound wave creates a periodic variation of the permittivity of the medium in time and space. Because it travels through the medium with a finite velocity, rather than remaining stationary in space, the acoustic wave Doppler-shifts the frequency of the diffracted light beam, as well as exciting it at an angle to the incident beam.

We first consider a simple physical model for *Raman–Nath diffraction*, when an optical wave of frequency ω_0 and propagation constant k_0 in the material propagates in the z direction. The optical beam interacts with a sound wave of frequency ω_a and propagation constant $k_a = 2\pi/\lambda_a$, propagating in the x direction. We determine the phase shift of the optical beam by using a simple integration of the phase shift across the acoustic beam. As we shall see later, the results we obtain are valid only when the Raman–Nath parameter Q, defined as

$$Q = \frac{k_a^2 w}{k_0} \qquad (4.9.14)$$

is such that $Q \ll \pi/2$, where w is the acoustic beam width [99]. Several diffraction spots, or orders of diffraction, occur, as illustrated in Fig. 4.9.1. We shall also derive the same results more rigorously and in a convenient normalized form as Eq. (4.9.61) in Sec. 4.9.5.

In Sec. 4.9.5 we shall consider *Bragg interaction*, which occurs when the angle between the direction of propagation of the input optical beam and the z axis is $\pm \theta_B$, where $\sin \theta_B = k_a/2k_0$ and θ_B is called the *Bragg angle*, in analogy to Bragg diffraction of x-rays from crystals. In this case, discussed in detail in Sec. 4.9.5, when $Q \gg \pi/2$, only one diffraction spot occurs.

It is convenient to write the variation in the dielectric constant in the form

$$\frac{\Delta \varepsilon}{\varepsilon} = \mathrm{Re}\left[\alpha e^{j(\omega_a t - k_a x)} \right] \qquad (4.9.15)$$

or

$$\frac{2\Delta n}{n_0} = \frac{\Delta \varepsilon}{\varepsilon} = \alpha \cos(\omega_a t - k_a x) \qquad (4.9.16)$$

where

$$\alpha = \frac{\varepsilon}{\varepsilon_0} p S_0 = n_0^2 p S_0 \qquad (4.9.17)$$

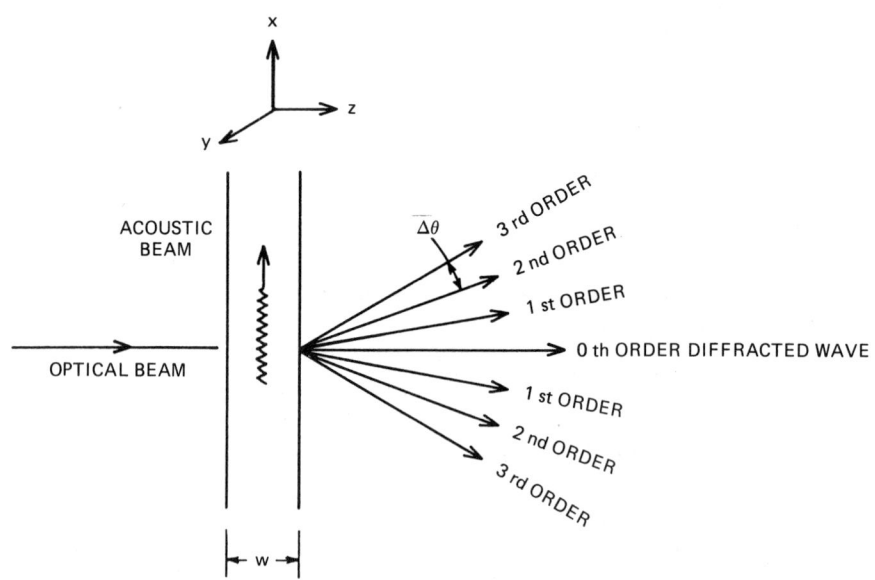

Figure 4.9.1 Raman–Nath diffraction.

and S_0 is the peak strain. As an example, since the acoustic wave velocity in water is 1.5 km/s, the wavelength λ_a at a frequency of 5 MHz is 300 μm or 0.3 mm. This is far larger than the optical wavelength, which is typically less than 1 μm.

We shall assume that the input optical wave in the medium is of frequency ω_0, with propagation constant k_0 in the z direction in the material, which has a refractive index n_0. A simple approximation to the phase change per unit length of the optical wave is therefore

$$\frac{\partial \phi}{\partial z} = -k_0\left(1 + \frac{\Delta\varepsilon}{2\varepsilon}\right) = -k_0\left(1 + \frac{\Delta n}{n_0}\right) \tag{4.9.18}$$

It follows that

$$\frac{\partial \phi}{\partial z} = -k_0\left[1 + \frac{\alpha}{2}\cos(\omega_a t - k_a x)\right] \tag{4.9.19}$$

where $k_0^2 = \omega^2\mu\varepsilon$. The phase change of a wave passing a distance z through the medium is therefore

$$\phi = -k_0 z - \frac{k_0}{2}\int_0^z \alpha \cos(\omega_a t - k_a x)\, dz \tag{4.9.20}$$

We define an amplitude parameter $v(z)$ by the relation

$$v(z) = \tfrac{1}{2}\int_0^z k_0 \alpha\, dz \tag{4.9.21}$$

This type of interaction between light and an acoustic beam depends on the integral of the amplitude across the acoustic beam.

If the acoustic beam is uniform over a width w, the phase change in a length w is

$$\phi = -k_0 w\left[1 + \frac{\alpha}{2}\cos(\omega_a t - k_a x)\right] \tag{4.9.22}$$

where

$$v(w) = \frac{k_0 \alpha w}{2} \tag{4.9.23}$$

Acousto-optic figure of merit. If A is the area of the acoustic beam, the power in the beam is

$$P = \tfrac{1}{2} A Z_0 V_a^2 S^2 \tag{4.9.24}$$
$$= \tfrac{1}{2} A \rho_{m0} V_a^3 S_0^2$$

where ρ_{m0} is the density of the material, V_a is the sound wave velocity, and Z_0 is the acoustic impedance.

It follows that we can write $v(w)$ in the form

$$v(w) = \frac{\pi w}{\Lambda}\left(\frac{2PM}{A}\right)^{1/2} = \frac{\pi w \alpha n_0}{\Lambda} \tag{4.9.25}$$

where Λ is the optical wavelength in free space. The parameter M is defined as the *acoustic-optic figure of merit* with

$$M = \frac{p^2 n_0^6}{\rho_{m0} V_a^3} \quad (4.9.26)$$

A table of the parameter M for various materials is given in Appendix B.

For most materials, M is less than the value for water, but for a few exceptional and technically very important materials like tellurium dioxide, M is larger than the value for water.

Example: Calculation of $v(w)$ for Water

For water, $p = 0.32$, $n_0 = 1.33$, $Z_0 = 1.5 \times 10^6$ kg/m²–s, $V_A = 1.5$ km/s, and $P = 1.69 \times 10^{12} AS_0^2$ watts. With $A = 10^{-5}$ m² and $P = 1$ W, it follows that $S_0 = 2.43 \times 10^{-4}$. Thus for $\Lambda = 0.63$ µm and $w = 3 \times 10^{-3}$ m, it follows that $\alpha = 1.38 \times 10^{-4}$, $v(w) = 2.75$.

Returning now to the calculation of phase, it follows that at a position z in the medium, the field $E(x, z)$ is of the form

$$E(x, z) = E_0 e^{j[\omega t - k_0 z - (\alpha k_0 z/2)\cos(\omega_a t - k_a x)]} \quad (4.9.27)$$

where E_0 is the entering plane wave field at $z = 0$. This is an expression for a phase-modulated wave which can be written in the form

$$E(x, z) = E_0 e^{j(\omega t - k_0 z)} \sum_{-\infty}^{\infty} (-j)^n J_{|n|}(v) e^{jn(\omega_a t - k_a x)} \quad (4.9.28)$$

where $J_n(v)$ is an nth-order Bessel function of the first kind and $v = \alpha k_0 z/2$.

We observe that the nth component of the wave has an effective propagation constant

$$\mathbf{k} = \mathbf{a}_z k_0 + \mathbf{a}_x n k_a \quad (4.9.29)$$

where $\mathbf{a}_x$ and $\mathbf{a}_z$ are unit vectors in the x and z directions, respectively. These waves propagate at angles $\pm n\Delta\theta$ to the axis, where $\Delta\theta \approx 2\theta_B$ and θ_B is the Bragg angle, which is defined as

$$\sin \theta_B = \frac{k_a}{2k_0} \quad (4.9.30)$$

Here we have assumed that $\Delta\theta$ is very small.

After a distance w, the intensity of the nth order scattered waves is

$$I_n = I_0 J_n^2(v) \quad (4.9.31)$$

where I_0 is the intensity of the entering optical beam and $v = \alpha k_0 w/2$. The nth-order diffracted beam propagates at an angle $n\Delta\theta$ to the axis. A plot of $[J_n(x)]^2$ for various orders of n is given in Fig. 4.9.2.

From Snell's law, we call the diffraction angle in air

$$\Delta\theta(\text{air}) = n_0 \Delta\theta \ (\text{medium}) \quad (4.9.32)$$

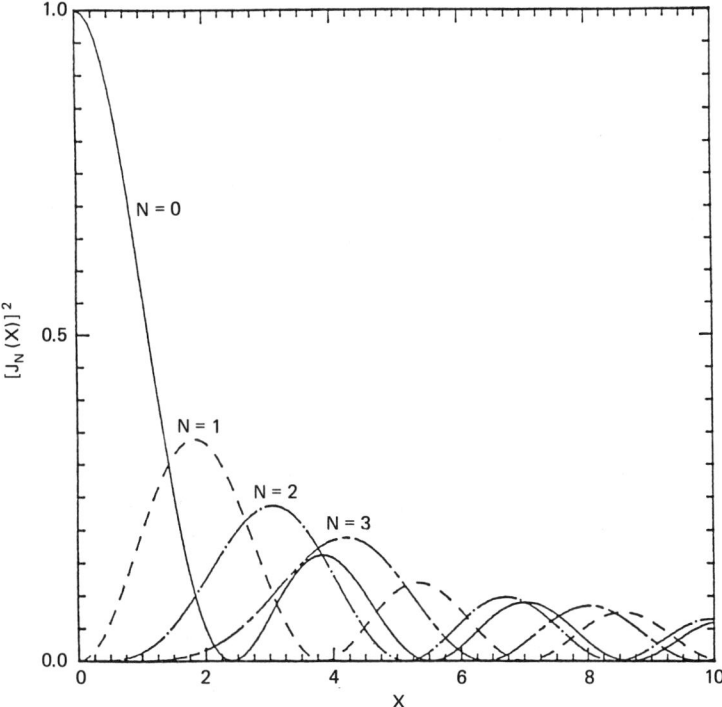

Figure 4.9.2 Plot of the function $[J_n(x)]^2$ for $n = 0, 1, 2,$ or 3 as a function of x.

where n_0 is the refractive index of the medium through which the acoustic wave is passing.

We expect to see a series of diffraction spots, the *Raman–Nath diffraction spots*, at angles $\theta_n = n\Delta\theta$ (air). The rays at these angles are Doppler-shifted to frequencies

$$\omega = \omega_0 + n\omega_a \qquad (4.9.33)$$

If the parameter v is such that $v \ll 1$, the first-order diffraction spot has an amplitude proportional to v, and the nth-order spot has an amplitude proportional to v^n. If v is larger, the amplitude of the zeroth-order spot decreases to become 0 when $v = 2.405$, while the amplitude of the first-order spot becomes maximum when $v = 1.84$. When v is sufficiently large, we see many diffraction spots, sometimes 20 or more.

We can observe the individual diffraction spots separately only if the beam diameter is somewhat smaller than the separation between spots. Suppose that the beam passes through a hole in a diaphragm of radius a. We can then show, from Fraunhofer diffraction theory (Sec. 3.2), that if the distance from the source is $z \gg a^2/\Lambda$, as illustrated in Fig. 4.9.3, the angle at which the beam amplitude becomes zero is

$$\theta_E(0) = \frac{0.61\Lambda}{a} \qquad (4.9.34)$$

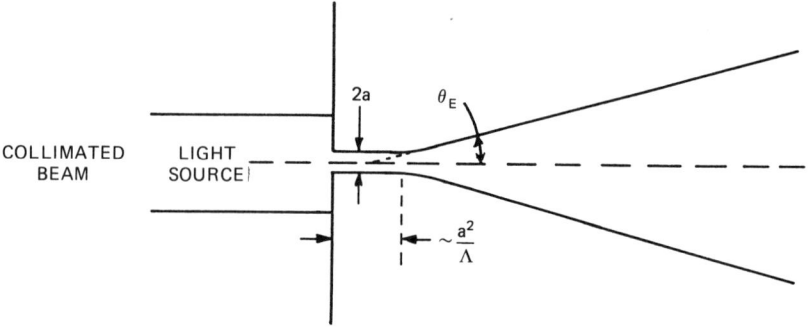

Figure 4.9.3 Collimated beam light source passed through a hole of radius a.

Each acoustically diffracted beam will behave the same way. If we place the maximum amplitude of the diffracted spot at the zero amplitude point of the neighboring spot, each spot should be distinguishable. This condition is the same as the Rayleigh criterion for the definition of a lens, where we require $\Delta\theta > \theta_E(0)$, which is equivalent to saying

$$\frac{k_a}{K} > \frac{0.61\Lambda}{a} \qquad (4.9.35)$$

or

$$\frac{a}{\lambda_a} > 0.61 \qquad (4.9.36)$$

where $K = 2\pi/\Lambda$. Such a condition is valid only at distances z from the diaphragm, where $z \gg a^2/\Lambda$, or the *Fresnel length*.

Suppose that $a = 0.5$ mm and $\Lambda = 6300$ Å. In this case, $a^2/\Lambda = 40$ cm. The Fresnel length varies as a^2, so the condition of Eq. (4.9.35) may be difficult to realize unless we use a lens to refocus the beam. In this case the condition is satisfied at the image plane of the lens.

On the other hand, if we observe the collimated beam at some plane $z < a^2/\Lambda$, the beam radius is approximately a and we require that

$$\frac{k_a}{K} = \Delta\theta \text{ (air)} > \frac{2a}{z} \qquad (4.9.37)$$

or

$$a < \frac{\Lambda z}{2\lambda_a} \qquad (4.9.38)$$

Limitations of Raman–Nath theory. We have assumed that $k_z \approx k_0$ throughout the analysis. The nth diffracted ray is diffracted at an angle $2n\theta_B$ so its phase change in the z direction becomes approximately $k_0 w \cos 2n\theta_B$. The

Raman–Nath theory neglects the difference between this phase shift and $k_0 w$. Therefore, this theory is valid only if

$$k_0 w (1 - \cos 2n\theta_B) \ll \frac{\pi}{4} \tag{4.9.39}$$

For $n\theta_B$ small, this relation becomes

$$2n^2 k_0 w \theta_B^2 \ll \frac{\pi}{4} \tag{4.9.40}$$

where

$$\theta_B \approx \frac{k_a}{2k_0} \approx \frac{\omega_a c}{2 V_a \omega_0} \tag{4.9.41}$$

and c is the velocity of light in the photoelastic medium. If $Q = 4 k_0 w \theta_B^2$, it follows that the criterion for Raman–Nath diffraction of the first diffraction spot is

$$Q \ll \frac{\pi}{2} \tag{4.9.42}$$

Example: Raman–Nath Criterion for Water

At $f = 5$ MHz in water, refractive index $n_0 = 1.33$, acoustic velocity $V_a = 1.5$ km/s, we define the maximum value of w for $Q = \pi/2$ as $w_{\max}$. With an optical wavelength $\Lambda = 0.63$ μm and $n = 1$, then $w_{\max} = 4.8$ cm. For $n = 2$, $w_{\max}$ decreases to 1.2 cm. For $f_a = 50$ MHz and $n = 1$, $w_{\max} = 0.5$ mm. For $f_a = 500$ MHz and $n = 1$, $w_{\max} = 5$ μm.

We can also see that at $f = 5$ MHz, with $\lambda_a = 0.3$ mm and the wavelength in air $\Lambda = 0.63$ μm, the Bragg angle is $\theta_B(\text{air}) = 0.06°$. At 500 MHz, $\theta_B(\text{air}) = 6°$. Thus, for low acoustic frequencies, θ_B is very small.

4.9.4 Schlieren Effect

The *Schlieren effect* is an important application of Raman–Nath diffraction used in measuring acoustic beam profiles. Consider what occurs when a wide optical beam of width L is passed through a rectangular acoustic beam of width w whose profile is nonuniform in the x and y directions, as illustrated in Fig. 4.9.4. It follows from Eqs. (4.9.28) and (4.9.31) that the acousto-optic interaction yields two first-order diffracted beams, each with an amplitude

$$\begin{aligned}|E(y, z)| &= |E_0 J_1[k_0 w \alpha(y, z)]| \\ &\approx \tfrac{1}{2} |E_0 k_0 w \alpha(y, z)| \quad (k_0 \alpha w \ll 1)\end{aligned} \tag{4.9.43}$$

We suppose that after the optical beam passes through the acoustic beam, it enters a convex lens. The undiffracted component of the beam passes through the focal point of the lens. We can remove this component with a stop consisting of a small dot on a glass plate, or with a knife edge, which removes not only the undiffracted beam but one of the first-order diffracted beams as well. It is sometimes convenient to keep both first-order diffracted components to maximize the available light; in this case we use the dot instead of the knife edge.

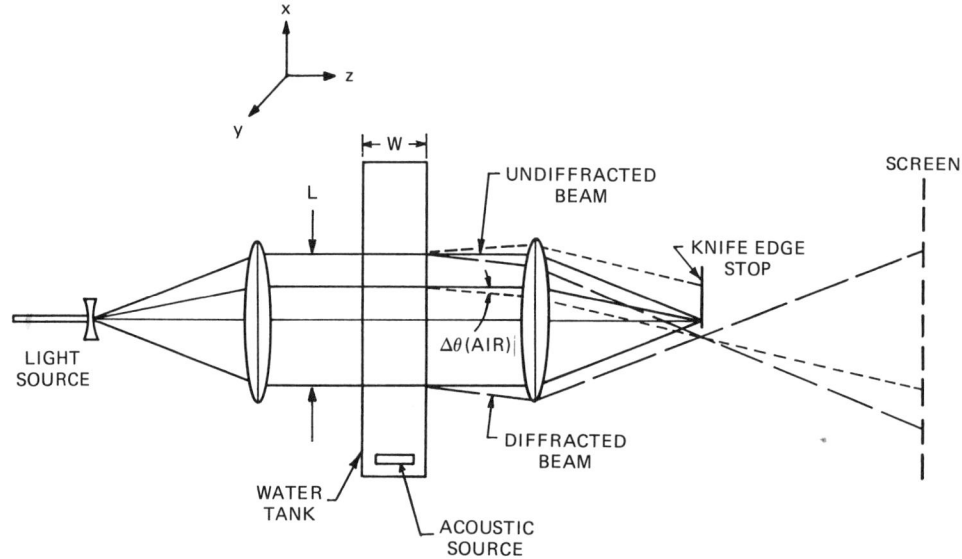

Figure 4.9.4 Schlieren optical system. The first-order diffracted rays are shown as dashed lines. Not all the *i*-order rays are shown; they may be eliminated by using an opaque dot or a knife edge as a stop.

We then display the diffracted beam on a distant screen. Each point on the screen x_I, y_I has a corresponding point in the acoustic beam x, y, where

$$x_I = x \frac{z - F}{F} \qquad (4.9.44)$$

$$y_I = y \frac{z - F}{F} \qquad (4.9.45)$$

F is the focal length of the lens and z is the distance of the screen from the focal plane of the lens.

This diffracted beam has each ray displayed by an angle $\Delta\theta(\text{air})$ from the corresponding undiffracted ray. If $\Delta\theta(\text{air})$ is small, we can display both diffracted rays without significantly blurring the intensity image of the acoustic beam. But if we want a highly defined image, we use a knife edge stop with only one diffracted beam, which yields an image corresponding to the intensity variation $\alpha^2(x, z)$ of the acoustic beam. An image of a square acoustic beam taken with a Schlieren optical system is shown in Fig. 4.9.5(a) and compared to a theoretical diffraction calculation for the same beam in Fig. 4.9.5(b).

If we use a high-intensity laser, an argon laser beam for instance, and pulse both the optical and acoustic source, we can observe the way in which an acoustic pulse propagates through a medium and even how longitudinal, shear wave, and Rayleigh wave modes are generated as a wave is reflected from a solid surface. (Other examples of the use of such a Schlieren optical system are shown in Figs. 2.5.8 and 4.9.15.)

(a)
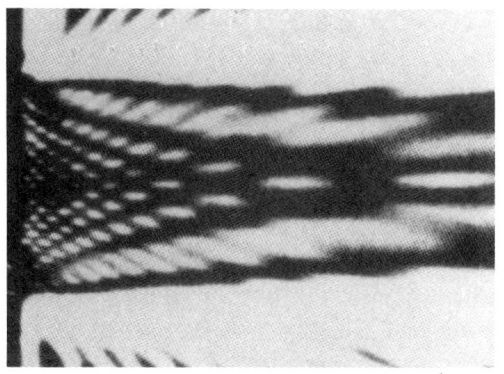

(b)
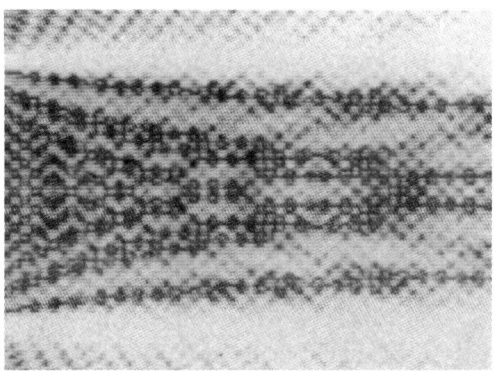

Figure 4.9.5 (a) Schlieren image of the near field of a square transducer with a ratio of width to acoustic wavelength of 14/37. (After Osterhammel, as noted by Cook et al. [100, 101].) (b) Schlieren image numerically calculated by Cook for the same transducer. (After Cook et al. [101].)

4.9.5 Bragg Diffraction by Sound

We now consider *Bragg diffraction*, a condition that occurs when the width of the acoustic beam is large, $Q \gg \pi/2$, and there are cumulative effects as the incident optical beam is injected at an angle θ_B to the wavefront of the acoustic beam. In this case, the wave is diffracted at an angle $-\theta_B$ to the acoustic wavefront and there is only one strong diffracted beam. The mathematical formalism we shall use is a rigorous one, suitable for dealing with Raman–Nath as well as Bragg diffraction.

We consider the configuration shown in Fig. 4.9.6, in which a sound wave passes in the x direction through an isotropic solid or liquid. Suppose that a light wave is incident on the material at an angle θ_i to the z direction, and diffracted by the sound wave at an angle $-\theta_R$ to the z direction, as shown in this example. We assume that the sound wave has frequency ω_a and propagation constant k_a, and we take the vector $\mathbf{k}_a$ to be in the x direction. If we want to see continuous and cumulative interaction over a finite length of the crystal, resulting from nonlinear interaction with the sound wave, all contributions to the diffracted wave must be added in phase. This implies that

$$\omega_0 \pm \omega_a = \omega_1 \qquad (4.9.46)$$

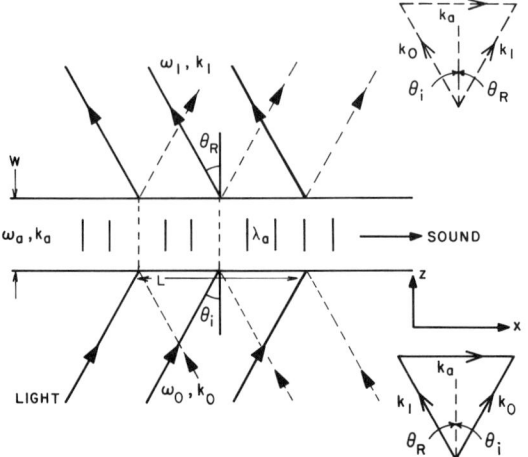

Figure 4.9.6 Bragg diffraction by sound. The solid-line vector diagram shows θ_i positive and $\mathbf{k}_a$ in the $+x$ direction; therefore, $\mathbf{k}_1 = \mathbf{k}_0 - \mathbf{k}_a$ and a diffracted optical beam travels in the $-z$ direction. The dashed lines illustrate Bragg diffraction with θ_i negative and the entering optical beam traveling in the $-x$ direction; $\mathbf{k}_1 = \mathbf{k}_0 + \mathbf{k}_a$ and a diffracted optical beam travels in the $+x$ direction. In the first case the frequency of the diffracted beam is downshifted; in the latter it is upshifted.

and
$$\mathbf{k}_0 \pm \mathbf{k}_a = \mathbf{k}_1 \quad (4.9.47)$$

where all parameters and angles are defined in the material where the interactions take place.

If we take the negative sign and use the angle convention given in the figure, it follows that

$$\omega_0 - \omega_a = \omega_1 \quad (4.9.48)$$

$$k_0 \sin \theta_i - k_a = -k_1 \sin \theta_R \quad (4.9.49)$$

and

$$k_0 \cos \theta_i = k_1 \cos \theta_R \quad (4.9.50)$$

When we consider diffraction by a sound wave in the UHF range, where $\omega_a \ll \omega_0$ or ω_1, we also see that $k_1 \approx k_0$, at least in an isotropic material, in which all light waves of the same frequency have the same propagation constant.† It follows that as $k_1 \approx k_0$, Eqs. (4.9.49) and (4.9.50) yield the results

$$\theta_i = \theta_R \quad (4.9.51)$$

and

$$\sin \theta_i = \sin \theta_R = \frac{k_a}{2k_0} = \frac{\lambda_0}{2\lambda_a} = \sin \theta_B \quad (4.9.52)$$

respectively. The latter condition is the familiar Bragg scattering relation, with the spacing d of the crystal lattice spacing replaced by the acoustic wavelength λ_a.

Note that because λ_0, the light-wave wavelength is typically much smaller than the acoustic wavelength λ_a and the diffraction angles are normally very small.

†The reader is referred to Appendix J for a discussion of Bragg diffraction in anisotropic materials.

We can also consider the solution using positive signs in Eqs. (4.9.46) and (4.9.47). In this case the frequency ω_1 is the upper, rather than the lower, sideband. Now $|\theta_i| = |\theta_R|$ as before, but the incident and diffracted rays, shown by the dashed lines, have positions on the other side of the z axis for those given in the first case.

Diffraction theory. We can now estimate how the amplitude of the diffracted wave increases with distance. The wave equation for the optical waves has the approximate form

$$\frac{\partial^2 \phi}{\partial x^2} + \frac{\partial^2 \phi}{\partial z^2} - \frac{\partial^2 \phi}{\partial t^2} \mu\varepsilon[1 + \alpha \cos(\omega_a t - k_a x)] = 0 \quad (4.9.53)$$

where ϕ is any component of the E field.

We look for a solution of the form

$$\phi = e^{j(\omega_0 t - k_0 z \cos\theta_i - k_0 x \sin\theta_i)} \sum_{n=-\infty}^{\infty} e^{jn(\omega_a t - k_a x)} \phi_n(z) \quad (4.9.54)$$

We assume ϕ_n to be a relatively slowly varying function. We write

$$\frac{\partial^2}{\partial z^2}\left(\phi_n(z)e^{-jkz}\right) \approx -\left(k^2\phi_n + 2jk\frac{\partial \phi_n}{\partial z} + \frac{\partial^2 \phi_n}{\partial z^2}\right)e^{-jkz} \quad (4.9.55)$$

Since ϕ_n is a slowly varying function, we neglect the last term, $\partial^2 \phi_n/\partial z^2$, in the parentheses on the right-hand side of Eq. (4.9.55) compared to $2jk(\partial \phi_n/\partial z)$.

We now consider the term that varies as

$$e^{j[(\omega_0 + n\omega_a)t - (k_0 \sin\theta_i + nk_a)x - k_0 z \cos\theta_i]}$$

We substitute Eqs. (4.9.54) and (4.9.55) into Eq. (4.9.53) to obtain, after some algebra, the following result:

$$\frac{\partial \phi_n}{\partial z} - j\beta_n \phi_n = -\frac{j\alpha k_0}{4\cos\theta_i}(\phi_{n+1} + \phi_{n-1}) \quad (4.9.56)$$

where

$$\beta_n = \frac{2nk_0 k_a \sin\theta_i + n^2 k_a^2}{2k_0 \cos\theta_i} \quad (4.9.57)$$

and we have assumed that $|n\omega_a| \ll \omega_0$.

Raman–Nath interaction: $Q \ll \pi/2$. Here we put $\theta_i = 0$ and assume that $wn^2 k_a^2/(2k_0 \cos\theta_i) \to 0$, or $Q \to 0$. Thus we can also assume that $\beta_n = 0$. Now consider the recurrence relation for the Bessel function:

$$\frac{\partial J_n}{\partial z} = -\tfrac{1}{2}[J_{n+1}(z) - J_{n-1}(z)] \quad (4.9.58)$$

and compare it to the simplified form of Eq. (4.9.56) with $\beta_n = 0$:

$$\frac{\partial \phi_n}{\partial z} = -\frac{j\alpha k_0}{4\cos\theta_i}(\phi_{n+1} + \phi_{n-1}) \quad (4.9.59)$$

If we write $\phi_n = (-j)^n \phi'_n$, and set $\theta_i = 0$, we find that

$$\frac{\partial \phi'_n}{\partial z} = \frac{-\alpha k_0}{4} (\phi'_{n+1} - \phi'_{n-1}) \tag{4.9.60}$$

Thus the solution to Eq. (4.9.60), if $\phi_0(0) = 1$, is

$$\phi'_n = J_n\left(\frac{k_0 \alpha z}{2}\right) = J_n\left[\frac{\pi z}{\Lambda}\left(\frac{PM}{2A}\right)^{1/2}\right] \tag{4.9.61}$$

This leads us to the Raman–Nath solution described by Eq. (4.9.28) and (4.9.31), with the same limitation, that $Q \ll \pi/2$.

Bragg regime: $Q \gg \pi/2$. We now assume that the only waves of importance are the $n = 0$, and either the $n = 1$ or $n = -1$ components. When only the $n = 0$ and $n = 1$ terms matter, Eq. (4.9.56) leads to the results

$$\frac{\partial \phi_1}{\partial z} - j\beta_1 \phi_1 = -\frac{j\alpha k_0}{4 \cos \theta_i} \phi_0 \tag{4.9.62}$$

and

$$\frac{\partial \phi_0}{\partial z} = -\frac{j\alpha k_0}{4 \cos \theta_i} \phi_1 \tag{4.9.63}$$

If $\beta_1 = 0$, it follows from Eq. (4.9.57) that

$$\sin \theta_i = -\frac{k_a}{2k_0} = -\sin \theta_B \tag{4.9.64}$$

The similar condition for $n = -1$ is exactly the same as the Bragg condition of Eq. (4.9.52), while Eq. (4.9.64) simply corresponds to the Bragg condition when θ_i is negative.

When the Bragg condition is satisfied, the interaction is cumulative. When α is constant, ϕ_1 initially increases linearly with distance. Equations (4.9.62) and (4.9.63) take the form of two simple coupled-mode equations. For α constant, Eqs. (4.9.62) and (4.9.63) give the solution

$$\phi_1(z) = -j\phi_0(0) \sin\left(\frac{\alpha k_0 z}{4 \cos \theta_i}\right) \tag{4.9.65}$$

So the ratio of the diffracted beam intensity I_1 to the incident beam intensity I_0 is

$$\frac{I_1}{I_0} = \sin^2\left(\frac{\pi z}{\Lambda} \sqrt{\frac{PM}{2A}}\right) \tag{4.9.66}$$

We observe that the intensity of the diffracted beam initially increases as the square of the distance, reaches a maximum value, and then decreases.

Response as a function of acoustic beam profile. In the general case when the Bragg condition is not exactly satisfied and we assume α to be nonuniform, we can obtain the solution to Eqs. (4.9.62) and (4.9.63). For simplicity, we shall assume from now on that the interaction length is short enough so that $\phi_0(z)$ is constant. We can then solve Eq. (4.9.62) by writing it in the form

$$e^{-j\beta_1 z}\frac{\partial}{\partial z}(\phi_1 e^{j\beta_1 z}) = -\frac{j\alpha k_0}{4\cos\theta_i}\phi_0 \quad (4.9.67)$$

with the result

$$\phi_1 = -\frac{jk_0 e^{-j\beta_1 z}\phi_0}{4\cos\theta_i}\int_0^z \alpha(z)e^{-j\beta_1 z}\,dz \quad (4.9.68)$$

It follows that, in this case, ϕ_1 is the Fourier transform of the acoustic beam profile.

Response as a function of input angle θ_i. When β_1 is finite and $\theta_1 \neq -\theta_B$, then it is convenient to write, for θ_B small,

$$\theta_i = -\theta_B + \Delta\theta \quad (4.9.69)$$

It follows from Eqs. (4.9.57) and (4.9.69) that $\beta_1 \approx k_a\,\Delta\theta$ and

$$\phi_1(\Delta\theta) = -\frac{jk_0 e^{-jk_a z \Delta\theta}\phi_0}{4\cos\theta_B}\int_0^z \alpha(z)e^{-jk_a z \Delta\theta}dz \quad (4.9.70)$$

If we rotate the crystal through an angle $\Delta\theta$ from the correct incident angle θ_B, Eq. (4.9.70) implies that the Fourier transform of the acoustic beam profile will be $\phi_1(\Delta\theta)$.

Response as a function of input frequency. We can find a similar relation for the response as a function of input frequency. From Eq. (4.9.54) we see that if the $n = 0$ term has a constant amplitude, the $n = 1$ diffracted wave varies as

$$\phi(x, z) = \phi_1 e^{j(\omega_0 + \omega_a)t}e^{-jk_0 z\cos\theta_i}e^{-j(k_0\sin\theta_i + k_a)x} \quad (4.9.71)$$

If θ_i is kept constant and ω_a is varied, the output angle is θ_R, where the propagation constant of the diffracted beam in the x direction is $k_x = k_0 \sin\theta_R$, and

$$\sin\theta_R = \frac{k_a + k_0\sin\theta_i}{k_0} \quad (4.9.72)$$

$$= \sin\theta_i + \frac{k_a}{k_0}$$

If we take the center frequency to be ω_{a0}, corresponding to a value of k_{a0}, where

$$-\sin\theta_{i0} = \frac{k_{a0}}{2k_0} = \sin\theta_{R0} = \sin\theta_B \quad (4.9.73)$$

and we assume that θ_R and θ_i are small, we find that

$$\theta_R \approx \frac{k_a - k_{a0}}{k_0} + \theta_{R0} \quad (4.9.74)$$

The angle of the emerging beam therefore varies linearly with the frequency of the acoustic wave. For general values of θ_R and θ_{R0}, the x position of the diffracted beam spot at the plane z is

$$x = z \sin \theta_R = z \sin \theta_{R0} + \frac{k_a - k_{a0}}{k_{a0}} z \quad (4.9.75)$$

Therefore, the x position of the diffracted spot varies linearly with frequency, and the amplitude variation of the diffracted beam with x position is the Fourier transform of the time response of the input acoustic signal.

When we vary the frequency, it follows from Eq. (4.9.57) that as $\beta_1 = 0$ when $k_a = k_{a0}$, then to first order in $(k_a - k_{a0})$,

$$\beta_1 = -(k_a - k_{a0}) \tan \theta_{i0} \quad (4.9.76)$$

If we substitute Eqs. (4.9.73) and (4.9.74) in Eq. (4.9.76), we see that for θ_{R0} and $|\theta_R - \theta_{R0}|$ small,

$$\beta_1 = \frac{(\theta_R - \theta_{R0})k_{a0}}{2} \quad (4.9.77)$$

We can then show from Eq. (4.9.68) that for a beam of width w, the variation of the diffracted beam amplitude with β_1 is

$$\left|\frac{\phi_1}{\phi_{10}}\right| = \left|\frac{\sin \beta_1 w/2}{\beta_1 w/2}\right| \quad (4.9.78)$$

where ϕ_{10} is the amplitude of the output beam when $\theta_i = \theta_{i0}$ and $\omega_a = \omega_{a0}$.

The angular range of the output beam to the 3-dB points is

$$\Delta\theta_R(3 \text{ dB}) = \frac{0.89\lambda_{a0}}{w} \quad (4.9.79)$$

Similarly, if we observe the output beam at a constant angle θ_{R0}, the bandwidth B between 3-dB points is

$$B = 0.89 \frac{V_a}{w \tan \theta_{i0}} \quad (4.9.80)$$

If the length of the acoustic beam is L, however, we can show that the diffracted optical beam has an angular extent to the 3-dB points of

$$\Delta\theta_E(3 \text{ dB}) = \frac{0.89\lambda_0}{L} \quad (4.9.81)$$

It follows that the number of resolvable spots N in the beam is

$$N = \frac{\Delta\theta_R}{\Delta\theta_E} = \frac{L}{w} \frac{\lambda_{a0}}{\lambda_0} \quad (4.9.82)$$

or

$$N = 1.12 BT \quad (4.9.83)$$

where $T = L/V_a$ is the time delay of the acoustic beam in a length L.

The basic physical principles governing each of the transversal filters we have examined in this chapter lead us to the same conclusion: *The number of resolvable points in any signal processing system is approximately equal to the time–bandwidth product of the device.*

4.9.6 Application of Bragg Deflection Systems

The first demonstrations of Bragg interaction used bulk waves and gave only limited angular and frequency bandwidths, as Eqs. (4.9.80) and (4.9.81) predicted. To circumvent this limitation in practical systems, several separate transducers, each with a different center frequency, are used. The individual transducers are placed at the optimum angle to the incident optical beam for efficient acousto-optic interaction. By this means, we can divide the total bandwidth into several frequency ranges to obtain a flat confined response over bandwidths of several hundred megahertz.

Surface acoustic waves are often used to deflect the optical beam. This makes it possible to employ a guided optical wave in place of the large optical components, attached to a stable table, that are usually required in a system using bulk acoustic waves. A layer of material is typically deposited on an SAW substrate, as illustrated in Fig. 4.9.7, and this waveguide layer is then arranged to propagate light waves with lower velocity than the substrate. As we saw earlier in our discussions of acoustic waveguides (Chapter 2 and Sec. 4.7), this implies that the total internal reflection of the light occurs at the top and bottom edges of the layer. Thus this layer acts as a waveguide, one that can consist, for instance, of a zinc oxide layer, like those used in the SAW devices described in Secs. 4.7 and 4.8. We can also form a high-quality, low-loss guide by diffusing titanium into lithium niobate.

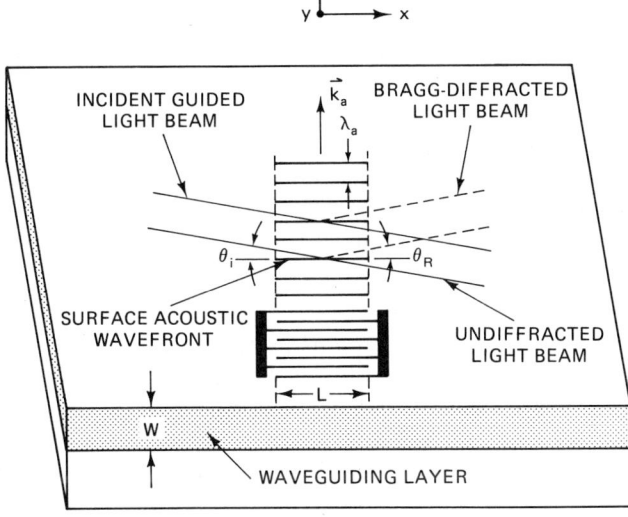

Figure 4.9.7 Guided-wave Bragg diffraction from single-surface acoustic wave. (After Tsai [102].)

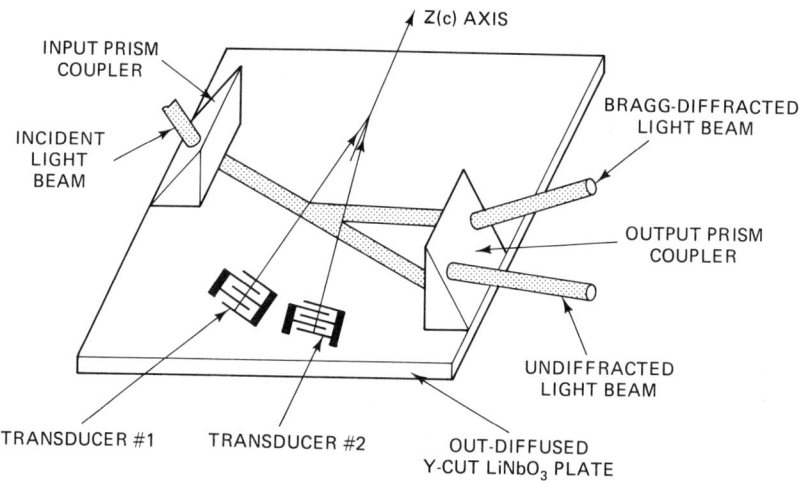

Figure 4.9.8 Experimental configuration for guided-wave acousto-optic Bragg deflection. (After Tsai [102].)

We introduce the optical beam into the guide by means of a *prism coupler*, illustrated in Fig. 4.9.8, a device operated on the same principle as the wedge coupler used with surface acoustic waves, which was described in Sec. 2.5.2. The prism is normally constructed of a material with a larger refractive index than the waveguide material. Thus if the propagation constant of a wave in the prism is k_p and the angle of the prism is θ, the effective propagation constant along the surface of the guide is k_p sin θ. If k_0 is the propagation constant of the guided wave and

$$k_p \sin \theta = k_0 \tag{4.9.84}$$

the two propagation constants will match and we can excite a waveguide mode. Using this method, optimum optical conversion efficiencies of the order of ~ 80% have been observed.

The Bragg deflector can be used for broadband operation if several SAW transducers, tilted with respect to each other, are employed, as illustrated in Fig. 4.9.9. Using three transducers in this arrangement, shown in Fig. 4.9.10, Tsai obtained a bandwidth of 385 MHz for a Bragg deflector.

Devices of this kind have been employed as wideband spectrum analyzers, operating in real time as acousto-optic modulators and acousto-optic phase shifters. Thus they provide a very useful component for optical waveguide systems.†

4.9.7 Acousto-Optic Convolver

An important type of signal processing device can be constructed using two acousto-optic deflectors in succession, as illustrated in Fig. 4.9.11. In such a device, known as the *acousto-optic convolver*, the optical beam enters at an angle θ_i to the normal

---

†An alternative technique which uses anisotropic materials to obtain broadband operation is discussed in Appendix J.

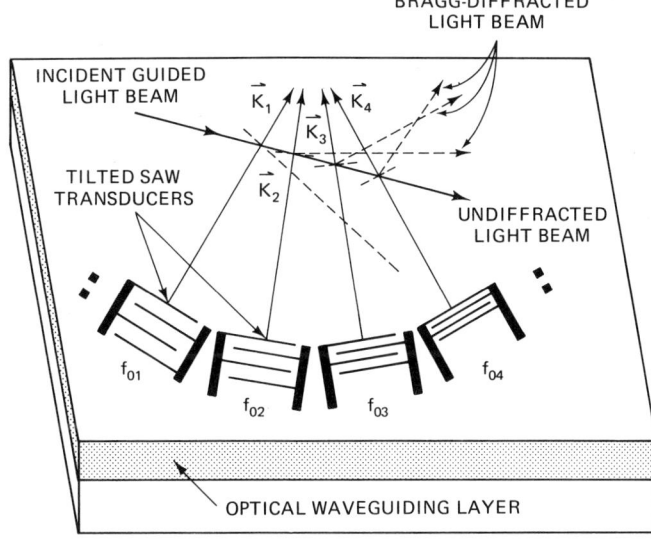

Figure 4.9.9 Guided-wave Bragg diffraction from multiple tilted surface acoustic waves. (After Tsai [102].)

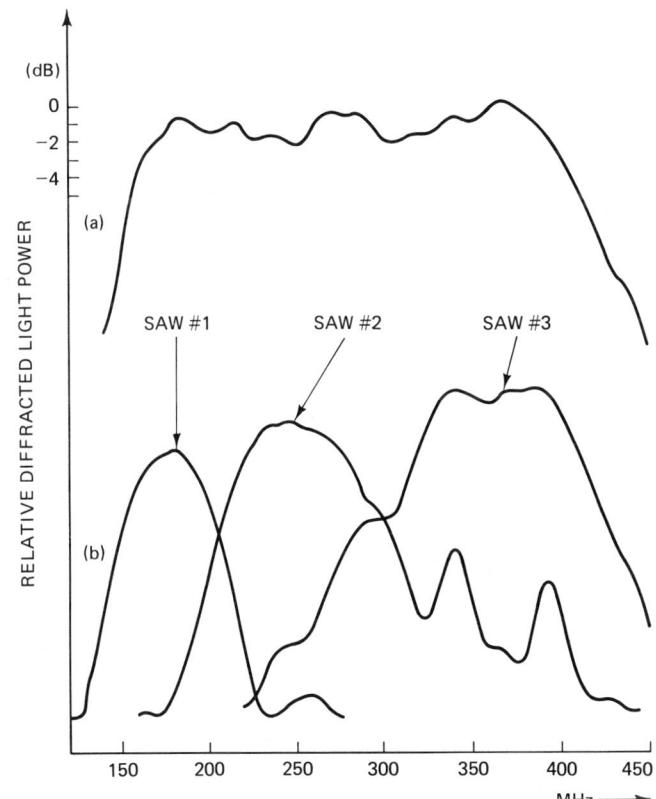

Figure 4.9.10 Frequency response of the Bragg diffracted light power from tilted surface acoustic waves (experimental): (a) three combined-surface acoustic waves; (b) individual-surface acoustic waves. (After Tsai [102].)

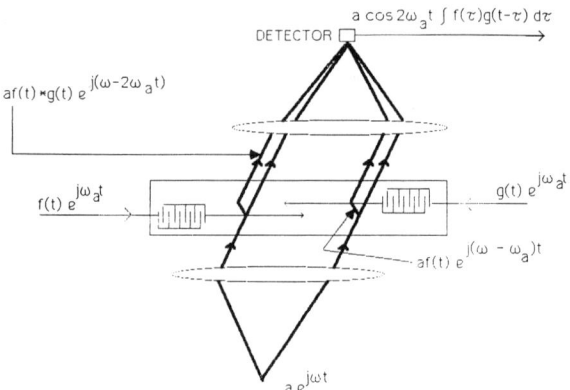

Figure 4.9.11 Waveform convolution using acousto-optic Bragg interaction. (From Lee et al. [103].)

acoustic beam axis. After diffraction by the first acoustic beam, the diffracted beam exits at an angle $\theta_R = -\theta_i$. This diffracted beam is diffracted once more by an acoustic beam traveling in the opposite direction and the rediffracted beam leaves the system at an angle θ_i. This beam, twice deflected by the acoustic waves, is modulated at a frequency $2\omega_a$. If we can separate this modulation component from the carrier and the component of frequency ω_a, we can detect the twice-deflected beam.

This system was first demonstrated by Lee et al. [103]. They employed a thin flat laser beam, focused by a cylindrical lens instead of an optical waveguide mode because the length of the acoustic beam (the aperture was 15 cm) and the lack of suitable lenses made it easier to produce a flat sheet beam from a 400-mW krypton laser ($\lambda = 0.657$ μm) than from a waveguide mode. A lens was used on the output to focus the diffracted beam to a small spot on a high-speed photodetector.

Suppose that the two acoustic wave signals entering the system in opposite directions are $f(t) \exp(j\omega_a t)$ and $g(t) \exp(j\omega_a t)$, respectively. The twice-diffracted beam is focused onto a photodetector. The signal received at the photodetector will be of the form

$$S(t) = e^{j(\omega + 2\omega_a t)} \int f\left(t - \frac{z}{v}\right) g\left(t + \frac{z}{v}\right) dz + \text{c.c.} \qquad (4.9.85)$$

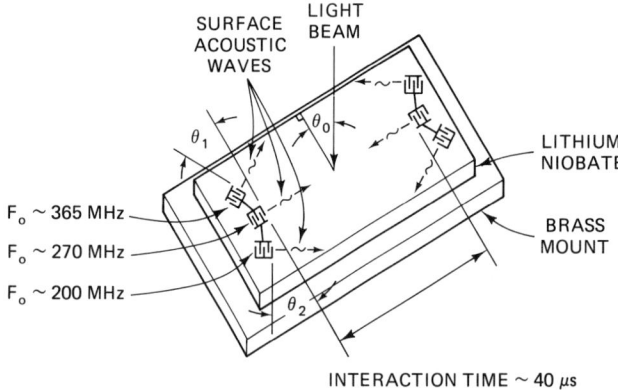

Figure 4.9.12 Large time-bandwidth, acousto-optic convolver. (After Berg and Udelson [104].)

where c.c. denotes the complex conjugate. In addition to this diffracted beam, there will be an undiffracted beam of frequency ω. When these two signals are mixed at the photodetector, the output at frequency $2\omega_a$ will be the convolution of the two input signals.

Another practical system, demonstrated by Berg and Udelson and illustrated in Fig. 4.9.12, employed a configuration with several input transducers of the type already described. Using three transducers in this manner, they obtained an input bandwidth of approximately 200 MHz and therefore a 400-MHz output bandwidth. Because the time delay in their device was of the order of 36 μs, we might expect it to be able to process signals with a time–bandwidth product of the order of 7000. Berg and Udelson obtained a dynamic range of 50 to 60 dB. They have used the system to carry out fast Fourier transforms and to process wide-bandwidth digital codes. Figures 4.9.13 and 4.9.14 show examples of phase-coded input and the corresponding outputs from the acousto-optic convolver. It is apparent that the acousto-optic convolver is a very powerful device for processing large-bandwidth signals.

4.9.8 Time-Integrating Correlator

Another interesting example of the many types of acousto-optic processors which have been developed is the time-integrating optical signal processor. This can make use of an incoherent optical source and, like the storage correlator, uses integration in time to obtain correlation between two signals. Because it employs optical signal processing, it is also possible to construct a two-dimensional correlator which can directly display real-time ambiguity functions.

We consider the system illustrated in Fig. 4.9.15. In this device, a laser diode or an LED is modulated by an input signal $f(t)$. A second signal $g(t)$ is inserted into an acousto-optic modulator and the output light beam is then passed through a Schlieren optical system of the type described in Sec. 4.9.4. A second lens is added to the system after the knife-edge stop and the output beam is imaged onto a CCD array. The output received at a point x on the array contains a term proportional to $f(t) g(t - x/V)$, where V is the effective velocity of a signal along the array. As we saw above, a Schlieren optical system basically images the amplitude of the acoustic wave at any point in the acoustic beam to a corresponding point on the image plane of the Schlieren optical system. By integrating the output of the photodetector as a function of time, we obtain the correlation function of the two input signals

$$R(x) = \int f(t) g\left(t - \frac{x}{V}\right) dt \qquad (4.9.86)$$

By reading out the image on the CCD array, we can obtain the correlation function directly.

The time–bandwidth product of the system, much as with the storage correlator, is dictated by the bandwidth of the input system and the integration time of the CCD array elements. The portion of the correlation function that can be read out is dictated by the delay time in the acousto-optic modulator. Thus the

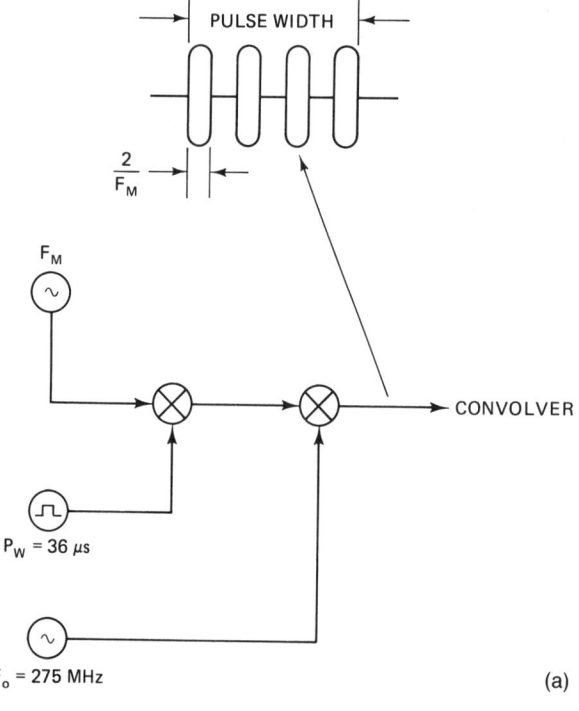

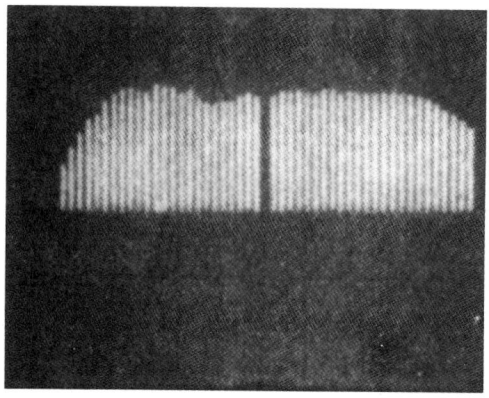

Figure 4.9.13 (a) Circuit generating phase-coded inputs for the acousto-optic convolver; (b) frequency-domain output from the convolver. (After Berg and Udelson [104].)

basic system has the same kinds of limitations on time bandwidth as the storage correlator, as well as a portion of the correlation function that can be read. The system itself has major advantages over many other kinds of acousto-optic correlators utilizing an optical reference to convert the acoustic phase modulation to amplitude modulation, because it does not require coherent optics with their attendant problems of speckle and noise.

We now analyze this system in more detail. The output of the photodetector is proportional to the image intensity $I(t, x)$, which can be written in the form

$$I(t, x) = I_1(t) I_2\left(t - \frac{x}{V}\right) \qquad (4.9.87)$$

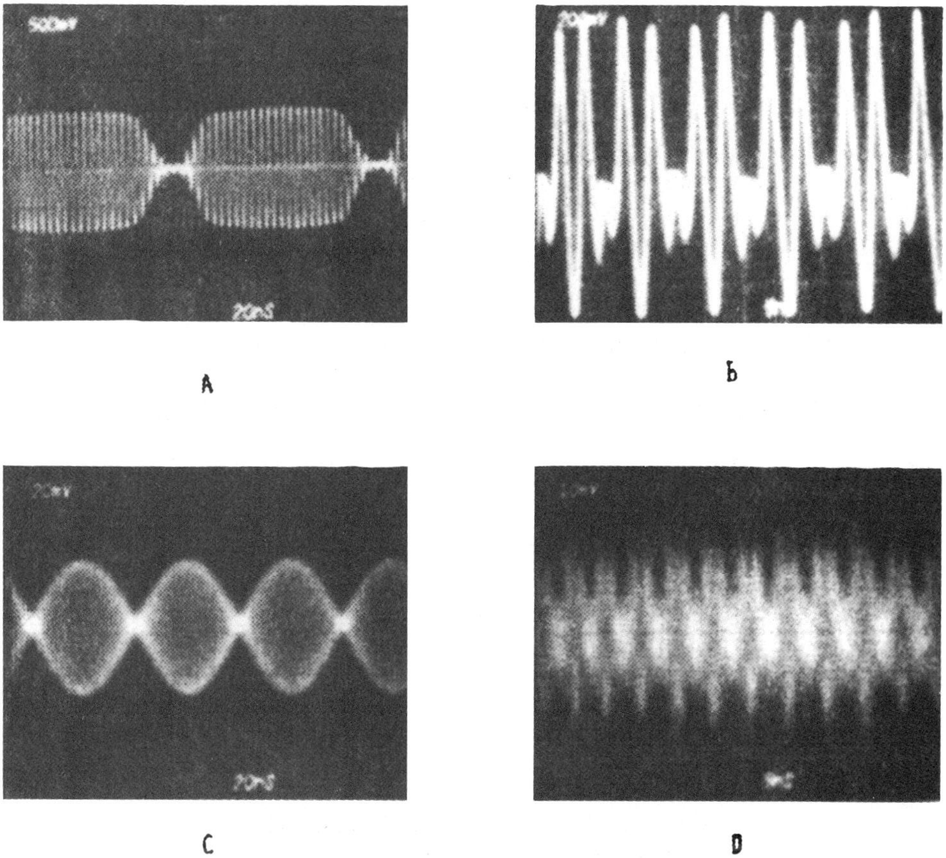

Figure 4.9.14 (a) Phase-coded inputs to the acousto-optic convolver; (b) corresponding time-domain outputs from the convolver. (After Berg and Udelson [104].)

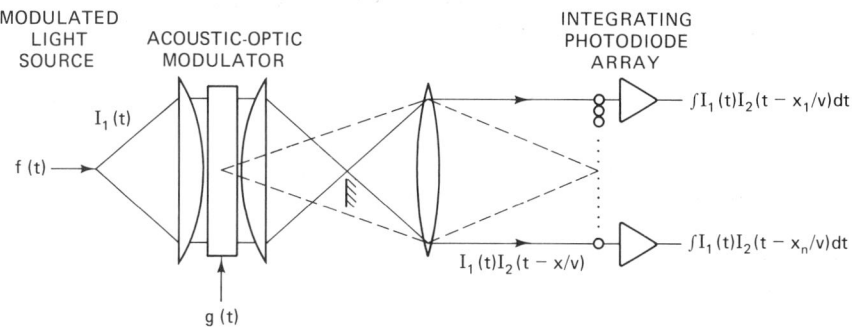

Figure 4.9.15 Time-integrating correlator. (From Kellman [105].)

Sec. 4.9 Acousto-Optic Filters

For double-sideband suppressed carrier modulation of the laser by a signal $f(t)$ cos $(2\pi f_0 t)$, it follows that

$$I_1(t) = A_1[1 + f(t) \cos(2\omega_0 t)] \quad (4.9.88)$$

where ω_0 is the carrier frequency of the signal and A_1 is a constant.

The complex electric field modulation due to the acousto-optic modulator is $E_2(t - x/V)$, where $I_2 = |E_2|^2$. For first-order diffraction,

$$E_2\left(t - \frac{x}{V}\right) = A_2^{1/2}\left[1 + g\left(t - \frac{x}{V}\right)e^{j\omega_0(t-x/V)}\right]e^{j\omega_c(t-x/V)} \quad (4.9.89)$$

where A_2 is a constant. In this case, the input acoustic signal is at a frequency $\omega_c \pm \omega_0$, but as the detector receives nothing but the first-order diffraction term, only the terms of frequencies ω_c and $\omega_c + \omega_0$ arrive at the detector. It follows that

$$I_2\left(t - \frac{x}{V}\right) = A_2\left[1 + g^2\left(t - \frac{x}{V}\right) + 2g\left(t - \frac{x}{V}\right)\cos\omega_0\left(t - \frac{x}{V}\right)\right] \quad (4.9.90)$$

If B is the bandwidth of the signals $f(t)$, $g(t)$, and $f_0 > 3B$, several cross-terms effectively integrate to zero and it can be shown that the total output from the detector is of the form

$$R\left(\frac{x}{V}\right) = TA_1A_2\left[1 + \frac{1}{T}\int_0^T g^2\left(t - \frac{x}{V}\right)dt \right.$$

$$\left. + \frac{1}{T}\cos\frac{\omega_0 x}{V}\int_0^T f(t)g\left(t - \frac{x}{V}\right)dt\right] \quad (4.9.91)$$

where T is the integration time of the system. The first two terms are essentially dc outputs; the last term is the correlation function we require. In practice, these unwanted terms can be eliminated by filtering.

One-dimensional transforms. One example of the use of this time-integrating system is for spectral analysis. In this case, the two input signals are FM chirps of $S_1(t) \cos(\omega_0 t + \mu t^2/2)$ and $\cos(\omega_0 t + \mu t^2/2)$, respectively. The input signal to the laser is a modulated chirp and the output is the Fourier transform of the modulation. The output is of the form

$$Y_{\text{out}}(x) = \text{Re}\left[e^{-j(\omega_0 x/V + \mu x^2/2V^2)}\int_0^T S_1(t)e^{-j\mu tx/V}dt\right] \quad (4.9.92)$$

Since the output occurs as a function of time $\tau = x/V$, we can multiply the output, as discussed in Sec. 4.7 by a chirp $\exp(j\mu\tau^2/2)$ to eliminate the chirp carrier and obtain phase information. The output with an input signal modulation $s_1(t) = \exp(j\Omega t)$ is

$$|Y_{\text{out}}(x)| = T\left|\frac{\sin[(\Omega - \mu x/V)T/2]}{(\Omega - \mu x/V)T/2}\right| \quad (4.9.93)$$

If the time delay in the acousto-optic delay line is $\tau_0 = L/V$, where L is the physical length of the array and there are N elements in the array, the minimum resolution is

$$\Delta f = \frac{\mu \tau_0}{2N\pi} \qquad (4.9.94)$$

The bandwidth of the chirp, however, is

$$B = \frac{\mu T}{2\pi} \qquad (4.9.95)$$

and the bandwidth of the chirp in a time τ_0 is

$$B' = \frac{\tau_0}{T} B \qquad (4.9.96)$$

Thus

$$\Delta f = \frac{B\tau_0}{NT} = \frac{B'}{N} \qquad (4.9.97)$$

The number of resolvable frequencies within the array is $B'/\Delta f = N$. It follows from Eq. (4.9.96) that the best possible resolution we could hope for is

$$\Delta f \approx \frac{1}{T} \qquad (4.9.98)$$

Thus for optimum resolution,

$$B\tau_0 = N \qquad (4.9.99)$$

which leads to an optimum choice of the chirp rate μ.

If we wish to examine a range of frequencies equal to the chirp bandwidth, we find, just as for the earlier cases examined in Sec. 4.7, that

$$B\tau_0 < \frac{N}{4} \qquad (4.9.100)$$

An example of a time-integrating spectrum analyzer output is shown in Fig. 4.9.16. In this example, $N = 1000$, $\tau_0 = 40$ μs, $B = 2.5$ MHz, $T = 2$ ms, and

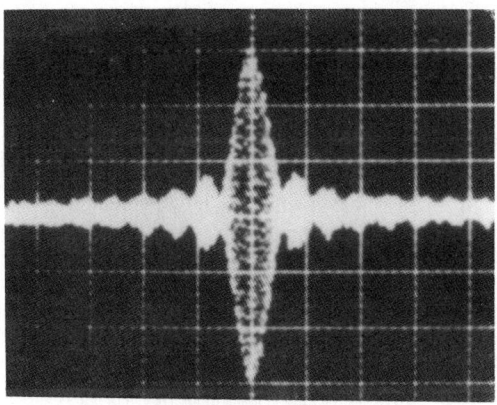

Figure 4.9.16 Time-integrating spectrum analyzer output. (After Kellman [105].)

$B\tau_0 = 100$. The possible resolution is 50 Hz. The dynamic range of the device was 40 dB.

Two-dimensional transforms. Two-dimensional transforms can also be carried out with another version of this system, as illustrated in Fig. 4.9.17. Two acousto-optic deflectors at right angles to each other are employed together with a modulated LED source. The output is now of the form

$$R(x, y) = \int f(t) g\left(t - \frac{x}{V}\right) h\left(t - \frac{y}{V}\right) dt \quad (4.9.101)$$

where the input signals are $f(t)$, $g(t)$, and $h(t)$, respectively.

Suppose that we wish to find the ambiguity function

$$\chi(\tau, f) = \int S_1(t) S_2^*(t - \tau) e^{-2\pi f t} dt \quad (4.9.102)$$

Then if we make the x dimension correspond to f and the y dimension to τ, we require the Fourier transform in the y direction of the function $S_1(t) S_2(t - \tau)$. To do this we must have complex input signals of the form

$$\begin{aligned} f(t) &= S_1(t) e^{j\mu t^2/2} \\ g(t) &= S_2^*(t) e^{-j\omega_2 t} \\ h(t) &= e^{j\mu t^2/2} e^{j\omega_1 t} \end{aligned} \quad (4.9.103)$$

The functions $f(t)$ and $g(t)$ without the $\exp(-j\mu t^2/2)$ term provide the required convolutions of $S_1(t)$ and $S_2(t)$. The combination of $h(t)$ with the $\exp(-j\mu t^2/2)$ term then provides the required Fourier transform. By substituting Eq. (4.9.103) in Eq. (4.9.101), it can be shown more directly that the output from the system is

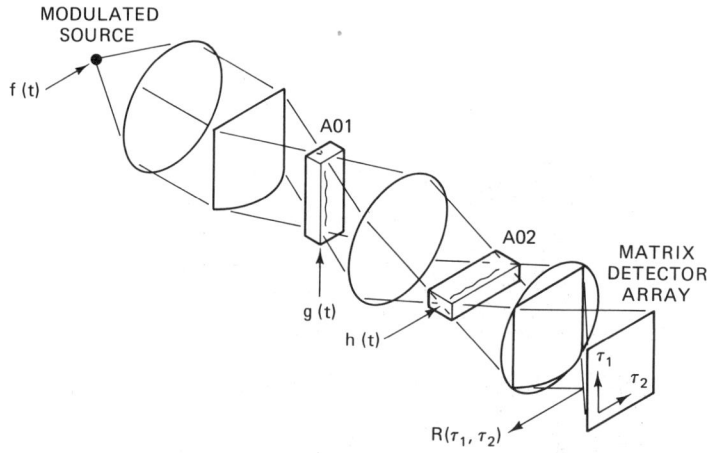

Figure 4.9.17 Two-dimensional time-integrating optical processor. (From Kellman [105].)

of the form

$$R(x, y) = \text{Re}\left[e^{-j(\omega_1 x/V - \omega_2 y/V)} e^{j\mu y^2/V^2} \int S_1(t) S_2^*\left(t - \frac{y}{V}\right) e^{j\mu t x/V} \, dt \right] \quad (4.9.104)$$

where the input signals to the two acousto-optic modulators have carrier frequencies of ω_1 and ω_2, respectively, and the integration time is the frame time of the system. The chirp itself must be of this length.

The modulation of the ambiguity function by the quadratic phase term can be canceled out through post-detection weighting. The variables x and y now correspond to Doppler shift and delay, respectively. The Doppler resolution is commensurate with the integration time and the time resolution is determined by the number of elements in the x dimension.

Kellman [105] has illustrated ambiguity function processing using a Hitachi HLP-20 light-emitting diode, a Fairchild SL62926 charge-coupled-device image sensor, and Isomet acousto-optic devices. The diode had a 30-MHz, 3-dB bandwidth and was biased at an average optical power of approximately 10 mW. The image sensor had 380 × 488 elements and was operated with an integration period of 33.34 ms. The acousto-optic devices had a 30-MHz, 1-dB bandwidth and a 50-μs delay. Chirp waveforms with a very large time–bandwidth product $BT = 524{,}288$ were synthesized digitally. The delay range was limited to 36 μs in the y direction and 27 μs in the x direction, to increase the bandwidth. The image plane sampling was therefore $380/36 \ \mu s^{-1} \approx 10$ MHz in one delay dimension and 18.1 MHz in the other.

A short pseudorandom code is shown in Fig. 4.9.18. The Doppler range was 1 to 3.8 kHz, determined by the chirp rate. The code-repetition period was 6.2 μs, so that four correlation peaks are evident in the ambiguity function. The Doppler shift in this example was 2 kHz. The time–bandwidth product was 5 MHz × 33.34 ms = 166,700. The sensor dynamic range was approximately 60 dB and the resultant input signal dynamic range was approximately 48 dB.

A similar system has been used for spectral analysis. This made it possible to obtain fine spectral information in one direction and coarse spectral information in the other. Thus the two-dimensional acousto-optic signal processor is an ex-

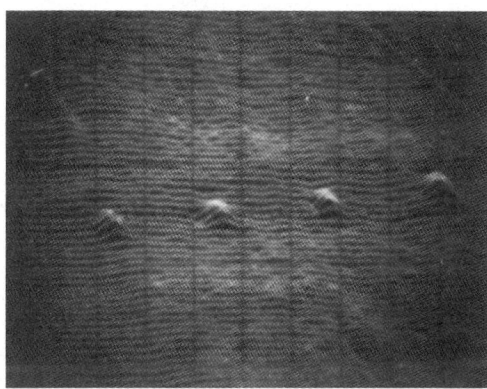

Figure 4.9.18 Ambiguity function for short code. (After Kellman [105].)

tremely powerful device and can provide a remarkable amount of real-time information. Demonstrations have been made of various other systems that can perform two-dimensional transforms. Examples of these are two-dimensional acoustic or acousto-optic convolvers. At the present time, the system described here seems to be the most practical.

PROBLEM SET 4.9

1. An isotropic material can be used for an acousto-optic deflector. Prove that interaction of a polarized optical beam with a longitudinal acoustic wave gives rise to an optical beam polarized in the same direction as the entering beam. Then prove that interaction of a polarized optical beam with an acoustic shear wave produces an optical beam polarized perpendicular to the polarization of the entering beam.

    Set up the problem by assuming that the perturbation of the dielectric constant due to the acoustic beam is very small, so that only first-order effects due to the acoustic wave need be considered. For a solid isotropic material, in a reduced coordinate system (see Sec. 1.5 and Appendix A), $p_{11} = p_{22} = p_{33}$, $p_{12} = p_{13} = p_{23} = p_{21} = p_{31} = p_{32}$, and $p_{44} = p_{55} = p_{66} = (p_{11} - p_{12})/2$; thus there are only two independent constants. For a liquid, in addition, $p_{11} = p_{12}$.

    We can write

    $$\varepsilon_0 E_i = B_{ij} D_j + \Delta B_{ij} D_j$$

    with

    $$\Delta B_{ij} = p_{ijkl} S_{kl}$$

    or, in reduced coordinates,

    $$\Delta B_I = p_{IJ} S_J$$

    (a) Consider a plane-polarized optical wave, polarized with the D field in the x direction and propagating in the z direction. Suppose that it is perturbed by a plane longitudinal acoustic wave propagating in the x direction (i.e., a wave with S_{xx} or S_1 finite). In the reduced coordinate system, what components of ΔB_I are finite? What components of E_i are finite? What is the polarization of the wave generated by the acousto-optic interaction?
    (b) What components of ΔB_I and E_i are finite for an incident optical beam, with the same intensity and interaction length as in part (a), when it interacts with a longitudinal acoustic wave propagating in the y direction?
    (c) If the interactions of parts (a) and (b) take place in a liquid instead of a solid, show that the respective output beam amplitudes are indentical.
    (d) What components of ΔB_I and E_i are finite, and what is the polarization of the generated optical beam, for interaction of an incident optical beam with a shear acoustic wave, propagating in the y direction, whose particle motion is in the x direction?
2. A Schlieren optical system (Fig. 4.9.4) is used to image a square acoustic beam in water, 1 cm on each side. The acoustic power in the beam is 50 mW and the operating frequency is 5 MHz. A knife edge is employed to pass only the first-order diffraction image.

A 5-mW helium–neon laser ($\Lambda = 6300$ Å in air) is used to produce the image, and the acoustic beam amplitude is uniform (although in practice, diffraction causes it to be nonuniform). Optical lenses of 4-cm diameter and focal length 50 cm are used to produce a uniform 4 cm diameter parallel optical beam and to focus it to a diffraction limited spot (see Sec. 3.3). Estimate the intensity of the Schlieren image and the width of the image of the acoustic beam at 3 m from the second lens.

3. A single-sideband acoustic modulator for a bimode fiber-optic delay line is constructed as shown below. The fiber's core is shaped slightly elliptically; this causes the fiber to have two possible modes of propagation, with fields E_1 and E_2 polarized at right angles. The fiber is bonded to an SAW delay line and a flat slab, and placed between them. The fiber is rotated so that the polarizations of the two orthogonal modes are at 45° to the surface of the slab. The displacement normal to the surface of the delay line, due to the surface acoustic wave, applies a stress component at 45° to the planes of polarization of the two optical waves, and couples them together.

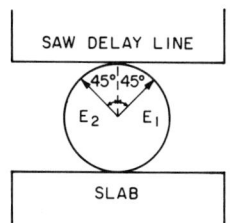

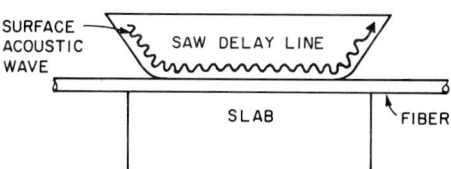

For optimum coupling of the two waves we know that

$$\omega_2 = \omega_1 \pm \omega_a \quad (1)$$

and

$$k_2 = k_1 \pm k_a \quad (2)$$

where ω_1 and ω_2 are the frequencies and k_1 and k_2 are the effective propagation constants of the two optical waves, and ω_a and k_a are the corresponding parameters of the acoustic wave. Typically, fibers can be obtained for which the beat wavelength λ_b is of the order of 1 mm, where $\lambda_b = 2\pi/(k_2 - k_1)$.

We may use a coupled mode theory, like that in Sec. 4.6.2, to write

$$\frac{\partial A_2}{\partial z} + jk_2 A_2 = j\kappa(z) A_1 \quad (3)$$

$$\frac{\partial A_1}{\partial z} + jk_1 A_1 = j\kappa(z) A_2 \quad (4)$$

where $\kappa(z)$ is the coupling coefficient between the optical waves, due to the presence of the acoustic wave, and $A_1(z, t)$ and $A_2(z, t)$ are the amplitudes of the two optical waves.

We may write

$$\kappa(z, t) = \kappa_0 \cos(\omega_a t - k_a z) \quad (5)$$

$$= \frac{\kappa_0}{2}\left[e^{j(\omega_a t - k_a z)} + e^{-j(\omega_a t - k_a z)}\right]$$

where

$$A_1(z, t) = a_1(z)e^{j(\omega_1 t - k_1 z)} \quad (6)$$

$$A_2(z, t) = a_2(z)e^{j(\omega_2 t - k_2 z)} \quad (7)$$

and in all cases Eq. (1) is satisfied.

(a) By putting $k_2 = k_a + k_1$, neglecting one of the terms in the brackets, and putting $A_2(0) = 0$ with $A_1(0)$ finite, find an expression determining the amplitudes of $A_2(z)$ and $A_1(z)$ along the fiber. Show that when k_a is positive (the acoustic wave is effectively propagating in the $+z$ direction), the output signal in mode 2 is upshifted in frequency.

(b) Show that if the effective propagation direction of the acoustic wave is reversed (k_a negative, $\theta > \pi/2$), the output signal is downshifted in frequency.

Note: A polarizer can easily separate out one signal from the other.

4. Following the assumptions and analysis of Prob. 3, assume that the coupling κ is weak ($\kappa \ll k_1$, $\kappa \ll k_2$) and, therefore, the amplitude of $a_1(z)$ is constant.

(a) Using the first of the coupled-mode equations [Eq. (3)], determine the 3-dB bandwidth for a coupler of length L, as ω_a is varied. Consider coupling from mode 1 to mode 2 and use only the first term in the brackets of Eq. (5). Put $k_a = \omega_a/V_a$ and assume V_a is constant. Then find $\Delta k_a/k_{a0}$ where k_{a0} is the acoustic propagation constant for optimum coupling.

(b) Using the second term in the brackets of Eq. (5), find how much signal is transferred into the lower sideband.

(c) Show that if the coupling coefficient κ is decreased and the length L is increased to obtain a given transfer efficiency to the upper sideband, both the bandwidth and the spurious signal level in the lower sideband are decreased.

5. An optical wave of the form $\phi_0(x) \exp[j(\omega_0 t - k_0 x)]$, traveling forward in the $+x$ direction, interacts with an acoustic wave traveling in the opposite direction that perturbs the effective dielectric constant of the medium by

$$\Delta \varepsilon = \alpha \varepsilon \cos(\omega_a t + k_a x)$$

The interaction of the two waves gives rise to an optical wave traveling backward in the $-x$ direction, of the form $\phi_1(x) \exp[j(\omega_1 t + k_1 x)]$, such that

$$\omega_1 = \omega_0 + \omega_a$$

and

$$k_1 = k_a - k_0$$

For strong interaction ($k_1 \approx k_0$), $k_a \approx 2k_0$ and the acoustic wave must have a frequency of the order of 25 GHz.

Write a wave equation similar to Eq. (4.9.53) for this interaction. Simplify this result in the same way as for Eq. (4.9.56), neglecting all terms but ϕ_0 and ϕ_1, to obtain two differential equations relating ϕ_0 and ϕ_1. Solve these equations for a system of length L with boundary conditions $\phi_0(0) = $ constant, $\phi_0(L) = 0$, and $k_1 = k_0 = k_a/2$, to find the amplitude of the excited wave at $x = 0$.*

Note: This effect has been used to detect thermal phonons. In addition, if two thin optical beams interact with an acoustic beam at the Bragg angle, the frequencies and propagation constants satisfy Eqs. (4.9.46) and (4.9.47); however, this type of analysis is better suited to thin optical beams than the one employed for broad beams in the text.

REFERENCES

1. C. E. Cook and M. Bernfeld, *Radar Signals: An Introduction to Theory and Application.* New York: Academic Press, Inc., 1967.
2. H. Matthews, ed., *Surface Wave Filters: Design, Construction, and Use.* New York: John Wiley & Sons, Inc., 1977.
3. "Special Issue on Surface Acoustic Wave Devices and Applications," *Proc. IEEE*, 64, No. 5 (May 1976), 577–832.
4. D. D. Buss, A. F. Tasch, Jr., and J. B. Barton, "Applications to Signal Processing," in *Charge-Coupled Devices and Systems*, M. J. Howes and D. V. Morgan, eds. Chichester, West Sussex, England: John Wiley & Sons Ltd., 1979, Chapter 3, pp. 103–75.
5. C. H. Séquin and M. F. Tompsett, *Charge Transfer Devices*, Supplement 8 for *Advances in Electronics and Electron Physics*, L. Marton, series ed. New York: Academic Press, Inc., 1975, Chapter VI, pp. 201–35.
6. R. Melen and D. Buss, eds., *Charge-Coupled Devices: Technology and Applications.* New York: IEEE Press, 1977.
7. R. W. Broderson, P. R. Gray, and D. A. Hodges, "MOS Switched-Capacitor Filters," *Proc. IEEE*, 67, No. 1 (Jan. 1979), 61–75.
8. F. L. J. Sangster and K. Teer, "Bucket-Brigade Electronics—New Possibilities for Delay, Time-Axis Conversion, and Scanning," *IEEE J. Solid-State Circuits*, SC-4, No. 3 (June 1969), 131–36.
9. S. A. Newton, K. P. Jackson, J. E. Bowers, C. C. Cutler, and H. J. Shaw, "Fiber-Optic Delay Line Devices for Gigahertz Signal Processing," *IEEE Int. Conf. Acoustics Speech Signal Process.*, Vol. 3 (Apr. 1983), 1204–7.
10. K. P. Jackson, S. A. Newton, and H. J. Shaw, "1-Gbit/s Code Generator and Matched Filter Using an Optical Fiber Tapped Delay Line," *Appl. Phys. Lett.*, 42, No. 7 (Apr. 1983), 556–58.
11. J. E. Bowers, S. A. Newton, and H. J. Shaw, "Fibre-Optic Variable Delay Lines," *Electron. Lett.*, 18, No. 23 (Nov. 11, 1982), 999–1000.
12. S. A. Reible, "Wideband Analog Signal Processing with Superconductive Circuits," *1982 Ultrason. Symp. Proc.* (IEEE), 82CH1823-4, Vol. 1, 190–201.

Reference: C. F. Quate, C. D. W. Wilkinson, and D. K. Winslow, *Proc. IEEE*, 53, No. 10 (Oct. 1965), 1604–23.

13. G. S. Kino and H. J. Shaw, "Acoustic Surface Waves," *Sci. Am.*, 227, No. 4 (Oct. 1972), 50–68.
14. M. G. Unkauf, "Surface Wave Devices in Spread Spectrum Systems," in *Surface Wave Filters: Design, Construction, and Use*, H. Matthews, ed. New York: John Wiley & Sons, Inc., 1977, Chapter 11, pp. 477–509.
15. C. S. Hartmann, D. T. Bell, Jr., and R. C. Rosenfeld, "Impulse Model Design of Acoustic Surface-Wave Filters," *IEEE Trans. Microwave Theory Tech.*, MTT-21, No. 4 (Apr. 1973), 162–75.
16. D. T. Bell, Jr., and R. C. M. Li, "Surface-Acoustic-Wave Resonators," *Proc. IEEE*, 64, No. 5 (May 1976), 711–21.
17. R. L. Rosenberg and L. A. Coldren, "Scattering Analysis and Design of SAW Resonator Filters," *IEEE Trans. Sonics Ultrason.*, SU-26, No. 3 (May 1979), 205–30.
18. R. C. Williamson, "Reflection Grating Filters," in *Surface Wave Filters: Design, Construction, and Use*, H. Matthews, ed. New York: John Wiley & Sons, Inc., 1977, Chapter 9, pp. 381–442.
19. C. Lardat, "Improved SAW Chirp Spectrum Analyser with 80 dB Dynamic Range," *1978 Ultrason. Symp. Proc.* (IEEE), 78 CH 1344-1SU, 518–21.
20. H. M. Gerard, O. W. Otto, and R. D. Weglein, "Development of a Broadband Reflective Array 10,000:1 Pulse Compression Filter," *1974 Ultrason. Symp. Proc.* (IEEE), 74 CHO 896-1SU, 197–201.
21. J. Crabb, M. F. Lewis, and J. D. Maines, "Surface-Acoustic-Wave Oscillators: Mode Selection and Frequency Modulation," *Electron. Lett.*, 9, No. 10 (May 17, 1973), 195–97.
22. A. J. Budreau, P. H. Carr, and K. R. Laker, "Frequency Synthesizer Using Acoustic Surface-Wave Filters," *Microwave J.*, 17, No. 3 (Mar. 1974), 65–69.
23. R. D. Weglein and O. W. Otto, "Microwave SAW Oscillators" *1977 Ultrason. Symp. Proc.* (IEEE), 77 CH 1264-1SU, 913–22.
24. Reticon Corporation, "Product Summary: Discrete Time Analog Signal Processing Devices." Sunnyvale, Calif.: Reticon Corp., 1977.
25. J. T. Walker and J. D. Meindl, "A Digitally Controlled CCD Dynamically Focused Phased Array," *1975 Ultrason. Symp. Proc.* (IEEE), 75 CHO 994-4SU, 80–83.
26. M. A. Jack, P. M. Grant, and J. H. Collins, "The Theory, Design, and Applications of Surface Acoustic Wave Fourier-Transform Processors," *Proc. IEEE*, 68, No. 4 (Apr. 1980), 450–68.
27. Reticon Corporation, "Preliminary Data Sheet: R5501-32 Parallel In/Serial Out (PISO)." Sunnyvale, Calif.: Reticon Corp., 1978.
28. Reticon Corporation, "S-Series Solid State Line Scanners, 512 and 1024 Elements." Sunnyvale, Calif.: Reticon Corp., 1978.
29. I. Deyhimy, R. C. Eden, R. J. Anderson, and J. S. Harris, Jr., "A 500-MHz GaAs Charge-Coupled Device," *Appl. Phys. Lett.*, 34, No. 2, (Jan. 15, 1980), 151–53.
30. R. D. Baertsch, W. E. Engeler, H. S. Goldberg, C. M. Puckette, and J. J. Tiemann, "Two Classes of Charge-Transfer Devices for Signal Processing," *Int. Conf. Technol. Applications Charge Coupled Devices* (Proc.), (Sept. 1974), 229–36.
31. R. D. Baertsch, W. E. Engeler, H. S. Goldberg, C. M. Puckette IV, and J. J. Tiemann, "The Design and Operation of Practical Charge-Transfer Transversal Filters," *IEEE J. Solid-State Circuits*, SC-11, No. 1 (Feb. 1976), 65–74.

32. R. Bracewell, *The Fourier Transform and Its Applications*, 2nd ed. New York: McGraw-Hill Book Company, 1978.
33. A. V. Oppenheim and R. W. Schafer, *Digital Signal Processing*. Englewood Cliffs, N.J.: Prentice-Hall, Inc., 1975.
34. P. M. Woodward, *Probability and Information Theory with Applications to Radar*, 2nd ed. Oxford: Pergamon Press Ltd., 1953.
35. D. Gabor, "Theory of Communication," *J. IEE* (Brit.), 93, Pt. III, No. 26 (Nov. 1946), 429–57.
36. H. Stark and F. B. Tuteur, *Modern Electrical Communications: Theory and Systems*. Englewood Cliffs, N.J.: Prentice-Hall, Inc., 1979.
37. M. Bernfeld, C. E. Cook, J. Paolillo, and C. A. Palmieri, "Matched Filtering, Pulse Compression and Waveform Design," Parts I-IV, *Microwave J.*, 7, No. 10 (Oct. 1964), 57–64, No. 11 (Nov. 1964), 81–90, No. 12 (Dec. 1964), 70–76, and 8, No.1 (Jan. 1965), 73–81.
38. J. R. Klauder, "The Design of Radar Signals Having Both High Range Resolution and High Velocity Resolution," *Bell Sys. Tech. J.*, XXXIX, No. 4 (July 1960), 809–20.
39. F. J. Harris, "On the Use of Windows for Harmonic Analysis with the Discrete Fourier Transform," *Proc. IEEE*, 66, No. 1 (Jan. 1978), 51–83.
40. J. F. Kaiser, "Digital Filters," in *System Analysis by Digital Computer*, F. F. Kuo and J. F. Kaiser, eds. New York: John Wiley & Sons, Inc., 1966, Chapter 7, pp. 218–85.
41. R. M. Hays, W. R. Shreve, D. T. Bell, Jr., L. T. Claiborne, and C. S. Hartmann, "Surface Wave Transform Adaptable Processor System," *1975 Ultrason. Symp. Proc.* (IEEE), 75 CHO 994-4SU, 363–70.
42. J. D. Maines, G. L. Moule, C. O. Newton, and E. G. S. Paige, "A Novel SAW Variable-Frequency Filter," *1975 Ultrason. Symp. Proc.* (IEEE), 75 CHO 994-4SU, 355–58.
43. C. Atzeni, G. Manes, and L. Masotti, "Programmable Signal Processing by Analog Chirp Transformation Using SAW Devices," *1975 Ultrason. Symp. Proc.* (IEEE), 75 CHO 994-4SU, 371–76.
44. J. D. Maines and E. G. S. Paige, "Surface-Acoustic-Wave Devices for Signal Processing Applications," *Proc. IEEE*, 64, No. 5 (May 1976), 639–52.
45. H. M. Gerard, P. S. Yao, and O. W. Otto, "Performance of a Programmable Radar Pulse Compression Filter Based on a Chirp Transformation with RAC Filters," *1977 Ultrason. Symp. Proc.* (IEEE), 77 CH 1264-1SU, 947–51.
46. G. R. Nudd and O. W. Otto, "Chirp Signal Processing Using Acoustic Surface Wave Filters," *1975 Ultrason. Symp. Proc.* (IEEE), 75 CHO 994-4SU, 350–54.
47. M. A. Jack and J. H. Collins, "Fast Fourier Transform Processor Based on the SAW Chirp Transform Algorithm," *1978 Ultrason. Symp. Proc.* (IEEE), 78CH 1344-1SU, 533–37.
48. M. A. Jack and E. G. S. Paige, "Fourier Transformation Processors Based on Surface Acoustic Wave Chirp Filters," *Wave Electron.*, 3, No. 3 (Nov. 1978), 229–47.
49. Reticon Corporation, "Preliminary Data Sheet: Quad Chirped Transversal Filter R5601." Sunnyvale, Calif.: Reticon Corp., 3/28/78.
50. F. G. Marshall, C. O. Newton, and E. G. S. Paige, "Theory and Design of the Surface

Acoustic Wave Multistrip Coupler," *IEEE Trans. Microwave Theory Tech.*, MTT-21, No. 4 (Apr. 1973), 206–15.

51. F. G. Marshall, C. O. Newton, and E. G. S. Paige, "Surface Acoustic Wave Multistrip Components and Their Applications," *IEEE Trans. Microwave Theory Tech.*, MTT-21, No. 4 (Apr. 1973), 216–25.
2. R. H. Tancrell and H. Engan, "Design Considerations for SAW Filters," *1973 Ultrason. Symp. Proc.* (IEEE), 73 CHO 807-8SU, 419–22.
53. C. Atzeni and L. Masotti, "Weighted Interdigital Transducers for Smoothing of Ripples in Acoustic-Surface-Wave Bandpass Filters," *Electron. Lett.*, 8, No. 19 (Sept. 21, 1972), 485–86.
54. T. Kodama, "Optimization Techniques for SAW Filter Design," *1979 Ultrason. Symp. Proc.* (IEEE), 79CH1482-9, 522–26.
55. C. S. Hartmann, W. S. Jones, and H. Vollers, "Wideband Unidirectional Interdigital Surface Wave Transducers," *IEEE Trans. Sonics Ultrason.*, SU-19, No. 3 (July 1972), 378–81.
56. A. J. DeVries, "A Design Method for Surface-Wave Filters Using Simple Structures as Building Blocks," *1973 Ultrason. Symp. Proc.* (IEEE), 73 CHO 807-8SU, 441–44.
57. A. J. DeVries and R. Adler, "Case History of a Surface-Wave TV IF Filter for Color Television Receivers," *Proc. IEEE*, 64, No. 5 (May 1976), 671–76.
58. A. J. DeVries, T. Sreenivasan, S. Subramanian, and T. J. Wojcik, "Detailed Description of a Commercial Surface-Wave TV IF Filter," *1974 Ultrason. Symp. Proc.* (IEEE), 74 CHO 896-1SU, 147–52.
59. K. Kishimoto, M. Ishigaki, K. Hazama, and S. Matsu-ura, "SAW Comb Filter for TV Frequency Synthesizing Tuning System," *1980 Ultrason. Symp. Proc.* (IEEE), 80CH1602-2, Vol. 1, 377–81.
60. S. Matsu-ura, K. Hazama, and T. Murata, "TV Tuning Systems with SAW Comb Filter," *IEEE Trans. Microwave Theory Tech.*, MTT-29, No. 5 (May 1981), 434–39.
61. R. A. Bergh, G. Kotler, and H. J. Shaw, "Single-Mode Fibre Optic Directional Coupler," *Electron. Lett.*, 16, No. 7 (Mar. 27, 1980), 260–61.
62. M. J. F. Digonnet and H. J. Shaw, "Analysis of a Tunable Single Mode Optical Fiber Coupler," *IEEE J. Quantum Electron.*, QE-18, No. 4 (Apr. 1982), 746–54.
63. G. S. Kino and H. Matthews, "Signal Processing in Acoustic Surface-Wave Devices," *IEEE Spectrum*, 8, No. 8 (Aug. 1971), 22–35.
64. G. S. Kino, "Acoustoelectric Interactions in Acoustic-Surface-Wave Devices," *Proc. IEEE*, 64, No. 5 (May 1976), 724–48.
65. G. S. Kino, S. Ludvik, J. H. Shaw, W. R. Shreve, J. M. White, and D. K. Winslow, "Signal Processing by Parametric Interactions in Delay-Line Devices," *IEEE Trans. Microwave Theory Tech.*, MTT-21, No. 4 (Apr. 1973), 244–55.
66. D. M. Bloom and G. C. Bjorklund, "Conjugate Wave-Front Generation and Image Reconstruction by Four-Wave Mixing," *Appl. Phys. Lett.*, 31, No. 9 (Nov. 1, 1977), 592–96.
67. D. P. Morgan and J. M. Hannah, "Correlation of Long Spread-Spectrum Wave-forms Using a SAW Convolver/Recirculation Loop Sub-system," *1976 Ultrason. Symp. Proc.* (IEEE), 76 CH 1120-5SU, 436–40.
68. D. P. Morgan, J. H. Hannah, and J. H. Collins, "Spread-Spectrum Synchronizer Using a SAW Convolver and Recirculation Loop," *Proc. IEEE*, 64, No. 5 (May 1976), 751–53.

69. P. K. Blair, "Receivers for the NAVSTAR Global Positioning System," *IEE Proc. (Brit.)*, 127, Pt. F, No. 2 (Apr. 1980), 163–67.

70. P. Defranould and C. Maerfeld, "A SAW Planar Piezoelectric Convolver," *Proc. IEEE*, 64, No. 5 (May 1976), 748–51.

71. I. Yao, "High Performance Elastic Convolver with Parabolic Horns," *1980 Ultrason. Symp. Proc.* (IEEE), 80CH1602-2, Vol. 1, 37–42.

72. J. B. Green and G. S. Kino, "SAW Convolvers Using Focused Interdigital Transducers," *IEEE Trans. Sonics Ultrason.*, SU-30, No. 1 (Jan. 1983), 43–50.

73. G. S. Kino and H. Gautier, "Convolution and Parametric Interaction with Semiconductors," *J. Appl. Phys.*, 44, No. 12 (Dec. 1973), 5219–21.

74. B. T. Khuri-Yakub and G. S. Kino, "A Detailed Theory of the Monolithic Zinc Oxide on Silicon Convolver," *IEEE Trans. Sonics Ultrason.*, SU-24, No. 1 (Jan. 1977), 34–43.

75. J. H. Cafarella, W. M. Brown, Jr., E. Stern, and J. A. Alusow, "Acousto-electric Convolvers for Programmable Matched Filtering in Spread-Spectrum Systems," *Proc. IEEE*, 64, No. 5 (May 1976), 756–59.

76. J. H. Goll and D. C. Malocha, "An Application of SAW Convolvers to High Bandwidth Spread Spectrum Communications," *IEEE Trans. Microwave Theory Tech.*, MTT-29, No. 5 (May 1981), 473–83.

77. S. A. Reible, "Acoustoelectric Convolver Technology for Spread-Spectrum Communications," *IEEE Trans. Sonics Ultrason.*, Su-28, No. 3 (May 1981), 185–95.

78. G. K. Montress and T. M. Reeder, "A High Performance SAW/Hybrid Coponent Fourier Transform Convolver," *1978 Ultrason. Symp. Proc.* (IEEE), 78CH 1344-1SU, 538–42.

79. H. C. Tuan, J. E. Bowers, and G. S. Kino, "Theoretical and Experimental Results for Monolithic SAW Memory Correlators," *IEEE Trans. Sonics Ultrason.*, SU-27, No. 6 (Nov. 1980), 360–69.

80. K. A. Ingebrigtsen, "The Schottky Diode Acoustoelectric Memory and Correlator—A Novel Programmable Signal Processor," *Proc. IEEE*, 64, No. 5 (May 1976), 764–69.

81. H. C. Tuan and G. S. Kino, "Large-Time-Bandwidth-Product Correlation and Holographic Storage with an S.A.W. Storage Correlator," *Electron. Lett.*, 13, No. 24 (Nov. 24, 1977), 709–10.

82. S. A. Reible and I. Yao, "An Acoustoelectric Burst Waveform Processor," *1980 Ultrasonics Symp. Proc.* (IEEE), 80CH1602-2, Vol. 1, 133–38.

83. H. C. Tuan, P. M. Grant, and G. S. Kino, "Theory and Application of Zinc-Oxide-on-Silicon Monolithic Storage Correlators," *1978 Ultrason. Symp. Proc.* (IEEE), 78 CH 1344-1SU, 38–43.

84. P. G. Borden and G. S. Kino, "Correlation with the Storage Convolver," *Appl. Phys. Lett.*, 29, No. 9 (Nov. 1, 1976), 527–29.

85. M. Cronin-Golomb, B. Fischer, J. O. White, and A. Yariv, "Passive (Self-Pumped) Phase Conjugate Mirror: Theoretical and Experimental Investigation," *Appl. Phys. Lett.*, 41, No. 8 (Oct. 15, 1982), 689–91.

86. G. L. Kerber, R. M. White, and J. R. Wright, "Surface-Wave Inverse Filter for Nondestructive Testing," *1976 Ultrason. Symp. Proc.* (IEEE), 76 CH 1120-5SU, 577–81.

87. Y. Murakami, B. T. Khuri-Yakub, G. S. Kino, J. M. Richardson, and A. G. Evans,

"An Application of Wiener Filtering to Nondestructive Evaluation," *Appl. Phys. Lett.*, 33, No. 8 (Oct. 15, 1978), 685–87.

88. D. Corl, "A C.T.D. Adaptive Inverse Filter," *Electron. Lett.*, 14, No. 3 (Feb. 2, 1978), 60–62.

89. R. W. Lucky, "Automatic Equalization for Digital Communication," *Bell Syst. Tech. J.*, XLIV, No. 4 (Apr. 1965), 547–88.

90. S. C. Tanaka, H. Tseng, L. T. Lin, and P. Chen, "An Integrated Real-Time Programmable Transversal Filter," *IEEE J. Solid-State Circuits*, SC-15, No. 6 (Dec. 1980), 978–83.

91. B. Widrow, "Adaptive Filters," in *Aspects of Network and System Theory*, R. E. Kalman and N. DeClaris, eds. New York: Holt, Rinehart and Winston, 1970, pp. 563–87.

92. J. E. Bowers, G. S. Kino, D. Behar, and H. Olaisen, "Adaptive Deconvolution Using a SAW Storage Correlator," *IEEE Trans. Microwave Theory Tech.*, MTT-29, No. 5 (May 1981), 491–98.

93. D. Behar, G. S. Kino, J. E. Bowers, and H. Olaisen, "Storage Correlator as an Adaptive Inverse Filter," *Electron. Lett.*, 16, No. 4 (Feb. 14, 1980), 130–31.

94. P. M. Grant and G. S. Kino, "Adaptive Filter Based on S.A.W. Monolithic Storage Correlator," *Electron. Lett.*, 14, No. 17 (Aug. 17, 1978), 562–64.

95. J. E. Bowers and G. S. Kino, "Adaptive Noise Cancellation with a SAW Storage Correlator," *Electron. Lett.*, 17, No. 13 (June 25, 1981), 460–61.

96. D. F. Barbe, W. D. Baker, and K. L. Davis, "Signal Processing with Charge-Coupled Devices," *IEEE J. Solid-State Circuits*, SC-13, No. 1 (Feb. 1978), 34–51.

97. B. K. Ahuja, M. A. Copeland, and C. H. Chan, "A Sampled Analog MOS LSI Adaptive Filter," *IEEE J. Solid-State Circuits*, SC-14, No. 1 (Feb. 1979), 148–54.

98. W. K. Masenten, "Adaptive Processing for Spread Spectrum Communication Systems," *Hughes Aircraft Company Report*, TP 77-14-22 (Sept. 27, 1977).

99. W. R. Klein and B. D. Cook, "Unified Approach to Ultrasonic Light Diffraction," *IEEE Trans. Sonics Ultrason.*, SU-14, No. 3 (July 1967), 123–34.

100. K. Osterhammel, "Optische Untersuchung des Schallfeldes Kolbenformig Schwinger Quarze," *Akust. Zh. Bd.*, 6 (1941), 73–86.

101. B. D. Cook, E. Cavanagh, and H. D. Dardy, "A Numerical Procedure for Calculating the Integrated Acoustooptic Effect," *IEEE Trans. Sonics Ultrason.*, SU-27, No. 4 (July 1980), 202–7.

102. C. S. Tsai, "Wideband Guided-Wave Acoustooptic Bragg-Devices and Applications," *1975 Ultrason. Symp. Proc.* (IEEE), 75 CHO 994-4SU, 120–25.

103. J. N. Lee, N. J. Berg, and B. J. Udelson, "Large Time-Bandwidth Acousto-Optic Signal Processors," *1977 Ultrason. Symp. Proc.* (IEEE), 77 CHU 1264-1SU, 451–55.

104. N. J. Berg and B. J. Udelson, "Large Time-Bandwidth Acousto-Optic Convolver," *1976 Ultrason. Symp. Proc.* (IEEE), 76 CH 1120-5SU, 183–88.

105. P. Kellman, "Time Integrating Optical Signal Processing," in *Acousto-Optic Bulk Wave Devices*, J. B. Houston, Jr., ed., *Proc. Soc. Photo-Opt. Instrum. Eng.*, 214 (1979), 63–73.

Appendix A

Stress, Strain, and the Reduced Notation

STRESS AND STRAIN VECTORS

In this appendix we derive the three-dimensional forms of the stress and strain tensors somewhat more fully and rigorously than in Secs. 1.1 and 1.5 of the text. We also describe the commonly used reduced notation based on the symmetry of the S and T tensors [1].

Strain Tensor

A point $\mathbf{r}$ in the material is displaced by stress to a point $\mathbf{r} + \mathbf{u}$, where $\mathbf{u}$ is the displacement vector. Suppose that we consider length l in the material between the point $\mathbf{r}$ and $\mathbf{r} + \delta\mathbf{r}$. After displacement, l changes to l' and, as illustrated in Fig. A.1, we can write

$$l^2 = (\delta r)^2 = (\delta x_1)^2 + (\delta x_2)^2 + (\delta x_3)^2 \tag{A.1}$$

and

$$l'^2 = (\delta\mathbf{r} + \delta\mathbf{u})^2 = l^2 + 2\delta\mathbf{u} \cdot \delta\mathbf{r} + (\delta\mathbf{u})^2 \tag{A.2}$$

We shall express δu_x in the form

$$\delta u_x = \frac{\partial u_x}{\partial x}\delta x + \frac{\partial u_x}{\partial y}\delta y + \frac{\partial u_x}{\partial z}\delta z \tag{A.3}$$

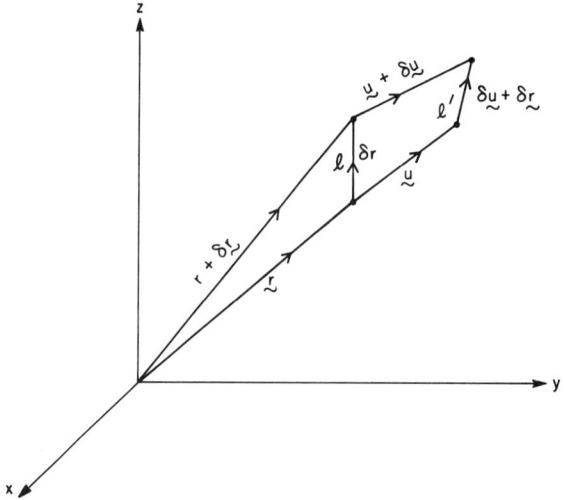

Figure A.1 Notation used in Eq. (A.1).

with similar notations for δu_y and δu_z. These relations can be summarized conveniently using tensor notation (see Sec. 1.5) and written in the form

$$\delta u_i = \frac{\partial u_i}{\partial x_j} \delta x_j \tag{A.4}$$

This is a shorthand notation for

$$\delta u_i = \sum_j \frac{\partial u_i}{\partial x_j} \delta x_j \tag{A.5}$$

where i can be x, y, or z, and where for a given i, the summation over the subscript j is understood.

We may write a Taylor expansion for Eq. (A.2) and keep terms to second order in δx_i. This procedure yields the result

$$l'^2 = l^2 + 2\frac{\partial u_i}{\partial x_j} \delta x_i \delta x_j + \frac{\partial u_i}{\partial x_j} \frac{\partial u_i}{\partial x_k} \delta x_j \delta x_k \tag{A.6}$$

where, as discussed in Sec. 1.5, the tensor notation now implies double summations over the three independent suffixes i, j, and k on the right-hand side of Eq. (A.6). Note that we have replaced the vector $\delta \mathbf{r}$ by δx_i in tensor notation. Similarly, a scalar product $\mathbf{A} \cdot \mathbf{B}$ is $A_i B_i$.

The change in l^2 is a true measure of the deformation of the material. If, instead, we were to use the change in $\delta \mathbf{r}$ as a criterion, this vector could be changed by a pure rotation of a rigid material without changing the length l; thus the change in $\delta \mathbf{r}$ would not be a measure of the deformation in this case.

We can now interchange the suffixes i and k in the third term of Eq. (A.6), and write the second term in the form

$$\frac{\partial u_i}{\partial x_j} \delta x_i \, \delta x_j = \frac{1}{2}\left(\frac{\partial u_i}{\partial x_j} + \frac{\partial u_j}{\partial x_i}\right) \delta x_i \, \delta x_j \qquad (A.7)$$

In this case, Eq. (A.6) can be written as

$$l'^2 = l^2 + 2S_{ij} \, \delta x_i \, \delta x_j \qquad (A.8)$$

where S_{ij} is known as the strain tensor and is defined as

$$S_{ij} = \frac{1}{2}\left(\frac{\partial u_i}{\partial x_j} + \frac{\partial u_j}{\partial x_i} + \frac{\partial u_k}{\partial x_i}\frac{\partial u_k}{\partial x_j}\right) \qquad (A.9)$$

We see from this definition that S_{ij} is a symmetric tensor. For small displacements, we can neglect the last term in Eq. (A.9) as being of second order. From now on, we shall write

$$S_{ij} = \frac{1}{2}\left(\frac{\partial u_i}{\partial x_j} + \frac{\partial u_j}{\partial x_i}\right) \qquad (A.10)$$

Alternatively, we can use a symbolic notation **S**, much like that for a vector, and define the strain S_{ij} in the form

$$\mathbf{S} = \begin{bmatrix} S_{11} & S_{12} & S_{13} \\ S_{21} & S_{22} & S_{23} \\ S_{31} & S_{32} & S_{33} \end{bmatrix} = \begin{bmatrix} S_{xx} & S_{xy} & S_{xz} \\ S_{yx} & S_{yy} & S_{yz} \\ S_{zx} & S_{zy} & S_{zz} \end{bmatrix} \qquad (A.11)$$

where now the subscripts 1, 2, and 3 are equivalent to x, y, and z, respectively, and are used interchangeably with them in the literature. It follows that

$$\frac{\partial S_{ij}}{\partial t} = \frac{1}{2}\left(\frac{\partial v_i}{\partial x_j} + \frac{\partial v_j}{\partial x_i}\right) \qquad (A.12)$$

where $\mathbf{v} = \partial \mathbf{u}/\partial t$ is the velocity of a particle in the material. This is equivalent to the one-dimensional equation of conservation of mass, given as Eq. (1.1.8), but it yields more information than just conservation of mass. If we take only the diagonal terms, we see that $\boldsymbol{\nabla} \cdot \mathbf{v} = \partial v_i/\partial x_i$ and that the equation of conservation of mass is

$$\rho_{m0} \boldsymbol{\nabla} \cdot \mathbf{v} + \frac{\partial \rho_{m1}}{\partial t} = 0 \qquad (A.13)$$

We use, from Eq. (A.12), the relation

$$\boldsymbol{\nabla} \cdot \mathbf{v} = \frac{\partial}{\partial t}(S_{11} + S_{22} + S_{33}) \qquad (A.14)$$

This is directly equivalent to Eq. (1.1.8) and will be derived in another way below. The diagonal terms are associated with longitudinal strain; the off-diagonal terms are associated with shear strain.

Stress and Strain Vectors

Change in Volume

It will be noted that the volume of a small portion δV of the material is $\delta x_1 \delta x_2 \delta x_3$. After deformation, it becomes $\delta V'$, where

$$\delta V' = (\delta x_1 + \delta u_1)(\delta x_2 + \delta u_2)(\delta x_3 + \delta u_3)$$

$$= \delta V \left(1 + \frac{\partial u_1}{\partial x_1}\right)\left(1 + \frac{\partial u_2}{\partial x_2}\right)\left(1 + \frac{\partial u_3}{\partial x_3}\right) \quad (A.15)$$

$$\approx \delta V \left(1 + \frac{\partial u_1}{\partial x_1} + \frac{\partial u_2}{\partial x_2} + \frac{\partial u_3}{\partial x_3}\right)$$

It follows that

$$\delta V' - \delta V = \delta V(S_{11} + S_{22} + S_{33}) \quad (A.16)$$

We see that the sum of the diagonal components of the strain tensor is the relative volume change $(\delta V' - \delta V)/\delta V$. *The shear terms do not contribute to a change in volume.* We can also see this result from substituting Eq. (A.14) in Eq. (A.13). This yields the result

$$\rho_{m1} = -\rho_{m0}(S_{11} + S_{22} + S_{33}) \quad (A.17)$$

which is identical to Eq. (A.15) for

$$\rho_{m1}/\rho_{m0} = -(\delta V' - \delta V)/\delta V \quad (A.18)$$

Stress Tensor

Here we shall give a different and more detailed derivation for stress than that given in Sec. 1.5. The force in the x direction on a body of volume V is $\int F_x \, dV$, where the force F_x is a scalar quantity. We can always write a scalar quantity as the divergence of a vector. Thus we put

$$F_x = \nabla \cdot \mathbf{A}$$
$$F_y = \nabla \cdot \mathbf{B} \quad (A.19)$$
$$F_z = \nabla \cdot \mathbf{C}$$

Then, from Gauss's theorem, we can write

$$\int_V F_x \, dV = \int_V \nabla \cdot \mathbf{A} \, dV = \int_S \mathbf{A} \cdot d\mathbf{s}, \quad \text{etc.} \quad (A.20)$$

where the surface integral is taken around the enclosing volume V.

It is apparent that we need nine components, $A_x, A_y, A_z, B_x, B_y, B_z, C_x, C_y,$ and C_z, to express $\int F_x \, dV$, $\int F_y \, dV$, and $\int F_z \, dV$. In tensor notation, we write

$$F_i = \frac{\partial T_{ij}}{\partial x_j} \quad (A.21)$$

which is shorthand for

$$F_i = \sum_j \frac{\partial T_{ij}}{\partial x_j} \tag{A.22}$$

or

$$F_x = \frac{\partial T_{xx}}{\partial x} + \frac{\partial T_{xy}}{\partial y} + \frac{\partial T_{xz}}{\partial z} \tag{A.23}$$

and so on. The quantity T_{ij} is called the *stress tensor*. In our previous notation, we see that $A_x = T_{xx} = T_{11}$, $A_y = T_{xy} = T_{12}$, $A_z = T_{xz} = T_{13}$, and so on.

It follows that the average force on an element of volume dV is

$$\frac{1}{dV} \int \nabla \cdot \mathbf{T}\, dV = \frac{1}{dV} \int \frac{\partial T_{ij}}{\partial x_j} dV = \frac{1}{dV} \int T_{ij}\, ds_j = \frac{1}{dV} \oint \mathbf{T} \cdot \mathbf{n}\, ds \tag{A.24}$$

where we define $\nabla \cdot \mathbf{T}$ as $\partial T_{ij}/\partial x_j$, and where ds_j is the surface element vector directed along the outward normal. The force on a surface in the z direction therefore has three components normal to the surface that comprise the vector $\mathbf{C}$ in Eq. (A.19); these are T_{xz}, T_{yz}, and T_{zz}. The first two terms are shear stresses that tend to distort the surface of an isotropic material, as shown in Fig. A.2(b).

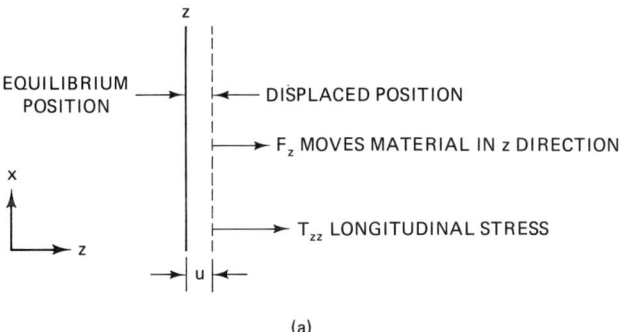

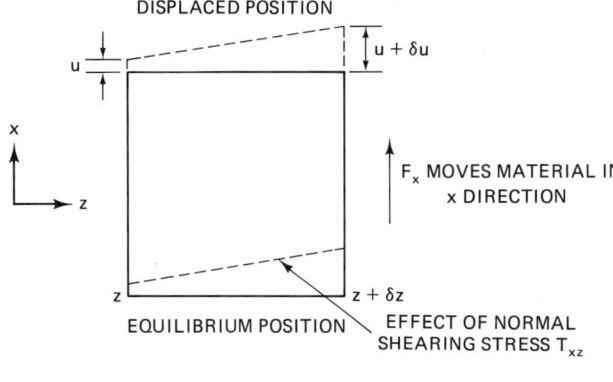

Figure A.2 Effect of normal longitudinal and shear stresses at a surface: (a) longitudinal stress; (b) shear stress.

Stress and Strain Vectors

The last term is a longitudinal stress, which acts as shown in Fig. A.2(a). All the stress components are applied to a cube, illustrated in Fig. 1.5.1. Because $\int T_{ij}\, ds_j = \int T_{ji}\, ds_i$, it can be shown that $T_{ij} = T_{ji}$ (i.e., **T** is a symmetric tensor).

Equation of Motion

The force on an element dV is $\int \mathbf{T} \cdot \mathbf{n}\, ds$ due to internal stresses. Thus, if only internal stresses are applied, we can write the equation of motion for first-order displacements as

$$\rho_{m0} \frac{\partial^2 \mathbf{u}}{\partial t^2} = \lim_{dV \to 0} \left[\frac{\int \mathbf{T} \cdot \mathbf{n}\, ds}{dV} \right] \tag{A.25}$$

or

$$\rho_{m0} \frac{\partial^2 \mathbf{u}}{\partial t^2} = \nabla \cdot \mathbf{T} \tag{A.26}$$

which is equivalent to

$$\rho_{m0} \frac{\partial^2 u_i}{\partial t^2} = \frac{\partial T_{ij}}{\partial x_j} \tag{A.27}$$

where ρ_{m0} is the mass density of the material.

SYMBOLIC NOTATION AND ABBREVIATED SUBSCRIPTS

Strain Tensor

To reduce the complexity of the stress and strain tensors, it is helpful to use symmetry and to work with an abbreviated subscript notation. Here we shall describe this abbreviated subscript notation and show how it is used.

We first consider the strain tensor S_{ij}, defined as

$$S_{ij} = \frac{1}{2}\left(\frac{\partial u_i}{\partial x_j} + \frac{\partial u_j}{\partial x_i} \right) \tag{A.28}$$

Because the strain is a symmetric tensor, we can replace S_{yx} with S_{xy}, and so on. Thus we can use a reduced notation with fewer subscripts. The standard reduced notation can be expressed in matrix form:

$$\mathbf{S} = \begin{bmatrix} S_{xx} & S_{xy} & S_{xz} \\ S_{yx} & S_{yy} & S_{yz} \\ S_{zx} & S_{zy} & S_{zz} \end{bmatrix} = \begin{bmatrix} S_1 & \dfrac{S_6}{2} & \dfrac{S_5}{2} \\ \dfrac{S_6}{2} & S_2 & \dfrac{S_4}{2} \\ \dfrac{S_5}{2} & \dfrac{S_4}{2} & S_3 \end{bmatrix} \tag{A.29}$$

Note that the notation follows a cyclic order, with the longitudinal strain terms corresponding to the subscripts 1, 2, and 3, respectively, and the shear strain terms corresponding to the subscripts 4, 5, and 6, respectively, as shown in the following table:

Normal tensor notation	Reduced notation	Corresponding strain
xx	1	Longitudinal in x direction
yy	2	Longitudinal in y direction
zz	3	Longitudinal in z direction
yz = zy	4	Shear y − z
zx = xz	5	Shear z − x
xy = yx	6	Shear x − y

Note that the off-diagonal terms are multiplied by $\frac{1}{2}$ so that we can write the strain in the form of a column matrix:

$$\begin{bmatrix} S_1 \\ S_2 \\ S_3 \\ S_4 \\ S_5 \\ S_6 \end{bmatrix} = \begin{bmatrix} \frac{\partial}{\partial x} & 0 & 0 \\ 0 & \frac{\partial}{\partial y} & 0 \\ 0 & 0 & \frac{\partial}{\partial z} \\ 0 & \frac{\partial}{\partial z} & \frac{\partial}{\partial y} \\ \frac{\partial}{\partial z} & 0 & \frac{\partial}{\partial x} \\ \frac{\partial}{\partial y} & \frac{\partial}{\partial x} & 0 \end{bmatrix} \begin{bmatrix} u_x \\ u_y \\ u_z \end{bmatrix} \qquad (A.30)$$

This can be done only because the $\frac{1}{2}$ was used in our definitions in Eq. (A.29). Following Auld [1], it is convenient to define the matrix in Eq. (A.30) with a symbolic notation, writing

$$\mathbf{S} = \nabla_s \mathbf{u} \qquad (A.31)$$

where $\nabla_s \mathbf{u}$ is defined as the symmetric part of $\nabla \mathbf{u}$. The symetric operator $\nabla_s u$ is defined by the first matrix on the right-hand side of Eq. (A.30). In the unreduced tensor notation, the symmetry of ∇_s is apparent because

$$S_{ij} = \frac{1}{2}\left(\frac{\partial u_i}{\partial x_j} + \frac{\partial u_j}{\partial x_i}\right) \qquad (A.32)$$

A simple example of longitudinal motion in the x direction with propagation in the x direction is defined by the relation $S_1 = \partial u_x / \partial x$; this follows from Eq. (A.30). On the other hand, a plane shear wave, in which propagation is in the z direction but particle displacement is only in the y direction, is defined by the relations $u_x = u_z = 0$ and $S_4 = \partial u_y / \partial z$. In this case, all other components of strain are on zero. The first case corresponds to a longitudinal wave passing through a flat plate; the second case corresponds to the flexural motion of a thin strip.

Stress Tensor

The stress tensor may be stated in terms of reduced subscripts just as the strain tensor was. Thus we write

$$\mathbf{T} = \begin{bmatrix} T_{xx} & T_{xy} & T_{xz} \\ T_{yx} & T_{yy} & T_{yz} \\ T_{zx} & T_{zy} & T_{zz} \end{bmatrix} = \begin{bmatrix} T_1 & T_6 & T_5 \\ T_6 & T_2 & T_4 \\ T_5 & T_4 & T_3 \end{bmatrix} \quad (A.33)$$

Note that the $\frac{1}{2}$ terms are not required here. The equation of motion for symbolic notation is

$$\nabla \cdot \mathbf{T} = \rho_{m0} \frac{\partial \mathbf{v}}{\partial t} \quad (A.34)$$

This can be put in reduced tensor form, by writing

$$\rho_{m0} \frac{\partial}{\partial t} \begin{bmatrix} v_x \\ v_y \\ v_z \end{bmatrix} = \nabla \cdot \mathbf{T} = \begin{bmatrix} \frac{\partial}{\partial x} & 0 & 0 & 0 & \frac{\partial}{\partial z} & \frac{\partial}{\partial y} \\ 0 & \frac{\partial}{\partial y} & 0 & \frac{\partial}{\partial z} & 0 & \frac{\partial}{\partial x} \\ 0 & 0 & \frac{\partial}{\partial z} & \frac{\partial}{\partial y} & \frac{\partial}{\partial x} & 0 \end{bmatrix} \begin{bmatrix} T_1 \\ T_2 \\ T_3 \\ T_4 \\ T_5 \\ T_6 \end{bmatrix} \quad (A.35)$$

For example, if the stress field has only one component, a shear stress $T_5 = T_{xz}$ propagating in the z direction, then $\nabla \cdot T$ becomes $\partial T_5 / \partial z$ and corresponds to an acceleration in the x direction.

Elasticity

Similarly, the elasticity tensor c_{ijkl} can be expressed in reduced notation. Because $S_{ij} = S_{ji}$ and $T_{ij} = T_{ji}$, it follows that $c_{ijkl} = c_{jikl} = c_{ijlk} = c_{jilk}$, which reduces the number of independent constants required from 81 to 36. Furthermore, because of symmetry, $c_{ijkl} = c_{klij}$. This further reduces the required number of independent constants in an arbitrary medium to 21. Thus we write

$$\begin{bmatrix} T_1 \\ T_2 \\ T_3 \\ T_4 \\ T_5 \\ T_6 \end{bmatrix} = \begin{bmatrix} c_{11} & c_{12} & c_{13} & c_{14} & c_{15} & c_{16} \\ c_{21} & c_{22} & c_{23} & c_{24} & c_{25} & c_{26} \\ c_{31} & c_{32} & c_{33} & c_{34} & c_{35} & c_{36} \\ c_{41} & c_{42} & c_{43} & c_{44} & c_{45} & c_{46} \\ c_{51} & c_{52} & c_{53} & c_{54} & c_{55} & c_{56} \\ c_{61} & c_{62} & c_{63} & c_{64} & c_{65} & c_{66} \end{bmatrix} \begin{bmatrix} S_1 \\ S_2 \\ S_3 \\ S_4 \\ S_5 \\ S_6 \end{bmatrix} \quad (A.36)$$

or
$$\mathbf{T} = \mathbf{c} : \mathbf{S} \tag{A.37}$$

where the general term is c_{IJ}; we use capital subscripts to denote the reduced notation and take $c_{IJ} = c_{JI}$.

Example: Cubic Crystal

Most crystals have certain symmetries that reduce the required number of constants. For instance, a cubic crystal looks the same in the x, $-x$, y, $-y$, z, and $-z$ directions. This implies that $c_{11} = c_{22} = c_{33}$, $c_{44} = c_{55} = c_{66}$, and $c_{12} = c_{13} = c_{23}$. All other diagonal terms are zero because of the mirror symmetry. Thus we find that for a cubic crystal,

$$\mathbf{c} = \begin{bmatrix} c_{11} & c_{12} & c_{12} & 0 & 0 & 0 \\ c_{12} & c_{11} & c_{12} & 0 & 0 & 0 \\ c_{12} & c_{12} & c_{11} & 0 & 0 & 0 \\ 0 & 0 & 0 & c_{44} & 0 & 0 \\ 0 & 0 & 0 & 0 & c_{44} & 0 \\ 0 & 0 & 0 & 0 & 0 & c_{44} \end{bmatrix} \tag{A.38}$$

When there is shear wave propagation along the z axis, with motion in the x direction, it follows from Eq. (A.30) that

$$S_5 = \frac{\partial u_x}{\partial z} \tag{A.39}$$

and from Eqs. (A.36) and (A.38) that

$$T_5 = c_{44} S_5 \tag{A.40}$$

Assuming that the RF components vary as $\exp(j\omega t)$, then $v_x = j\omega u_x$. It follows from Eq. (A.34) or Eq. (A.35) that

$$\frac{\partial T_5}{\partial z} = j\omega \rho_{m0} v_x \tag{A.41}$$

However, from Eqs. (A.39) and (A.40), we see that

$$c_{44} \frac{\partial v_x}{\partial z} = j\omega T_5 \tag{A.42}$$

Equations (A.41) and (A.42) are the transmission-line equations for shear wave propagation. Assuming that the waves propagate as $\exp(\pm j\beta_s z)$, we see that for shear waves in a cubic crystal,

$$\beta_s^2 = \omega^2 \left(\frac{\rho_{m0}}{c_{44}} \right) \tag{A.43}$$

If, on the other hand, we consider longitudinal motion in the z direction with only u_z or v_z finite, we find that the propagation constant β_l is given by the relation

$$\beta_l^2 = \omega^2 \left(\frac{\rho_{m0}}{c_{11}} \right) \tag{A.44}$$

Example: Isotropic Material

In this case, which is very much like that of the cubic crystal, the **c** tensor is of the same form as that of Eq. (A.38), with the additional condition that $c_{12} = c_{11} - 2c_{44}$. Note that the c_{12} term corresponds to the ratio of the longitudinal stress in the x direction to the longitudinal strain in the y direction. Such terms occur because when a material is compressed in one direction, it tends to expand in a perpendicular direction. The relation given follows from the requirement that the tensor **c** keeps the same form; however, the axes are rotated from their original position.

It follows that an isotropic medium has only two independent elastic constants. These are usually called the Lamé constants, defined as

$$\lambda = c_{12} \tag{A.45}$$
$$\mu = c_{44}$$

with

$$c_{11} = c_{12} + 2c_{44} = \lambda + 2\mu \tag{A.46}$$

The **c** matrices for different types of crystals are tabulated in Appendix A.2 of B. A. Auld's *Acoustic Fields and Waves in Solids* [1]. The similar **s** matrices, for which $\mathbf{s} = \mathbf{c}^{-1}$ or

$$\mathbf{S} = \mathbf{s} : \mathbf{T} \tag{A.47}$$

are also tabulated by Auld.

PIEZOELECTRIC TENSORS

The piezoelectric tensor e_{ijk} can also be expressed in reduced notation in the forms

$$T_I = c_{IJ}^E S_J - e_{Ij} E_j \tag{A.48}$$

and

$$D_i = \varepsilon_{ij}^S E_j + e_{iJ} S_J \tag{A.49}$$

These relations can also be written in the equivalent forms

$$\mathbf{T} = \mathbf{c}^S : \mathbf{S} - \mathbf{e} \cdot \mathbf{E} \tag{A.50}$$

and

$$\mathbf{D} = \boldsymbol{\varepsilon}^S \cdot \mathbf{E} + \mathbf{e} : \mathbf{S} \tag{A.51}$$

where, as before, we have used capital subscripts to express the reduced notation, and the notation i, j, k to represent the x, y, and z axes.† We define the general e_{iJ} tensor as follows:

$$e_{iJ} = \begin{bmatrix} e_{x1} & e_{x2} & e_{x3} & e_{x4} & e_{x5} & e_{x6} \\ e_{y1} & e_{y2} & e_{y3} & e_{y4} & e_{y5} & e_{y6} \\ e_{z1} & e_{z2} & e_{z3} & e_{z4} & e_{z5} & e_{z6} \end{bmatrix} \tag{A.52}$$

†Often, in the literature, the commonly used notation drops distinctions between uppercase and lowercase subscripts: $e_{x1} = e_{11}$, $e_{z6} = e_{36}$, $e_{z3} = e_{33}$, and so on.

The e_{Ij} tensor is defined similarly as

$$e_{Ij} = \begin{bmatrix} e_{1x} & e_{1y} & e_{1z} \\ e_{2x} & e_{2y} & e_{2z} \\ e_{3x} & e_{3y} & e_{3z} \\ e_{4x} & e_{4y} & e_{4z} \\ e_{5x} & e_{5y} & e_{5z} \\ e_{6x} & e_{6y} & e_{6z} \end{bmatrix} \quad (A.53)$$

Note that $e_{1x} = e_{x1}$, and so on.

Symmetry once more reduces the number of independent quantities required from a possible 27. Thus, for a 43-m class of cubic crystal (e.g., gallium arsenide or indium antimonide) only one constant, e_{x4}, is required, where

$$e_{iJ} = \begin{bmatrix} 0 & 0 & 0 & e_{x4} & 0 & 0 \\ 0 & 0 & 0 & 0 & e_{x4} & 0 \\ 0 & 0 & 0 & 0 & 0 & e_{x4} \end{bmatrix} \quad (A.54)$$

For shear wave propagation in the z direction, with motion in the x direction, e_{x4} is the constant required. The longitudinal wave piezoelectric coupling constant in this direction is zero. On the other hand, for propagation along a $\langle 111 \rangle$ axis, we need the form of e_{iJ} in a rotated system, with axes in the $\langle 111 \rangle$ direction. In this case, there is a finite longitudinal piezoelectric coupling constant in the $\langle 111 \rangle$ direction, which can be determined from e_{x4}.

An isotropic material, or a cubic crystal with a center of symmetry such as silicon, is not piezoelectric. Thus $e_{x4} = 0$.

REFERENCE

1. The notation and physics in this appendix are dealt with in more detail in B. A. Auld, *Acoustic Fields and Waves in Solids*, Vol. 1 (New York: John Wiley & Sons, Inc., 1973).

Appendix B

Acoustic Parameters of Common Materials

Tables of fundamental parameters of selected isotropic solids, single crystals, liquids, gases, piezoelectric materials, and acousto-optic materials are given in this appendix. The data have been taken from a large number of sources. More information is available for some materials than for others, so inevitably, there are blanks in the tables where the value of a parameter could not be found by the author.

A large number of parameters for isotropic solids, liquids, and gases have been tabulated recently by Selfridge [1]. In Tables B.1, B.2, and B.3 we give values for selected materials taken from Selfridge's work, supplementing his data from other sources [2–7].

Table B.4 tabulates the parameters of a number of common piezoelectric ceramics. Most of these data are taken from an article by Berlincourt et al. [8]. It has also been supplemented with data on Japanese piezoelectric ceramics, kindly supplied to the author by Fukumoto of the Matsushita Company [9].

Recently, Selfridge has kindly supplied the author with data on his measurements of certain new piezoelectric ceramics [10]. Some of the more important experimental data for these ceramics, together with data for certain piezoelectric single-crystal materials and one plastic piezoelectric material, polyvinylidene fluoride (PVF_2), are given in Table B.5.

Figure B.1 illustrates the parameters used in Tables B.4 and B.5.

Data on selected acousto-optic materials is given in Tables B.6 and B.7. These data are taken from the work of Dixon [11], the *CRC Handbook of Lasers* [12], and from tables given by Yariv [13].

TABLE B.1 BULK MATERIAL CONSTANTS FOR SELECTED SOLIDS

All materials are isotropic or polycrystalline unless otherwise noted. The notation is that used in Chapters 1, 2, and 3.

Material	V_l (km/s)	V_s (km/s × 10³)	ρ_{m0} (kg/m³ × 10⁶)	Z_l (kg/m²-s × 10⁶)	Z_s (kg/m²-s × 10⁶)	σ (dB/cm or $A = \alpha/f^2$ dB-s²/m × 10⁻¹⁵)	α
Aluminum	6.42	3.04	2.70	17.33	8.21	0.355	$A = 0.86$
Araldite 506/956	2.62		1.16	3.55			
Bakelite	1.59		1.40	3.63			
Beryllium	12.89		1.87	24.10	16.60	0.046	
Bismuth	2.20	1.10	9.80	21.5	10.75	0.33	
Boroncarbide	11.0		2.40	26.40			
Brass, yellow, 70%Cu, 30%Zn	4.70	2.10	8.64	40.6	18.14	0.38	
Butyl rubber	1.80		1.11	2.0			
Cadmium	2.80	1.50	8.60	24.0			
Carbon, Pyrolitic, variable properties	3.3		2.2	7.3			
Vitreous	4.26	2.68	1.47	6.26	3.82	0.17	
Chromium	6.65	4.03	7.0	46.6	28.21	0.21	
Copper, rolled	5.01	2.27	8.93	44.6	20.2	0.37	
Epoxy DER332, MPDA 15 parts per hundred by weight of resin, 60°C cure	2.68	1.15	1.21	3.25	1.39	0.37	6.7 at 2 MHz
Silver	1.9	0.98	2.71	5.14	2.65	0.32	16 at 2 MHz
Fused quartz	5.96	3.76	2.20	13.1	8.26	0.17	$A = 0.13$
Glass							
Corning sheet	5.66		2.49	14.1	6.26	0.28	
Crown	5.1	2.8	2.24	11.4	6.44	0.245	
Schott FK3	4.91	2.85	2.26	11.1	7.62	0.24	
Pyrex	5.64	3.28	2.24	13.1	23.6	0.42	
Gold, hard drawn	3.24	1.20	19.7	63.8			$A = 2.3$
Granite	6.5		4.1	26.8			
Hydrogen, solid at 4.2°K	2.19		.089	0.19			
Inconel	5.7	3.0	8.28	47.2	24.8	0.31	

TABLE B.1 (Continued)

Material	V_l (km/s)	V_s (km/s × 10³)	ρ_{m0} (kg/m³ × 10⁶)	Z_l (kg/m²-s × 10⁶)	Z_s (kg/m²-s × 10⁶)	σ (dB/cm or dB-s²/m)	α $A = \alpha/f^2$ × 10⁻¹⁵
Indium	2.56			18.7			
Iron	5.9	3.2	7.3	46.4	25.2	0.29	
Lead 2.2	0.7	11.2	7.69	7.83	0.44		
Lithium niobate LiNbO₃, crystal-type trigonal 3m, propagation along Z axis	7.33		24.6	34.0			$A = 0.0047$
Lucite or Plexiglas	2.7	1.1	4.70	3.1	1.26	0.40	3.2 at 5 MHz
Magnesium, drawn annealed	5.77	3.05	1.15	5.3	1.74	0.32	
Molybdenum	6.3	3.4	10.0	63.1	34.1	0.29	
Monel	5.4	2.7	10.0	47.6	23.8	0.33	
Mylar	2.54		8.82	3.0			
Nickel	5.6	3.0	1.18	49.5	26.5	0.30	$A = 92$
Niobium	4.92	2.10	8.84	42.2	18.0	0.39	
Nylon	2.6	1.1	8.57	2.9	1.23	0.39	2.9 at 5 MHz
Paraffin wax	1.5		1.12	2.3			
Platinum	3.26	1.73	1.5	69.8	37.0	0.32	
Polyethylene, low density	1.95	0.54	21.4	1.79	0.50	0.487	
Polypropylene	2.74		0.92	2.40			5.1 at 5 MHz
Polystyrene	2.40	1.15	0.88	2.52	1.21	0.35	1.8 at 5 MHz
Porcelain	5.9		1.05	13.5			
PVC, gray rod stock	2.38		2.3	3.27			11.2 at 5 MHz
Quartz, SiO₂ Propagation along Z axis	6.32		1.38	16.7			
Propagation along BZ axis			2.53	2.53			
RTV rubber			5.00			13.4	
RTV-11	1.05		1.18	1.24			2.5 at 0.8 MHz
RTV-577	1.08		1.35	1.46			3.8 at 0.8 MHz
Rubidium	1.26		1.53	1.93			
Rutile, TiO₂, crystal-type tetragonal 4/mmm, propagation along Z axis	7.90		4.26	33.6			

Material					long. wave	
Sapphire, Al$_2$O$_3$, crystal-type trigonal 3m, propagation along Z axis	11.1	6.04	3.99	25.2		$A = 0.0021$
Scotchtape	1.9		1.16			
Silicon nitride ceramic	11.0	6.25	3.27	20.5	0.26	
Silicone rubber (Sylgard 182)	1.027		1.05			Very lossy
Silly putty	1.0		1.0			
Silver	3.6	1.6	10.6	16.9	0.38	
Steel, mild	5.9	3.2	7.90	24.9	0.29	
Stycast 1267	2.57		1.16			4.6 at 3 MHz
Tantalum	4.10		16.6			
Teflon	1.39	2.90	2.14	38.8		3.9 at 5 MHz
Thorium	2.40	1.56	11.3	21.6	0.134	
Tin	3.3	1.7	7.3	12.5	0.31	
Titanium	6.1	3.1	4.48	13.9	0.32	
Tungsten	5.2	2.9	19.4	56.3	0.27	
Uranium	3.4	2.0	18.5	37.1	0.24	
Vanadium	6.0	2.78	6.03	16.8	0.36	
Vinyl, rigid	2.23		1.33			12.8 at 5 MHz
YAG[4], Y$_3$Al$_{15}$O$_{12}$, crystal-type cubic m3m, propagation along [001] direction	8.43		4.55			$A = 0.0034$
Zinc	4.2	2.4	7.0	16.9	0.26	
Zirconium	4.65	2.25	6.48	14.6	0.35	

TABLE B.2 MATERIAL CONSTANTS FOR LIQUIDS

The notation is that used in Chapters 1, 2, and 3. The parameter M is a quality factor of importance for microscopy [4]. It is defined by the relation $M = [V(H_2O)/V]\{[\alpha/f^2(H_2O)]/[\alpha/f^2]\}^{1/2}$, where the parameters for water correspond to a temperature of 30°C.

Material	V (km/s)	dV/dT (m/s-°C)	ρ_{m0} (10^3 kg/m³)	Z (10^6 kg/m²-s)	$A = \alpha/f^2$ (10^{-15} s²/m)	M
Acetone, CH₃OH at 25°C	1.174	−4.5	0.791	1.07	54	0.77
Alcohol C₂H₅OH at 25°C	1.207	−4.0	0.79	0.95	48.5	0.84
Alcohol, methanol CH₃OH at 25°C	1.103	−3.2	0.791	0.872	30.2	1.10
Argon, liquid at 87°K	0.840		1.43	1.20	15.2	2.01
Benzene C₆H₆ at 25°C	1.295	−4.65	0.87	1.12	873	
Fluorinert FC-40	0.640		1.86	1.19		
Gallium at 30°C	2.87		6.09	17.5	1.58	1.82
Glycol, ethylene at 25°C	1.658	−2.1	1.113	1.845	120	
Helium-4,						
Liquid at 0.4°K	0.238		0.147	0.035	1.73	
Liquid at 2°K	0.227		0.145	0.033	77	
Liquid at 4.2°K	0.183		0.126	0.023	226	
Honey, Sue Bee Orange	2.03		1.42	2.89		
Hydrogen, liquid at 20°K	1.19		0.07	0.08	5.6	2.34
Mercury at 23°C	1.45		13.53	19.6	5.8	1.89
Nitrogen, liquid at 77°K	0.860		0.85	0.68	120	2.2
Oil						
Castor, at 20.2°C	1.507	−3.6	0.942	1.420	10100	
Silicone Dow 710 at 20°C	1.352		1.11	1.50	8200	
Oxygen, liquid at 90°K	0.900		1.14	1.0	9.9	2.5
Sea water at 25°C	1.531	2.4	1.025	1.569		
Sonotrack couplant, Echo	1.62		1.04	1.68		
Water						
Liquid at 20°C	1.48		1.00	1.483		
Liquid at 25°C	1.497	2.4	1.00	1.494	22	0.94
Liquid at 30°C	1.509		1.00	1.509	19.1	1.00
Liquid at 60°C	1.55		1.00	1.55	10.2	1.29
Xenon, liquid at 166°K	0.630		2.86	1.80	22.0	

The reader is referred to Selfridge [1] for further information on solids, liquids, and gases. Auld [14] gives extensive tables of the parameters for crystalline solids, and the CRC handbook [12] has extensive tables on the parameters for acousto-optic diffraction. Some of the other references used in this appendix also give information on materials not tabulated here.

TABLE B.3 ACOUSTIC CONSTANTS FOR GASES

The notation is that used in Chapters 1, 2, and 3. The parameter M is a quality factor of importance for microscopy. It is defined by the relation $M = [V(H_2O)/V]\{[\alpha/f^2(H_2O)]/[\alpha/f_2]\}^{1/2}$, where the parameters for water are for a temperature of 30°C.

Material	V (m/s)	ρ_{m0} (kg/m³)	Z ($\times 10^2$ kg/m²-s)	$A = \alpha/f^2$ (dB $\times 10^{-12}$ s²/m)	M
Air, dry					
At 0°C	331	1.293	4.29		
At 20°C	344	1.24	4.27	1.64	
At 100°C	386	1.11	4.27		
At 500°C	553	0.77	4.28		
Ammonia, NH$_3$ at 0°C	415	0.771	3.20		
Argon					
At 0°C	319	1.78	5.67		
At 20°C, 40 bar	323	≈70.4	≈227	4.12×10^{-3}	1.03
At 20°C, 250 bar	323	≈440	≈1437	8.3×10^{-4}	2.28
Carbon dioxide, CO$_2$					
at 0°C	259	1.977	5.12		
Carbon monoxide, CO					
at 0°C	338	1.25	4.22		
Nitrogen, N$_2$ at 0°C	334		1.251	4.18	
Oxygen, O$_2$					
At 0°C	316	1.429	4.51		
At 20°C	328		1.32	4.33	
Xenon at 20°C, 40 bar	178			9.53×10^{-3}	1.23

Acoustic Parameters of Common Materials

TABLE B.4 PROPERTIES OF COMMONLY USED PIEZOELECTRIC CERAMICS

The parameters and elastic boundary conditions appropriate to this table are illustrated in Fig. B.1. Most of the elastic and piezoelectric parameters are defined in Sec. 1.5. The attenuation per unit length α of the material can be stated in terms of its Q as follows: $\alpha = \pi f/VQ$, where f is the frequency and V the acoustic velocity.
Some additional parameters not defined in Sec. 1.5 or the additional tables and diagrams are:

Q_M The mechanical Q of the material.
Q_E The electrical Q of the material.
N_1 Frequency constant of a thin resonant rod of length l. $N_1 = fl$.
N_{3t} Frequency constant of a resonant plate of thickness l. $N_{3t} = fl$.

The parameters given below are small signal values taken at 25°C.

Material	s^E_{33}	s^E_{11}	Q_M	s^E_{44}	s^E_{66}	s^D_{33}	s^D_{11}	s^D_{44}	c^E_{33}	c^E_{11}	c^D_{33}	c^D_{11}	N_1	N_{3t}	V^D_l	V^D_s	Density
		pm²/N							10¹⁰ N/m²				Hz-m		m/s		(10³ kg/m³)
PZT-4[a]	15.5	12.3	500	39.0	32.7	7.90	10.9	19.3	11.5	13.9	15.9	14.5	1650	2000	4600	2630	7.5
PZT-5A[a]	18.8	16.4	75	47.5	44.3	9.46	14.4	25.2	11.1	12.1	14.7	12.6	1400	1890	4350	2260	7.75
PZT-6H[a]	20.7	16.5	65	43.5	42.6	8.99	14.1	23.7	11.7	12.6	15.7	13.0	1420	2000	4560	2375	7.5
PZT-6A[a]	13.0	10.7	450	—	27.8	9.2	10.1	—	13.1	—	15.5	—	1770	2140	4570	—	7.45
PZT-6B[a]	9.35	9.0	1300	28.2	24.0	8.05	8.8	24.2	16.3	16.8	17.7	16.9	1920	2225	4820	2340	7.55
PZT-7A[a]	13.9	10.7	600	39.5	27.8	7.85	9.7	21.8	13.1	14.8	17.5	15.7	1750	2100	4800	2490	7.6
PZT-8[a]	13.5	11.5	1000	31.9	29.8	8.0	10.4	22.6	12.3	13.7	16.1	14.0	1700	2070	4580	2420	7.6
PZT-2[a]	14.8	11.6	680	45.0	29.9	9.0	10.7	22.9	11.3	13.5	14.8	13.6	1680	2090	4410	2400	7.6
BaTiO₃[a]	9.5	9.1	600	22.8	23.6	7.1	8.7	17.5	14.6	15.0	17.1	15.0	2200	2520	5470	3160	5.7
PbTiO₃	8.0					6.3			13.2		16.8						
95 w% BaTiO₃, 5 w% CaTiO₃	9.1	8.6	400	22.2	22.4	7.0	8.3	17.1	15.0	15.8	17.7		2290	2740	5640	3240	5.55
NRE-4[b]	—	8.1		—	—	—	—	—	—	—	—	—	2310	2760	—	—	5.7
PbNb₂O₆	25.4	—	11	—	—	21.8	—	—	—	—	—	—	—	—	—	—	6.0
Pb₀.₆Ba₀.₄Nb₂O₆[c]	—	11.5	250	—	—	—	10.9	—	—	—	—	—	1915	—	—	—	5.9
Na₀.₅K₀.₅NbO₃	10.1	8.2	240	27.0	—	6.4	7.6	15.8	16.8	—	21.4	—	2570	—	6940	3760	4.46
PCMUS-1[d]	12.9					8.5			13.1		15.3						
PCMUS-2[d]	14.5					7.6			11.8		16.0						
PCM-5A[d]	17.6					8.7			11.6		14.9						
PCM-33[d]	19.3					8.7			11.5		15.5						

	k'_{z3}	k_p	k_{z1}	k_{z3}	k_{x5}	k_T	$\varepsilon_{zz}^T/\varepsilon_0$	Q_E	$\varepsilon_{zz}^S/\varepsilon_0$	$\varepsilon_{xx}^T/\varepsilon_0$	$\varepsilon_{xx}^S/\varepsilon_0$
PZT-4[a]		−0.58	−0.33	0.70	0.71	0.51	1300	250	635	1475	730
PZT-5A[a]	0.66	−0.60	−0.34	0.705	0.685	0.49	1700	50	830	1730	916
PZT-5H[a]	0.70	−0.65	−0.39	0.75	0.675	0.505	3400	50	1470	3130	1700
PZT-6A[a]		−0.42	−0.25	0.54	—	0.39	1050	50	730	—	—
PZT-6B[a]		−0.25	−0.145	0.375	0.377	0.30	460	110	386	475	407
PZT-7A[a]	0.62	−0.51	−0.30	0.66	0.67	0.50	425	60	235	840	460
PZT-8[a]		−0.51	−0.30	0.64	0.55	0.48	1000	250	580	1290	900
PZT-2[a]		−0.47	−0.28	0.63	0.70	0.51	450	200	260	990	504
BaTiO$_3$[a]	0.47	−0.36	−0.21	0.50	0.48	0.38	1700	100	1260	1450	1115
PbTiO$_3$	0.46			.46		.46					
95 w% BaTiO$_3$, 5 w% CaTiO$_3$		−0.33	−0.19	0.48	0.48	0.38	1200	170	910	1300	1000
NRE-4[b]											
PbNb$_2$O$_6$		−0.31	−0.18	0.46	0.46	0.36	1420	200	1110	—	—
Pb$_{0.6}$Ba$_{0.4}$Nb$_2$O$_6$[c]		−0.07	−0.045	0.38	—	0.37	225	100	190	—	—
Na$_{0.5}$K$_{0.5}$NbO$_3$		−0.38	−0.22	0.55	—	—	1500	100	—	—	—
PCMUS-1[d]	0.66	−0.46	−0.27	0.605	0.645	0.46	496	70	306	938	545
PCMUS-2[d]	0.64					0.54	785		484		
PCM-5A[d]	0.66					0.55	734		369		
PCM-33[d]	0.69					0.50	1710		784		
						0.50	3530		1518		

Material	d_{z3}	d_{z1}	d_{x5}	g_{z3}	g_{z1}	g_{x5}	e_{z3}	e_{z1}	e_{x5}
	10^{-12} C/N			10^{-3} V-m/N			C/m^2		
PZT-4[a]	289	−123	496	25.1	−10.7	38.0	15.1	−5.2	12.7
PZT-5A[a]	374	−171	584	24.9	−11.4	38.0	15.8	−5.4	12.3
PZT-5H[a]	593	−274	741	1907	−9.1	26.8	23.3	−6.5	17.0
PZT-6A[a]	189	−80	—	20.4	−8.6	—	12.5	—	—
PZT-6B[a]	71	−27	130	1704	−6.6	31.0	7.1	−0.9	4.6
PZT-7A[a]	150	−60	362	3908	−15.9	48.8	9.5	−201	9.2
PZT-8[a]	225	−97	330	2504	−10.9	29.0	13.2	−4.0	10.4
PZT-2[a]	152	−60	440	3802	−15.1	50.1	9.0	−109	9.8
BaTiO$_3$[a]	190	−78	260	1206	−5.2	20.2	17.5	−4.3	11.4
95 w% BaTiO$_3$, 5 w% CaTiO$_3$	149	−58	242	14.0	−5.45	21.0	13.5	−3.1	10.9
NRE-4[b]	150	−59	—	1109	−4.7	—	—	—	—

TABLE B.4 (Continued)

Material	d_{z3}	d_{z1}	d_{x5}	g_{z3}	g_{z1}	g_{x5}	e_{z3}	e_{z1}	e_{x5}
		10^{-12} C/N			10^{-3} V-m/N			C/m^2	
PbNb$_2$O$_6$	85	~ -9	—	4205	~ -4.5	—	—	—	—
Pb$_{0.6}$Ba$_{0.4}$Nb$_2$O$_6$[c]	220	−90	—	1606	−6.8	—	—	—	—
Na$_{0.5}$K$_{0.5}$NbO$_3$	127	−51	306	29.0	−11.6	36.9	9.8	—	11.3
PCMUS-1[d]	176	−88							
PCMUS-2[d]	211	−93							
PCM-5A[d]	367	−186							
PCM-33[d]	575	−262							

[a]Trademark, Vernitron Piezoelectric Division.
[b]95 w% BaTiO$_3$, 5 w% CaTiO$_3$, plus 0.75 w% CoCO$_3$.
[c]General Electric Company.
[d]Trademark, Matsushita Electric Industrial Co. Ltd.

TABLE B.5 PROPERTIES OF SELECTED TRANSDUCER MATERIALS

The parameters in this table are defined in Fig. B.1 and in Chapters 1 and 2. The mechanical Q of the material is Q_M. The attenuation is $\alpha = \pi\omega/QV$. The Curie temperature of the material is T_C.

A. Longitudinal Waves

Material	V_l^p (10^3 m/s)	Z_l^p (10^6 kg/ms)	Q_M	$\varepsilon^S/\varepsilon_0$	ρ_{m0} (10^3 kg/m³)	k_T	k_p	tan δ	T_C (°C)
Aluminum nitride AlN thin film, hexagonal 6 mm, Z cut	10.4	34.0		8.5	3.27	0.17			
Cadmium sulfide CdS single crystal and thin film, hexagonal 6 mm, Z cut	4.46	36.0		9.5	5.68	0.15			
Lithium niobate LiNbO₃ single crystal, trigonal 3m, 36° Y cut	7.36	34.2		39.0	4.64	0.49		0.001	1150
Keramos K83 modified lead metaniobate	5.95	25.6	110	150	4.3	0.41			570
Murata P3 barium titanate	5.75	31.3	100	885	5.45	0.42		0.003	110
Murate P5 zirconate titanate	4.33	31.6	80	847	7.30	0.36		0.011	260
Murata P6 zirconate titanate	4.78	35.1	70	883	7.34	0.47		0.014	290
Murata P7 zirconate titanate	4.68	36.0	65	1000	7.69	0.51		0.019	320
Murata "surface wave material"	4.71	37.4	1000	240	7.95	0.48	0.25	0.0014	280
Pennwalt kynar polyvinylidene fluoride (PVF₂) plastic film $e_{z3} = -108 \times 10^{-3}$, $e_{z1} = 69 \times 10^{-3}$, $e_{z2} = 9 \times 10^{-3}$, $k_T = 0.11$, $k_{z1} = 0.072$, $k_{z2} = 9.4 \times 10^{-3}$ $Q \sim$ constant to 1 GHz	2.2	3.92	19	12	1.78	0.11		0.015 at 10^4 Hz	
Quartz, SiO₂ single crystal, trigonal 32, X cut	5.00	13.3		4.5	2.65	0.093			
Zinc oxide, ZnO single crystral and thin film, hexagonal 6mm, Z cut	6.33	36.0		8.8	5.68	0.28			

TABLE B.5 (Continued)

B. Shear Waves

Material	V_D^S (10^3 m/s)	Z_D^S (10^6 kg/m²-s)	$\varepsilon^S/\varepsilon_0$	k_T
Cadmium sulfide, X cut	1.76	8.5	9.0	0.19
Lithium niobate, 163°				
Y cut	4.44	20.6	58.1	0.55
X cut	4.80	22.6	44.0	0.68
Murata "surface wave material"	2.78	22.1	360	0.50
Quartz, X cut	3.80	10.1	4.5	0.14
Zinc oxide, X cut	2.72	15.5	8.3	0.32

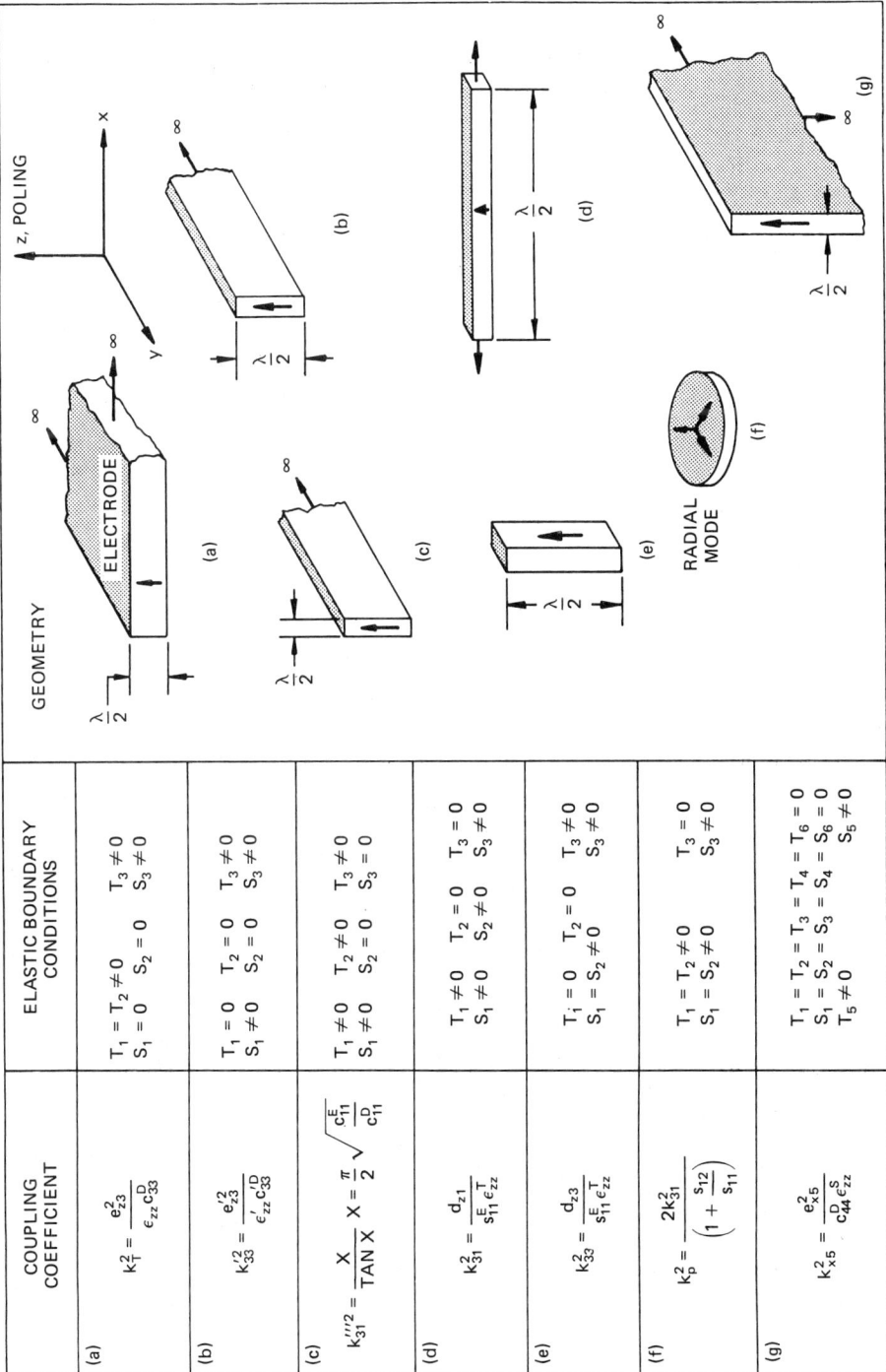

Figure B.1 Parameters used for defining piezoelectric ceramics in Tables B.4 and B.5.

TABLE B.6 PROPERTIES OF SOME COMMON ACOUSTO-OPTIC MATERIALS

The parameters used are defined in Sec. 4.9. The acousto-optic figure of merit $M = p^2 n_0^6 / \rho_{m0} V_a^3$ is defined in Eq. (4.9.26) [11]. Other related figures of merit are often used in the literature [11, 12]. The parameters used in this table are given by Yariv [13].

Material	ρ_{m0} (10^3 kg/m^3)	V_a (10^3 m/s)	n_0	p	$M/M(H_2O)$
Water	1.0	1.5	1.33	0.31	1.0
Extra-dense flint glass	6.3	3.1	1.92	0.25	0.12
Fused quartz (SiO$_2$)	2.2	5.97	1.46	0.20	0.006
Polystyrene	1.06	2.35	1.59	0.31	0.8
KRS-5	7.4	2.11	2.60	0.21	1.6
Lithium niobate (LiNbO$_3$)	4.7	7.40	2.25	0.15	0.012
Lithium fluoride (LiF)	2.6	6.00	1.39	0.13	0.001
Rutile (TiO$_2$)	4.26	10.30	2.60	0.05	0.001
Sapphire (Al$_2$O$_3$)	4.0	11.00	1.76	0.17	0.001
Lead molybdate (PbMO$_4$)	6.95	3.75	2.30	0.28	0.22
Alpha iodic acid (HIO$_3$)	4.63	2.44	1.90	0.41	0.5
Tellurium dioxide (TeO$_2$) (slow shear wave)	5.99	0.617	2.35	0.09	5.0

TABLE B.7 MATERIAL CONSTANTS OF SOME COMMON ACOUSTO-OPTIC MATERIALS

The parameters in this table are those used in Sec. 4.9. The parameter $M = p^2 n_0^6 / \rho_{m0} V_a^3$ is the acousto-optic figure of merit [11, 12]. The parameters used in this table are taken from Yariv [13].

Material	Optical wavelength (λ-μm)	n_0	ρ_{m0} (10^3 kg/m^3)	Acoustic wave polarization and direction	V_a (10^3 m/s)	Optical wave polarization and direction	M (10^{-15} s^3/kg)
Fused quartz	0.63	1.46	2.2	Long.	5.95	$\perp$	1.51
				Trans.	3.76	$\parallel$ or $\perp$	0.467
GaP	0.63	3.31	4.13	Long. in [110]	6.32	$\parallel$	44.6
				Trans. in [100]	4.13	$\parallel$ or $\perp$ in [010]	24.1
GaAs	1.15	3.37	5.34	Long. in [110]	5.15	$\parallel$	104
				Trans. in [100]	3.32	$\parallel$ or $\perp$ in [010]	46.3
TiO$_2$	0.63	2.58	4.6	Long. in [11–20]	7.86	$\perp$ in [001]	3.93
LiNbO$_3$	0.63	2.20	4.7	Long. in [11–20]	6.57	a	6.99
YAG	0.63	1.83	4.2	Long. in [100]	8.53	$\parallel$	0.012
				Long. in [110]	8.60	$\perp$	0.073
YIG	1.15	2.22	5.17	Long. in [100]	7.21	$\perp$	0.33
LiTaO$_3$	0.63	2.18	7.45	Long. in [001]	6.19	$\parallel$	1.37
As$_2$S$_3$	0.63	2.61	3.20	Long.	2.6	$\perp$	433
	1.15	2.46		Long.		$\parallel$	347
SF-4	0.63	1.616	3.59	Long.	3.63	$\perp$	4.51
β-ZnS	0.63	2.35	4.10	Long. in [110]	5.51	$\parallel$ in [001]	3.41
				Trans. in [110]	2.165	$\parallel$ or $\perp$ in [001]	0.57
α-Al$_2$O$_3$	0.63	1.76	4.0	Long. in [001]	11.15	$\parallel$ in [11–20]	0.34
CdS	0.63	2.44	4.82	Long. in [11–20]	4.17	$\parallel$	12.1
ADP	0.63	1.58	1.803	Long. in [100]	6.15	$\parallel$ in [010]	2.78
				Trans. in [100]	1.83	$\parallel$ or $\perp$ in [001]	6.43

TABLE B.7 Continued

Material	Optical wavelength (λ-μm)	n_0	ρ_{m0} (10^3 kg/m^3)	Acoustic wave polarization and direction	V_a (10^3 m/s)	Optical wave polarization and direction	M (10^{-15} s^3/kg)
KDP	0.63	1.51	2.34	Long. in [100]	5.50	$\parallel$ in [010]	1.91
				Trans. in [100]		$\parallel$ or $\perp$ in [001]	3.83
H$_2$O	0.63	1.33	1.0	Long.	1.5		160
Te	10.6	4.8	6.24	Long. in [11–20]	2.2	$\parallel$ in [0001]	4400
PbMoO$_4$	0.63	2.4		Long. $\parallel$ c axis	3.75	$\parallel$ or $\perp$	73

^aThe optical-beam direction actually differs from that indicated by the magnitude of the Bragg angle. The polarization is defined as parallel or perpendicular to the scattering plane formed by the acoustic and optical k vectors.

REFERENCES

1. A. R. Selfridge, "Approximate Material Properties in Isotropic Materials," *IEEE Trans. Sonics Ultrason.*, SU-32, No. 3 (May 1985), 381–94.
2. T. M. Reeder and D. K. Winslow, "Characteristics of Microwave Acoustic Transducers for Volume Wave Excitation," *IEEE Trans. Microwave Theory Tech.*, MTT-17, No. 11 (Nov. 1969), 927–41.
3. A. J. Slobodnik, Jr., R. T. Delmonico, and E. D. Conway, *Microwave Acoustics Handbook,* Vol. 3: *Bulk Wave Velocities*, in-house report RADC-TR-80-188 (May 1980), Rome Air Development Center, Air Force Systems Command, Griffiss Air Force Base, New York 13441.
4. R. A. Lemons and C. F. Quate, "Acoustic Microscopy," Chapter 1 in *Physical Acoustics: Principals and Methods*, Vol. XIV, W. P. Mason and R. N. Thurston, eds. New York: Academic Press, Inc. 1979, pp. 1–92.
5. J. Heiserman, D. Rugar, and C. F. Quate, "Cryogenic Acoustic Microscopy," *J. Acoust. Soc. Am.*, 67, No. 5 (May 1980), 1629–37.
6. H. K. Wickramsinghe and C. R. Petts, "Gas Medium Acoustic Microscopy," in *Scanned Image Microscopy*, E. A. Ash, ed. London: Academic Press, Inc. (London) Ltd., 1980.
7. M. Tone, T. Yano, A. Fukumoto, "High-Frequency Ultrasonic Transducer Operating in Air," *Jpn. J. Appl. Phys.*, Part 2, 23, No. 6 (June 1984), L436–38.
8. D. A. Berlincourt, D. R. Curran, and H. Jaffe, "Piezoelectric and Piezomagnetic Materials and Their Function in Transducers," in *Physical Acoustics*, Vol. I, W. P. Mason, ed. New York: Academic Press, Inc. 1964, Part A, Chapter 3, pp. 169–270.
9. A. Fukumoto, "The Application of Piezoelectric Ceramics in Diagnostic Ultrasound Transducers," private communication, unpublished.
10. A. R. Selfridge, private communication.
11. R. W. Dixon, "Photoelastic Properties of Selected Materials and Their Relevance for Applications to Acoustic Light Modulators and Scanners," *J. Appl. Phys.*, 38, No. 13 (Dec. 1967), 5149–53.

12. D. A. Pinnow, "Elastooptical Materials," in *CRC Handbook of Lasers with Selected Data on Optical Technology*, R. J. Pressley, ed. Cleveland, Ohio: CRC Press, Inc., 1971, pp. 478–88.
13. A. Yariv, *Quantum Electronics*, 2nd ed. New York: John Wiley & Sons, Inc., 1975.
14. B. A. Auld, *Acoustic Fields and Waves in Solids*, Vol. 1. New York: John Wiley & Sons, 1973.

Appendix C

Poynting's Theorem in Piezoelectric Media

We derive Poynting's theorem for piezoelectric media by generalizing the derivation of Eq. (1.1.21), the one-dimensional Poynting's theorem for nonpiezoelectric media. We shall show that Poynting's theorem can be generalized by adding the electromagnetic power flow in the medium to the mechanical power flow.

Appendix A shows that the three-dimensional equation of motion for an acoustic wave, in which all components vary as $\exp(j\omega t)$, is, in symbolic notation,

$$\nabla \cdot \mathbf{T} = j\omega \rho_{m0} \mathbf{v} \qquad (C.1)$$

or, in tensor notation,

$$\frac{\partial T_{ij}}{\partial x_j} = j\omega \rho_{m0} v_i \qquad (C.2)$$

Similarly, the relation between $\mathbf{v}$ and $\mathbf{S}$ is, in symbolic notation,

$$\nabla_s \mathbf{v} = j\omega \mathbf{S} \qquad (C.3)$$

or, in tensor notation,

$$\frac{\partial v_i}{\partial x_j} = j\omega S_{ij} \qquad (C.4)$$

It is convenient to consider the expression related to the Poynting vector

$$\frac{\partial}{\partial x_j}(v_i^* T_{ij}) = T_{ij}\frac{\partial v_i^*}{\partial x_j} + v_i^* \frac{\partial T_{ij}}{\partial x_j} \qquad (C.5)$$

Substituting from Eq. (C.2) and the complex conjugate form of Eq. (C.4) into Eq. (C.5), it follows that

$$\frac{\partial}{\partial x_j}(v_i^* T_{ij}) = j\omega(\rho_{m0} v_i v_i^* - T_{ij} S_{ij}^*) \qquad \text{(C.6)}$$

In reduced symbolic notation, Eq. (C.6) becomes

$$\nabla \cdot (\mathbf{v}^* \cdot \mathbf{T}) = j\omega(\rho_{m0}\mathbf{v} \cdot \mathbf{v}^* - \mathbf{T} : \mathbf{S}^*) \qquad \text{(C.7)}$$

To keep the notation simple, it will be more convenient from this point to use the symbolic tensor notation. Thus, from Eq. (C.7) and the piezoelectric constitutive relation

$$\mathbf{T} = \mathbf{c}^E : \mathbf{S} - \mathbf{e} \cdot \mathbf{E} \qquad \text{(C.8)}$$

we see that

$$\nabla \cdot (\mathbf{v}^* \cdot \mathbf{T}) = j\omega\rho_{m0}\mathbf{v} \cdot \mathbf{v}^* - j\omega\mathbf{S}^* : \mathbf{c}^E : \mathbf{S} + j\omega\mathbf{S}^* : \mathbf{e} \cdot \mathbf{E} \qquad \text{(C.9)}$$

It follows from symmetry that $\mathbf{S}^* : \mathbf{e} \cdot \mathbf{E} = \mathbf{E} \cdot \mathbf{e} : \mathbf{S}^*$. Hence we can substitute the piezoelectric constitutive relation

$$\mathbf{D} = \boldsymbol{\varepsilon}^S \cdot \mathbf{E} + \mathbf{e} : \mathbf{S} \qquad \text{(C.10)}$$

into Eq. (C.9), to write it in the form

$$\nabla \cdot (\mathbf{v}^* \cdot \mathbf{T}) = j\omega\rho_{m0}\mathbf{v} \cdot \mathbf{v}^* - j\omega\mathbf{S}^* : \mathbf{c}^E : \mathbf{S}$$
$$- j\omega\mathbf{E}^* \cdot \boldsymbol{\varepsilon}^S \cdot \mathbf{E} + j\omega\mathbf{D}^* \cdot \mathbf{E} \qquad \text{(C.11)}$$

Let us now consider the last term in Eq. (C.11). We use the complex conjugate form of Maxwell's equation,

$$\nabla \times \mathbf{H}^* = -j\omega\mathbf{D}^* + \mathbf{i}_c^* \qquad \text{(C.12)}$$

where $\mathbf{i}_c$ is the conduction current in the medium. After taking the dot product of Eq. (C.12) with $\mathbf{E}$, it follows that

$$j\omega\mathbf{D}^* \cdot \mathbf{E} = \mathbf{E} \cdot \mathbf{i}_c^* - \mathbf{E} \cdot (\nabla \times \mathbf{H}^*) \qquad \text{(C.13)}$$

We now use the vector relation

$$\nabla \cdot (\mathbf{A} \times \mathbf{B}) = \mathbf{B} \cdot (\nabla \times \mathbf{A}) - \mathbf{A} \cdot (\nabla \times \mathbf{B}) \qquad \text{(C.14)}$$

with Maxwell's equation,

$$\nabla \times \mathbf{E} = -j\omega\mu\mathbf{H} \qquad \text{(C.15)}$$

and it follows that Eq. (C.13) can be written in the form

$$j\omega\mathbf{D}^* \cdot \mathbf{E} = \mathbf{E} \cdot \mathbf{i}_c^* + \nabla \cdot (\mathbf{E} \times \mathbf{H}^*) - \mathbf{H}^* \cdot (\nabla \times \mathbf{E})$$
$$= \mathbf{E} \cdot \mathbf{i}_c^* + \nabla \cdot (\mathbf{E} \times \mathbf{H}^*) + j\omega\mu\mathbf{H} \cdot \mathbf{H}^* \qquad \text{(C.16)}$$

Finally, substituting Eq. (C.16) into Eq. (C.11) leads to the following relation:

$$\nabla \cdot (\mathbf{E} \times \mathbf{H}^* - \mathbf{v}^* \cdot \mathbf{T}) = j\omega \mathbf{S}^* : \mathbf{c}^E : \mathbf{S} - j\omega \rho_{m0} \mathbf{v} \cdot \mathbf{v}^*$$
$$+ j\omega \mathbf{E}^* \cdot \boldsymbol{\varepsilon}^s \cdot \mathbf{E} - j\omega \mu \mathbf{H}^* \cdot \mathbf{H} - \mathbf{i}_c^* \cdot \mathbf{E} \quad \quad (C.17)$$

Assuming that $\mathbf{c}^E$, ρ_{m0}, $\boldsymbol{\varepsilon}^s$, and μ are real, and adding Eq. (C.17) to its complex conjugate, it follows that

$$\text{Re}\,[\nabla \cdot (\mathbf{E} \times \mathbf{H}^* - \mathbf{v}^* \cdot \mathbf{T})] = -\text{Re}\,(\mathbf{i}_c^* \cdot \mathbf{E}) \quad \quad (C.18)$$

Suppose that we now integrate this relation over a region of volume V enclosed by a surface s. After using Gauss's theorem, we find that

$$\frac{1}{2} \text{Re} \int_s (\mathbf{E} \times \mathbf{H}^* - \mathbf{v}^* \cdot \mathbf{T}) \cdot \mathbf{n}\, ds = -\frac{1}{2} \text{Re} \int (\mathbf{i}_c^* \cdot \mathbf{E}\, dV) \quad \quad (C.19)$$

where $\mathbf{n}$ is the outward normal to the surface.

The left-hand side of this equation is the total power emitted from the volume V by the piezoelectric material. Therefore, the generalization of both the electromagnetic and mechanical forms of Poynting's theorem agree with intuition. We merely add the two contributions to the total power.

Appendix D

Determination of the Impedance Z_0 in Terms of $\Delta V/V$

We now determine the impedance Z_0 in Eq. (2.5.35) for a piezoelectric medium excited by a charge at its surface. We drop the subscript n where convenient and write Eq. (2.5.27) in the form

$$\frac{dA}{dz} + jkA = \frac{j\omega}{4}\int \rho_s \phi_0^* \, dl = \frac{j\omega}{4} \sigma \phi_0^* \tag{D.1}$$

where $\sigma = \rho_s w$, ρ_s is assumed to be uniform over the width w of the beam, and ϕ_0^* is the potential at the surface normalized to unit power.

We define a quantity Z_0 with the dimensions of impedance

$$Z_0 = \frac{\phi_0 \phi_0^*}{2} \tag{D.2}$$

Then, by assuming ϕ_0 is real because of the normalization employed, Eq. (D.1) can be written in the form

$$\frac{dA}{dz} + jkA = \frac{j\omega(2Z_0)^{1/2}\sigma}{4} \tag{D.3}$$

This coupling impedance is a measure of the strength of the electrical potential at the surface of the substrate and, hence, the coupling to an interdigital transducer placed on the substrate. It can be calculated directly by carrying out the field theory for Rayleigh wave propagation along the substrate. Alternatively, it can be related to K^2 or to $\Delta V/V$.

We shall now show how to relate the impedance Z_0 to $\Delta V/V$, the relative change in the velocity of the wave when a perfect conductor is laid down on the substrate.

When a perfect conductor is laid down on the substrate, the propagation constant of the wave changes from k to k'. We assume that $|k' - k| \ll |k|$. In this case, A now varies as $\exp(-jk'z)$, as does the surface charge on the metal $\sigma(z)$. We can therefore write Eq. (D.3) in the form

$$k - k' = \frac{\omega(2Z_0)^{1/2}\sigma}{4A} \tag{D.4}$$

We now need a further relation between $\sigma(z)$ and $A(z)$ to determine the change in k. This can be obtained by considering the potential at the metal conductor, which is made up of two parts: (1) the potential due to the surface wave ϕ_a, where

$$\phi_a = A\phi_0 \tag{D.5}$$

and (2) the electrostatic potential ϕ_s due to the charge σ itself. The total potential at the surface is

$$\phi = \phi_s + \phi_a \tag{D.6}$$

At a perfect conductor, we have the simple condition $\phi = 0$.

The solution of Poisson's or Laplace's equation creates a potential ϕ_s, the *electrostatic potential*, due to a charge distribution $\sigma(z)$, which varies as $\exp(-jk'z)$. Using the notation of Eq. (D.6), the solution of Laplace's equation in the region $y \geq 0$ is

$$\phi_s = Ce^{-k'y}e^{-jk'z}$$

$$E_{ys} = -\frac{\partial \phi_s}{\partial y} = Ck'e^{-k'y}e^{-jk'z} \tag{D.7}$$

$$E_{zs} = -\frac{\partial \phi_s}{\partial z} = jk'Ce^{-k'y}e^{-jk'z}$$

For the region $y < 0$ within the substrate, if the substrate is semi-infinite so that $\phi_s \to 0$ as $y \to -\infty$, then

$$\phi_s = De^{k'y}e^{jk'z}$$

$$E_{ys} = -Dk'e^{k'y}e^{-jk'z} \tag{D.8}$$

$$E_{zs} = jk'De^{k'y}e^{-jk'z}$$

where C and D are constants.

The electrostatic potential is continuous at $y = 0$, so that $C = D$. Furthermore, the value of D_y must obey the boundary condition

$$\varepsilon_0 E_{ys}^+ - \varepsilon E_{ys}^- = D_{ys}^+ - D_{ys}^- = \frac{\sigma}{w} \tag{D.9}$$

where the substrate is assumed to be isotropic, with a permittivity ε, and the medium above the substrate has a permittivity ε_0.

It then follows that

$$\phi_s(0) = \frac{\sigma}{k'w(\varepsilon_0 + \varepsilon)} \qquad (D.10)$$

This is the electrostatic potential at the surface of the substrate due to the surface charge in the conductor. We can now use Eqs. (D.4)–(D.6) and Eq. (D.10), with the condition that $\phi(0) = 0$, to obtain

$$\frac{\Delta V}{V} = \frac{k - k'}{k'} = \frac{-\omega w \phi_0^2}{4}(\varepsilon_0 + \varepsilon) \qquad (D.11)$$

It follows that

$$Z_0 = \frac{2}{\omega w(\varepsilon_0 + \varepsilon)} \left|\frac{\Delta V}{V}\right| \qquad (D.12)$$

Thus the impedance Z_0 is directly proportional to the quantity $\Delta V/V$, the relative change in acoustic velocity when an infinitesimally thin metal conductor is laid down on the piezoelectric substrate.

Appendix E

A Rigorous Derivation of Normal-Mode Theory

Here we will carry out a derivation of the normal-mode theory that is parallel to that of Sec. 2.5.2, using a rigorous approach based directly on the constitutive relations for the waves. To keep the technique as simple as possible, we will carry out the derivation for a nonpiezoelectric medium. This can be easily generalized to cover the case of a piezoelectric medium, yielding a result similar to Eq. (2.5.25).

We first obtain an orthogonality theorem for acoustic waves. This is needed to determine how an acoustic wave is excited by an external source or by a perturbation (i.e., to determine the coefficient α). It is shown in Appendix A that the three-dimensional equation of motion for an acoustic wave, in which all components vary as $\exp(j\omega t)$, is

$$\nabla \cdot \hat{\mathbf{T}} = j\omega\rho_{m0}\hat{\mathbf{v}} \tag{E.1}$$

or

$$\frac{\partial \hat{T}_{ij}}{\partial x_i} = j\omega\rho_{m0}\hat{v}_j \tag{E.2}$$

where we have used the sign $\hat{}$ to indicate parameters that vary with z. The relationship between $\hat{\mathbf{v}}$ and $\hat{\mathbf{S}}$ is, in reduced notation,

$$\nabla_s \hat{\mathbf{v}} = j\omega\hat{\mathbf{S}} \tag{E.3}$$

or, in tensor notation,

$$\frac{1}{2}\left(\frac{\partial \hat{v}_i}{\partial x_j} + \frac{\partial \hat{v}_j}{\partial x_i}\right) = j\omega\hat{S}_{ij} \tag{E.4}$$

Now, suppose that there are two modes of the system denoted by the subscripts m and n, such that

$$\hat{\mathbf{T}}_n(x, y, z) = \mathbf{T}_n(x, y)e^{-jk_n z}$$
$$\hat{\mathbf{T}}_m(x, y, z) = \mathbf{T}_m(x, y)e^{-jk_m z} \quad (E.5)$$

Similar definitions can be given for the other fields associated with these waves. By writing the appropriate equations [Eqs. (E.1) and (E.2)] for each mode, as well as their complex conjugates (i.e., reversing the sign of $j\omega$), we obtain the following relation:

$$\hat{X}_m^* \cdot \nabla \cdot \hat{\mathbf{T}}_n + \hat{\mathbf{T}}_n : \nabla_s \hat{\mathbf{v}}_m^* = j\omega\rho_{m0}\hat{\mathbf{v}}_m^* \cdot \hat{\mathbf{v}}_n - j\omega\hat{\mathbf{S}}_m^* : \hat{\mathbf{T}}_n \quad (E.6)$$

or

$$\hat{v}_{mi}^* \frac{\partial \hat{T}_{nij}}{\partial x_j} + \hat{T}_{nij} \frac{\partial \hat{v}_{mi}^*}{\partial x_j} = j\omega\rho_{m0}\hat{v}_{mi}^*\hat{v}_{ni} - j\omega\hat{S}_{mij}^*\hat{T}_{nij} \quad (E.7)$$

Using the tensor relation $\mathbf{a} \cdot \nabla \cdot \mathbf{B} + \mathbf{B} : \nabla_s \mathbf{a} = \nabla \cdot (\mathbf{a} \cdot \mathbf{B})$, Eq. (E.6) can be written in the form

$$\nabla \cdot (\hat{\mathbf{v}}_m^* \cdot \hat{\mathbf{T}}_n) = j\omega\rho_{m0}\hat{\mathbf{v}}_m^* \cdot \hat{\mathbf{v}}_n - j\omega\hat{\mathbf{S}}_m^* : \hat{\mathbf{T}}_n \quad (E.8)$$

or in the form

$$\frac{\partial}{\partial x_j}(\hat{v}_{mi}^*\hat{T}_{nij}) = j\omega\rho_{m0}\hat{v}_{mi}^*\hat{v}_{ni} - j\omega\hat{S}_{mij}^*\hat{T}_{nij} \quad (E.9)$$

It follows similarly that

$$\nabla \cdot (\hat{\mathbf{v}}_n \cdot \hat{\mathbf{T}}_m^*) = -j\omega\rho_{m0}\hat{\mathbf{v}}_m^* \cdot \hat{\mathbf{v}}_n + j\omega\hat{\mathbf{T}}_m^* : \hat{\mathbf{S}}_n \quad (E.10)$$

By using the constitutive relations

$$\hat{\mathbf{T}}_m = \mathbf{c} : \hat{\mathbf{S}}_m$$
$$\hat{\mathbf{T}}_n = \mathbf{c} : \hat{\mathbf{S}}_n \quad (E.11)$$

and the symmetry of the tensor $\mathbf{c}$, we can add Eqs. (E.8) and (E.10) to show that

$$\nabla \cdot (\hat{\mathbf{v}}_m^* \cdot \mathbf{T}_n + \hat{\mathbf{v}}_n \cdot \hat{\mathbf{T}}_m) = 0 \quad (E.12)$$

or

$$\frac{\partial}{\partial x_j}(\hat{v}_{mi}^*\hat{T}_{nij} + \hat{v}_{ni}\hat{T}_{mij}^*) = 0 \quad (E.13)$$

We can now take advantage of Gauss's theorem, which is commonly employed in electromagnetic (EM) theory, and use it for this acoustic problem. With the reduced notation, we can write, for a vector $\mathbf{B}$, $\int \nabla \cdot \mathbf{B}\, dV = \int \mathbf{B} \cdot \mathbf{n}\, ds$, where the surface integral is taken around the enclosing surface of the volume V and $\mathbf{n}$ is the outward vector normal to the enclosing surface.

We consider a region of length dz in a cylindrical system uniform in the z

direction, as illustrated in Fig. 2.5.1, and integrate through its volume. This yields the two-dimensional form of Gauss's theorem,

$$\int_s \nabla \cdot \mathbf{B}\, ds = \int_s \frac{\partial B_z}{\partial z}\, ds + \oint_l \mathbf{B} \cdot \mathbf{n}\, dl \tag{E.14}$$

where s is the cross section of the system and the line integral is taken around the circumference l of this area.

By using the relation in Eq. (E.14), it follows that

$$\frac{\partial}{\partial z}\int_s (\hat{\mathbf{v}}_m^* \cdot \hat{\mathbf{T}}_n + \hat{\mathbf{v}}_n \cdot \hat{\mathbf{T}}_m^*)_z\, ds + \oint_l (\hat{\mathbf{v}}_m^* \cdot \hat{\mathbf{T}}_n + \hat{\mathbf{v}}_n \cdot \hat{\mathbf{T}}_m^*) \cdot \mathbf{n}\, dl = 0 \tag{E.15}$$

We take the boundary condition at the surface of the cylinder to be either: (1) $\hat{\mathbf{T}} \cdot \mathbf{n} = 0$ (i.e., a stress-free boundary); or (2) $\mathbf{v} = 0$ (i.e., a rigid boundary). In this case, the line integral is zero and we find that

$$\frac{\partial}{\partial z}\int (\hat{\mathbf{v}}_m^* \cdot \hat{\mathbf{T}}_n + \hat{\mathbf{v}}_n \cdot \hat{\mathbf{T}}_m^*)_z\, ds = 0 \tag{E.16}$$

It then follows from Eq. (E.16) that

$$(k_m^* - k_n)\int (\hat{\mathbf{v}}_m^* \cdot \mathbf{T}_n + \mathbf{v}_n \cdot \mathbf{T}_m^*)_z\, ds = 0 \tag{E.17}$$

From this equation, either $k_n = k_m^*$ or

$$\int (\hat{\mathbf{v}}_m^* \cdot \mathbf{T}_n + \mathbf{v}_n \cdot \mathbf{T}_m^*)_z\, ds = 0 \tag{E.18}$$

which is known as the *orthogonality condition*. Note that $k_m^* = k_m$ for a propagating wave, but for a nonpropagating wave or cutoff mode, $k_m^* = -k_m$.

Normal-mode expansion. We now consider how to express the fields at any plane z of the system. We suppose that at any cross section, the total field is the sum of the fields of the individual modes of the system. Thus we write

$$\hat{\mathbf{T}}(x, y, z) = \sum_n A_n(z)\, \mathbf{T}_n(x, y) \tag{E.19}$$

and

$$\hat{\mathbf{v}}(x, y, z) = \sum_n A_n(z) \mathbf{v}_n(x, y) \tag{E.20}$$

Because of the orthogonality condition, it follows that the total power flow at the plane z is

$$p = -\tfrac{1}{2}\operatorname{Re}\int (\hat{\mathbf{T}} \cdot \hat{\mathbf{v}}^*)_z\, ds = -\tfrac{1}{4}\int (\hat{\mathbf{T}} \cdot \hat{\mathbf{v}}^* + \mathbf{T}^* \cdot \hat{\mathbf{v}})_z\, ds \tag{E.21}$$

Substituting from Eqs. (E.19) and (E.20), we see that

$$P = -\tfrac{1}{4}\sum A_n A_n^* \int (\mathbf{T}_n \cdot \mathbf{v}_n^* + \mathbf{T}_n^* \cdot \mathbf{v}_n)_z\, ds \tag{E.22}$$

or
$$P = \sum_n A_n A_n^* P_n \tag{E.23}$$

We shall define the amplitudes of T_n and v_n so that they correspond to unit power flow with $P_n = 1$ for forward waves and $P_n = -1$ for backward waves, that is, for a forward wave,

$$\tfrac{1}{4} \int_s (\mathbf{T}_n \cdot \mathbf{v}_n^* + \mathbf{T}_n^* \cdot \mathbf{v}_n)_z \, ds = -1 \tag{E.24}$$

It follows that the total power P is

$$P = \sum_n A_n A_n^* - B_n B_n^* \tag{E.25}$$

where B_n is the amplitude of the nth backward wave. Note that Eq. (E.24) is the generalization of Eq. (2.5.15).

The amplitude of the nth mode at the plane Z can be found by multiplying Eq. (E.19) by $v_n^*(x, y)$ and Eq. (E.20) by $T_n^*(x, y)$, adding the results, and integrating over the cross section. Because of the orthogonality condition, it follows that

$$A_n = -\tfrac{1}{4} \int (\hat{\mathbf{T}} \cdot \mathbf{v}_n^* + \mathbf{T}_n^* \cdot \hat{\mathbf{v}})_z \, ds \tag{E.26}$$

Thus if we know T and v at $z = 0$, we know T and v at all planes z.

We now consider how a wave is excited by an incident signal at the enclosing surface, or a perturbation on the enclosing surface. We take the unperturbed boundary condition to be $\mathbf{T}_n \cdot \mathbf{n} = 0$ at the enclosing surface (stress-free boundary conditions). The fields T and v obey Eqs. (E.1) and (E.2) within the volume of the system. However, $\mathbf{T}$ does not necessarily obey stress-free boundary conditions at the surface.

Following the same type of derivation that we made for Eq. (E.15), it follows that

$$\frac{\partial}{\partial z} \int (\hat{\mathbf{v}} \cdot \hat{\mathbf{T}}_n^* + \hat{\mathbf{v}}_n^* \cdot \hat{\mathbf{T}})_z \, ds = - \oint (\hat{\mathbf{v}} \cdot \hat{\mathbf{T}}_n^* + \hat{\mathbf{v}}_n^* \cdot \hat{\mathbf{T}}) \cdot \mathbf{n} \, dl \tag{E.27}$$

Substituting from Eq. (E.5), it follows that

$$\frac{\partial}{\partial z} e^{jk_n z} \int (\hat{\mathbf{v}} \cdot \mathbf{T}_n^* + \mathbf{v}_n^* \cdot \mathbf{T})_z \, ds = e^{jk_n z} \oint (\hat{\mathbf{v}} \cdot \mathbf{T}_n^* + \mathbf{v}_n^* \cdot \hat{\mathbf{T}}) \cdot \mathbf{n} \, dl \tag{E.28}$$

A further substitution from Eq. (E.26) and the use of the boundary condition $\mathbf{T}_n^* \cdot \mathbf{n} = 0$ leads to the result

$$\frac{dA_n}{dz} + jk_n A_n = \tfrac{1}{4} \oint (\mathbf{v}_n^* \cdot \hat{\mathbf{T}}) \cdot \mathbf{n} \, dl \tag{E.29}$$

Thus we have found a rigorously correct expression for the excitation of the nth mode by a stress field at the surface. This is the generalization for a nonpiezoelectric medium. With it, we can determine, for instance, how a perturbation at the surface of a substrate on which a Rayleigh wave can propagate will affect the amplitude of the wave.

A similar expression for the amplitude of the backward wave is

$$\frac{dB_n}{dz} - jk_n B_n = -\tfrac{1}{4} \int (\mathbf{v}^*_{-n} \cdot \hat{\mathbf{T}}) \cdot \mathbf{n}\, dl \tag{E.30}$$

where $\mathbf{v}_{-n}$ is the velocity field associated with the backward wave. Note that for any particular component in a planar system with the surface at $y = 0$, it follows by symmetry that if

$$v_{-nz} = -v_{nz} \tag{E.31}$$

then

$$v_{-ny} = v_{ny} \tag{E.32}$$

It follows, for example, that if the only term of importance at the surface $y = 0$ is T_2, then only the v_{ny} term matters, and Eq. (E.30) becomes

$$\frac{dB_n}{dz} - jk_n B_n = -\tfrac{1}{4} \int v^*_{ny} \hat{T}_2\, dx \tag{E.33}$$

Similar relations can also be derived for the cutoff modes. Such modes are required to form a complete set.

The same type of analysis can be carried through for acoustic waves in piezoelectric materials. Suppose that the potential associated with the nth wave is ϕ_n and the boundary condition at the surface is that of continuity of displacement density $\mathbf{D}_n \cdot \mathbf{n}$. Then, if a surface charge $\hat{\rho}_s = \hat{D}^+ - \hat{D}^-$ is introduced, as would be the case if electrodes were placed on the substrate, the equivalent of Eq. (E.29) is

$$\frac{dA_n}{dz} + jk_n A_n = \frac{j\omega}{4} \oint \hat{\rho}_s \phi^*_n\, dl \tag{E.34}$$

Alternatively, suppose that the boundary condition, initially, is $\phi_n = 0$ (short-circuit boundary condition) and the potential is changed to make ϕ finite, which occurs if a gap is cut in a metal film placed on the substrate. In such a case, the expression for the amplitude variation of the nth mode is

$$\frac{dA_n}{dz} + jk_n A_n = -\frac{j\omega}{4} \oint (\hat{\phi}\mathbf{D}^*_n) \cdot \mathbf{n}\, dl \tag{E.35}$$

A Rigorous Derivation of Normal-Mode Theory

Appendix F

Transducer Admittance Matrix

We follow the method of analysis given by Smith et al. [1]. We obtain the three-port admittance matrix of the transducer by finding the matrix of one interdigital period and then applying a cascading formalism. The admittance matrix for one section can easily be found by standard circuit analysis of the circuit shown in Fig. 2.4.7. Because this circuit is symmetric, the appropriate form of the admittance matrix is

$$[y] = \begin{bmatrix} y_{11} & y_{12} & y_{13} \\ y_{12} & y_{11} & -y_{13} \\ y_{13} & -y_{13} & y_{33} \end{bmatrix} \qquad (F.1)$$

where

$$\begin{bmatrix} i_{n-1} \\ -i_n \\ i_{3n} \end{bmatrix} = [y] \begin{bmatrix} e_{n-1} \\ e_n \\ e_{3n} \end{bmatrix} \qquad (F.2)$$

The values of the four independent elements differ accordingly as to whether the negative capacitor in Fig. 2.4.7 is short-circuited or included in the circuit. These two cases represent the *crossed-field* or *in-line* equivalent circuits, respectively (see Prob. 1.5.1).

1. The admittance parameters for the crossed-field model are

$$y_{11} = -jG_0 \cot \theta$$
$$y_{12} = jG_0 \csc \theta$$
$$y_{13} = -jG_0 \tan \frac{\theta}{4}$$
$$y_{33} = j\left(4G_0 \tan \frac{\theta}{4} + \omega C_s\right)$$
(F.3)

2. The admittance parameters for the in-line model are

$$y_{11} = -jG_0 \cot \frac{\theta}{4} \left(x - \cot \frac{\theta}{2}\right) 2 - \frac{[P - \csc(\theta/2)]^2}{[P - \cot(\theta/2)]^2}$$
$$y_{12} = jG_0 \frac{\cot \theta/4 \, [P - \csc(\theta/2)]^2}{2[2P - \cot(\theta/4)][P - \cot(\theta/2)]}$$
$$y_{13} = -jG_0 \frac{\tan(\theta/4)}{1 - 2P \tan(\theta/4)}$$
$$y_{33} = \frac{j\omega C_s}{1 - 2P \tan(\theta/4)}$$
(F.4)

where $G_0 = R_0$, $P = 2G_0/\omega C_s$, and $\theta = 2\pi\omega/\omega_0$.

The three-port matrix for the entire transducer is found by connecting the N periodic sections in cascade acoustically and in parallel electrically, as shown in Fig. 2.4.8.

The total transducer current is the sum of the currents flowing into the N sections. With the help of Eq. (F.1), we find that the total current into the electrical port is

$$I_3 = \sum_{n=1}^{N} i_{3n} = y_{13}(e_0 = e_N) + y_{33} \sum_{n=1}^{N} e_{3n}$$
(F.5)

We apply the boundary conditions ($e_0 = V_1$, $e_N = V_2$, and $e_{3n} = V_3$) and network symmetry to obtain the following results for the complete transducer, as illustrated in Fig. 2.4.8:

$$\begin{bmatrix} I_1 \\ I_2 \\ I_3 \end{bmatrix} = \begin{bmatrix} Y_{11} & Y_{12} & Y_{13} \\ Y_{12} & -Y_{11} & -Y_{13} \\ Y_{13} & -Y_{13} & Y_{33} \end{bmatrix} \begin{bmatrix} V_1 \\ V_2 \\ V_3 \end{bmatrix}$$
(F.6)

with

$$Y_{13} = Y_{31} = y_{13}$$
$$Y_{23} = Y_{32} = -y_{13}$$
$$Y_{33} = Ny_{33}$$
(F.7)

From Eq. (F.2), the following recursion relation may also be found:

$$\begin{bmatrix} e_n \\ i_n \end{bmatrix} = [R] \begin{bmatrix} e_{n-1} \\ i_{n-1} \end{bmatrix} + \begin{bmatrix} d_1 \\ d_2 \end{bmatrix} V_3 \tag{F.8}$$

where

$$[R] = \frac{1}{y_{12}} \begin{bmatrix} -y_{11} & 1 \\ y_{11}^2 - y_{12}^2 & -y_{11} \end{bmatrix} \tag{F.9}$$

The parameters d_1 and d_2 are also functions of y_{ij}, but will not be needed here. Applying Eq. (F.8) N times gives

$$\begin{bmatrix} e_N \\ i_N \end{bmatrix} = [R]^N \begin{bmatrix} e_0 \\ i_0 \end{bmatrix} + \sum_{n=0}^{N-1} [R]^n \begin{bmatrix} d_1 \\ d_2 \end{bmatrix} V_3 \tag{F.10}$$

Solving Eq. (F.10) for i_N and i_0, and again applying the boundary conditions, gives the result

$$Y_{11} = Y_{22} = \frac{-R_{N11}}{R_{N12}} \tag{F.11}$$

$$Y_{12} = Y_{21} = \frac{1}{R_{N12}}$$

where $[R_N] = [R]^N$. Equations (F.9) and (F.11) summarize the admittance parameters for the entire transducer. For the crossed-field model, the recursion matrix becomes

$$[R] = \begin{bmatrix} \cos\theta & -jR_0 \sin\theta \\ -jG_0 \sin\theta & \cos\theta \end{bmatrix} \tag{F.12}$$

This is the familiar circuit matrix for a transmission line of impedance R_0. Matrix $[S]$ is then obtained simply by replacing the θ in $[R]$ with $N\theta$.

In-Line Model

For frequencies near acoustic synchronism ($\theta = 2\pi\omega/\omega_0 \approx 2\pi$), the admittance matrix for both in-line and crossed-field models can be reduced to a much simpler form. In particular, for the in-line model at $\theta = 2\pi$, the matrix of one periodic section becomes

$$[y] = \frac{j\omega_0 C_s}{16} \begin{bmatrix} -1 & 1 & 4 \\ 1 & -1 & -4 \\ 4 & -4 & 0 \end{bmatrix} \tag{F.13}$$

It can be seen from Eq. (F.12) that $[R]$ is in canonical form, so that the transducer matrix is

$$[Y] = \frac{j\omega_0 C_s}{16} \begin{bmatrix} -\dfrac{1}{N} & \dfrac{1}{N} & 4 \\ \dfrac{1}{N} & -\dfrac{1}{N} & -4 \\ 4 & -4 & 0 \end{bmatrix} \tag{F.14}$$

Crossed-Field Model

For the crossed-field model, the matrix elements are infinite at $\theta = 2\pi$. However, the matrix can be simplified by writing $\theta = 2\pi + \delta$ and expanding the elements to first order in δ. The result is given in Eq. (2.4.24).

REFERENCE

1. W. R. Smith, H. M. Gerard, J. H. Collins, T. M. Reeder, and H. J. Shaw, "Analysis of Interdigital Surface Wave Transducers by Use of an Equivalent Circuit Model," *IEEE Trans. on Microwave Theory and Techniques*, MTT-17, No. 11 (Nov. 1969), 856–64.

Appendix G

Method of Stationary Phase

The method of minimum phase is a technique for evaluating exponential integrals in regions where the phase is changing rapidly. Suppose that we wish to evaluate

$$S = \int F(x) e^{j\phi(x)} \, dx \tag{G.1}$$

When the phase term $\phi(x)$ is changing rapidly with x, the exponential yields contributions that cancel each other out. Thus it is apparent that the main contribution to the integral is from the region where ϕ is changing least rapidly, that is, from the region around x_0 where

$$\left. \frac{\partial \phi}{\partial x} \right|_{x=x_0} = 0 \tag{G.2}$$

We may therefore write a Taylor expansion in the form

$$\phi(x) = \phi_0 + A(x - x_0)^2 \tag{G.3}$$

where

$$A = \frac{1}{2} \left. \frac{\partial^2 \phi}{\partial x^2} \right|_{x=x_0} \tag{G.4}$$

Thus we write

$$S \approx F(x_0) e^{j\phi_0} \int_{-\infty}^{\infty} e^{jA(x-x_0)^2} \, dx \tag{G.5}$$

We have taken the limits of the integral to be $-\infty$ and ∞, because the contributions from the regions where $A(x - x_0)^2 \gg 1$ are negligible. By substituting $z^2 = A(x - x_0)^2$, Eq. (G.5) becomes

$$S \approx \frac{F(x_0)e^{j\phi_0}}{\sqrt{A}} \int_{-\infty}^{\infty} e^{jz^2} \, dz \tag{G.6}$$

Equation (G.6) is in the form of a standard Fresnel integral, and yields the following result:

$$\sqrt{\frac{\pi}{2}} = \int_{-\infty}^{\infty} \cos x^2 \, dx \qquad \sqrt{\frac{\pi}{2}} = \int_{-\infty}^{\infty} \sin x^2 \, dx \tag{G.7}$$

Thus

$$S \approx (1 + j)F(x_0)e^{j\phi_0}\sqrt{\frac{\pi}{2A}} \tag{G.8}$$

with

$$A = \frac{1}{2}\frac{\partial^2 \phi}{\partial x^2}\bigg|_{x=x_0} \tag{G.9}$$

and

$$\frac{\partial \phi}{\partial x} = 0 \quad \text{at} \quad x = x_0 \tag{G.10}$$

Appendix H

Quasistatic Theory for Fields inside a Sphere

The scattering theory derived in Sec. 3.6 uses a quasistatic approximation to determine the fields inside a sphere. Here we will show how this quasistatic theory is derived.

We take the density of the sphere to be ρ'_{m0} and the density of the surrounding material to be ρ_{m0}. We assume that the applied pressure p_i is of the form

$$p_i = A_i e^{-jkz} \tag{H.1}$$

We take the origin of the coordinates of the center of a sphere of radius a and assume that $kz \ll 1$. In this case we can write

$$p_i = A_i(1 - jkz) = A_i(1 - jkr \cos \theta) \tag{H.2}$$

with

$$u_r = \frac{-jkr \cos \theta}{\omega^2 \rho_{m0}} A_i \tag{H.3}$$

The pressure, in general, obeys the relation

$$\nabla^2 p + k^2 p = 0 \tag{H.4}$$

In the neighborhood of the sphere, however, we expect relatively rapid changes in p. Thus if

$$\left|\frac{\partial^2 p}{\partial x^2}\right| \quad \text{or} \quad \left|\frac{\partial^2 p}{\partial y^2}\right| \quad \text{or} \quad \left|\frac{\partial^2 p}{\partial z^2}\right| \gg k^2$$

we can neglect the k^2 term in Eq. (H.4). This is the case if $ka \ll 1$, where a is the radius of the sphere. Within the sphere, the solution of Eq. (H.4) now has a variation identical to that given by a solution of Laplace's equation. We can use the quasistatic assumption to solve for the pressure in spherical coordinates by employing Laplace's equation,

$$\frac{1}{r^2}\frac{\partial}{\partial r}\left(r^2\frac{\partial \phi}{\partial r}\right) + \frac{1}{r^2 \sin\theta}\frac{\partial}{\partial \theta}\sin\theta\frac{\partial p}{\partial \theta} = 0 \tag{H.5}$$

This has solutions of the form

$$p = \sum_n P_n \cos(\theta)\left(A_n r^n + \frac{B_n}{r^{n+1}}\right) \tag{H.6}$$

where $P_n(\cos\theta)$ is the nth-order Legendre function. We will need only $P_0(\cos\theta) = 1$ and $P_1(\cos\theta) = \cos\theta$. As there are only terms independent of θ or that vary as $\cos\theta$ in the exciting pressure field inside the sphere, the pressure must take the form

$$p = A_0 + A_1 r \cos\theta \tag{H.7}$$

with

$$\omega^2 \rho'_{m0} u_r = A_1 \cos\theta \tag{H.8}$$

while outside the sphere the pressure is

$$p = A_i(1 - jkr\cos\theta) + \frac{B_0}{r} + \frac{B_1 \cos\theta}{r^2} \tag{H.9}$$

with

$$\omega^2 \rho_{m0} u_r = -jkA_i \cos\theta - \frac{B_0}{r^2} - \frac{2B_1 \cos\theta}{r^3} \tag{H.10}$$

The boundary conditions at the surface of the sphere $r = a$ require that u_r and p must be continuous. Thus it follows that

$$A_0 = A_i + \frac{B_0}{a} \tag{H.11}$$

$$0 = \frac{B_0}{a^2} \tag{H.12}$$

$$A_1 = -jkA_i + \frac{B_1}{a^3} \tag{H.13}$$

and

$$A_1 \frac{\rho_{m0}}{\rho_{m0'}} = -\frac{jkA_i}{\omega^2 \rho_{m0}} - \frac{2B_1}{a^3} \tag{H.14}$$

We conclude that

$$A_0 = A_i \tag{H.15}$$

and

$$A_1 = \frac{3jkA_i}{\omega^2 \rho_{m0}(2 + \rho_{m0}/\rho_{m0'})} \tag{H.16}$$

The internal symmetric pressure term is equal therefore to the external applied symmetric pressure term, while the internal value of u_r varies as $\cos \theta$. After dealing similarly with the u_θ component, it follows that

$$u_z = \frac{3u_{zi}}{2 + \rho_{m0}/\rho'_{m0}} \tag{H.17}$$

where u_{zi} is the external applied displacement field and u_z is the internal field.

A very similar but more complicated form of the derivation is used to solve for the exact theory for the scattering of a wave incident on a sphere. In this case, we follow the same procedure, but we employ solutions of the wave equation (H.4) instead of the quasistatic solution of Laplace's equation.

Appendix I

Rate of Movement of Charge in a CCD Due to Diffusion and Space Charge

A simple diffusion calculation of the rate of charge transfer through the CCD registers is made in this appendix. Three mechanisms contribute to the rate at which charge is transferred. These are: diffusion, due to the charge gradient present; self-induced drift, due to the fields associated with the carrier space charge; and drift associated with the fringing field between the electrodes. Numerical solutions of the relevant equations show that self-induced drift tends to dominate during short times, with diffusion becoming more prominent as the charge density drops; this is because the former mechanism depends on the carrier density. However, the fringe field at the gate is extremely important. In a well-designed, fast device, the fringe field effect is usually dominant. If surface state recombination effects are negligible, devices with 10-μm gates are operable at 10 MHz and devices with 4-μm gates are operable at frequencies as high as 100 MHz, with both operable down to extremely low frequencies of the order of a few hundred hertz.

Here we will calculate only the diffusion effects, and give some discussion of space-charge effects. For simplicity, we will make the calculations for a p-type carrier. The diffusion current density J_p for p-type carriers moving in the y direction is

$$J_p = -qD \frac{\partial p}{\partial y} \tag{I.1}$$

where p is the hole density per unit volume, q is the charge of a hole, and D is the diffusion coefficient of the holes. Using the continuity equation, we find that

$$\frac{\partial J_p}{\partial y} + q \frac{\partial p}{\partial t} = 0 \tag{I.2}$$

Hence the differential equation for the motion is

$$D \frac{\partial^2 p}{\partial y^2} = \frac{\partial p}{\partial t} \tag{I.3}$$

We suppose that the gate is of length L, and that the boundary conditions, referring to Fig. I.1, are zero carrier gradient, or current, at $y = 0$, and zero carrier concentration at $y = L$. We solve this equation by the method of separation of variables, that is, we use a product solution, writing

$$p = T(t)Y(y) \tag{I.4}$$

Inserting Eq. (I.4) into Eq. (I.3) gives the result

$$D \frac{T''}{T} = \frac{Y'}{Y} \tag{I.5}$$

This leads to a solution satisfying the boundary conditions of the form

$$p_k(y, t) = A_k \cos \lambda_k y \, e^{-\lambda_k^2 D t} \tag{I.6}$$

where

$$\lambda_k = \frac{\pi}{2L}(2k + 1) \tag{I.7}$$

If, in addition, we now require the initial carrier profile $p(y, 0)$ to be uniform, we can use a square-wave expansion to find the coefficients A_k in the series for $p_0 = p(y, 0)$, and write

$$p(y, t) = \sum_{k=0}^{\infty} A_k \cos \lambda_k y \, e^{-\lambda_k^2 D t} \tag{I.8}$$

with

$$A_k = \frac{4 p_0}{(2k + 1)\pi} \tag{I.9}$$

Each of the terms in this sum decays exponentially, but the first term decays least rapidly, so it is the dominant term in most cases. Thus, after some time,

$$p(y, t) \approx \frac{4 p_0}{\pi} \cos \frac{\pi y}{2L} e^{-(\pi^2 D / 4 L^2) t} \tag{I.10}$$

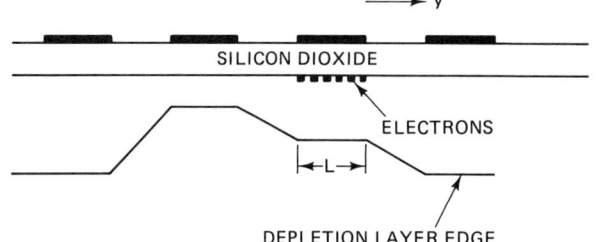

Figure I.1 Schematic of a CCD.

where the total charge $p_{tot} = \int_0^L p(y, t)\, dy$ remaining after time t is

$$p_{tot} = \frac{8}{\pi} p_{tot}(0) e^{-(\pi^2 D/4L^2)t} \tag{I.11}$$

Note that the decay of the total charge due to thermal diffusion can be considered to be exponential with a decay time constant of $\tau_0 = L^2/2.5D$. Thus it is highly advantageous to use as short a gate length as possible. It is also desirable to use as high a diffusion constant as possible, which implies that it is better to work with n-type carriers rather than with p-type carriers and, therefore, a p-type substrate.

As an example, with a gate length of 10 μm and $D = 20$ cm²/s, the decay time constant is $\tau_0 = 20$ ns. So with a clock rate of 5 MHz and a transfer time $t = 100$ ns, the remaining charge under the gate is 0.8%. Thus after 100 transfers, only 44% of the charge passes through the CCD. On the other hand, with $L = 5$ μm and $\tau_0 = 10$ ns, the remaining charge is 0.0045% and, after 100 transfers, 99.5% of the charge passes through the CCD.

The space-charge effects due to the carriers can be regarded as an enhanced diffusion mechanism. This will be shown in the following treatment. If we suppose that the charge per unit length is qp_s, the surface potential associated with this charge is

$$\phi_s = \frac{qp_s}{C_{0x}} + \text{constant} \tag{I.12}$$

where C_{0x} is the capacity per unit area of the gate oxide. The field associated with the surface potential is, in turn,

$$E_s = \frac{-\partial \phi_s}{\partial y} \tag{I.13}$$

It follows that the equation of motion of the carriers is

$$J_s = \frac{-q^2 \mu p_s (\partial p_s/\partial y)}{C_{0x}} \tag{I.14}$$

where μ is the mobility of the carriers. Thus the carrier density obeys the differential equation

$$\frac{q\mu}{2C_{0x}} \frac{\partial^2}{\partial y^2} (p_s^2) = \frac{\partial p_s}{\partial t} \tag{I.15}$$

It is as if the carriers have an enhanced diffusion coefficient D_{enh}, where

$$D_{enh} = \frac{q\mu p_s}{C_{0x}}$$
$$= D \frac{qp_s}{C_{0x}(kT/q)} \tag{I.16}$$

where from the Einstein relation $D = \mu kT/q$, T is the temperature and k is Boltzmann's constant.

When the carrier density is high, the voltage corresponding to the space charge $\phi_s = qp_s/C_{0x}$ can be of the order of 1 to 2 V, and much larger than kT/q. In this case, the space-charge effect is the strongest effect present. Thus, initially, when charge is released from a register, it moves relatively rapidly due to space-charge effect, but as the density drops, diffusion becomes the dominant effect.

Computed results of this kind are shown in Fig. I.2. At first the charge drops rapidly; it then drops less rapidly as normal diffusion effects become dominant.

It is advantageous to trap the carriers in a "buried layer" (i.e., in a potential well around a layer of negatively charged acceptors); then they will be some distance from the gates. Due to this effect, the carrier velocity is increased, which means that higher-speed devices can be made.

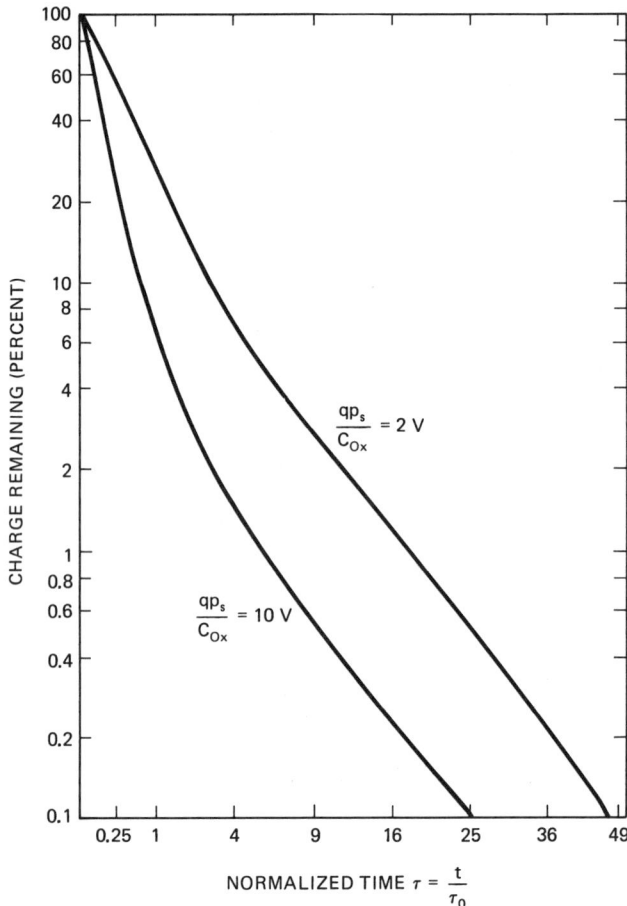

Figure I.2 Fraction of charge remaining under a plate as a function of the square root of the normalized time $\tau = t/\tau_0$. The space charge effect is strong in the region where $\tau < 1$. The curves are straight parallel lines over the diffusion portion of the characteristic. (From Séquin and Tompsett [1].)

REFERENCE

1. C. H. Séquin and M. F. Tompsett, *Charge Transfer Devices*, Suppplement 8 for *Advances in Electronics and Electron Physics*, L. Marton, series ed.   New York: Academic Press, Inc., 1975, Chapter VI, pp. 201–35.

Appendix J

Acoustooptic Effect in Anisotropic Crystals

We have only considered, in the main body of this text, the effect of acoustooptic diffraction in an isotropic crystal. We review here an important technique with major advantages, the use of shear or longitudinal acoustic waves to diffract an optical wave in an anisotropic crystal. We will demonstrate the important effects that occur in this case by using as an example shear wave diffraction in isotropic materials and an anisotropic material tellurium dioxide (TeO_2).

One advantage is the high value of the piezooptic coupling coefficient p and the acoustooptic factor of merit M, which for TeO_2 can be 515 times the value in fused quartz. A second reason for the use of TeO_2 is associated with the very slow acoustic shear wave velocity 0.617 km/sec in the [110] direction; this makes it possible to obtain relatively large Bragg angles at frequencies in the 50 to 100 MHz range. A third reason is associated with the fact that it is possible to obtain a diffracted wave with a constant amplitude over a broad acoustic input frequency range *without changing the input angle* θ_j.

We shall show, here, how this latter effect comes about when a polarized optical wave interacts with an acoustic shear wave. Tellurium dioxide is a tetragonal uniaxial crystal which is optically active. In an optically active material, a right hand circularly polarized light wave propagating along the optical axis, i.e., in the z direction or [001] direction has a refractive index n_0 different from the refractive index n_1 of a left hand circularly polarized wave propagating along the optical axis.

The general relation for the change in the dielectric impermeability due to interaction with an acoustic wave is

$$\Delta B_{ij} = p_{ijkl} S_{kl} \qquad (J.1)$$

Suppose we consider, first of all, an isotropic material through which a plane linearly polarized optical wave, polarized in the x direction, propagates in the z direction. If this wave interacts with an acoustic shear wave propagating in the x direction with its particle velocity in the y direction, it follows from Eq. (J.1) that there is a perturbation in the dielectric impermeability of the form

$$\Delta B_{xy} = P_{xyxy} S_{xy} \tag{J.2}$$

This expression, in turn, implies that there is a perturbation in the dielectric constant of the form $\Delta \varepsilon_{xy}$. So, it follows that the field E_x gives rise to a displacement density $D_y = \Delta \varepsilon_{xy} E_x$ or to a plane polarized wave in the medium with its polarization at right angles to the original incident wave. Therefore, one advantage of this type of interaction is that it is easy to use a polarizer to separate the diffracted wave from the incident wave.

For an anisotropic material like TeO_2, one possible mode of operation is to use an incident circularly polarized wave propagating along the optical axis, the [001] direction, which interacts with a pure acoustic shear wave propagating in the [100] direction with the particle velocity in the [001] direction. It may be shown by a similar argument to that given above that the diffracted wave will be one with the opposite circular polarization, at least when the diffraction angle is small.

We now consider how the Bragg diffraction equations change from those for an isotropic material. We follow the analysis of Dixon[1,2].

The Bragg diffraction equations are

$$\omega_0 \pm \omega_a = \omega_1 \tag{J.3}$$

and

$$\mathbf{k}_0 \pm \mathbf{k}_a = \mathbf{k}_1 \tag{J.4}$$

Here, the subscript 0 refers to the undiffracted wave with a refractive index n_0, and the subscript 1 to the diffracted wave with a refractive index n_1 and the opposite circular polarization.

Referring to Fig. 4.9.6, we can write

$$\omega_1 = \omega_0 - \omega_a \tag{J.5}$$

$$-k_1 \sin \theta_R = k_0 \sin \theta_i - k_a \tag{J.6}$$

and

$$k_1 \cos \theta_R = k_0 \cos \theta_i \tag{J.7}$$

when $k_1 \neq k_0$, it follows by squaring and adding Eqs. (J.6) and (J.7), that

$$\sin \theta_i = (k_a/2k_0) + (k_0^2 - k_1^2)/2k_a \tag{J.8}$$

with

$$\sin \theta_R = (k_a/2k_0) - (k_0^2 - k_1^2)/2k_a \tag{J.9}$$

Suppose that we assume that $k_0 > k_1$, and that for small deflection angles the variation of k_0 and k_1 with angle in an anisotropic material can be neglected compared to other effects. Then, as a function of frequency or k_a, it will be seen that the first term in Eq. (J.8) increases with frequency and the second term decreases with frequency. Here, we have assumed that k_0 and k_1 do not vary much for small diffraction angles. The second terms in Eqs. (J.8) and (J.9) would not exist in an isotropic medium. At a particular frequency ω_{a0}, the value of θ_i is minimum. For frequencies around this minimum value, θ_i hardly changes. In an isotropic medium, on the other hand, the optimum value of θ_i varies linearly with frequency. Therefore, operation of a TeO_2 acoustooptic Bragg cell can occur over a much larger bandwidth than in a simple isotropic medium. Because of the low phase velocity of the shear wave in the TeO_2 Bragg cell, and its highly anisotropic characteristics, the optimum frequency of operation is approximately 50 MHz.

By adding Eqs. (J.8) and (J.9), it will be seen that

$$\sin \theta_i + \sin \theta_R = k_a/k_0 \tag{J.10}$$

It follows that when the incident angle is kept fixed, the angle of diffraction varies linearly with frequency *in just the same manner as in an isotropic material*.

It has been shown by Yano et al.[3] that the operation of this type of Bragg cell can have certain problems due to nonlinear coupling at higher input powers from the diffracted wave to a second wave, a condition which occurs near 50 MHz, at the center of the band. This difficulty can be avoided by operation of the Bragg cell at angles slightly off axis (only a few degrees). Interestingly enough, in this case, the input beam can be linearly polarized; but the polarization direction is quite critical. With this type of Bragg cell, the diffraction efficiency can be as much as 95%, and operation with as much as 80% efficiency can be obtained over an octave bandwidth centered at 75 MHz.

REFERENCES

1. R. W. Dixon, "Acoustic Diffraction of Light in Anisotropic Media," IEEE J. Quant. Electron., 11, (Oct. 1972), pp. 85–93.
2. K. W. Warner, D. L. White and W. K. Bonner, "Acousto-optic Light Deflectors using Optical Activity in Paratellurite," J. Appl. Phys., 43, (Nov. 1972), pp. 4489–4495.
3. T. Yano, M. Kawabuchi, K. Fukumoto, and K. Watanabe, "TeO_2 Anisotropic Bragg Light Deflector Without Midband Degeneracy," Appl. Phys. Lett., 26, (June 15, 1975), pp. 689–691.

Index

A-scan, 220–22
   in medical diagnostic applications, 220, 222
   in nondestructive testing, 222
Abbreviated subscripts, 542–46
Acoustic attenuation, 9
Acoustic beam profile, 514–15
Acoustic convolver, 449–54
Acoustic emission transducer, 57
Acoustic holography, 279–91
   acoustic imaging intensity detector, 285
   holographic reconstruction, 279–85
   imaging of vibrating objects, 294–96
   scanned holographic imaging, 288–91
   scanning laser acoustic microscope (SLAM), 291–94
   Smith and Brenden technique, 286–88
   spherical wave sources, 281–82
Acoustic imaging intensity detector, 285
Acoustic impedance, 9–13
   input impedance, 12–13
   quarter-wave matching sections, 12
Acoustic losses, 8–9
   loss mechanisms, 9
   Poynting's theorem, 8
   viscous losses, 8–9
Acoustic parameters, of common materials, 548–61
Acoustic pulse echo system, 483
Acoustic transducers, 3
Acoustic waveguides, 104–14
   Lamb waves, 105, 108–9
   shear horizontal modes, 106–7
Acoustic wavelength:
   in response to elastic, 203
   in water, 197–98
Acousto-optic effect, in anisotrophic crystals, 588–90
Acousto-optic filters, 501–28
   Bragg deflection systems application, 517–18
   prism coupler, 518
   Bragg diffraction by sound, 511–17
   diffraction theory, 513
   Raman-Nath interaction, 513–14
   convolver, 518–21
   light diffraction by sound, 503–9
   acousto-optic figure of merit, 505–8
   Fresnel length, 508
   Raman-Nath theory, 508–9
   photoelastic effect, 501–3
   piezo-optic effect, 501–3
   Schlieren effect, 509–10
   time-integrating correlator, 521–28
   one-dimensional transform, 524–26
   two-dimensional transforms, 526–28
Acousto-optic materials:
   material constants for, 560–61
   properties of, 560
Adaptive filters, 480–500
   additional uses, 497–98
   eliminating an interfering signal, 495–98
   notch filter, 495
   inverse filter, 482–88
   zero-forcing algorithm, 485–86

Adaptive Filters (cont.)
  removing phase errors with, 481–82
  Wiener filter, 488–95
    pseudo-inverse filter, 491–92
    storage correlator as, 492–95
Air gap storage correlator, 471–74
Air-backed transducer, 50, 60–66
Air-gap convolver, 463
Ambiguity function:
  distributions, 378
  parabolic FM pulse waveforms, 381
Amplifiers, 3
Amplitude-weighted bandpass filter, 421–26
Anisotropic crystals, 588–90
Anisotropic elastic properties, 3
Apodization functions, examples of, 389
Apodized transducer, 330–32
Approximations:
  Born, 305–6
  Fraunhofer, 167–71
  Fresnel, 165, 182
  Kirchhoff, 308–9
  quasistatic, 306–7
Asymmetric atomic lattice, 17
Autoconvolution, 453
Axial spatial frequencies, 172

B-scan, 222–23
  in medical imaging, 223
B-scan frequency modulated chirp system, 257
B-scan pulse array system, 259
B-scan radial sector scan format, 226, 261
Bandpass filter, 329–32, 415–43
  amplitude-weighted, 421–26
  triple transitecho, 426
  building-block design technique, 427–31
  "comb" response, 436
  phase-weighted transducers, 426–27
  recursive comb filter, 436–37
  recursive filter, 432–43
  with feedback, 434–37
  fiber-optic, 439–43
  single-mode fiber, 439
  strip coupler, 416–21
Bandpass-bandstop filter, 395
Bandshape filter, 396

Bartlett weighting, 389
Biphase digital code, 324
Blackmann weighting, 389
Boltzmann's constant, 362
Born approximation, 305–6
Bragg cell, 293
Bragg condition, 514–15
Bragg deflection systems:
  application, 517–18
  prism coupler, 518
Bragg diffraction:
  by sound, 511–17
  diffraction theory, 513
  Raman-Nath interaction, 513–14
Bucket brigade device (BBD), 319, 339
  field-effect transistor (FET), 341
  performance limitation, 342–47
Bulk elastic modulus, 87
Buried channel, 350
Buried layer, 586

C-scan, 223–25
  in nondestructive testing, 223, 225
C-scan transmission system, 224
Capacitance, zero strain, 32
Cardiac imaging system, 261
Charge, rate of movement, 583–86
Charge-coupled device (CCD), 319, 339, 347–52
  burned channel, 350
  metal insulator semiconductor (MIS), 347
  split electrode method, 350
Charge-transfer device (CTD), 339–59
  bucket brigade devices (BBD), 339, 341–47
  field-effect transistor (FET), 341
  performance limitation, 342–47
  charged coupled devices (CCD), 347–52
  buried channel, 350
  metal insulator semiconductor (MIS), 347
  split-electrode method, 350
  imperfect charge-transfer efficiency, 352–56

single transfer device (STD), 356–58
switched capacitors, 356–58
fixed pattern noise, 356
Charge-transfer efficiency, imperfect, 352–56
Chip, 324
Chirp transform programmable matched filter, 401
Chirp z transform, 360
Chirp z transform implementation, 399–409
Chirp-focused systems, 251–58
Chirp-focused transmitter system, 254
Clamped transducer, 34–35
Clutter, 341
Comb filter, 436–37
  frequency characteristics, 438–39
  frequency response, 437
  fundamental structure, 436
Compressive receiver, 395
Computerized axial tomography (CAT) scanner, 275
Confocal scanned microscopy, 198
Conjugate filter, 325
Conservation of mass, 5
  in shear waves, 5
Constitutive relations, 19–20, 75–83
Conversion efficiency, of piezoelectric transducers, 1
Convolvers, 448–77
  acoustic, 449–54, 464–67
  airgap, 463
  airgap storage correlator, 471–74
  charge-coupled device, 467–68
  external mixer, 465
  input correlation mode, 474–75
  monolithic storage correlator, 471–74
  SAW storage correlator, 469–77
  semiconductor, 461–64
  spread spectrum communications, 454–59, 475–77
  frequency hopping modes, 458
  range finding, 459
  tapped-delay-line correlator, 470–71
  waveguide, 459–61

Correlation filter, 318
Correlators, 469–74
  airgap, 471–74
  diode memory, 474
  monolithic, 471–74
  SAW storage, 469–77
  spread spectrum, 456
  tapped-delay-line, 470–71
Creeping wave concept, 309–10
Cross-correlations, 398–99
Cubic crystal, 545–46
Curie point, 19

Deconvolution, of overlapping distorted pulses, 488
Delay lines:
  microwave, 3
  superconductive, 409–12
  tapped, 470–71
Delta-function model, 119–22
Depth of focus, 190–91
  geometrical concepts for, 191–94
Differential current integrator (DCI) output amplifier, 351
Diffraction:
  Green's function, 159–63
  spherical waves, 156–58
Diffusion, of charge, 583
Digitally coded devices, 322–25
Dilation, 87
Dirichlet weighting, 389
Displacement, 4, 75
Dolph-Tchebysheff weighting, 389
Doppler frequency shift, 368, 372, 527
Double-spiral dispersive delay line, 412

Edge-bonded transducer, 272
Elasticity, 1, 4, 77
  and Hooke's law, 4
Elasticity tensor, 544–46
Electrical impedance, of piezoelectric transducers, 1
Electrical input impedance, 32–33, 51–53
Electrical matching of a loaded transducer, 55–61
Electromagnetic resonator, 154
Electromechanical transducers, compared to piezoelectric transducers, 27

Electronic scanning, 225
Electrostatic field, 141
Electrostatic potential, 567
Energy, 87
Equation of motion, 4–5
  for solids and fluids, 90–92
Equivalent circuits:
  lumped, 37
  Mason, 33–35
  Redwood, 35, 41, 50
Excitation:
  by tone burst, 216–18
  nonuniform, 154–64
  pulsed, 212–18
Extensional waves, 14–15
  extensional wave velocity, 88–89
  Young's modulus, 14
External feedback loop, 336

Fabry-Pero interfesometer, 334
Far-field region (*See* Fraunhofer zone, 154)
Far-out sidelobes, causes of, 269
"Fat zero," 344
Ferroelectric materials, 17, 19–20
  domains in, 19–20
  poling in, 19
Fiber-optic coupler, construction of, 441
Fiber-optic recursive filter, 436–37
Field-effect transistor (FET), 341
Filter weighting, 387–91
  Bartlett, 389
  Blackmann, 389
  Dirichlet, 389
  Dolph-Tchebysheff, 389
  Gaussian, 388
  Hamming, 388–390
  Hanning, 388
  Kaiser, 389
Filters:
  acousto-optic, 501–28
  adaptive, 480–500
  amplitude-weighted bandpass, 421–26
  bandpass, 329–332, 415–43
  amplitude weighted, 421–26
  apodized transducer, 330–32
  building block design techniques, 427–31
  comb response, 436
  fiber-optic, 439–43

  phase-weighted transducers, 426–27
  recursive filter, 432–43
  single-mode fiber, 439
  strip coupler, 416–21
  uniform transducer, 329–30
  bandpass-bandstop, 395
  bandshape, 395
  comb filter, 436–37
  conjugate, 325
  correlation, 318
  FM chirp analog, 332–33
  chirp z transform implementation, 399–409
  filter weighting for sidelobe reduction, 387–91
  Fourier transform operations with, 391–99
  mathematical treatment, 384–87
  intermediate, 325
  inverse, 319, 360, 482–88
  zero-forcing algorithm, 485–86
  matched, 359–83
  ambiguity function, 376–83
  for complex signals, 367–68
  narrowband waveforms, 373–76
  pulse compression, 372
  radar system accuracy, 368
  theory, 361–67
  uncertainty relation, 372
  Wiener, 360
  notch, 495
  pseudo-inverse, 491–92
  RAC filters, 335–36
  recursive, 432–43
  comb, 436–37
  with feedback, 434
  fiber-optic, 439–43
  switched capacitor, 319
  fixed-pattern noise, 356
  time-delay, 396
  transversal, 117
  weighting for sidelobe reduction, 387–91
  Wiener, 360, 488–95
Finger weighting, 328
Finite exciting sources, 154–312
  diffraction and nonuniform excitation, 154
  focused transducers, 182–210
  lenless acoustic imaging, 218–96

Index

Finite exciting sources (cont.)
  plane piston transducers, 164–80
  pulsed excitation of transducers, 212–18
  reflection, 300–312
  scattering, 300–312
Fixed pattern noise, 356
Flat piston transducer, 182
Flexural mode, 109
FM chirp filter (*See* Frequency modulated chirp filter)
FM chirps (*See* Frequency-modulated chirps)
Focused transducers, 182–212
  3-dB definition, 185
  coherent imaging, 186–87
  depth of focus, 190–91
    geometrical concepts, 191–94
  F number, 185
  Gaussian beams, 206–10
    apodization, 209–10
  incoherent imaging, 186–87
  lens aperture, 185
  paraxial equation, 206–10
  Rayleigh two-point definition, 187–88
  reflection from a planar reflector, 194–96
  sidelobes, 185–86
  sparrow two-point definition, 189
  speckle, 189–90
  spherical, field of, 182–97
  transverse definition, 185
Focusing systems, acoustic imaging, 225–29
Fourier transform, 171–75
  Fourier transform convolver, 466
  Fourier transform output, 468
Fraunhofer approximation, 167–71
Fraunhofer region, 215
Fraunhofer zone, 154, 165–67
Frequency hopping modes, 458
Frequency modulated chirp analog filter, 332–33
Frequency modulated chirp filter, 383–99
  chirp *z* transform implementation, 399–409

with a charge-coupled device, 406–9
sliding transform, 408–9
with superconductive delay lines, 409–12
filter weighting for sidelobe reduction, 387–91
  Blackmann weighting, 389
  Dolph-Tchebysheff weighting, 389
  Gaussian weighting, 388
  Hamming weighting, 388–90
  Hanning weighting, 388
  Kaiser weighting, 389
  rectangular weighting, 388
Fourier transform operations with, 391–99
  bandpass-bandstop filter, 395
  bandshape filter, 395
  compressive receiver, 395
  correlation or convolution of two signals, 396–99
  nonlinear processing, 399
  spectral whitening, 399
  time-delay filter, 396
mathematical treatment, 384–87
main lobe width, 385
physics of, 386–87
sidelobes, 385–86
Frequency-modulated chirps, 319
Fresnel approximation, 165, 182
Fresnel length, 508
Fresnel lenses, 245–51
Fresnel limit, 166
Fresnel region, 214–15
Fresnel ripples, 174
Fresnel transducer, 228
Fresnel zone, 154, 165–67
Frionge field effect, of charge, 583

Gases, acoustic constants for, 553
Gauss theorem, 570
Gaussian beams, 206–10
Gaussian taper, 206, 209
Gaussian weighting, 388
Goos-Hanchen effect, 149
Green's function, 159–63
  Helmholtz's theorem, 159–60
  Kirchhoff formula, 160–61
  pressure release baffle, 162–63

Rayleigh-Sommerfeld formula, 161–62
rigid baffle, 162
Sommerfeld radiation condition, 160
transient source, 163
Guided-wave Bragg diffraction, 519

Half-wave resonator, 47
Hamming taper, 210
Hamming weighting, 388–90
Hankel transform, 171–75
Hanning weighting, 388
Helmholtz's theorem, 159–60
Hermite polynomials, 381
High-speed scanning, 225
Hilbert transform, 138
Holography, 226–27
  acoustic, 279–91
  scanned, 288–91
  Smith and Brenden technique, 286–88
Holosonics holographic ultrasound imaging system, 289
Hooke's law, 4, 14, 75, 77, 86–87, 109
  and elasticity, 4

Ideal fluid, 87–88
Image reconstruction, 282–85
Impedances:
  acoustic, 9–13
  electrical input, 32–33, 51–53
    at resonance, 52
    constructional techniques, 51–52
  input, 12–13
  input impedance, 12–13
  motional, 36
  motional impedance, 36
  quarter-wave matching sections, 12
  radiation, 32
  source, 58–59
  specific acoustic, 9
Imperfect charge-transfer efficiency, 352–56
Impulse model, 326
Interdigital transducers, 117–30, 322–29
  delta-function model, 119–22
  network theory, 123–30

normal-mode theory, 131–37
  leaky waves, 141–44
  perturbation theory, 137–41
Intermediate frequency filter, 322
Inverse filter, 319, 360, 482–88
  zero-forcing algorithm, 485–86
Isotropic media:
  acoustic wave guides, 104–14
    Lamb waves, 108–9
    shear horizontal modes, 106–7
    surface waves, 109–10
  equations of motion for solids and fluids, 90–92
  interdigital transducers, 117–30
    delta-function model, 119–22
    leaky waves, 141–44
    network theory, 123–30
    normal-mode theory, 131–37
    perturbation theory, 137–41
  mathematic formalism, 86–90
    dilation, 87
    energy, 87
    Hooke's law, 86–87
    ideal fluid, 87
    Lame constants, 86
    Poisson's ratio, 88–89
    wave velocities, 89–90
    Young's modulus, 86
  plane wave reflection, 94–105
    incident longitudinal wave, 96–98
  plane wave refraction, 100–105
  Rayleigh waves, 110–14
  waves in, 85–153
    longitudinal waves, 93–94
    shear waves, 94
    wave equation, 92–94
  wedge transducer, 144–52
    experimental results, 150
    finite attenuation, 146–47
    reflected wave, 149
Isotropic solids, 1

Kaiser weighting, 389
Kirchhoff approximation, 308–9
Kirchhoff formula, 160–61
KLM model, 41–47, 50

Lamb waves, 105, 108–9
  flexural mode, 109

Lame constants, 86
Leaky waves, 141–44
Least-mean-square (LMS) algorithm, 492
Leith-Upatnieks hologram, 280
Length expander bar, 83
Lens imaging, 278
Lens pupil function, 195
Lensless acoustic imaging, 218–96
  A-scan, 220–22
  acoustic holography, 279–91
    acoustic imaging intensity detector, 285
    holographic reconstruction, 279–85
    imaging of vibrating objects, 294–96
    scanned holographic imaging, 288–91
    scanning laser acoustic microscope (SLAM), 291–94
    Smith and Brenden technique, 286–88
    spherical wave sources, 281–82
  amplitude errors, 243–45
  applications, 218–20
    nondestructive testing (NDT) field, 219
  B-scan, 222–23
  C-scan, 223–25
  chirp-focused systems, 251–58
    basic system, 251–55
  focusing systems, 225–29
    basic imaging theory, 229–45
    elimination of physical lenses, 226–27
    holography, 226–27
    matched filter concepts, 229–31
    matched filter theory, 237–38
    paraxial approximation, 231–32
    phase-delay focusing, 228–29
    physical lens, 227
    radial sector scan system, 232–34
    range definition, 235–36
    Rayleigh criterion, 234
    sparrow criterion, 235
    time-delay focusing, 227–28

Fresnel lenses, 245–51
  basic system, 245
  phase sampling, 248–51
  sidelobe level, 249–50
  sidelobes, 248–51
  sidelobes, 238–39
  apodization, 241
  grating lobes, 239–41
  time-delay systems, 275–77
    dynamically focused system, 261
    grating lobes, 267–71
    range resolution, 266
    sampling lobes, 265–71
    sidelobes, 265–71
    surface acoustic wave imaging, 272–74
    synthetic aperture imaging system, 262, 270–74
    transverse definition, 266
    TV display, 265
  tomographic imaging systems, 275–77
Light diffraction by sound, 503–9
  acousto-optic figure of merit, 505, 505–8
  Bragg angle, 504
  Bragg interaction, 504
  Fresnel length, 508
  Raman-Nath diffraction, 503
  spots, 507
  Raman-Nath theory, 508–9
    limitations, 508–9
Linear passive surface wave devices, 322–39
  bandpass filter, 329–32
    apodized transducer, 330–32
    uniform transducer, 329–30
  frequency modulated chirp analog filter, 332–33
  interdigital transducers, 322–29
    digitally coded devices, 322–25
    mathematical model, 326–29
  resonators, 334
Liquid surface imaging system, 287
Liquids, material constants for, 552
Longitudinal waves, 2, 93
  longitudinal wave velocity, 88–89

Longitudinal-to-shear wave converter, 99
Loop output waveform, 457
Losses:
　acoustic, 8–9
　loss mechanisms, 9
　Poynting's theorem, 8
　viscous losses, 8–9
Lumped equivalent circuit, 37

Macrobend tapped delay line, 440
Mason equivalent circuit, 33–35
Matched filter, 359–83
　ambiguity function, 376–83
　for complex signals, 367–68
　pulse compression, 366
　radar system accuracy, 368
　　inequality proof, 373
　　narrowband waveforms, 373–76
　　uncertainty relation, 372
　theory, 361–67
　Wiener filter, 360
Material displacement, 4, 75
Maxwell's equation, 564
Medical applications:
　of acoustic sound waves, 194, 219–20, 222
　focused transducer for, 194
Metal insulator semiconductor (MIS), 347
Metal-oxide semiconductor (MOS), 339
Method of stationary phase, 179–80
Microwave delay lines, 3
Military radar systems, 461
Miller effect, 346
Monolithic storage correlator, 471–74
Motion tomography, 276
Motional impedances, 36
Moving target indicator (MTI) radar, 433

Narrowband waveforms, 373–76
NAVSTAR global navigation system, 459
Near-field region (*See* Fresnel zone, 154)
Near-in sidelobes, causes of, 269
Network theory, 123–30
　crossed-field model, 124–30
　normal mode formalism, 123

Nondispersive bandpass filter, 331
Noninvasive techniques in clinical testing, 219
Nonlinear interaction, between surface waves, 321
Nonpiezoelectric materials, 1–15, 75–82
　normal-mode theory, 569–73
　reduced subscript notation, 78–83
　soundwaves in, 2–15
　　acoustic impedance, 9–13
　　acoustic losses, 8–9
　　conservation of mass, 4–5
　　displacement, 4
　　elasticity, 4
　　energy, 6–7
　　equation of motion, 4–5
　　extensional waves, 14–15
　　Hooke's law, 4
　　Poynting's theorem, 7
　　propagation constant, 5–6
　　shear waves, 13–14
　　strain, 4
　　stress, 3–4
　tensor notation, 75, 77–78
Nonuniform excitation, 154–65
　Green's function, 159–63
　spherical waves, 156–58
Normal modes, 420
Normal-mode expansion, 571–73
Normal-mode theory, 131–37, 569–73
　power flow concepts, 133–34
Normalized line-spread function, 237
Notation, 542–46
　reduced subscript, 78–83
　tensor, 75, 77–78
Notch filter, 495

Open-circuited transducer, 35
Orthogonality, 571
Outward normal, 303

Paraxial rectilinear system, 232
Parseval's theorem, 363, 371, 377
Particle displacement, 1
Perfect plane reflector, 194
Perturbation theory, 137–41
Phase-delay focusing, 228–29
Phase-weighted transducers, 426–27

Photoacoustic microscopy, 205
Photoelastic effect, in acousto-optic filters, 501–3
Piezo-optic constant of water, 503
Piezo-optic effect, in acousto-optic filters, 501–3
Piezoelectric ceramics, properties of, 554–55
Piezoelectric constitutive relations, 19
Piezoelectric coupling constant, 1, 21
Piezoelectric materials, 1, 17–26, 75–83
　constitutive relations, 19–20, 75–83
　domains in ferroelectric materials, 19–20
　effect of coupling on wave propagation, 21–22
　　stiffened elastic content, 21
　electric field in, 17
　energy conservation in, 24–26
　ferroelectric materials, 17, 19
　poling in ferroelectric materials, 19–20
　　Curie point, 19
　Poynting's theorem, 563–65
　reduced subscript notation, 78–83
　stress-free dielectric constant, 22
　tensor notation, 75, 77–78
Piezoelectric tensor, 546–47
Piezoelectric transducers, 1, 27–75
　compared to electromechanical transducers, 27
　electrical input impedance, 51–53
　　at resonance, 52
　　constructional techniques, 51–52
　　terminated transducer, 52–53
　electrical matching of a loaded transducer, 55–61
　impedance of an unloaded transducer, 35–41
　　lumped equivalent circuits, 37
　　Mason equivalent circuit, 40
　　motional impedance, 36

KLM model, 41–47, 50
  effects of reflections on response, 45–47
  half-wave resonators, 47
  quarter-wave resonators, 47
Mason equivalent circuit, 33–35
power transfer to transducer, 53–54
radiation impedances, 32
reduced subscript notation, 78–83
Redwood equivalent circuit, 35, 41, 50
Reeder-Wislow design method, 61–66
tensor notation, 77–78
as a three-port network, 29–33
transducer acoustically matched at each end, 56
  acoustic load with electrical tuning, 58–59
  acoustic load with source impedance, 58–59
  acoustic load with source resistance, 58–59
  air-backed transducer matched at output, 56–57
  tuned air-backed transducer, 60–61
  untuned air-backed transducer, 60–61
  untuned transducer, 60
transducer design examples, 66–70
transducer pulse response with arbitrary terminations, 47–50
  air-backed transducer, 50, 60–66
  unmatched front surface, 50
transmitting constant, 31
velocity, 30
zero strain capacitance, 32
Piezoelectrically stiffened elastic content, 21
Piston transducer, 175
Pitch/catch imaging, 273, 275
Plane piston transducers, 164–82
  diffraction from rectangular transducers, 175–80
  fields on axis, 164
  Fourier transform, 171–75
  Fraunhofer approximation, 167–71

Fraunhofer zone, 165–67
Fresnel approximation, 165, 182
Fresnel limit, 166
Fresnel ripples, 174
Fresnel zone, 165–67
Hankel transform, 171–75
paraxial approximation, 165, 182
spatial frequency concepts, 171–75
Plane wave propagation, 1
Plane wave reflection, 94–105
  incident longitudinal wave, 96–98
Plane wave refraction, 100–105
PN-coded waveform, 455
Point spread function (PSF), 201, 232
Poisson's ratio, 88–89
Power flow per unit area, 7
Poynting's theorem, 7, 8, 24–25, 563, 563–65
Pressure release baffle, 162–63
Principal normal sections, 301
Prism coupler, 518
Processing gain, 325
Propagation of acoustic sound waves, 1–84
  acoustic impedances, 9–13
    input impedance, 12–13
    quarter-wave matching sections, 12
  acoustic losses, 8–9
    loss mechanisms, 9
    viscous losses, 8–9
  conservation of mass, 5
  constitutive relations, 75–83
  in crystalline materials, 2
  displacement, 4
  elasticity, 4
  energy, in a medium, 6–7
  equation of motion, 4–5
  extensional waves, 14–15
    Young's modulus, 14
  longitudinal wave, 2
  in nonpiezoelectric materials, 2–15
    conservation of mass, 4–5
    displacement of, 4
    equation of motion, 4–5
    Hooke's law, 4
    Poynting's theorem, 7
    propagation constant, 5–6

reduced subscript notation, 78–83
sound waves in, 2–15
tensor notation, 75, 77–78
one-dimensional theory, 2–15
piezoelectric materials, 17–25, 75–82
  constitutive relations, 19–20, 75–83
  Curie point, 19
  effect of coupling on wave propagation, 21–22
  electric field in, 17
  energy conservation in, 24–26
  stress-free dielectric constant, 22
piezoelectric transducers, 27–75
  electrical input impedance, 51–53
  electrical matching of a loaded transducer, 55–61
  impedance of an unloaded transducer, 35–41
  KLM model, 41–47, 50
  Mason equivalent circuit, 33–35
  power transfer to transducer, 53–54
  radiation impedance, 32
  reduced subscript notation, 78–83
  Redwood equivalent circuit, 35, 41, 50
  Reeder-Winslow design method, 61–66
  tensor notation, 77–78
  as a three-port network, 29–33
  transmitting constant, 31
  velocity, 30
  zero strain capacitance, 32
propagation constant, 5–6
shear wave, 2, 13–14
in solids, 1
strain, 4
stress, 3–4
tensor notation, 75–83
Propagation constant, 5–6
Proximity-tapped delay line, 412
Pseudo-inverse filter, 491–92
Pseudonoise code, 455
Pulse compression ratio, 325

Pulse response characteristics, of piezoelectric transducers, 1
Pulse-echo system, 483
Pulsed excitation:
　Fraunhofer region, 215
　Fresnel region, 214–15
　of transducers, 212–18
　transient response, 213–14

Quarter-wave resonator, 47
Quasistatic approximation, 306–7
Quasistatic theory, 580–82
　for fields inside a sphere, 580–82

Radar range accuracy, 360
Radial sector scan, 233
Radial spatial frequencies, 172
Radiation impedances, 32
Raman-Nath criterion for water, 509
Raman-Nath interaction, 513–14
Ray-tracing geometry, 200
Rayleigh criterion, 234
Rayleigh scattering limit, 306
Rayleigh waves, 105, 109, 110–14, 117
　detection, 323
　excitation, 323
　　by an acoustic microscope lens, 203
　Rayleigh root, 113
　velocity, 113–14
Rayleigh-Sommerfeld integral, 155
Receiver constant, 82
Rectangular weighting, 388
Recursive filters:
　comb, 436–37
　with feedback, 434
　fiber-optic, 439–43
Reduced form, 78
Reduced subscript notation, 78–83
Redwood equivalent circuit, 35, 41, 50
Reeder-Winslow design method, 61–66
Reflection in plane waves, 94–104
Reflective array compressor, 335
Refraction in plane waves, 94–104

Relative time of arrival determination, 401
Resonators, 334
Rigid baffle, 162

Sample-and-hold model, for a single transfer device (STD), 357
SAW devices (See Surface acoustic wave devices)
Scan imaging:
　A-scan, 220–22
　B-scan, 222–23
　C-scan, 223–25
Scanned holographic imaging, 288–91
Scanning laser acoustic microscope (SLAM), 291–94
Scattering:
　of acoustic waves, 300–313
　by large objects, 301–3
　general theory, 303–12
　　Born approximation, 305–6
　　Kirchhoff approximation, 308–9
　　quasistatic aproximation, 306–7
　　surface integral formulation, 304–5
　　volume integral formulation, 304–5
　Rayleigh scattering, 300
　　Born approximation, 305–6
　　general theory, 303–12
　　Kirchhoff approximation, 308–9
　　quasistatic approximation, 306–7
　　surface integral formulation, 303–4
　　volume integral formulation, 304–5
Schlieren effect, 509–10
Schlieren optical system, 509–10, 510, 521
Schwarz's inequality, 363, 366–67
Self-induced drift, of charge, 583
Semiconductor convolver, 461–64
Shear horizontal mode, 106–7
Shear horizontal wave, 100, 105
Shear modulus, 86

Shear wave, 2, 13–14, 94
Shear wave velocity, 88–89
Sidelobe reduction, 387–91
　Dirichlet weighting, 389
　Dolph-Tchebysheff weighting, 389
　Gaussian weighting, 388
　Hamming weighting, 388–90
　Kaiser weighting, 389
　rectangular weighting, 388
　triangular weighting, 389
Single transfer device (STD), 319, 339, 356–58
Single-mode fiber, 439
Single-pole recursive filter, 434
Sliding transform, 408–9
Smith and Brenden holographic technique, 286–88
Snell's law, 96, 460
Solids, bulk material constants for, 549–51
Sound wave propagation, 1–84
　acoustic impedances, 9–13
　　input impedance, 12–13
　　quarter-wave matching sections, 12
　acoustic losses, 8–9
　　loss mechanisms, 9
　　viscous losses, 8–9
　conservation of mass, 5
　constitutive relations, 75–83
　in crystalline materials, 2
　displacement of a material, 4
　energy, 6–7
　equation of motion, 4
　extensional waves, 14–15
　　Young's modulus, 14
　longitudinal wave, 2
　in nonpiezoelectric materials, 2–15
　　conservation of mass, 4–5
　　displacement of, 4
　　equation of motion, 4–5
　　Hooke's law, 4
　　Poynting's theorem, 7
　　propagation constant, 5–6
　　reduced subscript notation, 78–83
　　sound waves in, 2–15
　　tensor notation, 75, 77–78
　one-dimensional theory, 2–15
　in piezoelectric materials, 17–25, 75–82

constitutive relations, 19–20, 75–83
Curie point, 19
effect of coupling on wave propagation, 21–22
electric field in, 17
energy conservation in, 24–26
stress-free dielectric constant, 22
piezoelectric transducers, 27–75
electrical input impedance, 51–53
electrical matching of a loaded transducer, 55–61
impedance of an unloaded transducer, 35–41
KLM model, 41–47, 50
Mason equivalent circuit, 33–35
power transfer to transducer, 53–54
radiation impedance, 32
reduced subscript notation, 78–83
Redwood equivalent circuit, 35, 41, 50
Reeder-Winslow design method, 61–66
tensor notation, 77–78
as a three-port network, 29–33
transmitting constant, 31
velocity, 30
zero strain capacitance, 32
Poynting's theorem, 7, 8
propagation constant, 5–6
reduced subscript notation, 78–83
shear waves, 2, 13–14
in solids, 1
strain, 4
stress, 3–4
tensor notation, 75–83
Sparrow criterion, 235
Specific acoustic impedance, 9
Spectral whitening, 399
Specular reflection, 310
Spherical aberrations of lenses, 200
Spherical reference waves, 279–85

Spherical waves, 156–58
liquid medium, 157
vibrating spheres, 156–57
Split-gate sampled weighting, 408–9
Split-output electrode convolver, 464–65
Spread-spectrum communications, 454–59, 475–77
frequency hopping modes, 458
Spread-spectrum correlator, 456
Square pulse, matched filter for, 364–65
Stationery phase method, 578–79
Straight-crested wave, 332
Strain, 1, 4, 75
Strain tensor, 537–39, 542–44
Strain vectors, 537–42
strain tensor, 537–40
Stress, 1, 3–4, 75
in a solid, 3–4
Stress tensor, 540–41, 544
Stress vectors, 537–42
equation of motion, 542
stress tensor, 540–41
Stress-free dielectric constant, 22
Strip coupler, 416–21
Strip guide mode, 90
Superconductive delay line, 410
Superconductive dispersive filter, 413
Surface acoustic wave devices, 319
Surface acoustic wave excitation, 134–37
physical implications, 135
Surface acoustic wave storage correlator, 469–77
Surface integral formulation, 303–4
Surface waves, 109–10
Switched capacitors, 356–58
filter, 319
fixed pattern noise, 356
Symbolic notation, 542–46
elasticity tensor, 544–46
piezoelectric tensor, 546–47
strain tensor, 542–44
stress tensor, 544

Tapped systems, 346–47
BBD lines, 357–58
CCD line, 350, 357–58

Tapped-delay-line correlator, 470–71
Temperature drifts, 456
Tensor components, 78
Tensor notation, 77–78
Tensor quantities, 1
Tensors:
elasticity, 544–46
piezoelectric, 546–47
strain, 537–39, 542–44
stress, 540–41, 544
3-dB points, 180
Three-port network, transducer as, 29–33
Time-bandwidth product, 325
Time-delay filter, 396
Time-delay focusing, 227–28
Time-delay imaging systems, 261–75
Time-integrating correlator, 521–28
one-dimensional transforms, 524–26
two-dimensional transforms, 526–28
Tomographic imaging systems, 275–77
dynamically focused system, 261
grating lobes, 267–71
range resolution, 266
sampling lobes, 265–71
sidelobes, 265–71
surface acoustic wave imaging, 272–74
synthetic aperture imaging system, 262, 270–74
transverse definition, 266
TV display, 265
Total stored energy, 6–7
Transducer admittance matrix, 574–77
crossed-field model, 574, 577
in-line model, 574, 577
Transducer electrical input admittance, 128
Transducer materials, properties of, 557–58
Transducers:
acoustic, 3
acoustic emission, 57
air-backed, 50, 60–66
apodized, 330–32

Transducers (cont.)
  clamped, 34–35
  crossed-field, 83
  design examples, 66–70
  electromechanical, 27
  flat piston, 182
  focused, 182–212
  Fresnel, 228
  interdigital, 117–30, 322–29
  loaded, 55–61
  open-circuited, 35
  phase-weighted, 426–27
  piezoelectric, 1, 27–75
  plane piston, 164–82
  pulsed excitation of, 212–18
  rectangular, 175–80
  tuned, 60
  tuned air-backed, 60–61
  ultrasonic, 1
  uniform, 329–30
  unloaded, 35–41
  untuned, 60
  untuned air-backed, 60–61
  wedge, 144–52
  zinc oxide, 54
Transforms:
  chirp z transform, 360
  Fourier, 171–75
  Hankel, 171–75
  one-dimensional Bragg, 524–26
  sliding, 408–9
  two-dimensional Bragg, 526–28
Transmitting constant, 31, 82
Transversal filters, 117, 318–536
  acousto-optic filters, 501–28
    Bragg deflection systems applications, 517–18
    Bragg diffraction by sound, 511–17
    convolver, 518–21
    light diffraction by sound, 503–9
    photoelastic effect, 501–3
    piezo-optic effect, 501–3
    Schlieren effect, 509–10
    time-integrating correlator, 521–28
  adaptive filters, 480–500
    additional uses, 497–98
    eliminating an interfering signal, 495–98
    inverse filter, 482–88
    notch filter, 495
    removing phase errors with, 481–82
    Wiener filter, 488–95
  bandpass filters, 415–43
    amplitude-weighted, 421–26
    building block design technique, 427–31
    "comb" response, 436
    phase-weighted transducers, 426–27
    recursive comb filter, 436–37
    single-mode fiber, 439
    strip coupler, 416–21
  charge-transfer devices (CTD), 339–59
    bucket brigade device (BBD), 339, 341–47
    charge-coupled devices (CCD), 347–52
    imperfect charge-transfer efficiency, 352–56
    single transfer device (STD), 356–58
    switched capacitors, 356–58
  convolvers, 448–77
    acoustic, 449–54, 464–67
    airgap, 463
    air-gap storage correlator, 471–74
    charge-coupled device, 467–68
    external mixer, 465
    input correlation mode, 474–75
    monolithic storage correlator, 471–74
    SAW storage correlator, 469–77
    semiconductor, 461–64
    spread-spectrum communications, 454–59, 475–77
    tapped-delay-line correlator, 470–71
    waveguide, 459–61
  frequency modulated chirp filters, 383–99
    chirp z transform implementation, 399–409
    filter weighting for sidelobe reduction, 387–91
    Fourier transform operations with, 391–99
  linear passive surface wave devices, 322–39
    bandpass filter, 329–32
    frequency modulated chirp analog filter, 332–33
    interdigital transducers, 322–29
    reflective array compressor (RAC) filters, 335–36
    resonators, 334
    stabilized SAW oscillators, 336
  matched filters, 359–83
    ambiguity function, 376–83
    for complex signals, 367–68
    pulse compression, 366
    radar system accuracy, 368
    theory, 361–67
    Wiener filter, 360
Transverse spatial frequencies, 172
Triangular weighting, 389
Triple transit echo, 426
Tuned air-backed transducer, 60–61
Tuned transducer, 60

Ultra-high frequency (UHF) delay lines, 51
Ultrasonic transducer, 1
Uncertainty ellipse, 379
Unmatched front surface, 50
Untuned air-backed transducer, 60–61
Untuned transducer, 60

Variable delay line, 441
Vector components, 78
Vibrating objects, holographic imaging, 294–96
Volume integral formulation, 304–6

Waveguide convolver, 459–61
Wave guide convolver configuration, 461
Wave propagation with finite exciting sources, 154–317
  A-scan imaging, 220–22
  acoustic holography, 277–96
    imaging of vibrating objects, 294–96
    scanning laser acoustic microscope, 291–94
  acoustic imaging applications, 218–20
  amplitude errors, 243–45
  B-scan imaging, 222–23

basic imaging theory, 229–45
C-scan imaging, 223–25
chirp-focused systems, 251–77
  synthetic aperture imaging, 270–74
  tomographic imaging systems, 275–77
diffraction and nonuniform excitation, 154–64
digital sampling, 245–48
field of a focused spherical transducer, 182
fields on axis, 164
focused transducers, 182–212
focusing systems, 225–29
Fresnel lenses, 245–48
  sidelobes, 248–51
Gaussian taper, 206–10
grating lobes, 239–41
Green's function, 158–63
high-speed scanning, 225–29
lensless acoustic imaging, 218–96
matched fiber concepts, 229–31
plane piston transducers, 164–80
pulsed excitation of transducers, 212–18
radial sector scan system, 232
radial variation of the field, 171–75
rectangular transducer diffraction, 175–80
reflection by small and large objects, 300–312
scanned acoustic microscope, 197–206
scattering by small and large objects, 300–312

sidelobes, 238–39
spherical waves in a liquid or solid, 156–58
taper, 206–10
Wave(s):
  creeping, 309
  devices, 319
  equation, 92–94
  extensional, 14–15
  imaging, 272–74
  incident longitudinal, 96–98
  in isotropic media, 85–153
    acoustic waveguides, 104–14
    basic theory, 86–105
    equations of motion for solids and fluids, 90–92
    interdigital transducers, 117–31
    mathematic formalism, 86–90
    plane wave reflection and refraction, 94–104
    Rayleigh waves, 110–14
    wave equations, 92–94
    wedge transducers, 144–52
  Lamb, 108–9
  leaky, 141–44
  longitudinal, 2, 93–94
  plane, 94–105
  propagation, 1–84
    in crystalline materials, 2
    in nonpiezoelectric materials, 2–15
    in piezoelectric materials, 17–25, 75–82
  Rayleigh, 110–14
  reflected, 149
  scattering of, 300–313
  shear, 2, 13–14, 94, 100, 105
  shear horizontal, 100, 105

spherical, 156–58
  reference, 279–85
  sources, 281–82
  storage correlator, 469–77
  straight-crested, 332
  surface, 109–10
  velocities, 89–90
Wedge transducer, 144–52
  experimental results, 150
  finite attenuation, 146–47
  reflected wave, 149
Weighting:
  Bartlett, 389
  Blackmann, 389
  Dirichlet, 389
  Dolph-Tchebysheff, 389
  filter, 387–91
  finger, 328
  Gaussian, 388
  Hamming, 388–90
  Hanning, 388
  Kaiser, 389
  rectangular, 388
  triangular, 389
Weighting functions, properties of, 390
Wiener filter, 488–95
  pseudo-inverse filter, 491–92
  storage correlator as, 492–95
Windowing, 388
Windows, and their Fourier transforms, 389

X-rays, 219, 220

Young's modulus, 14, 88

Zero-forcing algorithm, 485–86
Zero-strain capacitance, 32
Zinc oxide transducer, 54

Index